Lecture Notes in Computer Science 4528

Commenced Publication in 1973
Founding and Former Series Editors:
Gerhard Goos, Juris Hartmanis, and Jan van Leeuwen

José Mira   José R. Álvarez (Eds.)

Nature Inspired Problem-Solving Methods in Knowledge Engineering

Second International Work-Conference on the Interplay
Between Natural and Artificial Computation, IWINAC 2007
La Manga del Mar Menor, Spain, June 18-21, 2007
Proceedings, Part II

Volume Editors

José Mira
José R. Álvarez
Universidad Nacional de Educación a Distancia
E.T.S. de Ingeniería Informática
Departamento de Inteligencia Artificial
Juan del Rosal, 16, 28040 Madrid, Spain
E-mail: {jmira, jras}@dia.uned.es

Library of Congress Control Number: 2007928351

CR Subject Classification (1998): F.1, F.2, I.2, G.2, I.4, I.5, J.3, J.4, J.1

LNCS Sublibrary: SL 1 – Theoretical Computer Science and General Issues

ISSN 0302-9743
ISBN-10 3-540-73054-0 Springer Berlin Heidelberg New York
ISBN-13 978-3-540-73054-5 Springer Berlin Heidelberg New York

Springer is a part of Springer Science+Business Media

springer.com

Printed in Germany

Typesetting: Camera-ready by author, data conversion by Scientific Publishing Services, Chennai, India
Printed on acid-free paper SPIN: 12076178 06/3180 5 4 3 2 1 0

Preface

The Semantic Gap

There is a set of recurrent problems in AI and neuroscience which have restricted their progress from the foundation times of cybernetics and bionics. These problems have to do with the enormous semantic leap that exists between the ontology of physical signals and that of meanings. Between physiology and cognition. Between natural language and computer hardware. We encounter this gap when we want to formulate computationally the cognitive processes associated with reasoning, planning and the control of action and, in fact, all the phenomenology associated with thought and language.

All "bio-inspired" and "interplay" movement between the natural and artificial, into which our workshop (IWINAC) fits, faces this same problem every two years. We know how to model and reproduce those biological processes that are associated with measurable physical magnitudes and, consequently, we know how to design and build robots that imitate the corresponding behaviors. On the other hand, we have enormous difficulties in understanding, modeling, formalizing and implementing all the phenomenology associated with the cognition field. We do not know the language of thought. We mask our ignorance of conscience with the term emergentism.

This very problem recurs in AI. We know how to process images, but we do not know how to represent the process for interpreting the meaning of behaviors that appear in a sequence of images computationally, for example. We know how to plan a robot's path, but we do not know how to model and build robots with conscience and intentions. When the scientific community can link signals and neuronal mechanisms with "cognitive magnitudes" causally we will have resolved at the same time the serious problems of bio-inspired engineering and AI. In other words, we will know how to synthesize "general intelligence in machines."

To attempt to solve this problem, for some time now we have defended the need to distinguish between own-domain descriptions of each level and those of the external observer domain. We also believe that it is necessary to stress conceptual and formal developments more. We are not sure that we have a reasonable theory of the brain or the appropriate mathematics to formalize cognition. Neither do we know how to escape classical physics to look for more appropriate paradigms.

The difficulty of building bridges over the semantic gap justifies the difficulties encountered up to now. We have been looking for some light at the end of the tunnel for many years and this has been the underlying spirit and intention of the organization of IWINAC 2007. In the various chapters of these two books of proceedings, the works of the invited speakers, Professors Monserrat and Paun, and the 126 works selected by the Scientific Committee, after the refereeing process, are included. In the first volume, entitled "*Bio-inspired Modeling of Cognitive Tasks*," we include all the contributions that are closer to the theoretical, conceptual and methodological aspects linking AI and knowledge engineering with neurophysiology, clinics and cognition. The second volume entitled "*Nature-Inspired Problem-Solving Methods in Knowledge Engineering*" contains all the contributions connected with biologically inspired methods and techniques for solving AI and knowledge engineering problems in different application domains.

An event of the nature of IWINAC 2007 cannot be organized without the collaboration of a group of institutions and people who we would like to thank now, starting with our university, *UNED*, and its Associate Center in Cartagena. The collaboration of the Universitat Politécnica de Cartagena and the Universitat de Murcia has been crucial, as has the enthusiastic and efficient work of José Manuel Ferrández and the rest of the Local Committee. In addition to our universities, we received financial support from the Spanish Ministerio de Educación y Ciencia, the Fundación SENECA-Agencia Regional de Ciencia y Tecnología de la Comunidad de Murcia, *DISTRON s.l.* and the Excelentísimo Ayuntamiento de Cartagena. Finally, we would also like to thank the authors for their interest in our call and the effort in preparing the papers, condition *sine qua non* for these proceedings, and to all the Scientific and Organizing Committees, in particular, the members of these committees who have acted as effective and efficient referees and as promoters and managers of pre-organized sessions on autonomous and relevant topics under the IWINAC global scope.

My debt of gratitude with José Ramón Alvarez and Félix de la Paz goes, as always, further than the limits of a preface. And the same is true concerning Springer and Alfred Hofmann and their collaborators Anna Kramer and Erika Siebert-Cole, for the continuous receptivity and collaboration in all our editorial joint ventures on the interplay between neuroscience and computation.

June 2007 José Mira

Organization

General Chairman

José Mira, UNED (Spain)

Organizing Committee

José Ramón Álvarez Sánchez, UNED (Spain)
Félix de la Paz López, UNED (Spain)

Local Organizing Committee

José Manuel Ferrández, Univ. Politécnica de Cartagena (Spain).
Roque L. Marín Morales, Univ. de Murcia (Spain).
Ramón Ruiz Merino, Univ. Politécnica de Cartagena (Spain).
Gonzalo Rubio Irigoyen, UNED (Spain).
Gines Doménech Asensi, Univ. Politécnica de Cartagena (Spain).
Vicente Garcerán Hernández, Univ. Politécnica de Cartagena (Spain).
Javier Garrigós Guerrero, Univ. Politécnica de Cartagena (Spain).
Javier Toledo Moreo, Univ. Politécnica de Cartagena (Spain).
José Javier Martínez Álvarez, Univ. Politécnica de Cartagena (Spain).

Invited Speakers

Gheorge Paun, Univ. de Sevilla (Spain)
Javier Monserrat, Univ. Autónoma de Madrid (Spain)
Álvaro Pascual-Leone, Harvard Medical School (USA)

Field Editors

Emilia I. Barakova, Eindhoven University of Technology (The Netherlands)
Eris Chinellato, Universitat Jaume-I (Spain)
Javier de Lope, Universitat Politécnica de Madrid (Spain)
Pedro J. García-Laencina, Universitat Politécnica de Cartagena (Spain)
Dario Maravall, Universitat Politécnica de Madrid (Spain)
José Manuel Molina López, Univ. Carlos III de Madrid (Spain)
Juan Morales Sánchez, Universitat Politécnica de Cartagena (Spain)
Miguel Angel Patricio Guisado, Universitat Carlos III de Madrid (Spain)
Mariano Rincón Zamorano, UNED (Spain)
Camino Rodríguez Vela, Universitat de Oviedo (Spain)

José Luis Sancho-Gómez, Universitat Politécnica de Cartagena (Spain)
Jesús Serrano, Universitat Politécnica de Cartagena (Spain)
Ramiro Varela Arias, Universitat de Oviedo (Spain)

Scientific Committee (Referees)

Ajith Abraham, Chung Ang University (South Korea)
Andy Adamatzky, University of the West of England (UK)
Michael Affenzeller, Upper Austrian University of Applied Sciences (Austria)
Igor Aleksander, Imperial College of Science Technology and Medicine (UK)
Amparo Alonso Betanzos, Universitate A Coruña (Spain)
José Ramón Álvarez Sánchez, UNED (Spain)
Shun-ichi Amari, RIKEN (Japan)
Razvan Andonie, Central Washington University (USA)
Davide Anguita, University of Genoa (Italy)
Margarita Bachiller Mayoral, UNED (Spain)
Antonio Bahamonde, Universitat de Oviedo (Spain)
Emilia I. Barakova, Eindhoven University of Technology (The Netherlands)
Alvaro Barreiro, Univ. A Coruña (Spain)
Josh Bongard, University of Vermont (USA)
Fiemke Both, Vrije Universiteit Amsterdam (The Netherlands)
François Brémond, INRIA (France)
Enrique J. Carmona Suárez, UNED (Spain)
Joaquín Cerdá Boluda, Univ. Politécnica de Valencia (Spain)
Enric Cervera Mateu, Universitat Jaume I (Spain)
Antonio Chella, Università degli Studi di Palermo (Italy)
Eris Chinellato, Universitat Jaume I (Spain)
Emilio S. Corchado, Universitat de Burgos (Spain)
Carlos Cotta, University of Málaga (Spain)
Erzsébet Csuhaj-Varjú, Hungarian Academy of Sciences (Hungary)
José Manuel Cuadra Troncoso, UNED (Spain)
Félix de la Paz López, UNED (Spain)
Ana E. Delgado García, UNED (Spain)
Javier de Lope, Universitat Politécnica de Madrid (Spain)
Ginés Doménech Asensi, Universitat Politécnica de Cartagena (Spain)
Jose Dorronsoro, Universitat Autónoma de Madrid (Spain)
Gérard Dreyfus, ESCPI (France)
Richard Duro, Universitate da Coruña (Spain)
Juan Pedro Febles Rodriguez, Centro Nacional de Bioinformática (Cuba)
Eduardo Fernández, University Miguel Hernandez (Spain)
Antonio Fernández-Caballero, Universitat de Castilla-La Mancha (Spain)
Jose Manuel Ferrández, Univ. Politécnica de Cartagena (Spain)
Kunihiko Fukushima, Kansai University (Japan)
Jose A. Gámez, Universitat de Castilla-La Mancha (Spain)
Vicente Garceran Hernández, Universitat Politécnica de Cartagena (Spain)

Pedro J. García-Laencina, Universitat Politécnica de Cartagena (Spain)
Juan Antonio García Madruga, UNED (Spain)
Francisco J. Garrigos Guerrero, Universitat Politécnica de Cartagena (Spain)
Tamás (Tom) D. Gedeon, The Australian National University (Australia)
Charlotte Gerritsen, Vrije Universiteit Amsterdam (The Netherlands)
Marian Gherghe, University of Sheffield (UK)
Pedro Gómez Vilda, Universitat Politécnica de Madrid (Spain)
Carlos G. Puntonet, Universitat de Granada (Spain)
Manuel Graña Romay, Universitat Pais Vasco (Spain)
Francisco Guil-Reyes, Universitat de Almería (Spain)
John Hallam, University of Southern Denmark (Denmark)
Juan Carlos Herrero, (Spain)
César Hervás Martínez, Universitat de Córdoba (Spain)
Tom Heskes, Radboud University Nijmegen (The Netherlands)
Fernando Jimenez Barrionuevo, Universitat de Murcia (Spain)
Jose M. Juarez, Universitat de Murcia (Spain)
Joost N. Kok, Leiden University (The Netherlands)
Yasuo Kuniyoshi, Univ. of Tokyo (Japan)
Petr Lánsky, Academy of Sciences of Czech Republic (Czech Republic)
Hod Lipson, Cornell University (USA)
Maria Longobardi, Università di Napoli Federico II (Italy)
Maria Teresa López Bonal, Universitat de Castilla-La Mancha (Spain)
Ramon López de Mántaras, CSIC (Spain)
Tino Lourens, Philips Medical Systems (The Netherlands)
Max Lungarella, University of Tokyo (Japan)
Manuel Luque Gallego, UNED (Spain)
Francisco Maciá Pérez, Universitat de Alicante (Spain)
george Maistros, The University of Edinburgh (UK)
Vincenzo Manca, University of Verona (Italy)
Riccardo Manzotti, IULM University (Italy)
Dario Maravall, Universitat Politécnica de Madrid (Spain)
Roque Marín, Universitat de Murcia (Spain)
Jose Javier Martinez Álvarez, Universitat Politécnica de Cartagena (Spain)
Rafael Martínez Tomás, UNED (Spain)
Jesus Medina Moreno, Univ. de Málaga (Spain)
Jose del R. Millan, IDIAP (Switzerland)
José Mira, UNED (Spain)
Victor Mitrana, Universitat Rovira i Virgili (Spain)
José Manuel Molina López, Univ. Carlos III de Madrid (Spain)
Juan Morales Sánchez, Universitat Politécnica de Cartagena (Spain)
Federico Morán, Universitat Complutense de Madrid (Spain)
Arminda Moreno Díaz, Universitat Politécnica de Madrid (Spain)
Ana Belén Moreno Díaz, Universitat Rey Juan Carlos (Spain)
Isabel Navarrete Sánchez, Universitat de Murcia (Spain)
Nadia Nedjah, State University of Rio de Janeiro (Brazil)

Taishin Y. Nishida, Toyama Prefectural University (Japan)
Richard A. Normann, University of Utah (USA)
Manuel Ojeda-Aciego, Univ. de Málaga (Spain)
Lucas Paletta, Joanneum Research (Austria)
José T. Palma Méndez, University of Murcia (Spain)
Miguel Angel Patricio Guisado, Universitat Carlos III de Madrid (Spain)
Gheorghe Paun, Universitat de Sevilla (Spain)
Juan Pazos Sierra, Universitat Politécnica de Madrid (Spain)
Mario J. Pérez Jiménez, Universitat de Sevilla (Spain)
José Manuel Pérez-Lorenzo, Universitat de Jaén (Spain)
Jose M. Puerta, University of Castilla-La Mancha (Spain)
Alexis Quesada Arencibia, Universitat de Las Palmas de Gran Canaria (Spain)
Günther R. Raidl, Vienna University of Technology (Austria)
Luigi M. Ricciardi, Università di Napoli Federico II (Italy)
Mariano Rincón Zamorano, UNED (Spain)
Victoria Rodellar, Universitat Politécnica de Madrid (Spain)
Camino Rodríguez Vela, Universitat de Oviedo (Spain)
Daniel Ruiz Fernández, Univ. de Alicante (Spain)
Ramón Ruiz Merino, Universitat Politécnica de Cartagena (Spain)
Ángel Sánchez Calle, Universitat Rey Juan Carlos (Spain)
Eduardo Sánchez Vila, Universitat de Santiago de Compostela (Spain)
José Luis Sancho-Gómez, Universitat Politécnica de Cartagena (Spain)
Gabriella Sanniti di Baja, CNR (Italy)
José Santos Reyes, Universitate da Coruña (Spain)
Shunsuke Sato, Aino University (Japan)
Andreas Schierwagen, Universität Leipzig (Germany)
Guido Sciavicco, Universitat de Murcia (Spain)
Radu Serban, Vrije Universiteit Amsterdam (The Netherlands)
Jesús Serrano, Universitat Politécnica de Cartagena (Spain)
Jordi Solé i Casals, Universitat de Vic (Spain)
Antonio Soriano Payá, Univ. de Alicante (Spain)
M^{a}. Jesus Taboada, Univ. Santiago de Compostela (Spain)
Settimo Termini, Università di Palermo (Italy)
Javier Toledo Moreo, Universitat Politécnica de Cartagena (Spain)
Jan Treur, Vrije Universiteit Amsterdam (The Netherlands)
Ramiro Varela Arias, Universitat de Oviedo (Spain)
Marley Vellasco, Pontifical Catholic University of Rio de Janeiro (Brazil)
Lipo Wang, Nanyang Technological University (Singapore)
Stefan Wermter, University of Sunderland (UK)
J. Gerard Wolff, Cognition Research (UK)
Hujun Yin, University of Manchester (UK)

Table of Contents

High Performance Implementation of an FPGA-Based Sequential DT-CNN

J. Javier Martínez-Alvarez, F. Javier Garrigós-Guerrero, F. Javier Toledo-Moreo, and J. Manuel Ferrández-Vicente

Dpto. Electrónica, Tecnología de Computadoras y Proyectos,
Universidad Politécnica de Cartagena, 30202 Cartagena, Spain
jjavier.martinez@upct.es

Abstract. In this paper an FPGA-based implementation of a sequential discrete time cellular neural network (DT-CNN) with 3×3 templates is described. The architecture is based on a single pipelined cell which is employed to emulate a CNN with larger number of neurons. This solution diminishes the use of hardware resources on the FPGA and allows the cell to process real time input data in a sequential mode. Highly efficient FPGA implementation has been achieved by manual design based on low level instantiation and placement of hardware primitives. The Intellectual Property Core offers an appropriate tradeoff between area and speed. Our architecture has been developed to assist designers implementing discrete CNN models with performance equivalent to hundreds or millions of neurons on low cost FPGA-based systems.

1 Introduction

Video and image processing algorithms demand an increasing computing performance, specially when they are targeted for real time applications. To accomplish the desired results, these systems are implemented using complex hardware solutions, which are time-consuming and difficult to develop. In this context, the Cellular Neural Network (CNN) paradigm is a general purpose tool that can be used to find commodity hardware solutions for these applications.

CNNs were proposed by Chua and Yang [1] like a bioinspired processor array able to solve complex signal processing problems. They have been used during decades for designing systems and applications that operate in a fully parallel manner. Some characteristics of CNNs make them different to other ANN models and especially useful for high-speed image and video processing applications [2]. They are useful also for bioinspired systems modeling, as the human retina [3,4]. One of the most important features is their local connectivity, any cell is connected only to other cells in a neighborhood, like biological neurons do. This allows the CNN to process information locally, while simultaneously been able to propagate information along the net, thus making at the end a global processing of all input data. Due to its character of universal Turing Machine [7], CNNs can be configured to perform different processing tasks, what makes them a useful and versatile tool for a great number of applications.

J. Mira and J.R. Álvarez (Eds.): IWINAC 2007, Part II, LNCS 4528, pp. 1–9, 2007.

CNNs have however important disadvantages that hinder their implementation as integrated circuits. Originally, CNNs were defined as a bidimensional array of processing elements working in a parallel manner. Due to its intrinsic parallelism, CNNs can process information at a very high speed. However, this structure prevents the implementation of large networks, as the circuit area is proportional to the network size.

Typically, parallel models of CNNs has been carried out as Application Specific Integrated Circuits (ASICs) due to their high speed and density, and low power consumption [5]. This approach has led to implementation of small CNNs [6], what limits their use in standard image and video applications. This solution has the additional disadvantages of high complexity, cost and development time. Looking for more flexibility and shorter design cycles, researchers have directed their efforts to FPGA-based solutions [8].

This is due to the fact that up to date FPGAs ofter not just an ever-increasing amount of logic, but also several internal resources like FIFOS, BlockRAMs, dedicated multipliers, fast carry chains, etc., that make them specifically useful for signal processing applications.

Due to the complexity of the calculations and the large number of processing elements (PE) involved, effective implementation of the CNN algorithm requires the best use of the FPGA internal resources. With this objective, in this paper we propose a novel sequential architecture for the DT-CNN model, particularized for a 3×3 neighborhood, which has been developed making use of low level FPGA primitives, relative placing constrains and manual routing of the PE to get the most efficient circuit in terms of both area and speed.

2 Design of a Sequential DT-CNN

The original CNN model is defined by means of an array of analogue neurons with local connectivity. In order to implement this algorithm on an FPGA, it is first necessary to digitize its analogue behavior. Several discretization methods (the Euler method, the differentia algorithm, the numeric integration algorithm and the response-invariant algorithm) have been tested [9] to determine the one that best fits the hardware resources on the FPGA while still having the smallest error. Simulation results showed that the Euler method-based model presents an excellent behavior with minimum use of hardware resources. Therefore, the Euler approximation method has been adopted to derive the recursive equations (1) and (2), and the model shown in figure 1.

$$X_{ij}[n] = \sum_{k,l \in Nr(ij)} A_{kl}[n-1]Y_{kl}[n-1] \ + \sum_{k,l \in Nr(ij)} B_{kl}[n-1]U_{kl} \ + \ I_{ij} \tag{1}$$

$$Y_{ij}[n] = \frac{1}{2}\left(|X_{ij}[n]+1| - |X_{ij}[n]-1|\right) \tag{2}$$

where I, U, Y and X denote input bias, input data, output data and state variable of each cell, respectively. The neighborhood distance r for cell (i, j) is given

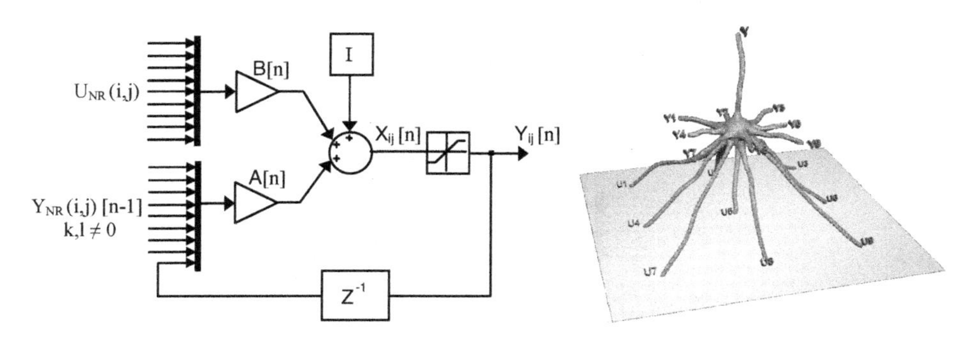

Fig. 1. (a) Discrete approximation of the original CNN model, obtained from the Euler's method, (b) 3D symbolical model

by $Nr(ij)$ function, where i and j denote the position of the cell in the network and k and l the position of the neighboring cells relative to the cell in consideration. B is the non-linear weights template for the inputs and A is the corresponding non-linear template for the outputs of the neighboring cells. Non-linearity means that templates can change over time. The templates are defined as space-invariant, so they are the same for each cell.

Despite this transformation, the direct implementation of the model over FPGAs presents two main disadvantages: the number of neurons that are necessary for most applications and the intrinsic parallelism of each neuron. For example, in a typical video processing task, a CNN for 640×480-pixel images running at 60 Hz would need an FPGA capable of accommodating more than 300,000 neurons, which is not feasible in practice.

Thus, the only solution is to develop a sequential architecture that multiplexes PE in time. In this approach, an N-cell CNN is folded in just one-recurrent-cell CNN keeping the area consumed to a minimum. Furthermore, this solution is the most appropriate for many applications, which exhibit a pure sequential behavior. This is the case of systems where the input data array is stored in a single memory, meaning that this data is read -mainly- sequentially, preventing their parallel processing. Some video applications have a similar behavior, because the video source provides data in a pixel-by-pixel scheme under rigid synchronization. This of course implies that the single neuron must be fast enough to process pixels at its input rate, but the solution is optimum in the sense that there is no need to use frame-buffers as intermediate image storage, that due to their size, are not possible to locate in the FPGA internal memory, and usually need expensive external devices. This would complicate the system design and add new constraints, like memory timing specifications.

With the last approach, the computation of a CNN is equivalent to a single neuron which shifts along the input array, similarly to the video signal generated by a camera. The area reduction is obtained at the cost of a loss in the computing performance. Following with the previous example, each cell in the parallel CNN has nearly 16 ms (1/60) to process one pixel. However, the single cell of

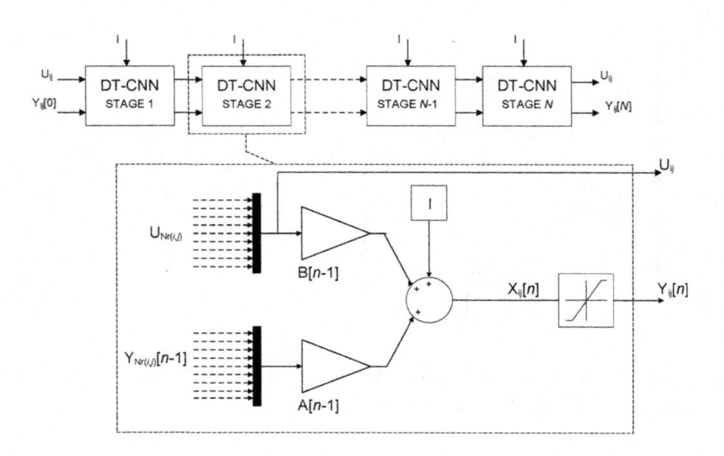

Fig. 2. The cell consists of N identical pipelined stages

the sequential approach should spend just 52 ns per pixel $(1/(640 * 480 * 60))$. Considering that a single iteration of the recurrent algorithm performed by the cell would need at least 40 ns to execute (as it will be shown in section 3.2), the parallel CNN could execute 400.000 iterations per pixel, for one iteration of the single neuron. This results in a very poor precision of the algorithm in the last case, what invalidates this approach for most of the real applications.

In order to overcome this problem, our proposal includes an architecture in which, instead of executing N iterations on the same recursive cell, the N iterations are executed by N different pipelined stages. Now, if 10 stages are considered, the area consumed by the cell is 10 times larger than the fully-folded version, but the computations are 10 times faster. Besides, the problem of the data storage between stages is overcome, since now the system can work on streaming data. This architecture is capable of processing real time input data with low area consumption and adjustable precision depending on the number of stages implemented. Simulations have revealed that 10 iterations, or stages, are enough in the vast majority of typical video processing applications to obtain data nearly identical to those of the original model. Details about the study can be found in [9]. In case of binary applications, however, just one or two iterations are enough.

As shown in figure 2, the stage has two input ports, the input vector U_{ij} and the outputs from previous stages $Y_{ij}[n{-}1]$; and two output ports, the corresponding output data $Y_{ij}[n]$ and the propagated input data, which is delayed as many cycles as the stage latency requires in order to ensure data synchronization. The propagation of the inputs makes it possible to pipeline a number of stages all working with the same input data, which enable the convergence of information and thus the possibility of configuring networks where one output can be function of multiple inputs, far more than the typical 3×3 template. Moreover, the inputs of each stage can be adjusted by different templates A and B, what enables the design of highly complex and powerful CNNs.

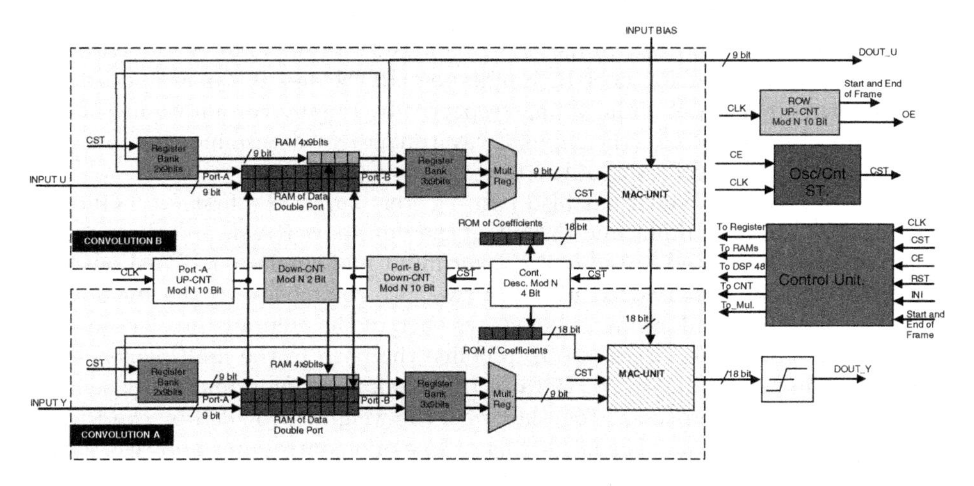

Fig. 3. RTL block diagram for the processing stage developed

3 Implementation of a 3×3 Stage

In this section the structure of every stage is described, particularized for a neighborhood radius of 1. The stage algorithm has been coded in VHDL, using low level structural descriptions and manual placement of primitives. Combination of both techniques granted the maximum exploitation of the FPGA internal resources while optimizing the performance.

Equations (3) and (4) show the DT-CNN model algorithm for this particular case. The corresponding RTL block diagram is shown in figure 3.

$$X_{ij}[n] = \sum_{k=1}^{-1}\sum_{l=1}^{-1} A_{kl}[n-1]Y_{kl}[n-1] + \sum_{k=1}^{-1}\sum_{l=1}^{-1} B_{kl}[n-1]U_{kl} + I_{ij} \quad (3)$$

$$Y_{ij}[n] = \frac{1}{2}\left(|X_{ij}[n]+1| - |X_{ij}[n]-1|\right) \quad (4)$$

From equation (3), a single stage is formed by two convolutional blocks of size 3×3, plus a bias. To calculate the convolutions, the coefficients of template A or B and information of 3 consecutive rows of the input data array must be previously stored. Each convolutional block has three memories for this purpose: a ROM for the 9 elements of its template, and 2 RAMs to store the last 3 rows. The first is a dual-port RAM with a size of 1024×18 bits to store the last 2 rows, using the 9 less-significant bits for the last row and the 9 most-significant bits for the second to last. The third row is just partially stored, as not all its data is used for the convolution, so a small single-port RAM of 4×9 bits does the work.

Sequential access to the stored data is granted by counters connected to the corresponding address buses. Because the two convolutional blocks work simultaneously, counters are shared between blocks to minimize resources. Double-port memories are connected to two modulo-N counters, an up-counter for the port-A and a down-counter for port-B. Single-port ROM and RAM are addressed by two down-counters of 4 and 2 bits respectively. Finally, an additional 10-bit counter checks the rows of the input array (image) to determine the end of a data frame and the beginning of the next.

Each convolutional block has also two register banks: the first, of 2×9 bits, connected to the data input and output of the dual-port RAM, and the second, with 3×9 bits, connected to the multiplexer input and single-port RAM output. The 2×9 bits bank is used to feed the two last rows back to the memories, while the 3×9 bits bank sequentially stores each of the columns involved in the convolution. This bank serves also to pipeline the path to the multiplexer, thus increasing the processing speed. Data in this bank are then sequentially sent to the MAC unit to be processed together with the weights provided by the ROM.

Circuit synchronization is achieved by means of a control unit and two clocks signals: the external clock (CLK) and a self-timed internal counter clock (CST). The period of CLK matches the input data rate, and represents the system clock. CST is of higher frequency, and is used to sequence the convolutional process in 17 cycles, for just one CLK period.

3.1 Resources Consumption

The sequential stage described has been implemented in a Xilinx XC4LX25-11 FPGA [11]. The VHDL description instantiates architecture specific low level primitives and was manually placed an routed to get the best area/speed tradeoff. As a consequence, the resultant circuit has a maximum deep of 4 logic levels, but a restrained area consumption.

The memory used to store the weights in each convolutional block has been implemented by means of a distributed 16×18 bits ROM with synchronous read operation. The size of this memory (and thus the precision of the weights) comes determined by the MAC input wide, which is fixed to 18 bits. A double-port Block-RAM configured as 1024×18bits keeps the consecutive image lines. Partial data from the third line is stored in a 4×9 bits distributed RAM with asynchronous read.

Every counter has been designed using platform-specific primitives (FDCPE, LUT4 and LUT3) and choosing for each particular case the up/down configuration more area-saving. Thanks to this manual design process the hard timing restrictions could be met.

Each multiplexer was also designed at a low level making use of just two primitives: LUT4 and MUXF5. As previously indicated, their output has been registered to increase pipelining and maximum working frequency. Register banks were implemented by means of FDCPE primitives, while each MAC unit required just a single DSP48. Finally, the control unit used just LUT primitives and a 5 bit up-counter connected to the CST clock.

To verify the proposed architecture, the circuit was synthesised and implemented using the Xilinx ISE 7.1.4 tools. Figure 1 shows the optimal manual placement figured out for the convolutional block. A summary of the resources consumed by the whole stage is depicted in Table 1.

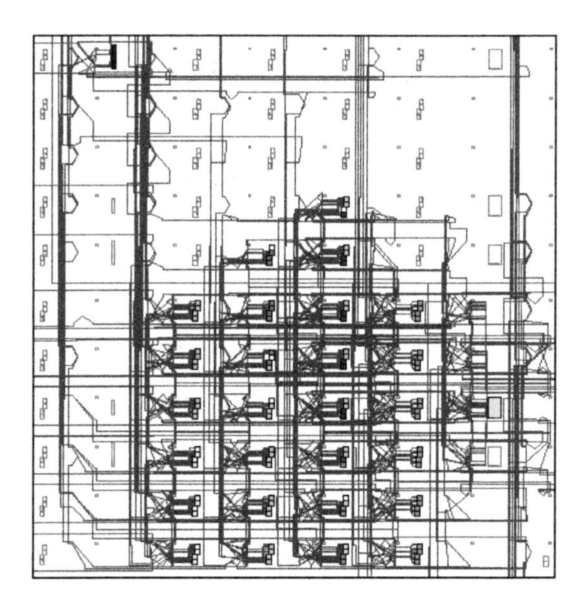

Fig. 4. Internal layout for the convolutional block on a Xilinx XC4LX15-11 part

Table 1. Summary of timing information and used resources for a single stage. Percentages are referred to a Xilinx XC4LX25-11 device.

Resources	Units	% Used
Area (slices)	150	1.40
Flip flops	172	0.80
DSP48	2	4.08
BlockRAM	2	2.78
Max. internal freq. (MHz)	410	-
Max. pixel Clk. (MHz)	24.11	-

3.2 Processing Speed

The circuit was simulated with ModelSim 6.0d and analyzed with the Timing Analyzer tools included in the Xilinx ISE 7.1.4. package to verify its correct behavior and determine the critical path. We made use of the last libraries provided by Xilinx, with the most accurate timing models for primitives. The critical path found was the signal from the BlockRAM to the register banks, formed by the BlockRAM propagation delay (1.830 ns), the net to the registers (0.808 ns) and the register's own set-up time (0.237 ns), for a total of 2.875 s for this device. This analysis shows that due to our optimization efforts, the circuit maximum working frequency comes now determined by the BlockRAM delay, a technological factor, and that there is no margin for improvement.

Experimental test were also carried out to set the real maximum working frequency and the circuit stability at this and lower frequencies. An Infinium 54830B oscilloscope with 4 Gsa/s and a XC4VLX25ff688-11 part from Xilinx were used for the test. The results shown that the circuit was able to run steady at an internal frequency (CST) of 415 Mhz. This means that a single stage can process more than 24 million inputs per second (period of 41.6 ns). Considering that the computing procedure consist of 35 operations (9 multiplications and 8 additions per convolutional block, followed by the additive bias) each stage can execute more than 840 MOPs. These results mean that our architecture can process a 640×480 pixels video signal at 78 fps. In other words, it can emulate the behavior or a 300,000 neurons CNN processing standard video in real time.

The complete stage uses around 4% of a XC4VLX25 part. Assuming that one stage is equivalent to an iteration of the recurrent model, and that it can be considered also a layer of a DT-CNN [10], using this device it would be possible to implement almost 25 layers of 300,000 neurons, that would be able to process real time standard video signal. With larger devices, for example the XC4VSX55 part, more than 160 layers could be implemented, meaning a total of almost 50 million equivalent-neurons in a single chip. Finally, it is worth mentioning that cascading new pipelined stages increments proportionally the processing capacity, but does not influence the processing speed.

4 Conclusions

A novel sequential DT-CNN architecture specifically tailored for hardware acceleration of the model on field programmable devices has been developed in this paper. A particularization of the general architecture has been implemented for the Virtex 4 family of FPGAs, that takes advantage of the low level primitives and advanced P&R options to get maximum efficiency and performance.

The resulting neuron stage, with a neighborhood radius of 1, has an outstanding area/speed tradeoff. The combination of an efficient architecture and a low level description/optimization procedure makes it a valuable solution for real time processing applications, where input data are provided sequentially. The stage has been designed to connect itself with other similar stages and so to set up large DT-CNN, without the need of using external memories or decreasing the processing performance. This way, our architecture supports cascading of multiple pipelined stages over several devices to form multi-FPGA computing platforms equivalent to hundreds or thousands of millions of neurons.

Acknowledgements

This research has been funded by MTyAS IMSERSO RETVIS 150/06.

References

1. L. O. Chua, L. Yang, "Cellular neural networks: theory", IEEE Trans. Circuits and Systems, CAS-35, 1988
2. K. R. Crounse, L. 0. Chua, "Methods for image processing and pattern formation in Cellular Neural Networks: a tutorial", IEEE Transactions on Circuits and Systems I, Vol. 42, Issue 10, pp. 583-601, 1995.
3. D. Balya, C. Rekeczky & T.Roska, "A realistic mammalian retinal model implemented on complex cell CNN universal machine", IEEE Intern. Symp. on Circuits and Systems, pp. 26-29, 2002
4. L.O. Chua, "The CNN: a brain-like computer Neural Networks", IEEE Intern. Joint Conference, Vol. 1, pp. 25-29, 2004.
5. T. Roska, A. Rodriguez-Vazquez, "Review of CMOS implementations of the CNN universal machine-type visual microprocessors", IEEE Int. Symp. on Circuits and Systems, ISCAS 2000, Geneva-Italia, vol. 2, pp. 120-123, 2000.
6. G. Linan, A. Rodriguez-Vazquez, S. Espejo, R. Dominguez-Castro, "ACE16K: a 128x128 focal plane analog processor with digital I/O", IEEE Int. Work. on Cellular Neural Networks and Their Applications, CNNA 2002, pp. 132-139, 2002.
7. K. R. Crounse, L. 0. Chua, "The CNN Universal Machine is as universal as a Turing Machine", IEEE Transactions on Circuits and Systems I. vol.43, Issue 4, pp. 353-355, 1996.
8. C. Hsin, H. Yung, C. Chang, L. Teh, C. Chun, "Image-processing algorithms realized by discrete-time cellular neural networks and their circuit implementations", Elsevier, Chaos Solitons and Fractals Journal, vol. 29, Issue 5, pp 1100-1108, 2006.
9. J.J. Martínez, F.J. Toledo, J.M. Ferrández, "Architecture Implementation of a Discrete Cellular Neuron Model (DT-CNN) on FPGA", The Int. Society for Optical Engineering, Bioengineered and bioinspired systems II, pp. 332-340, 2005.
10. Z. Nagy, P. Szolgay, "Configurable multilayer CNN-UM emulator on FPGA", IEEE Trans. on Circuits and Systems I, vol 50, Issue 6, pp.774-778, 2003.
11. Xilinx Inc, "Virtex-4 User Guide", data sheet(ug070), www.xilinx.com, 2004.

HANNA: A Tool for Hardware Prototyping and Benchmarking of ANNs

Javier Garrigós, José J. Martínez, Javier Toledo, and José M. Ferrández

Departamento de Electrónica, Tecnología de Computadoras y Proyectos
Universidad Politécnica de Cartagena, C/. Dr. Fleming s/n, 30202 Cartagena (MURCIA), Spain
Javier.Garrigos@upct.es
http://www.detcp.upct.es

Abstract. The continuous advances in VLSI technologies, computer architecture and software development tools make it difficult to find the adequate implementation platform of an ANN for a given application. This paper describes HANNA, a software tool designed to automate the generation of hardware prototypes of MLP and MLP-like neural networks over FPGA devices. Coupled with traditional Matlab®/Simulink® environments the generated model can be synthesized, downloaded to the FPGA and co-simulated with the software version to trade off area, speed and precision requirements. The tool and our design methodology are validated through two examples.

1 Introduction

Artificial Neural Networks (ANNs) have traditionally been realized by software modeling using general-purpose microprocessors. Looking for increased performance, designers have moved to other approaches, such as clusters of microprocessors or microcontrollers, application-specific microprocessors designed for intensive arithmetic computation (like PDSPs, Programmable Digital Signal Processors), and direct hardware implementations both over Application-Specific Integrated Circuits (ASICs) and Field Programmable Gate Arrays (FPGAs).

No matter of their flavor, software implementations execute sequentially the ANN algorithm (in essence) due to nature of the von Newmann architecture. On the other hand, hardware specific implementations are free to exploit the inherent parallelism of ANNs and thus are supposed to offer significant performance increments over software on conventional workstations. It has been demonstrated however that the validity of this assumption depends on many factors determining exactly the characteristics of the platforms been compared [1]. So far, results justifying performance of software and hardware implementations are difficult to compare and their leading usually stand just for particular applications designed over concrete platforms. This is partially due to the increased processing capabilities of current general-purpose processors, with clock rates over 3 GHz and architectural mechanisms (like pipelining and superscalar processing units) that provide high speed and certain degree of instruction-level parallelism. This is also

J. Mira and J.R. Álvarez (Eds.): IWINAC 2007, Part II, LNCS 4528, pp. 10–18, 2007.

the case of approaches based on PDSP devices, that now incorporate advanced characteristics such as multiple MAC (Multiply-and-Accumulate) units, fast A/D and D/A converters, barrel shifters, etc, with clock frequencies over 1 GHz.

In spite of these considerations it is clear that specific applications can benefit of an increased performance with massive-parallel execution over FPGA devices. Moreover, other FPGA advantages should be considered, like reduction in size and power consumption and better protection against unauthorized copies with respect to pure software approaches, and in-circuit reprogramability, shorter production time and lower NRE costs when compared to ASICs.

In some applications a designer must implement an ANN model over different platforms to meet performance specifications, a process still more painful when several hardware implementations have to be evaluated. In this paper we describe HANNA (Hardware ANN Architect), a tool to help designers in the generation of hardware implementations of MLP-like neural networks on FPGA devices. HANNA is a set of Matlab scripts that when combined with traditional Matlab/Simulink capabilities provides a rapid prototyping tool that permits creating a unique modeling of a ANN, but allows its implementation over different architectures with small changes. Moreover, the designer is able to co-simulate, characterize and compare the several implementations to verify both correctness and performance.

The rest of the paper is organized as follows: in Section 2 we expose the main advantages of the new system-level design methodologies and tools in which HANNA scripts are based, when compared to the traditional FPGA design tools. Section 3 depicts the neuron and network architecture implemented by HANNA, describing the main blocks and their behaviors. Sections 4 and 5 resume the results obtained when applying HANNA to generate hardware ANNs for two example applications: a simple XOR function and a speaker recognition problem. Finally, some concluding remarks are made in Section 6.

2 Software/Hardware Prototyping

To meet the challenges posed by the growing circuit complexity, new languages, tools and methodologies with greater modeling capabilities are been developed. In the area of design languages SystemC, SytemVerilog, Verilog 2005, Analogue and Mixed-Signal versions of Verilog and VHDL, or Vera are some of the new proposals [2]. Simultaneously, some companies are developing hardware modeling platforms based on standard simulation and software design tools to take advantage of well established design flows and engineer's back experience. Following this approach, Xilinx Inc. has developed System Generator™ for DSP[3], a plug-in to the Simulink® environment from The Mathworks Inc. designed to develop high-performance DSP systems for FPGAs. With the System Generator blockset, designers can make bit- and cycle-accurate simulations and HDL code generation for Xilinx FPGAs.

Moreover, System Generator provides co-simulation interfaces that make it possible to incorporate an FPGA directly into a Simulink simulation. When

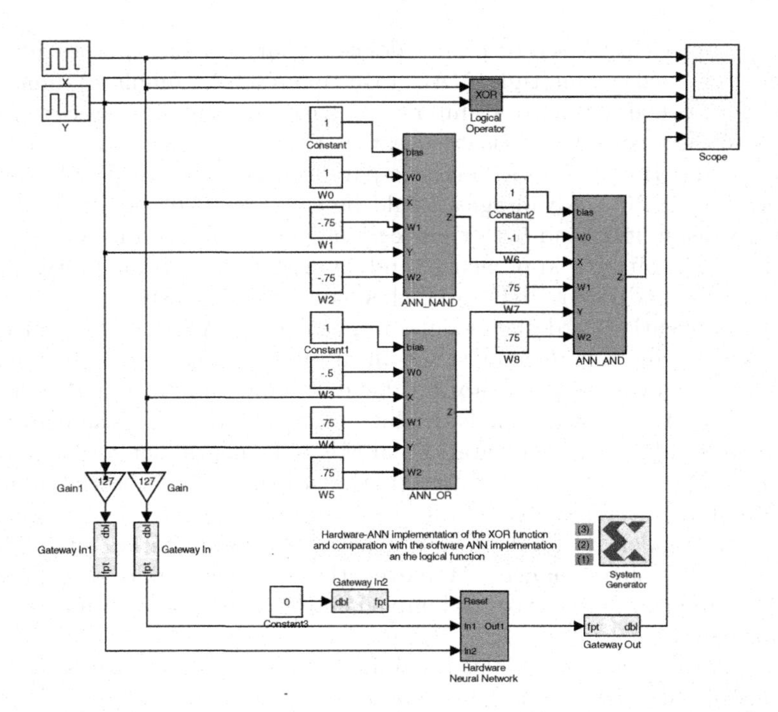

Fig. 1. Co-simulation of software and hardware models of an MLP implementing a simple XOR function

the design is simulated in Simulink, inputs to the hardware portion are sent by the communications interface to the real hardware, where the outputs are calculated and sent back to the Simulink environment. As advantages, the high-level schematic design flow, powerful Matlab visualization tools, the IP-based methodology and the hardware in the loop capability, are very effective in reducing design-time and risk.

With this workflow, successive refined software versions of an ANN can be developed with reduced additional effort, each one introducing more hardware-friendly characteristics, to observe the effect of latency, quantization, etc. Finally, the hardware version of the ANN can be developed and validated against the previous versions. As an example, Figure 1 demonstrate the implementation and co-simulation of three different versions of the XOR function: using the standard logical operator, a software (full-precision) ANN, and its hardware (quantized) version.

However, when using the System Generator blockset, the hardware model has to be defined at the Register Transfer Level (RTL), which requires certain knowledge of digital and logic design and the effort of generating a detailed description of the ANN algorithm. HANNA tools described in this paper are intended to help the designer in this task by automatically generating a synthesizable System Generator description starting from just a high level definition of the network. If the software version of the ANN has previously been trained and optimized, it takes minutes to have its hardware version downloaded and working on a FPGA prototyping board. Moreover, software and hardware outputs can be simultaneously verified with the same, unique, testbench.

3 Proposed ANN Hardware Architecture

In this section we expose the details of a hardware architecture well suited for ANN implementation over current high-end FPGA devices. The network model selected was the Multi-Layered Perceptron (MLP) and MLP-like networks, mainly for two reasons: on one hand it is one of the first and more studied ANN models proposed and by far the most used in industry; on the other hand, the model's mathematics are simple enough to make it attractive for hardware implementation when compared with more complex architectures.

Another remark is that HANNA inferences the models of trained MLPs, that is, the learning mechanism is not included in the hardware circuit. This limitation is coherent with our purpose of providing a tool to quickly prototype hardware versions of software ANNs, to guess whether it is practical or not to implement them over FPGAs when compared with their software counterparts. Moreover, faithful realizations of ANN learning algorithms are still too complex and usually consume more area than the ANN itself. Although several approaches have already been demonstrated, the development of FPGA-friendly learning algorithms is still an open subject [4].

3.1 Neuron Architecture

Each neuron in a MLP network calculates a sum of products (synaptic function) followed by a threshold nonlinearity (activation function). Its transfer function is

$$y = f(net) = f\left(\sum_{i=0}^{n} w_i x_i + b\right) \tag{1}$$

where n is the number of inputs, x_i are the inputs to the neuron, w_i are the inputs' weights and b is a bias term. The activation function f produces an excitatory or inhibitory signal at the output of the neuron, and usually takes the form of a step or sigmoid function.

This neuron model has been developed in VHDL using high-level behavioral descriptions, with parameterizable constructs that allow the designer to change the number of inputs, weights values and signals quantization. Each neuron receives its configuration when it is instantiated in the network by the HANNA tools. All operators are implemented using integer representation, as it has been demonstrated that it is possible to train ANN with integer weights, and it is still impractical to implement ANNs on FPGAs with floating-point precision weights [4].

A fully parallel implementation of each neuron including a multiplier for every input would be too costly in hardware resources, thus each neuron contains a

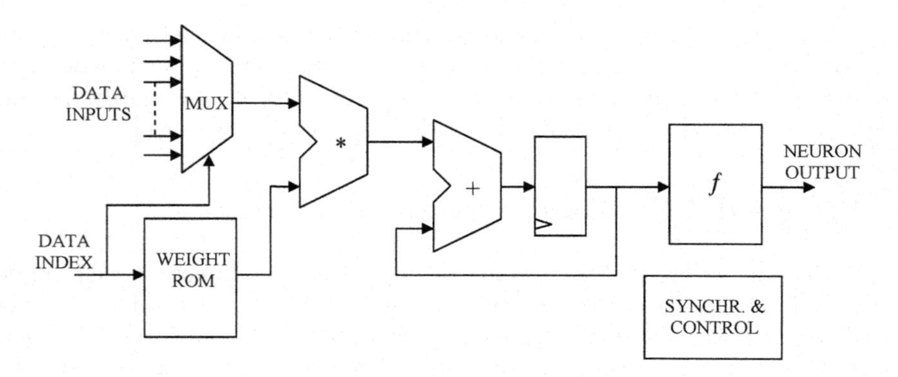

Fig. 2. Block diagram of the MPL neuron

unique MAC unit to calculate the synaptic function, as a trade-off between area and speed (Figure 2). When complex activation functions are selected (such as the sigmoid) they are implemented as a ROM table with the selected quantization. If a classical step function is selected a single comparator is inferred. In the first case, the ROM size can be reduced by one half if the function is symmetrical by implementing just the positive part of the function in the ROM and adding two comparators to determine the sign of the input and the output respectively. Of course, you can always select an option to manually edit the ROM table to implement any arbitrary function.

The MAC output is calculated using full-precision logic and then truncated to the appropriate bit number to be used as an input to the activation function. The operation of the previous components is controlled by a synchronization block that also updates the neuron output when the new data is ready for use.

3.2 Net Architecture

From the user viewpoint, after running HANNA a Simulink project file is created in which the specific neural network is just a new ANN Block (ANNB) that is ready to be connected to the testbench (input data generation blocks and verification blocks). The model generated includes also a synthesis block configured for the custom tools and hardware available for circuit implementation, as it is shown in the left side of Figure 3. Expanding the ANNB makes visible the net structure, formed by the Neuron Framework Blocks (NFBs) distributed in layers and the Network Controller (NCB) (Figure 3, middle side). Each NFB is composed of an input data preprocesing stage and the Neuron Block(NB) itself (Figure 3, right side).

Both the NCB and all NBs are the system primitives, and were defined using VHDL and highly-parameterizable RTL descriptions. The structure and behavior of the NBs was depicted in the previous section. The NCB's function is to control the parallel execution of the neurons and the pipeline synchronization. It is basically a state machine implemented in a ROM-fashioned style due to the

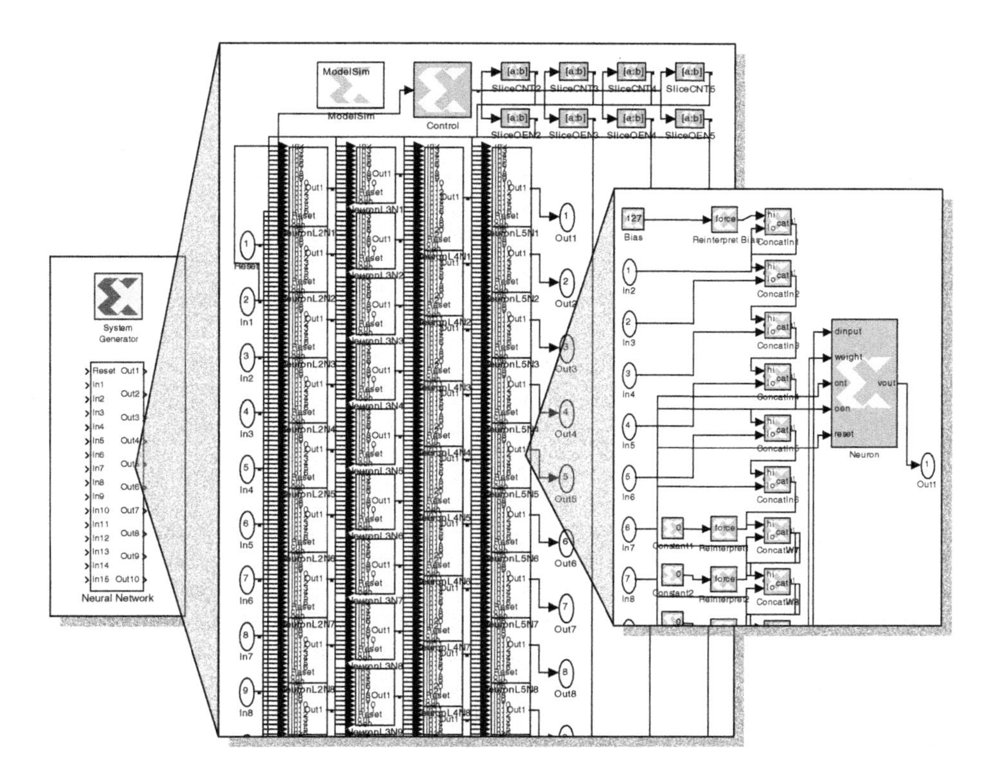

Fig. 3. Hierarchical view of a MLP model generated with HANNA (15 inputs, 10 outputs, 2 hidden layers, 10-20-15-10 neuron structure)

sequentiality of the control algorithm and its suitability for automated generation by HANNA scripts.

With the proposed architecture, the execution of a $n-$input neuron consumes $n+1$ clock cycles. Thus the latency of a $m-$layer network can be easily calculated as

$$L = \sum_{i=1}^{m} n_i + m \tag{2}$$

where n_i is the input number of neurons in the $i-$esim layer. The control ROM will then require a total of L words of $\sum_{i=1}^{m} \lceil \log(n_i) \rceil + m$ bits, which in practice results in insignificant ROM sizes compared to the network size.

4 Application Architecture

HANNA software is composed by a GUI (Figure 4), a set of matlab scripts and several complimentary files. The GUI is a front end tool for the network synthesis tools. It provides the same functionality than the scripts in a more convenient and easy to use interface, and it also shows the activation function

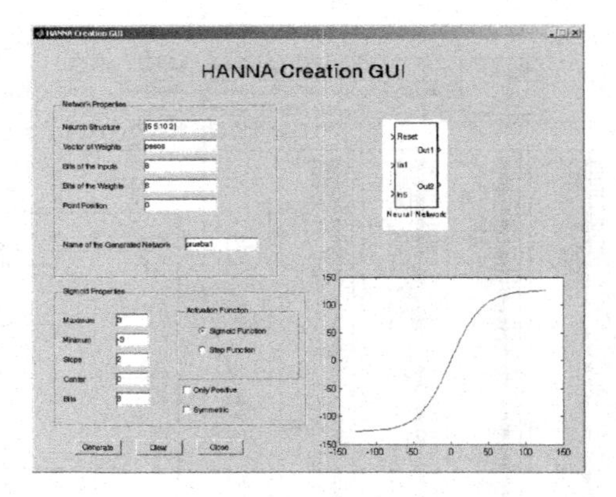

Fig. 4. HANNA graphics interface

selected for neurons. *Hanna.m* is the main script, which generates and connects the subsystems that compose the network. It uses two files, *control_config.m* and *neuron_wcte_config.m* that configure respectively the control system and the neuron parameters. VHDL code for control and neuron subsystems is generated by *hanna.m* in execution time, starting from proper templates, as their structure depends on the configuration selected by the user. Another file, template.m stores Simulink's primitive library blocks used to build the whole system.

After network generation, a Simulink model file is created which contains the network specification, and a second file with same name and sufix _vars is generated to store workspace and general configuration parameters.

5 Case Study: XOR Function

The first application seleccted to verify HANNA tools was the implementation of the XOR function. This is a simple problem to test the generation and performance of neurons with step-based activation functions. The network structure comprises two data inputs, one output and three neurons distributed in two layers with two and one neurons in the first and second layers respectively. The network structure and testbench were shown in Figure 1. The neurons in the first layer perform as a logical NAND and OR function respectively. The second layer neuron models a logical AND function. With these constrains, the corresponding weights and equations are:

$$\begin{aligned} Z_1(NAND) &= 1 - 0.75X - 0.75Y \\ Z_2(OR) &= -0.5 + 0.75X + 0.75Y \\ Z_3(AND) &= -1 + 0.75Z_1 + 0.75Z_2 \end{aligned}$$

where X and Y are the XOR input variables, and Z_i is the output of the $i-$esim neuron.

Table 1. Implementation and performance results obtained for the two benchmark applications

Parameter	XOR	Speaker Rec. 1	Speaker Rec. 2
I/O quantiz. (bits)	8	10	8
Weight quantiz. (bits)	8	10	8
Sigmoid quantiz. (bits)	8	10	8
Net structure[1]	[2,2,1]	[12,20,11]	[12,20,11]
FPGA Device (Xilinx)	s2s200e-6	xc2v1000-4	xc4vsx55-12
Clock rate (MHz)	46.762	37.115	60.22
Latency (clock cycles)	8	36	36
Throughput (Msa/s)	5.845	1.031	1.67
Slices (%)	8	82	2

The implementation results are summarized in Table 1. Eight bits were sufficient for input/output and activation function discretization. With these parameters, the network performed at 46.762 MHz over a SpartanII device from Xilinx, using just 192 slices (8% of the device area), and getting a throughput of 5.845 millions outputs per second.

6 Case Study: Speaker Recognition

In this application an ANN was trained to individually differentiate every member in a group of eleven -spanish- speakers. The identification process is based on the analysis of three vowels, *a*, *i* and *o*, that have the best clustering characteristics from a physiological point of view in the spanish language. In order to characterize spanish vowels starting from the vocal tract resonance frequencies, it is necessary to identify the first two formant frequencies (usually labeled F1 and F2), but better results have been reported if the first four formants are taken into account [5].

Therefore, in our case the ANN receives 12 data inputs corresponding with the four formants of the three vowels considered. The net was structured in two layers with 20 and 11 neurons respectively. The network was trained using a standard backpropagation algorithm and appropriated keywords so that only one of the second layer neurons will have a significant output for each speaker. The software model of the network was quantized to 10 bits for all weights, bias, inputs/outputs and sigmoid function. The backpropagation algorithm, implemented in floating point, was run over 300 iterations. To conclude, 11 test patterns (one for each speaker) were applied to the network that successfully identified all of them inverting 5.62 ms in the operation (Matlab software running on a 3.2GHz Pentium 4 with 2 GB RAM).

The hardware network model was generated by HANNA using, as in the software version, 10 bits for neurons I/O and weight quantization, and a 1024×1024 point sigmoidal activation function. We also generated a reduced version using just 8 bits. Both models were able to identify the 11 speakers without errors.

[1] Legend: [inputs, layer1_neurons, layer2_neurons, ...]

The 10 bits model was synthesized for Virtex-II part from Xilinx, while the 8 bit model targeted a Virtex-4 device, obtaining results shown in Table 1. In the first case, test patterns were evaluated in just 10.67 μs, implying that hardware model executed about 526 times faster than software for this application. In the second case, just 6.58 μs were spent, what means 854 times faster than software model. The 10 bit model consumes and 82% of the Virtex-2 part, while the smaller 8 bit model used just a 2% of the resources availabe on the -greater-Virtex-4 device.

7 Conclusions

In this work we have introduced HANNA, a tool for hardware prototyping of MLP-like ANNs over programmable devices. It has been demonstrated that the benefits offered by system level design tools, like the suite Matlab/Simulink/System Generator that allows targeting multiple implementation platforms, can be further extended with the HANNA tools in case of FPGA prototyping, by allowing a high level (neuron level) description of the ANN, instead of the standard lower-level RTL description of the network. The tool and our design methodology have been validated through two examples, resulting in hardware design times reduced to minutes and acceptable performance over current FPGA devices.

Future work involves increasing the capabilities of HANNA by including learning mechanisms, a microprocessor-based network controller for sophisticated execution schemes, and taking advantage of FPGAs dynamic reconfiguration features for on-line network evolution.

HANNA is available upon request for non-profit research institutions and organizations.

Acknowledgements

This research has been funded by the Spanish Government (MTyAS) under project number IMSERSO RETVIS 150/06.

References

1. A. F. Omondi, J. C. Rajapakse, "Neural Networks in FPGAs", Proc. of the 9th Int. Conf. on Neural Information Processing, Vol. 2, pp. 954–959, 2002.
2. D. Densmore, R. Passerone, "A Platform-Based Taxonomy for ESL Design", IEEE Design&Test of computers, Vol. 23, Issue 5, pp. 359–374, 2006.
3. Xilinx Inc., "Xilinx System Generator v7.1 User Guide", Xilinx Inc., www.xilinx.com, 2005.
4. J. Zhu, P. Sutton, "FPGA Implementations of Neural Networks - a Survey of a Decade of Progress", Proc. of 13*th* Int. Conf. on Field Programmable Logic and Applications, pp. 1062-1066, 2003.
5. J.L. Ramón, C. Sprekelsen, M.L. Marín, M. González-Ortín, M. Rubio, "Método de estudio de las frecuencias fundamentales y análisis espectrográfico de los dos primeros formantes de las vocales castellanas emitidas por cien sujetos normales", Acta Otorrinolaringológica Española, Vol. 30, pp. 399–414, 1986.

Improvement of ANNs Performance to Generate Fitting Surfaces for Analog CMOS Circuits

José Ángel Díaz-Madrid[1], Pedro Monsalve-Campillo[2], Juan Hinojosa[2], María Victoria Rodellar Biarge[3], and Ginés Doménech-Asensi[2]

[1] Fraunhofer Institute, Erlangen 91058, Germany
[2] Universidad Politcnica de Cartagena, Cartagena 30202, Spain
[3] Universidad Politcnica de Madrid, Campus de Montegancedo, Madrid 28660, Spain

Abstract. One of the typical applications of neural networks is based on their ability to generate fitting surfaces. However, for certain problems, error specifications are very restrictive, and so, the performance of these networks must be improved. This is the case of analog CMOS circuits, where models created must provide an accuracy which some times is difficult to achieve using classical techniques. In this paper we describe a modelling method for such circuits based on the combination of classical neural networks and electromagnetic techniques. This method improves the precision of the fitting surface generated by the neural network and keeps the training time within acceptable limits.

1 Introduction

Design of analog integrated circuits, require extensive simulations at different levels of analog design hierarchy. As we go deeper in details, down in the analog hierarchy, these simulations become more and more CPU time extensive, and so, the analog part of a typical mixed signal ASIC requires 90% of the design time while using only a small part of the silicon die.

In order to reduce the number of simulations at lower levels of the design hierarchy, the use of high level models with a good trade off between accuracy and simulation speed is becoming more and more important. With such models (also called macromodels) and the appropriate simulation tools, electronic designers can perform fast and accurate simulations which allow design parameters exploration in order to obtain optimal circuits for design specifications.

Macromodels are not new[1] and have been traditionally a way to override this problem and today extensive libraries of macromodels [2] exist for different applications. When these macromodels are properly fitted, they produce very accurate simulations, faster than those obtained with detailed device level SPICE like simulations. The parameters (gain, pole, input impedance, etc) of a macromodel can be obtained generating fitting surfaces from design parameters (device dimensions, biasing currents, etc).

There are several approaches to obtain parameters of macromodels from design parameters, like linear regression models [3], artificial neural networks [4],

J. Mira and J.R. Álvarez (Eds.): IWINAC 2007, Part II, LNCS 4528, pp. 19–27, 2007.

spline approximation [5], etc. One of the main drawbacks of these methods arises when the number of input design parameters increases. In this case, the error obtained increases also, producing inaccurate models which are not useful for macromodeling purposes. From the mentioned methods, those based on artificial neural networks are very precise and produce fitting surfaces with a remarkable precision. However their accuracy to fit a given surface needs to be improved when the number of input parameters increases, specially for analog CMOS modelisation problems.

One approach useful to solve this problem is to combine an inaccurate model of the circuit to be modelled with a classical neural network. This inaccurate model is very fast to compute and helps the neural network to reach a better solution for the fitting surface. So, the time needed to train the neural network is not considerably increased. This technique, which has been successfully addressed in RF/microwave devices modelling problems, is adapted here to the particular properties of CMOS circuits.

In this paper we use the mentioned techniques to improve efficiency of neural networks to generate fitting surfaces for CMOS circuits. In Sect. 2 we describe the modelling techniques. The inaccurate model of a CMOS transconductance amplifier is shown in Sect. 3. In Sect. 4, results obtained using these techniques are discussed and finally we conclude in Sect. 5.

2 Modelling Techniques

Fig. 1, shows the classical modelling approach using a neural network. In the classical approach using ANNs, the response (R_{ANN}) is such that:

$$R_{ANN}(x) \approx R_{FM}(x) \tag{1}$$

where x is the input parameter vector and R_{FM} is the fine model response obtained by device level SPICE type simulations. The response R_{ANN} is very

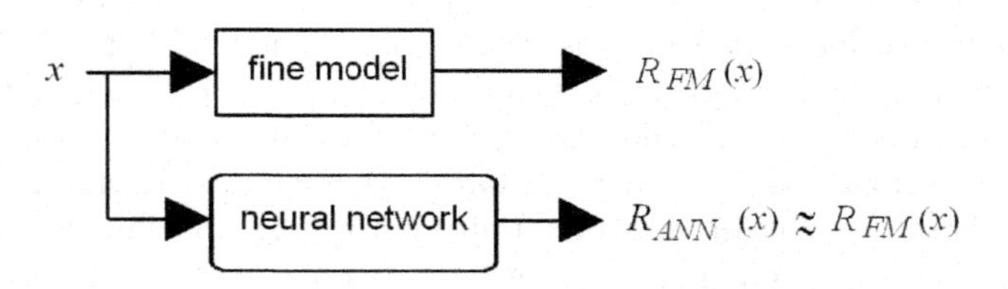

Fig. 1. Classical approach

precise for low number of parameters in x. However, when the size of this vector increases, R_{ANN} becomes more inaccurate making it unusable for CMOS modelling purposes. To solve this problem, alternative techniques must be used.

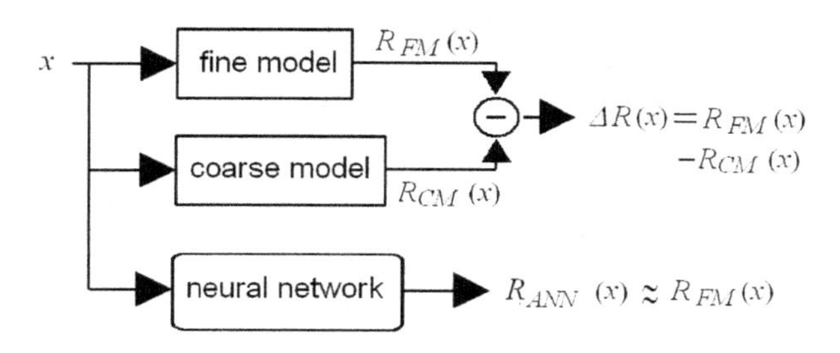

Fig. 2. EM approach

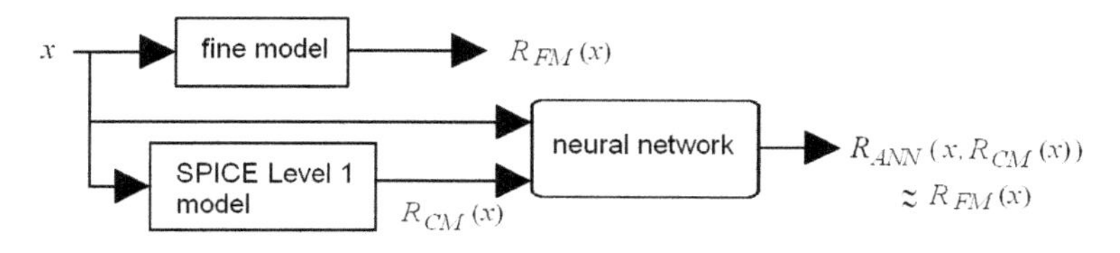

Fig. 3. PKI approach

Figs. 2 and 3 show the hybrid electromagnetic (EM) and the Prior Knowledge Input (PKI) modelling approaches.

In the EM method [6], the difference (ΔR) between the responses of both coarse and fine models is used to find the coefficients of the linear regression model (Fig. 2), reducing the number of fine model simulations or measurements due to a simpler input-output relationship:

$$\Delta R_(x) = R_{FM}(x) - R_{CM}(x) \tag{2}$$

Thus, the response of the EM model (R_{EM}) for a given input data is:

$$R_{EM}(x) = R_{CM}(x) + \Delta R(x) \approx R_{FM}(x) \tag{3}$$

In the PKI modelling approach [7], used for RF/Microwave devices, an initial model is required as coarse model (CM). For RF/microwave devices this initial model is an empirical model. For CMOS circuits we propose to use SPICE level 1 equations. In Fig. 3, vector x represents the design parameters, and R_{CM}, R_{FM} and R_{ANN} are the respective initial, fine and ANN model responses. R_{CM} is much faster to calculate, but less accurate than R_{FM}.

The CM output is used as input for the ANN model in addition to design parameters. The weights of the ANN are found such that its response is approximately equal to the fine model response. The aim of the PKI method is to find an appropriate response ($R_{PKI-ANN}$) such that:

$$R_{PKI}(x) = R_{ANN}(x, R_{CM}(x)) \approx R_{FM}(x) \tag{4}$$

3 Amplifier Coarse Model

The above modelling techniques have been applied to a fully differential telescopic transconductance amplifier (OTA) with gain enhancement [8] (Fig. 4). In this circuit, the aim is to obtain suitable models for gain, first pole (ω_0) and Slew-Rate (SR) from design parameters. A_1 and A_2 are folded cascade differential amplifiers and the common mode feedback block (CMFB) is a 40 MHz switched circuitry used to keep the common mode voltage at a given value. Technology used to develop this amplifier has been a CMOS 0.35μm from AustriaMicroSystem.

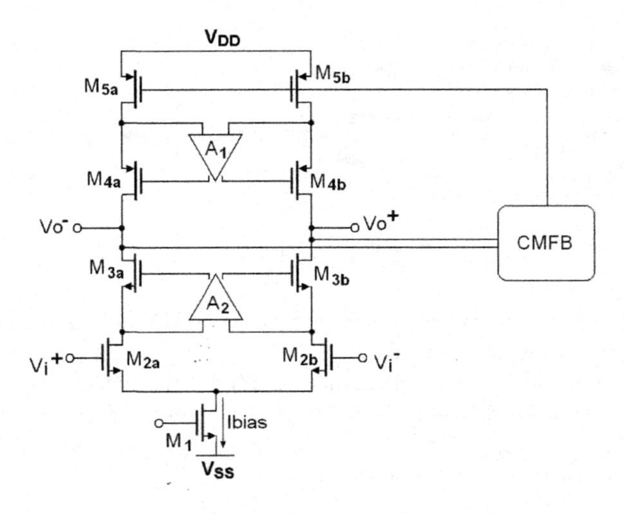

Fig. 4. Structure of the OTA

For this circuit, the coarse model was developed using SPICE level 1 equations, ignoring bulk effect, to obtain the gain and first pole:

$$gain = g_{m3} r_{out} \tag{5}$$

$$\omega_0 = \frac{1}{rout(C_L + C_a)} \tag{6}$$

$$r_{out} = g_{m3} r_{o2} r_{o3} A_1 || g m_4 r_{o4} r_{o5} A_2 \tag{7}$$

where g_{mi} , and r_{oi} are the small signal transconductance and output resistance of the MOS transistor, and C_L and C_a are, respectively, the load capacitor and the sum of internal MOS capacitances. The values of g_{mi} and r_{oi} are obtained through:

$$g_{mi} = \sqrt{2KI_{D_i}\frac{W_i}{L_i}} \tag{8}$$

$$r_{oi} = \frac{1}{\lambda I_{D_i}} \tag{9}$$

In these equations W_i and L_i are, respectively, the channel width and length of the i MOS transistor, I_{D_i} is the drain current, and K and λ are technology dependent parameters. For the Slew-Rate we have taken as coarse model the classical approximation given by:

$$SR = \frac{I_{bias}}{C_L} \quad (10)$$

From Fig. 4 is easy to see that I_D for transistor 2 to 5 is equal to $I_{bias}/2$. As K and λ are constant values for a given CMOS technology, the computation of (5) to (10) for given values of W and L is immediate.

Fig. 5 shows the macromodel used to model the fully differential OTA. The macromodel has two inputs, corresponding to V_i^+ and V_i^-, and two differential outputs. It includes also the possibility to define input and output common mode voltages through terminals V_{ICM} and V_{OCM} respectively.

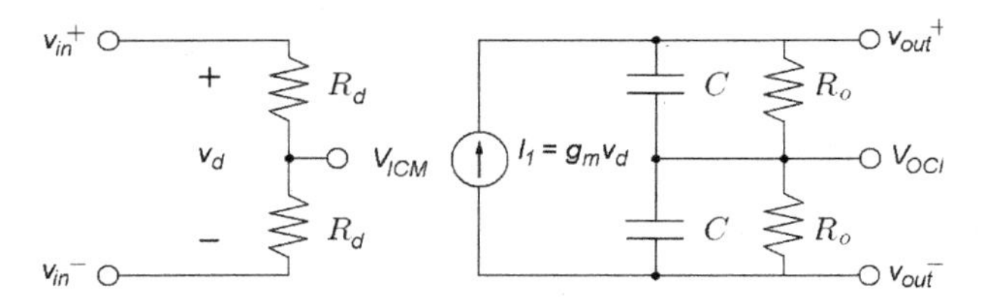

Fig. 5. Macromodel of the amplifier

In this macromodel, a single pole approximation, defined by R_o and C has been taken. The current source I_1 models the OTA gain and also the SR, limiting its maximum value to I_{bias}. The combination of SR and I_{bias} sets the value of the capacitor C. The output resistance R_o is computed from the pole value (ω_0) and the capacitor.

4 Results

ANN modelling techniques were applied to the OTA (Fig. 4) for the region of interest shown in Tab. 1 ($w_4 = 2.2w_3$, $l_4 = l_3$, and $w_5 = 75\mu m$, $l5 = 0.6\mu m$, $A_1 = A_2 = 40dB$, $C_L = 1.5pF$ fixed). Coarse and fine models data were provided, respectively, from SPICE level 1 equations and device level simulations using Cadence IC Package (BSIM3v3 models). Half of data exploring the whole of the region of interest (Tab. 1) were used for training and the other half for testing.

The ANN used has been a three-layer perceptron [9]. It has been implemented on Matlab and trained with Levenberg-Marquardt's learning algorithm [10] during 500 iterations. We have used 5 neurons for the input layer (6 for the PKI method) 30 neurons for the hidden layer and 3 neurons for the output layer. The inputs values to the neural network have been normalised.

Table 1. Region of interest for the transconductance amplifier

Parameters	Minimum value	Maximum value	Step
W_2	$400\mu m$	$2000\mu m$	$160\mu m$
L_2	$0.5\mu m$	$1.0\mu m$	$0.1\mu m$
W_3	$70\mu m$	$120\mu m$	$10\mu m$
L_3	$0.4\mu m$	$1.0\mu m$	$0.1\mu m$
I_{bias}	$500\mu A$	$1000\mu A$	$100\mu A$

Table 2. Relative errors between neural networks and fine model

	Gain		ω_0		SR	
	Mean	Max	Mean	Max	Mean	Max
ANN	0.1572	1.9075	0.2916	1.1018	0.0597	0.4679
EM-ANN	2.5284	4.1736	0.3351	1.2909	0.0423	0.3501
PKI-ANN	0.0902	0.8429	0.2240	0.9146	0.0450	0.3640

The mean and maximum relative errors over classical ANN, EM-ANN and PKI-ANN models for desired outputs (A_{DC}, ω_0 and SR) with respect to test data are shown in Tab. 2.

As we can see, the best global results are achieved using the PKI-ANN technique. For the gain value, the maximum error has been divided by two with respect to the classical ANN approximation, while the mean error has been reduced at 42%. For the ω_0 value, , the maximum error has been slightly reduced, but the mean error has clearly reduced (23% of improvement). Finally, the Slew-Rate mean error value has been improved as well as the error maximum value (about a 20% of reduction). For the Slew-Rate, however, the EM-ANN technique offers a better accuracy. On the other hand, the results obtained using the EM-ANN approach are not good at all for the gain and ω_0. So, for the transconductance amplifier we suggest to use the PKI-ANN technique to model both the gain and ω_0 and the EM-ANN technique to model the Slew-Rate.

We apply now these techniques to build the macromodel of Fig. 5. ω_0 and gain are those obtained with the PKI modelling technique for $w_2 = 1000\, mum$, $l_2 = 0.6\mu m$, $w_3 = 75\mu m$, $l_3 = 0.4\mu m$ and $I_{bias} = 450\mu A$, values not used in the training process. The output resistance R_o is computed from the pole ω_0 and the external capacitor C_L. The differential input resistance R_d is set to 1 $M\Omega$. The input and output common mode voltages have been set to $VICM = 1.35\,V$ and $VOCM = 1.8\,V$. The value of SR has been obtained for these same values using the EM modelling technique.

Figs. 6 and 7 show the open loop AC and closed loop transient response, respectively, of SPICE device level and macromodel simulations of the OTA.

As it can be seen in Fig. 6, AC response fits perfectly for the one pole simplification taken. At high frequencies, there are higher order effects which have been not modelled. Transient response (Fig. 7) exhibits small differences due to

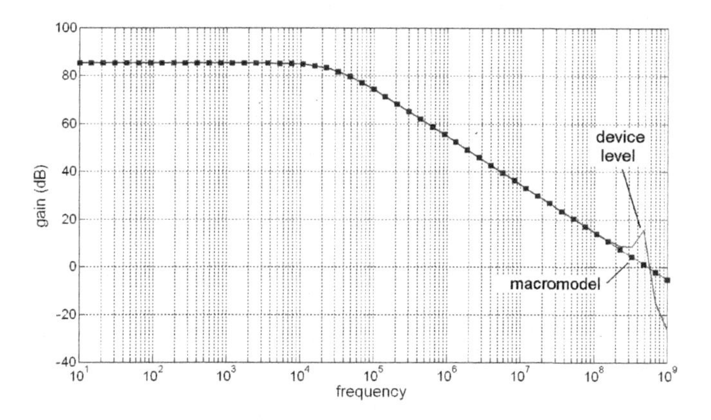

Fig. 6. AC analysis response

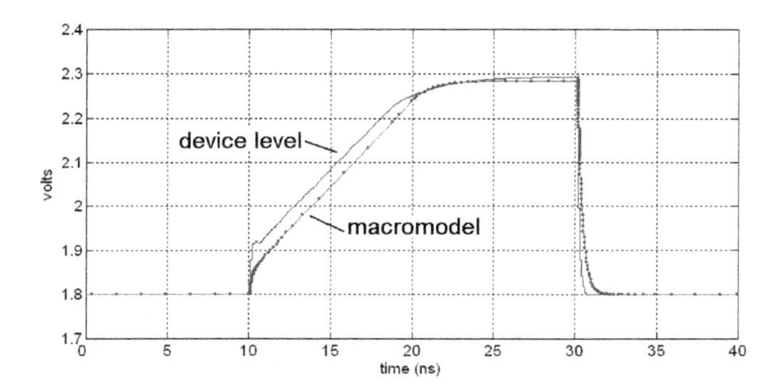

Fig. 7. transient analysis response

second order effects not included in the macromodel. However the rising slopes of both signals, due to SR effect, are similar. Device level AC and transient simulations took 150 ms and 7.34 s respectively while macromodel simulations took 10 ms and 700 ms, on a dual core twin 64-bit-processor computer running Suse Linux 9.1.

Although these differences are not remarkable in terms of absolute vales, there is a reduction of time of one order of magnitude using the macromodel. For a typical mixed signal ASIC, like a 12 bit pipeline analog to digital converter, a device level transient simulation of $3\mu s$ can take between 21 and 24 hours, while only 2.5 hours using a macromodel.

5 Conclusion

In this paper we have presented a modelling approach for analog CMOS circuits adding to the classical neural network approaches, the use coarse models, in order to improve the accuracy of those. This technique is used in RF/Microwave circuits in combination with empirical models, which have been substituted with SPICE level 1 equation for CMOS circuit applications. It has been successfully applied to model some parameters of a telescopic OTA. The main advantage of these modelling approaches is that they improve the accuracy of a classical neural network, keeping the simulation time within acceptable values.

From the results obtained, the PKI technique was the more precise for two of the parameters modelled, while the EM technique was the more accurate for the third parameter. This behaviour depends on the linearity and complexity of the fitting surface generated.

In order to test the validity of this approach, we have used a simple macromodel to include these parameters and we have performed AC and transient analysis on it. As expected, we have decreased the simulation time, which is now one tenth of the time expended using low level SPICE models.

Acknowledgment

This work has been partially supported by Fundación Séneca of Región de Murcia under grant 03094/PI/05 and Spanish Ministry of Education and Science (MEC) and the European Regional Development Fund of the European Commission (FEDER) under grant TIN2006-15460-C04-04

References

1. Boyle, G. R., Cohn, B. M., Pederson, D. O, Solomon, J. E.: Macromodeling of Operational Integrated Circuit Amplifiers. IEEE Journal of Solid–State Circuits **9** (1974) 353–364.
2. Wei, Y., Doboli, A.: Library of structural cell macromodels for design of continuous-time reconfigurable Delta-Sigma modulators. IEEE Intl. Symp. on Circuits and Systems (2006) 1764–1767.
3. Doménech-Asensi, G., Hinojosa, J., Martínez-Alajarín, J., Garrigós-Guerrero, J.: Empirical Model Generation Techniques for Planar Microwave Components using Electromagnetic Linear Regression Models. IEEE Trans. Microwave Theory and Tech. **53** (2005) 3305–3311.
4. Vancorenland, P., der Plas, G. V., Steyaert, M., Gielen, G., Sansen, W.: A Layout–aware Synthesis Methodology for RF Circuits. Proc. IEEE/ACM International Conference on Computer Aided Design (2001) 358–362.
5. McConaghy, T., Gielen, G.: Analysis of Simulation-Driven Numerical Performance Modeling Techniques for Application to Analog Circuit Optimization. IEEE Intl. Symp. on Circuits and Systems (2005) 1298–1301.
6. Watson, P. M., Gupta, K. C.: EM-ANN models for microstrip vias and interconnects in multilayer circuits. IEEE Trans. Microwave Theory Tech., **44**, (1996) 2495–2503.
7. Watson, P. M., Gupta, K. C., Mahajan, R. L.: Development of knowledge based artificial neural network models for microwave components. IEEE Int. Microwave Symp. Digest, (1998) 9–12.

8. Bult, K., Geelen, G.J.G.M.: A Fast-Settling CMOS Op Amp for SC Circuits with 90-dB DC Gain. IEEE Journal of Solid-State Circuits, **25**, (1998) 1379–1384.
9. Rumelhart, D. E., Inton, G. E., Williams, R. J.: Learning internal representations by error propagation. D. E. Rumelhart and J. L. McClelland, Eds. Parallel Data Processing, 1, Cambridge, MA: The MIT Press, (1986) 318–362
10. Hagan, M. T., Menhaj, M.: Training feedforward networks with the Marquardt algorithm. IEEE Trans. On Neural Networks, **5**, (1994) 989–993

Wavelet Network for Nonlinearities Reduction in Multicarrier Systems

Nibaldo Rodriguez, Claudio Cubillos, and Orlando Duran

Pontifical Catholic University of Valparaiso
Av. Brasil 2241, Chile
{nibaldo.rodriguez,claudio.cubillos,orlando.duran}@ucv.cl

Abstract. In this paper, we propose a wavelet neural network suitable for reducing nonlinear distortion introduced by a traveling wave tube amplifier (TWTA) over multicarrier systems. Parameters of the proposed network are identified using an hybrid training algorithm, which adapts the linear output parameters using the least square algorithm and the nonlinear parameters of the hidden nodes are trained using the gradient descent algorithm. Computer simulation results confirm that the proposed wavelet network achieves a bit error rate performance very close to the ideal case of linear amplification.

Keywords: Wavelet network, linearizing, multicarrier.

1 Introduction

Multicarrier systems (MC) based on orthogonal frequency division multiplexing (OFDM) have gained tremendous popularity thanks to their high spectral efficiency. However, OFDM shows a great sensibility to the nonlinear distortion introduced by the traveling wave tube amplifier, due to fluctuations of its non-constant envelope. Typically, a TWTA is modulated by nonlineal amplitude modulation to amplitude modulation (AM-AM conversion) and amplitude modulation to phase modulation (AM-PM conversion) functions in either polar or quadrature form [1]. To reduce both AM-AM and AM-PM distortions, it is necessary to operate the TWTA with a large power back off level. However, such amplification schemes posses low power efficiency since maximum power efficiency is only attained when the TWTA is operated near its saturation point. Hence, TWTA linearizing techniques are often necessary to achieve a trade-off between linear amplification and high power efficiency. In order to achieve the two key factors, several linearizing schemes for TWTA has been proposed in the literature. According to recent research in the field, artificial neural networks [2]-[6] and neuro-fuzzy systems [7]-[10] have been successfully used for linearizing schemes, since they can approximate any continuous functions arbitrarily well. In most existing models, complex input-output measured signals are initially converted to either a polar or rectangular representation and then two separate and uncouple real-valued models are used to estimate the output amplitude and phase as a function of the input power amplitude. The real parameters of the two

J. Mira and J.R. Álvarez (Eds.): IWINAC 2007, Part II, LNCS 4528, pp. 28–36, 2007.

models were obtained during a training procedure based on back-propagation algorithm. Therefore, the disadvantage of these linearizing techniques is their slow convergence speed and elevated requirements of computing resources.

Inspired by success of neural networks in various engineering areas, a linearizing scheme based on wavelet networks (WN) is introduced in the present study in order to create a transmitter operating with only one real-valued neural network instead of the two separate networks commonly used to achieve linear amplification and high spectral efficiency in OFDM based MC systems. WN was introduced in [11],[12] as an idea of combining wavelet theory with the adaptive learning scheme of feedforward neural networks. In this paper, WN has a feedforward structure consisting of two layers, a nonlinear hidden layer with wavelet activation function and a linear output layer. In order to optimize all of the WN parameters an hybrid training algorithm is used, which combines the gradient descent (GD) based search for the nonlinear parameters of the hidden wavelet nodes and the least squares (LS) estimation of the linear output parameters. In addition, the hybrid algorithm (LS-GD) is chosen here due to its fast convergence speed.

The remainder of this paper is organized as follows: In section 2, a brief description of the baseband transmission system is presented. The linearizing technique of the TWTA and an hybrid training algorithm for adjusting the WN parameters are presented in Section 3. The performance curves of the constellation warping effect and bit error rate (BER) of 16QAM-OFDM and 64QAM-OFDM signals are discussed in Section 4. Finally, the conclusions are drawn in the last section.

2 System Model

A simplified block diagram of the baseband-equivalent OFDM system is shown in Fig. 1. The M-ary quadrature amplitude modulation (MQAM) signals generator produces complex symbols q and after converts the MQAM stream into a block of N_c symbols. Through the inverse fast Fourier transform (IFFT) block, the baseband OFDM signal is generated as

$$x(n) = \sum_{k=0}^{N_c-1} q_k \exp\Big(j\frac{2\pi nk}{N_c}\Big) \qquad n = 0, 1, \ldots, N_c - 1 \tag{1}$$

where N_c is the number of subcarrier.

The modulated signal $x(n)$ is first passed through the wavelet linearizer (WL) and after nonlinearly amplified. The amplified signal $z(n)$ is propagated over an additive white Gaussian noise (AWGN) channel. In this paper, the amplified signal is represented as

$$z(n) = A(|y(n)|) \exp\{j\left[\theta_y + P(|y(n)|)\right]\} \tag{2}$$

where $|y(t)|$ and θ_y correspond respectively to the amplitude and phase of the linearized complex signal $y(t)$.

The functions $A(\cdot)$ and $P(\cdot)$ denote AM-AM conversion and AM-PM conversion; respectively. For a TWTA, the expresions for $A(\cdot)$ and $P(\cdot)$ are given by Saleh as [1]

$$A(|y(n)|) = \frac{\alpha_A |y(n)|}{1 + \beta_A |y(n)|^2} \tag{3}$$

$$P(|y(n)|) = \frac{\alpha_P |y(n)|^2}{1 + \beta_P |y(n)|^2} \tag{4}$$

with $\alpha_A = 2, \beta_A = 1, \alpha_P = \pi/3$ and $\beta_P = 1$.

The nonlinear distortion of a high power amplifier depends on the back off. The input back off (IBO) is defined as the ratio of the saturation input power, where the output power begins to saturate, to the average input power.

$$IBO(dB) = 10 \log_{10}\Big(\frac{P_{i,sat}}{P_{i,avg}}\Big) \tag{5}$$

where $P_{i,sat}$ is the saturation input power and $P_{i,avg}$ is the average power at the input of the TWTA.

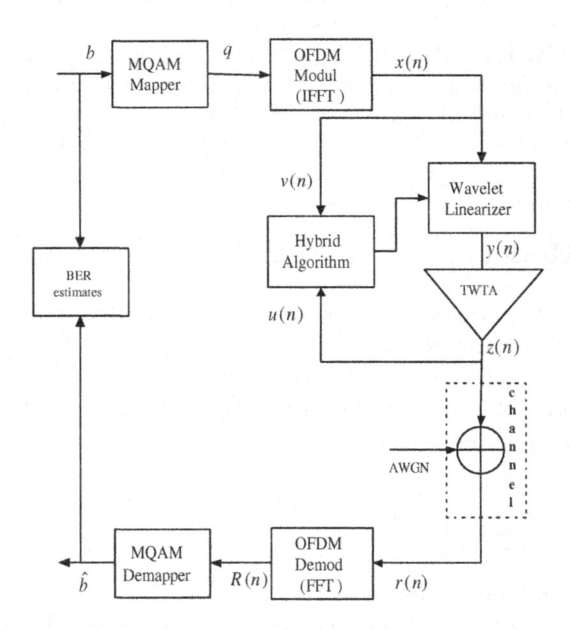

Fig. 1. Baseband OFDM transmission system

At time n, each fast Fourier transform (FFT) output at the receiver side is obtained as

$$R(n) = \sum_{k=0}^{N_c - 1} r_k \exp\Big(-j\frac{2\pi nk}{N_c}\Big) \qquad n = 0, 1, \ldots, N_c - 1 \tag{6}$$

where $r_k = z(k) + \zeta(k)$ and $\zeta(k)$ is the complex AWGN channel with two-sided spectral density.

The signal $R(n)$ is fed to the MQAM demapper device, which splits the complex symbols into quadrature and in-phase components; and puts them into a decision device, where they are demodulated independently against their respective decision boundaries. Finally, output bits stream $\hat{b}$ are estimated.

3 Wavelet Linearizing Scheme

A linearizer works by creating OFDM signal distortion that is the complement of the distortion inherent to the TWTA and its output signal is obtained as

$$y(n) = M(|x(n)|) \exp\{j [\theta_x + N(|x(n)|)]\} \tag{7}$$

where the functions $M(\cdot)$ and $N(\cdot)$ are used to invert the nonlinearities introduced by the TWTA.

Now, using (2) and (7) we obtain a complex signal envelope at the TWTA's output as

$$z(n) = A[M(|x(n)|)] \exp\left\{j \left[\theta_x + N(|x(n)|) + P[M(|x(n)|)]\right]\right\} \tag{8}$$

In order to achieve the ideal linearizing function, the signal $z(n)$ will be equivalent to the input signal $x(n)$. That is:

$$A[M(|x(n)|)] = \alpha |x(n)| \tag{9a}$$

$$N(|x(n)|) = -P[M(|x(n)|)] \tag{9b}$$

where $\alpha |x(n)|$ is the desired linear model. In this paper, desired linear gain was set to $\alpha = 1$, so that saturation power was reached at 0 dB. We therefore write ideal linearizing function as

$$y(n) = \frac{x(n)}{|x(n)|} A^{-1}[|x(n)|] \exp\{jP(A^{-1}[|x(n)|])\} \tag{10}$$

where $A^{-1}[\cdot]$ represent inverse AM-AM function of the TWTA.

Finally, in order to achieve (10) , it is necessary only to find the real-valued function $A^{-1}[\cdot]$, which can be approximated by using wavelet neural network and a finite number of samples of the AM-AM function.

3.1 Hybrid Training Algorithm

The linearizer's output signal $y(t)$ is approximated using a wavelet network and is obtained as

$$\hat{y} = Wh \tag{11}$$

with

$$W = [w_0 \; w_1 \; \ldots \; w_c \; w_{c+1}] \tag{12a}$$

$$h = [1 \; \phi_1(u) \; \ldots \; \phi_c(u) \; u] \tag{12b}$$

where w_j are the linear output parameters, c is the number of hidden wavelet nodes, u is the input value and $\phi_j(u)$ are hidden wavelet functions, which is derived from its mother wavelet ϕ through the following relation

$$\phi_j(u) = \frac{1}{\sqrt{d_j}}\phi\Big(\frac{u-t_j}{d_j}\Big) \tag{13}$$

where t_j and d_j are the translation and dilatation parameters of the wavelet functions; respectively. In the present paper, we choose the Mexican Hat wavelet function as a mother wavelet, which is defined as [11]-[12]

$$\phi(\lambda) = (1-\lambda^2)\exp\{-0.5\lambda^2\} \tag{14}$$

In order to estimate the linear parameters $\{w_j\}$ and nonlinear parameters $\{t_j, d_j\}$ of the wavelet linearizer an hybrid training algorithm is proposed, which is based on least square (LS) and gradient descent (GD) algorithms. The LS algorithm is used to estimate the parameters $\{w_j\}$ and the GD algorithm is used to adapts the nonlinear parameters $\{t_j, d_j\}$.

Now suppose a set of training input-output samples, denoted as $\{u_i, v_i, i = 1, \ldots, N_s)$. Then we can perform N_s equations of the form of (11) as follows

$$\hat{Y} = W\Phi \tag{15}$$

with the matrix Φ defined by

$$\Phi = \begin{pmatrix} 1 & \cdots & 1 \\ \phi_1(u_1) & \cdots & \phi_1(u_{N_s}) \\ \vdots & \vdots & \vdots \\ \phi_c(u_1) & \cdots & \phi_c(u_{N_s}) \\ u_1 & \cdots & u_{N_s} \end{pmatrix} \tag{16}$$

where the desired output v_i and input data u_i are obtained as

$$v_i = \frac{|x(n)|}{max\{|x(n)|\}}IBO \tag{17a}$$

$$u_i = |z(n)| \tag{17b}$$

For any given representation of the nonlinear parameters $\{t_j, d_j\}$, the optimal values of the linear parameters $\{\hat{w}_j\}$ are obtained using the LS algorithm as follows

$$\hat{W} = \Phi^{\dagger}V \tag{18}$$

where $V = [v_1\ v_2\ \cdots\ v_{N_s}]$ is the desired output and $\Phi^{\dagger}$ is the Moore-Penrose generalized inverse [13] of the wavelet function output matrix Φ.

Once linear parameters $\hat{W}$ are obtained, the gradient descent algorithm adapts the nonlinear parameters of the hidden wavelet functions minimizing mean square error, which is defined as

$$E(t,d) = \frac{1}{2}\left\|V - \hat{Y}\right\|^2 \tag{19a}$$

$$\hat{Y} = \hat{W}H \tag{19b}$$

Finally, the GD algorithm adapts the parameter $\{t_j, d_j\}$ according to the following equations

$$t_j = t_j + \mu(V - \hat{Y})\hat{w}_j \frac{\partial H}{\partial t_j} \qquad j = 1, \ldots, c \tag{20a}$$

$$d_j = d_j + \mu(V - \hat{Y})\hat{w}_j \frac{\partial H}{\partial d_j} \qquad j = 1, \ldots, c \tag{20b}$$

where μ is the step size of the GD algorithm.

4 Simulations Results

In this section, the performance evaluation of the wavelet linearizing scheme is presented. The parameters of the wavelet linearizer were estimated during the training process using two 16QAM-OFDM symbols with 64 subcarrier and the TWTA was operated at $IBO = 0$ dB. The wavelet linearizer was configured with one input node, one linear output node, two hidden wavelet nodes and two bias units. In the hybrid training process the initial nonlinear parameters $\{t_j, d_j\}$ were initialized by a Gaussian random process with a normal distribution $N(0, 1)$ and the step size was set to $\mu = 0.1$. Training was run with only 2 epoch and the normalized mean square error after convergence was approximately equal to -50 dB. In decision-direct mode, the wavelet linearizer is simply a copy of the wavelet network obtained during the hybrid training process.

To demonstrate the performance of the proposed wavelet linearizing scheme, we evaluated the constellation warping effect and the BER versus signal to noise (Eb/No) using 100 Monte Carlo run for an input data stream of length equal to 10,000 bits and the input back-off was set at $IBO = 0$dB for TWTA combined with wavelet linearizing (WLTWTA). Moreover, we also shows the performance for system without linearizing and system with ideal (linear) channel. The BER performance of the OFDM symbols without TWTA in the AWGN channel is very similar to the corresponding MQAM system and it is used here for benchmarking the performance of the MQAM-OFDM system with 64 carrier.

The effects of nonlinearities on the received 16QAM-OFDM and 64QAM-OFDM constellations in the absence of the AWGN channel are shown in Fig. 2 and 3, which correspond to the TWTA without and with linearizing operated at an input back off level of -9 dB, -14 dB; respectively. According to Fig. 2 and 3, it is observed that square 16QAM and 64QAM constellation are severely distorted by the nonlinear AM-AM and AM-PM characteristics of the TWTA without linearizing. This distortion is interpreted as in-band noise, and it is called constellation warping effect. From it can be seen that the proposed linearizing scheme significantly reduces the constellation warping effect on the received 16QAM-OFDM and 64QAM-OFDM symbols.

Fig. 4 shows the BER performance for 16QAM-OFDM with and without WL in presence of a TWTA. From Fig. 4 it can be seen that at $Eb/No = 16$ dB is achieved a $BER = -10$ dB when the TWTA without linearizing is operated at $IBO = -9$ dB and the same BER is achieved at $Eb/No = 3$ dB when

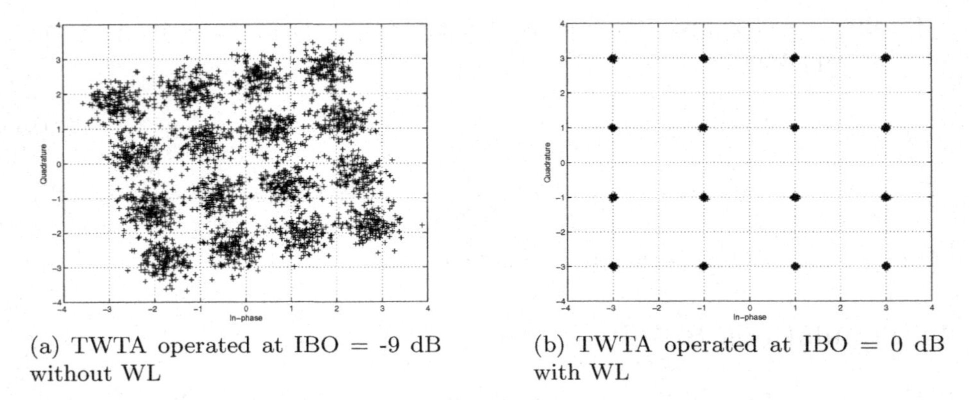

(a) TWTA operated at IBO = -9 dB without WL

(b) TWTA operated at IBO = 0 dB with WL

Fig. 2. Received 16QAM-OFDM Constellation in absence AWGN channel

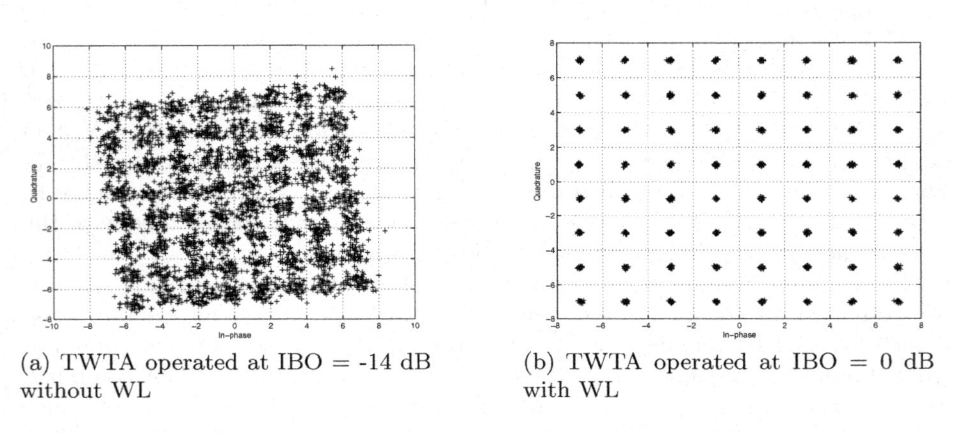

(a) TWTA operated at IBO = -14 dB without WL

(b) TWTA operated at IBO = 0 dB with WL

Fig. 3. Received 64QAM-OFDM Constellation in absence AWGN channel

the TWTA with wavelet linearizing (WLTWTA) is operated at $IBO = 0$ dB. Thus, the BER performance is very poor due to nonlinearities of the TWTA. Therefore, at $BER = -10$ dB the Eb/No reduction is 13 dB $((16-1)-(3-1))$ with the proposed linearizing scheme. In addition, also the linearizer achieves an $BER = -60$ dB at $Eb/No = 14.5$ dB, which is favorably compared to the BER corresponding to the linear amplification ideal case.

Fig. 5 shows the BER performance of 64QAM-OFDM signals for the TWTA without linearizing operated at $IBO = -14$ dB. From it can be seen that the Eb/No degradation at $BER = -10$ dB without linearizing is 17 dB (20-3), and with linearizing is obtained an improvement of Eb/No degradation equal to 13 dB $(17-(7-3))$. Furthermore, we can see that the BER performance is significantly improved with the linearizing scheme when is compared with the linear case, since at $BER = -60$ dB the Eb/No degradation is 0.25 dB $(19-18.75)$.

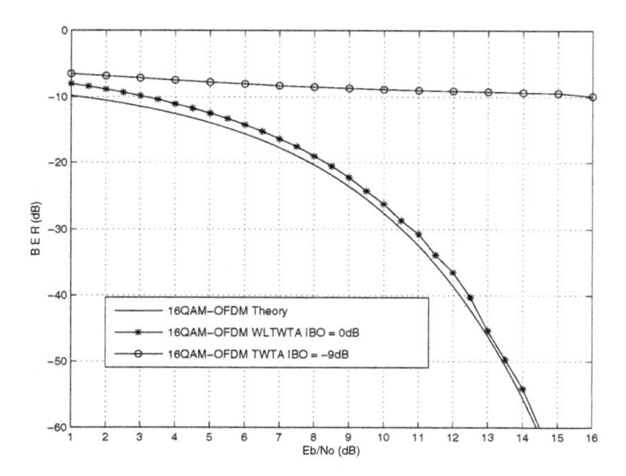

Fig. 4. BER performance of 16QAM-OFDM over a nonlinear AWGN channel

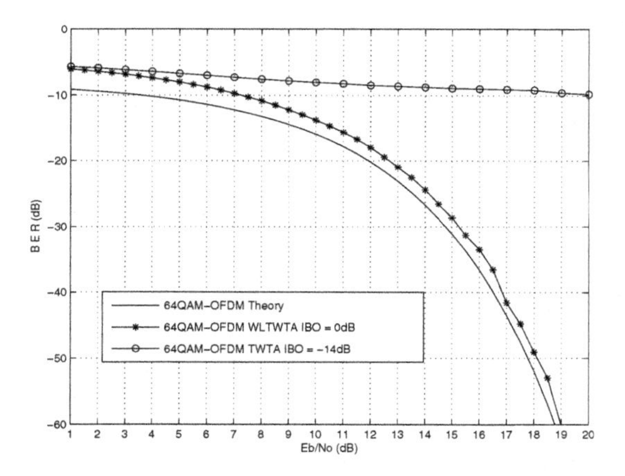

Fig. 5. BER performance of 64QAM-OFDM over a nonlinear AWGN channel

Finally, the proposed wavelet linearizing scheme achieves an improvement at the IBO level equal to 9 dB and 14 dB for 16QAM-OFDM and 64QAM-OFDM; respectively.

5 Conclusions

An adaptive baseband linearizing scheme based on a wavelet neural network to linearize a TWTA has been presented in this paper. The proposed linearizer uses only one real-valued wavelet network with 4 nonlinear parameters and 4 linear parameters to compensate nonlinear distortion in 16QAM-OFDM and 64QAM-OFDM systems. The linearizer coefficients adaptation was found by using 2 iterations of the (LS-GD) algorithm. Simulation results have shown that the proposed linearizing scheme offers significantly constellation warping effect and BER reduction. Moreover, linearizer achieves a BER very close with the BER corresponding to the ideal case of linear amplification.

Acknowledgments. This work has been partially supported by the Vicerectoria of the Catholic University of Valparaiso, .

References

1. A. M. Saleh: Frecuency-Independent and Frecuency-Dependent nolinear models TWT amplifiers, *IEEE Trans. Comm.*, vol. COM-29, pages. 1715-1719, November 1981.
2. B.E. Watkins and R. North: Predistortion of nonlinear amplifier using neural networks, *Proc. IEEE Military Comm. Conf.*, vol.1, pages 316-320, 1996.
3. M. Ibnkahla, J. Sombrin J., F. Castani and N.J. Bershad: Neural network for modeling non-linear memoryless communications channels, *IEEE Trans. Comm.*, vol. 45, no. 5,pages 768-771, July 1997.
4. M. Ibnkahla: Neural network modelling predistortion technique for digital satellite communications, *Proc. IEEE ICASSP*, vol.6, pages 3506-3509, 2000.
5. M. Ibnkahla: Natural gradient learning neural networks for adaptive inversion of Hammerstein systems, *IEEE Signal Processing Letters*, pages 315-317, October 2002.
6. F. Abdulkader, Langket, D. Roviras and F. Castanie: Natural gradient algorithm for neural networks applied to non-linear high power amplifiers, *Int. Journal of Adaptive Control and Signal Processing*, vol. 16, pages 557-576, 2002.
7. Ying Li,Po-Hsun Yang: Data predistortion with adaptive fuzzy systems, *IEEE Int. Conf. Syst., Man, and Cybern.*, vol. 6, pages 168-172, 1999.
8. D. Hong-min., H. Song-bai and Y. Jue-bang: An adaptive predistorter using modified neural networks combined with a fuzzy controller for nonlinear power amplifiers, *Int. Journal of RF and Microwave Computer-Aided Engineering*, vol. 14, no 1, pages 15-20, December 2003.
9. Lee, K. C. Gardner, P.: A Novel Digital Predistorter Technique Using an Adaptive Neuro-Fuzzy Inference System, *IEEE Comm. Letters*, vol.7, no 2, pages 55-57, Frebruary 2003.
10. Lee, K. C. Gardner, P.: Adaptive neuro-fuzzy inference system (ANFIS) digital predistorter for RF power amplifier linearization, *IEEE Trans. on Veh. Tech.*, vol.55, no 1, pages 43-51, Juanuary 2006.
11. Qianhua Zhang and A. Benvenist: Wavelet network, *IEEE Trans. Signal Processing*, vol.13, no 6, pages 889-898, 1992.
12. Qinghua Zhang: Using wavelet network in non-parameters estimation, *EEE Trans. Neural Networks*, vol. 8, no. 2, pages 227-236, 1997.
13. D. Serre, Matrices: Theory and applications.*New York: Springer-Verlag* , 2002

Improved Likelihood Ratio Test Detector Using a Jointly Gaussian Probability Distribution Function

O. Pernía, J.M. Górriz, J. Ramírez, C.G. Puntonet, and I. Turias

E.T.S.I.I., Universidad de Granada
C/ Periodista Daniel Saucedo, 18071 Granada, Spain
gorriz@ugr.es

Abstract. Currently, the accuracy of speech processing systems is strongly affected by the acoustic noise. This is a serious obstacle to meet the demands of modern applications and therefore these systems often needs a noise reduction algorithm working in combination with a precise voice activity detector (VAD). This paper presents a new voice activity detector (VAD) for improving speech detection robustness in noisy environments and the performance of speech recognition systems. The algorithm defines an optimum likelihood ratio test (LRT) involving Multiple and correlated Observations (MO). The so defined decision rule reports significant improvements in speech/non-speech discrimination accuracy over existing VAD methods with optimal performance when just a single observation is processed. The algorithm has an inherent delay in MO scenario that, for several applications including robust speech recognition, does not represent a serious implementation obstacle. An analysis of the methodology for a pair-wise observation dependence shows the improved robustness of the proposed approach by means of a clear reduction of the classification error as the number of observations is increased. The proposed strategy is also compared to different VAD methods including the G.729, AMR and AFE standards, as well as recently reported algorithms showing a sustained advantage in speech/non-speech detection accuracy and speech recognition performance.

1 Introduction

The emerging applications of speech communication are demanding increasing levels of performance in noise adverse environments. Examples of such systems are the new voice services including discontinuous speech transmission [1,2,3] or distributed speech recognition (DSR) over wireless and IP networks [4]. These systems often require a noise reduction scheme working in combination with a precise voice activity detector (VAD) [5] for estimating the noise spectrum during non-speech periods in order to compensate its harmful effect on the speech signal.

During the last decade numerous researchers have studied different strategies for detecting speech in noise and the influence of the VAD on the performance of speech processing systems [5]. Sohn *et al.* [6] proposed a robust VAD algorithm

J. Mira and J.R. Álvarez (Eds.): IWINAC 2007, Part II, LNCS 4528, pp. 37–44, 2007.

based on a statistical likelihood ratio test (LRT) involving a single observation vector. Later, Cho *et al* [7] suggested an improvement based on a smoothed LRT. Most VADs in use today normally consider hangover algorithms based on empirical models to smooth the VAD decision. It has been shown recently [8,9,10] that incorporating long-term speech information to the decision rule reports benefits for speech/pause discrimination in high noise environments, however an important assumption made on these previous works has to be revised: *the independence of overlapped observations.* In this work we propose a more realistic one: *the observations are jointly gaussian distributed with non-zero correlations.* In addition, important issues that need to be addressed are: *i*) the increased computational complexity mainly due to the definition of the decision rule over large data sets, and *ii*) the optimum criterion of the decision rule. This work advances in the field by defining a decision rule based on an optimum statistical LRT which involves multiple and *correlated* observations. The paper is organized as follows. Section 2 reviews the theoretical background on the LRT statistical decision theory. Section 3 considers its application to the problem of detecting speech in a noisy signal. Finally in Section 4 we discuss the suitability of the proposed approach for pair-wise correlated observations using the experimental data set AURORA 3 subset of the original Spanish SpeechDat-Car (SDC) database [11] and state some conclusions in section 5.

2 Multiple Observation Probability Ratio Test

Under a two hypothesis test, the optimal decision rule that minimizes the error probability is the Bayes classifier. Given an observation vector $\hat{\mathbf{y}}$ to be classified, the problem is reduced to selecting the hypothesis (H_0 or H_1) with the largest posterior probability $\mathrm{P}(H_i|\hat{\mathbf{y}})$. From the Bayes rule:

$$L(\hat{\mathbf{y}}) = \frac{p_{\mathbf{y}|H_1}(\hat{\mathbf{y}}|H_1)}{p_{\mathbf{y}|H_0}(\hat{\mathbf{y}}|H_0)} \begin{matrix} > \\ < \end{matrix} \frac{P[H_0]}{P[H_1]} \Rightarrow \begin{matrix} \hat{\mathbf{y}} \leftrightarrow H_1 \\ \hat{\mathbf{y}} \leftrightarrow H_0 \end{matrix} \tag{1}$$

In the LRT, it is assumed that the number of observations is fixed and represented by a vector $\hat{\mathbf{y}}$. The performance of the decision procedure can be improved by incorporating more observations to the statistical test. When N measurements $\hat{\mathbf{y}}_1, \hat{\mathbf{y}}_2, \ldots, \hat{\mathbf{y}}_N$ are available in a two-class classification problem, a multiple observation likelihood ratio test (MO-LRT) can be defined by:

$$L_N(\hat{\mathbf{y}}_1, \hat{\mathbf{y}}_2, ..., \hat{\mathbf{y}}_N) = \frac{p_{\mathbf{y}_1,\mathbf{y}_2,...,\mathbf{y}_N|H_1}(\hat{\mathbf{y}}_1, \hat{\mathbf{y}}_2, ..., \hat{\mathbf{y}}_N|H_1)}{p_{\mathbf{y}_1,\mathbf{y}_2,...,\mathbf{y}_N|H_0}(\hat{\mathbf{y}}_1, \hat{\mathbf{y}}_2, ..., \hat{\mathbf{y}}_N|H_0)} \tag{2}$$

This test involves the evaluation of an N-th order LRT which enables a computationally efficient evaluation when the individual measurements $\hat{\mathbf{y}}_k$ are independent. However, they are not since the windows used in the computation of the observation vectors $\mathbf{y}_k$ are usually overlapped. In order to evaluate the proposed MO-LRT VAD on an incoming signal, an adequate statistical model for the feature vectors in presence and absence of speech needs to be selected.

The joint probability distributions under both hypotheses are assumed to be jointly gaussian independently distributed in frequency and in each part (real and imaginary) of vector with correlation components between each pair of frequency observations:

$$L_N(\hat{\mathbf{y}}_1, \hat{\mathbf{y}}_2, ..., \hat{\mathbf{y}}_N) = \prod_{p \in \{R,I\}} \left(\prod_{\omega} \frac{p_{\mathrm{y}_1^\omega, \mathrm{y}_2^\omega, ..., \mathrm{y}_N^\omega | H_1}(\hat{\mathrm{y}}_1^\omega, \hat{\mathrm{y}}_2^\omega, ..., \hat{\mathrm{y}}_N^\omega | H_1)}{p_{\mathrm{y}_1^\omega, \mathrm{y}_2^\omega, ..., \mathrm{y}_N^\omega | H_0}(\hat{\mathrm{y}}_1^\omega, \hat{\mathrm{y}}_2^\omega, ..., \hat{\mathrm{y}}_N^\omega | H_0)} \right)_p \tag{3}$$

This is a more realistic approach that the one presented in [8] taking into account the overlap between adjacent observations. We use following joint gaussian probability density function for each part:

$$p_{\mathbf{y}_\omega | H_s}(\hat{\mathbf{y}}_\omega | H_s)) = K_{H_s,N} \cdot \exp\{-\frac{1}{2}(\hat{\mathbf{y}}_\omega^T (C_{\mathbf{y}_\omega, H_s}^N)^{-1} \hat{\mathbf{y}}_\omega)\} \tag{4}$$

for $s = 0, 1$, where $K_{H_s,N} = \frac{1}{(2\pi)^{N/2} |C_{\mathbf{y}_\omega, H_s}^N|^{1/2}}$, $\mathbf{y}_\omega = (\mathrm{y}_1^\omega, \mathrm{y}_2^\omega, \ldots, \mathrm{y}_N^\omega)^T$ is a zero-mean frequency observation vector, $C_{\mathbf{y}, H_s}^N$ is the N-order covariance matrix of the observation vector under hypothesis H_s and $|.|$ denotes determinant of a matrix. The covariance matrix will be modeled as a tridiagonal matrix, that is, we only consider the correlation between adjacent observations according to the number of samples (200) and window shift (80) that is usually selected to build the observation vector. This approach reduces the computational effort achieved by the algorithm with additional benefits from the symmetric tridiagonal matrix properties:

$$[C_{\mathbf{y}_\omega}^N]_{ij} = \begin{bmatrix} \sigma_{y_i}^2(\omega) \equiv E[|y_i^\omega|^2] & if & i = j \\ r_{ij}(\omega) \equiv E[y_i^\omega y_j^\omega] & if & j = i + 1 \\ 0 \quad other \quad case & & \end{bmatrix} \tag{5}$$

where $1 \leq i \leq j \leq N$ and $\sigma_{y_i}^2(\omega)$, $r_{ij}(\omega)$ are the variance and correlation frequency components of the observation vector $\mathbf{y}_\omega$ (denoted for clarity σ_i, r_i) which must be estimated using instantaneous values.

The model selected for the observation vector is similar to that used by Sohn *et al.* [6] that assumes the discrete Fourier transform (DFT) coefficients of the clean speech (S_j) and the noise (N_j) to be asymptotically independent Gaussian random variables. In our case the observation vector consist of the complex modulus of frequency DFT coefficient at frequency ω of the set of m observations.

3 Application to Voice Activity Detection

The use of the MO-LRT for voice activity detection is mainly motivated by two factors: *i*) the optimal behaviour of the so defined decision rule, and *ii*) a multiple observation vector for classification defines a reduced variance LRT reporting clear improvements in robustness against the acoustic noise present in the environment. The proposed MO-LRT VAD is described as follows. The

MO-LRT is defined over the observation vectors $\{\hat{\mathbf{y}}_{l-m}, \ldots, \hat{\mathbf{y}}_{l-1}, \hat{\mathbf{y}}_l, \hat{\mathbf{y}}_{l+1}, \ldots, \hat{\mathbf{y}}_{l+m}\}$ as follows:

$$\ell_{l,N} = \sum_{\omega} \frac{1}{2} \left\{ \mathbf{y}_{\omega}{}^T \Delta_N^{\omega} \mathbf{y}_{\omega} + ln \left(\frac{|C^N_{\mathbf{y}_\omega, H_0}|}{|C^N_{\mathbf{y}_\omega, H_1}|} \right) \right\} \tag{6}$$

where $\Delta_N^{\omega} = (C^N_{\mathbf{y}_\omega, H_0})^{-1} - (C^N_{\mathbf{y}_\omega, H_1})^{-1}$, $N = 2m + 1$ is the order of the model, l denotes the frame being classified as speech (H_1) or non-speech (H_0) and $\mathbf{y}_\omega$ is the previously defined frequency observation vector on the sliding window. Thus in the determination of the LRT the computation of the matrix inverses and determinants are required. Since the covariances matrices in $H_0 \& H_1$ are tridiagonal symmetric matrices, the inverses matrices can be computed as the following:

$$[C^{-1}_{\mathbf{y}_\omega}]_{mk} = \left[\frac{q_k}{p_k} - \frac{q_N}{p_N} \right] p_m p_k \quad N - 1 \geq m \geq k \geq 0 \tag{7}$$

where N is the order of the model and the set of real numbers q_n, p_n $n = 1 \ldots \infty$ satisfies the three-term recursion for $k \geq 1$:

$$0 = r_k(q_{k-1}, p_{k-1}) + \sigma_{k+1}(q_k, p_k) + r_{k+1}(q_{k+1}, p_{k+1}) \tag{8}$$

with initial values:

$$\begin{array}{l} p_0 = 1 \ \text{ and } p_1 = -\frac{\sigma_1}{r_1} \\ q_0 = 0 \ \text{ and } q_1 = \frac{1}{r_1} \end{array} \tag{9}$$

In general this set of coefficients are defined in terms of orthogonal complex polynomials which satisfy a Wronskian-like relation [12] and have the continued-fraction representation[13]:

$$\left[\frac{q_n(z)}{p_n(z)} \right] = \frac{1}{(z - \sigma_1) -} \ominus \frac{r_1^2}{(z - \sigma_2) -} \ominus \ldots \ominus \frac{r_{n-1}^2}{(z - \sigma_n)} \tag{10}$$

where $\ominus$ denotes the continuos fraction. This representation is used to compute the coefficients of the inverse matrices evaluated on $z = 0$. Since the decision rule is formulated over a sliding window consisting of $2m{+}1$ observation vectors around the frame for which the decision is being made (6), we can use the following relations to speed up the evaluation (computation of determinants and matrix inverses) of the future decision windows:

$$\left[\frac{q_n^l(z)}{p_n^l(z)} \right] = \frac{1}{(z - \sigma_1^l) -} \ominus \left[\frac{q_{n-1}^{l+1}(z)}{p_{n-1}^{l+1}(z)} \right] \tag{11}$$

$$\left[\frac{q_n^{l+1}(z)}{p_n^{l+1}(z)} \right] = \left[\frac{q_{n-1}^{l+1}(z)}{p_{n-1}^{l+1}(z)} \right] \ominus \frac{(r_{n-1}^{l+1})^2}{z - \sigma_n^{l+1}} \tag{12}$$

and

$$|C^N_{\mathbf{y}_\omega}| = \sigma_N |C^{N-1}_{\mathbf{y}_\omega}| - r^2_{N-1} |C^{N-2}_{\mathbf{y}_\omega}| \tag{13}$$

where super-index l denotes the current sliding window. In the next section we show a new VAD based on this methodology for $N = 2$, that is, this robust speech detector is intended for real time applications such us mobile communications. The decision function will be described in terms of the correlation and variance coefficients which constitute a correction to the previous LRT method [8] that assumed uncorrelated observation vectors in the MO.

4 Experimental Analysis of JGPDF Voice Activity Detector for $N = 2$

The improvement provided by the proposed methodology is evaluated in this section by studying the most simple case for $N = 2$. In this case, assuming that squared correlations ρ_1^2 under $H_0\&H_1$ and the correlation coefficients are negligible in H_0 (noise correlation coefficients $\rho_1^n \to 0$) vanish, the LRT can be evaluated according to:

$$\ell_{l,2} = \frac{1}{2}\sum_{\omega} L_1(\omega) + L_2(\omega) + 2\sqrt{\gamma_1\gamma_2}\left(\frac{\rho_1^s}{\sqrt{(1+\xi_1)(1+\xi_2)}}\right) \tag{14}$$

where $\rho_1^s = r_1^s(\omega)/(\sqrt{\sigma_1^s\sigma_2^s})$ is the correlation coefficient of the observations in H_1, $\gamma_i \equiv (y_i^\omega)^2/\sigma_i^n(\omega)$ and $\xi_i \equiv \sigma_i^s(\omega)/\sigma_i^n(\omega)$ are the SNRs a priori and a posteriori of the DFT coefficients, $L_{\{1,2\}}(\omega)) \equiv \frac{\gamma_{\{1,2\}}\xi_{\{1,2\}}}{1+\xi_{\{1,2\}}} - \ln(1+\xi_{\{1,2\}})$ are the independent LRT of the observations $\hat{\mathbf{y}}_1, \hat{\mathbf{y}}_2$ (connection with the previous MO-LRT [8]) which are corrected with the term depending on ρ_1^s, the new parameter to be modeled, and l indexes to the second observation.. At this point frequency ergodicity of the process must be assumed to estimate the new model parameter ρ_1^s. This means that the correlation coefficients are constant in frequency thus an ensemble average can be estimated using the sample mean correlation of the observations $\hat{\mathbf{y}}_1$ and $\hat{\mathbf{y}}_2$ included in the sliding window.

4.1 Experimental Framework

The ROC curves are frequently used to completely describe the VAD error rate. The AURORA 3 subset of the original Spanish SpeechDat-Car (SDC) database [11] was used in this analysis. The files are categorized into three noisy conditions: quiet, low noisy and highly noisy conditions, which represent different driving conditions with average SNR values between 25dB, and 5dB. The non-speech hit rate (HR0) and the false alarm rate (FAR0= 100-HR1) were determined in each noise condition.

Using the proposed decision function (equation 15) we obtain an almost binary decision rule as it is shown in figure 1(a) which accurately detects the beginnings of the voice periods. The detection of voice endings is improved using a hang-over scheme based on the decision of past frames as we will see in the following. Observe how this strategy cannot be applied to the independent MO-LRT [8] because of its hard decision rule and changing bias as it is shown

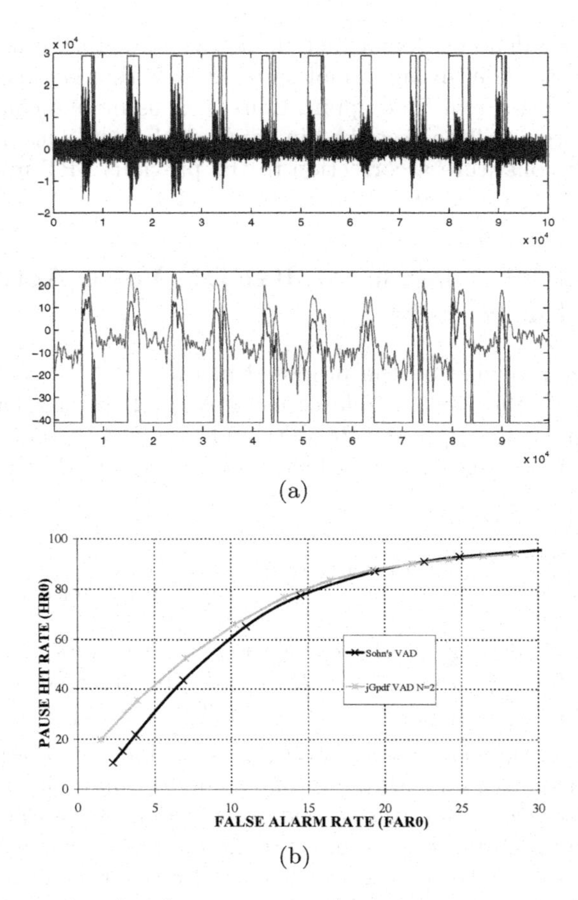

Fig. 1. a) JGPDF-VAD (blue) vs. MO-LRT (red) decision for $N = 2$. b) ROC curve for JGPDF VAD with $l_h = 8$ and Sohn's VAD [6] using a similar hang-over mechanism.

in the same figure. We implement a very simple hang-over mechanism based on contextual information of the previous frames, thus no delay obstacle is added to the algorithm:

$$\ell_{l,2}^{h} = \ell_{l,2} + \ell_{l-l_h,2} \tag{15}$$

where the parameter l_h is selected experimentally. The ROC curve analysis for this hang-over parameter is shown in figure 2(a) where the influence of hang-over in the zero hit rate is studied with variable detection threshold.

Finally, the benefits of contextual information [8] can be incorporated just averaging the decision rule over a set of jointly second order observations. A typical value for $m = 8$ produces increasing levels of detection accuracy as it is shown in the ROC curve in figure 2(b). Of course, these results are not the optimum ones since only pair-wise dependence is considered here however for a small number of observations the proposed VAD presents the best trade-off between detection accuracy and delay.

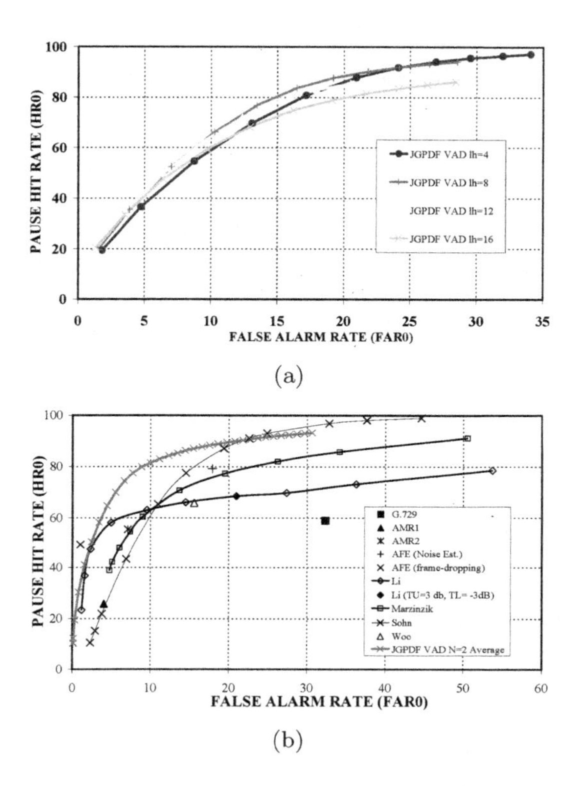

Fig. 2. a) ROC curve analysis of the JGPDF-VAD ($N = 2$) for the hang-over parameter l_h. b) ROC curves of the JGPDF-VAD using contextual information with $m = 8$ and standards and recently reported VADs.

5 Conclusion

This paper showed a new VAD for improving speech detection robustness in noisy environments. The proposed method is developed on the basis of previous proposals that incorporates long-term speech information to the decision rule. However, it is not based on the assumption of independence between observations since this hypothesis isn't realistic at all. It defines a statistically optimum likelihood ratio test based on multiple and correlated observation vectors which avoids the need of smoothing the VAD decision, thus reporting significant benefits for speech/pause detection in noisy environments. The algorithm has an optional inherent delay that, for several applications including robust speech recognition, does not represent a serious implementation obstacle. Several experiments were conducted for evaluating this approach for $N = 2$. An analysis based on the ROC curves unveiled a clear reduction of the classification error for two observations and also when the number of observations is increased. In this way, the proposed JGPDF-VAD outperformed, at the same conditions, the Sohn's VAD, that assumes a single observation in the decision rule and uses a HMM-based hangover, and other methods including the standardized G.729, AMR and AFE VADs, as well as other recently reported VAD methods in both speech/non-speech detection performance.

Acknowledgements

This work has received research funding from the EU 6th Framework Programme, under contract number IST-2002-507943 (HIWIRE, Human Input that Works in Real Environments) and SESIBONN project (TEC2004-06096-C03-00) from the Spanish government. The views expressed here are those of the authors only. The Community is not liable for any use that may be made of the information contained therein.

References

1. A. Benyassine, E. Shlomot, H. Su, D. Massaloux, C. Lamblin, and J. Petit, "ITU-T Recommendation G.729 Annex B: A silence compression scheme for use with G.729 optimized for V.70 digital simultaneous voice and data applications," *IEEE Communications Magazine*, vol. 35, no. 9, pp. 64–73, 1997.
2. ITU, "A silence compression scheme for G.729 optimized for terminals conforming to recommendation V.70," *ITU-T Recommendation G.729-Annex B*, 1996.
3. ETSI, "Voice activity detector (VAD) for Adaptive Multi-Rate (AMR) speech traffic channels," *ETSI EN 301 708 Recommendation*, 1999.
4. ——, "Speech processing, transmission and quality aspects (STQ); distributed speech recognition; advanced front-end feature extraction algorithm; compression algorithms," *ETSI ES 201 108 Recommendation*, 2002.
5. R. L. Bouquin-Jeannes and G. Faucon, "Study of a voice activity detector and its influence on a noise reduction system," *Speech Communication*, vol. 16, pp. 245–254, 1995.
6. J. Sohn, N. S. Kim, and W. Sung, "A statistical model-based voice activity detection," *IEEE Signal Processing Letters*, vol. 16, no. 1, pp. 1–3, 1999.
7. Y. D. Cho, K. Al-Naimi, and A. Kondoz, "Improved voice activity detection based on a smoothed statistical likelihood ratio," in *Proc. of the International Conference on Acoustics, Speech and Signal Processing (ICASSP)*, vol. 2, 2001, pp. 737–740.
8. J. Ramírez, J. C. Segura, C. Benítez, L. García, and A. Rubio, "Statistical voice activity detection using a multiple observation likelihood ratio test," *IEEE Signal Processing Letters*, vol. 12, no. 10, pp. 837–844, 2001.
9. J. M. Górriz, J. Ramírez, C. G. Puntonet, and J. C. Segura, "An effective cluster-based model for robust speech detection and speech recognition in noisy environments," *Journal of Acoustical Society of America*, vol. 120, no. 470, pp. 470–481, 2006.
10. J. M. Górriz, J. Ramirez, J. C. Segura, and C. G. Puntonet, "An improved mo-lrt vad based on a bispectra gaussian model," *Electronic Letters*, vol. 41, no. 15, pp. 877–879, 2005.
11. A. Moreno, L. Borge, D. Christoph, R. Gael, C. Khalid, E. Stephan, and A. Jeffrey, "SpeechDat-Car: A Large Speech Database for Automotive Environments," in *Proceedings of the II LREC Conference*, 2000.
12. N. Akhiezer, *The Classical Moment Problem.* Edimburgh: Oliver and Boyd, 1965.
13. H. Yamani and M. Abdelmonem, "The analytic inversion of any finite symmetric tridiagonal matrix," *J. Phys. A: Math Gen*, vol. 30, pp. 2889–2893, 1997.

Performance Monitoring of Closed-Loop Controlled Systems Using dFasArt

Jose Manuel Cano-Izquierdo, Julio Ibarrola, and Miguel Almonacid

Department of Systems Engineering and Automatic Control
Technical University of Cartagena
Campus Muralla del Mar, 30202 Cartagena, Murcia, Spain
{JoseM.Cano,Juliojose.Ibarrola,Miguel.Almonacid}@upct.es

Abstract. This paper analyzes the behaviour of closed-loop controlled systems. Starting from the measured data, the aim is to establish a classification of the system operation states. Digital Signal Processing is used to represent temporal signal with spatial patterns. A neuro-fuzzy scheme (dFasArt) is proposed to classify these patterns, in an on-line way, characterizing the state of controller performance. A real scale plant has been used to carry out several experiments with good results.

Keywords: dFasArt, neuro-fuzzy, temporal analysis, closed-loop systems.

1 Introduction

In control engineering, the closed-loop is the most frequently used control scheme [6]. Significant advances have been achieved in the design of controllers with new control strategies, but there are not many works dealing with the tracking of the controllers performance [5][7]. What does it happens during the working life of the controller? Is there a degradation on its performance or is it maintained similar? How does the controller react to unmeasured disturbances not taken into account at the design stage?

To answer these questions, tools are required to analyse the system behaviour in closed-loop and to represent it by means of a set of characteristics whose evolution could be compared and classified in a temporal way. Data coming from the Sensors/Actuators are used to study the system behaviour without affecting the process. The error signal, defined as the reference minus the output signal, is usually available in closed-loop systems. This error constitutes a temporal signal, often corrupted with noises of several natures.

In this work, a method to determine the controller performance is proposed. It is based on a first module to process the signal using Digital Signal Processing (DSP) techniques to estimate the spectral density function. The aim is to transform the temporal information contained into the error signal in a spatial vector, which has also a temporal character because it is calculated for several time intervals. The dFasArt model [3] is proposed to make a classification of the spatial-temporal vector. It allows distinguishing between different vector states.

J. Mira and J.R. Álvarez (Eds.): IWINAC 2007, Part II, LNCS 4528, pp. 45–53, 2007.

These states can be associated to different stationarities in the behaviour of the closed-loop system.

Section 2 describes the DSP method used to calculate the vector based in the spectral density function of the error signal. Section 3 describes the main equations of dFasArt model. In Section 4 some experimental results obtained from a real system are provided. Finally, Section 5 summarizes the main conclusions of the paper.

2 Digital Signal Processing

To represent a sample record, some statistical properties can be used: mean and mean square values, probability density functions, autocorrelation functions or autospectral density functions. Spectral density function is suitable to determine the properties of system [1]. In closed-loop systems this function can be applied to achieve a description of the control error nature. If a digital implementation is considered the components of the obtained vector will reflect the frequency composition of the control error energy. The analysis of the sliding windows illustrates the temporal evolution of the vector. If the process is stationary, the vector should remain time-constant and it could be concluded that the controller performance is invariant. Changes in the vector would indicate changes in the controller performance. For the data contained in the window, an approximation to the power spectral density function is calculated with the equation:

$$\tilde{G}_k = \tilde{G}_k\left(\frac{kf_c}{m}\right) = 2h\left[\hat{R}_0 + 2\sum_{r=1}^{m-1}\hat{R}_r\cos\left(\frac{\pi r k}{m}\right) + (-1)^k\hat{R}_m\right] \tag{1}$$

where $k = 0, 1 \ldots m$, h is the sample interval, $f_c = \frac{1}{2h}$ the Nyquist folding frequency and $\hat{R}_0 \ldots \hat{R}_m$ are the values of the approximation to the autocorrelation function coefficients carried out as follows:

$$\hat{R}_r = \hat{R}(rh) = \frac{1}{N}\sum_{n=1}^{N-r} x_n x_{n+r} \tag{2}$$

with $r = 0, 1, \ldots, m$ where m is the maximum lag number. This constant (m) must be small compared to the number of considered samples (N).

The Leakage reduction is used to achieve a smooth estimation of the power spectral density. It calculates the estimate harmonic $\hat{G}_k$ corresponding to the frequency $f = \frac{kf_c}{m}$ with $k = 0, 1, \ldots, m$.

$$\begin{aligned} \hat{G}_0 &= 0.5\tilde{G}_0 + 0.5\tilde{G}_1 \\ \hat{G}_k &= 0.25\tilde{G}_{k-1} + 0.5\tilde{G}_k + 0.25\tilde{G}_{k+1}\ k = 1, 2 \ldots, m-1 \\ \hat{G}_m &= 0.5\tilde{G}_{m-1} + 0.5\tilde{G}_m \end{aligned} \tag{3}$$

With this processing, the temporal signal for the considered window can be represented through a spatial vector. These vectors constitute a new temporal signal which can be classified by its variability. The result of the classification will determine the number of stationary states contained in the original signal.

3 dFasArt

FasArt model links the ART architecture with Fuzzy Logic Systems, establishing a relationship between the unit activation function and the membership function of a fuzzy set [2]. On the one hand, this allows interpreting each of the FasArt unit as a fuzzy class defined by the membership-activation function associated to the representing unit. On the other hand, the rules that relate the different classes are determined by the connection weights between the units.

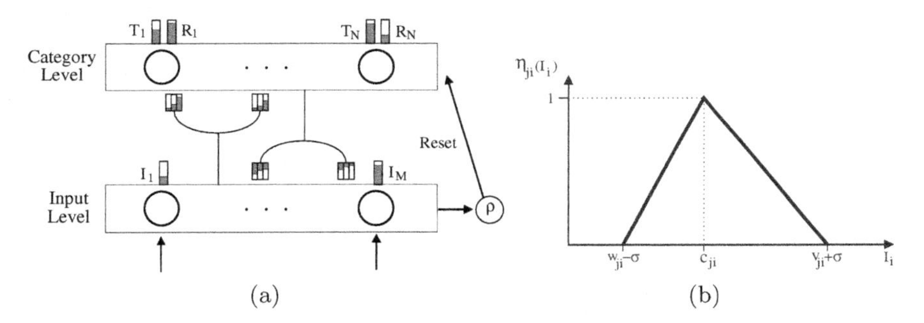

Fig. 1. (a) dFasArt model (b) Membership-activation function

dFasArt uses a dynamic activation function determined by the weights of the unit as the membership function of a fuzzy set. dFasArt model is represented in Figure 1(a). The signal activation is calculated as the AND of the activations of each one of the dimensions when a multidimensional signal is considered. This AND is implemented using the product as a T-norm. Hence, the activity T_j of unit j for a M-dimensional input $\boldsymbol{I} = (I_1 \dots I_M)$ is given by:

$$\frac{dT_j}{dt} = -A_T T_j + B_T \prod_{i=1}^{M} \eta_{ji}(I_i(t)) \tag{4}$$

where η_{ji} is the membership function associated to the *ith-dimension* of unit j, determined by the weights w_{ji}, c_{ji} and v_{ji}, as it is shown in Figure 3(b). The σ parameter determines the fuzziness of the class associated to the unit by:

$$\sigma = \sigma^* |2c_j| + \epsilon \tag{5}$$

The election of the winning unit J is carried out following the winner-takes-all rule:

$$T_J = \max_j \{T_j\} \tag{6}$$

The learning process starts when the winning unit meets a criterion. This criterion is associated to the size of the support of the fuzzy class that would contain the input if this was categorized in the unit. This value is calculated dynamically for each unit according to:

$$\frac{dR_j}{dt} = -A_R R_J + B_R \sum_{i=1}^{M} \left(\frac{\max(v_{Ji}, I_i) - \min(w_{Ji}, I_i)}{|2c_{Ji}| + \epsilon} \right) \tag{7}$$

The R_j value represents a measurement of the change needed on the class associated to the j unit to incorporate the input. To see if the J winning unit can generalize the input, it is compared with the design parameter ρ, so that:

- If:

$$R_J \geq \rho \tag{8}$$

the matching between the input and the weight vector of the unit is good, and the learning task starts.

- If:

$$R_J < \rho \tag{9}$$

there is not enough similarity, so the Reset mechanism is fired. This inhibits the activation of unit J, returning to the election of a new winning unit.

If the Reset mechanism is not fired, then the learning phase is activated and the unit modifies its weights. The *Fast-Commit Slow-Learning* concept is commonly used.

When the winning unit represents a class that had performed some other learning cycle (*committed unit*), the weights are *Slow-Learning* updated according to the equations:

$$\begin{aligned} \frac{d\boldsymbol{W}}{dt} &= -A_W \boldsymbol{W} + B_W \min(\boldsymbol{I}(t), \boldsymbol{W}) \\ \frac{d\boldsymbol{C}}{dt} &= A_C(\boldsymbol{I} - \boldsymbol{C}) \\ \frac{d\boldsymbol{V}}{dt} &= -A_V \boldsymbol{V} + B_V \max(\boldsymbol{I}(t), \boldsymbol{V}) \end{aligned} \tag{10}$$

For the case of the uncommitted units, the class is initialized with the first categorized value, hence *Fast-Commit*:

$$\boldsymbol{W}_J^{NEW} = \boldsymbol{C}_J^{NEW} = \boldsymbol{V}_J^{NEW} = \boldsymbol{I} \tag{11}$$

4 Experimental Study

To show the behaviour of the proposed method, a laboratory scale plant (Fig. 2) has been used. This plant has a light ball which moves vertically inside a tube, forced by the air flow generated by an electric fan installed at the bottom. The aim is to control the altitude of the ball by varying the voltage applied to the fan. The plant has a sensor to measure the ball altitude and a regulator to give a variable voltage to the fan. A PID controller, experimentally tuned, is used. Its output is generated by the following equation:

$$u[k] = K_p * e[k] + K_d * \Delta e[k] + K_i * \sum_j e[j] \tag{12}$$

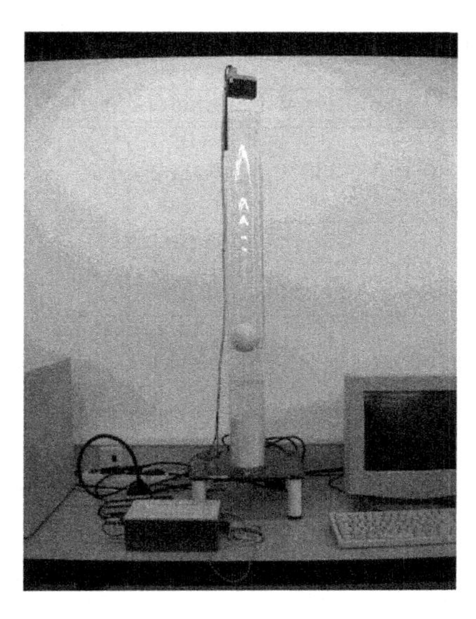

Fig. 2. Laboratory scale plant used

where $u[k]$ is the voltage applied to the fan at time $t = t_0 + k * \Delta t$ and $e[k] = ref - h[k]$, being $h[k]$ the altitude measured by the sensor and ref is the reference. The sample time is selected to $\Delta t = 0.33seg$.

The reference of the ball altitude is maintained constant while some changes have been done in the controller parameters, as shown in Table 1, where each set of parameters corresponds to an operating state of the plant. Fig. 3 shows the control error signal $e[k]$ along the experiment.

A sliding window of 200 samples is defined over the temporal error signal. With the selected sample time, the window corresponds to an interval of $66seg.$, as shown in Figure 4. The window is moved in steps of 10 samples ($3.3seg.$). The error signal for each window is processed with the digital processing method described in Section 2, considering 10 frequencies for the calculus of the autospectral density function. Thus, a 10-dimensional temporal signal is obtained and used as input to the dFasArt module.

For the learning task with dFasArt, the parameters of Table 2 are selected. Fig. 5 shows dFasArt output corresponding to the winning unit J. As it can be seen, four units have been committed, the units 1 and 2 correspond to the first state (first set of controller parameters), the unit 3 fits the second state and unit 4 corresponds to the third one.

In the second part of the experiment, the reference is also maintained constant with the same sets of controller parameters, but now the working order of these sets has been changed, as shown in Table 3. For this Table, the labels defined for the states of Table1 have remained the same.

In Fig. 6, the error data $e[k]$ gathered in this second test are shown. From these data a new set of patterns is generated using the DSP module. These patterns are presented to the dFasart module, which is able to recognise the

Table 1. Controller parameters for the learning test

Time interval	Parameters	State
$0 < t \leq 330$	$K_p = 0.2$ $K_i = 0.02$ $K_d = 1.4$	1
$330 < t \leq 660$	$K_p = 0.1$ $K_i = 0.03$ $K_d = 0.7$	2
$660 > t \leq 924$	$Kp = 0.2$ $Ki = 0.01$ $Kd = 0.7$	3

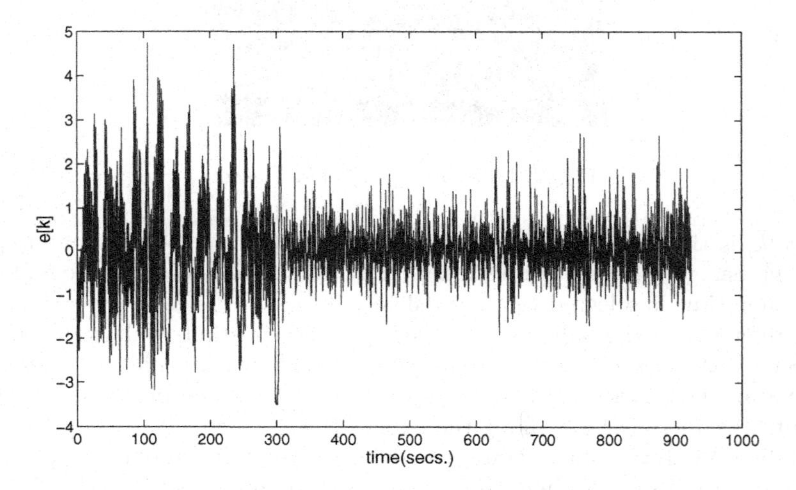

Fig. 3. Control error for the learning phase

Table 2. dFasAsrt parameters for the learning task

Parameters	Value
A_w	0.01
A_v	0.01
A_c	0.01
σ	5.0
α	0.0000001
ρ	0.075
A_r	0.0005
A_t	0.02

states codified in the previous phase. The reset and learning mechanisms have been disconnected to assure identification. The J winning unit, in this validation test, is represented in Fig. 7. Taken into account the relationship between units

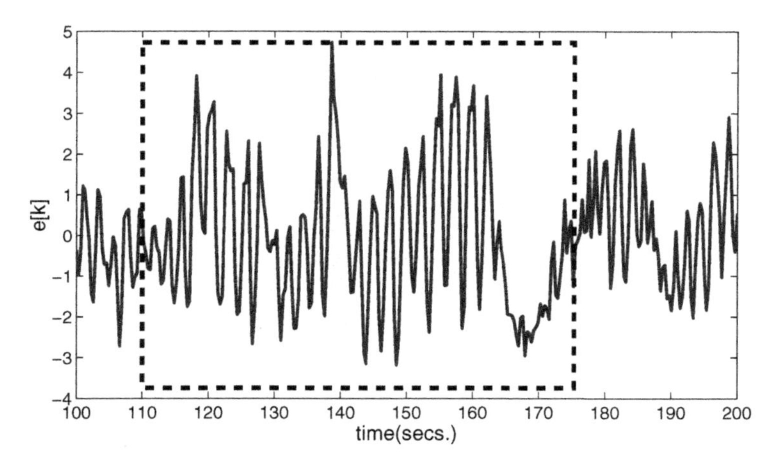

Fig. 4. Data window for error signal analysis

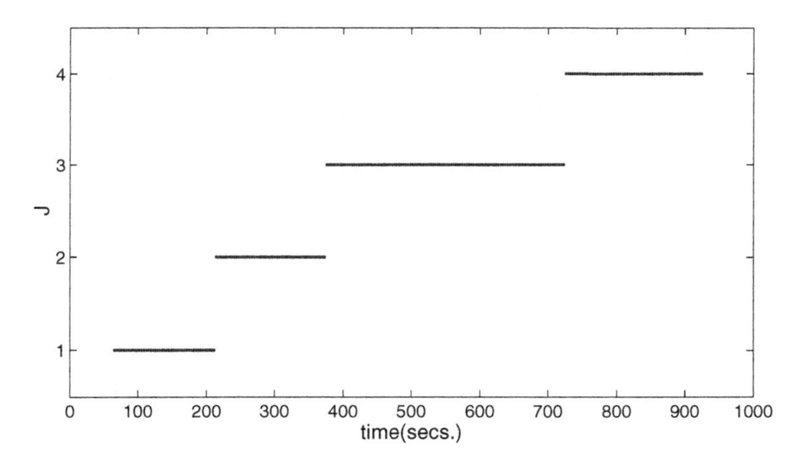

Fig. 5. Winning unit (J) in dFasArt module

and states, that could be defined from Fig. 5 (unit 4 $\equiv$ state 3, unit 2 $\equiv$ state 1, unit 3 $\equiv$ state 2) it can be seen how dFasArt recognise three different states and accurately predicts the order established in Table 3.

As it is shown in Fig. 7, dFasArt is able to identify the states of the new data set. A relevant characteristic is the fulfilment of the stability-plasticity dilemma that allows the network learning new states without damaging the stored information. So, dFasArt can quickly remember previous states without a new learning process.

It can be remarked that dFasArt works in an on-line and incremental way. To identify the operating states it only uses the last data generated by the plant. As a difference with other non-supervised clustering algorithms (i.e. k-means or

Table 3. Controller parameters for the validation test

Time interval	Parameters	State
$0 < t \leq 330$	$K_p = 0.2$ $K_i = 0.01$ $K_d = 0.7$	3
$330 < t \leq 660$	$K_p = 0.2$ $K_i = 0.02$ $K_d = 1.4$	1
$660 > t \leq 957$	$Kp = 0.1$ $Ki = 0.03$ $Kd = 0.7$	2

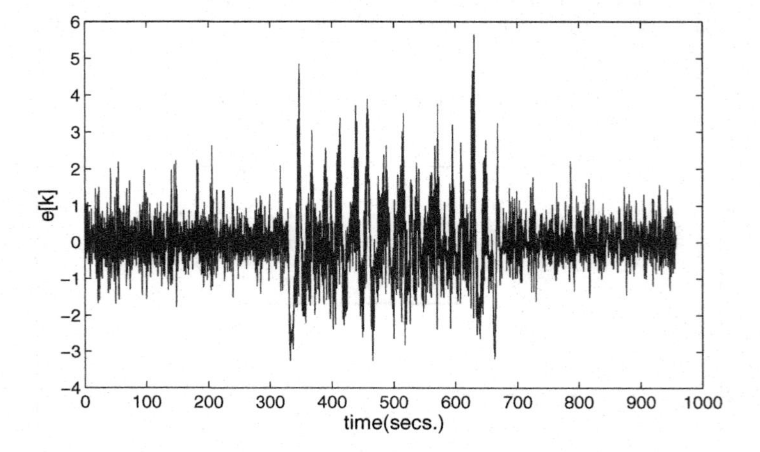

Fig. 6. Control error for the validation phase

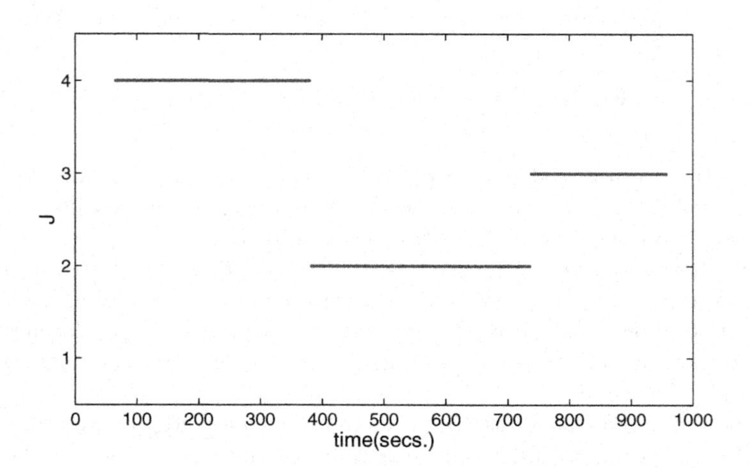

Fig. 7. Winning unit (J) in dFasArt module

fuzzy c-means [4].), dFasArt does not need neither the historic information nor future output data.

5 Conclusions

In this paper, the advantages of dFasArt model in the study of the operating behaviour of closed-loop systems have been presented. To this aim, a neuro-fuzzy scheme to process the information contained in the error signal is proposed. This signal is the minimum knowledge of the controlled process that can be measured in the worst case. A frequency analysis of the error is carried out to obtain a spatial vector based in the spectral density function representing the nature of the signal. This vector is categorized through a dFasArt model, identifying several stationary states. Thus, different operating points are detected in the closed-loop working.

The experimental results made with a laboratory scale plant show the feasibility of the proposed architecture. The analysis of the error signal with DSP and the non-supervised classification with dFasArt allows detecting the stationary states of a closed-loop controlled system starting from the error signal corrupted with noise.

Acknowledgments. We would like to thank the people of the NEUROCOR Group for their support in this work and specially to Miguel Pinzolas.

References

1. J. Bendat and A. Piersol. *Random Data: Analysis and Measurement Procedures.* Wiley-Interscience, 2000.
2. J. Cano, Y. Dimitriadis, E. Gómez, and J. Coronado. Learning from noisy information in FasArt and FasBack neuro-fuzzy systems. *Neural Networks*, 14:407–425, 2001.
3. J. Cano-Izquierdo, M. Almonacid, J. Ibarrola, and M. Pinzolas. Use of dynamic neuro fuzzy model dFasArt for identification of stationary states in closed-loop controlled systems. In *Proceedings of EUROFUSE 2007*, Jaen, Spain, 2007.
4. R. Duda, P. Hart, and D. Stork. *Pattern Classification.* John Wiley, 2000.
5. T. Harris, C. Seppala, and Desborough. A review of performance monitoring and assessment techniques for univariate and multivariate control systems. *Journal of Process Control*, 9:1–17, 1999.
6. I. Landau. Identification in closed loop: A powerful design tool (better design models, simpler controllers. *Control Engineering Practice*, 9:51–65, 2001.
7. M. Rossi and C. Scali. A comparison of techniques for automatic detection of stiction: Simulation and application to industrial data. *Journal of Process Control*, 15:505–514, 2005.

Normalising Brain PET Images

Elia Ferrando Juliá, Daniel Ruiz Fernández, and Antonio Soriano Payá

Department of Computer Technology. University of Alicante, PO 99, 03080 Alicante, Spain
{eferrando, druiz, soriano}@dtic.ua.es

Abstract. PET is a nuclear medical examination which constructs a three-dimensional image of metabolism inside the body; in this article in particular, images are taken from the brain. The high complexity inherent to the interpretation of the brain images makes that any help is important to the specialists in order to accurate the diagnostic. In order to reach reliable and good images, a normalization process is suggested in this paper, consisting of centring the brain in the three-dimensional image, scaling it according to a template brain and, finally, rotating the brain according to the inclination of the template. For not reducing the quality of the information the application works with PET image format and radioactivity measures instead of translate to an ordinary colour image.

Keywords: PET, brain, normalize, 3D images.

1 Introduction

In the last years the development of medical diagnosis task has been excellent. One of the improvements has been the use of advanced technological devices in diagnostic tasks. Specifically in medical image we find a great evolution from X-Ray to Positron Emission Tomography, doing necessary the use of computer algorithms to get better images because a clear image can be essential for doing an exact diagnostic.

Positron Emission Tomography (PET) is a technique based on the detection of radioactivity from a certain part of a living body. The radiation is caused by the emission of positrons emitted from a kind of fluorescent glucose injected to the bloodstream of the patient who has been fasting for at least 6 hours, and who has a suitably low blood sugar [1].

The tomography helps to visualize the biochemical changes that take place inside a body, like metabolism. The main advantage of this technique compare to other nuclear medicine tests, is that PET detects metabolism within body tissues, whereas other types just detect the amount of a radioactive substance collected in body tissue to examine the tissue's function [2]. The usually procedure is to administrate fluorodeoxyglucose[1] to the patients which is a kind of

[1] Its chemical formula is $C_6H_{11}FO_5$ and its chemical name is 2-Fluoro-2-Deoxy-D-Glucose abbreviated by FDG.

J. Mira and J.R. Álvarez (Eds.): IWINAC 2007, Part II, LNCS 4528, pp. 54–62, 2007.

sugar for fitting brain cells. Specifically it is used ^{18}F-FDG for assessment its glucose metabolism that has 109.8 minutes of half-life. The glucose is retained by cells with high metabolic activity, on the contrary, tissues with low or non-activity means damage regions [3]. These radioactive nuclei decays by emitting positrons that annihilate with electrons at the tissue. When a positron crashes against an electron, their collision produces two gamma rays having the same energy, but going in 180 directions. The gamma rays leave the patient's body and are detected by the PET scanner [4]. The specialized scanner emits light when it is hit by nuclear radiation from the patient, which is emitted from the ultra violet to the infrared range of wavelength (100 - 800 nm). Then the amount of data of radioactivity is collect to feed into a computer to be converted into a complex picture. There are several image formats that a scanner can make, nevertheless we have been working with ECAT 7 file format.

PET analysis is commonly used in oncology [5] because malignant tumours are identified by tissues with high metabolic activity as in neurological field [6] because high radioactivity is associated with brain activity and it helps to diagnose illnesses such as epilepsy, stroke or dementia. Besides, in cardiology [7], PET images are used for study atherosclerosis and vascular disease and finally, in pharmacology [8] it is used to test new drugs injected to animals.

The reason of working with radioactivity concentration instead of translate data image to a standard[2] colour image format is due to keep up all information about metabolism patient to be analyzed with accuracy. If translation to a standard image format is carried out, we can lost radioactivity information.

2 Objectives and Methods

The main objective of this work is to supply tools to prepare PET images to be analyzed using an automatic decision support system. The application has been made to process ranges of radioactivity from ECAT 7 files instead of some arbitrary colours [9]. These kinds of images are structured in three parts (Fig. 1):

1. Main header. It has information about the file, such as file type, patient's data, isotope name, isotope half-life, etc. In order to work with the files and to comply with the law of data protection, we deleted the patient name keeping his privacy.
2. Subheader. It stores specific information of the image matrix, for instants, X, Y and Z dimensions, data type, size of each edge of a voxel from each dimension, etc.
3. Image data. It is a three-dimensional array in which each cell contains the real quantitative value of the radioactivity detected in each point of the brain and the assessment unit is Bq/ml[3] or kBq/ml[4], which are the radioactivity

[2] TIFF, JPEG, GIFF or similar.
[3] Bequerel(s)/millilitre.
[4] Kilobequerel(s)/millilitre.

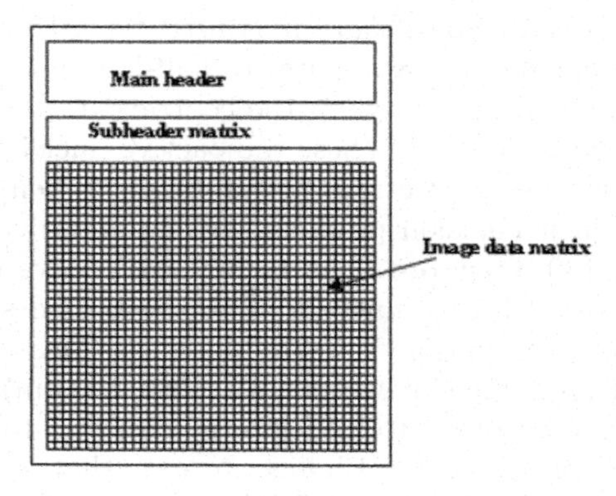

Fig. 1. Structure of an ECAT file

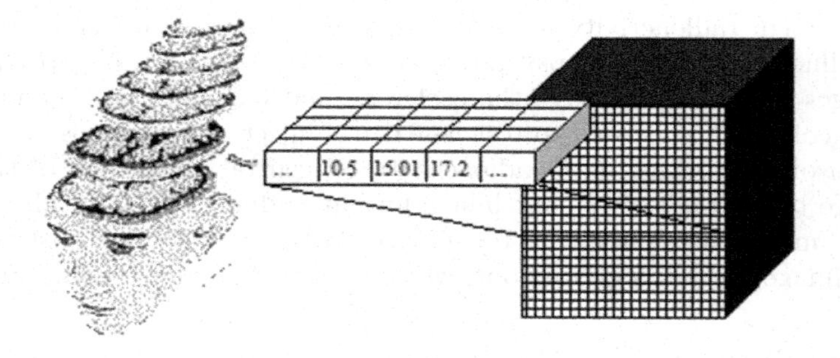

Fig. 2. The structure of PET raw data stored in the matrix image

concentration per the volume of the voxel[5]. There is one three-dimensional array for one frame of time, however, we just work with static images. One horizontal slice of a brain is stored in a two-dimensional array (Fig. 2).

When the patient is inside of a scanner, he has some possibility of movement, so he could nodded his head, additionally the most likely is that patient brain has different size comparing to template brain, hence we need to normalize the PET images to be compare between them and classify in health or damage brains.

To analyze PET images is necessary to have a model of reference to take from it the normalization parameters, like the degrees of inclination or the value of the area at the middle horizontal plane. In this work, we have a real human brain from which we know perfectly the areas, the inclination and its health estate. Normalization process could not be in other way, because all volume brains are different each one from another, so it is not possible to associate to a geometrical figure or something like that. This work is centred to correct three main characteristics: position of the brain in the image, the scale and the inclination.

[5] A cell represents a voxel when the image is printed.

3 Normalization Process

Every kind of image taken from a situation like a photograph, a TAC or a PET, all of them have noise. Then, the first step is to eliminate this noise using a filter. We identify noise with negative and very low values of radioactivity because it represents non-meaningful information.

It is also interesting to segment an image leaving just the brain and taking out cover tissues, because depending on the scan process it can take more or less regions out of the brain and the centre and volume could be unreliable [10]. So we have distributed radiation values in seven ranges, like rainbow colours because it is the best way to assign levels of metabolism (Fig. 3 Left). We recognized first and second level such cover tissues, which are not useful for medical test, consequently we just take into account the rest of levels (Fig. 3 Right).

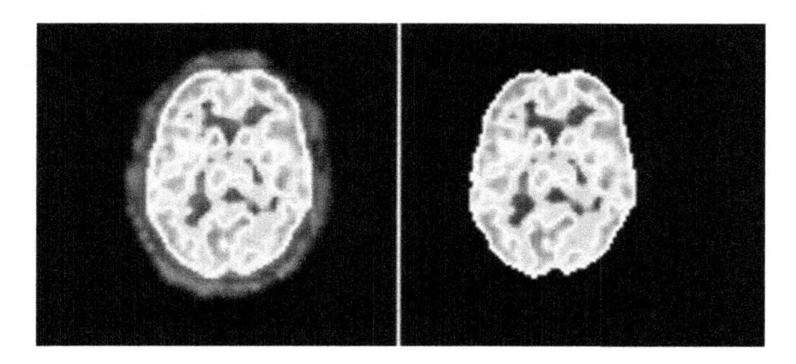

Fig. 3. (Left) Brain distributed in seven levels of radioactivity. (Right) Segmented brain.

3.1 Centring the Brain

To compare a PET image with the template, it is necessary to situate the brain at the centre of the image (CM) (Fig. 4). In order to do this operation we calculate the centre of mass (Eq. 1) of the brain and move it to the middle of three dimensions of the image.

$$CM = \frac{(\sum x_i + \sum y_i + \sum z_i)}{v} \qquad (1)$$

where x_i, y_i and z_i are voxel coordinates of the brain segmented and v is the volume of the entire brain segmented. CM obtained is the centre of the brain in the non-normalized image, therefore we have to move the brain until the middle of image and this is done adding the distance difference calculate in the next expressions:

$$Mov_x = \frac{dim_x}{2} - 1 - x_i; Mov_y = \frac{dim_y}{2} - 1 - y_i; Mov_z = \frac{dim_z}{2} - 1 - z_i \qquad (2)$$

where dim_x, dim_y and dim_z are values of dimension in X axis, Y axis and Z axis respectively and we subtract a unit because the image matrix data is numbered from 0 to dimension-1.

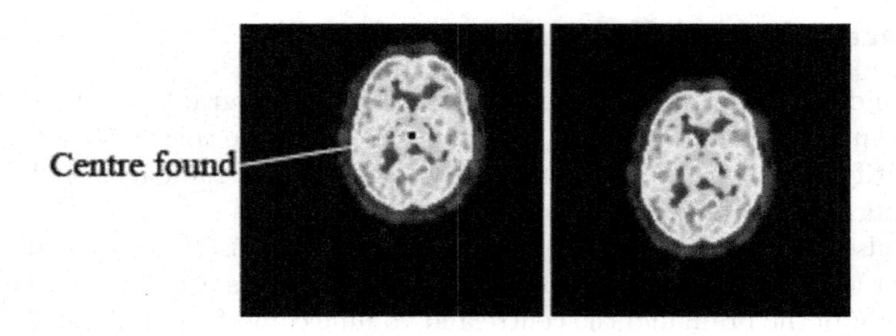

Fig. 4. (Left) Brain non-centred. (Right) Brain centred.

3.2 Scaling the Brain

Next function to use in normalization process is scaling the brain volume (Fig. 5). We have to estimate how bigger or smaller is the non-normalized image than the template and this is done using the area of the central horizontal plane of the PET image (Eq. 3).

$$p = \sqrt{\frac{areaImage}{areaTemplate}} \tag{3}$$

The value p obtained is the proportion to apply to each coordinate of the matrix image.

$$x_f = x_i \cdot p \quad y_f = y_i \cdot p \quad z_f = z_i \cdot p \tag{4}$$

The x_f, y_f and z_f obtained are destination coordinates and x_i, y_i and z_i are source coordinates. If p is more than a unit, we use interpolation (average of the neighbours values) to fill the empty values of the new voxels added due to the increase of the image. On the contrary, if p is less than a unit, interpolation is not necessary to be applied to the image because there is not empty voxels.

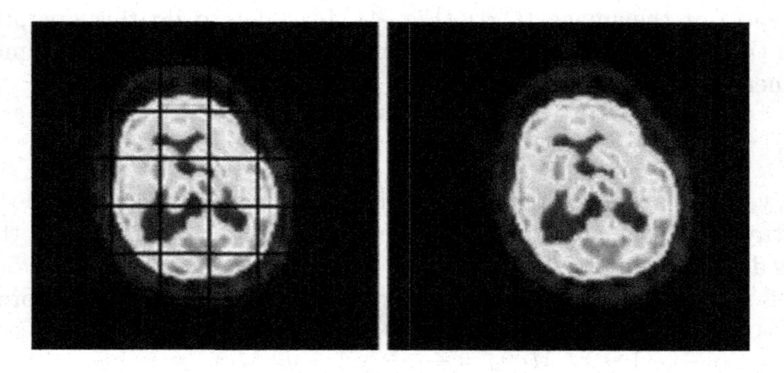

Fig. 5. (Left) Initial scale image. (Right) Final scale image.

3.3 Rotating the Brain

Finally, the brain must be at the same degrees inclination than the template brain. Our idea is based on to the comparison of the outline brains (Fig. 6). So, we implement a function to extract outline of a brain which is made up filtering third range of radiation frequency.

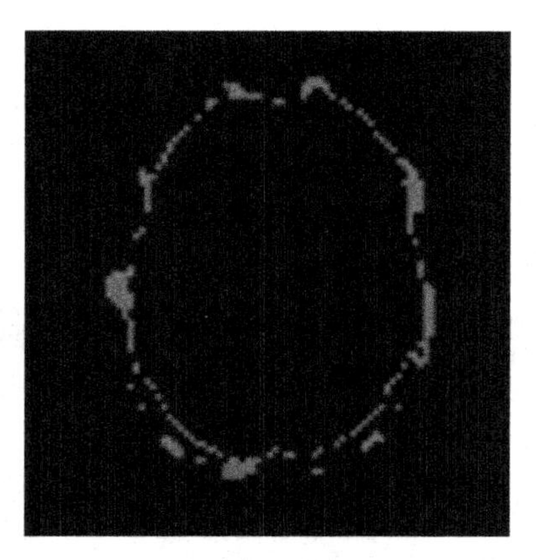

Fig. 6. Outline of the template brain

The rotation process has been approached from several points of view. The first one was oriented to find the maximum radius from template and from the patient brain. It works looking for the voxel which was situated at the highest distance. Then, the patient brain was rotated the degrees difference between the maximum radius of the template and from the patient. We discard this method because all brains are not identically, they have different outlines hence the maximum radius found was not the same for all the brains.

Secondly, we draw a rectangle rounding the brain in the middle planes at every dimensions and check tangent points between template and patient brain, after that, patient image was rotated the degrees difference. This mechanism was discarded due to the same problem as the previous method.

Finally we use the outline coordinates lists of the patient and template brain image, making up two mathematical series to analyze the correlation between them. We have used the coefficient of Pearson because it measures the correlation between two random linear series (Fig. 7) , looking for their tendency to increase or decrease together [11]. It was already implemented by open source. The method implemented follows these instructions:

X =*list of outline coordinates from template image*;
$for(degrees = 4\ until\ 24)\{$
Y = *list of outline coordinates from patient_brain*;
$r = Pearson\_Correlation\_Coefficient(X, Y)$;
$if(r > before\_r)\{$

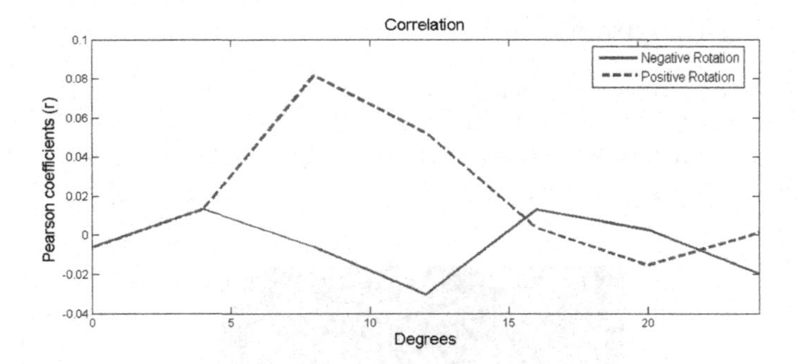

Fig. 7. Pearson coefficients of correlation: (Left) Negative rotation, (Right) Positive rotation

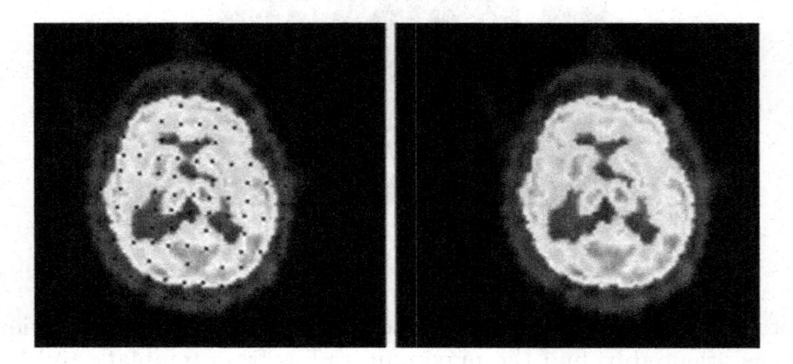

Fig. 8. (Left) Initial rotate image. (Right) Final rotate image.

```
            before_r = r;
            rotation = degrees;
        }
        Rotation(patient_brain, degrees);
        degrees+ = 4;
}
```

It is the same mechanism to rotate to opposite direction excepting that degrees are negative and when the process is finished, it rotate the patient image the degrees reached where the correlation coefficient was the biggest. This same process is executes for the three dimensions. The implementation to rotation coordinates is based on trigonometric functions (Eq. 5) where is moved each voxel to the position at degrees calculated before.

$$x_f = x_i \cdot cos\,(degrees) - y_i \cdot sin\,(degrees) \tag{5}$$
$$y_f = x_i \cdot sin\,(degrees) + y_i \cdot cos\,(degrees)$$

The x_f and y_f obtained are destination coordinates to where are going to move the source coordinates, x_i and y_i. In the construction of the image rotate, as we are not working in a discreet scope, there are some differents source coordinates which are rotate to the same coordinates destined, this is a problem because it leaves some empty voxels in the rotate image (Fig. 8 Left). For that reason, rotation method has a second phase in which it renders the rotate image looking for empty voxels. When a empty voxel is found, it is calculated the coordinate to the back position in the original image, that mean, to rotate negatively and the empty voxel takes the value of the voxel in the coordinate calculated at original image (Fig. 8 Right)[12].

4 Conclusions

We have implemented a set of tools which can do a preprocess of a PET image of the brain without lost of quality. This is possible because we operate always with the levels of radioactivity without convert the image to a graphics standard format. We use operations like centering, scaling and rotation to get an image that we can compare it with a template. Previously we have filtered the PET image eliminating noise and useless information.

We present this work as the previous one necessary to analise a PET image of the brain. Using the tools we have developed, nowadays we are implementing a diagnostic tool based in PET images to help radiologists and neurologists in the hard task of the diagnosis of brain diseases.

Acknowledgement. We appreciate the collaboration of Vesa Oikonen from the Turku PET Center helping us with the ECAT format and the libraries to read it. We are also grateful to Pedro Gonzalez Cabezas from PET Iberoa for lending us the PET images we have worked with.

References

1. Bailey, D.L., Townsend, D.W., Valk, P.E., Maisey, M.N.: Positron Emission Tomography. Publisher Springer (2005).
2. Silverman, D.H.S., Small, G.W., Chang, C.Y., Lu, C.S., Kung de Aburto, M.A., Chen, W., Czernin, J., Rapoport, S.I., Pietrini, P., Alexander, G.E., Schapiro, M.B., Jagust, W.J., Hoffman, J.M., Welsh-Bohmer, K.A., Alavi, A., Clark, C.M., Salmon, E., de Leon, M.J., Mielke, R., Cummings, J.L., Kowell, A.P., Gambhir, S.S., Hoh, C.K., Phelps, M.E.: Positron Emission Tomography in Evaluation of Dementia. Regional Brain Metabolism and Long-term Outcome. The Journal of the American Medical Association, Vol 286, 17 (2001).
3. Feng, D., Ho, D., Chen, K., Wu, L., Wang, J., Liu, R., Yeh, S.: An evaluation of the algorithms for determining local cerebral metabolic rates of glucose using positron emission tomography dynamic data. IEEE Xplore, 14 (1995) 697-710.
4. Camborde, M.L., Thompson, C.J., Togane, D.: A positron-decay triggered transmission source for positron emission tomography. IEEE Transactions on Nuclear Science, 51 (2004) 53-58.

5. Bengel, F.M., Ziegler, S.I., Avril, N., Weber, W., Laubenbacher, C., Schwaiger, M.: Whole-body positron emission tomography in clinical oncology, comparison between attenuation-corrected and uncorrected images. European Journal of Nuclear Medicine and Molecular Imaging, (1997) 1091-1098
6. Tai, Y.F., Piccini, P.: Applications of Positron Emission Tomography in Neurology. Journal of Neurology, Neurosurgery and Psychiatry, 75 (2004) 669-676
7. Johnson, T.R.C., Becker, C.R., Wintersperger, B.J., Herzog, P., Lenhard, M.S., Reiser, M.F.: Detection of Cardiac Metastasis by Positron-Emission Tomography-Computed Tomography. Circulation 112 (2005) 61-62.
8. Cunningham, V.J., Gunn, R.N., Matthews, J.C.: Quantification in positron emission tomography for research in pharmacology and drug development. Nuclear Medicine Communications, 25 (2004) 643-646.
9. Bockisch, A., Beyer, T., Antoch, G., Freudenberg, L.S., Kuhl, H, Debatin, J.F., Muller, S.P.: Positron Emission Tomography/Computed Tomography-Imaging Protocols, Artifacts and Pitfalls. Molecular Imaging and Biology, 6 (2004) 188-199.
10. Buchert, R., Wilke, F., Chakrabarti, B., Martin, B., Brenner, W., Mester, J., Clausen, M.: Adjusted Scaling of FDG Positron Emission Tomography Images for Statistical Evaluation in Patients with Suspected Alzheimer's Disease. Journal of Neuroimaging, 15 (2005) 348-355.
11. Allen, E.: An introduction to linear regression and correlation. WH Freeman (1976)
12. Semmlow, J.L.: Biosignal and Biomedical Image Processing, MATLAB-Based Applications. Marcel Dekker Ltd. 22 (2004)

Automatic Segmentation of Single and Multiple Neoplastic Hepatic Lesions in CT Images

Marcin Ciecholewski and Marek R. Ogiela

Institute of Automatics, AGH University of Science and Technology,
al. Mickiewicza 30, 30-059 Kraków, Poland
{ciechol, mogiela}@agh.edu.pl

Abstract. This paper describes an automatic method for segmenting single and multiple neoplastic hepatic lesions in computed-tomography (CT) images. The structure of the liver is first segmented using the approximate contour model. Then, the appropriate histogram transformations are performed to enhance neoplastic focal lesions in CT images. To segment neoplastic lesions, images are processed using binary morphological filtration operators with the application of a parameterized mean defining the distribution of gray-levels of pixels in the image. Then, the edges of neoplastic lesions situated inside the liver contour are localized. To assess the suitability of the suggested method, experiments have been carried out for two types of tumors: hemangiomas and hepatomas. The experiments were conducted on 60 cases of various patients. Thirty CT images showed single and multiple focal hepatic neoplastic lesions, and the remaining 30 images contained no disease symptoms. Experimental results confirmed that the method is a useful tool supporting image diagnosis of the normal and abnormal liver. The proposed algorithm is 78.3% accurate.

1 Introduction

Liver cancer is the fifth most widespread type of cancer in the world in terms of the number of cases, and the third most important as measured by the mortality rate [4]. The average incidence of liver cancer is 16 cases per 100,000 population worldwide, and 4 cases in Europe [4]. In order to improve the curability of liver cancer, early detection is critical. Liver cancer manifests itself in abnormal cells whose growth leads to the emergence of either single or multiple neoplastic lesion formations. If the hepatic tumor is detected early, the progress of the disease can be stopped and further treatment and therapy can be easier. Consequently, the design and development of computer-assisted image processing techniques to help doctors improve their diagnosis has made significant inroads into the clinical practice.

Generally, the literature of the subject includes several papers on segmenting neoplastic lesions in CT images. Papers [6, 9] describe how tumors can be located using statistical analyses. Park [6] determined a specific range of values called the statistical optimal threshold (SOT) which corresponds to the pixels

J. Mira and J.R. Álvarez (Eds.): IWINAC 2007, Part II, LNCS 4528, pp. 63–71, 2007.

representing neoplastic lesions. The SOT range is established using calculated values of the probability density of gray levels in the image, which is related to the Gaussian distribution, and determining the total value of the probability distribution error for pixels in the image. Values within the range (SOT) are used to locate neoplastic lesions. Seo [9] extracted blood vessels from CT images first. Then, just as in paper [6], a certain range of values called the Optimal Threshold (OT) was established to allow the neoplastic lesion to be located. The OT threshold is established on the basis of the probability density of gray levels in the image, which is related to the Gaussian distribution, and of the minimum error of the probability distribution for pixels in the image. In paper [10], the first step is to distinguish a convex hull containing the liver structure. Then the variance of grey levels is estimated for a single neoplastic lesion in the CT image. The authors of paper [1] used Kohonen neural networks to establish the features of the image which correspond to the area of neoplastic lesions. Papers [6, 9, 10] present interesting segmentation methods, albeit they can only be used for single tumors. Another limitation is that the research work was conducted on CT scans of no more than 10 patients. In radiology practice, there are as many cases of single tumors as there are of multiple neoplastic lesions (a 50/50 proportion).

This paper proposes a method which facilitates the automatic segmentation of single and multiple neoplastic lesions of various shapes and locations within the liver structure. Research was conducted on 60 cases of different patients. Thirty CT images contained single and multiple neoplastic lesions, while the remaining 30 showed healthy livers. The automatic method of neoplastic lesion segmentation in liver CT scans is presented in the next section. The section after that discusses the experiments conducted and research results. The last section summarises the study and presents research directions for the future.

2 Neoplastic Lesion Segmentation in Liver CT Images

This section presents the method for automatically segmenting single and multiple neoplastic lesions in liver CT images. First, the liver is extracted from the CT (computed tomography) image. Then, the histogram is transformed to enhance neoplastic lesions in CT images. In order to segment neoplastic lesions, the images are processed using binary morphological filtration operators with the application of a parametrized mean defining the distribution of gray levels of pixels in the image. The next step is to locate the edges of neoplastic lesions found inside the liver contour.

2.1 Liver Segmentation

The first step in the sequence of neoplastic lesion segmentation is to distinguish the liver structure, as this allows surplus elements to be removed from the image. The liver structure is segmented from the CT image by first finding the liver contour made up of a finite number of joint polylines which approximate particular fragments of the liver edge in the computed tomography image. This

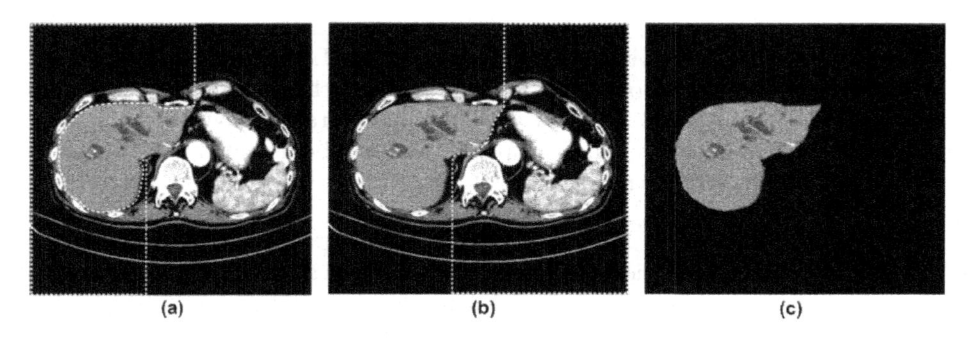

Fig. 1. Liver segmentation in a CT image of the abdominal cavity. (a) A CT image with a superimposed polygon containing the approximate liver edge and a fragment of the left side of the image. (b) A CT image with a superimposed polygon containing the approximate liver edge and a fragment of the right side of the image. (c) The segmented liver structure.

approach is presented in detail in papers [2, 3]. The area of the image located outside the liver contour is divided into two polygons which are eliminated from the image. Fig. 1.c shows an example of a CT image with the liver structure segmented using the approximate contour method [2, 3].

2.2 Histogram Transformations

Having segmented the liver structure in the CT image, the next step leading to the extraction of neoplastic foci is the transformation of the histogram by its equalization, which yields an even distribution of the number of pixels relative to the gray levels. The histogram is a single-dimensional statistical function obtained by counting the number of pixels corresponding to specific grey levels.

Let $g : M^2 \rightarrow Z$ be the gray-level CT image containing the liver structure and $(x, y) \in [0, M-1] \times [0, M-1]$ be the pixel coordinates. Then: $g(x, y) \in Z$. The histogram, $h(k) : Z \rightarrow Z$ is defined as the following set:

$$h(k) = \{(x, y) : g(x, y) = k\} \tag{1}$$

where k is the value of the grey level. Let $g_{HE_q} : M^2 \rightarrow Z$ be the histogram equalization transformation. It is represented by the following relationship:

$$\forall_{(x,y)\in M^2} \quad g_{HE_q}(x, y) = LUT(g(x, y)), LUT(k) = \frac{g_{max}}{x_{max} \cdot y_{max}} \sum_{j=0}^{k} h(j) \tag{2}$$

where LUT(look-up table) is the adjustment table allowing the gray levels of the input image to be changed in accordance with the values stored in the table. The g_{max} variable is the number of grey levels in the image, while the

$x_{max} \cdot y_{max}$ product defines the size of the image frame ($x_{max} \cdot y_{max} = M^2$), and $h(j)$ is the number of pixels with the j gray-level. Let $g_{GF} : M^2 \to Z$ be the Gaussian smoothing function [5] with the mask G with the dimensions of $5x5$, $g_{GF} = (g \times G)(x, y)$. As a result of the subsequent transformations g_{HE_q} and g_{GF}, we obtain an image in which neoplastic foci will be detected. The image is defined by the following formula:

$$g_{GF} = (g_{HE_q} \times G)(x, y) \tag{3}$$

The details of these operations are shown in Figures 2 and 3, which present the subsequent histogram transformations. Figures 2 (a) and 2 (b) contain the image of the liver structure g and its histogram $h(k)$. Figures 2 (c) and 2 (d) show the g_{HE_q} image after the histogram equalization transformation and the equalized histogram graph. Figures 3 (a) and 3 (b) contain the g_{GF} image after Gaussian smoothing and its histogram graph.

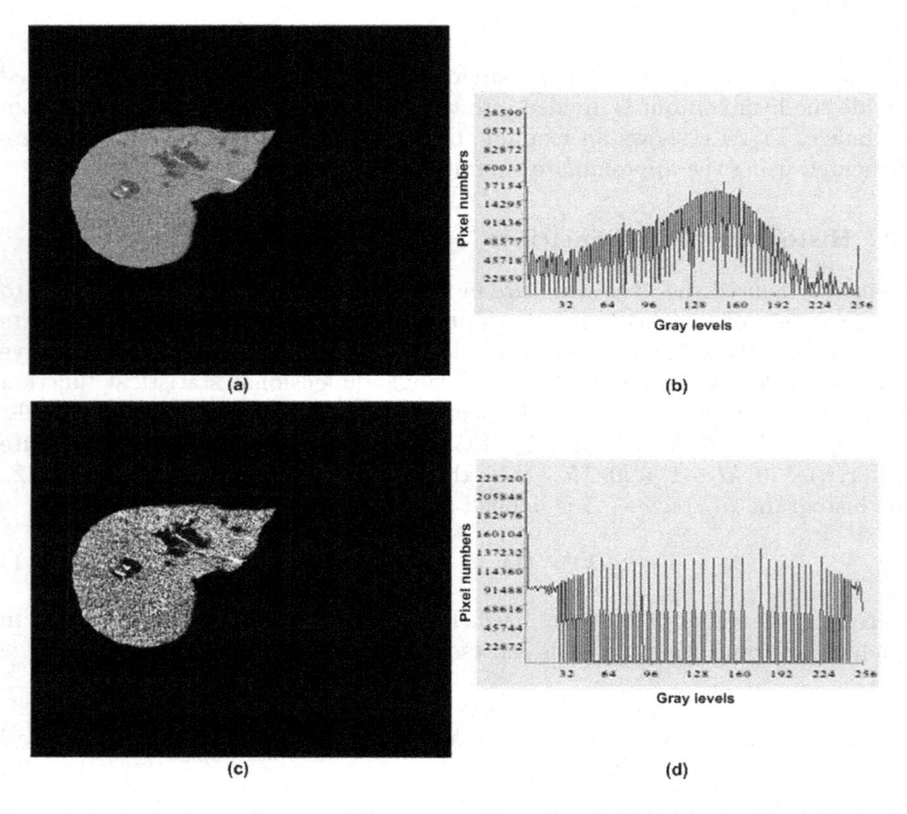

Fig. 2. Transformations in the CT image of the liver. (a) A CT image containing the liver structure (b) A histogram $h(k)$ (c) The image after the histogram equalization transformation (d) The histogram graph after the equalization.

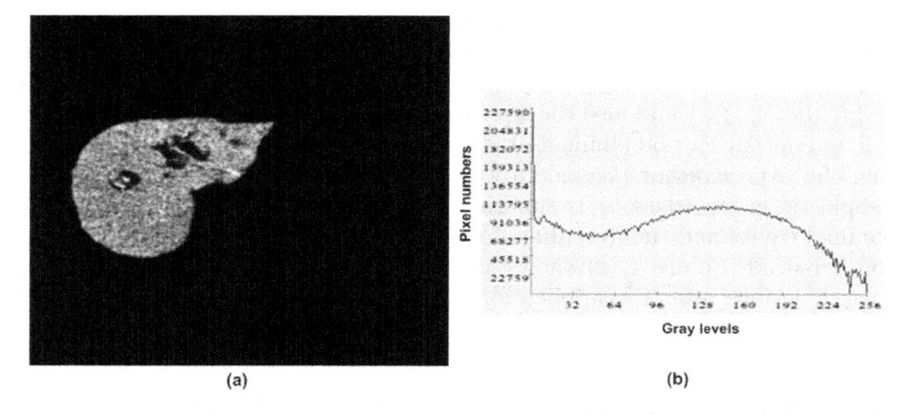

Fig. 3. Transformations in the CT image of the liver. (a) The image after Gaussian smoothing. (b) The histogram graph after Gaussian smoothing.

2.3 Segmenting Neoplastic Lesions

To segment neoplastic lesions, the image g_{GF} is first binarized using the function $b : M^2 \rightarrow M^2$, where

$$b = \begin{cases} 255 \text{ (white)} & \text{if } g_{GF}(x, y) > \mu \\ 0 \text{ (black)} & \text{if } g_{GF}(x, y) \leq \mu \end{cases} \tag{4}$$

The $\mu = \sum_{k \in Z} k \cdot \frac{h(k)}{M^2}$ parameter defines the mean calculated for the pixels of the g_{GF} image. Then, the binary image b is subjected to elementary filtration operations such as erosion, median filtration and edge detection.

Let the $e : M^2 \rightarrow M^2$ function define erosion and $e^i(b) : M^2 \rightarrow M^2$ be the iterative erosion (ie). Under the assumption that SE is the structural element with the dimensions of $3x3$, we can also define the median filtration operation. Let $m : M^2 \rightarrow M^2$ be the median filtration with the SE mask. Also, let $b_L : M^2 \rightarrow M^2$ define the linear filter using a Laplacian to detect the edges in the image. The Laplacian is represented by the L mask with the dimension of $3x3$. Every filtration is defined using image algebra operators for binary images [6, 8]. The definitions are as follows:

$$\text{Erosion and iterative erosion } e = \{b \ominus SE\}, e^i(b) = \{\ldots((b \ominus SE) \ominus SE) \ominus SE\} \tag{5}$$

$$\text{Median } m = \{b \circledm SE\} \tag{6}$$

$$\text{Laplacian } b_L = \{b \times L\} \tag{7}$$

The research has shown that the best method to improve neoplastic lesion segmentation is double erosion $e^2(b)$. As a result, the black image areas are magnified. Then, the image edges $e^2(b)$ are smoothed using median filtration. As a result of applying that filtration, we get an image defined by the following formula: $m = \{e^2(b) \circledm SE\}$. Figure 4 (a) shows a binarized specimen liver image, while Figure 4 (b) shows a liver image after double erosion and median filtration. As a result of using the linear filter $b_L = \{m \times L\}$, we get a set comprising a sub-set representing a specific number of edges of neoplastic lesions $b_f = \{f_1, f_2, \ldots, f_K\}$ and also the liver contour c. Thus $b_L = \{f_1, f_2, \ldots, f_K, c\}$. The liver contour can be eliminated, since its coordinates have been calculated during the liver structure segmentation, as presented in papers [2, 3]. The set of neoplastic lesion edges b_f is superimposed on the original CT image of the abdominal cavity and the liver, Fig. 4 (d). Thus is the neoplastic lesion segmentation achieved. Figure 4 (c) shows the contour of the liver and the edges of neoplastic lesions after the application of the linear filter which detects edges. Figure 4 (d) shows the liver image with the segmented neoplastic lesions.

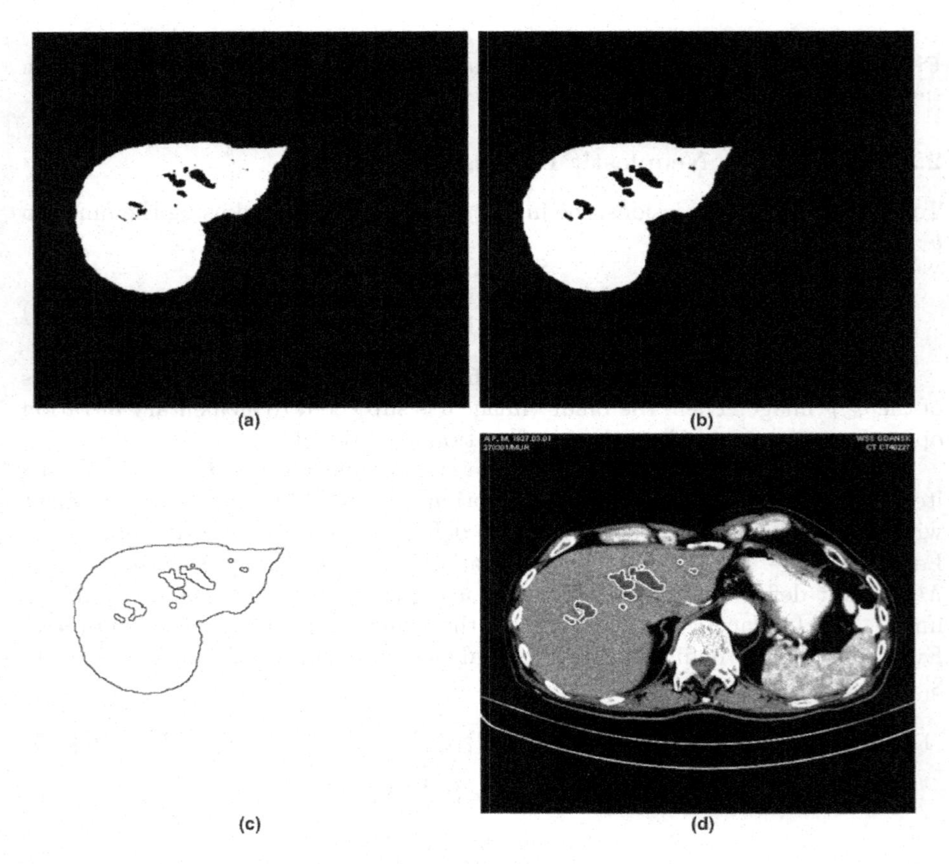

Fig. 4. The segmentation of hepatic neoplastic lesions in a CT image. (a) The binary image (b) The image after double erosion and median filtration (c) Edges of neoplastic lesions and the liver. (d) A CT image of the abdominal cavity and the liver with segmented neoplastic lesions.

3 Experiments Conducted and Research Results

During the studies on the analysis of CT images, the material from the Department of Image Diagnostics of the Provincial Specialist Hospital in Gdańsk, Poland, was used. The CT images were obtained with Somatom Emotion 6 (Siemens). CT images were performed with intravenous contrast enhancement. The parameters of exposure for the abdominal cavity were 130 kVp and 95 mAs for the venal phase, and 110 kVp and 90 mAs for the arterial phase, a 2 mm slice colimmation and a table speed of 15 mm/sec.

To verify the suitability of the proposed method, tests were run for two types of tumors: hemangiomas and hepatomas, and for images showing healthy livers. The tests were conducted on 60 cases of various patients. Thirty CT images contained single or multiple neoplastic lesions of various shapes and locations within the liver structure. Another thirty showed healthy livers. A radiologist participated in the tests to assess the segmentations of neoplastic lesions made. For every CT image, the neoplastic lesion segmentation was classified to one of the four basic possibilities: (TN) true negative, (FP) false positive, (FN) false negative and (TP) true positive [7]. Table 1 shows the results of classifying the processed images in relation to the number of cases.

Three ratios - sensitivity, specificity and accuracy - were calculated for the test data obtained. The sensitivity represents the number of patients who have neoplastic lesions:

$$Sensitivity = \frac{TP}{TP + FN} \cdot 100\% \tag{8}$$

Specificity defines the number of patients with no neoplastic lesions:

$$Specificity = \frac{TN}{TN + FP} \cdot 100\% \tag{9}$$

Accuracy is defined as:

$$Accuracy = \frac{TP + TN}{TP + TN + FP + FN} \cdot 100\% \tag{10}$$

The sensitivity of the method for test data amounted to 73.3%, its specificity to 83.3%, and accuracy to 78.3%. The ratios obtained confirm that the method is highly suitable for image diagnostics of the liver. However, it should be noted that the value of the FP parameter was obtained for images in which the falciform

Table 1. Test results for 60 CT images of the abdominal cavity and the liver

Lesion	Quantity	Number of TN	Number of FP	Number of FN	Number of TP
Hemangioma	15	0	0	3	12
Hepatoma	15	0	0	5	10
Normal Liver	30	25	5	0	0
Total Number	60	25	5	8	22

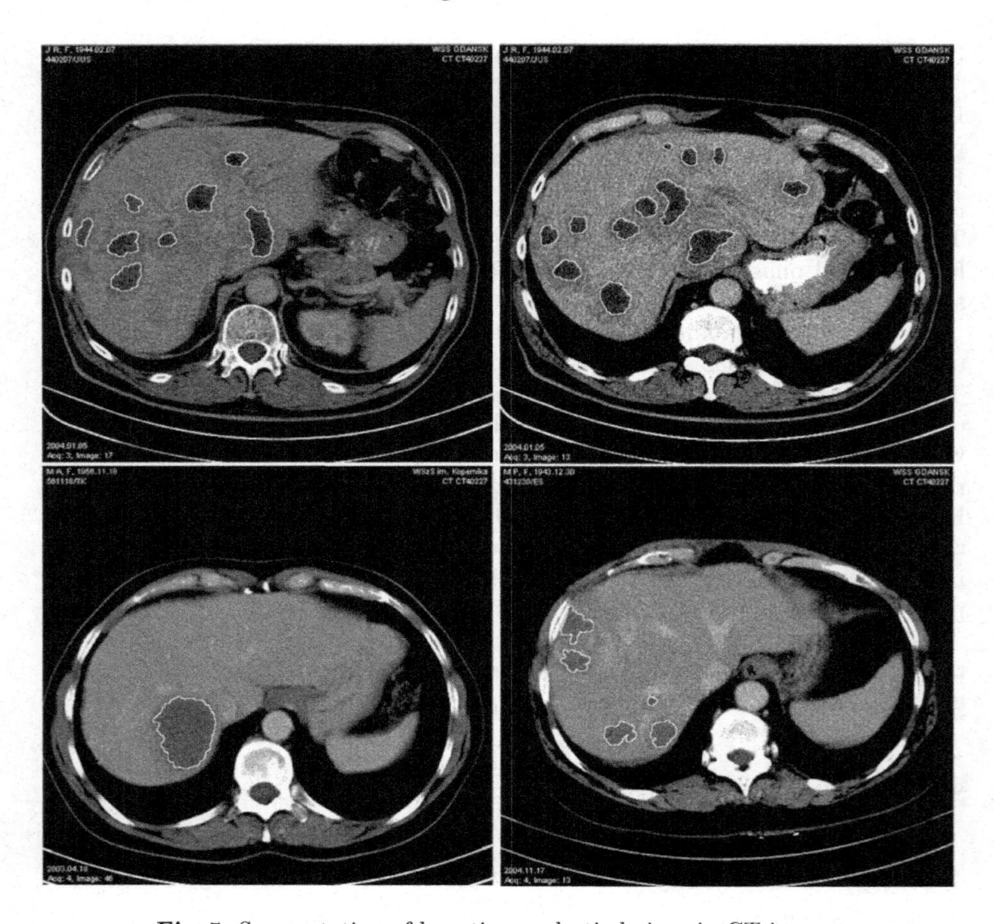

Fig. 5. Segmentation of hepatic neoplastic lesions in CT images

ligament which splits the liver into the right and left lobes was highly visible, which means that it could be interpreted as one of the possible foci. In the case of multiple neoplastic lesions whose foci had a diameter below 1 cm, FN values occurred for only some of them. Fig. 5 shows examples of hepatic lesions segmentation.

4 Summary and Further Research Directions

The article presents a method of automatic segmentation of single and multiple neoplastic lesions in CT images for a computer system assisting the early diagnostics of inflammation and neoplastic lesions of the liver. The segmentation of these lesions was achieved by first segmenting the liver from the computed tomography (CT) image. Then, the histogram was transformed to enhance the neoplastic lesions in CT images. In order to segment the neoplastic lesions, the images were processed using binary-morphological filtration operators with the application of a parametrized mean defining the distribution of grey levels of pixels in the image. The next step was to locate the edges of neoplastic lesions found inside the liver contour. To verify the suitability of the proposed method, tests were conducted for two types of tumors: hemangiomas and hepatomas, and also for images of healthy livers. The test data yielded the following ratio values: sensitivity 73.3%, specificity 83.3% and accuracy 78.3%. The ratio values obtained confirm the high suitability of the method for image diagnostics of the liver, both healthy and containing neoplastic lesions. In the future, research work will aim at developing a method which eliminates the falciform ligament as a lesion, because the current method classifies this ligament as a false positive (FP). Another research objective will be to raise the sensitivity of the method to reduce the occurrence of FNs and thus to improve the results of the segmentation of multiple neoplastic lesions below 1 cm in diameter.

Acknowledgements

This research project was financed with state budget funds for science in 2006-2007 as research project no. 3T11F 030 30 of the Ministry of Science and Higher Education.

References

1. Chen, E.L., Chung, P.C., Chen, C.L.,Tsai, H.M., Chang, C.I.: An automatic diagnostic system for CT liver image classification, IEEE Transactions on Biomedical Engineering, Vol. 45. No. 6.(1998) 783-794.
2. Ciecholewski M., Dębski K.: Automatic Segmentation of the Liver in CT Images Using a Model of Approximate Contour LNCS, Vol. 4263. (2006) 75-84.
3. Ciecholewski M., Dębski K.: Automatic detection of liver contour in CT images. Automatics, semi-annual journal of the AGH University of Science and Technology, Vol. 10. No. 2. (2006).
4. Danaei, G., Vander Hoorn S., Lopez A.D., Murray C.J., Ezzati M.: Causes of cancer in the world: comparative risk assessment of nine behavioural and environmental risk factors. Lancet, Vol. 366. (2005) 1874-1793.
5. Meyer- Bäse, A.: Pattern Recognition for medical imaging. Elsevier Academic Press (2004)
6. Park, S., Seo, K., Park, J.: Automatic Hepatic Tumor Segmentation Using Statistical Optimal Threshold. LNCS, Vol. 3514. (2005) 934-940.
7. Rangayan R.M.: Biomedical signal analysis. Wiley Computer Publishing, New York (1997)
8. Ritter, G.X., Wilson, J.N.: Computer Vision Algorithms in Image Algebra. CRC Press, Boca Raton, Florida (2000)
9. Seo, K.: Automatic Hepatic Tumor Segmentation Using Composite Hypotheses. LNCS, Vol. 3656. (2005) 922-929.
10. Seo, K., Chung, T.: Automatic Boundary Tumor Segmentation of a Liver. LNCS, Vol. 3483. (2005) 836-842.
11. Sonka, M., Fitzpatrick, J.M.: Handbook of Medical Imaging volume 2. Medical Image Processing and Analysis. SPIE Press, Bellingham WA (2000)
12. Tadeusiewicz, R., Ogiela, M.: Medical Image Understanding Technology. Springer, Berlin-Heidelberg (2004)

Biometric and Color Features Fusion for Face Detection and Tracking in Natural Video Sequences

Juan Zapata and Ramón Ruiz

Universidad Politécnica de Cartagena, Cartagena Murcia 30203, Spain
juan.zapata@upct.es
http://www.detcp.upct.es/~Personal/JZapata/index.html

Abstract. A system that performs the detection and tracking of a face in real-time in real video sequences is presented in this paper. The face is detected in a complex environment by a model of human colour skin. Very good results are obtained, since the colour segmentation removes almost all the complex background and it is realized to a very high-speed, making the system very robust. On the other hand, fast and stable real-time tracking is then achieved via biometric feature extraction of face using connected components labelling. Tracking does not require a precise initial fit of the model. Therefore, the system is initialised automatically using a very simple 2D face detector based on target ellipsoidal shape. Results are presented showing a significant improvement in detection rates when the whole sequence is used instead of a single image of the face. Experiments in tracking are reported.

1 Introduction

Real-time human face tracking in complex environment has many practical applications, such as visual surveillance and biometric identification, and is a challenging research topic in computer vision applications. Different techniques have been developed in the past for people detection/tracking attending to different parts of the human body: the face [2,3], the head [4,5], the entire body [6] or just the legs [7], as well as the human skin [8]. This work addresses the problem of face tracking in natural video sequences. Specifically, given a video stream in which a human face is in motion, the goal is to track the face across time varying images. In this framework, and depending on the examined application, many approaches have been proposed in the literature, focusing either in the highest accuracy, or the lowest computational complexity for real-time tracking. Among this last category, features invariant approaches has been successfully applied to track moving objects in cluttered environments, due to their performance in various problems concerning object tracking in natural video sequences. These algorithms aim to find structural features that exist even when the pose, viewpoint, or lighting conditions vary, and then use the these to locate faces. These methods are designed mainly for face localisation and detection.

J. Mira and J.R. Álvarez (Eds.): IWINAC 2007, Part II, LNCS 4528, pp. 72–80, 2007.

Accurate and real-time face tracking will improve the performance of face recognition, human activity analysis and high-level event understanding but, why face detection is a difficult task?

- Pose (Out-of-Plane Rotation): The images of a face vary due to the relative camera-face pose (frontal, 45 degree, profile, upside down).
- Presence or absence of structural components: Facial features such as beards, moustaches, and glasses may or may not be present and there is a great deal of variability among these components including shape, colour, and size.
- Facial expression: Face appearance is directly affected by a person's facial expression.
- Occlusion: Faces may be partially occluded by other objects. In an image with a group of people, some faces may partially occlude other faces.
- Orientation (In-Plane Rotation): Face images directly vary for different rotations about the camera's optical axis.
- Imaging conditions: When the image is formed, factors such as lighting (spectra, source distribution and intensity) and camera characteristics (sensor response, lenses) affect the appearance of a face.

Many methods have been proposed to build a skin colour model but skin colour alone is usually not sufficient to detect or track faces. A modular system using a combination of connected components labelling and colour segmentation for face tracking in a video sequence in real-time have been developed by us. Connected-component labelling is a method for identifying each object in a binary image. This procedure returns a matrix, called a label matrix. A label matrix is an image in which the objects in the input image are distinguished by different labels in the output matrix. Then, the procedure returns different features of the labelled objects, mainly the area and shape. The area is a measure of the size of the different objects of the image. The shape is a measure of the eccentricity of a object in comparison to a circle. Roughly speaking, the area is the number of pixels in the image and the shape is the number of likelihood to a circle. So, we use these features to detect, segment and track a face into the scene. We show this method and prove its feasibility in the task of tracking of a face in this paper.

2 Skin-Colour Modelling

The choice of colour space can be considered as the primary step in skin-colour classification for human face detection and tracking. The RGB colour space is the default colour space for most available image formats. Any other colour space can be obtained from a linear or non-linear transformation from RGB. The colour space transformation is assumed to decrease the overlap between skin and non-skin pixels thereby aiding skin-pixel classification and to provide robust parameters against varying illumination conditions. It has been observed that skin colours differ more in intensity than in chrominance. Hence, it has been a common practice to drop the luminance component for skin classification. Several colour spaces have been proposed and used for skin detection.

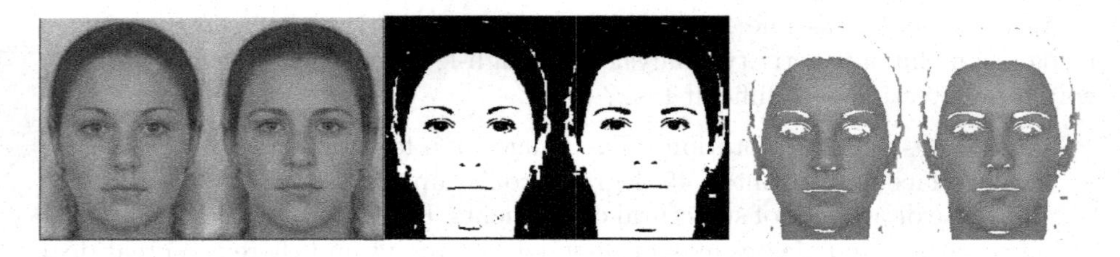

Fig. 1. Example results of the proposed skin-colour modelling to segment skin in still images

The orthogonal colour spaces reduce the redundancy present in RGB colour channels and represent the colour with statistically independent components (as independent as possible). As the luminance and chrominance components are explicitly separated, these spaces are a favourable choice for skin detection. The YCbCr space represents colour as luminance (Y) computed as a weighted sum of RGB values, and chrominance (Cb and Cr) computed by subtracting the luminance component from B and R values. The YCbCr space is one of the most popular choices for skin detection and has been used by Hsu et al. [10], Chai and Bouzerdoum [11] and Wong et al. [12].

The final goal of skin colour detection is to build a decision rule, that will discriminate between skin and non-skin pixels. This is usually accomplished by introducing a metric, which measures distance (in general sense) of the pixel colour to skin tone. The type of this metric is defined by the skin colour modelling method.

Our method in order to build a skin classifier is to define explicitly (through a number of rules) the boundaries skin cluster in our colour space.

In our case, the selected colour space is defined by a YCbCr colour components. Video acquisition is obtained by mean of a RGB video camera therefore a RGB to YCbCr transformation is required: (Y,Cb,Cr) is classified as skin if: $Y > 100$ and $Y < 140$ or $Cb > 128$ and $Cr < 128$

The simplicity of this method is a very important feature for the success of our procedure. The obvious advantage of this method is simplicity of skin detection rules that leads to construction of a very rapid classifier. The main difficulty achieving high recognition rates with this method is the need to find both good colour space and adequate decision rules empirically.

Colour information is an efficient tool for identifying facial areas for next detecting specific facial features. In Fig 1, we can observe how our skin-colour modelling can segment facial areas between another artifacts. In the next stage, we must segment only the face by means specific biometrical facial features (shape and size).

3 Connected Components Labelling

In a second phase, the main goal is to obtain connected components in the binary image. This binary image is the output image of our skin-colour model.

Fig. 2. Example results of three faces detected and segmented

Next, that objects must be labelled. The implement algorithm uses the general procedure proposed by Haralick and Shapiro [9]. The algorithm returns a matrix, the same size as the input image, containing labels for the 8-connected objects in the input image. The pixels labelled 0 are the background, the pixels labelled 1 represent one skin object, the pixels labelled 2 represent a second skin object, and so on. Each object is assigned an unique label to separate it from other objects. This assignation is as follows: all pixels within a object of spatially connected 1's are assigned the same label. In this way, it can be used to establish boundaries of objects in order to extract different features of components (objects) in an image. Typically, the procedure performs two passes through the image. In the first pass, the image is processed from left to right and top to bottom to generate labels for each pixel. These labels are stored in a pair of arrays. In the second pass, each label is replaced by the label assigned to its equivalence class.

Extracting and labelling of various disjoint and connected components in an image is central to many automated image analysis applications and, in binary images is a fundamental step in segmentation of an image objects.

Once all objects in the scene are labelled then each object can be processed independently. The binary object features include area, center of area, axis of least second moment, perimeter which give us information about where the object is located. Euler number, projections, thinness ratio, and aspect ratio tell us information about the shape.

Now, the size information is used in order to eliminate little objects and noise. For it, the size of all objects of the scene is obtained and a threshold value is calculated using Otsu's method [1]. Original Otsu's method is a threshold selection method from gray-level histograms, which chooses the threshold to minimise the intraclass variance of the thresholded black and white pixels.

Then, we use information of shape to eliminate hands, reflects due to illumination and other artifacts. A human face has a ellipsoid shape more o less. Hands and other artifacts has a very different shape. Now, all objects different to an ellipsoid are eliminated. The eccentricity is a key feature in order to find ellipsoids and circles between objects. The eccentricity is the ratio of the distance between the foci of the ellipse and its major axis length. An ellipse whose eccentricity is 0 is a circle, while an ellipse whose eccentricity is 1 is a line segment.

4 Proposed Tracking Scheme

In object tracking, the goal is to track object's contour along time; this actually means that, for each frame of a video sequence, we aim at separating the object from the background. The issue of object tracking in natural and cluttered sequences, involves various problems, especially when its most general confrontation is needed. The generality of a method is one of the most important goals, which means that we have to deal with as few as possible initial constraints, or even make the tracking problem independent from initial conditions

Fig. 3. Example results of the proposed method to track a face

or constraints. The difficulties that usually arise in real-world scenes are: (a) non-rigid (or articulated) moving objects, (b) moving objects with complicated contours, (c) object motions that are not simple translations, but also involve rotations, objects approaching or drawing away from the shooting camera (or camera zooming), (d) sequences with highly textured background, (e) existence of noise or abrupt/gradual environmental (external) changes, such as external lighting changes, (f) moving objects that get successively occluded by obstacles, that can be either static or moving. In this section, we describe the proposed tracking method that utilises the scheme model defined in previous sections, and exploits the results of our face detection scheme. Our method consists of three main steps: (a) definition of different zones (skin zones), in which the object (head/face) is supposed to be located, (b) labelling of connected components in each frame of video sequence and (c) calculation of the features of the objects and segmentation and tracking for feature analisys.

Usually the face shape can be chosen as ellipse or rectangle for fast tracking. In our system, the face is modelled as rectangle window and the size of the rectangle is changed online automatically. This change of dimensions is due to scale change of the face during the video sequence. The state of our tracker is defined as $S = x, y, w, h$, where (x, y) is the centre of the rectangle, (w, h) represents the width and height separately. Any lost of detection by the face tracker is defined by absence of rectangle. During the tracking, we pretend that some features like eyes and mouth keep inside this rectangle.

Shown in the figures of Fig 3 is the response of the tracking algorithm in a scene that collects several complex factors: variation on face scale, illumination conditions do not remain constant, the subject turns his face, and objects with skin-like colour appear on the background. With this method we are able to undergo real time people tracking in complex situations such as in unstructured backgrounds, varying illumination conditions and with rotations of the person face.

The main advantage of this scheme that use explicitly defined skin cluster boundaries is the simplicity and intuitiveness of the classification rules. However, the difficulty with them is the need to analyse some feature of detected objects, in our case labelled by means of connected components, and to obtain an adequate decision rule in order to segment the interest object in this case a human face frame to frame.

5 Results and Discussions

In order to test the reliability of the proposed methodology, a comparison between visual and automated evaluation was carried out. A set of 86 still images were qualified by four human inspectors as a function of their difficulty. Two different qualification schemes were used. In the first, the expert inspector qualified each of the images from its illumination quality into three categories, images of low, medium and high quality of illumination. In a second scheme, two quantitative measures were used to characterise the image database. The first measure was "contrast with background", whereby the database was ordered in to three groups; images of low, medium and high contrast. And lastly, from the appearance of the occluded face, the images were grouped in to another two categories; images with occluded or without occluded faces.

In order to compare the segmentation results, a modificated Pratt's quality measurement is computed [13]:

$$F = \frac{\sum_{A=1}^{A_A} \frac{1}{1+\alpha(d(i)^2)}}{max(A_A, A_I)} \tag{1}$$

where A_A, is the number of area pixels produced by computer-aided segmentation method, A_I, is the number of area pixels pointed by a human expert. $d(i)$ is the Euclidean distance between a pixel pointed by the technicians and the nearest pixel produced by computer-aided segmentation and α is a scaling constant, with a suggested value of 1/9 [13]. For our scheme computed by our procedure, $F = 0.866$. Our segmentation is very accurate.

In order to validate the proposed technique for detection of human faces, the area identified manually as face by an human was termed "real human face". Similarity is defined as the percentage of agreement between the face region detected by the system and the real face region.

$$\text{Similarity (\%)} = \frac{\text{concordance area}}{\text{real area}} \times 100$$

The test results indicate that the proposed methodology achieves a similarity of about 89%. The results are shown in Table 1 for each one of the categories.

In order to evaluate the performance of the system, a property termed sensibility was used which is defined in the following way:

$$\text{Sensibility(\%)} = \frac{VP}{VP + FN} \times 100$$

where VP is the number of True Positives and FN is the number of False Negatives. In this stage, our system is able to obtain a sensibility of 100%, i.e. the system detects all the faces observed by a human expert without occlusion. Only

Table 1. Face Detection Results

Illumination	
Low Quality	97
Medium Quality	98
High Quality	100
Contrast	
Low Contrast	88
Medium Contrast	89
High Contrast	100
Occlusion	
Occluded (partially and totally)	40
Not Occluded	100
Average Similarity	89%

a few misinterpretations occurred: in these cases, elliptical skin colour regions were falsely recognised as face. Skin colour misinterpretations are especially difficult to resolve since there is no strict definition of what a small face or artifact is.

The acquisition runs at 15 fps (320×240) on a 2.8 GHz pentium IV processor. The system was tested on different people using data collected over a period of various months. Among the test candidates, there were variations in sex, skin colour, hair colour, position of hair-line, hair style, amount of hair, head and face shape and clothing. The test data was collected in different room environments with a lot of variations in lighting conditions and background clutter.

To validate the proposed technique for the tracking of a human face, a set different movies was acquired in real time and processed by a our system. These were acquired because they contain single o multiple faces and face-free samples. A human inspector can perform the localisation of one o more faces on each image and to complete a short report on his, findings, the faces detected and their location. A set of movies were evaluated by the system. Example results of the proposed method to test one sequence are shown in Fig. 3.

The system is able to track the target face correctly from the start frame to the end frame in 90% of the videos (each video lasted 10 secs). The procedure is able to track the face correctly during partial occlusions and detect the face after temporary complete occlusion. In this sense, the face detection rate was improved when the whole sequence was considered instead of a single still image. The system partially fails only when the illumination has a very bad quality and the face is occluded totally for a long time or permanently.

Different techniques can be better than others depending on how "truth" is defined; majority vote, agreed opinion, expert opinion, retrospective revision. How does the performance of our system compare with previous studies? Unfortunately, no one knows the answer because different sets of images o movies were used in different studies. The collection of reference images and movies used in this work is not an accepted standard but it is a basis for comparison.

6 Conclusions

We have presented a real-time system for the detection and tracking of human face based on skin colour segmentation and connected components labelling. The core component of the system is based on a tracking face algorithm that involves low computational cost. Experimental results demonstrated the robust of the system to scale variations, relative fast subject movements and camera saccades, partial occlusion and rotations. There are still rooms to improve the proposed approach. In particular, our future research will be focused on addressing the following issues. Firstly, we are trying the detection and tracking of eyes and mouth using this scheme. Second, we are currently investigating the relation between colour spaces, in order to establish a reliable relation between them for a hybrid scheme. Also, the implementation of a more "intelligent" adaptive election of face or non face.

This paper presents a review of the generic techniques dedicated to implementing a system of detection and tracking of human faces: acquisition of the image, discriminate feature enhancement facing the interpretation, multi-level segmentation of the scene to isolate the areas of interest, classification in terms of individual and overall features, and finally, severity evaluation of the detection and cataloguing of the degree of acceptance in the detection and tracking. This paper presents and analyses the aspects in the design and implementation of a new methodology for the detection and tracking of human faces. After a test phase and updating for the specific proposed technique, an valuation of the relative benefits is presented. From the validation process developed with 86 set of images and different movies, it can be concluded that the proposed technique is capable of achieving very good results.

Acknowledgement

This work has been supported by Ministerio de Ciencia y Tecnología of Spain, under grant TIN2006-15460-C04-04.

References

1. Otsu, N.: A Threshold Selection Meted from Gray-Level Histograms. IEEE Transactions on Systems, Man and Cybernetics **9** (1979) 62–66
2. Erik Hjelmas and Boon Kee Low: Face detection: A survey. Computer Vision and Image Understanding **83(3)** (2001) 236–274.
3. Ming-Hsuan Yang, David Kriegman and Narendra Ahuja: Detecting faces in images: A survey. Transactions on Pattern Analysis and Machine Intelligence **24(1)** (2002) 34–58.
4. Stan Birchfield: Elliptical head tracking using intensity gradients and color histograms. Proceedings of the IEEE Conference on Computer Vision and Pattern Recognition (1998) 232–237.
5. Marco La Cascia and Stan Sclaro: Fast, reliable head tracking under varying illumination: An approach based on registration of texture-mapped 3d models. IEEE Trans. on Pattern Analysis and Machine Intelligence **22(4)** (2000) 322–336
6. Paul Viola, Michael J. Jones and Daniel Snow: Detecting pedestrians using patterns of motion and appearance. International Conference on Computer Vision **2** (2003) 734–741.
7. C. Papageorgiou, M. Oren, and T. Poggio: A general framework for object detection. Proceedings of the International Conference on Computer Vision (1998) 555–562.
8. Michael J. Jones and James M. Rehg: Statistical Color Models with Application to Skin Detection. Tech. Rep. Cambridge Research Laboratory (1998).
9. R.M. Haralick and L. Shapiro: Computer and Robot Vision, Addison Wesley **1** (1992) 28–48.
10. R.L. Hsu and M. Abdel-Mottaleb and A.K. Jain: Face detection in color images. IEEE Trans. Pattern Anal. Machine Intell **24(5)** (2002) 696–706.
11. D. Chai and A. Bouzerdoum: A Bayesian approach to skin color classification in YCbCr color space. IEEE TENCON00 **2** (2000) 421-424.
12. K.W. Wong and K.M. Lam and W.C. Siu. A robust scheme for live detection of human faces in color images. Signal Process. Image Commun **18(2)** (2003) 103–114
13. Pratt, W.K. Digital Image Processing (2e). John Wiley and Sons (1991)

Identifying Major Components of Pictures by Audio Encoding of Colours

Guido Bologna[1], Benoît Deville[2], Thierry Pun[2], and Michel Vinckenbosch[1]

[1] Laboratoire d'Informatique Industrielle, University of Applied Science HES-SO
Rue de la Prairie 4, 1202 Geneva, Switzerland
Guido.Bologna@hesge.ch, Michel.Vinckenbosch@hesge.ch
[2] Computer Science Center, University of Geneva
Rue Général Dufour 24, 1211 Geneva, Switzerland
Benoit.Deville@cui.unige.ch, Thierry.Pun@cui.unige.ch

Abstract. The goal of the See ColOr project is to achieve a non-invasive mobility aid for blind users that will use the auditory pathway to represent in real-time frontal image scenes. More particularly, we have developed a prototype which transforms HSL coloured pixels into spatialized classical instrument sounds lasting for 300 ms. Hue is sonified by the timbre of a musical instrument, saturation is one of four possible notes, and luminosity is represented by bass when luminosity is rather dark and singing voice when it is relatively bright. Our first experiments are devoted to static images on the computer screen. Six participants with their eyes covered by a dark tissue were trained to associate colours with musical instruments and then asked to determine on several pictures, objects with specific shapes and colours. In order to simplify the protocol of experiments, we used a tactile tablet, which took the place of the camera. Overall, experiment participants found that colour was helpful for the interpretation of image scenes.

1 Introduction

In this work we present *See ColOr* (**See**ing **Col**ours with an **Or**chestra), which is a multi-disciplinary project at the cross-road of Computer Vision, Audio Processing and Pattern Recognition. The long term goal is to achieve a non-invasive mobility aid for blind users that will use the auditory pathway to represent in real-time frontal image scenes. Ideally, our targeted system will allow visually impaired or blind subjects having already seen to build coherent mental images of their environment. Typical coloured objects (signposts, mailboxes, bus stops, cars, buildings, sky, trees, etc.) will be represented by sound sources in a three-dimensional sound space that will reflect the spatial position of the objects. Targeted applications are the search for objects that are of particular use for blind users, the manipulation of objects and the navigation in an unknown environment.

Sound spatialisation is the principle which consists of virtually creating a three-dimensional auditive environment, where sound sources can be positioned all around the listener. These environments can be simulated by means of loudspeakers or headphones. Among the precursors in the field, Ruff and Perret led a series of experiments on the space perception of auditive patterns [13]. Patterns were transmitted through a

J. Mira and J.R. Álvarez (Eds.): IWINAC 2007, Part II, LNCS 4528, pp. 81–89, 2007.

10x10 matrix of loudspeakers separated by 10 cm and located at a distance of 30 cm from the listener. Patterns were represented on the auditory display by sinusoidal waves on the corresponding loudspeakers. The experiments showed that 42% of the participants identified 6 simple geometrical patterns correctly (segment of lines, squares, etc). However, orientation was much more difficult to determine precisely. Other experiments carried out later by Lakatos taught that subjects were able to recognise with 60-90% accuracy ten alphanumeric characters [10].

Our See ColOr prototype for visual substitution presents a novelty compared to systems presented in the literature [9], [12], [4], [5] and [8]. More particularly, we propose the encoding of colours by musical instrument sounds, in order to emphasize coloured objects and textures that will contribute to build consistent mental images of the environment. Note also that at the perceptual level, colour is helpful to group the pixels of a mono-coloured object into a coherent entity. Think for instance when one looks on the ground and it "sounds" green, it will be very likely to be grass. The key idea behind See ColOr is to represent a pixel of an image as a sound source located at a particular azimuth and elevation angle. Depth is also an important parameter that we estimate by triangulation using stereo-vision. Finally, each emitted sound is assigned to a musical instrument, depending on the colour of the pixel.

In this work the purpose is to investigate whether individuals are able to learn associations between colours and musical instrument sounds and also to find out whether colour is beneficial to experiment participants. To the best of our knowledge this is the first study in the context of visual substitution for real time navigation in which colour is supplied to the user as musical instrument sounds.

2 Audio Encoding

This section illustrates audio encoding without 3D sound spatialization. Colour systems are defined by three distinct variables. For instance, the RGB cube is an additive colour model defined by mixing red, green and blue channels. We used the eight colours defined on the vertex of the RGB cube (red, green, blue, yellow, cyan, purple, black and white). In practice a pixel in the RGB cube was approximated with the colour corresponding to the nearest vertex. Our eight colours were played on two octaves : Do, Sol, Si, Re, Mi, Fa, La, Do. Note that each colour is both associated with an instrument and a unique note [2]. An important drawback of this model was that similar colours at the human perceptual level could result considerably further on the RGB cube and thus generated perceptually distant instrument sounds. Therefore, after preliminary experiments associating colours and instrument sounds we decided to discard the RGB model.

The second colour system we studied for audio encoding was HSV. The first variable represents hue from red to purple (red, orange, yellow, green, cyan, blue, purple), the second one is saturation, which represents the purity of the related colour and the third variable represents luminosity. HSV is a non-linear deformation of the RGB cube; it is also much more intuitive and it mimics the painter way of thinking. Usually, the artist adjusts the purity of the colour, in order to create different nuances. We decided to render hue with instrument timbre, because it is well accepted in the musical community that the colour of music lives in the timbre of performing instruments. This association

Table 1. Quantification of the hue variable by sounds of musical instruments

Hue value (H)	Instrument
red ($0 \leq H < 1/12$)	oboe
orange ($1/12 \leq H < 1/6$)	viola
yellow ($1/6 \leq H < 1/3$)	pizzicato violin
green ($1/3 \leq H < 1/2$)	flute
cyan ($1/2 \leq H < 2/3$)	trumpet
blue ($2/3 \leq H < 5/6$)	piano
purple ($5/6 \leq H < 1$)	saxophone

has been clearly done for centuries. For instance, think about the brilliant connotation of the Te Deum composed by Charpentier in the seventeenth century (the well known Eurovision jingle, before important sport events). Moreover, as sound frequency is a good perceptual feature, we decided to use it for the saturation variable. Finally, luminosity was represented by double bass when luminosity is rather dark and a singing voice when it is relatively bright.

The HSL colour system also called HLS or HSI is very similar to HSV. In practice, HSV is represented by a cone (the radial variable is H), while HSL is a symmetric double cone. Advantages of HSL are that it is symmetrical to lightness and darkness, which is not the case with HSV. In HSL, the Saturation component always goes from fully saturated colour to the equivalent gray (in HSV, with V at maximum, it goes from saturated color to white, which may be considered counterintuitive). The luminosity in HSL always spans the entire range from black through the chosen hue to white (in HSV, the V component only goes half that way, from black to the chosen hue). The symmetry of HSL represents an advantage with respect to HSV and is clearly more intuitive.

The audio encoding of hue corresponds to a process of quantification. As shown by table 1, the hue variable H is quantified for seven colours.

More particularly, the audio representation h_h of a hue pixel value h is

$$h_h = g \cdot h_a + (1 - g) \cdot h_b \quad (1)$$

with g representing the gain defined by

$$g = \frac{h_b - H}{h_b - h_a} \quad (2)$$

with $h_a \leq H << h_b$, and h_a, h_b representing two successive hue values among red, orange, yellow, green, cyan, blue, and purple (the successor of purple is red). In this way, the transition between two successive hues is smooth. For instance, when h is yellow then $h = h_a$, thus $g = 1$ and $(1 - g) = 0$; as a consequence, the resulting sound mix is only pizzicato violin. When h goes toward the hue value of green, which is the successor of yellow on the hue axis, the gain value g of the term h_a decreases, whereas the gain term of h_b ($(1 - g)$) increases, thus we progressively hear the flute appearing in the audio mix.

Table 2. Quantification of saturation by musical instrument notes

Saturation (S)	Note
$0 \leq S < 0.25$	Do
$0.25 \leq S < 0.5$	Sol
$0.5 \leq S < 0.75$	Sib
$0.75 \leq S \leq 1$	Mi

Table 3. Quantification of luminosity by double bass

Luminosity (L)	Double Bass Note
$0 \leq L < 0.125$	Do
$0.125 \leq L < 0.25$	Sol
$0.25 \leq L < 0.375$	Sib
$0.375 \leq L \leq 0.5$	Mi

Table 4. Quantification of luminosity by a singing voice

Luminosity (L)	Voice Note
$0.5 \leq L < 0.625$	Do
$0.625 \leq L < 0.75$	Sol
$0.75 \leq L < 0.875$	Sib
$0.875 \leq L \leq 1$	Mi

Once h_h has been determined, the second variable S of HSL corresponding to saturation is quantified into four possible notes, according to table 2.

Luminosity denoted as L is the third variable of HSL. When luminosity is rather dark, h_h is additionally mixed with double bass using the four notes depicted in table 3, while table 4 illustrates the quantification of bright luminosity by a singing voice. Note that the audio mixing of the sounds representing hue and luminosity is very similar to that described in equation 1. In this way, when luminosity is close to zero and thus the perceived colour is black, we hear in the final audio mix the double bass without the hue component. Similarly, when luminosity is close to one, the perceived colour is white and thus we hear the singing voice. Note that with luminosity at its half level, the final mix contains just the hue component.

3 Experiments

Our prototype is based on a sonified 17x9 sub-window pointed by the mouse on the screen which is sonified via a virtual Ambisonic 3D audio rendering system [3], [2], [6], [7], and [11]. In fact, the sound generated by a pixel is a monaural sound that is encoded into 9 Ambisonic channels; with parameters depending on azimuth and elevation angles. Then, the encoded Ambisonic signals are decoded for loudspeakers placed in a virtual cube layout. Finally, the physical sound is generated for headphones with

Fig. 1. The "Ducks" picture. Major components : sea (blue), sky (blue cyan), sun (yellow), clouds (white).

Fig. 2. The "Dolphin" picture. Major components : sea (blue), sky (blue cyan), dolphin (gray) and fish in the water (gray and cyan).

the use of HRTF functions related to the directions of virtual loudspeakers. The HRTF functions we use, are those included in the CIPIC database [1]. The orchestra used for the sonification is that described in section 2. The maximal time latency for generating a 17x9 sonified subwindow is 80 ms with the use of Matlab on a Pentium 4 at 3.0 GHz.

The purpose of this study was to investigate whether individuals are able to learn associations between colours and musical instrument sounds. Several experiments have been carried out by participants having their eyes enclosed by a dark tissue, and listening to the sounds via headphones [12]. In order to simplify the experiments, we used the T3 tactile tablet from the Royal National College for the Blind[1] (UK). Essentially, this device allows to point on a picture with the finger and to obtain the coordinates of the

[1] http://www.talktab.org/

Fig. 3. The "Tree" picture. Major components : grass (green), sky (blue cyan) and tree (dark green).

Fig. 4. The "Beach" picture. Major components : cliff (yellow), sky (blue and blue cyan) sea (blue) sand (bright-brown).

contact point. Moreover, we put on the T3 tablet a special paper with images including detected edges represented by palpable roughness.

Six participants were trained to associate colours with musical instruments and then asked to determine on several pictures, objects with specific shapes and colours. Experiments involved a learning phase with the use of elementary pictures. At the end of the training phase, a small test for scoring the performance of the participants was achieved. On the 15 heard sounds, the average number of correct colours among the six participants was 8.1 (standard deviation : 3.4). It is worth noting that the best score was reached by a musician who found 13 correct answers. Afterwards, participants were asked to explore and identify the major components of the pictures shown in figure 1.

Fig. 5. The "churchyard" picture. The goal in the experiments is to find one of the red doors.

Regarding the children draw picture illustrated in figure 1, all participants interpreted the major colours as the sky the sea and the sun; clouds were more difficult to infer (two individuals); instead of ducks, all the subjects found an island with yellow sand or a ship.

For the picture depicted in figure 2 all participants interpreted the major colours as the sky and the sea; an individual said that the dolphin is a "jumping animal", another said that it was a fish and the others determined a boat or a "round shape"; only a person found birds and no one was able to identify the small fish.

On the interpretation of real images, such as the picture shown in figure 3, four participants correctly identified the tree with the grass and the sky; a participant qualified the tree as a strange dark object and finally, the last individual inferred a nuclear explosion ! Concerning figure 4 , all subjects found major colours (blue and yellow); however no one made the distinction between the sky and the sea. Moreover, no one identified the yellow cliff.

The last assignment was to find a red door in figure 5. All participants found one of the red doors in a time range between 4 and 9 minutes.

4 Discussion

The first experiment concerning the recognition of 15 colours corresponding to 15 sounds exhibited that correct answers were given in a little bit more than half of the times, on average. Therefore, roughly speaking our group gave correct answers for five colours out of nine. That is clearly better than black and white identification. Thus, this experiment demonstrated that learning all colours is possible, but difficult in a short training time. It is worth noting that learning Braille is also complicated and requires a long period of training. Accordingly, the training phase with musical instrument sounds should be repeated a reasonable number of sessions.

The second experiment with children drawings demonstrated that the most important components of the pictures, such as the sky the sea and the sun were identified. Sometimes our participants were not completely sure, but with logical reasoning they inferred that if the top of the pictures is cyan and if the bottom is blue then the bottom is the sea and the top is the sky. Moreover, if something at the top of figure 1 is yellow and round shaped, this must be the sun. Another interesting observation is the difficulty to identify the three ducks. In fact, our common sense tells us that something yellow would be more likely to be the sand of an island or a yellow ship. Yellow ducks on the sea represent an unusual situation which is never considered by our participants.

The third experiment was performed with two real pictures. It is worth noting that figure 3 has three major components (sky, grass and tree), with a limited perspective view. Consequently, almost our participants gave a correct sketch of that picture. On the contrary, figure 4 presents noticeable perspective; as a result, the context of the picture was not determined by our six participants, although several individuals correctly identified the most important colours.

The fourth experiment consisted in finding one of the red doors of figure 5. All the people were successful, however the elapsed time was quite long. The first reason is that with A3 paper format on the T3 tablet, it takes a long time to explore the picture with a small sub-window of size 17x9 pixels. Moreover, the image scene is complicated with a high degree of perspective.

Five participants out of six said that colour was helpful for the interpretation of pictures. In fact, when one tries to identify a picture component, the presence of colour in the audio representation limits the number of possible interpretations. Finally, the experiments emphasised perspective as a major drawback for the understanding of two-dimensional figures.

5 Conclusion and Future Work

We presented the current state of the See ColOr project which aims at providing a mobility aid for visually impaired individuals. Because of real-time constraints, image simplification in our prototype was achieved by colour quantification of the HSL colour system translated into musical instrument sounds. With only a training session, the experiments on static pictures revealed that our participants were able to learn five out of nine principal colours, on average. Furthermore, colour was helpful for the interpretation of image scenes, as it lessened ambiguity. These experiments also demonstrated that the exploration time of pictures is quite long, probably because the sonified sub-window is small and should not expand too much for reasons related to the limits of human audio channel capacity. In the context of real time navigation, we foresee that a blind individual with a camera on her head would rely on image processing techniques, such as salient points determination, in order to quickly determine relevant parts of the environment.

Aknowledgements

We would like to thank the Hasler Foundation in Bern for financing the See ColOr research project under grant 1968. We also thank two internet sites[2] for making available the pictures used in the experiments.

References

1. Algazi, V.R., Duda, R.O., Thompson, D.P., Avendano (2001). The CIPIC HRTF Database. In IEEE Proc. Workshop on Applications of Signal Processing to Audio and Acoustics, Mohonk Mountain House, (WASPAA'01), New Paltz, NY.
2. Bologna, G., Vinckenbosch, M. (2005). Eye Tracking in Coloured Image Scenes Represented by Ambisonic Fields of Musical Instrument Sounds. In Proc. IWINAC (1) 327–337.
3. Bamford, J.S. (1995). An Analysis of Ambisonic Sound Systems of First and Second Order. Master Thesis, Waterloo, Ontario, Canada.
4. Capelle, C., Trullemans, C., Arno, P., Veraart, C. (1998). A Real Time Experimental Prototype for Enhancement of Vision Rehabilitation Using Auditory Substitution. IEEE T. Bio-Med Eng., 45, 1279–1293.
5. Cronly-Dillon, J., Persaud, K., Gregory, R.P.F. (1999). The Perception of Visual Images Encoded in Musical Form: a Study in Cross-Modality Information. In Proc. Biological Sciences, 266, 2427–2433.
6. Daniel, J. (2000). Acoustic Field Representation, Application to the Transmission and the Reproduction of Complex Sound Environments in a Multimedia Context. PhD thesis, University of Paris 6.
7. Gerzon, M.A. (1977). Design of Ambisonic Decoders for Multispeaker Surround Sound. Journal of the Audio Engineering Society (Abstracts), 25, 1064.
8. Gonzalez-Mora, J.L., Rodriguez-Hernandez, A., Rodriguez-Ramos, L.F., Dfaz-Saco, L., Sosa, N. (1999). Development of a New Space Perception System for Blind People, Based on the Creation of a Virtual Acoustic Space. In Proc. IWANN, 321–330.
9. Kay, L. (1974). A Sonar Aid to Enhance Spatial Perception of the Blind: Engineering Design and Evaluation. The Radio and Electronic Engineer, 44, 605–627.
10. Lakatos, S. (1993) Recognition of Complex Auditory-Spatial Patterns. Perception, 22, 363–374.
11. Malham, D.G., Myatt A. (1995). 3-D Sound Spatialisation using Ambisonic Techniques. Computer Music Journal, 19 (4), 58–70.
12. Meijer, P.B.L. (1992). An Experimental System for Auditory Image Representations. IEEE Transactions on Biomedical Engineering, 39 (2), 112–121.
13. Ruff, R.M., Perret, E. (1976). Auditory Spatial Pattern Perception Aided by Visual Choices. Psychological Research, 38, 369–377.

---

[2] http://www.freeimages.co.uk/galleries/buildings/country/index.htm
http://www.tuxpaint.org/gallery/

Towards a Semi-automatic Situation Diagnosis System in Surveillance Tasks

José Mira[1], Rafael Martínez[1], Mariano Rincón[1], Margarita Bachiller[1], and Antonio Fernández-Caballero[2,*]

[1] E.T.S.I. Informática - Univ. Nacional de Educación a Distancia, Madrid, Spain
{jmira,rmtomas,mrincon,marga}@dia.uned.es
[2] Escuela Politécnica Superior de Albacete & Instituto de Investigación en Informática de Albacete, Universidad Castilla-La Mancha, Albacete, Spain
caballer@dsi.uclm.es

Abstract. This paper describes an ongoing project that develops a set of generic components to help humans (semi-automatic system) in surveillance and security tasks in several scenarios. These components are based in the computational model of a set of selective and Active VI-Sual Attention mechanisms with learning capacity (*AVISA*) and in the superposition of an "intelligence" layer that incorporates the knowledge of human experts in security tasks. The project described integrates the responses of these alert mechanisms in the synthesis of the three basic subtasks present in any surveillance and security activity: real-time monitoring, situation diagnosing, and action planning and control. In order to augment the diversity of environments and situations where *AVISA* system may be used, as well as its efficiency as support to surveillance tasks, knowledge components derived from situating cameras on mobile platforms are also developed.

1 Introduction

Surveillance is a multidisciplinary task affecting an increasing number of scenarios, services and customers. It aims to detect threats by continually observing large and vulnerable areas of a scenario considered to be of economic, social or strategic value because it can suffer theft, fire, vandalism or attacks. The range of scenarios is very wide and of very different complexity, going from the mere detection of movement that sets off an alarm to an integral control system that monitors the scene with different sensors, diagnoses the situation and plans a series of consistent actions. In any case, it always implies the observation of mobile objects (people, vehicles, etc.) in a predetermined environment to provide a description of their actions and interactions. Hence, this implies the detection of moving objects and their tracking, the recognition of objects, the analysis of the

* The contribution to this paper from the other members of the AVISA project (A. Delgado, E. Carmona, J.R. Álvarez, F. de la Paz, M.T. López and M.A. Fernández) has been equal to the other authors. All of them should appear on the author list, but the rules of the congress stipulate a maximum of five recognized authors.

J. Mira and J.R. Álvarez (Eds.): IWINAC 2007, Part II, LNCS 4528, pp. 90–98, 2007.

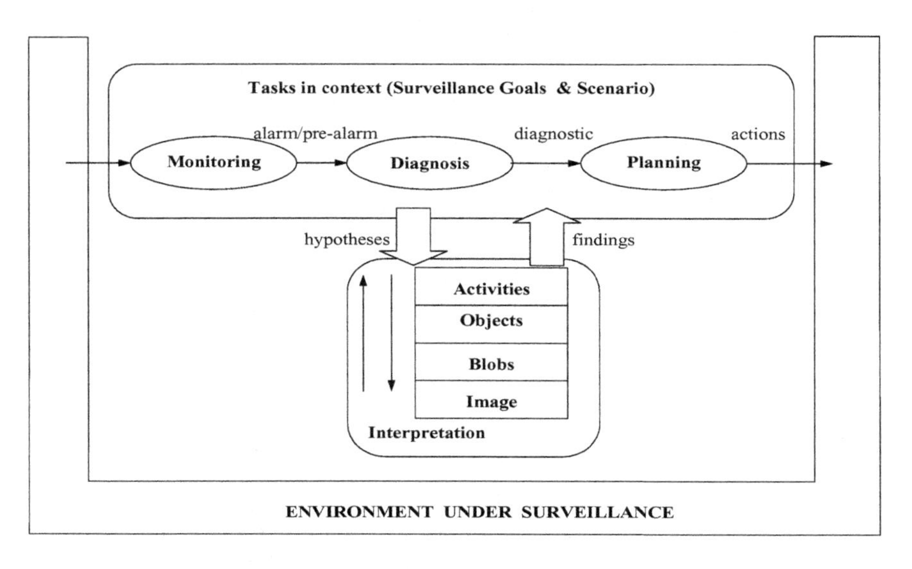

Fig. 1. Graphical Description of the Surveillance Task

specific movement for each type of object and the interpretation of the activity. Most approximations for activity analysis activity [4] [6] [15] [18] are posed as a classification problem, i.e., they are based on defining models for each type of activity in a specific application domain, and they are highly dependent on the results obtained during tracking. As you may observe in Fig. 1 the surveillance task follows a common structure with general control tasks: (1) monitoring of a number of critical variables, whose deviation from normality is a sign of some disfunction, (2) diagnosis of the problem, and (3) planning of relevant actions for solving the problem.

2 The *AVISA* Approach to Surveillance

In our surveillance approach a semi-automatic approximation is considered; at the end it is the operator of an alarms power station who takes the decisions. However, one of the fundamental problems in this kind of projects is the enormous semantic gap between the physical signal level and the knowledge level. One of the ways of overcoming this gap is by partitioning it. There is consensus in the area, although with some variety and dispersion in the nomenclature, in accepting different description levels with an increasing degree of semantics [13] [14], which facilitates the injection of domain knowledge. In our work, four description levels are considered: image, blobs, objects and activities/behaviours (see again Fig. 1). Each of these levels is modular and independent, and the information handled comes from the ontology of the proper level and from the lower level.

In the blob level, the entities are associated with the visible part of objects of interest or with parts of them, the blobs. In the object level the information associated with blobs for producing a description of the objects of interest on the scene is reorganised. We move from a frame-oriented description to an object-oriented description. The models of the objects of interest are described here, which contain: 1) the visual characterisation of the object and its spatial-temporal evolution; 2) the composition relations used to describe complex objects; and 3) the relations between objects for generating the geometric description of the task-focused scene. At object level, it is necessary to inject additional domain knowledge, in terms of models of the objects in any description level where they are used. At activity level, the identification of a set pattern of elementary activities (events), each of them basically normal, can be interpreted as an alarming activity together.

The aims of the *"selective and Active VISual Attention mechanisms with learning capacity"* (*AVISA*) project correspond to modelling and formalization of the tasks associated to each level and their integration. This purpose is divided in *AVISA* two subprojects: (a) "Distributed monitoring of scenarios with different kinds of moving non-rigid objects", with focus on the processes associated to pixels, blobs and static and dynamic objects considered for monitoring the scene. The monitoring output is a warning in the sense that *AVISA* system has detected some anomaly. Thus it is the alarm that activates a diagnosis process. (b) "A set of diagnosis, planning and control agents with capacity of learning and cooperation with humans", which shares the object level but focuses on the situation diagnosis task, which starts from the results of the monitoring task. The following sections introduce the tasks related to *AVISA* project in detail.

3 Monitoring and Pre-diagnosis

The aim of the visual monitoring and scene pre-diagnosis tasks is to collect visual information from diverse coordinates of the scene and to pre-process them according to the selective visual attention (*SVA)* mechanisms: *Accumulative computation* and *algorithmic lateral inhibition* (*ALI*). The global purpose is to focus attention of the cameras (static and mobile) on those elements of the scene that better fulfil the criteria specified by the human expert as "objects of interest". The processes corresponding to these tasks are related to the image, blob and object levels. The decision on the proper monitoring message is taken in terms of the value of a set of parameters of motion, size and shape, firstly obtaining the blobs, and then injecting the semantic needed to get the objects of interest. It has been possible to work on a real outdoor scenario monitored by the cooperating company *SECISA* (see image at Fig. 2). This process has been developed in a sequential and incremental manner, starting with *AC* and *ALI* [9], formulated firstly at physical level and then in terms of multi-agent systems, where a totalizing process on the individual opinions of a set of working memories as agent coordination mechanism has been used. Next both mechanism have been combined [5] to face the dynamic *SVA* problem [7], that is to say, "where to look to" and "what to look to". Our strategy has been the integration of the bottom-up (connectionist) and top-down (symbolic) organizations usually accepted in Neuroscience.

Fig. 2. Outdoor scenario. Monitoring of a test case. SECISA video sequence.

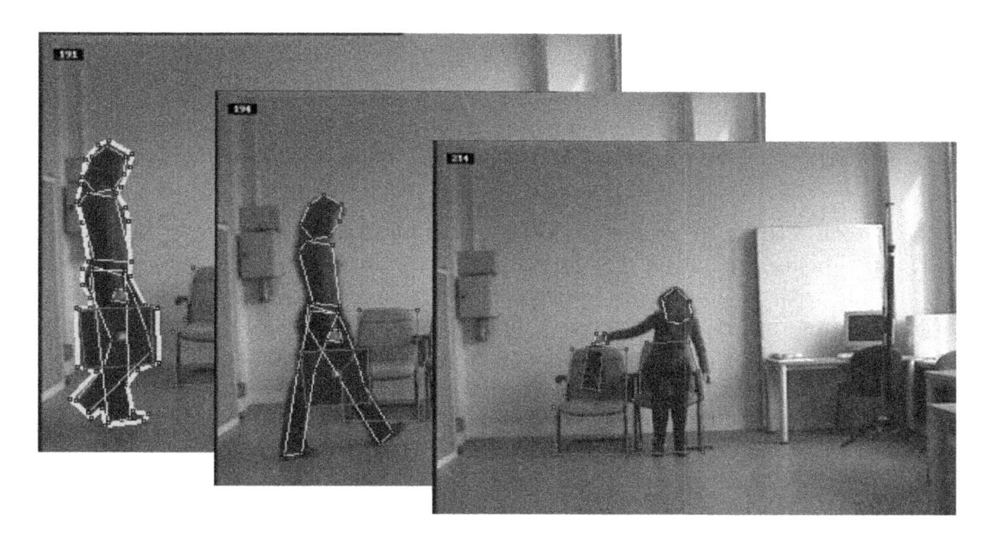

Fig. 3. Use of an annotation tool for video sequences (the indoor scenario in the figure), which allows the representation of time, position, attributes, relations, states, and events as visual objects

To exemplify the methodological proposal developed, initially a very simple scenario was defined (Fig. 3). It is an indoor space that is a pass-through area for humans. Humans can move freely, come in and go out of the observation area, sit down, carry a briefcase, leave it, pick it up, etc. This scenario simplifies the problem (there are no dogs, trees or cars, etc.) and it helps us specify the other two components of a context, the aims and sensors. An alarm would go off when someone leaves a briefcase (a package) in the area under surveillance.

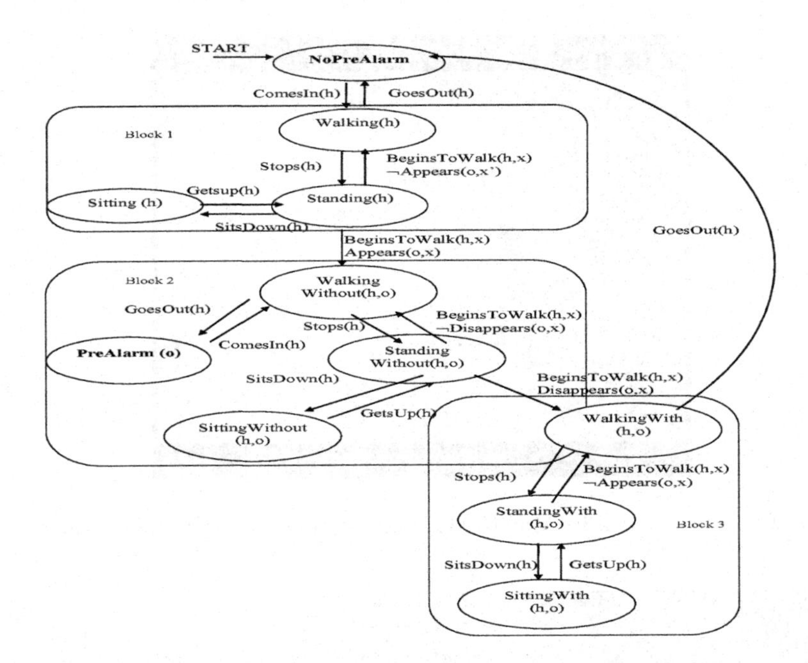

Fig. 4. State transition diagram used to model the activity level in the monitoring and pre-diagnosis steps according with the reconocibles events in the video sequence

If the person returns to his initial position and picks up the briefcase, the alarm will stop. This is a situation that forms part of the most thorough surveillance of a situation, with no going back, where it is identified that a person who leaves a larger site (an airport, hospital, etc.), first causing an alarm of this sort because he has left a package, which would require a response action. Fig. 4 shows the state transition diagram for the automaton describing the scene at activity leve in the context of the monitoring task.

The main difficulties found are related to the interaction between moving objects (e.g., a human) and static ones (for instance, a suitcase) and to the complexity of working in real-time in real scenarios (camera movement, noise, 2D vs. 3D, calibration, stereovision, and so on).

4 Diagnosis of the Activity Level

The aim of the diagnosis task is to model the knowledge of the human expert in surveillance tasks. The main problem faced in this task is that, when interpreting a scene the external knowledge necessary to fill the enormous semantic gap between the physical signal level and the knowledge level has to be injected, i.e., the AVISA "intelligence" level needs to adapt the sensory information to the abstraction level.

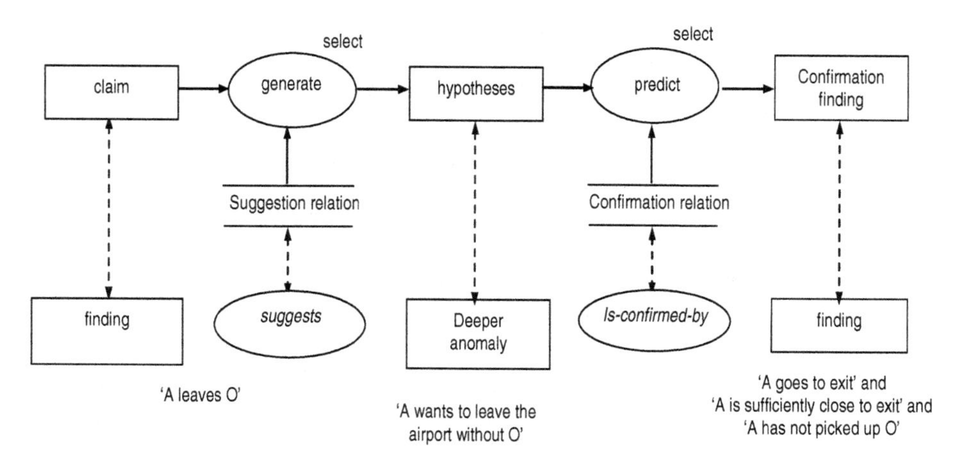

Fig. 5. The "generate and predict" inferential scheme. A first inference "generate" selects the hypotheses from a claim (from monitoring) requiring confirmation. This inference is based on the "suggestion" relation between anomalies. From this hypothesis, the inference "predict" selects the confirmation findings. In this instance the inference is based on the confirmation relation between the anomaly of a specific depth level requiring confirmation and the relevant findings.

If the monitoring step focuses on the bottom-up interpretation, the logical continuation would be to include a "diagnosis" stage. The use of models is a way of introducing domain knowledge in a top-down control which makes it possible to improve segmentation by aiding detection of the constituent parts of the object, detecting regions of potential confusion between object and background, and improving tracking with specific methods depending on the object type.

From the firing of the alarm by the monitoring module, there are two possible relations with the diagnosis which determine its meaning in the global process:

a) The alarm triggers the "diagnosis of what has happened" process. In other words, there is an analysis of what has happened from the recorded images (and data from other sensors).

b) The alarm triggers the "diagnosis of what is happening" process. This process is focused and guided by hypotheses of what could be happening. The diagnosis ends with confirmation of hypotheses that require a response that the planning phase will determine, although it could also conclude that the situation is not in fact alarming.

Figure 5 shows the inferential schema of the problem solving method (PSM) "generate-predict" used to solve the diagnosis task in AVISA. Certain behaviors suggest individual causal relations and that those relations have visual consequences that can be verified. So, for example, the event "A leaves the object O" suggests two hypotheses: "A picks up O" and "A wants to leave the airport without O". The confirmation of this last one demands the obtaining of the findings: "A goes to Exit", "A has not picked up O" and "A is sufficiently close to Exit". They are findings that directly do not emerge from the interpretation scene, but that there is to look for them.

In the following table, the generetad hypothesis is associated to the confirmation finding and this one is associated to its decomposition in more elementary events and their space-temporary relation.

Claim	Hypothesis	Confirmation finding Decomposition (looking for)
Leaves(A, O, t_1, x)	Takes(A, O, t_2, x)	Picks(A, O, t_2, x)
	Wants-to-leave-airport (A, O, t_3)	Is-going-to(A, Exit, t_3) $\neg$Picks-up(A, O, t_4) $\forall t_4 \in [t_1, t_3]$ Is-close-to(A, Exit, t_4)

5 Learning

The objectives of learning task are (1) to guide the search process in the *SVA* mechanisms by reinforcing the descriptions of the objects that look more like the predefined patterns, (2) to guide the robots in their pathways towards the selected coordinates, and, (3) to accumulate the *AVISA* experience in its collaboration with the human operator. Of these three functions the first one has been fully accomplished [11], with the complementation of reinforcement learning with other evolutionary models. The second function is in development phase; to date an autonomous navigation system based on the calculation of the centre of area in open space is available [1]. The third function, associated to the interaction of the prototype with the human operator, is currently being developed.

Fig. 6. Use of learning by reinforcement to select those parameters that contribute to a better discrimination (diagnosis) of different situations (video sequences) in accordance with the intention of the observer

The main problems of the learning-related tasks are given by the difficulty of constructing valid training sets for supervised learning in different scenarios. The solutions explored so far are the construction of a case base for each concrete scenario and the use of evolutionary procedures to enhance the segmentation task. Those parameters that contribute to a better discrimination (diagnosis) of different situations in accordance with the intention of the observer have to be reinforced [11].

6 Conclusions

This paper has presented an ongoing project denominated *AVISA* that develops a set of generic components to help humans in surveillance and security tasks in several scenarios. The paper has introduced the tasks related to the project, highlighting the effort taken, the difficulties found and the main contributions offered so far.

The greatest difficulty found arises when connecting blobs to activities in different real scenarios. To solve this difficulty, at least partially, the following measures have been adopted: (1) Modelling and labelling of all families of sequences of interest in the proposed real scenarios and at the levels of objects, events and behaviours (construction of the ontologies of each level for each scenario). (2) Establishment of a clear frontier between automatic and human interpretation for each context (scenario, surveillance intention, information sources, robot accessibility, and so on). (3) Emphasizing in alarm pre-diagnosis as a complement to monitoring, leaving the decision on the action at each situation to the human operator. (4) Dedicating an additional effort to the construction of a case base that enables reasoning at each scenario by analogy to previous situations, in cooperation with any learning-related task. (5) Exploring new ways of representing activities that enable limiting the combinatory explosion of the state transition diagrams of the deterministic finite automata. (6) Making explicit the difficulty of the situation diagnosis task and trying to integrate the efforts of this project with some complimentary ones (face recognition, etc.).

In spite of the difficulties found, some preliminary results has been obtained: (1) Modelling and implementation of the motion detection task [12]. (2) Modelling of the image understanding process [17]. (3) Generation of a tool for annotating images [16] and its later use in learning.

Acknowledgements

This work is supported in part by the Spanish coordinated TIN2004-07661-C02-01 and TIN2004-07661-C02-02 grants.

References

1. Álvarez J.R., De la Paz F. & Mira J. (2005). A robotics inspired method of modeling accessible open space to help blind people in the orientation and traveling tasks. Lecture Notes in Computer Science, 3561, pp. 405-415.

2. Bolles B. & Nevatia R. (2004). A hierarchical video event ontology in OWL. Final Report 2004, ARDA Project.
3. Bremond F., Maillot N., Thonnat M. & Vu V.T. (2004). Ontologies for video events. INRIA, Research Report, 5189.
4. Collins R.T., Lipton A.J. & Kanade T. (eds.) (2000). Special Issue on Video Surveillance, IEEE Transactions on Pattern Analysis and Machine Intelligence, 22:8.
5. Fernández-Caballero A., López M.T., Mira J., Delgado A.E., López-Valles J.M. & Fernández M.A. (2007). Modelling the stereovision-correspondence-analysis task by lateral inhibition in accumulative computation problem-solving method. Expert Systems with Applications, 34 (4).
6. Haritaoglu I., Harwood D. & Davis L.S. (2000). W4: Real-time surveillance of people and their activities. IEEE Transactions on Pattern Analysis and Machine Intelligence, 22(8), pp. 809-830.
7. López M.T., Fernández M.A., Fernández-Caballero A., Mira J. & Delgado A.E. (2007). Dynamic visual attention model in image sequences. Image and Vision Computing, 25 (5), pp. 597-613.
8. López M.T., Fernández-Caballero A., Fernández M.A., Mira J. & Delgado A.E. (2007). Real-time motion detection from ALI model. Eleventh International Conference on Computer Aided Systems Theory, EUROCAST 2007.
9. López M.T., Fernández-Caballero A., Mira J., Delgado A.E. & Fernández M.A. (2006). Algorithmic lateral inhibition method in dynamic and selective visual attention task: Application to moving objects detection and labelling. Expert Systems with Applications, 31 (3), pp. 570-594.
10. López M.T., Fernández-Caballero A., Fernández M.A., Mira J. & Delgado A.E. (2006). Visual surveillance by dynamic visual attention method. Pattern Recognition, 39 (11), pp. 2194-2211.
11. López M.T., Fernández-Caballero A., Fernández M.A., Mira J. & Delgado A.E. (2006). Motion features to enhance scene segmentation in active visual attention. Pattern Recognition Letters, 27 (5), pp. 469-478.
12. Mira J., Delgado A.E., Fernández-Caballero A. & Fernández M.A. (2004). Knowledge modelling for the motion detection task: The algorithmic lateral inhibition method. Expert Systems with Applications, 27 (2), pp. 169-185.
13. Nagel H.H. (2004). Steps towards a cognitive vision system. AI Magazine, 25 (2), pp. 31-50.
14. Neuman B. & Weiss T. (2003). Navigating through logic-based scene models for high-level scene interpretations. 3rd International Conference on Computer Vision Systems, ICVS-2003. Lecture Notes in Computer Science, 2626, pp. 212-222.
15. Regazzoni, C.S., Fabri, G. & Breñaza, G. (eds.) (1999). Advanced Video-Based Surveillance Systems, Kluwer Academic Publishers, Dordrecht.
16. Rincón M. & Martínez-Cantos J. (2007). An annotation tool for video understanding. Eleventh International Conference on Computer Aided Systems Theory, EUROCAST 2007, 12-16 February 2007, Las Palmas de Gran Canaria (Spain).
17. Rincón M., Bachiller M. & Mira J. (2005). Knowledge modeling for the image understanding task as a design task. Expert Systems with Applications, 29 (1), pp. 207-217.
18. Robertson N. & Reid I. (2006). A general method for human activity recognition in video. Computer Vision and Image Understanding, 104, pp. 232-248.
19. Taboada M., Des J., Martínez D. & Mira J. (2005). Aligning reference terminologies and knowledge bases in the health care domain. Lecture Notes in Computer Science, 3561, pp. 437-446.

An Implementation of a General Purpose Attentional Mechanism for Artificial Organisms

J.L. Crespo, A. Faiña, and R.J. Duro

Integrated Group for Engineering Research
Universidade da Coruña (Spain)
{jcm,afaina,richard}@udc.es
http://www.gii.udc.es

Abstract. Attention is a mechanism present in most of the more complex and developed living beings. It is responsible for much of their ability to operate in real time in unstructured and dynamic environments with a limited amount of processing resources. In this paper, an architecture for developing attentional functions in agents is presented. This architecture is based on the concept of attentor and it allows for the real time adaptation to the environment and tasks to be performed in a natural manner. One of the main requirements of the system was the ability to handle different sensorial varieties and attentional streams in a transparent manner while, at the same time, being able to progressively create more complex attentional structures. The main characteristics of the architecture are presented through its implementation in a real robot.

Keywords: Attention, Autonomous robots, Sensorial processing.

1 Introduction: Attentional Phenomena

Attention could be described as a type of sensorial spaces categorization by means of both physically salient and task-related information. The usefulness of this categorization is to allow natural organisms to consider the varied circumstances that could appear in their usual operation.

The study of natural attention phenomena has led to the development of artificial attentional systems. Through this type of approximations more efficient use of the detection, processing and cognitive capabilities of an agent or robot can be achieved. Looking at the literature regarding attentional systems in living beings [1], most authors postulate the existence of two information streams that may guide the attentional process: one of them is related with the search for physically salient stimuli (especially bright or colored regions, certain oriented edges, motion, etc.). This is called the "bottom up" modality. Examples of this approach are the well known papers like [2][3] who present a bioinspired model for determining the most interesting positions in a visual scene from color, intensity and orientation maps.

On the other hand, some type of higher level information about the task to be carried out or the objective may be known and thus used. This approach

J. Mira and J.R. Álvarez (Eds.): IWINAC 2007, Part II, LNCS 4528, pp. 99–108, 2007.

is usually called the "top-down" approach. For example, if someone is looking for fire extinguishers in a corridor, attention will be focused on walls and the subject will be looking for red objects. Some work on this approach is the classic reference [4] who follows a usual trend in the area that uses a bottom-up approximation (often called preattentional stage) to modulate the operation of certain prespecified detectors.

Even though most authors have adhered to one or the other approach, it is becoming clear that the two streams must be considered when designing an efficient real system as most attentional functions in the real world involve both types of information and, even if this were not so in natural attentional systems, we could find it desirable for an artificial agent to consider both types of information, and in this case for it to be intrinsic to the attentional mechanism and transparent.

Additionally, most implementations of artificial attentional systems presented in the literature imply just one sensorial modality, being, in general, quite difficult to add other sensorial modalities (vision, sound, haptics, etc.) in the same global attentional mechanism. In fact, the most extended sensorial modalities are the visual and auditive ones, while very few references could be found that considered other types, such as haptics [5]. What is more important is that even those that considered different modalities did not naturally integrate them into a unit. If we want to develop attentional functions of any kind, it would be desirable to handle different sensorial modalities in a homogeneous way.

In this paper, we present an approximation for developing attentional functions in a way that is independent both from the nature of information (physical or task-related salience) and the sensorial modalities involved. This approximation has the form of a modular architecture that permits building ad-hoc attentional functions. This architecture has been called AMADIS (Attention Modular Architecture with Distribution of Information and Sensing). The rest of this paper is structured as follows: First, a novel approximation for developing attentional functions is presented, and conceptually compared to other approximations. In the second section, we comment the most important characteristics of the architecture through the description of the results of some experiments in a real robot. Finally, some conclusions and comments are presented.

2 Attentors and AMADIS

The most important concept introduced in this paper is an element we have called "Attentor". An attentor is a structure that constructs, during the lifetime of an agent, the relation between a sensing-related space and an attentional-related space. For example, an attentor dedicated to the sound frequencies that must be scanned in order to detect some musical instrument receives as input a time-domain signal, and has as its output a probability histogram of the frequency domain indicating the areas where more search effort should be devoted. The particular form of this function is learnt throughout the life of the attentor or introduced directly when the information is known by the designer. Obviously,

Fig. 1. Representation of AMADIS attentor values (a) A sample of image, as seen by the agent. (b) The graph shows the relevance values for the image (there are two regions of maximum values, at left and right sides of the central region). (c) pseudocolor and 8-level quantized version of "b".

an attentor is directly related to a given detector as it represents how the efforts devoted to finding the things this detector is trained to find should be spent.

As an example in the visual domain, which is the most intuitive, in the image 1 the graphical representation of an attentor corresponding to a chair detector using a camera in a meeting room is shown in the form of a 3-D graph, where the X and Y axes correspond to the dimensions of the visual space and the Z axis shows, for every point of the space, the level of attention that must be paid after this attentor was constructed by exploring the images using the corresponding detector as explained later.

The main problem is how to construct a given attentor. This problem has been solved in this work through the implementation of a very simple mechanism that is guided by the detections provided by the detector or detectors associated with a given attentor in the architecture. Each position of the portion of the sensing space the attentor considers with a given level of discretization has an associated relevance value. Initially, when there is no information, the attentor relevance surface is basically flat. The detector is applied over the sensing space using a random function that is weighed by the relevance values, that is, the higher the relevance, the higher the probability of that area inputting the detector. Consequently, at the beginning the detector is randomly applied over the sensing space and, as the detector produces positive detections, the location of these detections in the sensing space (and its surroundings according to some type of neighbourhood function such as a Gaussian) increase their relevance value. As the interaction between the agent and the environment is more prolonged, more detections are made and the relevance of the different positions in the sensing space will be modified accordingly, leading to an attentor in the form of a representation of the relevance of the different areas of the sensing space and an application of the detector more often in the areas of high relevance and more seldom in those with a low relevance. As this mechanism is probabilistic, it intrinsically provides for the whole sensing space having a probability of being input to the detector thus allowing for adaptation to changes and preventing the mechanism from getting stuck in certain areas as is the case of [2] where they have had to implement a combination of a "winner take all" (for determining the selected location) and an additional inhibition of return mechanism to prevent this by ignoring the previously selected positions.

There is, obviously, a close relation between attentors and detectors, as detectors provide the inputs that are going to be used for establishing attentor parameter values. However, this relation, which is also present, although from a different perspective, in other attentional approximations, does not imply exclusiveness in the case of AMADIS. This means that, unlike other mechanisms, the detectors could be changed, without modifying the parameter values or form of the attentor.

In order to provide the basic building blocks for an attentional architecture that considers different sets of detectors over different sensorial modalities a classification of attentors can be obtained. The attentor types in this classification differ in the way in which their values are acquired and the source of the input data:

- Primary attentors: they are fed directly from sensorial detectors, and their values are established during the life of the agent in which they are implemented. Examples of this type of attentor are "attentor for red color", "attentor for chairs", "attentor for characteristic frequencies of a violin", etc.
- Secondary attentors: they may take their inputs both from sensorial detectors or from primary attentors, and their values also vary with the passing of time. An example of this type of attentor is a fire extinguisher attentor based on the results of the "red color attentor" and an attentor for cylindrical shapes.
- Instinctive and reactive attentors: Both of them are used for implementing attentional behaviours that imply a sudden change of actuation (for example, what to do if an emergency occurs). The difference between them is that instinctive attentors have predefined and non-changing values, while reactive attentor values are acquired during the life of the agent.

Attentors are not the only component of the architecture. It is obviously necessary to establish ways in which to integrate or combine information coming from different attentors in order to construct the current attentional space. This can be achieved through different procedures: The most usual way is to establish attentor combination values where the type of combination (linear, quadratic, etc.), the nature of its implementation (polynomial, neural network, etc) and the characteristic coefficients could be either predefined by the designer or through some automatic learning or adaptation stage.

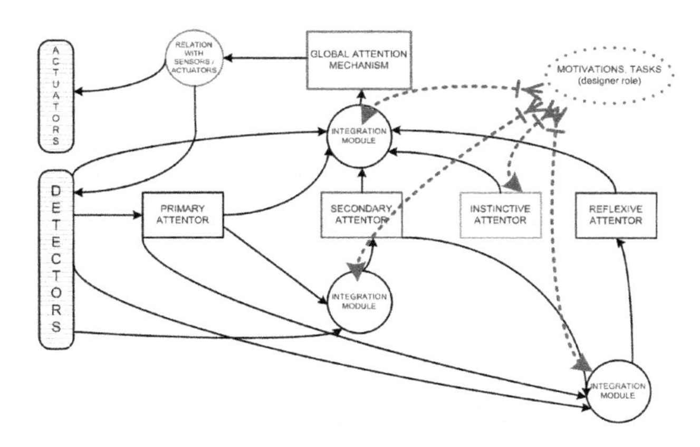

Fig. 2. A schematic representation of AMADIS and its components. Only one instance of each type of element is shown.

Figure 2 displays a schematic representation of the different elements that are defined in the architecture and the possible relations among them. In a particular implementation of AMADIS, it is not necessary for all the elements to be present.

In order to understand how a current global attention map is created, the following example can be considered: Imagine a corridor where a robot with a movable camera with a very narrow field of view moves. The robot is interested in chairs, lights and open space and it has a camera it has to move around in order to apply the corresponding detectors so as to find these features. As shown in figure 3, three individual detectors for each of these categories is implemented. During its lifetime the robot has created, depending on its previous experience, attentors determining where it was more efficient to point the cameras in order to find these objects (lights around the ceiling, chairs are usually in the bottom of the field of view to its left and the open space is usually in front of it). Now, as it must move taking these three types of objects into account, the global attentor is created by means of an integration module (in this case, the module function is just a linear combination with pre-fixed weights) and this module is the one that is going to determine each moment of time how the camera pointing resource is going to be used. The weights in the combination modulate the relative importance for the attention mechanism of each of the individual elements.

As stated before, the mechanism just establishes the relevance of each area in the sensing space, leading to a mechanism that makes the camera look more often in the direction of the locations with the highest relevance and seldom to those with very low relevance. In AMADIS no area is left out of the exploration, leading to a natural implementation of what some authors call the curiosity and exploration mechanism present in natural attentional systems. It consists in the exploration of certain regions of the sensorial spaces even when it seems a priori useless (very low relevance) or even when the experimental subjects have been advised or forbidden about it. Some psychological researchers claim that this actually improves the survivability of individuals by enhancing their ability for early detection of threats [6].

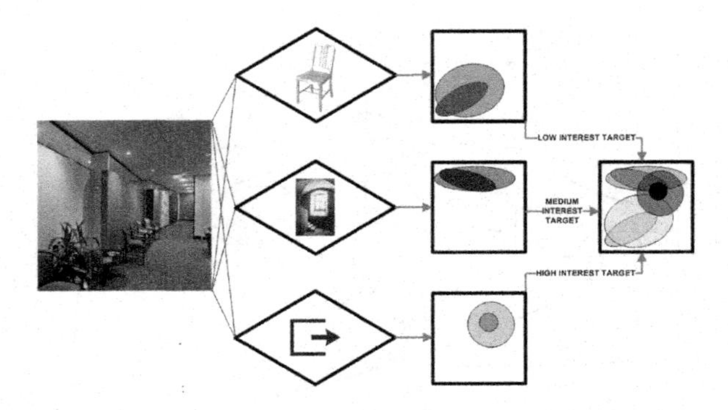

Fig. 3. An example of the operation of the architecture using visual stimuli where three attentors are integrated into a global attentor

3 Experimental Results

This section provides some results that permit presenting the main features in the use of AMADIS. The experiments have been designed to show both particular and combined properties of the architecture.

3.1 Experiment 1: Formation of Disjointed Foveal Regions

The foveal region is a part of the animal retina characterized by a higher resolution than the rest. By moving the foveal region, living beings can analyze in a more detailed way parts of the scene. Concepts that are similar to visual fovea can be found in other sensorial modalities (people have the ability of attend to certain sound frequency distributions -for example, follow a musical instrument among a group-). Although some references (see [1]) identify foveal regions with the term focus of attention, this is not absolutely true. However, in the context of attentional system research for applications, this identity may be accepted. Thus high interest regions can be identified with foveal regions.

One of the properties of AMADIS is the possibility to establish, in a coherent manner, more than one region of interest. This allows the agent to divide its attention between disjoint regions.

To show the process of building disjoint high interest regions, we have designed the following experiment: We have an AIBO robot that explores an environment with its vision system. For the sake of clarity, we have reduced the difficulties associated with the detection stage (that has been left to the AIBO image processing system), as our purpose is to display the attentional capabilities, not the sensorial ones.

In this hemispherical environment, two positive (from the point of view of the task) stimuli are found. The robot will explore the environment and the attentor associated with this stimulus (through its detector) will evolve. It must be noticed that, due to the field of view of the robot, there are several head coordinates that correspond to a detection, so the situation is similar to a moving target being detected by a narrow field robot camera.

Fig. 4. The experimental environment for the first experiment

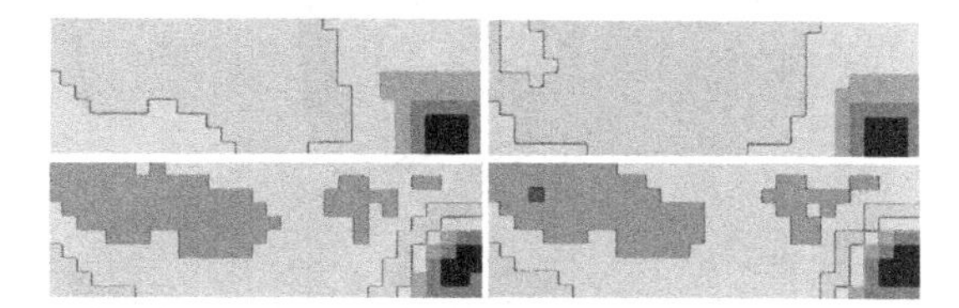

Fig. 5. A series of global attention space representations for experiment number 1. Top images correspond to the beginning of the experiment, when there is no high interest valued region. Bottom images correspond to stable stages, in which two disjoint attentional regions (red or dark) can be appreciated.

Due to the simplicity of the task (involving a single detector), the general attentor and the attentor associated with the detection of blue blobs will have exactly the same values, even when they are not the same from a conceptual point of view.

The experiment starts with a robot that has no prior knowledge about the environment; the robot has only been instructed with the relation between attentional positive value and blue stimulus detection.

The series of images in figure 5 display the evolution of the attentor corresponding to the detection of blue blobs in the visual sensorial space (in this case the whole field of view achievable by rotating the head of the robot which contains the camera) as the robot interacts with the world. Red (dark in the bw image) indicates high relevance, blue low. The first image corresponds to the beginning of the experiment, the robot has no experience and no region of interest is defined yet. As time passes, the attentor starts to create two disjoint attention regions corresponding to the positions in which the stimuli appeared (usually separated to the left and right of the robot). Finally, the robot has two well defined attention regions which is where it explores the most. This was achieved without any previous knowledge on the number of attention regions that could be present. The fact that the attention regions appear with a different size even though the stimuli are symmetrical can be explained by the statistical nature of the system, that is by the number of detections for each of the stimulus at that time which may be different. If a longer time period is allowed, the sizes will become similar.

3.2 Experiment 2: Considerations About Sensorial Modalities

As has been said before, one of the properties of AMADIS is that it is able to handle different sensorial modalities as a response to any attentional request. Usually, the most usual relation between sensorial modalities involves sound and visual stimuli, and, in most of cases, in a hierarchized manner, that is, one (usually sound) modulates the behaviour of the other.

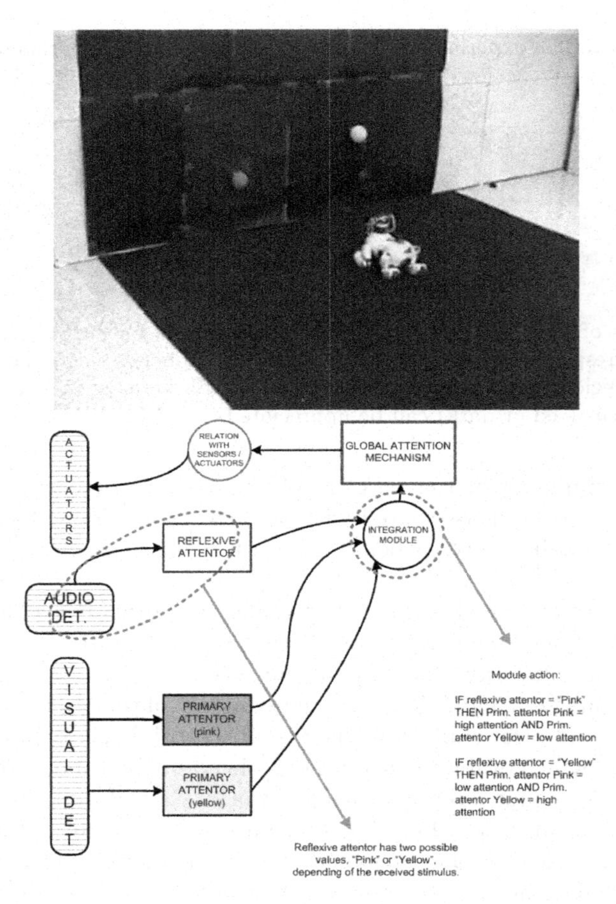

Fig. 6. (top) Experimental environment for experiment number 2. (bottom) AMADIS configuration for this experiment: there are two visual attentors that map the attentional spatial distribution for each colored stimuli, there is another attentor having to do with speech commands.

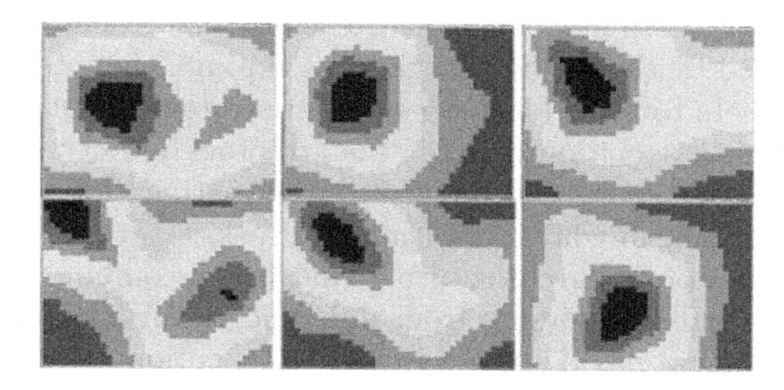

Fig. 7. (from left to right and top to bottom) In the first stages (a) of the experiment there are no high attention regions, but after some interaction with the environment (b) the presence of the stimulus builds an attention surface (to the right of the robot). Then a voice order changes the target and causes another unstable period (c), because, for the sake of stability, the global attentor uses both present and past data. After some iterations, the new attention surfaces appears (bottom left of the robot) (d). We change back the focus of attention (e & f).

In the following experiment we are going to establish a hierarchized version of these sensorial modalities with the objective of showing how the global attention space changes in its relevance values and, thus, determines how the AIBO employs its camera resources in the search for a given stimulus that must be detected. Two differently colored stimuli are presented to the robot in different regions of its visual space, and there is a voice command that indicates which should be the focus of attention.

In the experiment, the robot begins to establish its attention map based on one of the color stimulus. When attention has been focused on the corresponding area, a voice order switches the stimulus that is going to guide the attentional process to the other. The process is repeated. The evolution of the attentors in the experiment is shown in image 7 (they must be read in usual reading direction). At the beginning of the experiment there are no defined high interest regions; as time passes, stabilization is achieved providing a higher attention or relvance value to the right of the robot, where yellow balls are usually present. Then a voice order switches the object of attention to the search for pink balls which are always presented to the bottom left of the visual field of the robot. This results in another transitory stage until a new stable situation is achieved. Finally, we switch back to the first objective.

This experiment shows how the global attention mechanism can operate in a smooth and stable manner even when the goal directed focus of attention is changed.

4 Conclusions

In this paper we have proposed a mechanism that allows an adaptive implementation of an attentional mechanism that is based on the concept of attentor as a learnable relevance map associated with each detector. The values of the attentors are obtained by the artificial organisms through the interaction with the environment and help to provide a more efficient strategy for utilizing the organisms resources. This association between the attentors and detectors is the basis for a distributed architecture that integrates information and helps to guide the behaviour of the organism while performing its tasks.

The main features of this architecture is that it can combine sensorial or task based information streams coming from different sensors while preserving the most important natural based characteristics of attention such as the existence both of the chance for any spatial region to be explored while preserving the concentration of sensorial and cognitive resources on the most promising areas.

References

1. Pashler, H.E.: The psychology of attention. MIT Press, Cambridge, MA (1998)
2. Itti, L., Koch, C., & Niebur, E.: A model of saliency-based visual attention for rapid scene analysis. IEEE Transactions on Pattern Analysis and Machine Intelligence. (20(11)).(1998) 1254–1259
3. Itti, L. and Koch, C.: A saliency-based search mechanism for overt and covert shifts of visual attention. Vision Research. (40(10-12)). (2000) 1489–1506
4. Yarbus, A. L.: Eye movements and vision. Plenum Press, NY (1967)
5. Terada, K., Nakamura, T., Takeda, H., and Nishida, T.: A cognitive robot architecture based on tactile and visual information. Advanced Robotics. Vol. 13, (8). (2000) 767-778
6. Sokolov, Y. N.: Perception and the Conditioned Reflex. MacMillan, New York (1963)

Optimal Cue Combination for Saliency Computation: A Comparison with Human Vision

Alexandre Bur and Heinz Hügli

Institute of Microtechnology, University of Neuchâtel
Rue A.-L. Breguet 2, CH-2000 Neuchâtel, Switzerland
alexandre.bur@unine.ch, heinz.hugli@unine.ch

Abstract. The computer model of visual attention derives an interest or saliency map from an input image in a process that encompasses several data combination steps. While several combination strategies are possible, not all perform equally well. This paper compares main cue combination strategies by measuring the performance of the considered models with respect to human eye movements. Six main combination methods are compared in experiments involving the viewing of 40 images by 20 observers. Similarity is evaluated qualitatively by visual tests and quantitatively by use of a similarity score. The study provides insight into the map combination mechanisms and proposes in this respect an overall optimal strategy for a computer saliency model.

1 Introduction

It is generally admitted today that the human vision system makes extensive use of visual attention in order to select relevant visual information and speed up the vision process. Visual attention represents also a fundamental mechanism for computer vision where similar speed up of the processing can be envisaged. Thus, the paradigm of computational visual attention has been widely investigated during the last two decades. Today, computational models of visual attention are available in numerous software and hardware implementations [1,2] and possible application fields include color image segmentation [3] and object recognition [4].

First presented in [5], the saliency-based model of visual attention generates, for each visual cue (color, intensity, orientation, etc), a conspicuity map, i.e. a map that highlights the scene locations that differ from their surroundings according to the specific visual cue. Then, the computed maps are integrated into a unique map, the saliency map which encodes the saliency of each scene location. Depending on the scene, visual cues may contribute differently to the final saliency and of course, some scene locations may have higher saliency values than others. Therefore, the cue combination process should account optimally for these two aspects.

Note that the map integration process, described here for the purpose of combining cues, is also available at earlier steps of the computational model, namely

J. Mira and J.R. Álvarez (Eds.): IWINAC 2007, Part II, LNCS 4528, pp. 109–118, 2007.

for the integration of multi-scale maps or integration of different features. Omnipresent in the model, the competitive map integration process plays an important role and deserves careful design. The question which of the cue combination model performs better in comparison to human eye movements motivated this research.

In [6] four methods are considered for performing the competitive map integration and the methods were evaluated with respect to the capability to detect reference locations, but no comparison with eye movements is performed. Specifically, the authors propose an interesting weighting method as well as a so-called iterative method performing a non-linear transform of a map. Both will be considered here. Another feature integration scheme which comprises several masking mechanisms was also proposed in [7]. In [8], the authors propose an alternative non-linear integration scheme that shows quite superior to the more traditional linear scheme and will therefore be considered here.

Another aspect of the cue integration strategy refers to the way each cue contribution is weighted with respect to the others. The long-term normalization proposed in [9] will be considered along with the more traditional instantaneous peak-to-peak normalization approach.

The comparison of saliency maps with human eye fixations for the purpose of model evaluation has been performed previously. In [10] the authors propose the notion of chance-adjusted saliency for measuring the similarity of eye fixations and saliency. This requires the sampling of the saliency map at the points of fixations. In [11] the authors propose the reconstruction of a human saliency map or human saliency map from the fixations and perform the comparison by evaluating the correlation coefficient between fixations and saliency maps. This method was also used in [7]. The similarity score, used in [8], expresses the chance-adjusted saliency in a relative way that makes it independent of the map scale; it will therefore also be used here.

The remainder of this paper is organized as follows. Section 2 gives a brief description of the saliency-based model of visual attention. Section 3 defines the tools used for comparing saliency and fixations. Section 4 is devoted to the selection and definition of the six map integration methods that are evaluated by experiments described in section 5. Finally, section 6 concludes the paper.

2 The Saliency-Based Model of Visual Attention

The saliency-based model of visual attention was proposed by Koch and Ullman in [5]. It is based on three major principles: visual attention acts on a multi-featured input; saliency of locations is influenced by the surrounding context; the saliency of locations is represented on a scalar saliency map. Several works have dealt with the realization of this model i.e. [1]. In this paper, the saliency map results from 3 cues (intensity, contrast, orientation and chromaticity) and the cues stem from 7 features. The different steps of the model are detailed below.

First, 7 features (1..j..7) are extracted from the scene by computing the so-called feature maps from an RGB color image. The features are: (a) Intensity feature F_1, (b) Four local orientation features $F_{2..5}$ and (c) Two chromatic features based on the two color opponency filters red-green F_6 and blue-yellow F_7.

In a second step, each feature map is transformed into its conspicuity map: the multiscale analysis decomposes each feature F_j in a set of components $F_{j,k}$ for resolution levels k=1...6; the center-surround mechanism produces the multiscale conspicuity maps $\mathcal{M}_{j,k}$ to be combined, in a competitive way, into a single ***feature conspicuity map*** C_j in accordance with:

$$C_j = \sum_{k=1}^{K} \mathcal{N}(\mathcal{M}_{j,k}) \tag{1}$$

where $\mathcal{N}(.)$ is a normalization function that simulates both intra-map competition and inter-map competition among the different scale maps.

In the third step, using the same competitive map integration scheme as above, the seven (j=1...7) features are then grouped, according to their nature, into the three cues intensity, color and orientation. Formally, the ***cue conspicuity maps*** are thus:

$$C_{int} = C_1; \quad C_{orient} = \sum_{j \epsilon \{2,3,4,5\}} \mathcal{N}(C_j); \quad C_{chrom} = \sum_{j \epsilon \{6,7\}} \mathcal{N}(C_j) \tag{2}$$

In the final step of the attention model, the cue conspicuity maps are integrated, by using the scheme as above, into a ***saliency map*** S, defined as:

$$S = \sum_{cue \epsilon \{int, orient, chrom\}} \mathcal{N}(C_{cue}) \tag{3}$$

3 Comparing Fixations and a Saliency Map

The idea is to design a computer model which is close to human visual attention and, here, our basic assumption is that human visual attention is tightly linked to eye movements. Thus, eye movement recording is a suitable means for studying the spatial deployment of human visual attention. More specifically, while the observer watches at the given image, the K successive fixation locations of his eyes

$$\mathbf{X}^i = (\mathbf{x}_1^i, \mathbf{x}_2^i, \mathbf{x}_3^i, ...) \tag{4}$$

are recorded and then compared to the computer generated saliency map.

The degree of similarity of a set of successive fixations with the saliency map is evaluated qualitatively and quantitatively. For the qualitative comparison, the fixations are transformed in a so-called human saliency map which resembles the saliency map and the similarity is evaluated by comparing them visually. For the quantitative comparison, a similarity score is used.

3.1 Human Saliency Map

The human saliency map $H(\mathbf{x})$ is computed under the assumption that it is an integral of gaussian point spread functions $h(\mathbf{x}_k)$ located at the position of the successive fixations. It is assumed that each fixation $\mathbf{x}_k$ gives rise to a gaussian distributed activity. The width of the gaussian is chosen to approximate the size of the fovea. Formally, the human saliency map $H(\mathbf{x})$ is:

$$S_{human} = H(\mathbf{x}) = \frac{1}{K}\sum_{k=1}^{K} h(\mathbf{x}_k) \tag{5}$$

3.2 Score

In order to compare a computational saliency map and human fixation patterns quantitatively, we compute a score s, similar to the chance-adjusted saliency used in [10]. The idea is to define the score as the difference of average saliency $\overline{s}_{fix}$ obtained when sampling the saliency map S at the fixations points with respect to the average $\overline{s}$ obtained by a random sampling of S. In addition, the score used here is normalized and thus independent of the scale of the saliency map, as argued in [8,12]. Formally, the score s is thus defined as:

$$s = \frac{\overline{s}_{fix} - \overline{s}}{\overline{s}}, \quad with \quad \overline{s}_{fix} = \frac{1}{K}\sum_{k=1}^{K} S(\mathbf{x}_k) \tag{6}$$

4 Six Cue Integration Methods

Transforming initial features into a final saliency map, the model of visual attention includes several map integration steps described by eq. 1, 2 and 3 in which, the function $\mathcal{N}(.)$ formally determines the competitive map integration. Two different normalizations, a weighting scheme and three different map transforms will now be defined and used for describing six different cue integration methods $\mathcal{M}_{1...6}$ to be compared.

4.1 Peak-to-Peak vs. Long-Term Normalization

When the integration concerns maps issued from similar features, their value range is similar and the maps can be combined directly. This is the case for integration of multiscale maps (eq. 2) and also for the integration of similar features into the cue maps (eq. 3). However, the case of the integration of several cues into the saliency map (eq. 4) is different because the channels intensity, chrominance and orientation have different nature and may exhibit completely different value ranges. Here a map normalization step is mandatory.

Most of the previous works dealing with saliency-based visual attention [1] normalize therefore the channels to be integrated using a ***peak-to-peak normalization***, as follows:

$$\mathcal{N}_{PP}(C) = \frac{C - C_{min}}{C_{max} - C_{min}} \tag{7}$$

where C_{max} and C_{min} are respectively the maximum and the minimum values of the conspicuity map C. This peak-to-peak normalization has however an undesirable drawback. It maps each channel to its full range, regardless of the effective amplitude of the map. An alternative normalization procedure proposed in [9] tends to escape it.

The idea is to normalize each channel with respect to a maximum value which has universal meaning. The procedure, named ***long-term normalization***, scales the cue map with respect to a universal or long-term cue specific maximum $\overline{M}_{cue}$ by

$$\mathcal{N}_{LT}(C) = \frac{C}{\overline{M}_{cue}} \tag{8}$$

Practically, the long-term cue maximum can be estimated for instance by learning from a large set of images. The current procedure computes it from the cue maps $C_{cue}(n)$ of a large set of more than 500 images of various types (lanscapes, traffic, fractals, art, ...) by setting it equal to the average of the cue map maxima.

4.2 Weighting Scheme

Inter-map competition can be implemented by a map weighting scheme. The basic idea is to assign to each map C a scalar weight w that holds for its individual contribution. Most of the previous works dealing with saliency-based visual attention use such a competition-based scheme for map combination based on a weight. The weight is computed from the conspicuity map itself and tends to catch the global interest of that map. We consider following weight definitions:

$$w_1(C) = (M - \overline{m})^2 \quad or \quad w_2(C) = \frac{C_{max}}{\overline{C}} \tag{9}$$

The first weight expression w1 stems from [1]. In it, M is the maximum value of the normalized conspicuity map $N_{pp}(C)$ and $\overline{m}$ is the mean value of its local maxima. This weight function w_1 tends to promote maps with few dissimilar peaks and to demote maps with a lot of same peaks. The second weight expression w_2, proposed in [8], derives also the weight from the conspicuity map C. The values C_{max} and $\overline{C}$ are respectively the maximum and mean values of the conspicuity map. This weight tends to promote maps with few large peaks and demote maps with a lot of similar peaks.

4.3 Map Transform

The first map transform adopts a linear mapping and a weighting scheme for inter-map competition. The corresponding mapping is:

$$\mathcal{N}_{lin}(C) = w(C) \cdot C \tag{10}$$

where $w(C)$ is any of the weights above.

The second map transform corresponds to the iterative scheme proposed in [6]. Here, the function $\mathcal{N}(.)$ consists in an iterative filtering of the conspicuity map by a difference-of-Gaussians-filter (DoG) according to:

$$\mathcal{N}_{iter}(C) = C_n \quad with \quad C_n = |C_{n-1} + C_{n-1} * DoG - \varepsilon|_{\geq 0} \tag{11}$$

where the filtering, initiated by $C_0 = C$, iterates n times and produces the iterative mapping $\mathcal{N}_{iter}(C)$. The iterative mapping tends to decrease lower values and increases higher values of the map.

The third map transform, proposed in [8], consists in an exponential mapping of the conspicuities and a weighting scheme for inter-map competition, defined as:

$$\mathcal{N}_{exp}(C) = w(C_\gamma) \cdot C_\gamma \quad with \quad C_\gamma = (C_{max} - C_{min})(\frac{C - C_{min}}{C_{max} - C_{min}})^\gamma \tag{12}$$

the mapping has exponential character imposed by $\gamma > 1$: it promotes the higher conspicuity values and demotes the lower values; it therefore tends to suppress the lesser important values forming the background.

4.4 Comparison of Six Methods

The considered methods combine the 3 map transforms ($\mathcal{N}_{lin}$ $\mathcal{N}_{iter}$ and $\mathcal{N}_{exp}$) with one of the two normalization schemes ($\mathcal{N}_{PP}$ and $\mathcal{N}_{LT}$), resulting in six cue integration methods $\mathcal{M}$:

$$\begin{array}{llll}
\mathcal{M}_1: & \mathcal{N}_{lin/PP}(C) = \mathcal{N}_{lin}(\mathcal{N}_{PP}(C)) & \mathcal{M}_4: & \mathcal{N}_{lin/LT}(C) = \mathcal{N}_{lin}(\mathcal{N}_{LT}(C)) \\
\mathcal{M}_2: & \mathcal{N}_{iter/PP}(C) = \mathcal{N}_{iter}(\mathcal{N}_{PP}(C)) & \mathcal{M}_5: & \mathcal{N}_{iter/LT}(C) = \mathcal{N}_{iter}(\mathcal{N}_{LT}(C)) \\
\mathcal{M}_3: & \mathcal{N}_{exp/PP}(C) = \mathcal{N}_{exp}(\mathcal{N}_{PP}(C)) & \mathcal{M}_6: & \mathcal{N}_{exp/LT}(C) = \mathcal{N}_{exp}(\mathcal{N}_{LT}(C))
\end{array}$$

In [8], the performance of the weighting schemes w_1 and w_2 were similar. Here only the weight w_2 defined eq. 9 is used.

5 Comparison Results

5.1 Experiments

The experimental image data set consists in 40 color images of various types like natural scenes, fractals, and abstract art images. The images were shown to 20 human subjects. Eye movements were recorded with a infrared video-based tracking system (EyeLinkTM, SensoMotoric Instruments GmbH, Teltow/Berlin). The images were presented, in a dimly lit room on a 19″ CRT display with a resolution of 800×600, 24 bit color depth, and a refresh rate of 85 Hz. Image viewing was embedded in a recognition task. The images were presented to the subjects in blocks of 10, for a duration of 5 seconds per image, resulting in an average of 290 fixations per image.

5.2 Qualitative Results

Figure 1 provides a qualitative comparison of the peak-to-peak ($\mathcal{N}_{PP}$) and the long-term ($\mathcal{N}_{LT}$) normalization schemes. Two examples are given. The first one (1) compares $\mathcal{M}_1$ and $\mathcal{M}_4$ methods for image #28 (flower). More specifically, the images provide a comparison of the saliency maps (D) obtained from $\mathcal{M}_1$ and $\mathcal{M}_4$ with the human saliency map (E). We notice that $\mathcal{M}_4$ is more suitable than $\mathcal{M}_1$. An explanation is given by analyzing the cue contributions: color contribution (A), intensity contribution (B) and orientation contribution (C). We notice that with $\mathcal{M}_1$, which applies the peak-to-peak normalization, all cues contribute in a similar way to saliency, although the intensity contrast clearly dominates in the image. The performance is poor. $\mathcal{M}_4$ however, which applies the long-term normalization, has the advantage to take into account the relative contribution of the cues. Thus, color and orientation are suppressed while intensity is promoted. To summarize, this example illustrates the higher suitability of the long-term normalization $\mathcal{N}_{LT}$, compared to the peak-to-peak $\mathcal{N}_{PP}$. Example (2) compares $\mathcal{M}_3$ with $\mathcal{M}_6$ for image #37 (blue traffic sign). It also illustrates the higher suitability of the $\mathcal{N}_{LT}$ compared to the $\mathcal{N}_{PP}$.

In Figure 2, we discuss the performances of the linear ($\mathcal{N}_{lin}$), iterative ($\mathcal{N}_{iter}$) and exponential ($\mathcal{N}_{exp}$) map transforms by comparing the saliency map issued

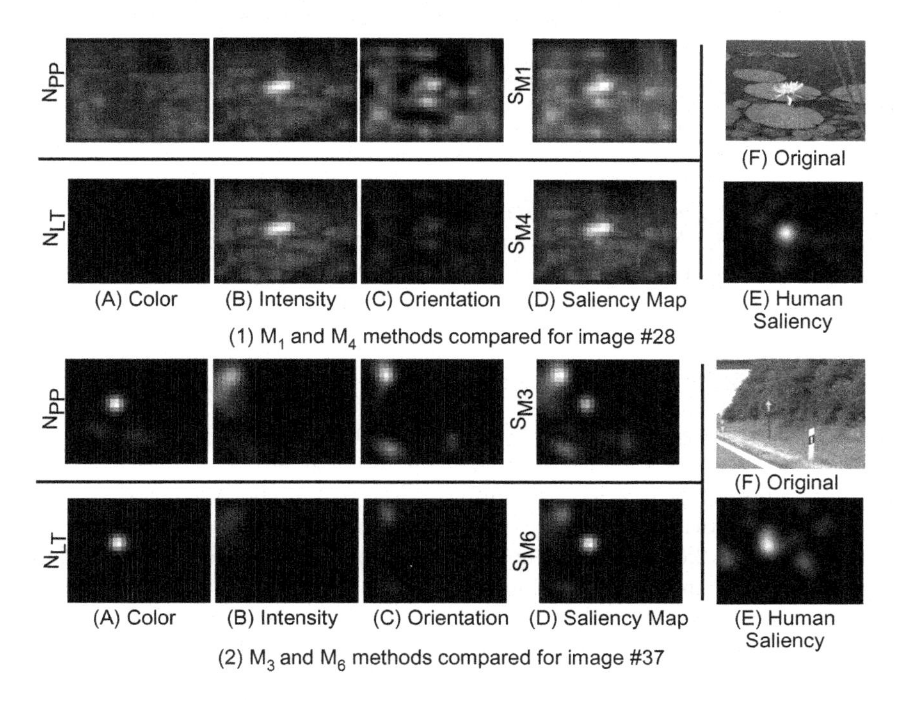

Fig. 1. Peak-to-peak $\mathcal{N}_{PP}$ versus long-term $\mathcal{N}_{LT}$ normalization scheme: (1) $\mathcal{M}_1$ compared to $\mathcal{M}_4$, (2) $\mathcal{M}_3$ to $\mathcal{M}_6$

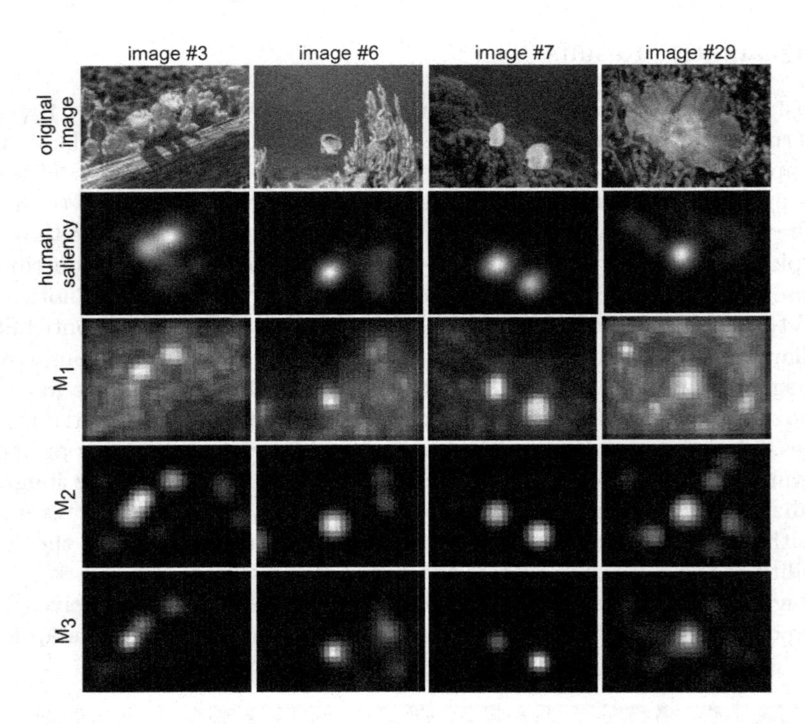

Fig. 2. A comparison of the map transforms $\mathcal{N}_{lin}$, $\mathcal{N}_{iter}$ and $\mathcal{N}_{exp}$, by comparing the saliency map issued from methods $\mathcal{M}_1$, $\mathcal{M}_2$ and $\mathcal{M}_3$ with the human saliency map

from methods $\mathcal{M}_1$, $\mathcal{M}_2$ and $\mathcal{M}_3$. Here, the results illustrate the highest suitability of the iterative $\mathcal{N}_{iter}$ and exponential $\mathcal{N}_{exp}$ methods, compared to the linear model $\mathcal{N}_{lin}$. Hence, $\mathcal{N}_{iter}$ and $\mathcal{N}_{exp}$ are non-linear map transform, which tend to promote the higher conspicuity values and demotes the lower values. It therefore tends to suppress the low-level values formed by the background.

5.3 Quantitative Results

Table 1 shows the average scores for the six methods, computed over all the images and issued from the five first fixations over all subjects. The results with a different number of fixation are similar. Concerning the $\mathcal{N}_{iter}$ and $\mathcal{N}_{exp}$, the evaluation is performed by testing different number of iterations (3, 5 and 10) and γ value (1.5, 2.0, 2.5, 3.0).The quantitative results presented here confirm the qualitative results. First, the long-term normalization ($\mathcal{N}_{LT}$) provides higher scores than the peak-to-peak ($\mathcal{N}_{PP}$) over all map transforms ($\mathcal{N}_{lin}$, $\mathcal{N}_{iter}$ and $\mathcal{N}_{exp}$). Second, both non-linear map transforms ($\mathcal{N}_{iter}$ and $\mathcal{N}_{exp}$) perform equally well and provide higher scores than the linear $\mathcal{N}_{lin}$.

Table 2 reports the t and p-value obtained with a paired t-test in order to verify that the $\mathcal{N}_{LT}$ scores are statistically higher than the $\mathcal{N}_{PP}$ scores over all the images, and also $\mathcal{N}_{iter}$ and $\mathcal{N}_{exp}$ scores higher than $\mathcal{N}_{lin}$ scores. Here, the

Table 1. An overview of the average scores for the six methods

normalization methods		$\mathcal{N}_{PP}$	$\mathcal{N}_{LT}$
$\mathcal{N}_{lin}$		$\mathcal{M}_1$: 0.49	$\mathcal{M}_4$: 0.65
$\mathcal{N}_{iter}$	n=3	1.49	1.88
	n=5	$\mathcal{M}_2$: 1.92	$\mathcal{M}_5$: 2.40
	n=10	2.36	2.94
$\mathcal{N}_{exp}$	γ=1.5	0.84	1.22
	γ=2.0	1.46	2.33
	γ=2.5	$\mathcal{M}_3$: 2.06	$\mathcal{M}_6$: 3.39
	γ=3.0	2.55	4.24

Table 2. t-value and p-value obtained with a paired t-test for comparing the different normalization schemes

	$\mathcal{N}_{PP}$ vs $\mathcal{N}_{LT}$			$\mathcal{N}_{lin}$ vs $\mathcal{N}_{iter}$ vs $\mathcal{N}_{exp}$		
Paired t-test	$\mathcal{M}_1$ vs $\mathcal{M}_4$	$\mathcal{M}_2$ vs $\mathcal{M}_5$	$\mathcal{M}_3$ vs $\mathcal{M}_6$	$\mathcal{M}_4$ vs $\mathcal{M}_5$	$\mathcal{M}_4$ vs $\mathcal{M}_6$	$\mathcal{M}_5$ vs $\mathcal{M}_6$
t-value	3.09	2.92	2.46	4.52	3.37	0.25
p-value	< 0.005	< 0.01	< 0.025	< 0.005	< 0.005	-

presented values are computed for $\mathcal{N}_{exp}(\gamma = 2.0)$ and for $\mathcal{N}_{iter}(n = 5)$ iteration, the p-values confirm both statements above.

Finally, this study suggests that an optimal combination strategy for saliency computation uses long-term normalization ($\mathcal{N}_{LT}$) combined with one of the non-linear transforms ($\mathcal{N}_{iter}$, $\mathcal{N}_{exp}$). If computation costs are to be considered as additional criteria of selection for a non-linear map transform, the less complex exponential $\mathcal{N}_{exp}$ would probably be preferred.

6 Conclusions

This paper presents main cue combination strategies in the design of computer model of visual attention and analyzes the performance in comparison to human eye movements. Two normalization schemes (peak-to-peak $\mathcal{N}_{PP}$ and long-term $\mathcal{N}_{LT}$) and three map transforms (linear $\mathcal{N}_{lin}$, iterative $\mathcal{N}_{iter}$ and exponential $\mathcal{N}_{exp}$) are considered, resulting in six cue integration methods. The experiments conducted for evaluating the methods involve the viewing of 40 images by 20 human subjects.

The qualitative and quantitative results conclude two principal points: first, the long-term normalization scheme $\mathcal{N}_{LT}$ is more appropriate than the peak-to-peak $\mathcal{N}_{PP}$. The main difference between both schemes is that $\mathcal{N}_{LT}$ has the advantage to take into account the relative contribution of the cues which is not the case of $\mathcal{N}_{PP}$. Second point, both non-linear map transforms $\mathcal{N}_{iter}$ and $\mathcal{N}_{exp}$ perform equally well and are more suitable than the linear $\mathcal{N}_{lin}$.

From this study, we can state that the optimal cue combination scheme for computing a saliency close to a collective human visual attention is the long-term

$\mathcal{N}_{LT}$ normalization scheme combined with one of the non-linear map transforms $\mathcal{N}_{iter}$ or $\mathcal{N}_{exp}$, with a possible preference for the later method for its lesser computation costs.

Acknowledgments

The presented work was supported by the Swiss National Science Foundation under project number FN-108060 and was done in collaboration with the Perception and Eye Movement Laboratory (PEML), Dept. of Neurology and Dept. of Clinical Research, university of Bern, Switzerland.

References

1. L. Itti, Ch. Koch, and E. Niebur. A model of saliency-based visual attention for rapid scene analysis. *IEEE Transactions on Pattern Analysis and Machine Intelligence (PAMI), Vol. 20, No. 11, pp. 1254-1259*, 1998.
2. N. Ouerhani and H. Hugli. Real-time visual attention on a massively parallel SIMD architecture. *International Journal of Real Time Imaging, Vol. 9, No. 3, pp. 189-196*, 2003.
3. N. Ouerhani and H. Hugli. MAPS: Multiscale attention-based presegmentation of color images. *4th International Conference on Scale-Space theories in Computer Vision, Springer Verlag, LNCS, Vol. 2695, pp. 537-549*, 2003.
4. D. Walther, U. Rutishauser, Ch. Koch, and P. Perona. Selective visual attention enables learning and recognition of multiple objects in cluttered scenes. *Computer Vision and Image Understanding, Vol. 100 (1-2), pp. 41-63*, 2005.
5. Ch. Koch and S. Ullman. Shifts in selective visual attention: Towards the underlying neural circuitry. *Human Neurobiology, Vol. 4, pp. 219-227*, 1985.
6. L. Itti and Ch. Koch. A comparison of feature combination strategies for saliency-based visual attention systems. *SPIE Human Vision and Electronic Imaging IV (HVEI'99), Vol. 3644, pp. 373-382*, 1999.
7. O. Le Meur, P. Le Callet, D. Barba, and D. Thoreau. A coherent computational approach to model bottom-up visual attention. *PAMI, Vol. 28, No. 5*, 2006.
8. N. Ouerhani, A. Bur, and H. Hügli. Linear vs. nonlinear feature combination for saliency computation: A comparison with human vision. volume 4174 of *Lecture Notes in Computer Science*, pages 314–323, Springer, 2006.
9. N. Ouerhani, T. Jost, A. Bur, and H. Hugli. Cue normalization schemes in saliency-based visual attention models. *International Cognitive Vision Workshop, Graz, Austria*, 2006.
10. D. Parkhurst, K. Law, and E. Niebur. Modeling the role of salience in the allocation of overt visual attention. *Vision Research, Vol. 42, No. 1, pp. 107-123*, 2002.
11. N. Ouerhani, R. von Wartburg, H. Hugli, and R. Mueri. Empirical validation of the saliency-based model of visual attention. *Electronic Letters on Computer Vision and Image Analysis (ELCVIA), Vol. 3 (1), pp. 13-24*, 2004.
12. L. Itti. Quantitative modeling of perceptual salience at human eye position. *Visual Cognition, in press*, 2005.

The Underlying Formal Model of Algorithmic Lateral Inhibition in Motion Detection

José Mira[1], Ana E. Delgado[1], Antonio Fernández-Caballero[2], María T. López[2], and Miguel A. Fernández[2]

[1] Universidad Nacional de Educación a Distancia
E.T.S.I. Informática, 28040 - Madrid, Spain
{jmira,adelgado}@dia.uned.es
[2] Universidad de Castilla-La Mancha
Instituto de Investigación en Informática (I3A) and
Escuela Politécnica Superior de Albacete, 02071 - Albacete, Spain
{caballer,mlopez,miki}@dsi.uclm.es

Abstract. Many researchers have explored the relationship between recurrent neural networks and finite state machines. Finite state machines constitute the best characterized computational model, whereas artificial neural networks have become a very successful tool for modeling and problem solving. Recently, the neurally-inspired algorithmic lateral inhibition (ALI) method and its application to the motion detection task have been introduced. The article shows how to implement the tasks directly related to ALI in motion detection by means of a formal model described as finite state machines. Automata modeling is the first step towards real-time implementation by FPGAs and programming of "intelligent" camera processors.

1 Introduction

Recently the algorithmic lateral inhibition (ALI) method and its application to the motion detection task have been introduced [1]-[5]. And, currently our research team is involved in implementing the method into real-time in order to provide efficient response time in visual surveillance applications [6]-[7].

In recent years, many researchers have explored the relation between discrete-time recurrent neural networks and finite state machines, either by showing their computational equivalence or by training them to perform as finite state recognizers from example [8]. The relationship between discrete-time recurrent neural networks and finite state machines has very deep roots [9]-[11]. The early papers mentioned show the equivalence of these neural networks with threshold linear units, having step-like transfer functions, and some classes of finite state machines. More recently, some researchers have studied the close relationships more in detail [12]-[13], as well as the combination of connectionist and finite state models into hybrid techniques [14]-[15]. During the last decades specialized algorithms even have extracted finite state machines from the dynamics of discrete-time recurrent neural networks [16]-[19].

J. Mira and J.R. Álvarez (Eds.): IWINAC 2007, Part II, LNCS 4528, pp. 119–129, 2007.

The article shows how to implement the tasks directly related to ALI in motion detection and introduced by means of a formal model described finite state machines, and the subsequent implementation in hardware, as automata modeling may be considered as the first step towards real-time implementation by field programmable gate arrays (FPGAs) [20] and programming of "intelligent" camera processors.

2 ALI in Motion Detection

The operationalization of the ALI method for the motion detection application has led to the so-called lateral interaction in accumulative computation [2],[4]. From [2],[4] we cite and reformulate the most important concepts and equations.

Temporal Motion Detection firstly covers the need to segment each input image I into a preset group of gray level bands (N), according to equation 1.

$$x_k(i,j;t) = \begin{cases} 1, \text{if } I(i,j;t) \in [\frac{256}{N} \cdot k, \frac{256}{N} \cdot (k+1) - 1] \\ 0, \text{otherwise} \end{cases} \quad (1)$$

This formula assigns pixel (i,j) to gray level band k. Then, the accumulated charge value related to motion detection at each input image pixel is obtained, as shown in formula 2:

$$y_k(i,j;t) = \begin{cases} v_{dis}, \text{if } x_k(i,j;t) = 0 \\ v_{sat}, \text{if } (x_k(i,j;t) = 1) \cap (x_k(i,j;t-\Delta t) = 0) \\ \max[x_k(i,j;t-\Delta t) - v_{dm}, v_{dis}], \\ \qquad \text{if } (x_k(i,j;t) = 1) \cap (x_k(i,j;t-\Delta t) = 1) \end{cases} \quad (2)$$

The charge value at pixel (i,j) is discharged down to v_{dis} when no motion is detected, is saturated to v_{sat} when motion is detected at t, and, is decremented by a value v_{dm} when motion goes on being detected in consecutive intervals t and $t - \Delta t$.

Spatial-Temporal Recharging is thought to reactivate the charge values of those pixels partially loaded (charge different from v_{dis} and v_{sat}) and that are directly or indirectly connected to saturated pixels (whose charge is equal to v_{sat}). Values z_k are initialized to y_k. Formula 3 explains these issues, where v_{rv} is precisely the recharge value.

$$z_k(i,j;t+l\cdot\Delta\tau) = \begin{cases} v_{dis}, & \text{if } z_k(i,j;t+(l-1)\cdot\Delta\tau) = v_{dis} \\ v_{sat}, & \text{if } z_k(i,j;t+(l-1)\cdot\Delta\tau) = v_{sat} \\ \min[z_k(i,j;t+(l-1)\cdot\Delta\tau) + v_{rv}, v_{sat}], \\ \qquad \text{if } v_{dis} < z_k(i,j;t+(l-1)\cdot\Delta\tau) < v_{sat} \end{cases} \quad (3)$$

This step occurs in an iterative way in a different space of time $\tau \ll t$. The value of $\Delta\tau$ will determine the number of times the mean value is calculated.

Spatial-Temporal Homogenization aims in distributing the charge among all connected neighbors holding a minimum charge (greater than v_{dis}) - now, O_k is initialized to z_k. This occurs according to the following equation.

$$\begin{aligned} O_k(i,j;t+m\cdot\Delta\tau) = & \frac{1}{1+\delta_{i-1,j}+\delta_{i+1,j}+\delta_{i,j-1}+\delta_{i,j+1}} \\ & \times[O_k(i,j;t+(m-1)\cdot\Delta\tau)+ \\ & \delta_{i-1,j}\cdot O_k(i-1,j;t+(m-1)\cdot\Delta\tau)+ \\ & \delta_{i+1,j}\cdot O_k(i+1,j;t+(m-1)\cdot\Delta\tau)+ \\ & \delta_{i,j-1}\cdot O_k(i,j-1;t+(m-1)\cdot\Delta\tau)+ \\ & \delta_{i,j+1}\cdot O_k(i,j+1;t+(m-1)\cdot\Delta\tau)] \end{aligned} \tag{4}$$

where

$$\forall(\alpha,\beta)\in[i\pm 1,j\pm 1], \delta_{\alpha,\beta} = \begin{cases} 1, & \text{if } O_k(\alpha,\beta;t+(m-1)\cdot\Delta\tau) > v_{dis} \\ 0, & \text{otherwise} \end{cases} \tag{5}$$

Lastly, we take the maximum value of all outputs of the k gray level bands to show the silhouette of a moving object:

$$O(i,j;t) = \arg\max_k O_k(i,j;t) \tag{6}$$

3 Formal Model for ALI in Motion Detection

The control knowledge is described in extensive by means of a finite automaton in which the state space is constituted from the set of distinguishable situations in the state of accumulated charge in a local memory [5]. Thus, we distinguish $N+1$ states $S_0, S_1, ..., S_N$, where S_0 is the state corresponding to the totally discharged local memory (v_{dis}; in general $v_{dis}=0$), S_N is the state of complete charge ($v_{sat}=7$) and the rest are the $N-1$ intermediate charge states between v_{dis} and v_{sat}.

Let us suppose, without loss of generality, that it is enough to distinguish eight levels of accumulated charge ($N=8$) and, consequently, that we can use as a formal model of the control underlying the inferential scheme that describes the data flow corresponding to the calculation of this subtask an 8 states automaton ($S_0, S_1, ..., S_7$), where S_0 corresponds to v_{dis} and S_7 to v_{sat}. Let us also suppose that discharge ($v_{dm}=2$) and recharge ($v_{rv}=1$) initially take the values corresponding to the descent of two states and to the ascent of one state. This way, the state transition diagram corresponds to a particular kind of reversible counter ("up-down") controlled by the result of the lateral inhibition (dialogue among neighbors).

To complete the description of the states, together with the accumulated charge value, v ($v_{dis}\leq v\leq v_{sat}$), it is necessary to include come binary signals, A_P and $A_C=0,1$. When $A_P=1$, a pixel tells its neighbors that it has detected a mobile, or that some neighbor has told him to have detected a mobile. A_C indicates that motion has been detected on the pixel.

3.1 ALI Temporal Motion Detecting

The task firstly gets as input data the values of the 256 gray level input pixels and generates $N=8$ binary images, $x_k(i,j;t)$. The output space has a FIFO memory structure with two levels, one for the current value and another one for the previous instant value. Thus, for N bands, there are $2N = 16$ binary values for each input pixel; at each band there is the current value $x_k(i,j;t)$ and the previous value $x_k(i,j;t-\Delta t)$, such that:

$$x_k(i,j;t) = \begin{cases} 1, & \text{if } I(i,j;t) \in [32 \cdot k, 32 \cdot (k+1) - 1] \\ 0, & \text{otherwise} \end{cases} \quad (7)$$

Thus, a pair of binarised values at each band, $x_k(i,j;t)$ and $x_k(i,j;t-\Delta t)$, constitutes the input space of the temporal non recurrent ALI. The output space is the result of the individual calculus phase in each calculus element. The inputs are observables $x_k(i,j;t)$ and $x_k(i,j;t-\Delta t)$ and the current charge value that initially is at state S_0. It also receives the comparison rule and the numerical coding of the different discrepancy classes ($D1, D2, D3$). The output is the class of discrepancy selected at this time, $D(t)$.

$$D(t) = \begin{cases} D2, & \text{if } (x_k(i,j;t) = 1) \cap (x_k(i,j;t-\Delta t) = 0) \\ D3, & \text{if } (x_k(i,j;t) = 1) \cap (x_k(i,j;t-\Delta t) = 1) \\ D1, & \text{otherwise} \end{cases} \quad (8)$$

This class is now an input in charge of filtering a specific charge value (before dialogue) from a set of potential values. These potential values are v_{dis}, v_{sat} and $max[x_k(i,j;t-\Delta t) - v_{dm}, v_{dis}]$. The output of subtask ALI Temporal Motion Detecting constitutes the accumulated charge value, $y_k(i,j;t)$, complemented by label A_C. Remember that $A_C = 1$ denotes the fact that a movement has been locally detected by this pixel.

$$A_C = \begin{cases} 1, & \text{if } D(t) = D2 \\ 0, & \text{otherwise} \end{cases} \quad (9)$$

$$y_k(i,j;t) = \begin{cases} v_{dis}, & \text{if } D(t) = D1 \\ v_{sat}, & \text{if } D(t) = D2 \\ \max[x_k(i,j;t-\Delta t) - v_{dm}, v_{dis}], & \text{if } D(t) = D3 \end{cases} \quad (10)$$

The following situations can be observed in Fig. 1 (see discrepancy class D):

1. $x_k(i,j;t-\Delta t) = 0, 1, x_k(i,j;t) = 0$ (corresponding to discrepancy class $D1$) In this case the calculation element (i,j) has not detected any contrast with respect to the input of a mobile in that band ($x_k(i,j;t) = 0$). It may have detected it (or not) in the previous interval ($x_k(i,j;t-\Delta t) = 1, x_k(i,j;t) = 0$). In any case, the element passes to state $S_0[v = v_{dis}, A_C = 0]$, the state of complete discharge, independently of which was the initial state.
2. $x_k(i,j;t-\Delta t) = 0, x_k(i,j;t) = 1$ (corresponding to discrepancy class $D2$) The calculation element has detected in t a contrast in its band ($x_k(i,j;t) = 1$), and it did not in the previous interval ($x_k(i,j;t-\Delta t) = 0$). It passes to state $S_7[v = v_{sat}, A_C = 1]$, the state of total charge, independently of which was the previous state. Also A_C passes to 1, in order to tell its potential dialogue neighbors that this pixel has detected a mobile.
3. $x_k(i,j;t-\Delta t) = 1, x_k(i,j;t) = 1$ (corresponding to discrepancy class $D3$). The calculation element has detected the presence of an object in its band ($x_k(i,j;t) = 1$), and it had also detected it in the previous interval ($x_k(i,j;t-\Delta t) = 1$). In this case, it diminishes its charge value

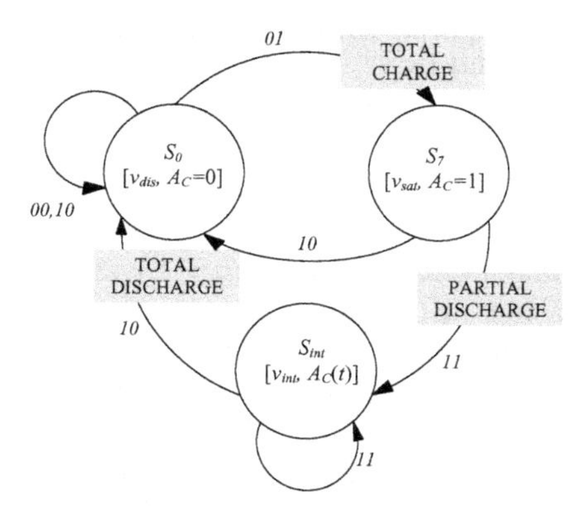

Fig. 1. Control automaton that receives inputs $x_k(i,j;t-\Delta t)$ and $x_k(i,j;t)$, and produces three outputs, coincident with its three distinguishable charge states ($S_0 = v_{dis}$, $S_7 = v_{sat}$, and v_{int})

in a certain value, v_{dm}. This discharge - partial discharge - can proceed from an initial state of saturation $S_7[v_{sat}, A_C = 1]$, or from some intermediate state $(S_6, ..., S_1)$. This partial discharge due to the persistence of the object in that position and in that band, is described by means of a transition from S_7 to an intermediate state, $S_{int}[v_{int}, A_C = 0, 1]$, without arriving to the discharge, $S_0[v_{dis}, A_C = 0]$. The descent in the element's state is equivalent to the descent in the pixel's charge, such that (as you may appreciate on Fig. 1) only the following transitions are allowed: $S_7 \rightarrow S_5, S_6 \rightarrow S_4, S_5 \rightarrow S_3, S_4 \rightarrow S_2, S_3 \rightarrow S_1, S_2 \rightarrow S_0$, and $S_1 \rightarrow S_0$.

3.2 ALI Spatial-Temporal Recharging

In the previous task the individual "opinion" of each computation element has been obtained. But, our aim is also to consider the "opinions" of the neighbors. The reason is that an element individually should stop paying attention to motion detected in the past, but before making that decision there has to be a communication in form of lateral inhibition with its neighbors to see if any of them is in state S_7 (v_{sat}, maximum charge). Otherwise, it will be discharging down to S_0 (v_{dis}, minimum charge), because that pixel is not bound to a pixel that has detected motion. The output is formed after dialogue processing with neighboring pixels by the so called permanency value, $z_k(i,j;t)$.

The values of charge accumulated before dialogue are written in the central part of the output space of each pixel (C^*) that now enters in the dialogue phase. The data in the periphery of receptive field in the output space of each pixel (P^*) contains now the individual calculi of the neighbors. Let $v_C(t) = y_k(i,j;t)$ be the initial charge value at this subtask. Each pixel takes into account the set of individual calculus, $v_C(t + k \cdot \Delta\tau), A_j$, by means of the logical union of the labels:

$$A_{P*}(\tau) = \bigcup_j A_j(\tau) \tag{11}$$

This result, A_{P*}, is now compared with A_C, giving rise to one of two discrepancy classes (recharge or stand-by).

$$D(t+l\cdot\Delta\tau)=\begin{cases} stand-by(v_{dis}), \text{ if } v_C(t+l\cdot\Delta\tau)=v_{dis} \\ stand-by(v_{sat}), \text{ if } v_C(t+l\cdot\Delta\tau)=v_{sat} \\ recharge, \text{ if } (v_{dis}<v_C(t+l\cdot\Delta\tau)<v_{sat})\cap(A_{P*}=1) \end{cases} \tag{12}$$

Subsequently, the class activated plays the role of selection criteria to output the new consensus charge value after dialogue, $z_k(i,j;t+\Delta t)$, with $\Delta t=k\cdot\Delta\tau$, being k the number of iterations in the dialogue phase, a function of the size of the receptive field. Notice that τ is a parameter that only depends on the size of the objects we want to detect from their motion. So, the purpose of this inference is to fix a minimum object size in each gray level band. The whole dialogue process is executed with clock τ, during k intervals $\Delta\tau$. It starts when clock t detects the configuration $y_k(i,j;t-\Delta t)=y_k(i,j;t)=1$ and ends at the end of t, when a new image appears.

$$A_C=\begin{cases} 1, \text{ if } D(t+l\cdot\Delta\tau)=\{stand-by(v_{sat})\cup recharge\} \\ 0, \text{ otherwise} \end{cases} \tag{13}$$

$$v(t+l\cdot\Delta\tau)=\begin{cases} v_{dis}, \text{ if } D(t+l\cdot\Delta\tau)=stand-by(v_{dis}) \\ v_{sat}, \text{ if } D(t+l\cdot\Delta\tau)=stand-by(v_{sat}) \\ \min[v(t+(l-1)\cdot\Delta\tau)+v_{rv},v_{sat}], \\ \qquad\qquad \text{if } (D(t+l\cdot\Delta\tau)=recharge \end{cases} \tag{14}$$

$$A_C=0, \text{ if } D(t+(l-1)\cdot\Delta\tau)=\{stand-by(v_{sat})\cup recharge\} \tag{15}$$

In each dialogue phase (in other words, in each interval of clock $\Delta\tau$), the calculation element only takes into account values $y_k(i,j;t-\Delta t)$, $y_k(i,j;t)$ and $A_C(t)$ present in that moment in its receptive field. To diffuse or to use more distant information, new dialogue phases are necessary. That is to say, new inhibitions in $l\cdot\Delta\tau$ $(1<l\leq k)$ are required. This only affects to state variable $A_C(\tau)$, as $y_k(i,j;t-\Delta t)$ and $y_k(i,j;t)$ values remain constant during the intervals used to diffuse τ and to consensus the different partial results obtained by the calculation elements. Notice that the recharge may only be performed once during the whole dialogue phase. That is why $A_C=0$, when a recharge takes place. Lastly, the output will be:

$$z_k(i,j;t+\Delta t)=v_C(t+\Delta t) \tag{16}$$

In the corresponding state transition diagram the following situations have to be distinguished:

1. $y_k(i,j;t-\Delta t)=0,1, y_k(i,j;t)=0$
   In any case, independently of the pixel's dialogue with the neighbors, at the end of Δt the pixel passes to state $S_0[v=v_{dis},A_C=0]$.
2. $y_k(i,j;t-\Delta t)=0, y_k(i,j;t)=1$
   Again, independently of the dialogue phase, the pixel's state will be $S_7[v=v_{sat},A_C=1]$.
3. $y_k(i,j;t-\Delta t)=1, y_k(i,j;t)=1$
   (a) Local memory is in $S_0[v_{dis},A_C=0]$. Pixels in state S_0 are not affected by lateral recharge due motion detection in their periphery. Thus, the pixel maintains the same state S_0.

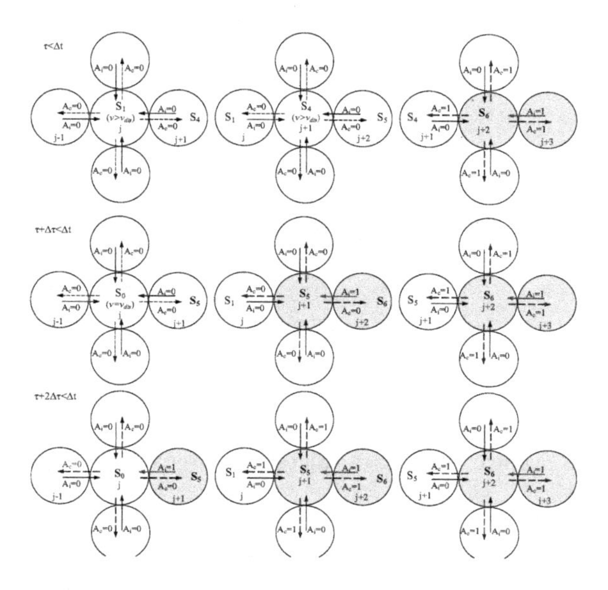

Fig. 2. Detail of the dialogue where diffusion of motion detection is shown through "transparent" pixels ($j+2$ and $j+1$), while pixel j deserves an "opaque" behavior

(b) Local memory is in $S_7[v_{sat}, A_C = 1]$. Pixels in state S_7 are maximally charged. So, they can not be recharged. They also maintain the state.

(c) Local memory is in $S_{int}[v_{int}, A_C(t)]$. Depending on their four neighbors' charge values, it can stay in S_{int} if all neighbors have variable $A_j = 0$ or transit up to S_7 if it finds some neighbor with variable $A_j = 1$.

   i. Transit from S_i to S_{i+1}. After recharge, the calculation element is now in S_{i+1}. It sends $A_C = 1$ and waits up to the end of Δt. In a second clock cycle $\Delta\tau$, $A_C = 1$ is potentially used by its neighbors to increment their charge values. Thus, the dialogue extends in steps of size the receptive field. Pixels with are said to be "transparent" if they allow information on motion detection by some neighbor (in state S_7) of their receptive field to cross them.

   ii. Remain in S_i. If none of its neighbors has transmitted $A_j = 1$, the pixel stays in S_i, without recharging in the first $\Delta\tau$. In this case, it maintains its proper $A_{C^*} = 0$, and its behavior is called "opaque". However, if in a later $\Delta\tau$ and inside the dialogue interval it does receive any $A_j = 1$, it will pass to S_{i+1}. Fig. 2 illustrates this diffusion mechanism through "opaque" and "transparent" pixels of the receptive field.

3.3 ALI Spatial-Temporal Homogenization

Now, the aim of this task is to obtain all moving patches present in the scene. The subtask considers the union of pixels that are physically together and at a

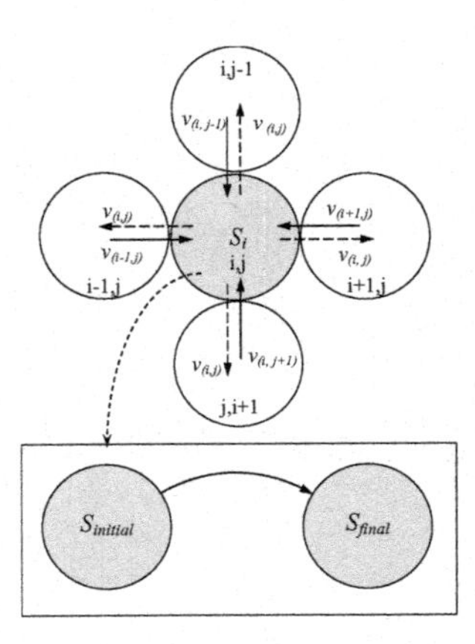

Fig. 3. Dialogue to average the charge values that overcome a threshold inside each gray level band

same gray level band to be a component of an object. A set of recurrent lateral inhibition processes are performed to distribute the charge among all neighbors that possess a certain minimum charge ("permanency value", $z_k(i, j; t)$, of previous task), that is to say, those pixels in states S_1 to S_7, and are physically connected. A double objective is aimed:

1. To dilute the charge due to the image background motion among other points of the own background, so that only moving objects are detected. To dilute the charge due to the image background motion does not mean that we are dealing with moving cameras. Instead of it, we are facing the problem of false motion detected where moving objects are just leaving pixels that now pertain to the background.
2. To obtain a parameter common to all pixels of the object those belong to the same gray level band (simple classification task).

Charge values, $z_k(i, j; t + \Delta t)$ are now evaluated in the center and in the periphery. Now, let v_C be the initial charge value. The result of the individual value (C) is compared with the mean value in (P) - notice that in P^* we have the average of those neighbors that have charge values different from θ_{min}, the so called "permanency threshold value" - and produces a discrepancy class according with threshold, θ_{min}, and passes the mean charge values that overcome that threshold. After this, the result is again compared with a second threshold, namely θ_{max}, eliminating noisy pixels pertaining to non-moving object.

$$D(t + \Delta t) = \begin{cases} D1, \text{ if } v_C = \theta_{min} \\ D2, \text{ if } (\theta_{min} < v_C < v_{sat}) \cap (\theta_{min} < v_P < v_{sat}) \\ D3, \text{ if } (\theta_{min} < v_C < v_{sat}) \cap (v_P = \theta_{min}) \end{cases} \tag{17}$$

$$O_k(i, j; t + \Delta t) = \begin{cases} \theta_{min}, \text{ if } v_C = \theta_{min} \\ (v_C + v_P)/2, \\ \text{ if } (\theta_{min} < v_C < v_{sat}) \cap (\theta_{min} < v_P < v_{sat}) \\ v_C, \text{ if } (\theta_{min} < v_C < v_{sat}) \cap (v_P = \theta_{min}) \end{cases} \tag{18}$$

$$O_k(i, j; t + \Delta t) = v_{dis}, \text{ if } O_k(i, j; t + \Delta t) > \theta_{max} \tag{19}$$

Fig. 3 illustrates the dialogue scheme and the description of the control automaton where the transitions among the initial state $S_i(t)$ (whenever $S_i(t)$ different from S_0) and the final state $S_i(t+\Delta t)$ state are carried out in agreement with rule:

$$S_{i_{final}} = 1/N_{k+1}(S_{i_{initial}} + \sum_{RF_k} v_j) \tag{20}$$

where the sum on sub-index j extends to all neighbors, v_j, belonging to the subset of the receptive field, RF_k, such that its state is different from S_0, and N_k is the number of neighbors with state different from S_0.

4 Data and Results

In order to test the validity of our proposal, in this section the result of applying 8∗8 ALI modules on specific areas of a well-known benchmark image sequences is shown. Figure 4 shows the first and the last images of the Ettlinger-Tor sequence. The treated $64 * 64$-pixel zone of the benchmark is marked.

Fig. 4. Ettlinger Tor sequence. (a) Frame number 1. (b) Frame number 40.

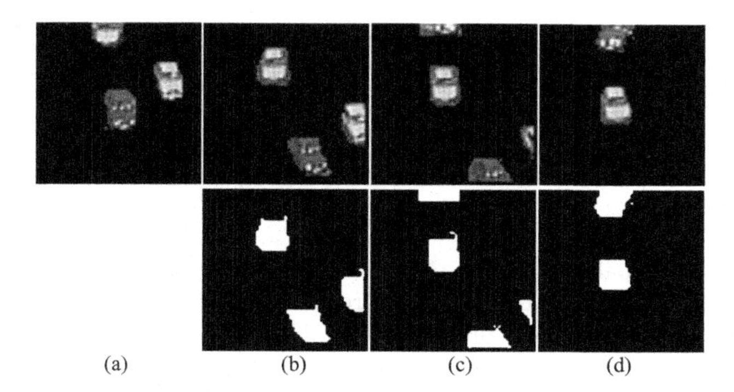

Fig. 5. Ettlinger Tor sequence ground truth and result. (a) After frame number number 1. (b) After frame number 20. (c) After frame number 30. (d) After frame number 40.

Figure 5 shows the result on some frames. After a few frames, the cars are perfectly segmented, as you may appreciate by comparing with the ground truth provided. Again, like in the previous sequence, you may appreciate that there must be enough motion to detect the moving objects. And, concerning the searched real-time performance, let us highlight that the results for $8*8$ modules have been obtained at a frequency of 0.966 MHz (1.035 μs). When extrapolating to usual $512 * 512$ pixel images, which need 4096 $8 * 8$ ALI modules, the results should be obtained after 4.24 ms. This performance may be considered as excellent, as in order to work in real-time we have up to 33 ms per image frame.

5 Conclusions

This paper starts from previous works in computer vision, where our algorithmic lateral inhibition method applied to motion detection has proven to be quite efficient. We have shown in this article how the algorithmic lateral inhibition model, based in recurrent neural networks, has been modeled by means of finite state automata, seeking for real-time through an implementation in FPGA-based reconfigurable hardware.

The design by means of programmable logic enables the systematic and efficient crossing from the descriptions of the functional specifications of a sequential system to the equivalent formal description in terms of a Q-states finite state automata or a N-recurrent-neurons neuronal network, where $Q \leq 2^N$. Starting from this point, a hardware implementation by means of programmable logic is very easy to perform. This kind of design is especially interesting in those application domains where the response time is crucial (e.g. monitoring and diagnosing tasks in visual surveillance and security).

Acknowledgments

This work is supported in part by the Spanish CICYT TIN2004-07661-C02-01 and TIN2004-07661-C02-02 grants, and the Junta de Comunidades de Castilla-La Mancha PBI06-0099 grant.

References

1. A. Fernández-Caballero, J. Mira, M.A. Fernández, and M.T. López, Segmentation from motion of non-rigid objects by neuronal lateral interaction, Pattern Recognition Letters, vol. 22, no. 14, pp. 1517-1524, 2001.
2. A. Fernández-Caballero, J. Mira, A.E. Delgado, and M.A. Fernández, Lateral interaction in accumulative computation: A model for motion detection, Neurocomputing, vol. 50C, pp. 341-364, 2003.
3. A. Fernández-Caballero, M.A. Fernández, J. Mira, and A.E. Delgado, Spatio-temporal shape building from image sequences using lateral interaction in accumulative computation, Pattern Recognition, vol. 36, no. 5, pp. 1131-1142, 2003.

4. A. Fernández-Caballero, J. Mira, M.A. Fernández, and A.E. Delgado, On motion detection through a multi-layer neural network architecture, Neural Networks, vol. 16, no. 2, pp. 205-222, 2003.
5. J. Mira, A.E. Delgado, A. Fernández-Caballero, and M.A. Fernández, Knowledge modelling for the motion detection task: The lateral inhibition method, Expert Systems with Applications, vol. 7, no. 2, pp. 169-185, 2004.
6. M.T. López, A. Fernández-Caballero, M.A. Fernández, J. Mira, A.E. Delgado, Visual surveillance by dynamic visual attention method, Pattern Recognition, vol. 39, no, 11, pp. 2194-2211, 2006.
7. M.T. López, A. Fernández-Caballero, M.A. Fernández, J. Mira, A.E. Delgado, Motion features to enhance scene segmentation in active visual attention, Pattern Recognition Letters, vol. 27, no. 5, pp. 469-478, 2006.
8. R.P. Ñeco, and M.L. Forcada, Asynchronous translations with recurrent neural nets, in Proceedings of the International Conference on Neural Networks, ICNN'97, vol. 4, pp. 2535-2540, 1997.
9. W.S. McCulloch, and W.H. Pitts, A logical calculus of the ideas immanent in nervous activity, Bulletin of Mathematical Biophysics, vol. 5, pp. 115-133, 1943.
10. S.C. Kleene, Representation of events in nerve nets and finite automata, in Automata Studies, Princeton University Press, 1956.
11. M.L. Minsky, Computation: Finite and Infinite Machines, Englewood Cliffs, Prentice-Hall Inc., 1967.
12. R.C. Carrasco, J. Oncina, and M.L. Forcada, Efficient encoding of finite automata in discrete-time recurrent neural networks, in Proceedings of the Ninth International Conference on Artificial Neural Networks, ICANN'99, vol. 2, pp. 673-677, 1999.
13. R.C. Carrasco, and M.L. Forcada, Finite state computation in analog neural networks: Steps towards biologically plausible models, in Lecture Notes in Computer Science, vol. 2036, pp. 482-486, 2001.
14. F. Prat, F. Casacuberta, and M.J. Castro, Machine translation with grammar association: Combining neural networks and finite state models, in Proceedings of the Second Workshop on Natural Language Processing and Neural Networks, pp. 53-60, 2001.
15. G.Z. Sun, C.L. Giles, and H.H. Chen, The neural network pushdown automaton: Architecture, dynamics and training, in Lecture Notes in Artificial Intelligence, vol. 1387, pp. 296-345, 1998.
16. A. Cleeremans, D. Servan-Schreiber, and J.L. McClelland, Finite state automata and simple recurrent networks, Neural Computation, vol. 1, no. 3, pp. 372-381, 1989.
17. C.L. Giles, C.B. Miller, D. Chen, H.H. Chen, G.Z. Sun, and Y.C. Lee, Learning and extracted finite state automata with second-order recurrent neural networks, Neural Computation, vol. 4, no. 3, pp. 393-405, 1992.
18. P. Manolios, and R. Fanelli, First order recurrent neural networks and deterministic finite state automata, Neural Computation, vol. 6, no. 6, pp. 1154-1172, 1994.
19. M. Gori, M. Maggini, E. Martinelli, and G. Soda, Inductive inference from noisy examples using the hybrid finite state filter, IEEE Transactions on Neural Networks, vol. 9, no. 3, p. 571-575, 1998.
20. F. Bensaali, and A. Amira, Accelerating colour space conversion on reconfigurable hardware, Image and Vision Computing, vol. 23, pp. 935-942, 2005.

Novel Strategies for the Optimal Registration of Biomedical Images*

Jorge Larrey-Ruiz, Juan Morales-Sánchez, and Rafael Verdú-Monedero

Departamento de las Tecnologías de la Información y las Comunicaciones,
Universidad Politécnica de Cartagena, Cartagena (Murcia) 30202, Spain
{jorge.larrey,juan.morales,rafael.verdu}@upct.es

Abstract. This paper is intended to address three of the most common problems arising in the field of non-rigid biomedical image registration. Firstly, a regularization term based on fractional-order derivatives is proposed. It can be seen as a generalization of the bio-inspired diffusion and curvature smoothing terms but, since it incorporates features of both regularizers, with this approach it is possible to obtain better registration results from a variational point of view. Next, a frequency-domain formulation for the image registration problem is presented. It provides efficient and stable implementations for the considered registration techniques. Finally, a two-step method is proposed for obtaining the optimal values of the registration parameters, because in the literature there is no agreement about which are the optimal values for these parameters, leading the authors to arbitrarily fix them. The resulting registration scheme, after the incorporation of these three strategies, is tested on two biomedical imaging scenarios.

1 Introduction

Image registration is the process of finding an optimal geometric transformation that aligns points in one view of an object with corresponding points in another view of the same object [1]. Particularly, in biomedical imaging there are several applications that require registration (e.g. image fusion, atlas matching, pathological diagnosis) [2]. In these applications, a non-linear transformation is necessary to correct the local differences between the images. This non-rigid registration is an ill-posed problem, because uniqueness of the solution, if one exists, is not guaranteed. A common method to solve this issue is to add some prior knowledge of the displacement field (e.g. to assume that a physical plausible deformation should be regular and smooth). A regularization term is then used to preferentially obtain more likely solutions. The most popular regularizers used in the literature are the *diffusion* [3], [4] (inspired by the cells with semi-permeable membranes), and *curvature* [5], [6] (inspired by the cells with

* This work is partially supported by the Spanish Ministerio de Ciencia y Tecnología, under grant TEC2006-13338/TCM, and by the Consejería de Educación y Cultura de Murcia, under grant 03122/PI/05.

J. Mira and J.R. Álvarez (Eds.): IWINAC 2007, Part II, LNCS 4528, pp. 130–141, 2007.

elastic-type membranes) smoothing terms, which are based respectively on first and second order derivatives. During the past two years, a great interest in finding new regularization terms suitable for non-parametric biomedical image registration has arisen. Some recent examples are the *minimal curvature* [7], the *membrane energy* [8] or the *consistent symmetric* [9] approaches, all of them based on the diffusion and/or curvature smoothing terms. In this paper, a new regularization term based on fractional order derivatives is proposed. It can also be seen as a generalization of the diffusion and curvature smoothing terms.

The computation of a numerical solution for a non-parametric registration scheme is not straightforward [10], [11], [7]. This work aims to contribute to the implementation of non-parametric image registration methods in the frequency domain. The corresponding Euler-Lagrange equations are translated into the frequency domain by means of the Fourier transform and serve as a starting point for all implementations.

In contrast to many other ill-posed problems, where efficient strategies are available to automatically estimate the regularization parameters, for image registration such satisfactory approaches are yet missing [12], leading the authors of previous works to arbitrarily fix these parameters. This setting may become critical when the computed transformation is applied to clinical data [13]. This paper is also intended to address this problem by providing the guidelines on how to choose the registration and simulation parameters for non-parametric image registration methods.

2 The Variational Framework

Given two images, a reference $R \equiv R(\mathbf{x})$ and a template $T \equiv T(\mathbf{x})$, with $\mathbf{x} \in \mathbf{\Psi} \equiv]0,1[^d$, the aim of image registration is to find a transformation from T onto R in such a way that the transformed template matches the reference. Then the purpose of the registration is to determine a displacement field $\mathbf{u} \equiv \mathbf{u}(\mathbf{x})$ such that $T_{\mathbf{u}} \equiv T\left(\mathbf{x} - \mathbf{u}(\mathbf{x})\right)$ is similar to $R(\mathbf{x})$ in the geometrical sense. It turns out that this problem may be formulated in terms of a variational approach [3], [11], [9]. To this end we introduce the joint energy functional to be minimized:

$$\mathcal{J}[\mathbf{u}] = \mathcal{D}[R, T; \mathbf{u}] + \alpha\, \mathcal{S}[\mathbf{u}]\,, \tag{1}$$

where $\mathcal{D}$ represents a distance measure and $\mathcal{S}$ is a penalty term that determines the smoothness of $\mathbf{u}$. The parameter α is used to control the strength of the smoothness of the displacement vectors versus the similarity of the images.

Mutual Information (MI) is typically used to measure the distance between biomedical images with different intensity levels. Assuming that the transformation class is sufficient to align the images, the MI is maximum if $T_{\mathbf{u}}$ matches R (i.e., if the images are perfectly aligned) [14]. $\mathcal{D}$ is obtained by writing the MI in terms of the Kullback-Leibler divergence [15].

The bio-inspired diffusion and curvature smoothers are given by (2) and (3):

$$\mathcal{S}^{\mathrm{diff}}[\mathbf{u}] = \frac{1}{2}\, a[\mathbf{u}, \mathbf{u}] = \frac{1}{2} \sum_{l=1}^{d} \int_{\mathbf{\Psi}} \langle \nabla u_l, \nabla u_l \rangle_{\mathbb{R}^d}\, d\mathbf{x} = \frac{1}{2} \sum_{l=1}^{d} \int_{\mathbf{\Psi}} \|\nabla u_l\|^2\, d\mathbf{x}\,, \tag{2}$$

where $a[\mathbf{u},\mathbf{u}] = \int_\Psi \langle \mathcal{B}[\mathbf{u}], \mathcal{B}[\mathbf{u}]\rangle_{\mathbb{R}^d}\, d\mathbf{x} = \int_\Psi \langle \mathcal{A}[\mathbf{u}], \mathbf{u}\rangle_{\mathbb{R}^d}\, d\mathbf{x}$ is a bilinear form, and $\langle \cdot,\cdot\rangle_{\mathbb{R}^d}$ denotes the dot (or inner) product in $\mathbb{R}^d$.

$$\mathcal{S}^{\text{curv}}[\mathbf{u}] = \frac{1}{2}\, a[\mathbf{u},\mathbf{u}] = \frac{1}{2}\sum_{l=1}^{d} \int_\Psi \langle \nabla^2 u_l, \nabla^2 u_l\rangle_{\mathbb{R}^d}\, d\mathbf{x} = \frac{1}{2}\sum_{l=1}^{d} \int_\Psi (\Delta u_l)^2\, d\mathbf{x}\,, \quad (3)$$

where $\Delta \equiv \nabla^2 \equiv \nabla^\top \nabla$ is the d-dimensional Laplace operator.

According to the calculus of variations, the first variation of (1) in any direction (also know as the Gâteaux derivative) must be zero, i.e.,

$$d\mathcal{J}[\mathbf{u};\mathbf{z}] = d\mathcal{D}[R,T;\mathbf{u};\mathbf{z}] + \alpha\, d\mathcal{S}[\mathbf{u};\mathbf{z}] = 0\,,\ \forall \mathbf{z}\,, \quad (4)$$

or, in other words, a displacement field $\mathbf{u}$ minimizing equation (1) necessarily must be a solution for the Euler-Lagrange equation:

$$\mathbf{f}(\mathbf{x};\mathbf{u}) + \alpha\, \mathcal{A}[\mathbf{u}](\mathbf{x}) = \mathbf{0}\,, \quad (5)$$

subject to appropriate boundary conditions [6]. The force field $\mathbf{f}(\mathbf{x};\mathbf{u})$ (related to the distance measure) is used to drive the deformation. $\mathcal{A}[\mathbf{u}]$ is a partial differential operator related to the smoother [4], [5]:

$$\mathcal{A}^{\text{diff}}[\mathbf{u}](\mathbf{x}) = -\Delta \mathbf{u}(\mathbf{x})\,, \quad (6)$$

$$\mathcal{A}^{\text{curv}}[\mathbf{u}](\mathbf{x}) = \Delta^2 \mathbf{u}(\mathbf{x})\,. \quad (7)$$

The resulting non-linear partial differential equations (PDE's) can be solved numerically using a finite difference approximation of $\mathcal{A}[\mathbf{u}]$ (Appendix contains the kernels of the discrete approximations used for the derivatives), and a fixed-point or time-marching scheme. This provides an iterative procedure, whose matrix is large, sparse and often ill-conditioned, and has to be inverted with special techniques [10]. Therefore the resulting scenario in the spatial domain implies considerable computational load and memory requirements.

3 Proposed Strategies

3.1 Generalized Regularization Term

At this point, the novel regularization term proposed in this work is introduced:

$$\mathcal{S}^{\text{fract}}[\mathbf{u}] = \frac{1}{2}\, a[\mathbf{u},\mathbf{u}] = \frac{1}{2}\sum_{l=1}^{d} \int_\Psi \langle \nabla^\sigma u_l, \nabla^\sigma u_l\rangle_{\mathbb{R}^d}\, d\mathbf{x} = \frac{1}{2}\sum_{l=1}^{d} \int_\Psi \|\nabla^\sigma u_l\|^2\, d\mathbf{x} \quad (8)$$

with $\sigma \in [1,2]$. Note that (8) includes the diffusion (2) and curvature (3) cases. The Gâteaux derivative of (8), $d\mathcal{S}^{\text{fract}}[\mathbf{u};\mathbf{z}]$, is given by:

$$d\mathcal{S}^{\text{fract}}[\mathbf{u};\mathbf{z}] = a[\mathbf{u},\mathbf{z}] = \sum_{l=1}^{d} \int_\Psi \langle \nabla^\sigma u_l, \nabla^\sigma z_l\rangle_{\mathbb{R}^d}\, d\mathbf{x} = \int_\Psi \langle \mathcal{A}^{\text{fract}}[\mathbf{u}], \mathbf{z}\rangle_{\mathbb{R}^d}\, d\mathbf{x}\,. \quad (9)$$

Using the fact that $\nabla^\sigma u_l$ is a locally summable function with respect to the Lebesgue measure in $\mathbb{R}^d$, i.e. $\int_\psi \|\nabla^\sigma u_l\| \, d\mathbf{x} < +\infty$ for every bounded domain $\psi \subset \mathbf{\Psi}$, and imposing Neumann boundary conditions, i.e. $\nabla^\sigma u_l|_{\partial\mathbf{\Psi}} = \mathbf{0}$, where $\partial\mathbf{\Psi}$ is the (Lipschitz) boundary of $\mathbf{\Psi}$, we can use the Formula of Partial Integration (see e.g. [16]) to get the equation of the generalized derivative of order σ, in the Sobolev sense, of $\nabla^\sigma u_l$:

$$\int_{\mathbf{\Psi}} \langle \nabla^\sigma u_l, \nabla^\sigma z_l \rangle_{\mathbb{R}^d} \, d\mathbf{x} = (-1)^\sigma \int_{\mathbf{\Psi}} \nabla^{2\sigma} u_l \, z_l \, d\mathbf{x} \,. \tag{10}$$

Finally, from (9) and (10) we can obtain the partial differential operator $\mathcal{A}^{\mathrm{fract}}[\mathbf{u}]$:

$$\mathcal{A}^{\mathrm{fract}}[\mathbf{u}](\mathbf{x}) = (-\Delta)^\sigma \mathbf{u}(\mathbf{x}) = \Big(- \sum_{m=1}^{d} \partial_{x_m x_m} \Big)^\sigma \mathbf{u}(\mathbf{x}) \,. \tag{11}$$

Note that equation (11) holds (6) and (7), and that it cannot be implemented directly in the spatial domain if $\sigma \notin \mathbb{N}$.

3.2 Efficient Frequency Implementation

This paper proposes a new implementation for solving (5) in the frequency domain by using the d-dimensional Fourier transform (d-$\mathcal{FT}$) over the discrete spatial variable $\mathbf{x}$. Translating (5) into the frequency domain results in:

$$\tilde{\mathbf{f}}(\boldsymbol{\omega}) + \alpha \, \tilde{\mathcal{A}}(\boldsymbol{\omega}) \, \tilde{\mathbf{u}}(\boldsymbol{\omega}) = \mathbf{0} \,, \tag{12}$$

where $\boldsymbol{\omega} = (\omega_1, \cdots, \omega_d)$ is the d-dimensional variable in the frequency domain associated to the discrete space $\mathbf{x}$, and $\tilde{\mathbf{f}}(\boldsymbol{\omega})$ is the d-$\mathcal{FT}$ of the external forces. The d-$\mathcal{FT}$ of the regularization forces can be written as $\tilde{\mathcal{A}}(\boldsymbol{\omega}) \, \tilde{\mathbf{u}}(\boldsymbol{\omega})$, being $\tilde{\mathcal{A}}(\boldsymbol{\omega})$ an operator which performs the spatial derivatives in the frequency domain by means of products. When considering $d = 2$, (12) can be written as:

$$\begin{bmatrix} \tilde{f}_1(\boldsymbol{\omega}) \\ \tilde{f}_2(\boldsymbol{\omega}) \end{bmatrix} + \alpha \begin{bmatrix} \tilde{\mathcal{A}}_{11}(\boldsymbol{\omega}) \, \tilde{\mathcal{A}}_{12}(\boldsymbol{\omega}) \\ \tilde{\mathcal{A}}_{21}(\boldsymbol{\omega}) \, \tilde{\mathcal{A}}_{22}(\boldsymbol{\omega}) \end{bmatrix} \begin{bmatrix} \tilde{u}_1(\boldsymbol{\omega}) \\ \tilde{u}_2(\boldsymbol{\omega}) \end{bmatrix} = \begin{bmatrix} \mathbf{0} \\ \mathbf{0} \end{bmatrix} \,. \tag{13}$$

For the registration schemes considered in this work, the components of the displacement field are independent and are not coupled, i.e., $\tilde{\mathcal{A}}_{12}(\boldsymbol{\omega}) = \tilde{\mathcal{A}}_{21}(\boldsymbol{\omega}) = \mathbf{0}$, and $\tilde{\mathcal{A}}_{11}(\boldsymbol{\omega}) = \tilde{\mathcal{A}}_{22}(\boldsymbol{\omega})$ is given (see Appendix) by:

$$\tilde{\mathcal{A}}_{11}^{\mathrm{diff}}(\boldsymbol{\omega}) = 2\big((1 - \cos\omega_1) + (1 - \cos\omega_2)\big) \,, \tag{14}$$

$$\tilde{\mathcal{A}}_{11}^{\mathrm{curv}}(\boldsymbol{\omega}) = \big(2\big((1 - \cos\omega_1) + (1 - \cos\omega_2)\big)\big)^2 \,, \tag{15}$$

$$\tilde{\mathcal{A}}_{11}^{\mathrm{fract}}(\boldsymbol{\omega}) = \big(2\big((1 - \cos\omega_1) + (1 - \cos\omega_2)\big)\big)^\sigma \,. \tag{16}$$

It should be noted that (16) holds (14) and (15) if $\sigma \in [1, 2]$.

Using a time-marching scheme to solve (12) gives rise to the equation:

$$\partial_t \tilde{\mathbf{u}}(\boldsymbol{\omega}, t) + \tilde{\mathbf{f}}(\boldsymbol{\omega}, t) + \alpha \, \tilde{\mathcal{A}}(\boldsymbol{\omega}) \, \tilde{\mathbf{u}}(\boldsymbol{\omega}, t) = \mathbf{0} \,, \tag{17}$$

where $\partial_t \tilde{\mathbf{u}}(\boldsymbol{\omega}, t) = [\partial_t \tilde{u}_1(\boldsymbol{\omega}, t)\ \ \partial_t \tilde{u}_2(\boldsymbol{\omega}, t)]^\top$ (in the steady-state $\partial_t \tilde{\mathbf{u}}(\boldsymbol{\omega}, t) = \mathbf{0}$ and (17) holds (12)). In order to solve (17), the time t is discretized, $t = \xi\,\tau$, being τ the time step and $\xi \in \mathbb{N}$ the iteration index, and the time derivative of $\tilde{\mathbf{u}}(\boldsymbol{\omega}, t)$ is replaced by its discrete approximation[1]. Using the notation $\tilde{\mathbf{u}}(\boldsymbol{\omega}, \xi\,\tau) = \tilde{\mathbf{u}}^{(\xi)}(\boldsymbol{\omega})$, the resulting iterative scheme is the following:

$$\tilde{\mathbf{u}}^{(\xi)}(\boldsymbol{\omega}) = \left(\mathbf{I} + \tau\,\alpha\,\tilde{\mathcal{A}}(\boldsymbol{\omega})\right)^{-1} \left(\tilde{\mathbf{u}}^{(\xi-1)}(\boldsymbol{\omega}) - \tau\,\tilde{\mathbf{f}}^{(\xi-1)}(\boldsymbol{\omega})\right), \tag{18}$$

where $\tilde{\mathbf{u}}^{(\xi)}(\boldsymbol{\omega})$ is usually initialized to zero, $\tilde{\mathbf{u}}^{(0)}(\boldsymbol{\omega}) = \mathbf{0}$.

For the registration techniques under consideration, matrix $\tilde{\mathcal{A}}(\boldsymbol{\omega})$ can be written as $\tilde{\mathcal{A}}(\boldsymbol{\omega}) = \mathbf{I}_2 \otimes \tilde{\mathcal{A}}_{11}(\boldsymbol{\omega})$, where $\mathbf{I}_2$ is the 2×2 identity matrix and $\otimes$ denotes the Kronecker product. In these cases, matrix inversions have disappeared due to the fact that the multiplication of a circulant matrix and a column vector becomes a pointwise product of their respective spectra in the frequency domain [17]. Then, the iteration for the l-th component is given by:

$$\tilde{u}_l^{(\xi)}(\boldsymbol{\omega}) = \frac{1}{1 + \tau\,\alpha\,\tilde{\mathcal{A}}_{11}(\boldsymbol{\omega})} \left(\tilde{u}_l^{(\xi-1)}(\boldsymbol{\omega}) - \tau\,\tilde{f}_l^{(\xi-1)}(\boldsymbol{\omega})\right). \tag{19}$$

Direct solution schemes in the spatial domain are available for the diffusion and curvature registration techniques. However, these schemes require, for $n_1 \times n_2$ images, the inversion of an ill-conditioned matrix of size $n_1 n_2 \times n_1 n_2$, and therefore they are not applicable for images of more than 64×64 pixels (assuming 1 GByte RAM, see below), since the computer runs out of memory. There exists the possibility of fast implementations based on a DCT-type factorization [18], without any matrix inversion involved in the procedure. The maximum image size is in this case 2048×2048 pixels. Particularly, for the diffusion registration technique, an efficient implementation based on an additive operator splitting (AOS) scheme is also available [4], but it is outperformed by the DCT-based scheme, which is faster due to the excellent implementation of the FFT routine provided by MATLAB[2] (note that the DCT routine calls the FFT). However, this scheme cannot be applied to the proposed registration method, because the DCT is a real transform, and using it with complex numbers (e.g. $(-1)^\sigma$, with $\sigma \notin \mathbb{N}$) may lead to non-real displacement fields. Then the frequency approach is in this case the only possibility for its implementation.

Table 1 shows the timings for the solution of one linear system of equations arising in a generic non-parametric registration scenario, as well as the actual and theoretical ratios between the timings of the implementations under discussion. These timings were obtained on a PC with Intel Pentium IV, 2.8 GHz, 1 GByte RAM, and the computations were performed under MATLAB 6.5 (R13). Note that the complexity of a 2D-DCT of size $n_1 \times n_2$ is approximately four times the complexity of a 2D-FFT of size $n_1 \times n_2$. Therefore the estimation of the complexity is $\mathcal{O}(8\,N \log_2 N + 2\,N)$ (where $N = n_1 n_2$) for the DCT-based scheme and $\mathcal{O}(2\,N \log_2 N + 4\,N)$ for the proposed frequency-domain implementation.

[1] First backward difference $\partial_t \tilde{\mathbf{u}}(\boldsymbol{\omega}, t) \approx (\tilde{\mathbf{u}}(\boldsymbol{\omega}, \xi\,\tau) - \tilde{\mathbf{u}}(\boldsymbol{\omega}, \xi\,\tau - \tau))\,/\tau$.

[2] The FFT functions are based on a library called FFTW [19].

Table 1. Timings and ratios for one iteration of the DCT-based (valid only for the diffusion and curvature cases) and the proposed frequency domain implementations

Image size (pixels)	DCT timings (s)	Frequency timings (s)	actual ratio	theoretical ratio
128 × 128	0.058	0.016	3.596	3.563
256 × 256	0.081	0.022	3.625	3.611
512 × 512	0.842	0.228	3.688	3.650
1024 × 1024	3.394	0.913	3.719	3.682
2048 × 2048	32.247	8.341	3.866	3.701

3.3 Optimal Parameters Selection

The methodology exposed in this paper consists of two sequential steps:

1. *Initial estimation of the parameters proportionality.* For a small number of iterations $\hat{\xi}$ (typically between 100 and 200), the value of the regularization parameter $\hat{\alpha}$ that minimizes the joint energy functional (1) is obtained, as seen in figure Fig.1(a) (in this example, where a diffusion registration is performed, $\hat{\xi} = 100$ and $\hat{\alpha} = 30$). Note that due to the small value of $\hat{\xi}$, the computational load of this step is relatively light. The proportionality ρ between $\hat{\xi}$ and $\hat{\alpha}$ can be calculated as:

$$\rho = \frac{\hat{\xi}}{\hat{\alpha}} \,. \tag{20}$$

   The displacement field resulting from these parameters is the optimal in terms of the best trade-off, according to the variational approach, between $\mathcal{D}$ and $\mathcal{S}$. As addressed in [8], a scaling factor Γ must be computed so that $\mathcal{D} \sim \Gamma \, \alpha \, \mathcal{S}$, since these energies do not have the same scale and therefore they cannot be compared. The scaling factor is given by the following expression:

$$\Gamma = \left| \frac{\mathcal{D}_0 - \mathcal{D}_\infty}{\alpha_0 \, \mathcal{S}_0 - \alpha_\infty \, \mathcal{S}_\infty} \right| \,, \tag{21}$$

   where $\mathcal{D}_0$ and $\mathcal{S}_0$ are respectively the similarity energy and the regularization energy without any regularization (i.e. $\alpha_0 \simeq 0$), and $\mathcal{D}_\infty$ and $\mathcal{S}_\infty$ are the values of these energies for a large enough regularization parameter which makes the registration not appreciable (typically, $\alpha_\infty > 10\,\hat{\alpha}$).

   At this point, it should be noted that the heuristic choice of $\hat{\xi}$ is not optimal: it is almost certainly too small (i.e., the algorithm has not converged yet) or too large (i.e., the optimal registration could be reached sooner).
2. *Optimal parameters computation.* Our experiments over different types of images show that if the computed proportionality ρ between $\hat{\xi}$ and $\hat{\alpha}$ is kept constant, the energies of the joint functional (1) show the same behavior as in figure Fig.1(a) for a large enough value of the regularization parameter (typically, $\alpha > \hat{\alpha}/10$). To show the validity of the previous reasoning, figures

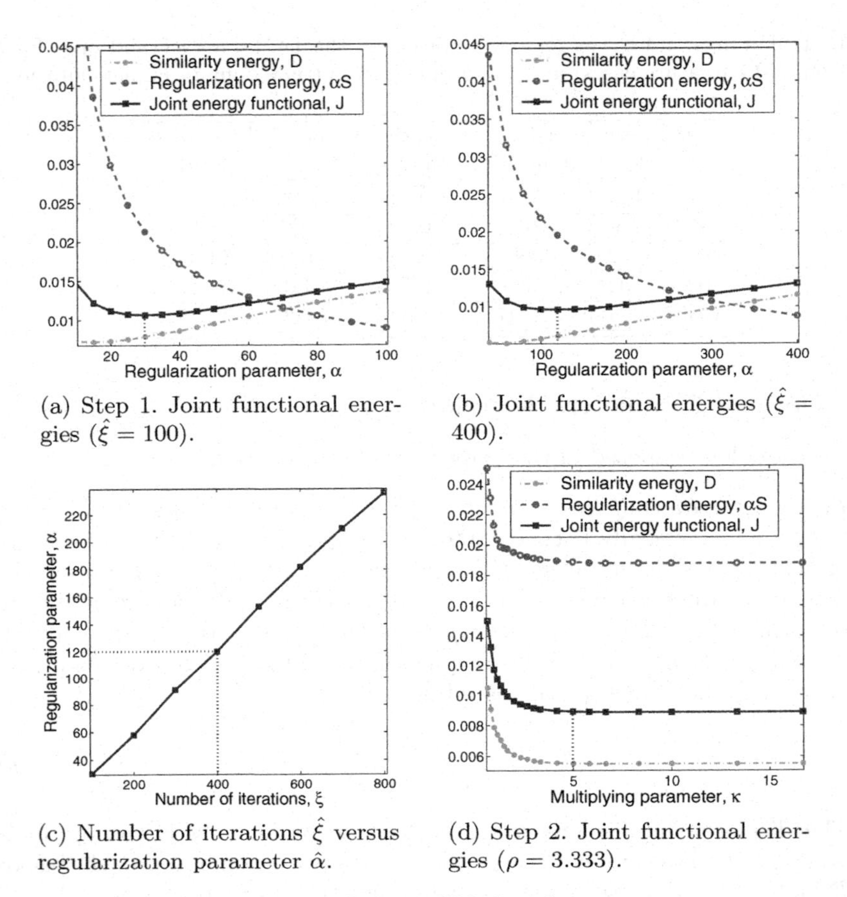

(a) Step 1. Joint functional energies ($\hat{\xi} = 100$).

(b) Joint functional energies ($\hat{\xi} = 400$).

(c) Number of iterations $\hat{\xi}$ versus regularization parameter $\hat{\alpha}$.

(d) Step 2. Joint functional energies ($\rho = 3.333$).

Fig. 1. Proposed Methodology

Fig.1(b) and Fig.1(c) are presented. In the figure Fig.1(b) the number of iterations is four times higher than in figure Fig.1(a) (in this example, $\rho = \frac{100}{30} = \frac{400}{120} = 3.333$). The figure Fig.1(c) illustrates that ρ remains constant for every pair $(\hat{\xi}, \hat{\alpha})$ that minimize (1). The idea is then to find a multiplying constant κ_o that allows for an optimal registration from the variational point of view and minimizes the number of iterations of the algorithm, i.e.,

$$\alpha_o \simeq \kappa_o \, \hat{\alpha} \,, \tag{22}$$

$$\xi_o \simeq \kappa_o \, \hat{\xi} = \kappa_o \, \rho \, \hat{\alpha} = \rho \, \alpha_o \,. \tag{23}$$

This parameter κ_o is obtained as the minimum value of the multiplying parameter κ from which the slope of the joint functional (1) is close to zero ($\sim 10^{-5}$), i.e., the convergence has been reached, see figure Fig.1(d) (in the example, $\kappa_o \simeq 5$). Thus, the optimal parameters α_o and ξ_o can be calculated by employing equations (22) and (23).

4 Registration Results

In this section, the proposed regularization term is tested on two typical biomedical imaging applications: the atlas matching (or scene-to-model) scenario and the multimodal (or intra-subject) registration scenario. In the first case, the aim is to register an axial slice obtained from a computed tomography (CT, Fig.2(b)) and an image corresponding to an atlas or model (Fig.2(a)). The mutual information before the registration process is MI $= 1.14$. In the second experiment, the objective is to register an axial slice obtained from a positron emission tomography (PET) scan (Fig.3(b)) and an axial slice from a CT of a human thorax (Fig.3(a)). The mutual information before the registration is in this case MI $= 0.98$. In both simulations, exists at least one value of σ in the interval $]1, 2[$ that outperforms, in a lower number of iterations of the registration algorithm, the optimal results obtained with the diffusion and curvature schemes in terms of both similarity of the images and smoothness of the computed transformation. These values of σ were obtained by sweeping the fractional order of derivation in steps of 0.05.

Table 2. Summary of the parameters involved in the first experiment (Fig.2)

Registration scheme	*Diffusion*	*Curvature*	$\sigma = 1.85$
Number of iterations, $\hat{\xi}$ (step 1)	100	100	100
Regularization parameter, $\hat{\alpha}$ (step 1)	50	5×10^3	5×10^3
Computed proportionality, ρ (step 1)	2	0.02	0.02
Multiplying parameter, κ_o (step 2)	6.5	4	3.2
Optimal number of iterations, ξ_o (step 2)	650	400	320
Optimal regularization parameter, α_o (step 2)	325	20×10^3	16×10^3
Mutual Information after the registration	1.39	1.42	1.44
Regularization energy, $\mathcal{S}$	0.046	0.002	0.001

Table 3. Summary of the parameters involved in the second experiment (Fig.3)

Registration scheme	*Diffusion*	*Curvature*	$\sigma = 1.15$
Number of iterations, $\hat{\xi}$ (step 1)	100	100	100
Regularization parameter, $\hat{\alpha}$ (step 1)	60	50×10^3	100
Computed proportionality, ρ (step 1)	1.667	0.002	1
Multiplying parameter, κ_o (step 2)	8	3.5	3
Optimal number of iterations, ξ_o (step 2)	800	350	300
Optimal regularization parameter, α_o (step 2)	480	175×10^3	300
Mutual Information after the registration	1.60	1.26	1.62
Regularization energy, $\mathcal{S}$	0.079	0.081	0.076

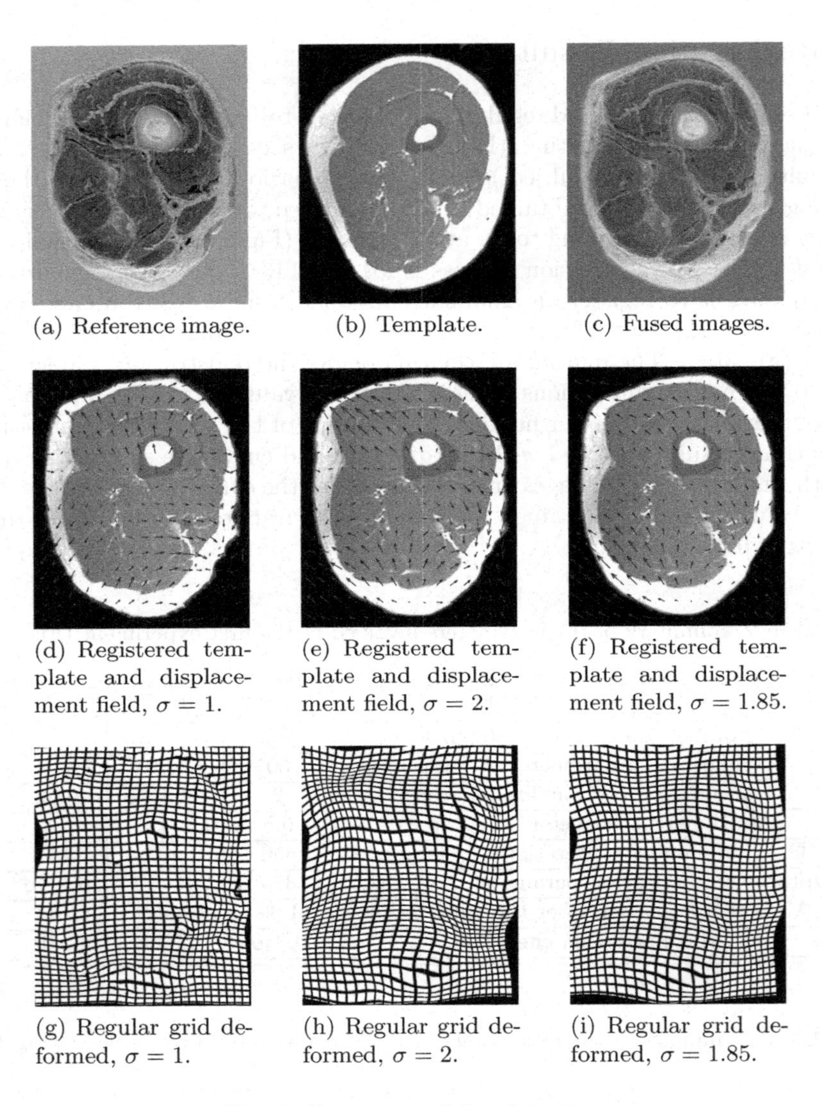

(a) Reference image. (b) Template. (c) Fused images.

(d) Registered template and displacement field, $\sigma = 1$. (e) Registered template and displacement field, $\sigma = 2$. (f) Registered template and displacement field, $\sigma = 1.85$.

(g) Regular grid deformed, $\sigma = 1$. (h) Regular grid deformed, $\sigma = 2$. (i) Regular grid deformed, $\sigma = 1.85$.

Fig. 2. Scene-to-model registration

Table 2 and Table 3 summarize the simulation parameters α_o and ξ_o, as well as the computed mutual information and the regularization energy $\mathcal{S}$ for each registration scheme. In these simulations, the highest value of the similarity measure and the lowest value of the regularization energy correspond to the novel approach. As can be appreciated in Fig.3(i) and Fig.3(g), an advantage of using $\sigma = 1.15$ versus $\sigma = 1$ is that in this way global rigid transformations are partially corrected (and thus the resulting grid is smoother), because the regularizer includes a slight component of curvature registration. In the same way, the use of $\sigma = 1.85$ instead of $\sigma = 2$ avoids local curvatures that could be unlikely for

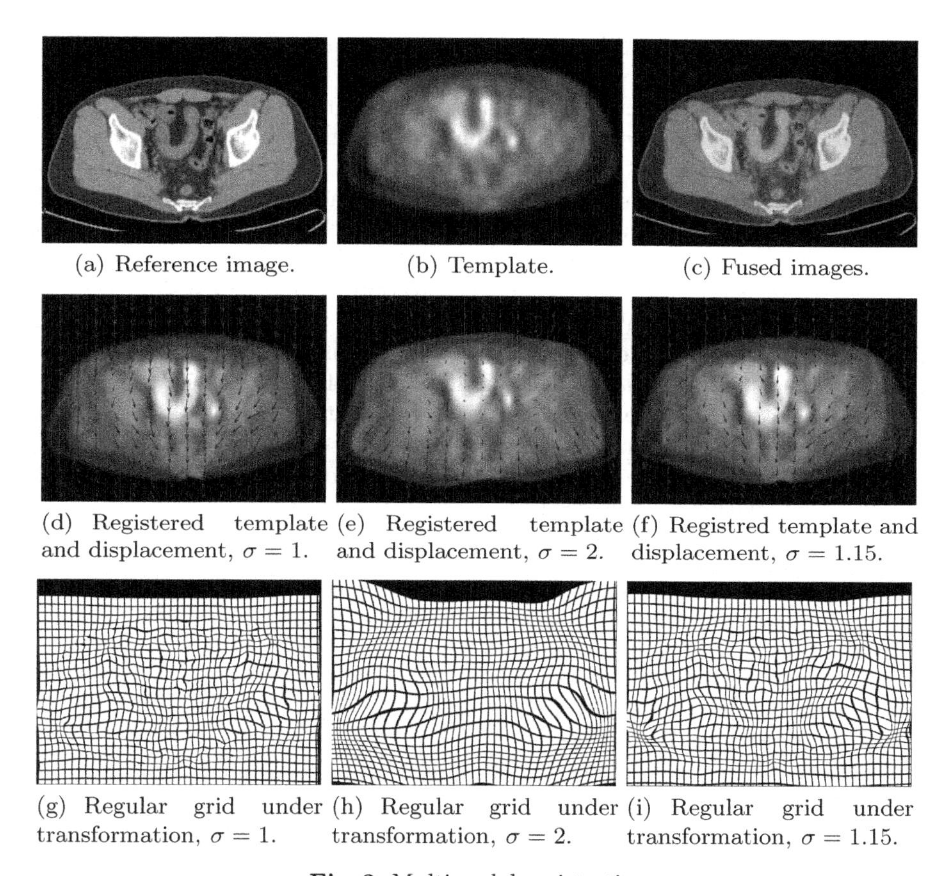

(a) Reference image. (b) Template. (c) Fused images.

(d) Registered template and displacement, $\sigma = 1$. (e) Registered template and displacement, $\sigma = 2$. (f) Registred template and displacement, $\sigma = 1.15$.

(g) Regular grid under transformation, $\sigma = 1$. (h) Regular grid under transformation, $\sigma = 2$. (i) Regular grid under transformation, $\sigma = 1.15$.

Fig. 3. Multimodal registration

the tissues under consideration, because in this case the regularizer includes a small component of diffusion registration, see Fig.2(i) versus Fig.2(h). Finally, it should be noted that the new approach can perform the optimal registration in the lowest number of iterations ξ_o, i.e., the proposed hybrid diffusion-curvature technique allows for a faster convergence of the registration algorithm.

5 Conclusion

In this paper, three novel strategies are presented in order to address some of the most common problems which arise in the field of non-rigid biomedical image registration. Firstly, we propose a regularization term based on fractional order derivatives, which can be seen as a generalization of the diffusion and curvature smoothing terms, since it includes features of both regularizers, thus providing better registration results in terms of both similarity of the images and smoothness of the transformation. However, it should be noted that no constraint is appropriate *everywhere* in the images, because these typically may represent tissues with very different properties, which may therefore need to be treated differently.

Next, the minimization of the joint energy functional is accomplished by the translation of the Euler-Lagrange equations into the frequency domain, yielding non-linear PDE's which are solved by means of a time-marching algorithm, thus providing an iterative scheme. The presented frequency domain formulation allows for an efficient, FFT-based implementation, which reduces considerably the numerical complexity and memory requirements of the overall iterative process.

Finally, a two-step procedure is proposed to obtain, for the considered registration schemes, the optimal parameters that achieve the best trade-off between similarity of the images and smoothness of the transformation. However, it is possible to obtain better *subjective* results, in terms of visual quality (e.g. higher MI) of the registered image, by relaxing the registration parameters (i.e. lower α), but always at the expense of a loss of smoothness and/or continuity in the computed mapping. In this case, the resulting transformation would not be optimal from a variational point of view.

In summary, the main goal of this work is to offer an objective upper limit of registration quality and to provide the design guidelines to efficiently reach it.

References

1. Zitová, B., Flusser, J.: Image registration methods: a survey. Image and Vision Computing **21** (2003) 997–1000
2. Maintz, J., Viergever, M.: A survey of medical image registration. Medical Image Analysis **2**(1) (1998) 1–36
3. Amit, Y.: A nonlinear variational problem for image matching. SIAM Journal of Scientific Computing **15**(1) (1994) 207–224
4. Fischer, B., Modersitzki, J.: Fast diffusion registration. M.Z. Nashed, O. Scherzer (eds), Contemporary Mathematics 313, Inverse Problems, Image Analysis, and Medical Imaging, AMS (2002) 117–129
5. Fischer, B., Modersitzki, J.: Curvature based image registration. Journal of Mathematical Imaging and Vision **18**(1) (2003) 81–85
6. Braumann, U.D., Kuska, J.P.: Influence of the boundary conditions on the results of non-linear image registration. IEEE International Conference on Image Processing **I** (2005) 1129–1132
7. Henn, S., Witsch, K.: Image registration based on multiscale energy information. Multiscale Modelling and Simulation **4**(2) (2005) 584–609
8. Noblet, V., Heinrich, C., Heitz, F., Armspach, J.P.: Retrospective evaluation of a topology preserving non-rigid registration method. Medical Image Analysis **10** (2006) 366–384
9. Zhang, Z., Jiang, Y., Tsui, H.: Consistent multi-modal non-rigid registration based on a variational approach. Pattern Recognition Letters **27** (2006) 715–725
10. Fischer, B., Modersitzki, J.: Fast inversion of matrices arising in image processing. Numerical Algorithms **22** (1999) 1–11
11. Fischer, B., Modersitzki, J.: Fast image registration - a variational approach. Proceedings of the International Conference on Numerical Analysis & Computational Mathematics, G. Psihoyios (ed.), Wiley (2003) 69–74

12. Fischer, B., Modersitzki, J.: Large scale problems arising from image registration. GAMM Mitteilungen **27**(2) (2004) 104–120
13. Ue, H., Haneishi, H., Iwanaga, H., Suga, K.: Nonlinear motion correction of respiratory-gated lung SPECT images. IEEE Transactions of Medical Imaging **25**(4) (2006) 486–495
14. Viola, P., Wells, W.M.: Alignment by maximization of mutual information. International Journal Computer Vision **24** (1997) 137–154
15. Hermosillo, G., Chefd'Hotel, C., Faugeras, O.: A variational approach to multimodal image matching. Technical Report 4117, INRIA (February 2001)
16. Bronshtein, I.N., Semendyayev, K.A., Musiol, G., Muehlig, H.: Handbook of mathematics. Springer (2004)
17. Davis, P.J.: Circulant matrices. Wiley-Interscience, NY (1979)
18. Fischer, B., Modersitzki, J.: A unified approach to fast image registration and a new curvature based registration technique. Linear Algebra and its Applications **308** (2004) 107–124
19. Frigo, M., Johnson, S.G.: The design and implementation of FFTW3. Proceedings IEEE **93**(2) (2005) 216–231

Appendix

In order to compute the discrete approximations of first and second order spatial derivatives, the following operators are defined:

$$\mathrm{d}^{-}[n] = \delta[n] - \delta[n-1]\,,$$

$$\mathrm{d}^{+}[n] = \delta[n+1] - \delta[n]\,,$$

$$\mathrm{d}^{2}[n] = \mathrm{d}^{-}[n] * \mathrm{d}^{+}[n]\,,$$

where d^{-}, d^{+} and d^{2} perform the backward difference, forward difference and second order difference, respectively, δ is the Kronecker delta and $*$ denotes the linear convolution. Then, the partial differential operators for the bidimensional case are defined by:

$$\Delta \mathbf{u}(\mathbf{x}) = \begin{bmatrix} \partial_{x_1x_1} u_1(\mathbf{x}) + \partial_{x_2x_2} u_1(\mathbf{x}) \\ \partial_{x_1x_1} u_2(\mathbf{x}) + \partial_{x_2x_2} u_2(\mathbf{x}) \end{bmatrix} \approx \begin{bmatrix} \left(\mathrm{d}^2[n_1] + \mathrm{d}^2[n_2]\right) * u_1(\mathbf{x}) \\ \left(\mathrm{d}^2[n_1] + \mathrm{d}^2[n_2]\right) * u_2(\mathbf{x}) \end{bmatrix},$$

$$\Delta^2 \mathbf{u}(\mathbf{x}) = \begin{bmatrix} \left(\partial_{x_1x_1} + \partial_{x_2x_2}\right)^2 u_1(\mathbf{x}) \\ \left(\partial_{x_1x_1} + \partial_{x_2x_2}\right)^2 u_2(\mathbf{x}) \end{bmatrix} \approx \begin{bmatrix} \left(\mathrm{d}^2[n_1] + \mathrm{d}^2[n_2]\right) * \left(\mathrm{d}^2[n_1] + \mathrm{d}^2[n_2]\right) * u_1(\mathbf{x}) \\ \left(\mathrm{d}^2[n_1] + \mathrm{d}^2[n_2]\right) * \left(\mathrm{d}^2[n_1] + \mathrm{d}^2[n_2]\right) * u_2(\mathbf{x}) \end{bmatrix}.$$

Finally, in order to obtain (14), (15) and (16), it should be noted that

$$\mathcal{FT}\{\mathrm{d}^2[n]\} = -\left|1 - e^{-j\omega}\right|^2 = -2\left(1 - \cos\omega\right).$$

Characterization of Artificial Muscles Using Image Processing

Rafael Berenguer Vidal[1], Rafael Verdú Monedero[2], Juan Morales Sánchez[2], and Jorge Larrey Ruiz[2,⋆]

[1] Department of Technical Sciences, Catholic University of Murcia, 30107, Murcia, Spain
rberenguer@pdi.ucam.edu
[2] Department of Information Technologies and Communications, Technical University of Cartagena, 30202, Cartagena, Spain
rafael.verdu@upct.es

Abstract. Artificial muscles are bio-inspired devices formed by several layers of conducting polymers. These devices have the ability of transform electrical energy into mechanical energy through an electrochemical reaction, which is produced by an oxidation or reduction of the polymer due to an electric current. Since the device have a strip shape, this reaction results in a macroscopic swelling and shrinking movement. This movement is similar to the biological muscles and it has several applications as motor prostheses and as part of complex biomaterials. In this paper we describe a computer vision system developed to analyze and characterize these devices through their cycle of life. The method includes cameras for tracking the movement of the muscle from different angles and a set of algorithms to characterize the motion of the device through its use. By means of active contours it is determined the instantaneous position of the muscle in the space. From these contours other parameters like the parametric motion and energy of curvature are calculated. These data are compared with the physical parameters of the device, like the tension and energy consumption, providing a way for performing automatic testing on the research of artificial muscles.

1 Introduction

The devices known as artificial muscles have been developed since 1992 [1]. They consist of a thin strip composed by a set of several layers of conducting and non-conducting polymer. Their main interest is focused on their parallelism with the natural muscles: applying an electric current on the muscle in an electrolytic media, a macroscopic angular bending movement is described by the free end of the muscle. The movement of these devices is based on the changes of volume of the conducting polymer associated to the oxidation and reduction processes.

⋆ This work is partially supported by *Ministerio de Ciencia y Tecnología*, under grant TEC2006-13338/TCM, and by *Fundación Séneca*, project 03122/PI/05.

J. Mira and J.R. Álvarez (Eds.): IWINAC 2007, Part II, LNCS 4528, pp. 142–151, 2007.

Therefore, the angular movement is caused by the gradient of tension inside the device due to the volume changes in the conducting polymer film.

The velocity and efficiency of the devices depend on different issues like the number, type and thickness of the polymers constituent of the device; as well as the characteristics of the polymer itself. Thus, it is necessary the use of a system that allow the characterization of the device in an efficient and precise way.

This paper shows the last approach for the analyzing procedure based on a computer vision system and deformable models [2] for analyzing the muscle contour in the space. This system overcome the restrictions of previous releases [3] due to its ability of analyze the movement in the 3D space and a faster implementation of the algorithm in the frequency domain. Additionally the evolution of the efficiency of the device through the cycle of life can be characterized by the parameters derived from the model.

The structure of the paper is as follows: In Section 2 the architecture and characteristics of the artificial muscle are explained. Afterwards, Section 3 describes the parameters and the system used for the characterization. Section 4 shows the experimental results obtained from the study and discusses their importance. Conclusions and further work are presented in Section 5.

2 Artificial Muscle Architecture

The design and operation of the devices known as artificial muscles are based on their constituent materials: conducting polymers. By using these polymers as wet materials, we can obtain interesting electrochemical properties. In a wet polymer film there is a presence of polymer, solvent, and ionic components. Each of these components has a specific task. If we locate the electronic conductor and ionic conductor in the same interface, the first material is able to store electrons whereas the second one stores the electroactive substance. The cooperative effort of the different parts in the film, causes its swelling and shrinking during oxidation or reduction respectively, producing a macroscopic change in volume able to produce controlled movements and mechanical work.

A multilayer polymeric device was constructed and patented [1]. Its architecture is based on a triple-layer device formed by a conducting polymer, a two-sided tape, and another conducting polymer. When one of the polymer acts as anode, the second acts as a cathode. The flow of an anodic current oxidizes the conducting polymer and the film swells. If the current is inverted, it appears a reduction in the polymer, shrinking the film. These processes of oxidation and reduction cause reverse both micro- and macroscopic effects. As one of the ends of the device is fixed and the other one can move freely, the macroscopic effect of the entire oxidation-reduction cycle is the film torsion or bending.

The macroscopic motion of the device has been analyzed by using the so-called chronopotentiometry technique. A constant current density is applied to the working electrode. This current can be either positive (causing the oxidation of the polymer) or negative (causing its reduction). In both cases, conducting polymers induce a constant voltage increase or decrease. In order to avoid an excessive muscle degradation the experiment is finished when a certain voltage is reached. The actuator response to this experiment is registered in the chronopotentiogram. This response is characterized by the following physical parameters:

Electric Energy. The electrical energy, E, to move the muscle is defined by $E = \int V(t)i(t)dt$, where i is the electrical current, V is the electrical potential with respect to the voltage of the equilibrium state, and t is time.

Electric Charge. The electric charge consumed by the muscle, Q, is defined as the integral of the current that flows through the working electrode, $Q = \int I(t)dt$, where I is the electrical current. This consumed charge is directly related to the actual oxidation state.

Since the film movement is produced by oxidation or reduction of the conducting polymer, control over the movement rate is easily accomplished by stopping the current flow and the movement is reversed by reversing the direction of the current. Likewise, the motion speed is increased by raising the current intensity.

3 Mechanical Characterization with a Vision System

In order to characterize the mechanical properties of the polymeric device through its use, we execute multiple cycles of swelling-shrinking movements of the device. All these cycles constitute the known *cycle of life* of the artificial muscle. This experiment is analyzed by means of an specifically developed artificial vision system. Its basic system architecture is composed of a grabbing device connected to a workstation, which is responsible of the semi-automatic processing and characterization. The following subsections describe the entire process.

3.1 Parameters Under Study

The main interest of the experiment is to analyze the relationship between the current flow applied to the device and the resultant macroscopic movement. Therefore, we should make use of some parameters that allow us to characterize this mechanical behavior of the device.

In a first approach, we consider the artificial muscle as a lineal device with a contour shape, which can be described as a parametric function:

$$\mathbf{v}(s,t) = (x(s,t), y(s,t), z(s,t)) \tag{1}$$

where the parameter s is the length index $s \in [0, L]$ and t is the time. This expression determines the position of the device for each value of length and time. The movement of a device like this is limited to a swelling-shrinking. Later we can consider it as a flexible strip, described by the length and width indexes and with an additional degree of liberty: a twist movement.

The parameters considered for describing the mechanical properties of the muscle are the parametric motion and the energy of curvature.

Parametric Motion. This expression represents the motion of the device for each value of the parameters length s and time t. We consider that the device length keeps constant during the movement and the function $\mathbf{v}(s,t)$ describes the muscle in every frame. Then, the displacement occurred is defined by

$$\boldsymbol{d}(s,t) = (x(s,t+\Delta t) - x(s,t), y(s,t+\Delta t) - y(s,t), z(s,t+\Delta t) - z(s,t)) \quad (2)$$

which indicates that each displacement vector is the difference between each position of the muscle in two correlative frames. Velocity or parametric motion is obtained easily as the rate of those position changes, by multiplying the displacement vectors by the frame rate,

$$\boldsymbol{vel}(s,t) = \mathbf{d}(s,t)/\Delta t \quad (3)$$

where $\Delta t = 1/fps$, and fps is the frame rate.

Energy of Curvature. The curvature of a contour at a point can be defined as the inverse of the radius of the arc that overlays the contour at that point [4]. A continuous representation of curvature for a parametric contour element is given by the following expression:

$$C(s) = \frac{n \cdot v_{ss}}{|v_s|^3} = \frac{\hat{n} \cdot v_{ss}}{|v_s|^2} \quad (4)$$

where $\hat{n}$ is a unitary vector normal to the contour. Thus, curvature is equal to the magnitude of the normal component of the term $v_{ss} = (x_{ss}, y_{ss}, z_{ss})$, where i_{ss} is the second derivative of the i component respect to the parameter s, divided by the square of the incremental arc length. Since curvature is a function of radius, this measurement is not scale-invariant. A scale-invariant version of curvature is defined by multiplying equation (4) by the incremental contour length, as the following equation shows:

$$E^e_{curv}(s) = |C(s)||v_s|/\pi. \quad (5)$$

The total sum energy for the entire contour is obtained by integrating $E^e_{curv}(s)$ over each element in the contour and summing the resulting integrals. In order to avoid the cancellation of negative with positive curvature magnitudes, the modulus of $C(s)$ is used.

3.2 Stereoscopic Computer Vision System

A computer vision system has been used for analyzing and modelling the device. The used system architecture consists of a dual-camera grabbing system linked to a processing station. The cameras, which are controllable remotely, are placed orthogonally both in the XY plane in front of the moving muscle, following the scheme of Fig. 1. Assuming that both cams work with the same tilt angle α and the distance d between each cam and the target is equal, the XZ and YZ projection of the polymeric film can be easily obtained.

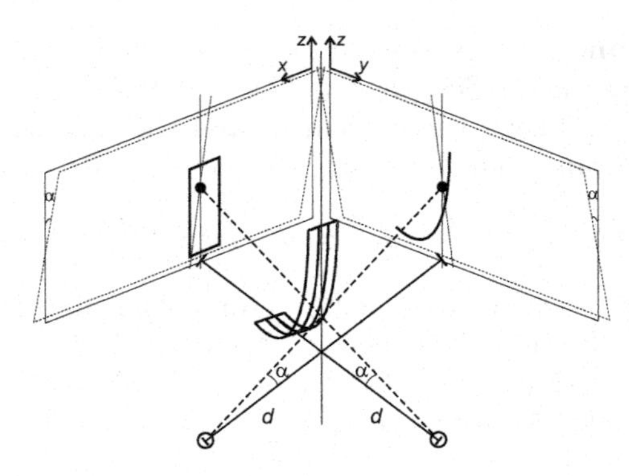

Fig. 1. Location of cams, muscle and respective projections

The artificial muscle movement is registered by the grabbing system and processed later by the workstation by means of techniques of image processing. The two projections of the device are the input data for this characterization algorithm. As the polymeric stripe experiences a bending movement which is clearly non-rigid, the task of tracking non-rigid motion is better accomplished with deformable models [5].

An application is specially developed to perform all the process. At the set up of this application, it is performed a system initialization necessary for improving the quality of the parameter estimation. It consists in a calibration of the system, including both the cams subsystem and the chronopotentiometry equipment. Additionally the system identifies the target, maximizing the view of the moving muscle in both cams. For this purpose the cam parameters are adjusted using the tilt and zoom movements. This process is performed only once at the beginning of the cycles of swelling-shrinking movements.

Secondly, it begins the loop of muscle movements. The application grabs a stereo-video with the moving polymeric film for each cycle. The requested parameters are estimated afterwards. These parameters are provided from each stereo-frame and cams parameters, by means of active contours (whose implementation is detailed in Appendix A). As is described in the following section, the characterization of the muscle is performed considering the parameters obtained from the two different sources: equipment and the computer vision.

4 Experimental Results

Three-layer actuator can generate a movement of up to 360 degrees when they are subjected to an electrical current, although in this case a smaller range is analyzed. This experiment starts with the muscle in vertical position (0 degrees). This means that the device is in electrical equilibrium (the reference electrode is short-circuited). Then a flow of a constant current of +5mA for a time interval of 30 s is applied, producing an angular movement of 90 degrees of the free end of the muscle. After that a current of -5 mA is applied for 30 s in order to

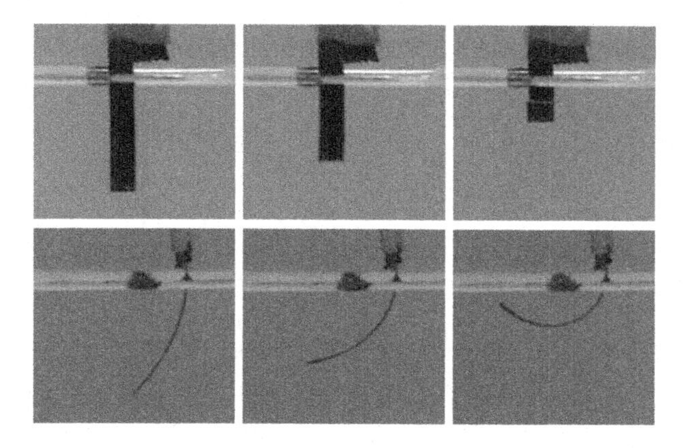

Fig. 2. XZ (top) and YZ (bottom) projections of the muscle movement

recover the original position, closing a cycle. This process is executed again a fixed number of cycles. The tension and energy consumption of the muscle are measured directly from the device through the chronopotentiometry equipment. Fig. 2 shows a group of the stereo-frames registered by the system.

4.1 Frame-to-Frame Motion Estimation

Over each acquired frame and before the characterizing of the device, simple automatic preprocessing tasks are performed: a linear projection with angle α is applied to obtain the projection over the XZ and YZ plane (see Fig. 1); a segmentation with a region competition based algorithm [6] is executed and the resulting images are refined by means of morphological operations. This process leads to a thin 2D curve that defines the XZ or YZ projections of the inner film contour, providing the external forces in the later algorithm.

The second phase includes the device tracking by means of a 3D active contour $\mathbf{v}(s,t) = (x(s,t), y(s,t), z(s,t))^T$ and the extraction of the parameters derived from the curve. For the tracking of the muscle for each stereo-frame, we applied the iterative algorithm described in the Appendix A. Note that in the equation (11) it appears the $\mathbf{q}_\xi$ term, which represents the external forces vector for each time ξ. As our interest is to obtain a 3D active contour, we consider both projections simultaneously to estimate this vector. Consequently, for each stereo-frame we obtain a 3D parametric curve that defines the position of the device in the space. Left part of Fig. 3 shows the set of contours (equally time-spaced) that describe the trajectory of the film during one cycle.

From the contour $\mathbf{v}(s,t)$ we can calculate easily the parameters described in Section 3.1, parametric motion and energy of curvature. In the middle of Fig. 3, it is shown the motion occurred between two consecutive frames. This motion throughout the film length in terms of direction and magnitude is represented by displacement vectors. The right side of same figure shows the resulting distribution of the curvature along the muscle for the corresponding frames.

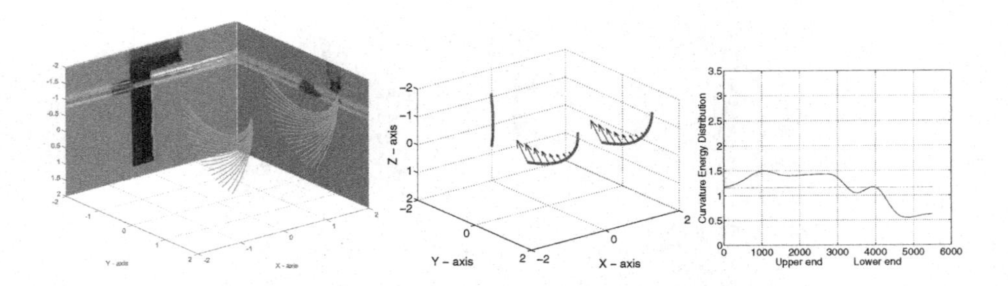

Fig. 3. Motion and curvature energy distribution

4.2 Evolution of Parameters Through the Life Cycle of the Muscle

Our interest is focused in characterizing the artificial muscle and its parameters through its cycle of life. So we consider the parameters described in Section 3.1 to determine the efficiency of the muscle. The efficiency is defined as the amount of movement obtained for a fixed consumption of energy. For this purpose we apply a process of 500 cycles of swelling-shrinking movements, each of them obtained by the application of a flow of a constant current of +5mA for 30 s followed by a flow of negative current with identical magnitude and time interval.

The tension and energy consumption of the muscle are measured directly from the device through the equipment laboratory. Left part of Fig. 4 shows the evolution of the muscle voltage during current flow for cycles 1, 250 and 500. An almost linear evolution is observed. Both, the slope of the line and the initial voltage rise for increasing cycles of the muscle life. For each cycle, the total energy consumption of the device can be calculated as the integral of these curves. Right part of Fig. 4 displays these values fitted by a quadratic curve. Therefore, from the result it can be concluded that the energy used by the muscle for the bending movement rises with the cycles.

On the other hand, we can calculate the energy of curvature of the muscle for each cycle by means of the computer vision method. Fig. 5 shows the curves for cycles 1, 250 and 500 and their linear fitting. It can be seen that the slope of the lines decreases with the cycles of movement of the muscle. Since the energy of curvature represents the average of resulting movement, it can be deduced that this movement is reduced as the number of cycles increases. On the left, the evolution of total sum energy of curvature is shown for different cycles of the experiment. This figure is calculated as the integral of energy of curvature for each cycle. A decreasing exponential fits this progression.

These figures prove that through the cycle of life of the device, the consumption of electrical energy increases lineally whereas the curvature energy achieved by the device falls with a decreasing exponential shape. Therefore, from the experimental results we can conclude that the efficiency of the artificial muscles is reduced with their use. Using the device by successive cycles under the same current flow, it is produced a greater energy consumption whereas a smaller movement is achieved.

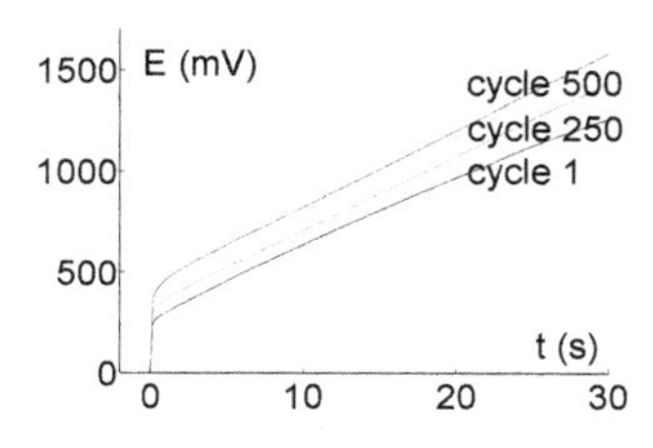

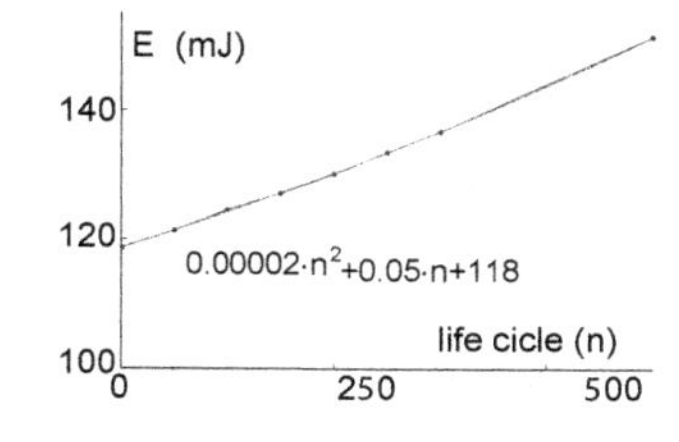

Fig. 4. Tension and energy measured in the device

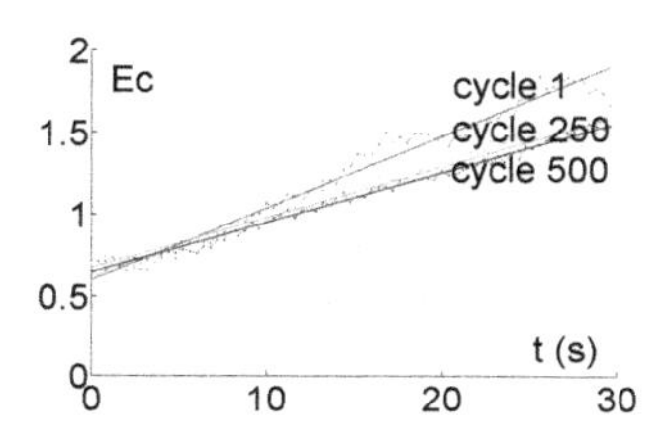

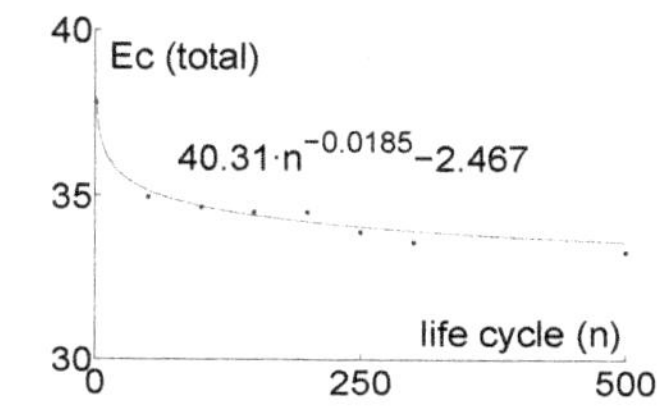

Fig. 5. Energy of curvature in cycle 1, 250 and 500, and during life cycle

5 Conclusions and Further Work

In this paper, it has been described an improved computer vision-based method to quantify the mechanical properties of artificial muscles. The method is based on active contours algorithms for tracking non-rigid moving objects. The system supplies the parameters related to the motion and curvature distribution of the polymeric film in any direction of the space. These values are compared with the electrochemical magnitudes resulting during the experimental process, such as consumed energy, voltage and current applied. Performing the analysis during repetitive cycles of swelling-shrinking movements, it can be drawn new and relevant conclusions about the behavior of the device in cycles of intense work.

The study performed on a wet polymeric device reveals interesting results. A slight degradation of the performance of the muscle with the increasing number of movements can be observed. The energy consumed is directly related to the curvature suffered by the muscle; nevertheless, the energy consumption increases lineally with the number of cycles previously performed. In addition, the results show that the energy of curvature increases with the curvature of the muscle although this curvature decreases with the cycle of life of the muscle. Both results prove that the efficiency of the current devices falls with the use of the device. By using this system, the performance and characteristics of artificial muscles can be analyzed in order to improve their efficiency in future approaches.

Finally, the current work lines consist in an improved tracking algorithm able to consider the width of the polymer strip. To achieve this feature, it must be used an active surface instead of a active contour in the modelling of the muscle.

This enables the ability to analyze not only the bending movement but also the torsion movement, which appears on some implementations of the devices.

References

1. Otero, T.F.: Patent EP-9200095 and EP-9202628 (1992)
2. Weruaga, L., Verdú, R., Morales, J.: Frequency domain formulation of active parametric deformable models. IEEE PAMI, **26**(12) (2004) 1568–1578
3. Verdú, et al. : Mechanical characterization of artificial muscles with computer vision. SPIE Annual Int. Symposium on Smart Structures and Materials (2002)
4. Stroud, K.A.: Engineering Mathematics. McMillan (1987)
5. Terzopoulos, D.: Deformable models: classic, topology-adaptive and generalized formulations. In Osher, S., Paragios, N., eds.: Geometric Level Set Methods in Imaging, Vision and Graphics. Springer-Verlag New York, Inc. (2003) 21–40
6. Zhu, S.C., Yuille, A.L.: Region competition: Unifying snakes, region growing, and bayes/MDL for multiband image segmentation. IEEE PAMI, **18**(9) (1996) 884–900
7. Liang, J., McInerney, T., Terzopoulos, D.: United snakes. Medical Image Analysis **10**(2) (2006) 215–233
8. Davis, P.J.: Circulant Matrices. Wiley-Interscience, NY (1979)

A Appendix. Implementation of Active Contours in the Frequency Domain

A dynamic active contour or snake is a time-variant parametric curve $\mathbf{v} \equiv \mathbf{v}(s,t) = (x(s,t), y(s,t), z(s,t))^T$ in the space $(x,y,z) \in \mathbb{R}^3$, where x, y and z are the coordinate functions, $s \in [0, L]$ is the spatial domain parameter and t is time. The shape of the contour outlining an object is governed by an energy functional with internal and external components [7], $\mathcal{E}(\mathbf{v}) = \mathcal{S}(\mathbf{v}) + \mathcal{P}(\mathbf{v})$. The internal energy of deformation, $\mathcal{S}(\mathbf{v})$, is described by $\alpha(s)$ and $\beta(s)$ which are the physical parameters of elasticity and rigidity, respectively. The external energy, $\mathcal{P}(\mathbf{v})$, couples the contour to the target in the image with the internal and external constraints.

In a practical implementation, the contour is divided into N elements using a finite elements formulation, $\mathbf{v} = \sum_{n=0}^{N-1} \mathbf{v}_n$. Each element of each component x, y and z is constructed geometrically by a shape function and its parameters. Since the coordinate functions are independent, the following formulation describes only the x component, $\mathbf{v}_n^x(s,t) = \mathbf{u}_n^x(t)\mathbf{N}(s)$, being the same for all components. Gathering the shape parameters of the N elements in the vector $\mathbf{u}(t)$[1], the equations that minimice the energy functional are the following [7]:

$$\mathbf{M}\frac{d^2\mathbf{u}(t)}{dt^2} + \mathbf{C}\frac{d\,\mathbf{u}(t)}{dt} + \mathbf{K}\mathbf{u}(t) = \mathbf{q}(s,t), \tag{6}$$

where $\mathbf{M}$ is the mass matrix, $\mathbf{C}$ is the damping matrix, $\mathbf{K}$ is the stiffness matrix and $\mathbf{q}$ is the external forces vector which depends on the position of the curve

[1] In order to make easier the notation, the super-index x has been removed.

at time t. These matrices are banded (few elements out of the main diagonals are different from zero), and are assembled overlapping the submatrices of each element weighted by the value of the parameters of elasticity, rigidity, mass and damping of each element.

In order to solve the equations of motion (6), the time is discretized, $t = \xi \Delta t$, being Δt the time step and $\xi \in \mathbb{N}$ the iteration index (the notation $\mathbf{u}(\xi \Delta t) = \mathbf{u}_\xi$ is used), the time derivatives of $\mathbf{u}(t)$ are replaced by their discrete approximations producing the following second order iterative system

$$\mathbf{u}_\xi = \mathbf{A}^{-1}\mathbf{A}_1\mathbf{u}_{\xi-1} + \mathbf{A}^{-1}\mathbf{A}_2\mathbf{u}_{\xi-2} + \mathbf{A}^{-1}\mathbf{q}_{\xi-1}, \tag{7}$$

where $\mathbf{A} = \mathbf{M}/(\Delta t)^2 + \mathbf{C}/\Delta t + \mathbf{K}$, $\mathbf{A}_1 = 2\mathbf{M}/(\Delta t)^2 + \mathbf{C}/\Delta t$, $\mathbf{A}_2 = -\mathbf{M}/(\Delta t)^2$ and vector $\mathbf{q}_{\xi-1}$ contains the external forces at time $t = (\xi - 1)\Delta t$.

In practice it is usual to assume that mass and damping are constant inside each element. Also, in many applications, mass and damping do not vary and keep constant for all elements of the contour, $m_n = m$ and $c_n = c$. Then,

$$\left(\left(\frac{m}{\Delta t^2} + \frac{c}{\Delta t}\right)\mathbf{F} + \mathbf{K}\right)\mathbf{u}_\xi = \left(\frac{2m}{\Delta t^2} + \frac{c}{\Delta t}\right)\mathbf{F}\,\mathbf{u}_{\xi-1} + \left(\frac{-m}{\Delta t^2}\right)\mathbf{F}\,\mathbf{u}_{\xi-2} + \mathbf{q}_{\xi-1}, \tag{8}$$

where $\mathbf{F}$ is the $N \times N$ shape function matrix, which is a circulant matrix defined by its first row $\boldsymbol{f}$ (using finite difference shape function, this $1 \times N$ vector is $\boldsymbol{f} = [1\ 0\ 0\ \cdots\ 0\ 0]$). Assuming also that the contour is closed and the static parameters, elasticity and rigidity, do not vary along the curve, that is, $\alpha_n = \alpha$ and $\beta_n = \beta$, the stiffness matrix $\mathbf{K}$ is a circulant matrix and it is completely defined by its first row $\mathbf{k}$, called the stiffness spatial kernel [2].

The result of multiplying a circulant matrix with a vector is equivalent to the circular convolution of the first row of the matrix with the aforementioned vector [8]. Then, defining the following variables $\eta = m/\Delta t^2 + c/\Delta t$, $\gamma = \Delta t\ c/m$, $a_1 = 1 + (1+\gamma)^{-1}$, $a_2 = -(1+\gamma)^{-1}$, the equation (8) can be written as:

$$\eta^{-1}(\eta\mathbf{f} + \mathbf{k}) \otimes \mathbf{u}_\xi = a_1\,\mathbf{f} \otimes \mathbf{u}_{\xi-1} + a_2\,\mathbf{f} \otimes \mathbf{u}_{\xi-2} + \eta^{-1}\mathbf{q}_{\xi-1}, \tag{9}$$

where $\otimes$ denotes the circular convolution operation between sequences of length N. The discrete Fourier transform (DFT) of the circular convolution of two sequences implies in the frequency domain the multiplication of their DFT's,

$$\eta^{-1}(\eta + \hat{\mathbf{k}})\,\hat{\mathbf{u}}_\xi = a_1\,\hat{\mathbf{u}}_{\xi-1} + a_2\,\hat{\mathbf{u}}_{\xi-2} + \eta^{-1}\hat{\mathbf{q}}_{\xi-1}, \tag{10}$$

where $\hat{\mathbf{k}}$, $\hat{\mathbf{u}}_\xi$, $\hat{\mathbf{u}}_{\xi-1}$, $\hat{\mathbf{u}}_{\xi-2}$ and $\hat{\mathbf{q}}_{\xi-1}$ are the N points DFT's of their respective spatial sequences. Now, the matrix operations have become the result of multiplying the frequency spectrum of signals, and it is possible to isolate $\hat{\mathbf{u}}_\xi$ on the left part of equation (10),

$$\hat{\mathbf{u}}_\xi = \hat{\mathbf{h}}\left(a_1\hat{\mathbf{u}}_{\xi-1} + a_2\,\hat{\mathbf{u}}_{\xi-2} + (\eta\,\hat{\mathbf{f}})^{-1}\hat{\mathbf{q}}_{f\xi-1}\right). \tag{11}$$

Vector $\hat{\mathbf{h}} = \eta\,(\eta + \hat{\mathbf{k}})^{-1}$ represents the inverse filter of $\hat{\mathbf{k}}$, η is the global mass, related with the bandwidth of the filter $\hat{\mathbf{h}}$, γ represents the relationship between damping and mass, and the constants a_1 and a_2 are coefficients of the second order difference equation.

Segmentation of Sequences of Stereoscopic Images for Modelling Artificial Muscles

Santiago González-Benítez, Rafael Verdú-Monedero[1], Rafael Berenguer-Vidal[2], and Pedro García-Laencina[1,*]

[1] Department of Information Technologies and Communications, Technical University of Cartagena, 30202, Cartagena, Spain
rafael.verdu@upct.es

[2] Department of Technical Sciences, Catholic University of Murcia, 30107, Murcia, Spain
rberenguer@pdi.ucam.edu

Abstract. In this paper, an implementation of the *Region Competition* algorithm for segmenting stereoscopic video sequences is shown. This algorithm is an essential task in the method in order to obtain a 3D characterization of artificial muscles. Image sequences are acquired by a two-cam computer vision system. Optimal and efficient segmentation of these images is our goal; information obtained from the segmented first frame of the video sequence is used for segmenting the next frame and so on. Redundancy between stereoscopic pairs of images is also used to optimize the segmentation. In this paper, the *Region Competition* algorithm is described and our own specific implementation is addressed. Particular problems of stereoscopic video segmentation are shown and how they are solved. Finally, results yielded from simulations are presented and conclusions close the paper.

1 Introduction

The algorithm described in this paper is specifically designed for segmentation of stereoscopic image sequences. The images to segment are sequential pictures of artificial muscles in motion. Artificial muscles [1] are built with polymer conductors which are able to transform electrical energy into mechanic work. When exposed to an electrical current, artificial muscles can bend or stretch themselves depending on the sign of that current.

From human prosthesis to motor systems for small robots, there's a wide range of applications for artificial muscles and new ones appear every day. Inspired in natural muscles, polymer-made muscles are excited by electrical currents and their movement is due to dimensional variations of the polymers caused by electrochemical reactions. This situation creates the need of a system to measure small variations of the artificial muscles and their behaviour in different conditions. Observation and characterization of these devices is essential for its investigation and improvement.

* This work is partially supported by *Ministerio de Ciencia y Tecnología*, under grant TEC2006-13338/TCM, and by *Fundación Séneca*, project 03122/PI/05.

J. Mira and J.R. Álvarez (Eds.): IWINAC 2007, Part II, LNCS 4528, pp. 152–161, 2007.

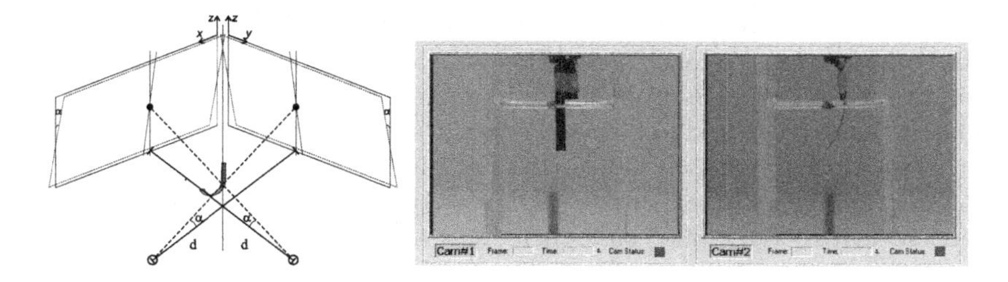

Fig. 1. Cameras arrangement for image acquisition and acquired stereo-frames

In [2] a new method is developed in order to obtain the 3D characterization of the muscles. Two digital cameras take orthogonal pictures of the muscle in motion. Using stereoscopic vision techniques and digital image processing we get an automatic way to extract the parameters we're looking for. This method is based in artificial vision and image processing, the different stages are:

- Camera control and stereoscopic images acquisition (see Fig. 1).
- Image Processing. Which includes image segmentation (detection of the muscle in the images and separation from the background) and tracking using a 3D active contour model [3].
- Data mining: parameters of movement, curvature energy, etc.

Our work focuses on image processing which is a key stage for the whole process since it's impossible to characterize the muscle if the obtained parameters are inaccurate. In this paper we describe the proposed algorithm that carries out the optimized segmentation. The starting point for our segmentation algorithm was Zhu and Yuille's *Region Competition* [4] algorithm.

2 Segmentation Techniques

In the analysis of the objects in images it is essential that we can distinguish between the objects of interest (target) and "the rest" (also referred to as the background). Before developing our algorithm, a wide study of several segmentation techniques has been carried out: edge detection techniques, thresholding, region growing, active contours, etc. All these techniques have something in common, they formulate some hypotheses about the image, test features, and make decisions by applying thresholds explicitly or implicitly. The main difference between these different approaches lies in the domains on which the hypotheses, tests, and decisions are based. Edge based techniques only make use of local information and cannot guarantee continuous closed edge contours. Active contour models make use only of information along the boundary and require good initial estimates to yield correct convergence. An advantage of region growing is that it tests the statistics inside the region, however it often generates irregular boundaries and small holes. In addition, all these three methods lack a global criterion for segmenting the entire image. By contrast, global optimization approaches based on energy functions or Bayesian and MDL (Minimum Description Length) criteria often have problems to find their minima.

Zhu and Yuille presented in [4] a statistical framework for image segmentation using a novel algorithm which they called Region Competition (RC). It is derived by minimizing a generalized Bayes/MDL criterion and combines the attractive geometrical features of snake/balloon models and the statistical techniques of region growing.

2.1 The Region Competition Algorithm

The goal of image segmentation is to partition the image into subregions with homogeneous intensity (color or texture) properties which will hopefully correspond to objects or object parts. Now suppose that the entire image domain R has been initially segmented into M piecewise "homogeneous" underlying regions R_i where $i = 1, 2, \ldots, M$ and $R = \cup_{i=1}^{M} R_i$, $R_i \cap R_j = \emptyset$ if $i \neq j$. Let ∂R_i be the boundary of region R_i, where we define the direction of ∂R_i, to be counter-clockwise. Let $\Gamma = \cup_{i=1}^{M} \Gamma_i$ be the edges or segmentation boundaries of the entire image with $\Gamma_i = \partial R_i$.

Now consider an MDL criterion (a global energy functional). This gives:

$$E\left[\Gamma, \{\alpha_i\}\right] = \sum_{i=1}^{M} \{ \frac{\mu}{2} \oint_{\partial R_i} ds - \log P\left(\left\{I_{(x,y)} : (x,y) \in R_i\right\} |\alpha_i\right) + \lambda \} \; , \qquad (1)$$

where the first term within the braces is the length of the boundary curve ∂R_i, for region R_i. The second term is the sum of the cost for coding the intensity of every pixel (x, y) inside region R_i, according to a distribution $P(\{I_{(x,y)} : (x,y) \in R_i\} |\alpha_i)$. λ is the code length needed to describe the distribution and code system for region R_i, and we simply assume λ is common to all regions.

Because the energy E in (1) depends on two groups of variables (the segmentation Γ and the parameters α_i) Zhu and Yuille proposed a greedy algorithm which consists of two alternating stages. The first stage locally minimizes the energy with the number of regions fixed. It proceeds by iterating two steps both of which cause the energy to decrease. The second stage merges regions provided this decreases the energy. The first stage includes two steps. In the first step, we fix Γ, and $I(x, y)$, and we solve for the α_i to minimize the description cost for each region. In other words, the α_i are estimated by maximizing the conditional probabilities. In the second step, we fix the α_i and do steepest descent with respect to Γ.

Both steps in this first stage of the algorithm cause the energy function to decrease. In addition the function is bounded below and so the algorithm is guaranteed to converge to a local minimum. This two step process, however, does not allow us to alter the number of regions. Thus a second stage is added where adjacent regions are merged if this causes the energy to decrease. This is followed by the two step iteration stage again, and so on. Overall each operation reduces the energy and so a local minimum is reached. The whole process, as described in [4], goes as follows:

1. Initialize the segmentation, put N seeds randomly across the image, all background is treated as a single region with uniform probability distribution.
2. Fix the boundary Γ, compute the parameters $\{\alpha_i\}$ by maximizing $P(I : \alpha_i)$.
3. Fix $\{\alpha_i\}$, move the boundary Γ by minimizing the energy function. When two seed regions meet, edge is formed at their common boundary, then these two regions compete along this boundary.
4. Execute step 2,3 iteratively until the motion of boundary Γ converges. Then goto step 5.
5. If there is background region not occupied by any seed regions, then put a new seed in the background, and goto step 2; else goto step 6.
6. Merge two adjacent regions so that the merging causes the largest energy decrease, goto step 2. If no merge can decrease the energy, then goto step 7.
7. Stop.

3 RC Based Segmentation of Stereoscopic Sequences of Artificial Muscles

We describe in this Section our adaptation of RC algorithm for artificial muscles segmentation. As shown in the introduction of this article, our segmentation work is a key stage in the method designed to obtain a characterization of artificial muscles. The input for our application are pairs of orthogonal pictures of an artificial muscle in motion (see Fig. 1). Our algorithm is responsible of segmenting these images along the stereoscopic sequence. The output generated will be used by active contours to characterize the muscle movements.

In the following, we'll see a general overview of how our algorithm works and how it deals with important issues as initial segmentation, use of redundancy in stereoscopic pairs of frames and tracking of the target along the sequence.

3.1 General Overview of Our Algorithm Operation

The easiest way to understand our algorithm is to describe its operation from the first pair of frames acquired. Let $A(t)$ be the frame captured in the XZ plane and $B(t)$ the frame captured in the YZ plane at the same instant t. Fig. 2 shows a sequence of three stereo frames. As shown in the figure, differences between consecutive frames are small. This will help the target tracking later. Also, both cameras are calibrated so that relative position of the muscle is the same in both frames $A(t)$ and $B(t)$.

Our application operates with pairs of frames $A(t), B(t)$ simultaneously. The first pair $A(0), B(0)$ is segmented without any "a priori" information. RC algorithm is applied to $A(0)$ with a number of initial region seeds chosen by the user (any number of seeds will yield a correct segmentation). The process follows the algorithm described in the previous section until the segmentation Γ is complete. Once the initial segmentation of $A(0)$ is done (the whole image $I(x, y)$ is partitioned into N regions R_i), the user is prompted to choose the region of interest (the artificial muscle). Now the algorithm "knows" what to look for and

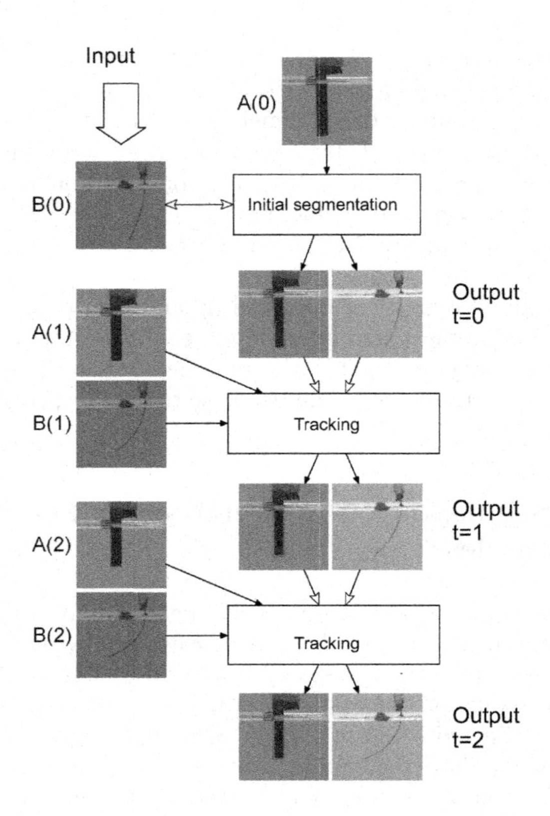

Fig. 2. Sequence segmentation flow. White arrows show information transfer.

uses that information to segment $B(0)$ much faster. This initial segmentation could be accelerated by using the knowledge we already have about the images but this way we ensure that the algorithm will work with images of any kind.

At this point, we have the segmentation Γ of both initial frames $A(0)$ and $B(0)$. From now on, the algorithm will work in a "stationary" mode until the end of the whole sequence. On this stage, we do have information about previous frames and we can use it to segment consecutive images much faster. On this "stationary" mode, RC algorithm is not fully used as described in the previous section. Instead of that, the number of regions is limited to *target* and *background* and the merging step almost disappears so the computational cost is reduced. Target tracking through the stereoscopic video is done by change detection in consecutive frames. Frame difference is calculated between frames $A(t)$ and $A(t-1)$ and the region competition algorithm will only take care about the pixels that have changed.

3.2 Initial Segmentation

As seen in the previous section, segmentation of the first frame $A(0)$ is carried out by the RC algorithm described before. This first segmentation is important because the next frames ($B(0)$, $A(1)$) are segmented much faster thanks to the information obtained. Moreover, if the muscle segmentation in the first frame is wrong our application will probably not be able to track its movements. In this section we'll explain some important issues related to this part of the process.

First step is to put N seeds randomly across the image (see Fig. 3(b)). The size of these initial seeds is 1 pixel but the initial descriptors α_i are calculated with a 5×5 pixels sampling window. This is to ensure that initial descriptors do not correspond to noisy peaks in the image. If one initial seed falls into one of these peaks, the statistical forces wouldn't let it grow as neighbor pixels won't pass the likelihood test. On the other hand, the sampling window cannot be too big because if it covers a significant edge it may be ignored. In fact, the optimal size of the sampling window depends on the image signal to noise ratio, in [4] a discussion about seeds and sampling windows can be found.

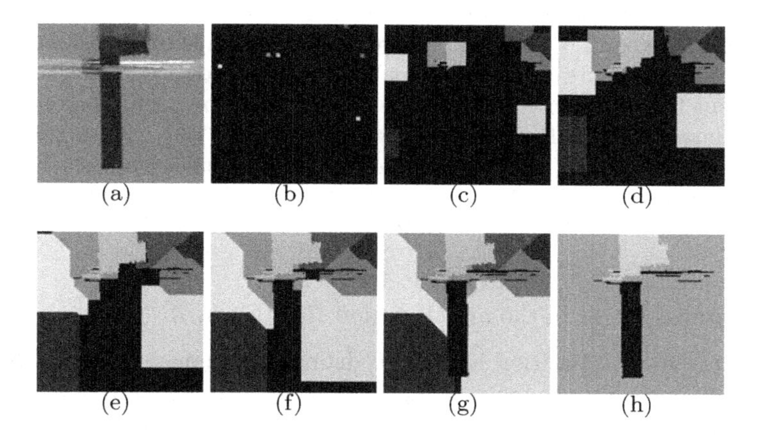

Fig. 3. Initial segmentation $A(0)$ with $N = 8$ random seeds (a) Frame $A(0)$. (b) Segmentation $\Gamma_{A(0)}$ after first growing iteration.; (c), (d), (e), (f), (g) show $\Gamma_{A(0)}$ evolution every 10 iterations. (h) Final segmentation $\Gamma_{A(0)}$ after 160 iterations.

Now we have an approximation of the final segmentation $\Gamma_{A(0)}$, this is a $m \times n$ matrix of labels, which is the same size of the frame. This matrix has zeros in every position except in those N pixels with initial seeds R_i. Every seed region has a descriptor α_i associated which is used to control its growing. Next step is to make those N regions grow. This is done by morphologic operators that dilate the initial regions with a 3×3 pixels structuring element. New pixels are tested against the region descriptors and only become part of the region if they fit the distribution $P(I|\alpha_i)$. Using gray level images and one descriptor per region, this decision is taken by comparing the pixel value with the mean of the pixels in the region. If the difference is not higher than the growing threshold T_g, the new pixel is finally added. Descriptors α_i are recalculated in every growing iteration to keep them representative of their regions R_i. When the values of $\Gamma_{A(0)}$ converge (regions can't grow anymore) the algorithm looks for pixels without any region assigned and, if there are any, it puts a new seed and the previous steps are repeated. Once every pixel in the image has been labeled next step is to merge regions if this causes the energy to decrease, using a new threshold that we call T_m (see Fig. 3(g) and (h)).

If the values chosen for T_g and T_m are too tight, regions will have problems to grow and new seeds will be added to complete $\Gamma_{A(0)}$. This means more regions, more descriptors and much more time to obtain an oversegmented image as final result. On the other hand, if T_g and T_m are too relaxed our algorithm will be less sensitive to details and we may loose some regions of interest. However, most artificial muscle pictures have a good contrast between target and background which makes easy to tune both thresholds properly.

3.3 Redundancy Between Stereoscopic Pairs of Images

As we said before, our application only uses the information obtained from $\Gamma_{A(t)}$ to segment $B(t)$ when $t = 0$. In other words, information transfer between stereoscopic pairs only occurs for the first pair of images (see Fig. 2). It is possible to implement this transfer for every frame and it may be useful for images with different characteristics. However, experiments carried out with artificial muscle sequences don't show any improvement. In this section we describe how we use the information obtained from $\Gamma_{A(t)}$ to speed up the segmentation of $B(t)$. Assuming that $\Gamma_{A(t)}$ is correct, the information we can use to segment $B(t)$ is:

- Muscle position $R_{interest}^{A(t)}$ in frame $A(t)$.
- Descriptor $\alpha_{interest}^{A(t)}$ of the muscle region $R_{interest}^{A(t)}$ in $A(t)$.

We use the muscle position in $A(t)$ to delimit the range of pixels where the algorithm will search for the muscle in $B(t)$. Cameras are calibrated so that both views of the muscle are in the same relative position and scale. Therefore we can assume that muscle coordinates in z axis are the same in $A(t)$ and $B(t)$. Pixels lying out of this range are labeled as background (see Fig. 4(c)).

The value of descriptor $\alpha_{interest}^{A(t)}$ is used to find the y axis coordinates of the muscle in $B(t)$. The algorithm looks for pixels with values similar to $\alpha_{interest}^{A(t)}$ and labels the rest as background (see Fig. 4(d)). At this point, the segmentation of frame $B(t)$ starts by applying the RC algorithm. However, as more than 90% of the image is already segmented (labeled as background) there's no competition between regions except in some special cases. Part of our target gets occluded when the muscle's end reaches the water surface. In this situation our application needs the region competition strength to complete the segmentation correctly. Once we've delimited the muscle position in $B(t)$, segmentation starts putting one seed at the center of the delimited zone and we let it grow until segmentation is finished. More specifically, z axis (vertical) coordinate of the initial seed is found by calculating $R_{interest}^{A(t)}$ gravity center. The y axis (horizontal) coordinate is chosen by finding the more similar pixel to $\alpha_{interes}^{A(t)}$. Initial seed position is carefully chosen because we don't want regions to compete when it's not necessary.

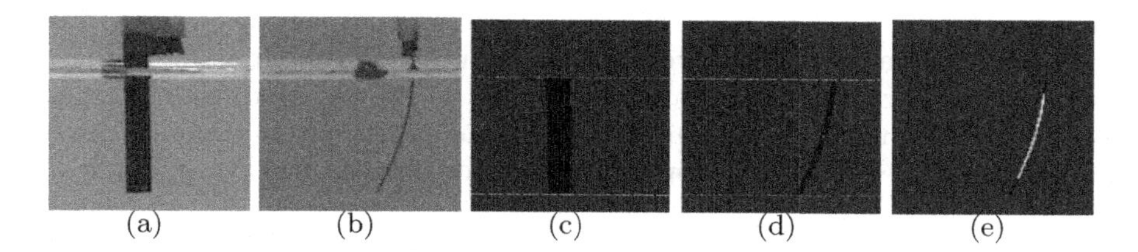

Fig. 4. Information transfer between stereoscopic pairs. (a) Frame $A(t)$. (b) Frame $B(t)$. (c) Segmentation $\Gamma_{A(t)}$. (d) Search range in $B(t)$. One seed is put in the z coordinate of the gravity center of $R^{A(t)}_{interes}$. y coordinate is found looking for similar pixels to $\alpha^{A(t)}_{interest}$. (e) shows the regions growing.

3.4 Video Segmentation: Muscle Tracking

In this section we show how $A(t-1)$ and $B(t-1)$ frames are segmented using the segmentation of previous frames. The following explanation is referred to YZ plane frames ($B(t)$ and $B(t-1)$) but segmentation of $A(t)$ frames is analogous.

The technique applied to track the muscle through the video sequence is based in change detection between consecutive frames. These changes are usually detected calculating the difference $I(t) - I(t-1)$ between consecutive frames. In our case, we use the knowledge that we already have about the image and its movement to optimize the segmentation. As the muscle speed is limited, we assume that its position will be only slightly different in $B(t)$. Therefore, we can estimate the muscle position in $B(t)$ if we know $\Gamma_{B(t-1)}$. In order to do this, the muscle position in $B(t-1)$ is dilated morphologically which gives us a limited space where the muscle might be in $B(t)$ (see Fig. 5(b)).

Once more, most pixels in $B(t)$ are already labeled as background before starting RC algorithm. Now we apply change detection technique calculating the difference between $B(t-1)$ and $B(t)$ but only in the estimated zone. Operation time is reduced this way because most pixels of both frames are ignored. Besides, we prevent the algorithm from being distracted by slight changes is water surface.

Next step is to find the optimal position to place our initial seed. We use $\alpha^{B(t-1)}_{interest}$ descriptor to test those pixels not excluded yet by our estimation. New seed is put at the more similar pixel found in $B(t)$ and it starts growing

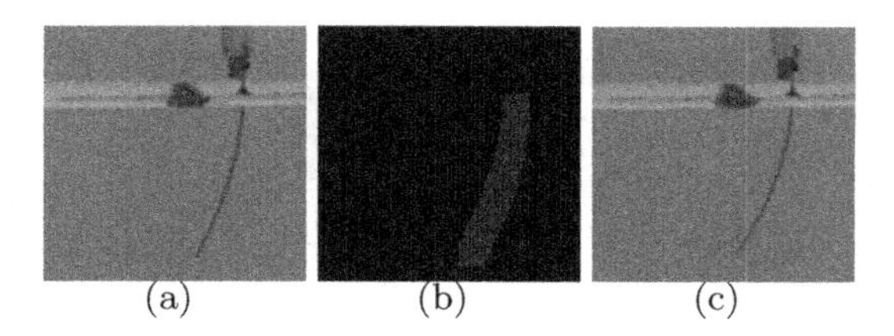

Fig. 5. Estimated target position using previous knowledge. (a) Frame $B(t-1)$. (b) Target estimation dilating previous segmentation. (c) Frame $B(t)$.

(see Fig. 6(b)). In RC algorithm all regions grow alternatively to ensure a "fair" competition but at this point we let the region of interest grow first. This is a way to avoid problems with possible changes in illumination (due to muscle movement) that may mislead our algorithm. Once the muscle region is grown, the background region starts growing until segmentation of $B(t)$ is finished.

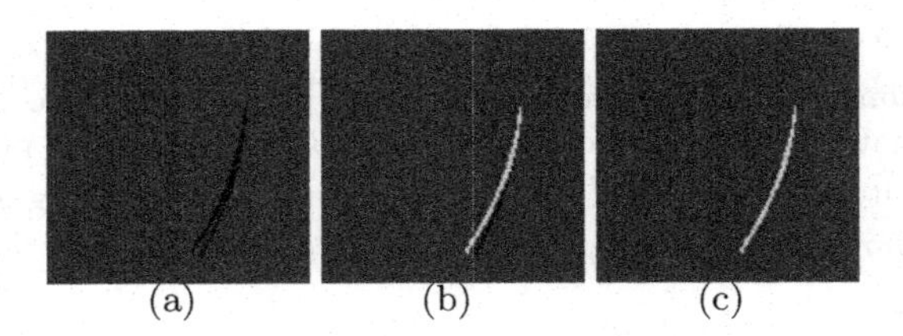

Fig. 6. (a) Change detection between $B(t)$ and $B(t-1)$. (b) Initial seed is placed using $\alpha_{interest}^{B(t-1)}$ and it starts growing in first place. (c) Final segmentation $\Gamma_{B(t)}$.

Experiments show that tracking process is remarkably accelerated thanks to this techniques because region competition only occurs in special cases like the occlusion problem mentioned before. Nevertheless, segmentation yielded is correct in every case.

4 Segmentation Results

Several tests have been run with different sequences and different types of artificial muscles and the results yielded are satisfactory. The segmentations obtained represent faithfully the muscle contour and are suitable to extract its parameters. Fig. 7 shows the results obtained in the first frames of a 60 pairs sequence

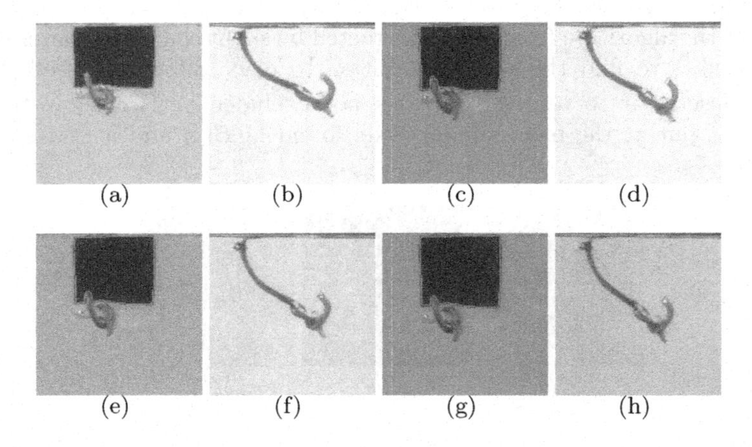

Fig. 7. Stereoscopic video segmentation. The muscle is loaded with a metal ring.

of 150×150 pixels. The muscle has been loaded with a piece of metal in order to test its strength. XZ plane segmentation is correct as it doesn't add the metal ring to the target region. Notice the changes in lighting between frames $A(t)$ and $B(t)$ and the presence of hard noise. Results obtained are also satisfactory in this situation.

5 Conclusion

This paper shows the application developed for segmenting stereoscopic video sequences based in the region competition algorithm. The original RC algorithm has been tuned and optimized for artificial muscle video segmentation in gray level images and results yielded are satisfactory. Parameters of muscle behaviour can be obtained from the output of our algorithm even in adverse situations of noise and lighting. The process can be run without any human participation after thresholds are adjusted. Redundancy between stereoscopic pairs of images has been used to optimize the segmentation with satisfactory results. Muscle tracking has also been optimized by estimating target position in consecutive frames. As a future work it could be interesting to extend the method by adding some topological model to ensure coherent segmentations in both planes.

References

1. Otero, T.F., Sansiñena, J.M.: Bilayer dimensions and movement of artificial muscles. Bioelectrochem. Bioenergetics **47** (1997)
2. Verdú, R., Morales, J., Fernandez-Romero, A.J., Cortés, M.T., Otero, T.F., Weruaga, L.: Mechanical characterization of artificial muscles with computer vision. Proc. of SPIE Annual Int. Symposium on Smart Structures and Materials (2002)
3. Liang, J., McInerney, T., Terzopoulos, D.: United snakes. Proc. Int. Conf. Computer Vision (1999) 933–940
4. Zhu, S.C., Yuille, A.L.: Region competition: Unifying snakes, region growing, and bayes/MDL for multiband image segmentation. IEEE Trans. Pattern Anal. Machine Intell. **18**(9) (1996) 884–900

A Support Vector Method for Estimating Joint Density of Medical Images

Jesús Serrano, Pedro J. García-Laencina,
Jorge Larrey-Ruiz, and José-Luis Sancho-Gómez

Dpto. Tecnologías de la Información y las Comunicaciones
Universidad Politécnica de Cartagena,
Plaza del hospital 1, 30202 Cartagena (Murcia), Spain
jsg@alu.upct.es

Abstract. Human learning inspires a large amount of algorithms and techniques to solve problems in image understanding. Supervised learning algorithms based on support vector machines are currently one of the most effective methods in machine learning. A support vector approach is used in this paper[1] to solve a typical problem in image registration, this is, the joint probability density function estimation needed in the image registration by maximization of mutual information. Results estimating the joint probability density function for two CT and PET images demonstrate the proposed approach advantages over the classical histogram estimation.

1 Introduction

Human learning can be defined as the change in a subjet behaviour as a result of the experience, by the stablishment of associations between stimulus and responses by means of the practice. This practice implies an iterative *trial-and-error* proccess in which better responses are used before known stimulus. Even new responses are generated in front of unknown stimulus, as a result of evaluating similar known ones previously learned. These learning processes are not limited to the humankind, in the other hand, they are shared with other living beings. It is possible to extrapolate all this conceps concerning with biological learning to the field of the artificial intelligence, being the result the research area named *machine learning* [1]. This broad subfield, inspired in these biological learning processes, is in charge of develop algorithms and techniques that make the computers be able to "learn".

One of the most relevant type of algorithm in machine learning is the *supervised training*, in which the aim is to create a function from a training data, consisting in pairs of input objects and desired outputs. Support Vector Machines (SVM) are powerfull supervised training methods [2], which are "cousins" of

[1] This work is partially supported by Ministerio de Educación y Ciencia under grant TEC2006-13338/TCM, and by Consejería de Educación y Cultura de Murcia under grant 03122/PI/05.

J. Mira and J.R. Álvarez (Eds.): IWINAC 2007, Part II, LNCS 4528, pp. 162–170, 2007.

the artificial neural networks. SVM are able to solve classification and regression problems from a training data set, supporting the solution in a small subset of it.

Probability density function estimation is a particular approach of SVM regresion. Density approximation is a common task in many image understanding problems, particularly in those that need to compute the Mutual Information (MI) between images as a similarity measure. This is the case of the medical image registration by Maximization of Mutual Information (MMI) [3,4]. This method requires to estimate the joint probability density function of two or more medical images, which usually are 3-D digital ones. Normally joint histogram is the used method to estimate the joint density, but this is an innacurate estimator. In this work a support vector density estimator is used, based on the support vector density estimation method decribed in [5], which provides a smooth and sparse solution.

This paper structures as follows: Section 2 describes the support vector density estimation method; in Section 3 the application of this method in medical images is presented with a brief introduction of the image registration problem and the medical image registration using MMI in subsections 3.1 and 3.2 respectively, and the implementation of the support vector joint density estimation in subsection 3.3; the Section 4 shows the described method results; finally the conclusions and references close the paper.

2 Support Vector Density Estimation

Let $p(x)$ denote the density function from a data set $\{x_1, ..., x_l\}$ which is tried to be estimated. The distribution function is

$$F(x) = P(X \leq x) = \int_{\infty}^{x} p(t)dt \tag{1}$$

Thus, finding density requires solving a linear operator equation $Ap = F$, where A is the linear operator.

The empirical distribution function is computed from the data set using

$$F_l(x) = \frac{1}{l}\sum_{i=1}^{l} \theta(x - x_i) \tag{2}$$

with

$$\theta(x) = \begin{cases} 1, & x > 0 \\ 0, & \text{otherwise} \end{cases} \tag{3}$$

The obtained distribution is introduced in (1). The Support Vector (SV) method can be used to solve linear operator equations $Ap(t) = F(x)$ [6], using $F_l(x_i)$ as the desired output y_i. For any fixed point x, $F_l(x)$ is a unbiased estimation of $F(x)$ and its standard deviation is

$$\sigma = \sqrt{\frac{1}{l}F(x)((1 - F(x)))} \tag{4}$$

so the accuracy of the approximation can be characterized as

$$\varepsilon_i = \sigma_i = \sqrt{\frac{1}{l} F_l(x_i)((1 - F_l(x_i)))} \tag{5}$$

thus, in order to solve the regression problem we consider the triples$(x_i, F_l(x_i), \varepsilon_i)$.

This method does not guarantee the obtained distribution will always be positive due to the used kernel may be non-monotonic. It is possible to choose monotonic kernels, but the drawback is the monotonic functions expressed with Mercer kernels [2] have not the desired shapes to perform an accurate regression. It can be obtained using classical density estimation kernels, although these kernels not satisfies the Mercer's condition, by a linear programming approach for solving the SV regression [7].

If approximating the density from a mixture of gaussian shapes is desired, the regression in the image space can be a mixture of sigmoids [5]. Then the kernels may have the form

$$K(x_{sv}, x) = \frac{1}{1 + e^{\gamma(x_{sv} - x)}} \tag{6}$$

where x_{sv} is the kernel centre and γ the kernel width, which is a pre-fixed parameter, and the *cross-kernel* [5] derived from the kernel K is

$$\mathcal{K}(x_{sv}, x) = \frac{\gamma}{2 + e^{\gamma(x_{sv} - x)} + e^{-\gamma(x_{sv} - x)}} \tag{7}$$

Instead of having a fixed parameter γ, an adaptative kernel width can be used. This is achieved by means of a dictionary of κ kernels for each SV centre, resulting in an approximation of the distribution function as the following expansion of support vectors:

$$F(x) = \sum_{i=1}^{l} \sum_{k=0}^{\kappa} \alpha_i^k K_k(x_i, t) \tag{8}$$

and, from these support vectors is possible to obtain the desired density

$$p(t) = \sum_{i=1}^{l} \sum_{k=0}^{\kappa} \alpha_i^k \mathcal{K}_k(x_i, t) \tag{9}$$

where each K_i and $\mathcal{K}_i$ has a width γ_i. Thus, generalizing the linear programming SV density estimation, we obtain the following optimization problem [5]:

$$\min \left(\sum_{i=1}^{l} \sum_{k=0}^{\kappa} \alpha_i^k + C \sum_{i=1}^{l} \xi_i + C \sum_{i=1}^{l} \xi_i^* \right) \tag{10}$$

under the constraints

$$y_i - \varepsilon_i - \xi_i \leq \textstyle\sum_{j=1}^{l} \sum_{k=0}^{\kappa} \alpha_j^k K_k(x_j, x_i) \leq y_i + \varepsilon_i + \xi_i^*, \; i = 1, \ldots, l, \tag{11}$$

$$\textstyle\sum_{i=1}^{l} \sum_{k=0}^{\kappa} \alpha_i^k = 1, \tag{12}$$

$$\alpha_i \geq 0, \quad \xi_i \geq 0, \quad \xi_i^* \geq 0 \qquad i = 1, \ldots, l \tag{13}$$

where the regularizer $\Omega(\alpha) = \sum_{i=1}^{l} \sum_{k=0}^{\kappa} \alpha_i^k$ can be changed for other ones that provide better results, like a weighted sum of the coefficients

$$\sum_{i=1}^{l} \sum_{k=0}^{\kappa} w_k \alpha_i^k \tag{14}$$

3 Support Vector Joint Density Estimation of Medical Images

3.1 Image Registration Problem

It is common that medical images are represented as sequences of 2-D cross-sectional slices, which are used to construct a 3-D volume by the geometrical relationships between the slices, and there are not information about the relative positions of the patients in the scanners. The problem generally consists in the prospective registration of one or several 3-D images to another one, obtaining the parameters involved in the transformation. This process is desirable to be automated and accurate [8]. It is not a trivial problem, because of the large number of variables to take into account, such as the different positioning of the patient in the scanners, the different images resolution, distortions introduced in the images which are specific in each modality, and so on. There are many approaches for prospectively image registering, but one of them has acquired special importance in the last years because of its robustness, simplicity and mathematical elegance. This method is based in MMI, and defends that the Mutal Information (MI) of the images to be registered becomes maximal if they are geometrically aligned [3,4].

3.2 Medical Image Registration Using MMI

Let $\mathcal{R}$ (*reference* image) and $\mathcal{F}$ (*floating* image) denote two images related by a registration transformation $\mathbf{T}_{\boldsymbol{\alpha}}$ with parameters $\boldsymbol{\alpha}$ such that voxels $\mathbf{p}$ in $\mathcal{R}$ with intensity r physically correspond to voxels $\mathbf{T}_{\boldsymbol{\alpha}}(\mathbf{p})$ in $\mathcal{F}$ with intensity f [9]. The information that a value contains about the other is measured as the mutual information $I(F, R)$ of the variables $F = \{f\}$ and $R = \{r\}$

$$\begin{aligned} r &= \mathcal{R}(\mathbf{p}) \\ f &= \mathcal{F}(\mathbf{T}_{\boldsymbol{\alpha}}(\mathbf{p})) \\ I(F, R) &= \sum_{f,r} p_{FR}(f, r) \log \frac{p_{RF}(f, r)}{p_F(f) \cdot p_R(r)} \end{aligned} \tag{15}$$

where $p_{RF}(r, f)$, $p_R(r)$ and $p_F(f)$ are the joint and marginal densities respectively.These distributions are usually computed by simple normalization of the joint histogram. Since the density $p_{RF}(r, f)$, and, in general, the marginal distributions $p_R(r)$ and $p_F(f)$, depend on the mapping $\mathbf{T}_{\boldsymbol{\alpha}}(\mathbf{p})$, also the mutual information $I(R,F)$ does. The mutual information criterion postulates that the images are geometrically aligned by the transformation $\mathbf{T}_{\boldsymbol{\alpha}^*}(\mathbf{p})$ for which $I(R,F)$ is maximal

$$\boldsymbol{\alpha}^* = \arg\max_{\boldsymbol{\alpha}} I(R,F) \tag{16}$$

3.3 SV Joint Density Estimation Method

The density estimation method previously described is applied to one-dimensional densities. This section extends it to a multi-dimensional problem, such as the joint density estimation of medical images. This is a 2-dimensional problem, in which we estimate the density $p(\mathbf{x})$ with its corresponding distribution function

$$F(\mathbf{x}) = P(\mathbf{X} \le \mathbf{x}) = \int_{-\infty}^{x_1} \int_{-\infty}^{x_2} p(t)dtdt$$

where x_1 and x_2 are the images intensity values.

Let $\mathcal{R}$ and $\mathcal{F}$ denote the reference and floating images, which have a corresponding fixed resolutions $\rho_{\mathcal{R}}$ and $\rho_{\mathcal{F}}$, and generally, $\rho_{\mathcal{R}} \neq \rho_{\mathcal{F}}$. First, the intensity values are linearly rescaled into the continuous range $(0, n_{\mathcal{R}} - 1)$ and $(0, n_{\mathcal{F}} - 1)$ with $n_{\mathcal{R}} = n_{\mathcal{F}} = 256$. Instead of being discrete ranges, these ranges must be continuos because rounding to the nearest integer may cause a loss of information. The next step is the computation of the empirical joint distribution function, in which is taken into account the different images resolution $\rho_{\mathcal{R}}$ and $\rho_{\mathcal{F}}$. Due to $\mathbf{T}_{\boldsymbol{\alpha}}(s)$ will not coincide with a grid point of $\mathcal{R}$, a interpolation is needed. The classical interpolation schemes (Nearest Neighbour, Trilinear or Partial Volume) only extend the influence of the voxel sample s to the nearest neighbors of $\mathbf{T}_{\boldsymbol{\alpha}}(s)$, without considering that the images may have so different resolutions. In this case the influence region of a voxel in one of the images affects in the another one either a larger or a smaller number of voxels than the number of nearest neighbors. In this work an interpolation scheme based on the volume overlap of the voxels is proposed, assuming that each voxel represents a parallelepiped centered on it, which dimensions and orientation depend respectively on the slice separation and pixel spacing in each image, and on the grid positions and transformation parameters $\boldsymbol{\alpha}$. Figure 1 shows a 2-D projection of this concept. Each voxel of the reference and floating images has a fixed volume $V_{\mathcal{R}}$ and $V_{\mathcal{F}}$, and each pair of voxels in $\mathbf{T}_{\boldsymbol{\alpha}}(s)$ and n_i has a overlap volume $\mathcal{V}_{s,n_i}$. The joint distribution function is computed using the ratio of the overlap volume over the total volume V

$$\left.\begin{array}{c} \forall f \ge f(s), f \in \mathbf{F} \\ \wedge \\ \forall r \ge r(n_i), r \in \mathbf{R} \end{array}\right\} \Rightarrow F(f,r) += \frac{\mathcal{V}_{s,n_i}}{V} \tag{17}$$

where $\mathbf{F}$, $\mathbf{R}$ are discrete ranges between 0 and $(n_{\mathcal{F}} - 1)$ or $(n_{\mathcal{R}} - 1)$, respectively. The number of points in these ranges is a parameter to be chosen, if either a coarse or a fine estimation is desired.

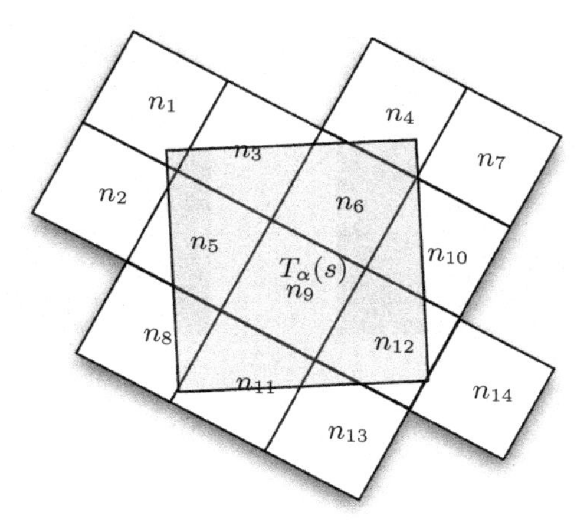

Fig. 1. 2-D projection of volume overlap based interpolation

Since the problem is 2-dimensional, then a 2-dimensional kernel is used. In this work the bi-dimensional kernel is chosen to be a tensor product of one dimensional gaussian-like kernel shown in section 2

$$K(\mathbf{x}_{sv}, \mathbf{x}) = \frac{1}{1 + e^{\gamma_1(x_{sv,1} - x_1)}} \times \frac{1}{1 + e^{\gamma_2(x_{sv,2} - x_2)}} \tag{18}$$

where $\mathbf{x}_{sv} = (x_{sv,1}, x_{sv,2})$ is the centre of the kernel, and $\boldsymbol{\gamma} = (\gamma_1, \gamma_2)$ is a vector containing the kernel widths in each one of the two dimensions. Therefore, the cross-kernel results:

$$\mathcal{K}(\mathbf{x}_{sv}, \mathbf{x}) = \frac{\gamma_1}{2 + e^{\gamma_1(x_{sv,1} - x_1)} + e^{-\gamma_1(x_{sv,1} - x_1)}} \times \frac{\gamma_2}{2 + e^{\gamma_2(x_{sv,2} - x_2)} + e^{-\gamma_2(x_{sv,2} - x_2)}} \tag{19}$$

The used regularizing term in (10) is the same as in (14), i.e., a weighted sum of the SV coefficients. These weights are chosen to penalize the kernels with small width, in order to perform a smooth solution. In addition, since a highly accuracy is desired, the ε_i used is choosen equal to $0.1\sigma_i$.

4 Results

The described method is tested in two thorax images from a dual CT-PET scanner, obtained from the OsiriX open source site[2]. The CT image is consisted of 41 slices with a resolution equal to 512 × 512, and the PET image has the same number of slices than the CT image and resolution 128 × 128.

[2] `http://homepage.mac.com/rossetantoine/osirix/Index2.html`

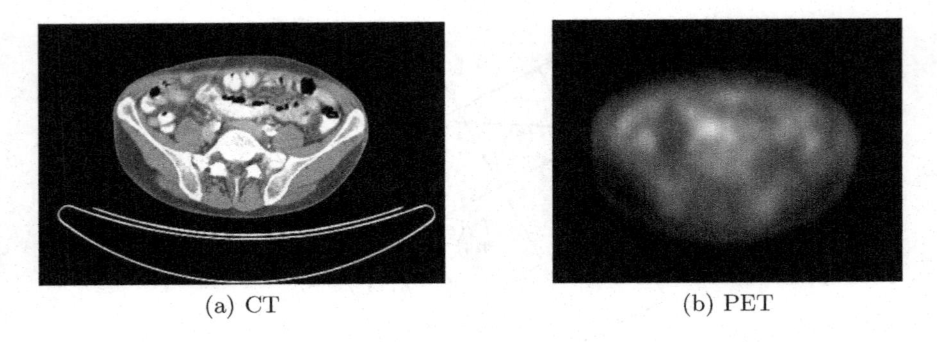

(a) CT (b) PET

Fig. 2. Cross slices of the CT (a) and PET (b) images used

Two cross slices of both images are shown in Figure 2. After computing the empirical distribution, the SV joint density estimation is carried out, obtaining 249 support vectors, which means a 16.45% over the whole data set. Therefore, the SV density estimation results in a data compression, which allows working over fewer points than the 256×256 histogram points. We compare our method with the classical and simple estimation procedure based on the histogram.

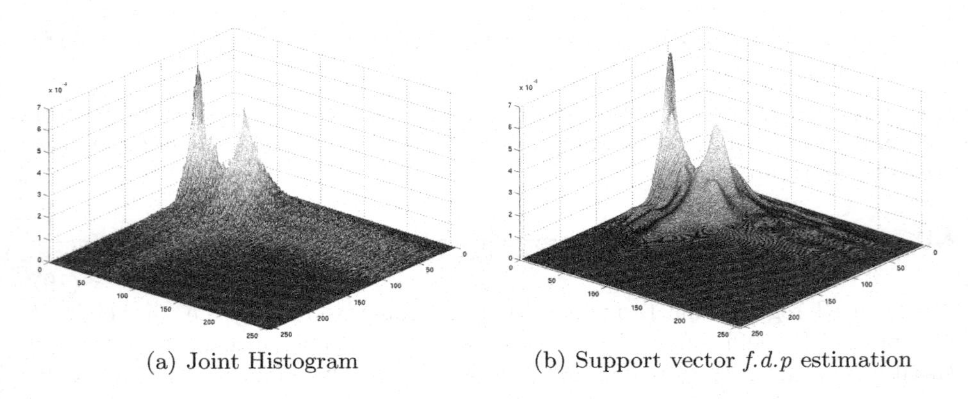

(a) Joint Histogram (b) Support vector *f.d.p* estimation

Fig. 3. CT-PET joint histogram (a) and SV probability joint density estimation (b)

Figures 3 and 4 show that the obtained result looks like the histogram shape, but it is smoother. An important advantage of our approach is that, since the density obtained is like a sum of Gaussians, analytic expresions can be derived for stochastic approximations of mutual information and its gradient. Therefore, the mutual information derivatives can be easy obtained in order to optimize it by means of a gradient-based or similar procedure.

(a) 2D joint histogram (b) 2D $f.d.p$ estimation

Fig. 4. Joint histogram (a) and support vector density estimation (b) 2D projections

5 Conclusions

This paper describes a SV method for estimating joint densities of medical images. The method makes an empirical distribution regression to find the support vectors, and after that, the linear operator that relates the distribution to the density is applied to the kernels. In addition, the empirical distribution is obtained using a novel interpolation method that takes into acount the superposition volume amount between voxels of both images. In this approach, a sigmoidal kernel is needed to achieve a correct destribution regression, which implies obtaining a density as a sum of Gaussians. Therefore, due to the obtained result is a kernel mixture, analytic expressions of the mutual information gradient can be derived to perform gradient maximization based approaches, in order to implement an efficient MMI based image registration. The resulting density is smooth and sparse, in contrast to the estimation based on histogram or other methods such as parzen windows.

This work will stimulate future works in many directions. Some of them would be to include the SV estimation method in a MMI registration procedure, and a comparative study of the developed volume overlap interpolation approach with the most common interpolation schemes used in the literature.

References

1. Mitchel, T.M.: Machine Learning. McGraw-Hill, New York (1997)
2. Vapnik, V.: The Nature of Statistical Learning Theory. Spinger, New York (1995)
3. Collignon, A., Maes, F., Delaere, D., Vandermeulen, D., Suetens, P., Marchal, G.: Automated Multi-modality Image Registration based on Information Theory. In Bizais, Y., Barillot, C., Di Paola, R., eds.: Proceedings XIVth International Conference on Information Processing in Medical Imaging – IPMI'95. Volume 3 of Computational Imaging and Vision., Ile de Berder, France, Kluwer Academic Publishers (June 1995) 263–274

4. Viola, P., Wells, W.M.: Alignement by Maximization of Mutual Information. In: Proceedings of the 5th International Conference on Computer Vision, Cambridge, MA (1995) 16–23
5. Weston, J.A.E.: Extensions to the Support Vector Method. PhD thesis, University of London (1999)
6. Vapnik, V., Golowich, S.E., Smola, A.: Support Vector Method for Function Approximation, Regression Estimation and Signal Processing. In Mozer, M.C., Jordan, M.I., Petsche, T., eds.: Advances in Neural Information Processing Systems. Volume 9., Cambridge, MA, MIT Press (1997) 281–287
7. Vapnik, V.: Statistical Learning Theory. Wiley, New York (1998)
8. Maintz, J.B.A., Viergever, M.A.: A survey of medical image registration. Medical Image Analisys **2**(1) (1998) 1–36
9. Maes, F., Vandermeulen, D., Suetens, P.: Medical Image Registration Using Mutual Information. In: Proceedings of the IEEE – Special Issue on Emerging Medical Imaging Technology. Volume 91. (2003) 1699–1722 (invited paper).

Segmentation of Moving Objects with Information Feedback Between Description Levels

M. Rincón, E.J. Carmona, M. Bachiller, and E. Folgado

Dpto. de Inteligencia Artificial. ETSI Informatica. UNED.
Juan del Rosal 16, 28040 Madrid, Spain
{mrincon, ecarmona, marga, efolgado}@dia.uned.es

Abstract. In real sequences, one of the factors that most negatively affects the segmentation process result is the existence of scene noise. This impairs object segmentation which has to be corrected if we wish to have some minimum guarantees of success in the following tracking or classification stages. In this work we propose a generic knowledge-based model to improve the segmentation process. Specifically, the model uses a decomposition strategy in description levels to enable the feedback of information between adjacent levels. Finally, two case studies are proposed that instantiate the model proposed for detecting humans.

1 Introduction

In recent years, many researchers have focused their attention on detecting and tracking moving objects in video sequences given that it is the first significant step for many machine vision applications, such as semantic video annotation, pattern recognition, video surveillance, traffic control, detection and tracking of people, perceptual interfaces, etc. Obviously, depending on the application, it will be necessary to describe the scene with a different degree of detail, which implies applying different techniques to extract image information and a different degree of precision in segmenting the objects of interest. Thus, in some video surveillance applications it is necessary to distinguish the movement of the different parts of the body to recognise the specific action that the human is doing, for example, to determine whether he is carrying a briefcase or some dangerous object in his hands or not. By contrast, in other applications, the human can be treated as a rigid body, since only the system needs to detect his presence in a room, his passing through a specific area or, generally, the analysis of his path. Therefore, in the first instance a more precise segmentation is required than in the second one. This work focuses on obtaining a robust segmentation with enough degree of precision for the application.

The segmentation algorithm outputs, especially if we work with real scenes, generally contain noise. Noises are primarily due to the intrinsic noise of the video camera, to unwanted reflections, to objects that have a colour that matches the background totally or partially and the existence of sudden shadows and artificial

J. Mira and J.R. Álvarez (Eds.): IWINAC 2007, Part II, LNCS 4528, pp. 171–181, 2007.

or natural changes in the lighting. The total effect of these factors is twofold: first, it may mean that areas that do not belong to the moving objects are incorporated into the foreground (foreground noise), and second, that certain areas, which belong to the objects, do not appear in the foreground (background noise).

There are a number of methods for segmenting moving objects present in a video sequence. These are based, for example, on statistical methods [1][2], the subtraction of consecutive frames [3], optic flow [4], genetic algorithms [5] or on hybrid methods [6][7][8][9] that combine some of these techniques. However, due to the speed and ease of implementation, one of the most frequently used methods, with a fixed camera, is the one based on background subtraction and its many variants [10][11][12][13]. In all these works, a segmentation is generated whose goodness depends on adjusting the method parameter configuration for a specific type of scene, but there is no resegmentation of the scene in the event of error.

In real scenes it is difficult to obtain a precise segmentation in a first approach. Although, previously, knowledge on the type of objects was used to refeed the segmentation process [14][15][16], it was only used on static images and using basic generic characteristics of the objects of interest (continuity or smoothness properties of the contours, local uniformity of movement, etc).

Generally, the main problem in interpreting images is the huge semantic gap that exists between the physical signal level and the knowledge level. To facilitate this gap it is necessary to insert new levels and inject the knowledge available. Following the proposal by Nagel et al. [17], in this work we distinguish different description levels with an increasing degree of semantics: Pixel level, Blob level, Object level, and Activity-behaviour level. Specifically, the final aim is to show how the segmentation (blob level) results can be improved when there is an exchange of information between the object level and blob level assuming that models related to the type of objects of interest exist.

This paper is organized as follows. Section 2 describes in the first instance the generic segmentation model proposed, and after the specific model for human segmentation. Section 3 analyses two case studies that highlight the instantiation of the model in two different situations where the segmentation is improved by applying different operators. Finally, section 4 analyses the results obtained and the future work proposed for improving the proposal.

2 Description of the Segmentation Model

Figure 1 shows the segmentation model proposed where the feedback cycle between levels is evident. The segmentation process begins by taking a video sequence frame as input. The result of this operation will be an initial proposal of the set of blobs associated with moving scene objects. This set of blobs, from the blob level, plays the role of findings at object level and is the input to the diagnosis task. The approach used at object level to refeed the blob level is based on the well-known strategy of diagnosis and planning of therapies used in medicine. Here, the quality of the segmentation is diagnosed based on normality

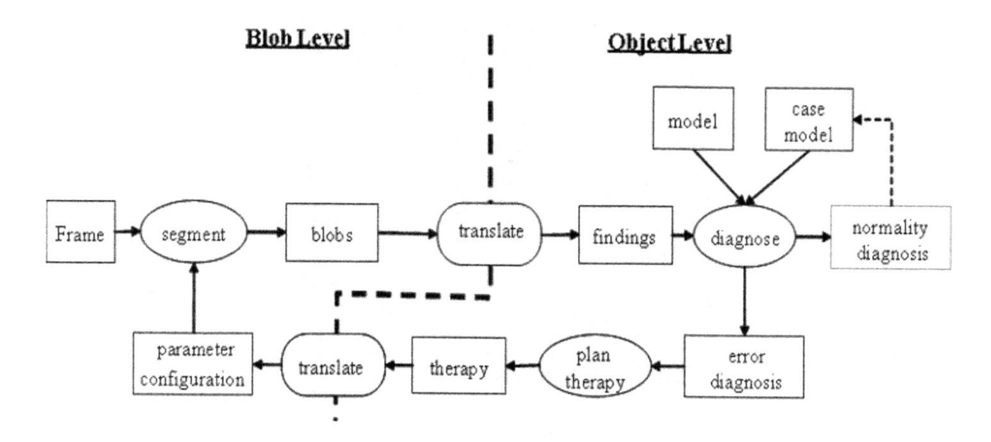

Fig. 1. Feedback structure proposed

models of the objects being recognized, distinguishing between normal situations (normality diagnosis) and abnormal situations (error diagnosis).

If a normality diagnosis is obtained, the resulting information updates the description of the object to that moment (case model). Conversely, if an abnormal situation has been detected, the feedback process is used to solve it by applying the appropriate therapy. This therapy is translated, at blob level, into a parameter configuration that affects the segmentation process. The feedback cycle is thus completed.

The separation into description levels means that it is necessary to introduce translation operations to adapt the entities that play the roles between adjacent levels. Thus, whereas at blob level, we speak of blobs associated with an object, these very blobs play the role of findings at object level. Similarly, the therapy planned to solve the segmentation problem becomes a new parameter configuration that will affect the segmentation process.

2.1 The Initial Segmentation

The process begins with an initial segmentation of the frame i of the video sequence using a method independent of the domain and tuned for the type of scene captured by the camera (context dependent). Specifically, in this work, as an initial segmentation method, the approach explained in [13] characterised by its ease of implementation and low computational cost was used. This approach is inspired by the subtraction background method and, therefore, it will be assumed that at every moment we will work with scenes taken with a fixed camera.

2.2 The Human Model

In this work we applied the segmentation model to detect humans. Consequently, a human model is required as a reference model for the diagnosis. The human

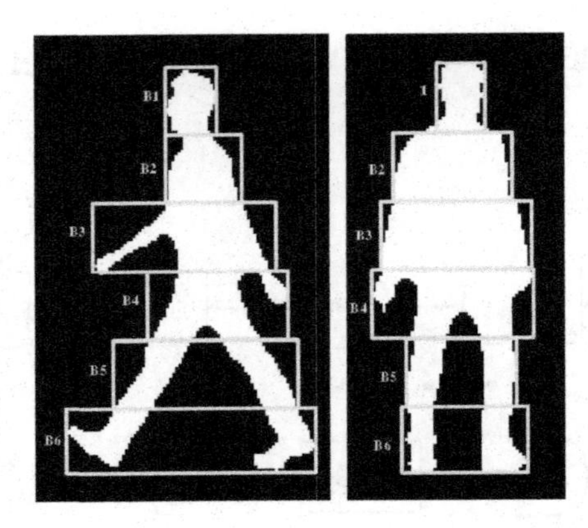

Fig. 2. A human blob in frontal and lateral position divided into blocks

model used here [18] consists on a block model. Basically it consists of dividing the blob corresponding to the human vertically into six regions the same height (figure 2). Each of these regions is defined by the rectangle that circumscribes it and that we will call block. Conceptually, the blocks of this division correspond to zones related to the physical position of specific parts of the body when usual actions are done with normal human movement (head, hands, feet, trunk). The main advantage of this division is that it enables us to study the human in parts. Thus, for example, if we are analysing the hands we know that in a normal situation these are between blocks B3 and B4, otherwise we would detect an abnormal situation. Another advantage of this model is that it is also possible to handle frontal and lateral human views homogeneously.

2.3 Therapy Diagnosis and Planning

In the diagnosis stage, the aim is to evaluate whether the result obtained in the segmentation is coherent with the human model described in the previous section. In our case, with this model it must be possible to detect, for example, the absence of some significant part of the body, the division of the body in several unconnected blobs, an unjustified change in the segmentation from one frame to the next, the presence of parts not related to the human (due to foreground noise), etc.

After the problem has been detected, it is necessary to generate a plan to correct it. As the problems raised are related with the segmentation of an object, either because all the corresponding blobs have not been assigned to the object, or because more blobs have been assigned than necessary, operators must be applied to modify this segmentation. Obviously, if we tried to improve the segmentation globally on all the scene, we would encounter numerous problems, since segmentation operator behaviour is not usually homogeneous (whereas in some regions of the image the segmentation detects the objects precisely, in others it may introduce foreground noise, for example). Conversely, if we focus on the problem region and reduce the analysis region, operator behaviour is probably more effective. In fact, the smaller the analysis region, the more homogeneous segmentation operator behaviour has to be.

2.4 Segmentation Operators

To improve the segmentation we exemplify here two operators that work at blob level: combination operator and restoration operator.

- The combination operator consists of modifying the set of blobs associated with an object, i.e., assigning or withdrawing some blob to/from the set under certain determining factors imposed by the domain and controlled from the object level via the therapy proposed.
- The restoration operator restores those human pixels associated with background noise. For example, let us assume that, following the previous example, the blobs corresponding to the head, arms and legs of the human have been located, but that one part of the body is so like the background that it has not been detected in the first instance by the segmentation algorithm. In this case, the restoration operator can be applied so that the missing part emerges focusing exclusively on the specific region at blob level of the human model and its associated blobs.

3 Case Studies

This section shows two segmentation examples of moving humans, using the model shown in figure 1 and each of the two segmentation operators described above.

Case 1

One example of combination operator use is that shown in figure 3. The human on the right has been correctly detected as one blob. This will happen as long as the human is notably different from the background. However, the human on the left has been divided into two blobs because his shirt has not been detected. In this instance, it is at object level where this situation has to be detected, i.e., the blob on the right will be recognised as human, but the same will not occur with the other two blobs when they are analysed separately and recognised as not being consistent with a complete human (error diagnosis). In this last instance, the operator will seek to match one of these blobs with different parts of the body, for example, the blob situated in the upper left part (figure 3b) has a high probability of corresponding to a head. Therefore, it is logical to think that the rest of the body must be within a region that has some dimensions matching the scale of a human in the type of scene considered and this region is in the centre of the lower part of the head. Thus, the combination operator will assign all the blobs found in that region to the set of blobs belonging to the human, thereby achieving an initial improvement of the object segmentation.

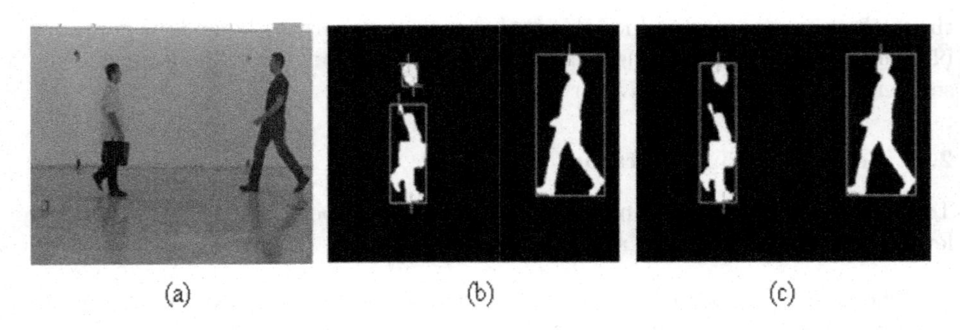

Fig. 3. Example of combination operator use: a) original frame b) the human on the left has been segmented into two very different blobs that have been initially assigned to two different objects c) using a combination operator both blobs have been grouped, and the bounding box has been obtained corresponding to the complete object

This implies passing, as information to the blob level, the reference blob (the one recognised as a head) and a region of interest or ROI where the blobs must be sought that have to be joined to this reference blob. The result of the process is shown in figure 3c.

Case 2

When the background subtraction method is used, the main causes of a background noise are usually associated with those pixels of the moving object whose RGB value is very like that of the background. Indeed, in this instance, the difference of the RGB values of this type of pixels with their background equivalents is not large enough to exceed the threshold and, thus, they are classified as not belonging to the foreground, when in fact they do belong there. Thus, the aim in this example is to use the feedback process to restore those parts of the human which, due to the presence of the background noise, were not detected in the first instance during the segmentation stage and were indeed detected as missing at object level from the human model. Observe that now, when making those parts of the human emerge which were missing, not only do we manage to restore the human silhouette, but also, as in the case analysed before, we are associating all those blobs belonging to the same individual explicitly.

The exchange of information between the different levels can also be described following figure 1. At the object level, using the human model described in secton 2.2 as a reference and the set of blobs resulting from the segmentation process as input (findings), all those blobs that may belong to the human are recognised. Using the human model, for example, the aim is to find whether some block exists with a significant absence of pixels (error diagnosis). If this is the case, this region of the image will be proposed as the region of interest where the missing blobs should be sought (therapy). Another error diagnosis could be the detection of unconnected blobs, which suggests the need to connect them by obtaining new regions belonging to the object (therapy). In either of these two cases, the position of the box and its dimensions are used as information feedback (parameter configuration). Already at blob level, the restoration operator focuses its analysis on the region of interest (ROI) of the image to locate new blobs not detected before. The aim is to make the largest amount of pixels belonging to the human and only to the human emerge which were not detected in the initial segmentation process. Then all the blobs associated with the object are grouped to generate a new set of blobs. These once more pass to the object level, where the diagnosis process will again check whether the degree of restoration of the human silhouette is sufficiently acceptable for the application needs or, on the contrary, whether it is necessary to repeat the cycle. Observe that, although the process description was done based on only one block of the human model with the need for restoration, really there is no limit to the number of boxes that can be refed at lower levels.

Restoration Operator

An important element in the description of all the previous process exposed in case 2 is the restoration operator of the ROIs where it is hypothesised that part of the object must be detected. The characteristics of this operator are based on the truncated cones method [13] for background substraction. Basically, the idea is the following, as can be seen in Figure 4a, for each point p of the background model, a revolution cone can be built using the straight line containing its associated RGB vector, B_t^p, as the axis and another straight line as a generator which, passing through the origin, forms an angle, ω, with the previous straight line. If we now trace three perpendicular planes to the vector B_t^p, one containing the point p (reference plane), and the other two, situated above and below this, at a distance h_1 and h_2, respectively, these planes will delimit, together with the cone surface, two adjacent regions of interest: a truncated cone situated in the upper part of the reference plane and another in the lower part. Since the pixels associated with parts of the object not detected in the first segmentation present RGB values very close to the background RGB values, it is obvious that if sufficiently small h_1, h_2 and ω values are chosen, these parameter values will delimit a region where this type of pixels are contained. However, the disadvantage is that in this region all the pixels belonging to the fluctuation noise are also included. Indeed, the RGB level of the image pixels that do not belong to moving objects is not exactly the same as the RGB level of the background model pixels, but rather the RGB level of the image pixels presents small fluctuations around the RGB level of the background model pixels.

Therefore, the problem is how to appropriately choose the value of the parameters that define the truncated cone region to separate, if it is possible, the two types of pixels mentioned. For this, we unfold each of the three parameters defined before into two to divide the original truncated cone volume into new subregions. Thus, as is indicated in figure 4b, $h_{21}$and h_{22} (together with ω) will make it possible to delimit a truncated cone volume in the upper part of the original volume. Similarly, h_{11} and h_{12}(together with ω) will make it possible to

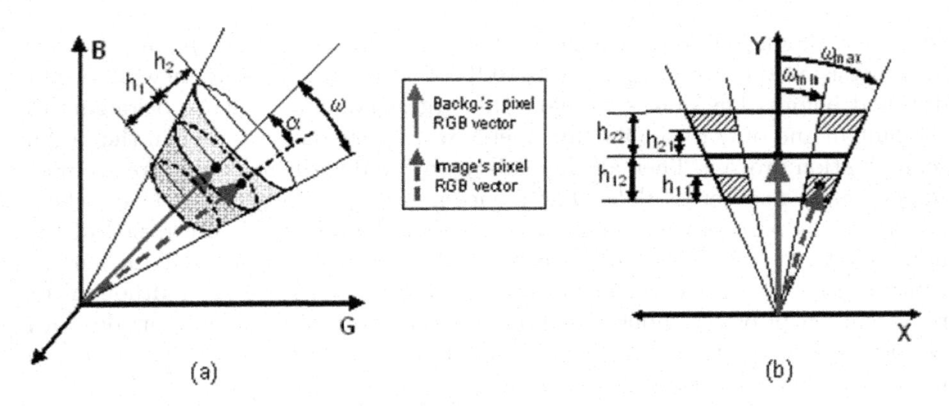

Fig. 4. Background truncated cones associated with a background point (a) in the RGB space and (b) as a projection on the XY plane (the Y axis is made to coincide with the vector RGB of the background point)

delimit another truncated cone volume included in the lower part of the original volume. Analogically, ω_{min} and ω_{max} (together with h_1 and h_2, in figure 4a) will make it possible to delimit a truncated cone crown also included in the original volume. Finally, if we calculate the intersection of these three volumes, we obtain two truncated cone crowns, whose section (striped area) can be seen in figure 4b. The idea is that if the value of these 6 parameters is appropriately chosen, a pixel belonging to a fluctuation noise, by the extreme nearness of its RGB value to the background, will not have much probability of being confined in these two crowns and will have much more probability of being in the remaining original volume. On the other hand, if we admit as a working hypothesis that the number of object pixels that are extremely similar to the background is very low, then we can affirm that if a pixel belongs to the object, there is a high probability that this pixel belongs to one of the two truncated cone crowns defined above.

To tune the value of the six parameters that characterise the operator four probability distributions will be used. On the one hand, in order to estimate the background noise, we will consider how all those pixels are distributed which according to the initial segmentation method do not belong to the foreground of the whole frame under analysis. In the first place, we will calculate this distribution for different values of angle α (see figure 4a), i.e., the angle existing between the vector RGB of a point of the image and the vector RGB equivalent to the background model. Similarly, on the other hand, we will calculate the same distribution, but for the pixels of the ROI that we want to restore (see figure 5a). The two remaining distributions are obtained in the same way but depending on different h values normalised according to the background vector module (see figure 5b).

Comparing the curve slopes in figure 5a enable us to establish that approximately both curve slopes are conserved from 0° to the angle value 0.7^o. This means that most of the ROI pixels that are within this range belong to the

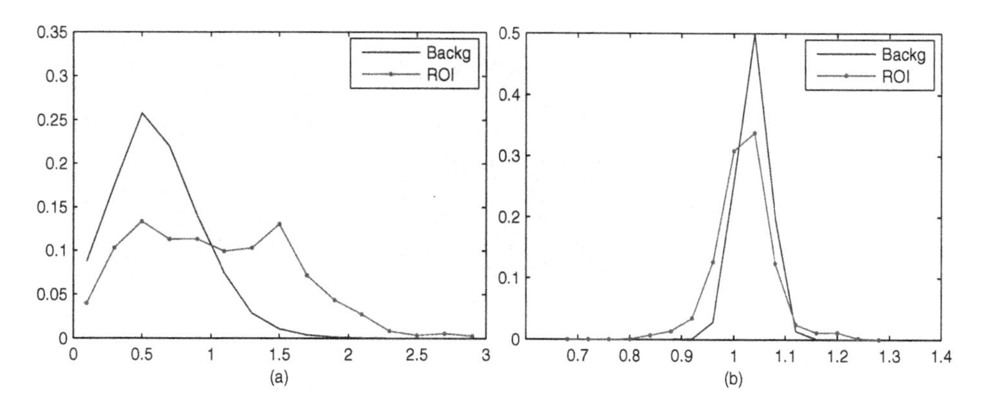

Fig. 5. Probability distributions of the number of points that according to the segmentation stage do not belong to the foreground, (a) depending on the value of angle α and (b) depending on the offset h

fluctuation pixel category. From 0.7^o the trend for both curves is different, thereby indicating the presence of points associated to the object in the ROI. This frontier value enables us to initialise the ω_{min} value. Observe that the ω_{max} value is not critical, a value will simply be chosen that is large enough to guarantee that no object pixel remains outside, for example, $\omega_{max} = 10^o$. The h parameter values are determined by inspecting figure 5b. In this figure it is observed that in the range $(0, 1.05)$, because of the disparity of slopes between the two curves, there is a greater probability of finding an object pixel than a fluctuation pixel in the ROI. Conversely, in the interval $(1.05, 1.09)$, this probability is inverted. Finally, in the interval $(1.09, 2)$ the probability of finding object pixels in the ROI increases once more. In the light of this analysis, for the frame under analysis, $(h_{11}, h_{12}, h_{21}, h_{22}) = (0, 1.05, 1.09, 2)$ will be taken. Observe that the critical h values are now h_{12} and h_{21} because they are the ones that mark the frontier between fluctuation and object pixels. On the other hand, the h_{11} and h_{22} parameters are no longer as critical because they do not mark the frontier limits between both types of pixels, suffice it be to assign them a value small and large enough so as not to leave any object pixel outside.

Finally, figure 6 shows the result of applying all the steps indicated to one of the frames of a scene with a human which presents parts of her body with RGB values very similar to the background and whose distribution curves were those represented in figure 5. Thus, taking the frame indicated in figure 6a as input, a segmentation is done that produces as a result a set of blobs indicated in figure 6b. An analysis at object level of the blobs obtained reveals that there are several boxes associated with the human model with no pixels. The application of the restoration operator to the ROIs associated with these boxes, together with the blobs already existing in the segmentation stage, produces the result shown in figure 6c.

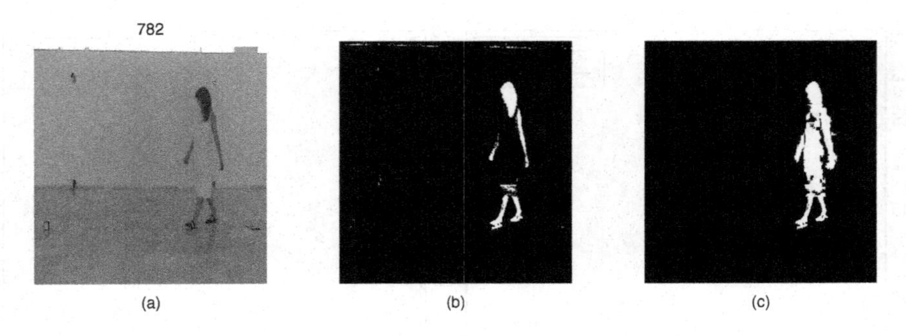

Fig. 6. Result of the ROI restoration process: (a) current frame (b) set of blobs obtained in the segmentation stage (c) set of segmentation blobs after the restoration

4 Conclusions and Future Works

This work presents a knowledge-based segmentation model based on different description levels, which makes it possible to feedback information from the more abstract object level to the blob level to improve the segmentation process results. In order to test the viability of the model, it is instantiated in the two case studies, each of which uses a different resegmentation operator and whose results support the validity of this model.

In future works, the study will focus on developing tasks to diagnose and plan therapy belonging to the object level. Thus, for example, the task of diagnosing implies recognising humans from the human model and also parts of their body. Similarly, at blob level, it will be necessary to refine the already existing operators and develop new operators that make it possible to do the segmentation therapies proposed at object level at this level.

Acknowledgments

The authors would like to thank the CiCYT for financial support via project TIN-2004-07661-C0201 and the UNED project call 2006.

References

[1] T. Horprasert, D. Harwood, L. S. Davis. *A statistical approach for realtime robust background subtraction and shadow detection.* In Proc. of IEEE Frame Rate Workshop, pp 1-19, Kerkyra, Greece, 1999.

[2] C. Stauffer, W. Grimson. *Adaptive background mixture models for real-time tracking.* In Proc. of the IEEE Computer SocietyConference on Computer Vision and Pattern Recognition, pp.246-252, 1999.

[3] A. J. Lipton, H. Fujiyoshi, R. S. Patil. *Moving target classification and tracking from real-time video.* In Proc. of WorkshopApplications of Computer Vision, pages 129-136, 1998.

[4] L. Wang, W. Hu, T. Tan. *Recent developments in human motion analysis.* Pattern Recognition. vol. 36(3), pp. 585-601, March 2003.

[5] E. Y. Kim, S. H. Park. *Automatic video segmentation using genetic algorithms.* Pattern Recognition Letters 27 (11), pp. 1252-1265, 2006.

[6] E. J. Carmona, J. Martínez-Cantos and J. Mira. *Posprocesamiento morfológico adaptativo basado en algoritmos genéticos y orientado a la detección robusta de humanos.* Campus Multidisciplinary in Perception and Intelligence, CMPI-2006, pp. 249-261, Albacete (Spain), July 2006.

[7] J. Martínez-Cantos, E.J. Carmona, A. Fernández-Caballero, M.T. López. *Mejora paramétrica de la interacción lateral en computación acumulativa.* Campus Multidisciplinary in Perception and Intelligence, CMPI-2006, pp. 262-273, Albacete (Spain), July 2006.

[8] R. T. Collins, A. J. Lipton, A. J, T. Kanade. *Special Issue on Video Surveillance.* IEEE Trans. Pattern Anal. Mach. Intell. 22 (8) 2000.

[9] Y. Dedeoglu. *Moving Object Detection, Tracking and Classification for Smart Video Surveillance.* Ph.D. Thesis, 2004.

[10] I. Haritaoglu, D. Harwood, L.S. Davis. *W4: Real-time Surveillance of People and Their Activities.* PAMI, 22(8), pp. 809-830, Aug. 2000.

[11] C. Stauffer, W. Grimson. *Learning Patterns of Activity Using Real-Time Tracking.* IEEE Trans. Pattern Analysis and MachineIntelligence, vol. 22, no. 8, pp. 747-757, 2000.

[12] R. Cucchiara, M. Piccardi, A. Prati. *Detecting Moving Objects, Ghost and Shadows in Video Streams.* IEEE Trans. Pattern Analysis and Machine Intelligence, vol. 25, no. 10, pp. 1337-1342, 2003.

[13] E. J. Carmona, J. Martínez-Cantos and J. Mira. *A new video segmentation method of mobile objects based on blob-level knowledge.* Pattern Recognition Letters. (under review), 2007.

[14] C. Dillon, T. Caelli. *Learning Image Annotation: the CITE system.* Videre , 1, 2, pp. 90-123, 1998.

[15] S. Sing and A. Sowmya. *RAIL: Road Recognition from Aerial Images Using Inductive Learning.* In International Archives of Photogrammetry and Remote Sensing, volume (32) 3/1, pp. 367-378, 1998.

[16] J. Tani. *Model-based learning for mobile robot navigation from the dynamical systems perspective.* IEEE Transactions on Systems, Man and Cybernetics, Part B, pp. 421-436, Vol : 26, Issue: 3, Jun 1996.

[17] H. H. Nagel. *Steps toward a cognitive vision system.* AI Mag. 25(2), pp.31-50,2004.

[18] E. Folgado, M. Rincón, E.J. Carmona, M. Bachiller. *A block-based model for monitoring of human activity.* IEEE trans. on Pattern Analysis and Machine Intelligence. (under review). 2007.

Knowledge-Based Road Traffic Monitoring

Antonio Fernández-Caballero, Francisco J. Gómez, and Juan López-López

Instituto de Investigación en Informática de Albacete (I3A) and
Escuela Politécnica Superior de Albacete
Universidad de Castilla-La Mancha, 02071-Albacete, España
{caballer,fgomez}@dsi.uclm.es

Abstract. This article presents a knowledge-based application to study and analyze traffic behavior on major roads, using as the main surveillance artefact a video camera mounted on a relatively high place with a significant image analysis field. The system described presents something new which is the combination of both traditional traffic monitoring systems, that is, monitoring to get information on different traffic parameters and monitoring to detect accidents automatically. Therefore, we present a system in charge of compiling information on different traffic parameters. It also has a surveillance module, which can detect a wide range of the most significant incidents on a freeway or highway.

1 Introduction

Until a few years ago, surveillance amounted to no more than the presence of policemen on the roads who informed of any road violations. In the last years, there has been a growing interest in the use of automatic mechanisms capable of providing information about the behavior of automobiles on highways and city roads [14] [7] [13]. The most attractive of these, without a doubt, is video image detection (e.g. [6] [8] [9]). A survey of video processing techniques for traffic applications [10] demonstrates the great importance of the topic addressed. Notice that every video camera-based traffic control system is classified into two types: (a) Traffic monitoring, which includes: (1) Monitoring to obtain information on different traffic parameters, such as: number of vehicles per unit of time, vehicle classification, average speed, individual speed of each vehicle, etc. (2) Monitoring to detect accidents automatically, also called AID (Automatic Incident Detection). These focus on finding irregularities, such as: stopped traffic, slow traffic, traffic jams, etc. (b) Traffic control for toll purposes or sanctions. The ones focusing on sanctions control traffic violations and detect vehicles unmistakably. Those focusing on tolls analyze traffic at the toll booths, studying the number of vehicles in each line, average waiting time in line, etc.

This article presents an application which allows a study and analysis of traffic behavior on major roads (more specifically freeways and highways), using as the main surveillance artefact a video camera mounted on a relatively high place (such as a bridge) with a significant image analysis field. It is a traffic monitoring system which allows the gathering of some traffic parameters and the detection of some of the most significant and frequent incidents

J. Mira and J.R. Álvarez (Eds.): IWINAC 2007, Part II, LNCS 4528, pp. 182–191, 2007.

that can take place on a freeway or highway. Our approach is similar to [1] in the sense that image-processing modules extract visual data from the scene by spatio-temporal analysis [11] and high-level modules are designed to work on vehicles and their attributes and to exploit expert knowledge on traffic conditions.

2 Road Traffic Monitoring Architecture

The system is capable of identifying each vehicle that appears in each image and it assigns them the type of vehicle they belong to and the real situation they occupy on the road. Once a vehicle is detected [12] [2], it is followed closely through the images captured by the video camera, so that each vehicle is tracked down from the time it enters the scene until it leaves it. Through this tracking, we get information from that vehicle regarding traffic parameters, such as speed, number of vehicles of each type that have entered the scene, location and speed of each vehicle, as well as the detection of any incidents on the road. Incident detection is carried out by two modules integrated into the system. These modules of the *Road Traffic Monitoring* architecture (see Fig. 1) are called *Static Analysis* and *Dynamic Analysis*, respectively, depending on how they analyze the current image or the comparison of that image with the previous one. Another feature of the application is letting the user determine which incidents he/she wants to control at all times.

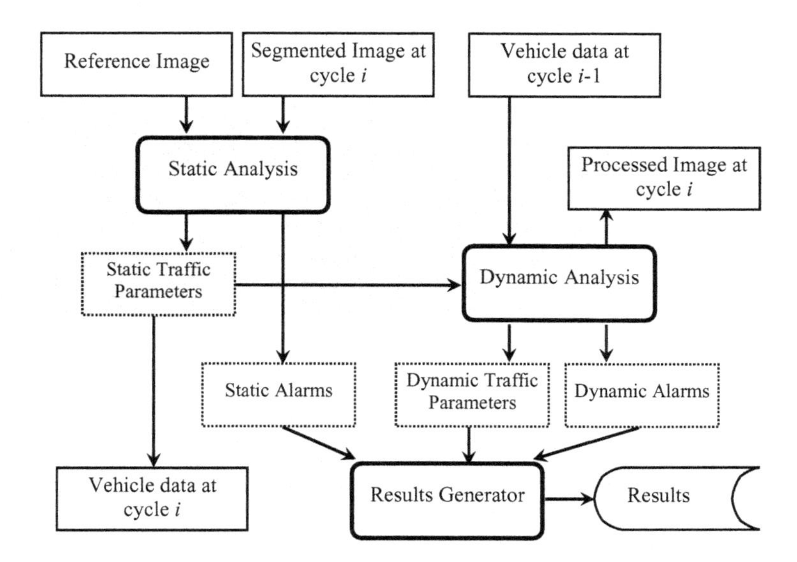

Fig. 1. Road Traffic Monitoring architecture

At the end of the image sequence processing, the system is in the position to offer the following detailed list of parameters:

- Total number of motorcycles, cars, vans and heavy vehicles (trucks and buses) traveling on the road during the processing period.
- Total number of vehicles traveling in the right and left hand lanes and middle lane, if there is one.
- Total number of vehicles traveling in the entrance/exit ramp, if there is one.
- Total number of vehicles traveling on the road.
- Average speed of traveling motorcycles, cars, vans and heavy vehicles.
- Average speed of vehicles traveling in the right and left hand lanes and middle lane, if there is one.
- Average speed of vehicles traveling in the entrance/exit ramp, if there is one.
- Average speed of all vehicles traveling on the road.

The system uses a camera mounted on a bridge which captures real traffic images. These images, which are processed in a 256-color-gray scale format, are segmented according to [3] [4] [5] model in such a way that each object's movement captured in the real image may be observed. The segmented image shows the image's background in black, whereas the motion detected in the vehicles is represented by pixels in different gray scales (see Fig. 2b). A great number of pixels can be seen in gray scales which do not represent moving objects; those pixels constitute the image's noise. This application analyzes the sequence of segmented images, giving rise to a sequence of processed images, where each vehicle is shown through an identifying color which is maintained while the vehicle travels along the area of analysis. The shape of the vehicle is rectangular (see Fig. 2c). This whole process is carried out periodically with a 0.5 second frequency.

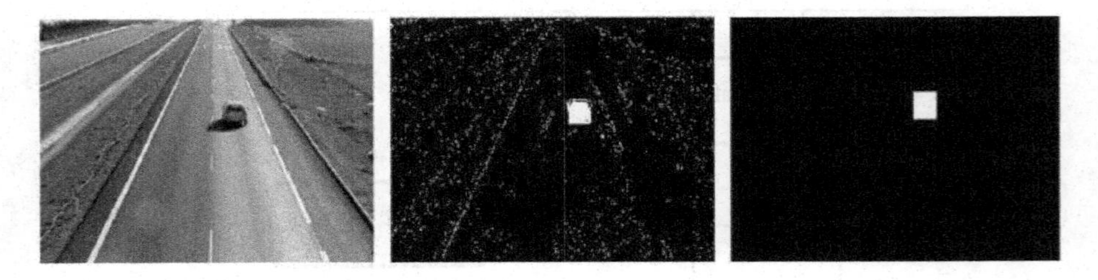

Fig. 2. Road traffic monitoring images. (a) Real image in gray scales. (b) Segmented image. (c) Processed image.

3 Static Analysis for Road Traffic Monitoring

Static Analysis uses "Segmented Image at cycle *i*" and "Reference Image" as inputs and "Static Traffic Parameters" as output. To detect moving objects in the scene, the algorithms described in [5] are performed. Once a moving object has been detected, the dimensions of the object are established, its center is calculated and the coordinates of its center are extrapolated to the "Reference Image", which returns the value of the colors of those coordinates. Those colors are related to the position the object occupies, according to Table 3, and are associated to the "Reference Image" (see example in Fig. 3b). Notice that the coordinates of the center of the object are the lower boundary and the half point between the left and right boundaries. This is done this way so that the center can have better correspondence with the real situation of the vehicle.

Table 1. Colors table

Color	Color (byte)	Area that it represents
White	255	Reserved
Yellow	252	Entrance/exit ramp
Red	224	Shoulder
Dark Gray	146	Right Lane
Light Blue	31	Left Lane
Green	28	Middle Lane
Dark Blue	3	"Out of bounds" (it does not belong to the road)
Black	0	Lines and dividing lines

Fig. 3. Example of Reference Image. (a) Road traffic image. (b) Reference Image. (c) Distance zones.

Thus, through moving object detection, the number of moving objects in the area of analysis at the time is determined, as well as their dimensions, and position on the road. A vehicle classification process determines the objects whose established position is not "Out of bounds". A vehicle which inherits the characteristics of the object is created and the object is eliminated. The vehicle is classified, according to its type. The type of vehicles established as valid vehicles are: "Motorcycle", "Car", "Van" or "Heavy Vehicle" (both trucks and buses are included). If the object does not belong to any type, it is labeled as a "Foreign object". Once the vehicle is classified, an estimate of its minimum and maximum position is calculated in the following image, depending on its minimum or maximum speed, respectively. All this information is called "Static Traffic Parameters" and it goes on to the *Dynamic Analysis* as such and to the following execution cycle under the name of "Vehicle data at cycle i".

To know a vehicle's dimensions and the distance between vehicles, three "Distance Zones" have been defined which indicate three areas where the value of the resolution of each pixel, length and crosswise, is going to be the same for that whole area but different from the other areas. The user indicates this area so that the pixels in each distance zone represent a number of centimeters which is lower than the pixels from an area further away from the camera. An example of distance zones for the "Reference Image" is found in Fig. 3c.

Now, a static surveillance step is capable of controlling three incidents through the study of the object's/vehicle's dimension and its position, as shown next:

- Vehicle traveling along shoulder: The coordinates in the center of the vehicle are extrapolated to the reference image. Depending on the color of the pixel corresponding to those coordinates, the vehicle will be in one area or another. If the pixel is red, the vehicle is traveling on the shoulder or median.
- Restricted traffic: During those dates when traffic is restricted, the traveling of vehicles in the "heavy vehicles" category is controlled. It is necessary for the user to have previously defined the restriction dates. The task "Static Surveillance" is in charge of comparing those dates with the current date and whether any of the objects detected has "heavy vehicle" characteristics or not.
- Foreign objects: The existence of objects detected that do not belong to any valid type of vehicle is checked.

Hence, the information called "Static Alarms" is generated. For each vehicle detected: (a) the position that the vehicle occupies is checked. If it is on the shoulder, a record is kept informing that the vehicle is traveling on the shoulder; (b) the type of vehicle is checked. If it is a heavy vehicle, the date is checked to see if it is traveling on restricted dates; if it is a "Foreign object", a record is kept indicating that an unidentified object remains in the area of analysis.

4 Dynamic Analysis for Road Traffic Monitoring

An image comparison step is the fundamental part of the *Dynamic Analysis*. It is in charge of finding the differences between an image and the previous one. This is done by comparing the information obtained from the *Static Analysis* of the vehicles in the current segmented image, located in "Static Traffic Parameters", and information "Vehicle data at cycle $i-1$", which contains static information from vehicles in the segmented image included in the previous cycle. The first thing that is calculated is the combinatory relationship of each vehicle in the previous image with each vehicle in the current image. This process is based on weight classification for each relationship, according to the following criteria. It is essential for the current vehicle to be further ahead than the other one, in the direction of traffic. If this is so, then:

- The weight established in parameter Advanced Position is established in the total weight of this relationship. This weight is expressed as W_{AP}.

- If they are in the same lane, the weight established in characteristic Same Lane is added to the total weight of the relationship. The weight for this argument is called W_{SL}.
- If both vehicles belong to the same type, the weight established in Same Type is added to the total weight of the relationship. This weight is called W_{ST}.
- If the current vehicle is within the position estimates done for the previous vehicle, and it was within the speed limits, the weight established in parameter Within Speed Limits is added to the total weight of the relationship. This weight is called W_{WSL}.

If the characteristics of any parameter are not fulfilled, that weight is, naturally, not added to the total weight of the relationship. Therefore, the total weight of a combinatory relationship is called TW_{CR} and it is defined as:

$$TW_{CR} = W_{AP} + W_{SL} + W_{ST} + W_{WSL} \tag{1}$$

Once all possible combinatory relationships have been established and the weight of those relationships has been determined, we go on to the final classification between vehicles from the previous and current images. In this process, the combinatory relationships established are compared and a vehicle from the previous image is matched with one from the current image through the selection of the relationship with the greatest possible weight. This is done in such a way that if a match has been established between a previous and a current vehicle and a subsequent relationship containing the same current vehicle with greater weight is found: (a) the previous match is undone, (b) a new match is established for the current vehicle with the previous vehicle from the relationship with the greatest weight, and, (c) another match is tried for the previous vehicle through the search of another relationship which will contain it.

Once this process is completed, the final classifications between vehicles from the previous and current images are obtained. This means that the previous and current vehicles which are matched represent the same real vehicle. There might be previous or current vehicles unable to be matched. This condition can correspond to five different situations, as listed next:

- Current vehicle entering the area of analysis.
- Previous vehicle leaving the area of analysis.
- Previous vehicle which stops on the road after suddenly slowing down.
- Previous vehicle concealed or overlapped by another vehicle.
- Current vehicle coming out of concealment or overlapping or a stopped vehicle that has resumed speed.

When the final matching is done and we know which vehicle from the previous image corresponds to which from the current image, the following operations - related to vehicle tracking - are carried out:

1. A color tag is assigned to the current vehicle, depending on:
   (a) if the current vehicle is matched to a previous vehicle, the tag from the previous vehicle is assigned to the current one.
   (b) if the current vehicle is not matched to a previous vehicle, it is assigned a new color tag.
2. The vehicle is drawn in the processed image.

Afterwards, the speed of the vehicles is calculated and the traffic parameter counters are updated, according to the vehicles that have entered the scene. For every final match, information about the coordinates of the center of the previous and current vehicles of the match is extracted. The distance traveled by the vehicle is calculated, according to the compromise area where it is situated and where it was situated before, taking into account the resolution of the pixels in the distance zones. Since the distance traveled between two images and the actual period of time between those two images are known, the speed at which the vehicle has been traveling during that time is calculated. The result of all these operations are the "Dynamic Traffic Parameters".

Using the previous information, a dynamic surveillance procedure can start for detecting irregularities linked to object motion. This way information "Dynamic Alarms", which contains all the incidents detected in this execution cycle, is created. That information contains the following items:

- Date and time of the incident.
- Type of incident.
- Type of vehicle involved, as well as the assigned tag.
- Place on the road where the incident has taken place.

The incidents detected by this task, as well as the way to control them, are the following ones:

- Traveling too slowly or too fast: If the speed of the current vehicle exceeds the minimum between the highest speed limit for that type of vehicle and the highest limit for the lane where the vehicle is currently traveling in, an incident involving excessive speed is detected, that is, if the speed is below the maximum between the lowest speed limit for that type of vehicle and the lowest limit for the lane the vehicle is currently traveling in, an incident involving a vehicle traveling too slowly is detected.
- Overstepping a solid line: The reference image is checked to see that there is no solid line established between the position of the previous and current vehicles. To do this, we go through the middle row between the centers of both vehicles and from the column of one of them to that of the other. We also check to see if there are three consecutive black pixels.
- Stopped vehicle: A previous vehicle that is not matched to a current one, that is, a vehicle that has disappeared from the scene and whose speed had slowed down or had traveled at a lower speed that the minimum speed limit.
- Traffic congestion: When the number of stopped vehicles is greater that the threshold set up, it is considered a traffic jam.

- Strange movement: An object moves in such a way that its transverse movement is greater than its longitudinal one. The longitudinal movement is the number of rows of pixels which separates the previous from the current vehicle. The transverse movement is the number of columns which separates the two.

To this purpose, first the value of the speed of each current vehicle is extracted. If that speed is less than the minimum corresponding speed, an indication that there has been an incident involving lack of speed is stored in a data structure. If the speed is greater than the maximum corresponding limit, information about an incident involving excessive speed is stored. Later, all the final matches are checked over, and:

- The coordinates from the center of the previous and current vehicle implicated in the match are recorded. The longitudinal distance and the transverse length traveled by the vehicle are calculated. If the vehicle's movement is mainly transverse in the scene, an incident involving a strange movement is indicated.
- The segment from the intermediate line which separates the coordinates from both centers is checked over. If three consecutive black pixels are found, an incident of solid line overstepping is reported.

All previous unmatched vehicles are gone through. If any of them is stopped, an incident involving a stopped vehicle is reported. This means that the counter for the number of stopped vehicles increases by one and if that counter is equal to or greater than the threshold established, information about an incident involving a traffic jam is stored.

5 Data and Results

Firstly, the performance of the application is studied by taking a controlled vehicle traveling at a constant speed of 100 km/hour through the area of analysis (see Fig. 4a). A vehicle belonging to the heavy vehicle type, traveling in the right lane at 100 km/hour, has been detected. The type of vehicle and the lane where it was traveling has been detected correctly. The vehicle has not committed any traffic violations. The next situation is similar to that of the previous example. In this case, the controlled vehicle traveled at a constant speed of 120 km/hour (see Fig. 4b). It is detected in the right lane traveling at 122 km/hour. These results practically agree with reality, since the real speed was 120 km/hour. The deviation in the speed calculation is around 1.6%. As for incidents, the truck is detected to be speeding. The incident is only recorded in the image where it is detected.

In the last study case presented in Fig. 4c, a typical traffic scene with several types of vehicles is analyzed. The weather conditions were partly cloudy with strong winds. Furthermore, the date in which the images were processed is specified as a restricted-traffic date in order to see how the incident detection tool works. The traffic parameters obtained in this set are: four vehicles traveling in the right lane at an average speed of 121 km/hour and only one traveling in the left lane at 129 km/hour. According to the type of vehicle: three vehicles traveled at an average speed of 141 km/hour and two heavy vehicles at an average speed of 96 km/hour. Five vehicles traveled altogether at an average speed of 123 km/hour. Two heavy vehicles are detected on a restricted date. Moreover, the following vehicles were speeding: the heavy vehicle matched with the yellow one in the processed image, which was traveling in the right lane and the cars matched with the red, gray and blue ones in the processed images.

Fig. 4. Some road traffic monitoring scenes

6 Conclusions

The application proposed implements a knowledge-driven system capable of controlling traffic on highways and freeways in one direction of traffic. At maximum, the application controls three lanes of regular traffic (right, middle and left) and an entrance / exit ramp. The images are captured by a video camera mounted on top of a bridge. Through segmentation, this system provides a sequence of images where the movements of the objects in the scene are shown. The position of each object on the road is calculated and the objects are classified according to the categories of valid type of vehicle or to strange object category. Up to this point in the processing, the application can detect the following traffic incidents: presence of a strange object on the road, heavy vehicle traveling on restricted dates and vehicle traveling on the shoulder.

By analyzing the current and previous segmented images, we can determine the movement of each vehicle in the scene. The application has the capability of matching these vehicles in such a way that vehicles in the previous segmented image are matched to those in the current one. Likewise, by analyzing motion, the application can detect the following incidents: a vehicle speeding or driving too slowly, a stopped vehicle, congestion on the road, a vehicle overstepping the solid line and a vehicle's strange movement.

Acknowledgements

This work is supported in part by the Spanish CICYT TIN2004-07661-C02-02 grant and the Junta de Comunidades de Castilla-La Mancha PBI06-0099 grant.

References

1. Cucchiara R., Piccardi M., Mello, P. (2000). Image analysis and rule-based reasoning for a traffic monitoringsystem. IEEE Transactions on Intelligent Transportation Systems, 1 (2), 119-130.
2. Fernández M.A., Fernández-Caballero A., López M.T., Mira J. (2003). Length-Speed Ratio (LSR) as a characteristic for moving elements real-time classification. Real-Time Imaging, 9 (1), 49-59.
3. Fernández-Caballero A., Mira J., Fernández M.A., Lopez M.T. (2001). Segmentation from motion of non-rigid objects by neuronal lateral interaction. Pattern Recognition Letters, 22 (14), 1517-1524.
4. Fernández-Caballero A., Mira J., Delgado A.E., Fernández M.A. (2003). Lateral interaction in accumulative computation: A model for motion detection. Neurocomputing 50, 341-364.
5. Fernández-Caballero, A., Fernández M.A., Mira J., Delgado A.E. (2003). Spatio-temporal shape building from image sequences using lateral interaction in accumulative computation. Pattern Recognition 36 (5), 1131-1142.
6. Gupte S., Masoud O., Martin R.F.K., Papanikiolopoulos N.P. (2002). Detection, and classification of vehicles. IEEE Transactions on Intelligent Transportation Systems, 3 (1), 37-47.
7. Ha D.M., Lee J.M., Kim Y.D. (2004). Neural-edge-based vehicle detection and traffic parameter extraction. Image and Vision Computing, 22 (11), 899-907.
8. Hsieh J.W., Yu S.H., Chen Y.S., Hu W.F. (2006). Automatic traffic surveillance system for vehicle tracking and classification. IEEE Transactions on Intelligent Transportation Systems, 7 (2), 175-187.
9. Ji X., Wei Z., Feng Y. (2006). Effective vehicle detection technique for traffic surveillance systems. Journal of Visual Communication and Image Representation, 17 (3), 647-658.
10. Kastrinaki V., Zervakis M., Kalaitzakis K. (2003). A survey of video processing techniques for traffic applications. Image and Vision Computing, 21 (4), 359-381.
11. López M.T., Fernández-Caballero A., Fernández M.A., Mira J., Delgado A.E. (2006). Visual surveillance by dynamic visual attention method. Pattern Recognition, 39 (11), 2194-2211.
12. Mira J., Delgado A.E., Fernández-Caballero A., Fernández M.A. (2004). Knowledge modelling for the motion detection task: The algorithmic lateral inhibition method. Expert Systems with Applications, 27 (2), 169-185.
13. Rad R., Jamzad M. (2005). Real time classification and tracking of multiple vehicles in highways. Pattern Recognition Letters, 26 (10), 1597-1607.
14. Tai J.C., Tseng S.T., Lin C.P., Song K.T. (2004). Real-time image tracking for automatic traffic monitoring and enforcement applications. Image and Vision Computing, 22 (6), 485-501.

Comparison of Classifiers for Human Activity Recognition*

Óscar Pérez[1], Massimo Piccardi[2], Jesús García[1], and José M. Molina[1]

[1] Departamento de Informática. Universidad Carlos III de Madrid.
Avenida de la Universidad Carlos III, 22 Colmenarejo 28270. Madrid. Spain
oscar.perez.concha@uc3m.es, jgherrer@inf.uc3m.es, molina@ia.uc3m.es
[2] Faculty of Information Technology University of Technology, Sydney PO Box 123.
Broadway NSW 2007
massimo@it.uts.edu.au

Abstract. The human activity recognition in video sequences is a field where many types of classifiers have been used as well as a wide range of input features that feed these classifiers. This work has a double goal. First of all, we extracted the most relevant features for the activity recognition by only utilizing motion features provided by a simple tracker based on the 2D centroid coordinates and the height and width of each person's blob. Second, we present a performance comparison among seven different classifiers (two Hidden Markov Models (HMM), a J.48 tree, two Bayesian classifiers, a classifier based on rules and a Neuro-Fuzzy system). The video sequences under study present four human activities (inactive, active, walking and running) that have been manual labeled previously. The results show that the classifiers reveal different performance according to the number of features employed and the set of classes to sort. Moreover, the basic motion features are not enough to have a complete description of the problem and obtain a good classification.

1 Introduction

The topic of human activity recognition is an open and challenging problem to solve by the research community. Usually, the analysis is carried out by extracting motion features [1] with which to compute the motion patterns to recognise a set of activities. For example, the works [2]-[4] show how to compute and then select the most relevant features for the final classification. There are other works that employ non motion features, like for instance the gait energy image [5] or silhouettes [6].

As it is pointed out in [7], many of the activity recognition works use body components features, like the position of head [8], hands and feet. The reality is that these features may not be found or located in many circumstances.

* Funded by CICYT TEC2005-07186, CAM 15 MADRINET S- 0505/TIC/0255, FOMENTO SINPROB.

J. Mira and J.R. Álvarez (Eds.): IWINAC 2007, Part II, LNCS 4528, pp. 192–201, 2007.

On the other hand, one of the main issues in this kind of works is the definition of the activities. They can be inferred from clusters of the extracted features [9] or manually assigning action labels to the video sequences [10]. In this case, we have to take into account the subjectivity and ill-definition of the manual labelling.

Thus, this work proposes the comparison of different classifiers of activity recognition by using only motion features described and calculated by a sequence of displacements of the 2D centroid and the height and width of each person's blob. This motion features can be provided by a simple tracker with no need of locating other parts of the body like head or hands. The CAVIAR [10] sequences are utilised in this paper to recognize the set of activities corresponding to 'Inactive' (IN), 'Active' (AC), 'Walking' (WK) and 'Running' (R).

The first step was to extract the most relevant motion features for the activity recognition and feeding the classifiers. Our approach consists of using a wrapper method that produces empirical evaluations for a classifier. The wrapper system is based on a machine learning (ML) method and combines a search method with a machine learning algorithm - the search is driven by the performance of the induced concepts [11]-[12]. In our case J.48, and a Genetic Algorithm are the selected ML algorithm and the search method respectively. To carry out this wrapper we employed WEKA 3.5.2 [13].

Subsequently, we decided to select HMM (Hidden Markov Models) since they classify each scene as a function of the future, actual and previous frames and they have been used efficiently for this task in previous works. As it was suggested in [14], the Baum-Welch method and a Genetic Algorithm (GA) are employed to adjust the parameters of the mixtures of Gaussians that in our work define each state of the HMM. In addition, we selected a J.48 tree, two Bayesian classifiers (Bayesian Network and Naive Simple), a classifier based on rules (PART)(all of them included in the WEKA software [13]) and finally a Neuro-Fuzzy classifier [15] so that we have a wide variety of different methods for classification.

This work presents in Section 2 the computation and selection of the motion features which will feed the classifiers. Section 3 displays the parameters to adjust the classifiers, the way in which the experiments are carried out and the main results of this experiments. Then, Section 4 elaborates an analysis of the results and the most important conclusions of our work.

2 Data Base and Extraction of Features

The data base used for our work is the one built for the CAVIAR project [10]. We selected four videos with the criterion of having the maximum number of activities: *Fight_RunAway1.mpg*, *Fight_RunAway2.mpg*, *Fight_OneManDown.mpg* and *Fight_Chase.mpg*. These videos were recorded by the CAVIAR team with a wide angle camera lens in the entrance lobby of the INRIA Labs at Grenoble, France. The sequences have half-resolution PAL standard (384 x 288 pixels, 25 frames per second) and were compressed using MPEG2.

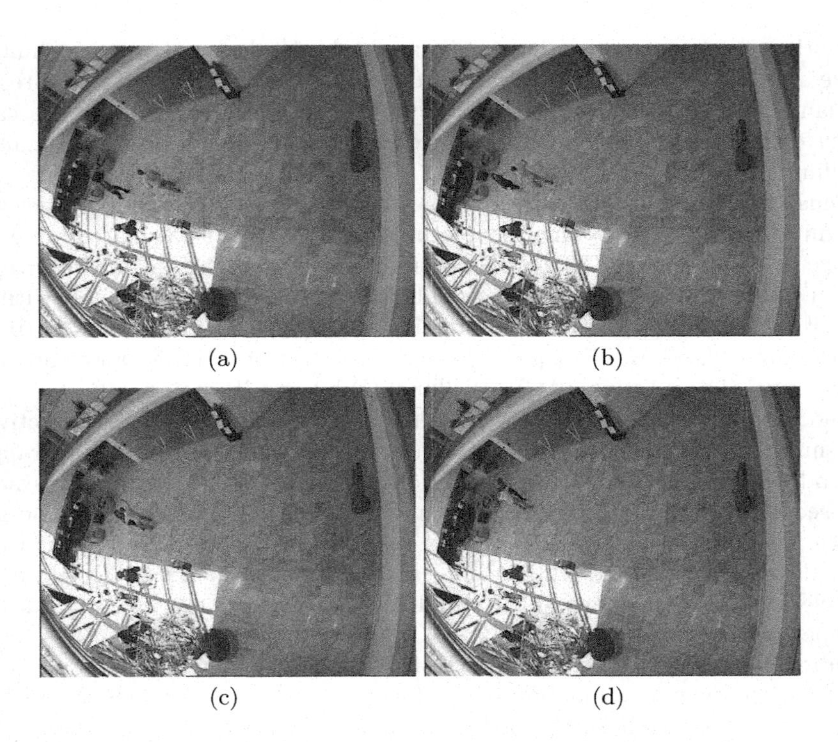

Fig. 1. Frames 277 (a), 289 (b), 302 (c) and 368 (d) extracted from the video sequence *Fight_Chase.mpg*

Among all the measurements provided as ground truth, we selected the basic data that any simple tracker can provide: the 2D centroid position *(x,y)*, height and width of each person's surrounding box *(h,w)*. Then, we compute a set of 40 features that are divided into two groups:

1. Velocity and Speed for a frame window (f) of 3, 5, 15 and 25 frames:
   - velocity for the x-axis and y-axis:

$$v^x_{f,i} = (x_i - x_{i-f}) \tag{1}$$

$$v^y_{f,i} = (y_i - y_{i-f}) \tag{2}$$

   - speed:

$$speed_{f,i} = \frac{1}{f}\sqrt{(x_i - x_{i-f})^2 + (y_i - y_{i-f})^2} \tag{3}$$

   - mean speed:

$$mean\_speed_{f,i} = \frac{1}{f}\sum_{j=0}^{f-1}\sqrt{(x_{i-j} - x_{i-(j+1)})^2 + (y_{i-j} - y_{i-(j+1)})^2} \tag{4}$$

2. Height, Width and Area for a frame window (f) of 3, 5, 15 and 25 frames:
   - difference of height, width and area:

$$diff\_height_{f,i} = (h_i - h_{i-f}) \tag{5}$$

$$diff\_width_{f,i} = (w_i - w_{i-f}) \tag{6}$$

$$diff\_area_{f,i} = |diff\_height_f \cdot diff\_width_f| \tag{7}$$

   - mean difference of height, width and area:

$$mean\_diff\_height_{f,i} = \frac{1}{f}\sum_{j=0}^{f-1}(h_{i-j} - h_{i-(j+1)}) \tag{8}$$

$$mean\_diff\_width_{f,i} = \frac{1}{f}\sum_{j=0}^{f-1}(w_{i-j} - w_{i-(j+1)}) \tag{9}$$

$$mean\_diff\_area_{f,i} = \frac{1}{f}\sum_{j=0}^{f-1}|((h_{i-j} \cdot w_{i-j}) - (h_{i-(j+1)} \cdot w_{i-(j+1)}))| \tag{10}$$

where i is the current frame and i=*1..N* being N the number of samples. The number of samples for the training and validation groups are 4344 and 2679 samples.

Subsequently, the most important of the 40 motion features defined above are selected, discarding the ones not relevant. This selection of features is carried out by means of a wrapper that employs a J.48 and a GA as a classifier and search method respectively with 10 cross fold validation. The outcome of this filter (called *importance weight* from now on) is the number of times that each feature is used in each of the rounds of the 10 cross fold validation. We decided to extract and choose the features like that: a first group with the 13 features above or equal 70% of importance and a more restricted one with the 4 features equal 100% weight. The table 1 show the selected features and their correspondent weight in pertecentage:

The histograms of these features (Figure 2) show the difficulty of this problem due to the subjectivity in the manual process carried out when labeling the activities. The challenging goal of the classifiers is to sort the classes taking these features with such a low level of separability.

3 Experiments

Subsequently, we had to choose the set of the parameters for each of the classifiers.

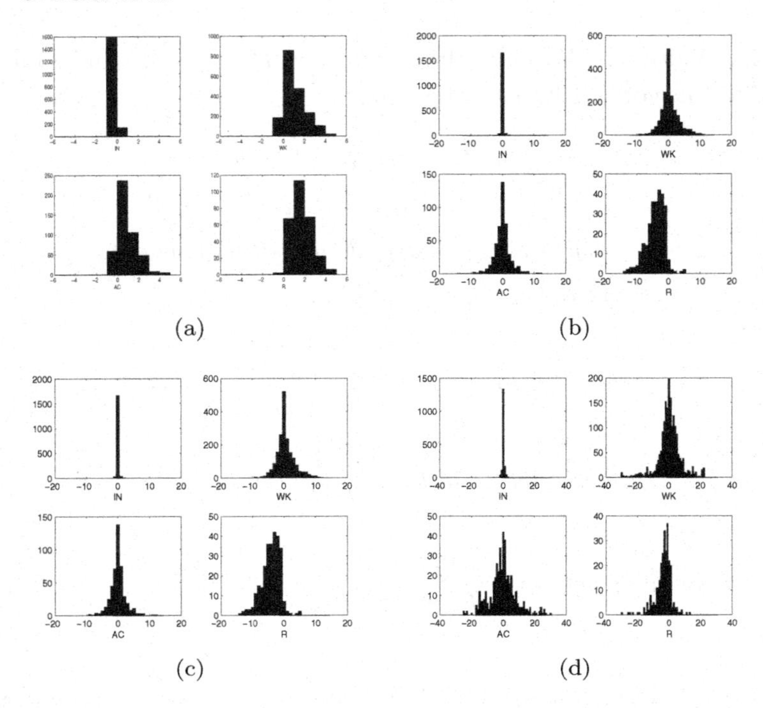

Fig. 2. *Speed_3, i* (a) $v^x_{3,i}$ (b), $v^y_{3,i}$ (c) and $diff\_height_{15,i}$ (d) for inactive (IN), walking (WK), active (AC) and running (R)

The parameters for the different classifiers are adjusted as follows:

- HMM (Baum-Welch):
  - The initial means, $\mu_0 = (0, 1, 2, 7.0)$, for 4 activities and $\mu_0 = (0, 1, 7.0)$ for 3 activities.
  - The terms of the initial covariance matrix $\sum_0$ are randomly selected in an interval of [-2..2]. Moreover, the covariance matrix have to be positive semidefinite.
  - The transition matrix A is initialized randomly.
  - Prior probability: $prior_0$ = (IN=1.0, WK=0.0, AC=0.0, R=0.0).
- HMM (GA):
  - The initial means, μ_0, the initial terms of the covariance matrix, $\sum_0$, the transition matrix A and the initial probability of state, $prior_0$, are chosen randomly. The covariance matrix have the restriction of being positive semidefinite.
  - The typical variables of the GA are initialised as in [14].
- Neuro-fuzzy:
  - Number of variables in each fuzzy set: 2.
  - Form of the set: triangular.
  - Size of the rule base: automatically determine.
  - Rule learning Procedure: best per class
  - Learning Rate: 0.01

  - Maximum Number of epochs: 500
  - Minimum Number of epochs: 0
  - Number of Epochs after optimum: 100
  - Admissible Classification errors: 0
- Rest of classifiers: The default parameters of WEKA.

Table 1. Features selected by the feature selection algorithm: 13 Features above the 70% weight; 4 features equal the 100% weight

$v^x_{3,i}$	100%
$v^y_{3,i}$	100%
$v^y_{5,i}$	80%
$v^x_{15,i}$	100%
$v^x_{25,i}$	80%
$v^y_{25,i}$	90%
$speed_{3,i}$	100%
$speed_{5,i}$	70%
$speed_{15,i}$	90%
$speed_{25,i}$	90%
$diff\_height_{15,i}$	80%
$diff\_height_{25,i}$	70%
$mean\_diff\_area_{25,i}$	90%

Table 2. Results for 13 features and 3 activities (inactive, walking, running)

	HMM (Baum-Welch)	HMM (GA)	J.48	Bayes Net	Naive Bayes	PART	Neuro-Fuzzy
Training	63.90%	70%	99.44%	92%	90.63%	99.47%	87.91%
Validation	48.81%	64%	74.79%	82.82%	80.13%	76.29%	67.74%

Table 3. Results for 4 features and 3 activities (inactive, walking, running)

	HMM (Baum-Welch)	HMM (GA)	J.48	Bayes Net	Naive Bayes	PART	Neuro-Fuzzy
Training	86.49%	89.92%	90.93%	87.00%	88.81%	91.34%	64.79%
Validation	76.89%	88.61%	77.74%	75.76%	79.43%	78.38%	49.40%

Table 4. Results for 13 features and 4 activities (inactive, walking, active, running)

	HMM (Baum-Welch)	HMM (GA)	J.48	Bayes Net	Naive Bayes	PART	Neuro-Fuzzy
Training	54.11%	56.27%	98.87%	83.03%	70.53%	98.71%	52.13%
Validation	32.68%	39.32%	54.56%	50.71%	39.25%	55.68%	26.03%

Table 5. Results for 4 features and 4 activities (inactive, walking, active, running)

	HMM (Baum-Welch)	HMM (GA)	J.48	Bayes Net	Naive Bayes	PART	Neuro-Fuzzy
Training	63.23%	80.57%	85.84%	77.62%	79.30%	86.21%	61.55%
Validation	56.53%	78.31%	60.00%	65.01%	73.15%	62.14%	46.90%

Table 6. Results for 4 features and 3 activities (inactive, walking, running) for the HMM (Baum-Welch) and 76.89% of correct classifications

HMM (Baum-Welch)		Predicted		
Actual		IN	WK	R
	IN	619	14	0
	WK	437	1440	0
	R	1	167	0

Table 7. Results for 4 features and 3 activities (inactive, walking, running) for the HMM (GA) and 88.61% of correct classifications

HMM (GA)		Predicted		
Actual		IN	WK	R
	IN	599	34	0
	WK	95	1774	8
	R	0	168	0

Table 8. Results for 4 features and 3 activities (inactive, walking, running) for the Naive Bayes and 79.43% of correct classifications

Naive Bayes		Predicted		
Actual		IN	WK	R
	IN	572	61	0
	WK	803	1553	119
	R	0	164	0

Table 9. Results for 4 features and 4 activities (inactive, walking, active, running) for the HMM (GA) and 78.31% of correct classifications

HMM (GA)		Predicted			
Actual		IN	WK	AC	R
	IN	600	99	0	0
	WK	190	1497	12	8
	AC	0	170	0	0
	R	0	168	0	0

Table 10. Results for 4 features and 4 activities (inactive, walking, active, running) for the Naive Bayes and 73.15% of correct classifications

Naive Bayes		Predicted			
Actual		IN	WK	AC	R
	IN	576	57	0	0
	WK	202	1377	0	128
	AC	15	155	0	0
	R	2	160	0	0

Then, we divided the data in training and validation data by following the criterion of having a representative number of active and running activities in both groups (since they are the less numerous activities). Thus, the first and the third videos are used as training data, whereas the second and the forth constitute the validation data.

The next step was to employ the training and validation data in the different classifiers. In a first approach, we trained the classifiers for three activities (inactive, walking and running) as a reduced case of the general one and subsequently the training is carried out with the four activities.

Thus, we carried out four rounds for the comparison of performance:

1. 3 activities (IN, WK and R) and 13 features (those above or equal 70% of weight)
2. 3 activities (IN, WK and R) and 4 features (those equal 100% of weight)
3. 4 activities (IN, WK, AC and R) and 13 features (those above or equal 70% of weight)
4. 4 activities (IN, WK, AC and R) and 4 features (those equal 100% of weight)

The next tables (Tables 2-5) show the percentage of correct classifications for each classifier in each trial.

It is interesting to show some of the confusion matrix to check how the classification is carried out. We can see (Tables 6-8) the confusion matrix corresponding to the HMM adjusted with Baum-Welch and GA and the Naive Bayes for the case of 4 features and 3 activities. Furthermore, we included the confusion matrix of the HMM (GA) and Naive Bayes for 4 features and 4 activities (Tables 9-10).

4 Discussion

In this section we analysed the results and subsequently extracted the most relevant conclusions.

4.1 Analysis of the Results

The outcome of the classification of 3 activities shows that the results are very similar for 4 and 13 features in the case of the J.48, PART and the two Bayesian classifiers, being above the 74% of correct classification in the validation set of data. Nevertheless, the classification differs a lot for the HMMs, which present a good performance in the training for 4 features and a poor performance in the case of 13 features. This can be due to the adjustment of the HMM's states by means of a mixture of Gaussians, that is, the more dimensions the more difficult is for the mixture of Gaussian to model the probability landscape of each activity. Moreover, since HMMs are sequential classifiers, they may benefit more from the reduction of the size of the feature set. On the validation set, HMMs (adjusted by GA) outperform all the other classifiers in the low dimensional case.

On the other hand, the Neuro-fuzzy classifier has the opposite behaviour and it is able to build a more robust classifier by taking a high number of features.

In the case of 4 features, the performance for all classifiers is lower than in the previous case. The reason why this happens is that it is very difficult to distinguish between active and walking with the only information of centroid and surrounding boxes. On the validation set, the relative performance of the classifiers is similar to that with three classes. We can highlight the good working of the Naive Bayes and HMM (GA), with 73.15% and 78.31%.

4.2 Conclusions

We can infer very interesting conclusions by analysing the confusion matrix for each classifier.

For example, the good result obtained with the HMMs (Baum-Welch and GA) in the case of 4 features and only 3 activities (Tables 6-7) informs that the running case is ignored, and the total performance is still good since the running samples are much less than the remaining classes. In general, this phenomenon happened for all the classifiers as we can observe in the tables 6-10.

In order to avoid this, we carried out several experiments where the training data had the same number of samples for each activity and surprisingly the results did not improve. This can be explained as the difficulty in distinguishing the activities due to the ill definition of the labels in the ground truth. Thus, the good results are partly obtained for the good classification of the most common class.

This can be proven by considering the confusion matrix for 4 activities (Tables 9-10). We can check that the active (AC) and running (R) activities are misclassified almost always.

In general, the results show a good performance (above the 74% of correct classification) for all classifiers in the case of using only three clearly different classes: IN, WK and R (Tables 2-3). In the case of utilising the four activities (Tables 4-5), the classifiers present a low performance, except for the HMM which states have been adjusted by means of a Genetic Algorithm (GA) and the Naive Bayes classifier. The only exception is the Neuro-fuzzy classifier which can not exceed the 68% of correct classification in the best of the cases.

Thus, we can inferred that the HMM (GA) and the Naive Bayes are good classifiers even under not very separable classes. As a general rule, the J.48, the Bayesian classifiers and PART are good classifiers for this task, whereas Neuro-fuzzy does need further analysis and research for improvement.

References

1. J.C. Nascimento,M. A. T. Figueiredo and J. S. Marques: Segmentation and Classification of Human Activities, HAREM 2005: International Workshop on Human Activity Recognition and Modelling, Oxford, UK, September 2005
2. Pedro Canotilho Ribeiro and José Santos-Victor. Human Activity Recognition from Video: modeling, feature, selection and classification arquitecture, HAREM 2005: International Workshop on Human Activity Recognition and Modelling, Oxford, UK, September 2005

3. Sebastian Brännström. Extraction, Evaluation and Selection of Motion Features for Human Activity Recognition Purposes, Master's Thesis in Computer Science at the School of Engineering Physics Royal Institute of Technology 2006
4. Osama Masoud and Nikos Papanikolopoulos. A Method For Human Action Recognition, Department of Computer Science and Engineering University of Minnesota 2003
5. Xiaotao Zou and Bir Bhanu. Human Activity Classification Based on Gait Energy Image and Coevolutionary Genetic Programming Pattern Recognition, 2006. ICPR 2006. 18th International Conference on Volume 3, 20-24 Aug. 2006 Page(s):556 - 559
6. Mohiuddin Ahmad and Seong-Whan Lee HMM-based Human Action Recognition Using Multiview Image Sequences, Pattern Recognition, 2006. ICPR 2006. 18th International Conference on Volume 1, 20-24 Aug. 2006 Page(s):263 - 266 .
7. M. Leo, T. D'Orazio, I. Gnoni, P. Spagnolo and A. Distante: Complex Human Activity Recognition for Monitoring Wide Outdoor Environments, Pattern Recognition, 2004. ICPR 2004. Proceedings of the 17th International Conference on Volume 4, 23-26 Aug. 2004 Page(s):913 - 916 Vol.4
8. Paul E. Rybski and Manuela M. Veloso: Human Activity Recognition from Video: modeling, feature selection and classification architecture, HAREM 2005 - International Workshop on Human Activity Recognition and Modeling. Oxford, UK, September 2005
9. Zhou Feng and Tat-Jen Cham. Video-based Human Action Classi.cation with Ambiguous Correspondences, Computer Vision and Pattern Recognition, 2005 IEEE Computer Society. 20-26 June 2005 Volume: 3, On page(s): 82- 82
10. "CAVIAR PROJECT", http://homepages.inf.ed.ac.uk/rbf/CAVIAR
11. D.M. Santoro, E.R. Hruschska Jr and M. do Carmo Nicoletti. Selecting feature subsets for inducing classifiers using a committee of heterogeneous methods, Systems, Man and Cybernetics, 2005 IEEE International Conference on Volume 1, 10-12 Oct. 2005 Page(s):375 - 380 Vol. 1
12. Ron Kohavi and Dan Sommerfield. Feature Subset Selection Using the Wrapper Method: Overfitting and Dynamic Search Space Topology, First International Conference on Knowledge Discovery and Data Mining (KDD-95).
13. "WEKA", http://www.cs.waikato.ac.nz/ml/weka/
14. Óscar Pérez, Massimo Piccardi, Jesús García, M.A. Patricio, J.M. Molina. Comparison between Genetic Algorithms and the Baum-Welch Algorithm in Learning HMMs for Human Activity Classification Proceedings of EvoIASP2007: Ninth European Workshop on Evolutionary Computation in Image Analysis and Signal Processing. Valencia, Spain, 11-13 April 2007
15. "NEFCLASS", http://fuzzy.cs.uni-magdeburg.de/nefclass/

A Multi-robot Surveillance System Simulated in Gazebo

E. Folgado, M. Rincón, J.R. Álvarez, and J. Mira

Dpto. de Inteligencia Artificial. ETSI Informatica. UNED.
Juan del Rosal 16, 28040 Madrid, Spain
{efolgado, mrincon, jras, jmira}@dia.uned.es

Abstract. A special kind of surveillance problem is the monitoring of wide enclosed areas with difficult access and changing environments. The characteristics of this kind of problem recommend a solution based on a multi-agent system, where several robots cooperate to solve the problem. A multi-robot system for surveillance in these kinds of environments has been designed and simulated on the Gazebo 3D simulator. Typical surveillance tasks are simulated and experimental results are shown.

1 Introduction

Nowadays it is very usual for public and private institutions to have sophisticated surveillance systems to monitor the state of their "business" in order to detect anomalous situations and avoid adverse or undesirable situations. A special kind of surveillance problem is surveillance of wide enclosed areas with difficult access and changing environments. The characteristics of this kind of problem require a solution based on a multi-agent system, where several robots cooperate to solve the problem.

Robots in these multi-agent systems are generally simple in design and control, homogeneous and low cost. They are well suited for distributed problem solving, where just one robot is not enough or is very costly in terms of time and design. However, when more than one robot moves in a common workspace, each robot becomes a mobile obstacle for the others. For this reason, the robots need to be coordinated.

Numerous surveillance projects currently exist which use robots in different environments. Some of these projects simply focus on the navigation of robots within enclosed environments, while others go further and add recognition tasks in determined situations. The AUVs project [1] aims to perform underwater surveillance tasks, primarily for oceanographic issues. The basic problem that arises when performing autonomous navigation is to avoid obstacles in the water. To exchange information on the position and state of the robots, distributed intelligence has been used [2]. NRaD and MSSM [3] is a project which uses a robot vehicle called "cypher" to perform surveillance tasks by remote control plane. It's equipped with visible light and infrared cameras and laser and audio sensors. Cyberguard [4] is a mobile robot for security, whose aim is to patrol

J. Mira and J.R. Álvarez (Eds.): IWINAC 2007, Part II, LNCS 4528, pp. 202–211, 2007.

enclosed areas in order to detect intruders, fires, etc. The robot has a large number of sensors that allow it to navigate and detect events of interest. Millibot is a project on recognition and surveillance tasks, whose aim is to construct distributed teams of robots between 5 and 10 cm, with cameras, thermometers and movement sensors. One of the main lines, in addition to the design and mobility of robots, is the use of coordination architectures to control the location of all the team, construct maps and detect objects of interest.

Some of the video-surveillance projects are: VSAM [5], which studies a group of video analysis techniques for surveillance applications in an urban environment and indicates incidents detected by a large number of sensors distributed at different points to an operator; CAVIAR [6], which studies video analysis techniques, in order to improve the performance of surveillance systems with applications in urban environments and shopping centres; and a large number of projects related to traffic surveillance, among others, [7][8].

The common feature of all these works is cooperation between the agents participating in the surveillance task. Some of them require an operator to control them, while others, which do not require an operator, inform him. The present design does not require any operator, the robots move autonomously and warn a central alarm system of an anomalous situation. When a robot detects an anomalous situation, it focuses the camera on it and warns the other robots of its position so that they go to the point where the alarm has occurred.

In this paper, typical surveillance tasks are simulated on the Gazebo 3D simulator (http://playerstage.sourceforge.net). This makes it possible to check the behaviour of the robots in different enclosed areas in quite a real way, since its configuration can be changed so that it can act on simulated or real robots. With this purpose in mind, the Gazebo simulator was used on Pioneer3AT robot.

The rest of the paper is structured as follows. The second section describes the multi-robot surveillance system proposed. The third section gives the details of its simulation on the Gazebo environment. The fourth section evaluates the system. Finally, the fifth section presents the conclusions and further work.

2 Description of the Multi-robot Surveillance System

The surveillance system proposed (figure 1) consists of a number of robots that can be configured by the user and with a maximum of 10, due to the limitations of the Gazebo simulator. The robots will set off from an initial position and, from there, will wander around the enclosed area and watch it by analysing the information from their cameras. In order to simulate the intruder in Gazebo another robot has been introduced, which will wander around the enclosed area avoiding the obstacles.

If something moves and it is not one of the surveillance robots, then an alarm situation will be detected. In this case, an alarm signal will be sent to the central alarm system (operator) and to the other robots, indicating the position and, consequently, where they must go.

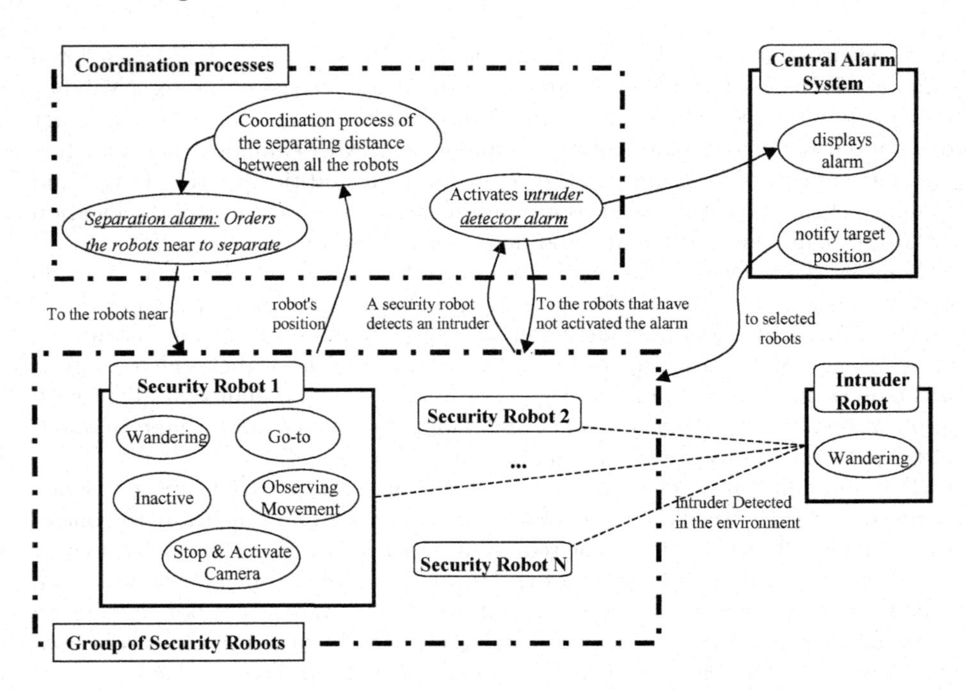

Fig. 1. Multi-agent Surveillance System Architecture

The robots move randomly, so to avoid collisions it is necessary to test the separation between them. Accordingly, a coordination process has been created, which will be in charge of evaluating the distance between robots at time intervals (10 s). If a robot is too near to one another (1.5m), the coordination process will send an alarm to the robots nearby, which will stop and change direction 90° left or right in order to move away from each other.

Usually, the robots will wander around the defined space, stopping at set intervals (every 20 s). When the robots are stationary, the cameras will activate and any movement will be detected. It is assumed that a stationary robot must not detect any movement unless something around him moves. Currently, to detect movement, a change in the image segmentation captured by the camera is detected. If movement is detected, the robot will send signals to the other robots to ensure that it is not one of them. If it is an intruder, the robot that has detected it will warn the other security robots to go to him. At the same time, it sends a warning to the central alarm system to warn the operator.

Once the robots reach their destination they stop and focus on the intruder. The maximum separating distance allowed between a security robot and the intruder is 2 metres. At this distance, in the simulator, the camera has range enough for the robots to focus on the intruder. In order to focus on him, when the security robot detects the intruder it sends the other robots a parameter with its global coordinates and orientation. This position allows the other robots when

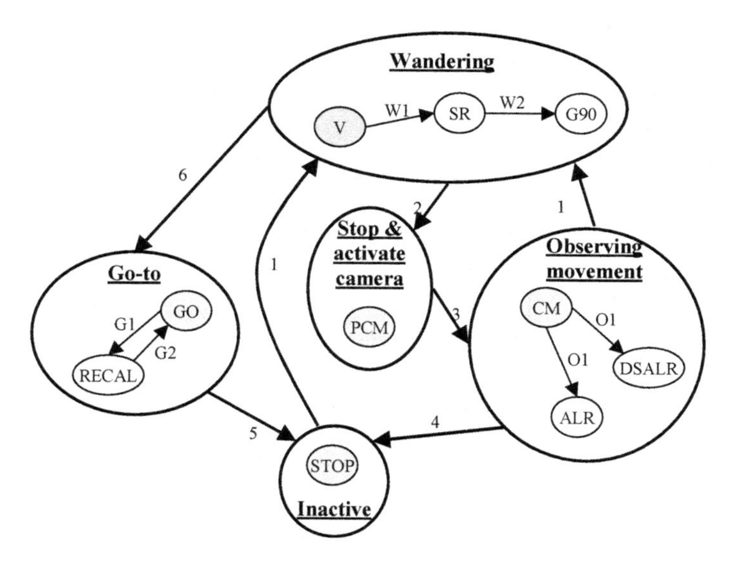

Fig. 2. Transition diagram of robot's behaviour modes

they arrive at their destination to stop and make the necessary turns to be in the right direction to point their cameras.

In this architecture, the activities of the security robots has been organized in behaviour modes which may consist of various parallel processes. The transition diagram shown in figure 2 describes the relations between behaviour modes and processes. Tables 2 and 1 describe its corresponding internal processes and transition signals.

3 System Simulation

For the system simulation the Player/Stage/Gazebo (PSG) [9] [10], platform was used, which was created at the University of South California, in a research study on systems with numerous robots. The aim was to create a tool to program, interconnect and even simulate this kind of systems. PSG consists of the Stage (2D) and Gazebo (3D) simulators and the Player server, to which the application program is connected to collect the sensory data or send orders to the actuators. The support to a wide variety of robots and the simulators that it incorporates, make it a very complete and extensively used platform.

PSG has a client/server design: the applications establish a dialogue by TCP/ IP with the Player server, which is responsible for providing sensory readings and implementing the action commands. As well as permitting remote access, this design provides PSG applications, great independence of language and minimum architecture impositions. Stage supports enormous colonies of cooperating robots in 2D simulations, while Gazebo can manage up to 10 robots in 3D simulations.

The configuration used for the system implementation is shown in Figure 3. The robot chosen is Pioneer 3AT. The surveillance system is going to be

Table 1. Robot's behaviour modes and internal processes

Behaviour mode	Process	Description
Wandering	V	Wandering around the enclosed area
	SR	It checks the distance to the other robots
	G90	It turns 90° left or ritht in order to move away from another near robot
Stop & Activate the camera	PCM	It stops and activates the camera readings
Observing movemen	CM	Stationary observing the possible movement in the image
	ALR	Activate the warning alarm with the position for the other robots, when an intruder has been found
	DSALR	Deactivate all the approach alarms
Go-to	GO	Going towards the target position
	RECAL	Recalculate Target Position
Inactive	STOP	Stop robot movement if required and deactivate all its sensors

Table 2. Description of the Signals between robot's processes

Signals	Description
1	Activate wandering around the enclosed area
2	Stop & activate camera
3	Camera control activated
4	Stop observing movement
5	Stop robot movement because it has found the target position
6	Signal received from the central alarm system to send the robot to a specific position
W1	Near robot alarm
W2	Turn
G1	Reached position is not the intruder's one
G2	The recalculation indicates a new target position (odometric coordenates)
O1	Intruder detected alarm

controlled by just one client program, capable of managing different robots concurrently. Thus, it is thus possible to program as if we were doing it with one robot. Communication between the different robots will be via the "Player" program, provided by the simulator, which we will access from our client program.

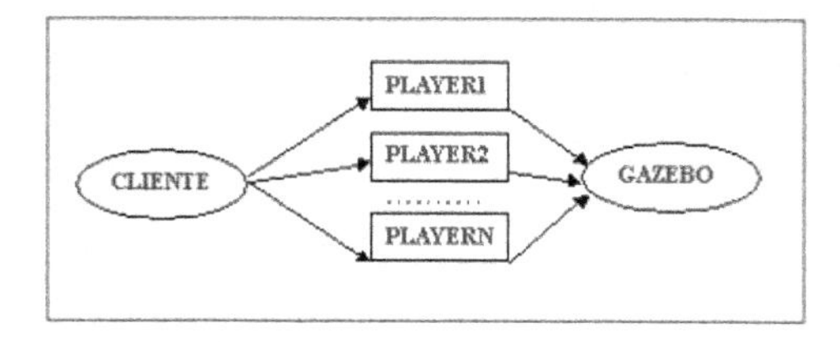

Fig. 3. Diagram of communication between the different elements

For the navigation, the security robots use the VFH (*Vector Field Histogram*) method, for which a driver in Gazebo already exists. The intruder robot, on the other hand, does not use this method but simply wanders around the enclosed area using its sensors to avoid obstacles.

To detect movement the change in image segmentation is analysed. Specifically, access to the "`BlobfinderProxy`" class of the Gazebo "`cmvision`" driver is used , which segments the colours predefined in the "`colors.txt`" configuration file in the image.

Recalculation Algorithm

Each time the robot stops, the local reference system is updated, so that the odometric coordinate origin (0,0) becomes this new position. This means that the odometric coordinate origin changes each time the robot stops. Its calculation is based on translations and rotations and the simulator driver that operates it is: "position".

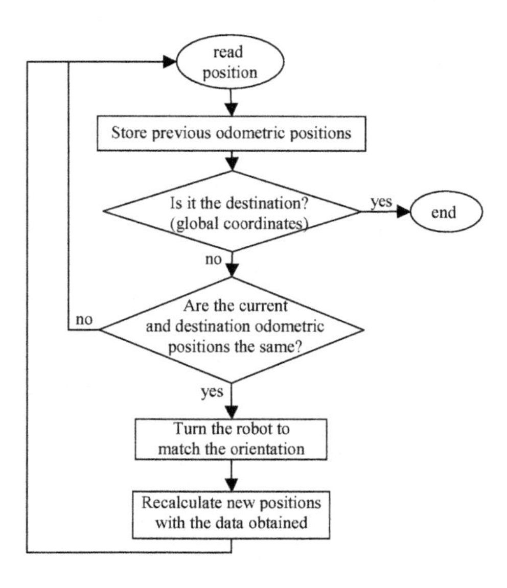

Fig. 4. Recalculation algorithm functioning scheme

There is a problem, known as "dead-reckoning", which is due to location errors that accumulate during navigation because of robot rotations and translations. To solve it, a recalculation algorithm has been done, which uses the global coordinates on the map of the enclosed area to recalculate the target position in terms of odometric coordinates. In a real system, the global coordinates on the map of the enclosed area can be provided by a global positioning device like, for example, a GPS installed in each robot, landmarks settled in the environment, etc. This global position is simulated via the simulator driver: "`truth`".

This algorithm recalculates the target position when a robot thinks that it has reached its destination based on odometric coordinates. Figure 4 shows the recalculation algorithm.

4 Evaluation

This section presents the results of different tests done to check and evaluate the system's functionality. Navigation tests were done first, followed by intruder detection tests and lastly, recalculation algorithm evaluation tests. All the tests were done on the same scenario.

4.1 Test 1: Navigation

Figure 5 shows the initial situation and the path followed by two robots wandering around the enclosed area. It is checked that robots wander normally and avoid the obstacles. Also, the coordination process calculates the distance between all the robots at 10-second intervals and a robot turns 90° if it is near to another. Every 20 s, the robot is stationary to observe the environment and warn of the possible movement of an intruder robot.

4.2 Test 2: Camera Reading and Intruder Detection

This test tries to observe the response of each robot to the camera readings. Movement is detected via the change in the image segmentation captured by the camera.

Once one robot has observed movement, it checks that it is not one of the other security robots by exchanging information on their positions. When one of the robots detects an intruder, it remains stationary and informs the other security robots of the global position where it is (global coordinates and orientation). At the same time it sends a message to inform the central alarm system.

From this moment, the robots must go to the robot that has sounded the alarm. Figure 6 shows the paths of the three robots. The robot labelled number 1 is the one that has found the intruder and focuses on it with its camera. It sends messages to the other robots to change their path to go there. Points A and B on the figure mark the points where robots 2 and 3 received the signal from robot 1 to go to the position of robot 1.

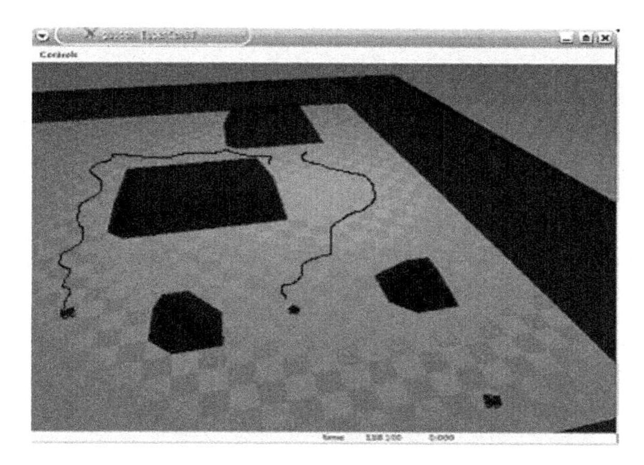

Fig. 5. Two robots wander around the enclosed area while their cameras capture information on the environment

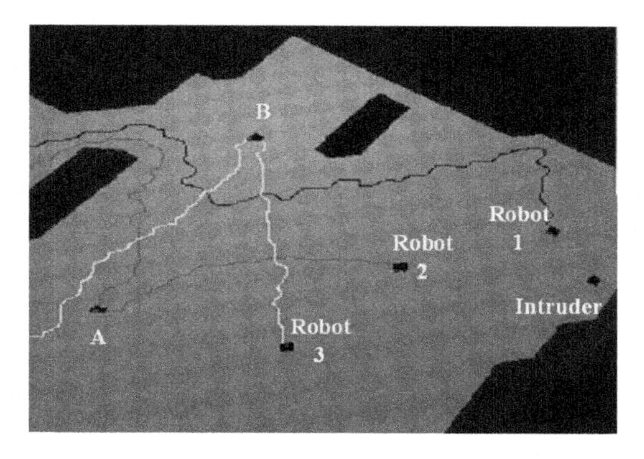

Fig. 6. Tracing of the routes of the robots when they go to the position where the intruder has been detected by robot 1

4.3 Test 3: Warning from the Central Alarm System to Go to the Intruder

In this case, the operator in charge of watching the enclosed area, has detected something or someone suspicious. In this situation, he decides to send the security robots to the suspected area. Accordingly, the operator notifies the robots chosen, using the warning program called "central alarm system", which move towards the point where the suspicious situation has occurred. It is assumed that the operator has a scale map to select the appropriate coordinates. Therefore, in this case, it is the central alarm system that communicates with the robots via the controller. A specific instance of this is that the central alarm system warns all the robots to go to the exit in case of danger.

4.4 Evaluation of the Recalculation Algorithm

Figure 7 compares the dead-reckoning error before and after introducing the recalculation algorithm. The error points are calculated as the difference in metres between the target position and the real position when target is located at different distances. The line shows the mean value. An improvement is clearly seen from the recalculation algorithm, since it limits the error to 2 m.

Finally, figure 8 shows how the time to find the intruder increases with distance due to the cost of the recalculation algorithm.

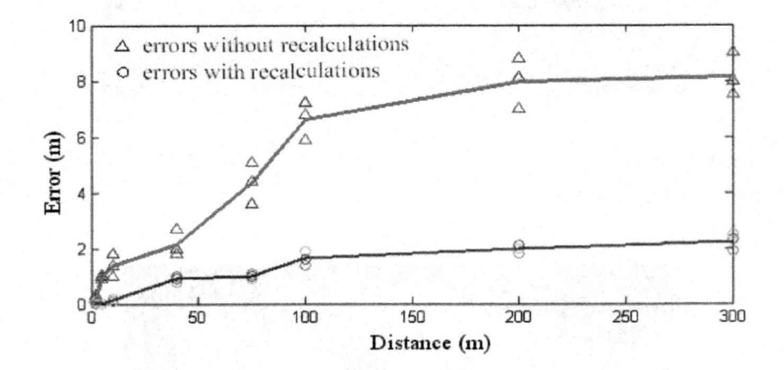

Fig. 7. Error made with and without the recalculation algorithm when trying to reach a target position depending on the distance. The points correspond to the different tests done for the statistical evaluation.

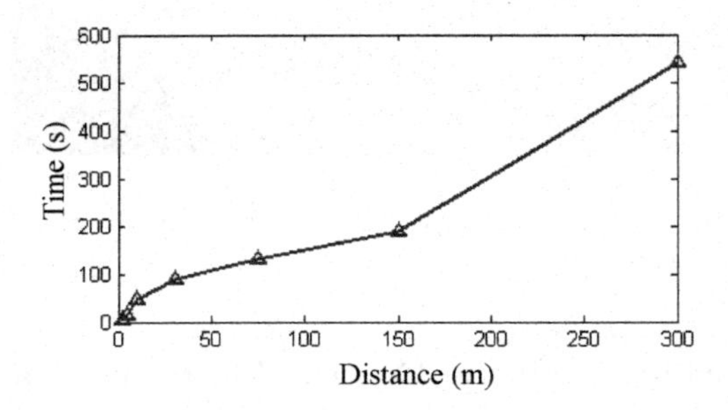

Fig. 8. Time that the robot takes to reach a target position depending on the distance

5 Conclusions

In this paper, a multi-robot system for performing surveillance tasks within a predetermined wide enclosed area has been defined. Surveillance tasks are simulated on the Gazebo 3D simulator and experimental results are shown. The surveillance system allows to control up to 10 robots and it is composed of global coordination processes and security robots, where processes are organized by behaviour modes. Intruder detection is based on movement detection from video signals captured by a robot's camera.

The proposed system is a basic surveillance system, our goal here has been to evaluate the use of Gazebo to simulate multi-robot systems for surveillance applications. As further work, we plan to improve coordination system efficiency by means of more efficient navigation algorithms. Also, the intruder detection system can be improved by allowing more complex situation management, i.e., allowing intruder detection while the security robot is wandering (not still) or recognizing specific moving objects as no intruders.

Acknowledgments

The authors would like to thank the CiCYT for financial support via project TIN-2004-07661-C0201.

References

[1] J.G Bellingham. New oceanographic uses of autonomous underwater vehicles. Marine Technology Society Journal, 31:34-47, 1997.

[2] J.H. Kim, B.A. Moran, J.J. Leonard, J.G. Bellingham, and S.T. Tuohy. Experiments in remote monitoring and control of autonomous underwater vehicles. In Oceans'96 MTS/IEEE, Ft. Lauderdale, Florida, 1996.

[3] D.W. Murphy, J.P. Bott, W.D. Bryan, J.L. Coleman, D.W. Cage,H.G. Nguyen,and M.P. Cheatham. MSSMP: No place to hide. In Association for Unmanned Vehicle Systems International 1997 Conference (AUVSI'97), Baltimore, MD, June 1997.

[4] H.R. Everett R.T. Laird, G.A. Gilbreath, R.S Inderieden, and K.J.Grant. Early user appraisal of the mdars interior robot. In American Nuclear Society, 8th International Topical Meeting on Robotics and Remote Sensing(ANS'99), Pittsburgh, PA, April 1999.

[5] R. Collins, A. Lipton, T. Kanade, H. Fujiyoshi, D. Duggins, Y. Tsin,D.Tolliver, N. Enomoto, and O. Hasegawa. A system for video surveillance and monitoring. Technical Report CMU-RI-TR-00-12, Robotics Institute, Carnegie Mellon University,Pittsburgh, PA, May2000.

[6] Pedro M. Jorge, Arnaldo J. Abrantes, Jorge S. Marques, Estimation of the Bayesian Network Architecture for Object Tracking in Video Sequences. ICPR,Cambridge, August 2004.

[7] T. Huang and S. Russell. Object identification: A bayesian analysis with application to track surveillance. Artificial Intelligence, 103:117,1998.

[8] H. Pasula, S. Russell, M. Ostland, and Y. Ritov. Tracking many objects with many sensors. In IJCAI'99, August 1999.

[9] Gerkey, Brian P., Richard T. Vaughan and Andrew Howard (2003). The Player/Stage project: tools for multi-robot and distributed sensor systems. In: Proceedings of the 11th International Conference on Advanced Robotics ICAR'2003. Coimbra (Portugal).

[10] Nathan Koenig, Andrew Howard. Design and Use Paradigms for Gazebo, An Open-Source Multi-Robot Simulator. Robotics Research Labs, University of Southern California Los Angeles, CA 90089-0721, USA.

Context Data to Improve Association in Visual Tracking Systems*

A.M. Sánchez, M.A. Patricio, J. García, and J.M. Molina

Universidad Carlos III de Madrid
Computer Science Department
Applied Artificial Intelligence Group
Avda. Universidad Carlos III 22, 28270 Colmenarejo (Madrid)
{amsmonte,mpatrici,jgherrer}@inf.uc3m.com,molina@ia.uc3m.es

Abstract. A key aspect in visual surveillance systems is robust movement segmentation, which is still a difficult and unresolved problem. In this paper, we propose an architecture based on a two-layer image-processing modules: General Tracking Layer (GTL) and Context Layer (CL). GTL describe a generic multipurpose tracking process for video-surveillance systems. CL is designed as a symbolic reasoning system that manages the symbolic interface data between GTL modules in order to asses a specific situation and take the appropriate decision about visual data association. Our architecture has been used to improve the association process of a tracking system and tested in two different scenarios to show the advantages in improved performance and output continuity.

1 Introduction

One of the most widely explored topics in computer vision is the development of tracking systems based on video sequences [1]. Generically, we can say this process follows the structure of a Multi-Target Tracking application (MTT): extracting detections, gating detections with existing tracks, matching detections to tracks and update them, and tracks initialization.

However, the particular conditions for operating this type of systems require from quite specialized solutions. They must handle complex situations and interactions such as passing, occlusions and often with sudden and extraneous manoeuvres. Each real object may generate multiple observations and collide with other projected objects, so the generation of a continuously reliable and stable output requires a deep effort in design and adjustment. The tracking algorithms are usually the most flexible and parametrical part of vision systems, and practically all of them exploit external information to model the objects and their context. The configuration is done aiming at a trade-off between computational resources, needs of robustness and behavior.

In spite of the complex situation mentioned above, an expected feature of these systems is the capability to track and maintain identity (ID) of all detected

* Funded by CICYT TEC2005-07186, CAM 15 MADRINET S- 0505/TIC/0255 and IMSERSO AUTOPIA.

J. Mira and J.R. Álvarez (Eds.): IWINAC 2007, Part II, LNCS 4528, pp. 212–221, 2007.

objects over time, even if their detected images split, get occluded or merge with other targets moving within the Field of View (FoV) of the camera [2]. For instance, in order to recognize behaviours, the detection of moving objects is not enough. An analysis of trajectory and interaction with the scene context (other objects and elements) is necessary to infer the behaviour type [3].

An additional problem of shadows is usually present too [4], since large shadows of moving objects are also detected and produce important difficulties in the segmentation process. Several attempts to remove shadows from objects boundaries have been proposed, such as an adapted morphological processing [4].

The performance of a multi-target video tracking system critically depends on the data association method used for assigning detected regions (blobs) to the tracks. Only in the case that a single target appears, with no clutter, there is no association problem. However, in the general case it is necessary to determine which observations correspond to which objects before updating their tracks. Although visual tracking has been extensively studied, most works have assumed that the motion correspondence problem has been previously solved in the image segmentation phase or it is trivial, so that a nearest neighbor (NN) strategy assigning blobs to closest track is enough. NN strategy is indeed adequate in many applications, but its performance quickly degrades in a dense environments [5].

Recently, data association algorithms have received wider attention by the computer vision community. The full problem of data association in the visual context is a correspondence of multiple blobs to multiple tracks. Multiply fragmented or merged blobs may appear as candidates to update each track. Besides, the evaluation function, such as a distance considering both residuals and structural information is necessary to evaluate the assignment hypotheses.

The problem of split and merging has been considered from different points of view. Some proposals do not consider explicit hypotheses and use lower-level image information to address the problem. For instance, w4 system for tracking people [6], is based on low-level correlation operators to resolve occlusions and people-group tracking. Kim et al[7] propose object-tracking through a "natural correspondence" derived from the use of labels retained from a frame to the next one, created by a distributed genetic algorithm for spatial segmentation. The segmentation result from previous frame is evolved to next one, so labels are propagated from one site to the next by selection. The authors of this work [8] have previously proposed a method using a fuzzy system which evaluates heuristics computed from overlapping blobs and tracks. It computes "confidence levels" that are used to weight each gated blob's contribution to update the target track and its associated shape. These proposals build a sub-optimal solution avoiding the combinatorial complexity of enumerating the possible data association alternatives.

In most systems, contextual information and specific knowledge is additionally used to improve performance. In [9], contextual information is represented by means of 2D polygons on the image, each of them having a list of attributes: IN-OUT zone, NOISY zone, OCCLUSION zone, AREA-OF-INTEREST zone, etc. This type of information helps for taking suitable decisions. So, specific rules are typically applied to protect the systems against occlusions. In the case of track-based occlusions (those between visual objects) the problem is to assign shared points to the correct track. Instead, in the case of background object-based occlusions, the track model must not be updated in the occluded part. In [10] the goal is keeping the track history consistent before and after the occlusion, even trying to separate the group of people into individuals during the occlusion also. The concept of macro-object is introduced there to handle this situation with adhoc rules takes into account possible errors due to visual overlaps.

Therefore, it is very usual to develop visual tracking systems making use of contextual information to address specific problems and improve their performance. It can be seen as an enrichment of unconstrained output resulting of a general-purpose trackers, in which observations are associated with targets, which are updated, created and deleted accordingly. In this contribution we propose a general architecture to define and exploit contextual information of visual trackers to apply them in specific operational conditions.

Next section gives an overview of the tracking system and the architecture we propose to improve the association process of the tracking system, although this architecture can be extended to other modules of the tracking system. The third section introduces the association process and some solutions given by other researchers. After the association introduction detail is given about the Context Layer (CL) proposed in this paper and the type of problems it can solve. This description of CL is followed by experimentation and evaluation of the system propose. Finally, some conclusions are presented.

2 Tracking System Overview

In this section we describe our approach to design video-tracking systems. In our proposal, we aimed to overcome the limitations of a real tracking system, where we have to endow it with generality and flexibility, while at the same time coping with the efficiency requirements mandatory in real-time systems.

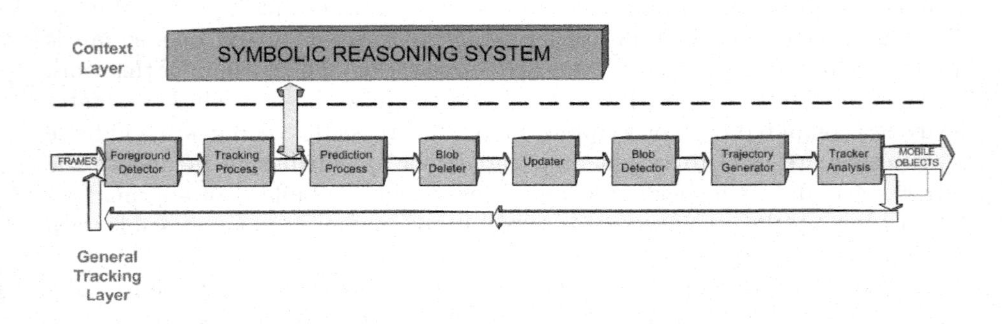

Fig. 1. Tracking System Architecture

Our architecture is depicted in Figure 1. This architecture is based on a two-layer image-processing modules: General Tracking Layer (GTL) and Context Layer (CL). GTL describe a generic multipurpose tracking process for video-surveillance systems. GTL is arranged in a pipe-line structure of several modules, shown in Figure 1. GTL directly interface with the image stream coming from a camera and extract the track information of the mobile objects in the current frame. The interface between adjacent modules in GTL is symbolic data and it is set up, so that for each module different algorithms are interchangeable.

The implementation and the algorithm used by each of the GTL modules does not affect the CL. Within the tracking process and receiving a list of blobs from the previous module, the tracking process will solve a problem of blob-to-track multi-assignment, where several (or none) blobs may be assigned to the same track and simultaneously several tracks could overlap and share common blobs. So the association problem to solve is the decision of the most proper grouping of blobs and assignation to each track for each frame processed, next section gets into more detail in the association process .

GTL modules depend on many parameters that should be adjusted for a specific implementation, but that's not enough, specifics scenarios give special conflicts that the tracking system does not know how to solve properly. CL performs symbolic reasoning based on the context data of a specific tracking scenario. CL manages the symbolic interface data between GTL modules with the goal of assessing a specific tracking scenario and trigger the appropriate action to treat that situation. CL is designed as a symbolic reasoning system. One of the more employed reasoning systems in industry and services are knowledge-based system (universally known as expert systems). Knowledge-based systems embed a large component of domain-specific knowledge but, differently from other heuristic-based systems, knowledge is represented in an identifiable separate part of the system rather than being dispersed throughout the whole program.

3 Association

The data association can be dealt as a sequential decision process which takes, for each frame, current detections and the result of assignments up to that time, represented with sufficient statistics contained in the track state vectors. The assignment problem is so addressed as maximizing the total distance among tracks vectors and blobs.

The association problem is formalized with the assignment matrix, $\mathbf{A}$[k], defined as $A_{ij}[k] = 1$ if blob $b_i[k]$ is assigned to object o_j; and $A_{ij}[k] = 0$ otherwise. The blobs extracted in the k-th frame are $b[k] = \{b_1[k], ..., b_{N_k}[k]\}$ and the objects tracked up to now (last assignment of blobs was at frame k-1) are $o[k-1] = \{o_1[k], ..., o_{M_k}[k-1]\}$.

The size of matrix $\mathbf{A}$[k], $N_k x M_k$, changes with time, since i=1,...,N_k represents the blobs extracted from the k-th frame, whose number depends on the variable effects mentioned above during the detection process, and j=0,...,M_k represent the objects in the scene, whose number may also dynamically change when objects appear and disappear in the scene. Special index j=0 is kept to represent assignment of blobs to "null track" at current time, which are used to initialize new objects or are discarded.

So, the association decision A[k] will consider the distances between sets of blobs and tracks, computed through the usual Mahalanobis formula:

$$d_{ij} = \left[\hat{x}_j - \bar{f}_i\right]^t (\mathbf{P}_j^{-1}) \left[\hat{x}_j - \bar{f}_i\right] ; i = 1, ..., N_k, j = 1, ..., M_k$$

The features vector $\bar{f}_i$ are extracted from the sets of blobs corresponding to j-th track ($b_i[k]$ such that A_{ij}[k]=1). The estimated state vectors, $\hat{x}_j$, containing state information about objects, and associated covariance $\mathbf{P}_j$, are recursively updated with the sequence of assigned observations, using motion and observation models, by means of a Kalman filter. The optimal decision would be that combination in which the sum of distances between current detections and predicted tracks is minimized:

$$\underset{\mathbf{A}[k]}{min} \sum_{i=1,j=1}^{N_k,M_k} d_{ij}$$

Under complex circumstances (several close objects with very fragmented images), the number of possible ways to combine their image regions into tracks grows at exponential rate. The number of possible solutions is $s^{N_k * M_m}$. The data association task would search for the optimal (or near-optimal) hypothesis, according to the previous metric to assess the quality of assignments. In the general case, it may be impractical to find the optimal decision through exhaustive enumeration of all hypotheses because of the high computational load required.

The procedure proposed here is the use of a simplified association technique (analogous to a nearest neighbour approach, requiring low computation load), complemented with a knowledge-based system. This last system takes the appropriate decisions under certain conditions, predefined by a number of variables describing the situation in which the decision is taken. Otherwise, the default decision will be based on the conventional association method. Furthermore, given the modular architecture, the intervention of the knowledge-based system is independent of the specific technique applied for data association.

4 Context Layer

Context Layer adopts the general-purpose model for knowledge representation based on a production-system model with forward chaining reasoning. The major two components of the expert system are the knowledge base and the inference engine. The knowledge base contains facts and rules about the subject at hand i.e. the domain the expert system is functioning in. Rules consists of heuristics that enable the human expert to make educated guesses when necessary, to recognize promising approaches to problems, and to deal effectively with erroneous or incomplete data. Expert systems is data-driven reasoning system which uses `'IF THEN'` rules to deduce a problem solution from initial data.

Therefore, in order to design the Context Layer of our architecture, the aim is to set up properly the rules and facts involved in a specific scenario. As we mentioned earlier, there are complex circumstances where we need context information to achieve blob-to-track association suitably. In Figure 2, the knowledge-based reasoning architecture for Context Layer is depicted. So, we must analyze each particular scenario and identify rules and facts which are useful in setting up the proper blob-to-track association from mobile objects. These rules and facts take into account that mobile objects are imprecisely perceived, partially overlapped, missed, or cluttered by shadow.

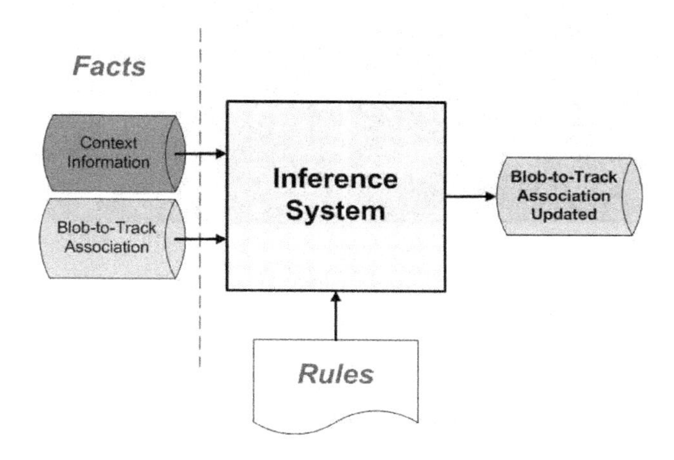

Fig. 2. Knowledge-based architecture for Context Layer

Next we are going to show an implementation example of part of the Context Layer for a squash match scenario. For instance, Figure 3 shows a sequence (from 1 to 4) of foreground frames where two squash players (the white spots) run toward each other and then separate. Before the collision (frame 1 and 2) the tracker was able to track the two persons without any problem. The problem arises when the two players become a unique blob (frame 3), at this moment the tracker eliminates the two tracks to later on create a new and only track for the two players (frame 4).

In this scenario, we are able to establish some of the context information and the blob-to-track association as the following facts:

```
(Number-of-tracks  2)
(Player-Size 20)
(Player-Aspect-Ratio 0.5)
(Track Player1)
(Track Player2)
(Association A)
```

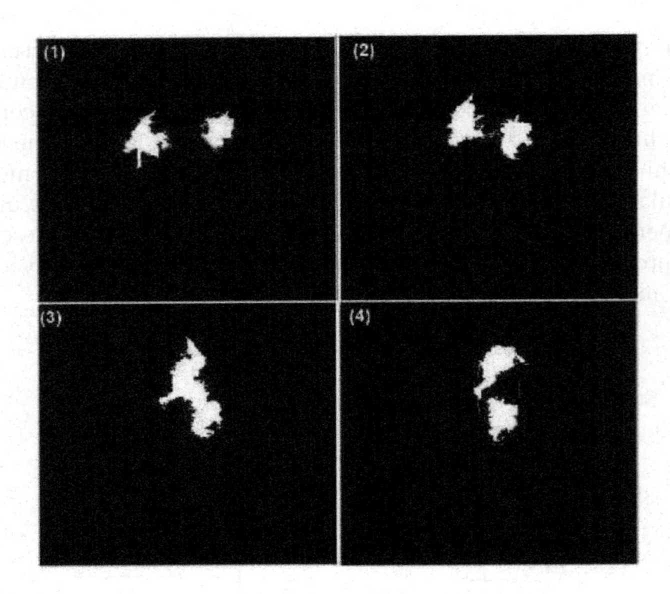

Fig. 3. Sequence of frames (from 1 to 4) where two squash players run toward each other and separate

where in `Player1` and `Player2` is represented the position, size and kinematics of the players; and `A` is the assignment matrix.

Collisions between players are troublesome as they became one big blob (Figure 3.3) creating an error in data association with the consequent loss of target tracks. To prevent such problems, we need some rules based on the actual facts, that try to improve the assignment established by matrix `A`. So, the following rules are incorporated to the knowledge-based system:

```
( p match-player1
  p match-player2
  p match-A
  if {Conflict player1 player2}
     {All-Assignment A player1 player2}
     {Empty-Assignment A}
  -> modify( A ^ distribute-players(player1, player2))
)
```

where predicate `p` looks for facts that match `player1`, `player2` and assignment matrix `A`; `Conflict` is function that returns `true` whether there exists a conflict between `player1` and `player2` (see Figure 4); `All-Assignment` indicates if two players are assigned to the same one blob; `Empty-Assignment` is `true` if there is one blob that has not been assigned to any player (see Figure ??); and `distribute-players` reassign matrix `A` based on the players' information (position, size and kinematics).

Fig. 4. Collision between two tracks

5 Performance Evaluation

In this section simulation experiments, carried out in two different realistic scenarios, are discussed in order to demonstrate the superior performance of a knowledge-based tracking system with respect to a standard tracking system that does not exploit the knowledge-based. To evaluate tracking performance the following metrics are adopted:

1. standard deviation of number of tracks;
2. standard deviation of track size;

To evaluate the performance of our symbolic reasoning system, we captured information about the tracks during the tracking process and analyze it. Ground truth was not available and it is inconvenient and labored to extract the ground truth in the scenarios used. The time overload produced by the use of the CL was barely noticeable for example for one of the scenarios it was 5.50%.

The two scenarios used are a squash game, and a handball game. The first one is 4 a simpler scenario, where there are only two players, while the handball game (Figure 4) is a more complicated scenario due to the number of players and their size.

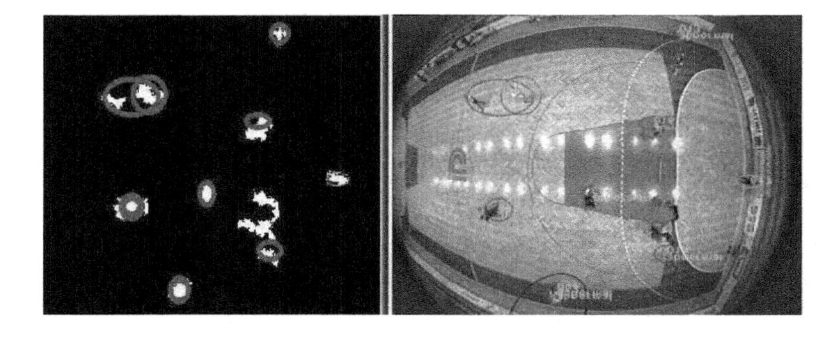

Fig. 5. Handball scenario

An important measurement to assess tracking continuity is the number of tracks. Table 1 shows the mean and the standard deviation of the number of tracks that the system maintained at all times. For the squash game the real number of players is 2, while for the handball game there should be 14 players, however these 14 are not in the camera's field of view at all times. In Table 1 it is possible to see how for both scenarios the system with the CL gave a closer mean of the number of tracks to the number of players, and lower standard deviation, showing that the CL improved the system.

Table 1. Number of Targets

Scenario	Data	Without CL	With CL
Squash	*Mean*	1.49	1.81
	Standard Deviation	0.51	0.47
Handball	*Mean*	3.82	8.48
	Standard Deviation	2.67	1.74

The other metric chosen to evaluate the system is the standard deviation of the track's size. This metric was selected because the tracking system without the CL grouped several players into one track as shown in the bottom right image of Figure 3. Therefore by looking at the track's size's standard deviation it is possible to see if the tracking system has improved by not grouping several players in one track and thus creating a larger track. Table 2 shows the standard deviation of the track's sizes . It is possible to see how the standard deviation in both cases has been reduced considerably. This means that the system did not group several players in one only track.

Table 2. Standard deviation of target's size

Scenario	Without CL	With CL
Squash	40.60%	29.62%
Handball	45.59 %	34.49%

6 Conclusions

The association process in Target Tracking is a complicated task due to occlusions, collisions and missed detections of targets that may occur during the tracking process. To improve association we propose a Context Layer, containing facts and rules about the scenario that is going to be track, and statistics about how it is behaving. The Context Layer modifies if necessary the blobs to track association made by the association algorithm, in order to improve the tracking.

The design and deployment of a Context Layer to support a tracking system has proved to soften the errors that the tracking system can make, without slowing down the system in a noticeable way. This Context Layer should be specific for different scenarios.

In this work we have concentrated in solving the associations problems, however the tracking system is far from being perfect, and this solution can be extended to other problems caused at the initialization, actualization or at any other level of the tracking system.

References

1. M. A. Patricio, J. Carbó, O. Pérez, J. García, and J. M. Molina. Multi-agent framework in visual sensor networks. *EURASIP Journal on Advances in Signal Processing*, 2007:Article ID 98639, 21 pages, 2007. doi:10.1155/2007/98639.
2. Pankaj Kumar, Surendra Ranganath, Kuntal Sengupta, and Weimin Huang. Cooperative multi-target tracking and classification. In *ECCV 2004, 8th European Conference on Computer Vision*, volume 3021 of *Lecture Notes in Computer Science*, pages 376–389. Springer, 2004.
3. O. Pérez, M. Piccardi, J. García, M A. Patricio, and J M. Molina. Comparison between genetic algorithms and baum-welch algorithm for the adjustment of hmm to classify human activities. In *Applications of Evolutionary Computing, EvoWorkshops 2007*, Valencia, España, April 2007.
4. O. Pérez, M A. Patricio, J. García, and J M. Molina. Improving the segmentation stage of a pedestrian tracking video-based system by means of evolution strategies. In *8th European Workshop on Evolutionary Computation in Image Analysis and Signal Processing. EvoIASP 2006*, Budapest, Hungary, April 2006.
5. Ingemar J. Cox and Sunita L. Hingorani. An efficient implementation of reid's multiple hypothesis tracking algorithm and its evaluation for the purpose of visual tracking. *IEEE Trans. Pattern Anal. Mach. Intell.*, 18(2):138–150, 1996.
6. Ismail Haritaoglu, Davis Harwood, and Larry S. David. W4: Real-time surveillance of people and their activities. *IEEE Trans. Pattern Anal. Mach. Intell.*, 22(8):809–830, 2000.
7. Eun Yi Kim and Se Hyun Park. Automatic video segmentation using genetic algorithms. *Pattern Recogn. Lett.*, 27(11):1252–1265, 2006.
8. J. Garcia, J. M. Molina, J. A. Besada, and J. I. Portillo. A multitarget tracking video system based on fuzzy and neuro-fuzzy techniques. *EURASIP Journal on Applied Signal Processing*, 14:2341–2358, 2005.
9. B. Lienard, X. Desurmont, B. Barrie, and J. Delaigle. Real-time high-level video understanding using data warehouse. In *Real-Time Image Processing 2006*, Valencia, España, February 2006.
10. Rita Cucchiara. Multimedia surveillance systems. In *VSSN '05: Proceedings of the third ACM international workshop on Video surveillance & sensor networks*, pages 3–10, New York, NY, USA, 2005. ACM Press.

Automatic Control of Video Surveillance Camera Sabotage

P. Gil-Jiménez, R. López-Sastre, P. Siegmann, J. Acevedo-Rodríguez, and S. Maldonado-Bascón

Dpto. de Teoría de la Senal y Comunicaciones
Universidad de Alcalá, Alcalá de Henares, 28805 (Madrid), Spain
{pedro.gil,robertoj.lopez,philip.siegmann, javier.acevedo,saturnino.maldonado}@uah.es

Abstract. One of the main characteristics of a video surveillance system is its reliability. To this end, it is needed that the images captured by the videocameras are an accurate representation of the scene. Unfortunately, some activities can make the proper operation of the cameras fail, distorting in some way the images which are going to be processed. When these activities are voluntary, they are usually called sabotage, which include partial o total occlusion of the lens, image defocus or change of the field of view.

In this paper, we will analyze the different kinds of sabotage that could be done to a video surveillance system, and some algorithms to detect these inconveniences will be developed. The experimental results show good performance in the detection of sabotage situations, while keeping a very low false alarm probability.

1 Introduction

Automatic video surveillance systems are typically used to monitor particular places, trying to detect suspicious activities such as vandalism or intrusion [1,2,3,4,5]. These kinds of systems are normally composed of a number of videocameras and an image processing block, which is generally implemented over a computer or specialized hardware. The robustness of a video surveillance system depends heavily on both the image processing algorithms and the quality of the images received.

When it comes to the video quality, some extern parameters can influence it, such as illumination, weather conditions, or voluntary actions. In systems where these are crucial aspects, some dispositions must be taken to prevent, or at least, detect these situations. In this work, we focus in voluntary actions, which are typically called sabotage, and some algorithms have been designed to detect them. There are not many works in the literature that deal with this problem. In [6], for instance, a camera dysfunction detector is implemented, designed mainly for surveillance systems inside vehicles.

Figure 1 shows the block diagram of the system implemented. Essentially, the system is composed of a videocamera, where the images captured are processed

J. Mira and J.R. Álvarez (Eds.): IWINAC 2007, Part II, LNCS 4528, pp. 222–231, 2007.

by a group of image processing algorithms, such as motion detection, or object tracking, running over a PC. To perform the sabotage detection, another block has been added, which analyzes the images before the video surveillance block does. If no sabotage is detected, the system works as usual. However, if the block detects any kind of sabotage, an alarm is triggered to warn the surveillance staff. In this state, it makes no sense the operation of the video surveillance block, so it can be deactivated until the normal state is reached again.

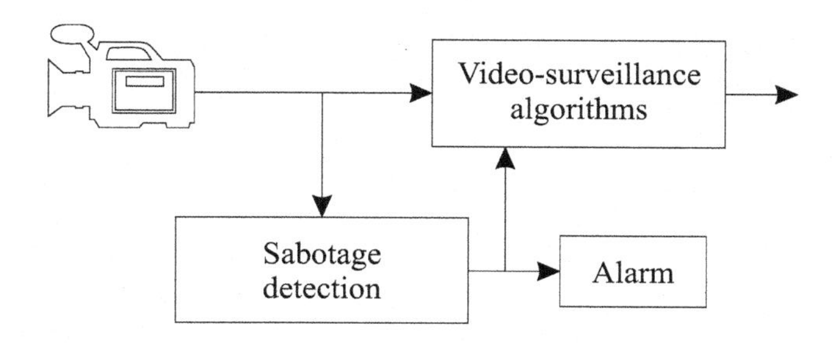

Fig. 1. Block diagram of a video surveillance system with sabotage detection

2 Camera Sabotage

In this paper, we will consider sabotage as any voluntary action on a videocamera that changes the physical parameters of the camera, affecting, therefore, the images received and processed by the video surveillance algorithms. These actions include:

- **Image occlusion:** The lens of the videocamera can be occluded with an opaque object, preventing the system to analyze the surveyed scene.
- **Image defocus:** The focus setting of a videocamera can be altered, changing the image captured, specially the edges of the objects in the scene.
- **Change of the field of view:** The field of view can be changed easily displacing the camera, in such a way that the new image is partially or totally distinct to the previous one.

In Fig. 2 we can see some examples. Figure 2(a) shows an image of a scene without any kind of sabotage. In Fig. 2(b), the lens has been partially occluded with a blank sheet of paper, hiding part of the scene to the camera. In Fig. 2(c), the lens setting of the camera has been changed to defocus the image, distorting specially the edges of the objects. Finally, in Fig. 2(d) the camera has been shifted to the right, to partially change the field of view. In any of these situations, common video surveillance algorithms, such as background subtraction, or background modeling, would not be able to operate correctly, and so, the output of the system would no longer be valid. Although the video surveillance system will not be able to solve this problem by itself, some algorithms can be designed to detect these situations and alert the surveillance staff, who can finally analyze the situation, and perform the corresponding actions.

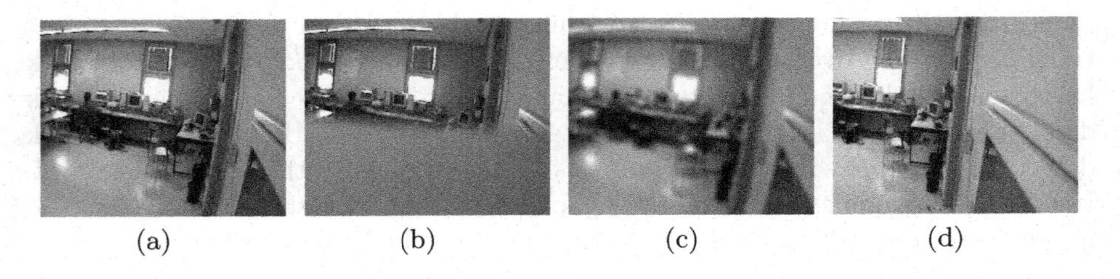

Fig. 2. Some examples of video camera sabotage. (a) No-sabotaged image. (b) Partial occlusion example. (c) Defocus example. (d) Displacement example.

3 Sabotage Detection

The algorithm which has to be designed for the detection of sabotage activities must fulfill two different constrains. Firstly, it must be robust enough to ensure a high detection probability, while maintaining the false alarm probability as low as possible. Secondly, the algorithm should be computationally simple to carry out with the real time requirements.

We have to consider also the scenario where the system is working. As in many other video surveillance algorithms, the fundamental concept is the comparison between the current frame and some model built from previous images, which is usually called the background model. In the sabotage case, two different situations must be distinguished. On the one hand, when the cameras are not altered, normal motion of objects or people in the scene must not be considered as sabotage, as it must be interpreted by the video surveillance algorithms themselves. On the other hand, voluntary alterations of the parameters of the cameras do must be considered as sabotage. Although both situations are completely different to each other, from the algorithm's point of view they can be quite similar, since both are reflected as a change between the current frame and the background model.

To allow the anti-sabotage system to discern between each situation, we compute a model of the background using a supervised sequence of the scene where no sabotage exists. Although many techniques have been proposed for the computation of background models which comprise the whole image, in this work we use the edges of the scene instead of the whole image. This yields us two advantages. Edges are more robust against illumination changes than smooth parts of the image when comparing a frame with the computed model of the background. But also, by using only the edges in the comparisons to perform the sabotage detection, a smaller number of operations are done by the algorithm, reducing the computational complexity, and so, the time required to perform the sabotage detection.

3.1 Background Model

As it has been explained above, the background model only includes the information at the edges of the objects belonging to the background. The background model is a binary matrix of the same size than the original images, where pixels set to true belong to the edges of the background, and vice versa. To ensure that an edge belongs to the background, and not to a moving object, the edges are computed for M consecutive frames,

$$B_n = \sum_{i=nM}^{(n+1)M-1} D_i \ , \tag{1}$$

where D_i is the edge matrix computed for frame i, whose pixels take the value 1 if it belongs to an image edge, and 0 otherwise. Furthermore, to allow the system for slow changes of the background, the matrix B is computed for each M new frames, as it is indicated in (1), and the background model updated according to:

$$P_n = \alpha P_{n-1} + (1-\alpha)\, B_n \ , \tag{2}$$

where α is a parameter between 0 and 1 that establishes the influence of the new frames with respect to the last background model. Finally, only those pixels which have been computed as edges for an enough number of times (in our case we chose $P_n \geq M/2$) are considered to belong to the background model.

In Fig. 3 we can see an example of this computation. Figure 3(a) shows a frame of the sequence recorded, while from (b) to (d) it is shown the background model state for $M = 30$ and $\alpha = 0.3$ after 30 ($n = 1$ iteration), 60 ($n = 2$) and 120 ($n = 4$) frames respectively. It is obvious that the model becomes more stable as the number of frames used in the computation increases, and after about 3 iterations the model remains almost unchanged as long as the background remains stationary. From now on, we will use this model with the above mentioned settings to perform the sabotage detections, as it is explained in the following sections.

3.2 Occlusion Detection

When an object covers partly or totally the lens of the camera, part of the image is no longer visible, and a number of pixels, or even all the pixels, of the background model disappear as well. This occlusion must be differentiate from partial hidings due to moving objects. To this end, the entropy of the pixels belonging to the background model is computed:

$$E = -\sum_k p_k \cdot \ln(p_k) \ , \tag{3}$$

where p_k is the probability of appearance of the gray level k in the image. Since the model is not altered until new M consecutive frames arrive, causing its update, this computation is performed only once each M frames. For each new frame, the same entropy is calculated again. Let suppose now that an object is placed near the lens, so that we can occlude part of the image. In this situation, a loss of information will occur in the zone of the image that has been occluded, resulting in a decrease of its entropy.

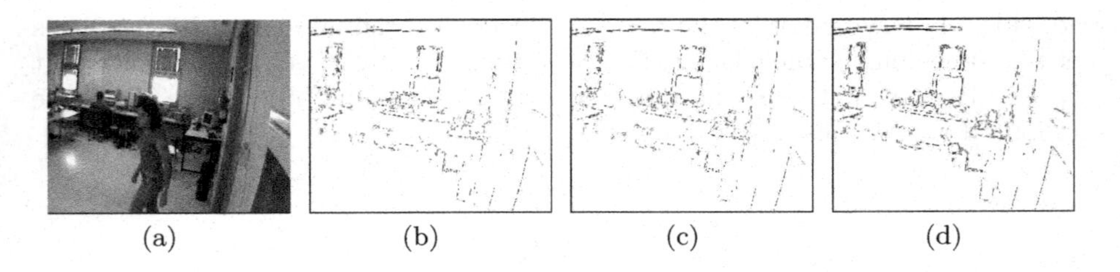

(a) (b) (c) (d)

Fig. 3. Computation of the background model. (a): An image of the original sequence. (b-d) Background model after (b) 30 frames (1 iteration), (c) 60 frames (2 iterations), (d) 120 frames (4 iterations).

Although this would be enough in the case of a total occlusion for the detection of the sabotage, in partial occlusions this effect can be masked by the entropy of the rest of the image, which probably remains unchanged. To allow the system to detect partial occlusion as well as total ones, the image is divided into several sub-images, or blocks, of equal size, and the process is performed over each sub-image independently. A sabotage will be detected when the current entropy for at least one of these blocks is lower than a given threshold.

3.3 Defocus Detection

The defocus of the lens of a camera implies the degradation of the edges, so that most of them disappear from the image [7]. Therefore, we can easily count the number of pixels in the background model and, for each new frame, the same parameter can be computed. If the camera has been defocused, this number for the current frame will be much lower than that of the background model, and this reduction can be used to detect that the camera is not correctly focused.

3.4 Displacement Detection

The detection of shifts on the field of view is computed using the zero-mean normalized cross correlation ($ZNCC$), as in classical block matching algorithms, but in this case, to speed up the algorithm, only for the pixels of the background model:

$$ZNCC_i(m,n) = \frac{\sum_{x,y} \left(I_{i-1}(x,y) - \mu_{I_{i-1}}\right)\left(I_i(x+m,y+n) - \mu_{I_i}\right)}{\sigma_{I_{i-1}}\sigma_{I_i}}, \quad (4)$$

where I is the image captured by the camera. In the computation, only the pixels belonging to the background model are considered. A block matching algorithm returns two parameters [8], the optimal value in both axes (m,n) of the translation between the current frame and the background model, and the value of the $ZNCC$ for that displacement, and both parameters are used to detect

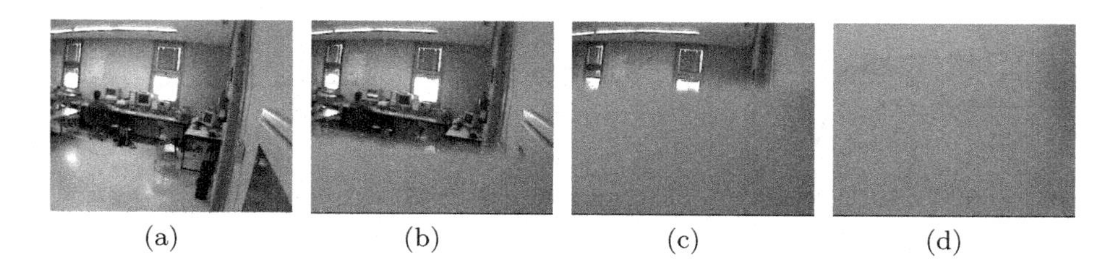

(a) (b) (c) (d)

Fig. 4. Different frames of the occlusion example

a possible sabotage. The second parameter can be used to estimate the reliability of the matching. If this value is lower than a given threshold, we interpret that as an error in the matching, which can be caused for many reasons, such as a partial occlusion, too large a displacement, noise, etc. In this case, the algorithm waits for the next frames to come, to see if the problem still exists.

The first parameter must be analyzed more carefully. A small displacement, of the order of about 5 pixels, could be caused by vibrations or blows, and should not be considered as sabotage. Bigger displacements however do must be considered as sabotage. A problem can arise if the camera is displaced slowly so that the system could consider it as a consequence of vibrations or noise for each frame, and no sabotage will be detected. To solve this inconvenience, the system computes both the relative displacement and the accumulated displacement, and whenever one parameter surpasses the threshold a sabotage is detected.

4 Experimental Results

The proposed system has been implemented using C++ language as an optional block to the video surveillance system our group is currently developing, so that the changes in the original system needed to incorporate the algorithms described in this paper are minimal. The system processes 5 gray level images per second, running in real time over a PC.

The system has been tested in indoor and outdoor environments, with images that include fast (lights on and off) and gradual (different hours of the day) illumination changes, and background motion (specially people and cars). During the recording of the sequence, the camera has been intentionally sabotaged to test the response of the system. These actions include the three kinds of sabotage analyzed in this work, as has been described previously.

Figure 4 shows some images of an occlusion example. These images have been extracted from a longer sequence, which comprises enough time for the background model to stabilize, and a few more frames to register the sabotage action. In Fig. 5 the evolution of the entropy for the last 8 frames of the sequence is displayed. As it has been described before, the whole image has been divided into 9 blocks, and the 9 graphics of Fig. 5 correspond with these 9 blocks. In the graphics it is also marked in dotted lines the reference entropy computed from

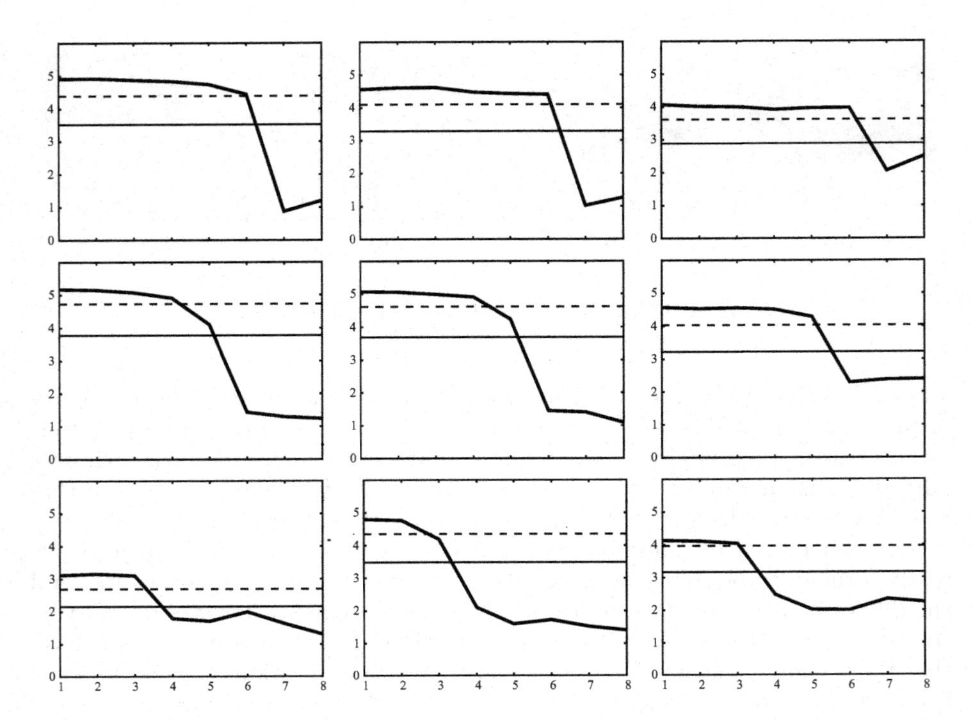

Fig. 5. Occlusion detection for the 9 blocks of the image. Horizontal axis shows the number of sequence from 1 to 8. Vertical axis measures the entropy of the corresponding frame and block.

the background model, and in solid lines the threshold for each block, which is computed as 0.8 times the reference entropy. From a look at this figure, it can be seen that, between frames 3 to 7 the entropy of all blocks go below its corresponding threshold, and so, an occlusion has been detected. We can also see that, since the lens of the camera has been occluded from bottom to top, that is, the sheet of paper have been moved upwards, the blocks in the bottom line reach the threshold, and detect the occlusion, before the top line's do.

Some frames of a defocus sabotage example is shown in Fig. 6. Again, these images correspond to a sequence longer enough to allow the background to stabilize. For each sub-image, the original frame is displayed at the left, and its corresponding edge mask at its right. The results of the defocus sabotage detector are displayed in Fig. 8(a), along with the reference in dotted line, which is the number of pixels that compose the background model, and the threshold in solid line, which is computed as 0.4 times the reference described above. In this case, the threshold is reached at frame 7, and at this point, a defocus sabotage is then detected.

Finally, and example of a displacement sabotage can be seen in Fig. 7, and the results of the sabotage detector in Fig. 8(b). The measured parameter is the value of the normalized correlation. While there is no camera motion, this

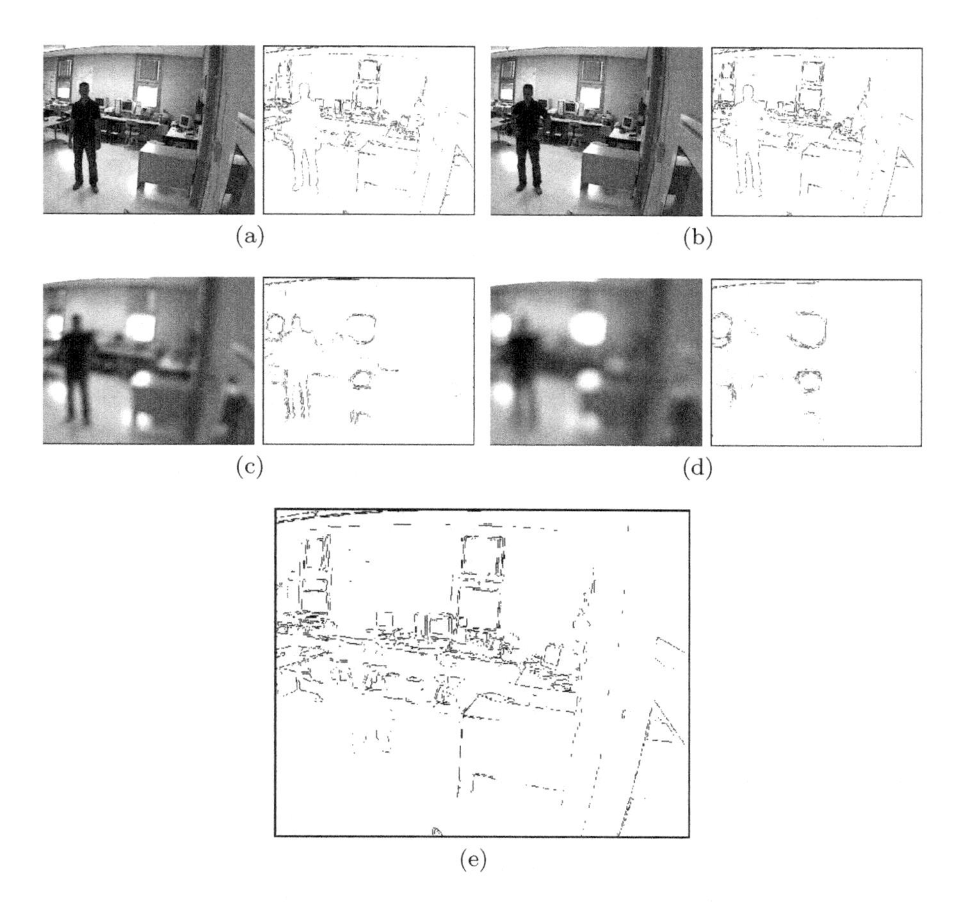

Fig. 6. (a-d) Different frames of the defocus example. For each sub-image the original frame is displayed on the left, and the computed edge mask on the right. (e) Background model.

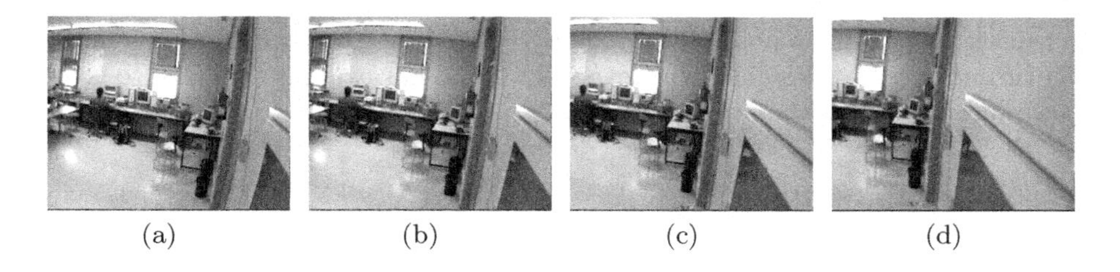

Fig. 7. Different frames of the displacement example. (a) Original position of the camera. (b-d) Successive displacement of the camera.

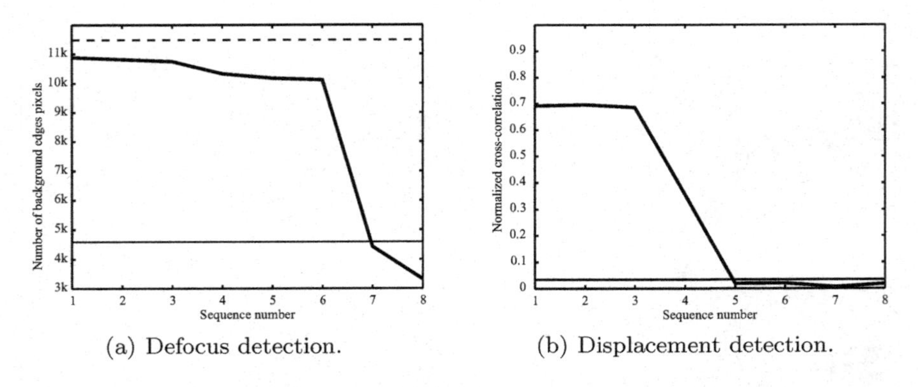

(a) Defocus detection. (b) Displacement detection.

Fig. 8. Defocus and displacement detection examples

parameter takes values around 0.7, which is mainly due to camera noise and vibrations. However, when the camera in shifted to change its point of view, the correlation decreases to values under 0.01. The threshold for the detection of camera displacements has been set to 0.03. This value has been drawn in Fig. 8(b), that allows us to see that, in this case, the displacement sabotage is detected in frame 5.

5 Conclusion

This paper describes the main kinds of sabotage that could be done to a video surveillance system, and how this is reflected into the images captured, and to what extent could the operation of a video surveillance system be affected. The paper also proposes some algorithms to trigger an alarm when a sabotage has been detected. The algorithms fulfill the typical video surveillance constrains, that is, robustness to ensure a high detection and low false alarm probabilities, and simplicity to accomplish with the real time requirements.

The future work is basically related with the testing of the system on more complex sequences, such as adverse outdoor conditions (night, rain, fog or wind among others), crowed scenarios, etc. Although the basic idea of the system is to keep it as simple as possible to reduce the computation time, more complex algorithms could be designed whenever additional hardware is available to go on maintaining the real time requirements.

Acknowledgments

This work was supported by the project of the Ministerio de Educación y Ciencia de España number TEC2004/03511/TCM.

References

1. Foresti, G.L., Mhnen, P., Regazzoni, C.S.: Multimedia video-based surveillance systems. Requirements, issues and solutions. 1 edn. Kluwer Academic Publishers, 101 Philip Drive, Assinippi Park, Norwell, Massachusetts 02061 USA (2000)
2. Di Stefano, L., Neri, G., Viarani, E.: Analysis of pixel-level algorithms for video surveillance applications". In: CIAP01. (2001) 541–546
3. Toyama, K., Krumm, J., Brumitt, B., Meyers, B.: Wallflower: Principles and practice of background maintenance. In: ICCV (1). (1999) 255–261
4. Boult, T., Micheals, R., Gao, X., Eckmann, M.: Into the woods: Visual surveillance of noncooperative and camouflaged targets in complex outdoor settings. In: Proceedings of the IEEE. Volume 89. (2001) 1382–1402
5. Dawson-Howe, K.: Active surveillance using dynamic background subtraction. Technical Report TCD-CS-96-06, Dept of Computer Science, Trinity College, Dublin, Ireland (1996)
6. Harasse, S., Bonnaud, L., Caplier, A., Desvignes, M.: Automated camera dysfunctions detection. In: Image Analysis and Interpretation, IEEE (2004) 36 – 40
7. Pentland, A.: A new sense for depth of field. IEEE Transactions on Pattern Analysis and Machine Intelligence **9** (1987) 523–531
8. Roma, N., Santos-Victor, J., Tom, J.: A comprative analysis of cross-correlation matching algorithms using a pyramidal resolution approach. World Scientific (2002)

Complex Permittivity Estimation by Bio-inspired Algorithms for Target Identification Improvement

David Poyatos[1], David Escot[1], Ignacio Montiel[1], and Ignacio Olmeda[2]

[1] Laboratorio de Detectabilidad
Instituto Nacional de Técnica Aeroespacial (INTA)
Ctra. Ajalvir Km. 4, 28850, Torrejón de Ardoz, Spain
{poyatosmd,escotbd,montielsi}@inta.es
http://www.inta.es

[2] Laboratorio Sun Microsystems de Finanzas Computacionales
Universidad de Alcalá, Spain
josei.olmeda@uah.es
http://lfc.uah.es

Abstract. Identification of aircrafts by means of radar when no cooperation exists (Non-Cooperative Target Identification, NCTI) tends to be based on simulations. To improve them, and hence the probability of correct identification, right values of permittivity and permeability need to be used. This paper describes a method for the estimation of the electromagnetic properties of materials as a part of the NCTI problem. Different heuristic optimization algorithms such as Genetic Algorithms (GA) and Particle Swarm Optimization (PSO), as well as other approaches like Artificial Neural Networks (ANN), are applied to the reflection coefficient obtained via free-space measurements in an anechoic chamber. Prior to the comparison with real samples, artificial synthetic materials are generated to test the performance of these bio-inspired algorithms.

Keywords: NCTI, ANN, GA, PSO, permittivity, permeability, free-space measurements.

1 Introduction

For many years, a great effort has been done in different NATO countries to develop an identification system capable to make a reliable classification of aircrafts into different groups (friendly, hostile or neutral) at the maximum surveillance and weapons systems range. The traditional way to identify a target is through "cooperative" techniques, like IFF (Identification, Friend or Foe) systems that need the collaboration of the unknown aircraft, which sends a predefined code to the interrogating system. By this method, only friendly targets with properly working systems are recognized. No positive hostile or neutral identification can be performed. The problem arises when an incorrect identification occurs and results in fratricide or engagement of civilian aircrafts. To prevent this kind

J. Mira and J.R. Álvarez (Eds.): IWINAC 2007, Part II, LNCS 4528, pp. 232–240, 2007.

of problems, Non-Cooperative Target Identification (NCTI) systems are being investigated [1].

These systems operate by comparing data provided by different sensors (surveillance radars, infrared, etc) against a database of known aircraft signatures. Among them, radar based NCTI (Fig. 1) is one of the most promising techniques due to its long range and recent improvements in resolution capabilities and signal processing techniques [2].

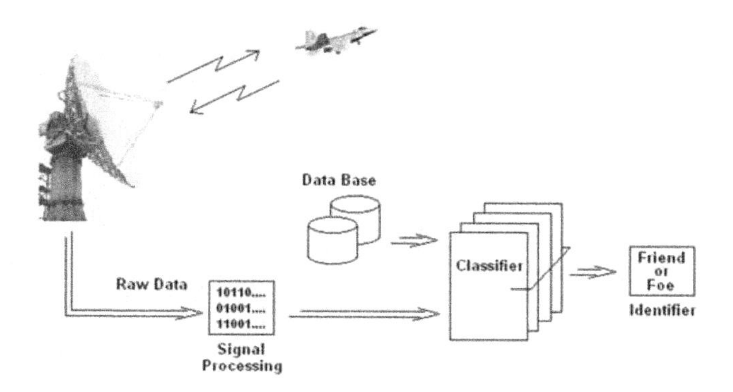

Fig. 1. Radar-Based NCTI

Depending on the selected radar technique, the generation of the database may or not be an issue. JEM techniques have shown their potentiality and the associated database generation simplicity, but the problems arising when the S/N ratio is not good enough enforce the exploitation of other techniques. Nowadays, imaging techniques like ISAR or HRRPs [3] have shown more completeness regarding aspect angles but the database seems very difficult to be generated.

There are different possibilities to fulfill this task. Measurement campaigns of flying aircrafts are very expensive and recording all the aspects is too difficult to be performed. Subscale model measurements are expensive too because they need complex scale models and specific facilities with a very high cost to be maintained and run. Finally, predictions by software seem to be the most feasible way to obtain those desired data [4].

As prediction tools solve Maxwell's equations, a right estimation of the electromagnetic properties of the materials constituting the target is mandatory. Moreover, recent NCTI techniques tend to examine the effect of resonant parts of aircrafts, such as antennas, inlet ducts, exhaust pipes, cockpit, etc, [5] where different structures with diverse materials have to be defined. This context-based information together with auxiliary data from other sources (electronic support measurements, flight/mission plans, flight profile and target behaviour, etc.) will be the input in an identification system.

Hence, it becomes clear that accurate estimation of dielectric constants of materials plays an important role in recent NCTI techniques. There are many ways to measure the complex electromagnetic properties of samples in the time and frequency domain and they all basically fall into two categories: either destructive methods, in which sample preparation is needed for accurate evaluation, or nondestructive methods, which require very little or no sample preparation. The measured quantity(s) of the sample will enable the computation of its permittivity and permeability.

The probe method offers the advantages of being a broadband and nondestructive method, but requires perfect contact between the probe and the sample. The surface roughness of the sample seriously limits the accuracy of the measurement [6]. The perturbation of a resonant cavity by the introduction of a dielectric sample can be used to compute electrical properties by measuring the change in resonant frequency and its quality factor [7]. In the waveguide and cavity methods, the sample should fit exactly into the sample holder, and small misalignments can cause significant measurement errors.

In the free-space method, the antennas focus microwave energy at the measurement plane, and the sample is fixed at the common focal plane between the two antennas [8]. Since the sample is at the focal plane of the antenna and is not in contact with the applicator, it can be adapted easily for measurements at high or low temperatures and hostile environments.

INTA's Detectability Lab has the capability and know-how in free-space reflection measurements [9] and in this paper presents its first approach to the estimation of electromagnetic characteristics of materials via the application of bio-inspired algorithms. There are numerous examples of this kind of techniques focused on electromagnetics [10]-[13]. In the first part of this paper, the measurement method is described and the problem is presented. Then, Genetic Algorithms (GA) and Particle Swarm Optimization (PSO) are used to obtain the complex permittivity constant from artificial, simulated materials. After that, Artificial Neural Networks (ANN) are applied to the same set of materials to compare performance. Finally, these techniques are used on real measurements obtained from INTA's anechoic chamber.

2 Problem Formulation

The free-space method employed at INTA follows the configuration shown in Fig. 2.a, where a PC controls the positioner and a Vector Network Analyzer (VNA), which is also connected to a transmitting and receiving antenna. This method allows broadband and contactless measurements. Using this setup, S_{11} parameter is measured and reflection coefficient is obtained for a metal-backed sample (Fig. 2.b). From transmission line theory, reflection coefficient is related to complex permittivity and permeability via the following general equations (no assumptions or approximations for low losses materials have been made):

$$\Gamma_{\perp} = \frac{\sqrt{\frac{\mu_r^*}{\epsilon_r^*}} cos(\theta_i) tanh(jk_0 d\sqrt{\mu_r^* \epsilon_r^*} cos(\theta_t)) - cos(\theta_t)}{\sqrt{\frac{\mu_r^*}{\epsilon_r^*}} cos(\theta_i) tanh(jk_0 d\sqrt{\mu_r^* \epsilon_r^*} cos(\theta_t)) + cos(\theta_t)} . \quad (1)$$

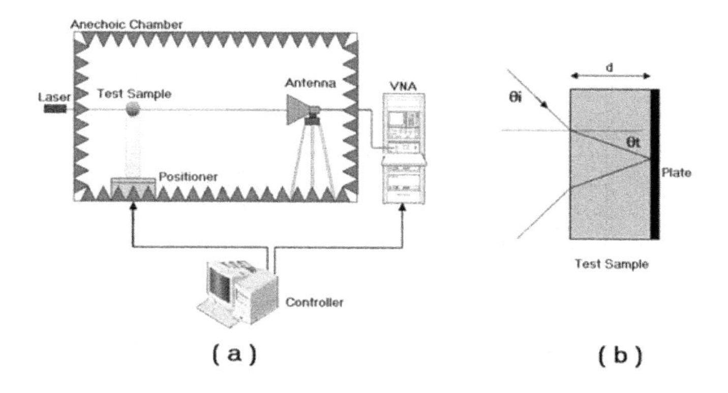

Fig. 2. Measures in anechoic chamber

$$\Gamma_{\|} = \frac{\sqrt{\frac{\mu_r^*}{\epsilon_r^*}} cos(\theta_t) tanh(jk_0 d\sqrt{\mu_r^* \epsilon_r^*} cos(\theta_t)) - cos(\theta_i)}{\sqrt{\frac{\mu_r^*}{\epsilon_r^*}} cos(\theta_t) tanh(jk_0 d\sqrt{\mu_r^* \epsilon_r^*} cos(\theta_t)) + cos(\theta_i)} . \quad (2)$$

where $\Gamma_{\perp}$ y $\Gamma_{\|}$ are perpendicular and parallel reflection coefficients, d is the sample thickness, $k_0 = \frac{2\pi}{\lambda}$ is the free-space wavenumber, θ_i is the incidence angle, θ_t the transmitted angle (Fig. 2.b) and ϵ_r^* y μ_r^* are relative complex permittivity and permeability:

$$\epsilon_r^* = \epsilon_r^{'} - j\epsilon_r^{''} . \quad (3)$$

$$\mu_r^* = \mu_r^{'} - j\mu_r^{''} . \quad (4)$$

Because ϵ_r^* and μ_r^* cannot be easily expressed in terms of the reflection coefficients and d, this paper propose to find them by bio-inspired processes.

3 Results

In this first approach by INTA, both simulated and real materials are used to test the algorithms proposed. As materials under test are non-magnetic, there is no need to measure off-normal so $\theta_i = 0$ and (1) and (2) are the same, so one of them is enough to extract real ($\epsilon_r^{'}$) and imaginary parts ($\epsilon_r^{''}$) of ϵ_r^*. For this reason, the synthetic materials created are non-magnetic and the problem is reduced to the estimation of the complex permittivity. This simplification diminish the complexity of the problem but doesn't limit its utility as the conclusions can be easily extrapolated to oblique incidence and μ_r^* determination.

3.1 Genetic Algorithms and Particle Swarm Optimization

Given electric permittivity, thickness, and frequency, reflection coefficients can be calculated from (1). Table 1 shows the fifteen different artificial materials

Table 1. Artificial materials

Material	$\epsilon_r^{'}$	$\epsilon_r^{''}$	$d(mm)$	Frequency (GHz)
AR1	2	0	2	8-12.4
AR2	5	0	2	8-12.4
AR3	10	0	2	8-12.4
AR4	10	1	2	8-12.4
AR5	10	10	2	8-12.4
AR6	2	0	1	8-12.4
AR7	5	0	1	8-12.4
AR8	10	0	1	8-12.4
AR9	10	1	1	8-12.4
AR10	10	10	1	8-12.4
AR11	2.45	0	0.796	8-12.4
AR12	2.55	0	1.589	8-12.4
AR13	2.01	0	1.539	8-12.4
AR14	9.8	0	1.234	8-12.4
AR15	2.2	5.2	1.2	8-12.4

generated. For each of these materials, GA and PSO are applied separately using the `MATLAB` *Genetic Toolbox* and a specifically developed code respectively. The main parameters of these algorithms are presented in Table 2. The output of both algorithms is excellent, as they can match the desired real and imaginary parts of the complex permittivity for all the cases at all the frequencies with negligible error. As an example, Fig. 3 shows the results obtained for the last

Table 2. GA and PSO parameters

GA		PSO	
Parameter	Value	Parameter	Value
PopInitRange	[1 0;20 20]	Range	[1 0;20 20]
PopulationSize	50	Population	20
EliteCount	4	C1, C2	2
Generations	100	Iterations	140
SelectionFcn	@selectionroulette	Initial inertia	0.9
CrossoverFcn	@crossoverintermediate,0.5	Final inertia	0.4

artificial material (AR15). To emulate the measurement error and evaluate its influence in the determination of $\epsilon_r^{'}$ and $\epsilon_r^{''}$, the reflection coefficient related to this material is contaminated with a gaussian error (zero mean and a variance of 0.5 dB in modulus and 0.1° in phase). The results obtained are nearly the same for GA and PSO (Fig. 4), and the influence of this errors becomes clear, deriving in an incorrect estimation of ϵ_r^*.

In the next step, a 20x20 cm real sample of Arlon© CuClad 250GX-0620 55 11 is measured in the anechoic chamber, and the reflection coefficient is treated

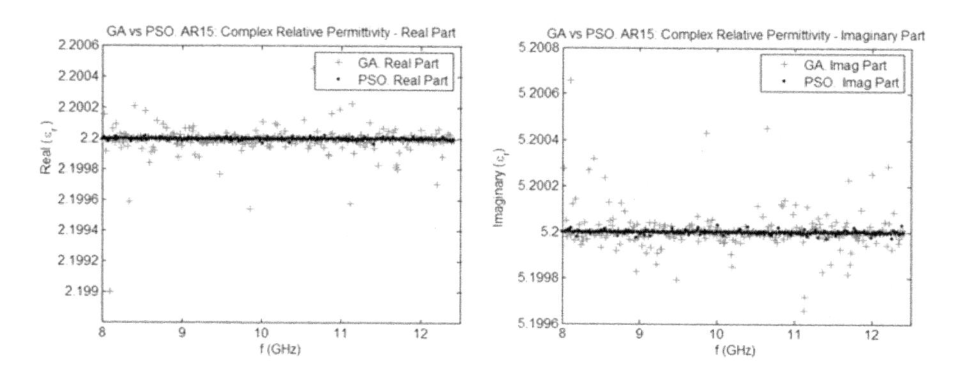

Fig. 3. Estimated real and imaginary part of AR15. GA vs PSO.

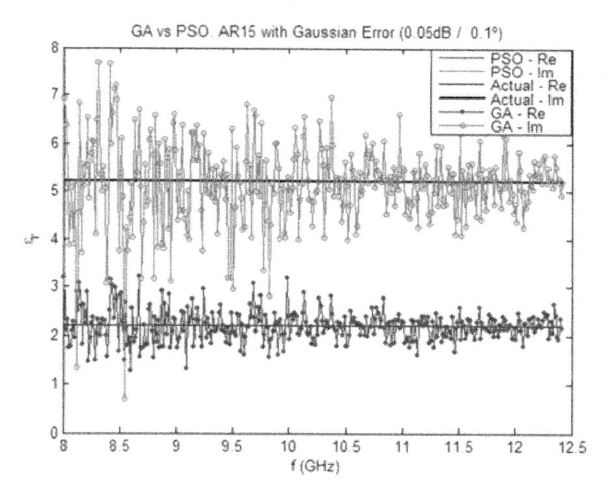

Fig. 4. GA and PSO results for AR15 reflection coefficient contaminated with noise

with GA and PSO to obtain the dielectric constant. The sample has a thickness of $d = 1.70\ mm$ and the manufacturer asserts that its nominal real part and loss tangent ($tan\delta = \frac{\epsilon_r{''}}{\epsilon_r{'}}$) are 2.55 and 0.0022 respectively, with minimum variations over a wide frequency band[1]. Comparisons with results obtained with the estimation presented in this paper are shown in Fig. 5, proving that bio-inspired optimization are a good and easy-to-implement alternative. This good results are supported by the fact that the actual measurement error is lower than the proposed for AR15.

[1] The measurement method followed by Arlon© is the accepted industry standard IPC TM-650 2.5.5.5, a stripline resonator test for permittivity and loss tangent (dielectric constant and dissipation factor) at X-Band–3/98.

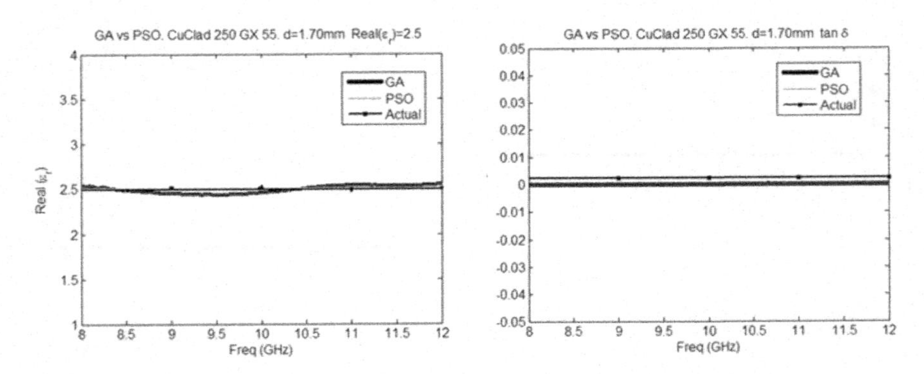

Fig. 5. Real part and loss tangent estimation with GA and PSO of CuClad 250GX

3.2 ANN

Using the *Neural Network Toolbox* provided by MATLAB, different multilayer feed-forward backprop networks are designed. All of them have four inputs (real and imaginary parts of the reflection coefficient, frequency and thickness), two outputs (real and imaginary parts of complex relative permittivity), and use, for training, validation and test, a set of twenty artificial materials where random values have been chosen: $\epsilon_r^{'} \in [1, 10]$, $\epsilon_r^{''} \in [0, 10]$ and $d \in [0.5, 2]$ mm.

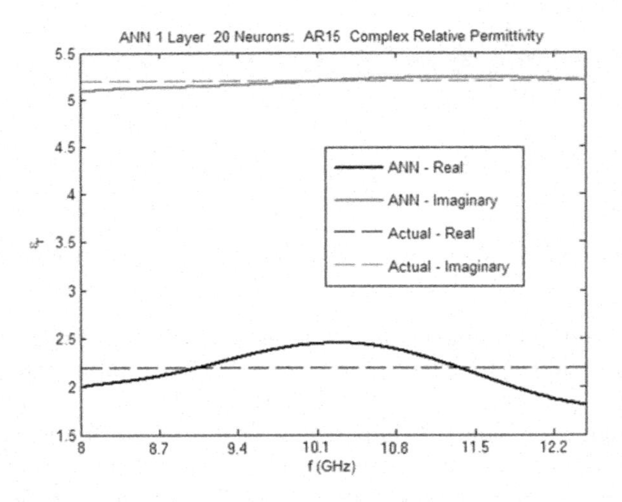

Fig. 6. ANN applied to artificial material AR15

Minmax normalization is applied for input and output parameters and `tansig`, `purelin` and `trainlm` are used as hidden layers activation function, output layer activation function and training function respectively. After grouping the materials by frequency, 1/2 of data is used for training, 1/4 for validation

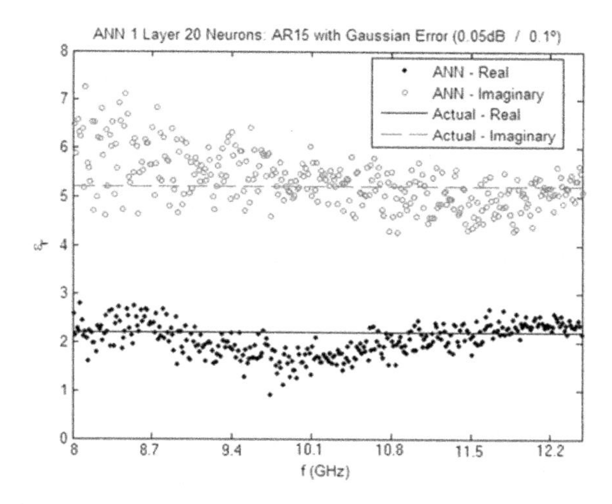

Fig. 7. ANN applied to contaminated artificial material AR15

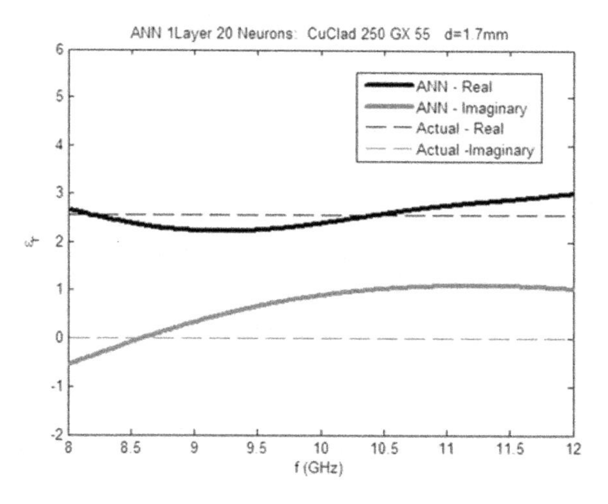

Fig. 8. ANN applied to measured Arlon© CuClad 250GX-0620 55 11

and 1/4 for test. Different architectures have been designed, namely with one hidden layer and 10, 15 or 20 neurons and with two hidden layers with 20-10 neurons or 25-15 neurons. The best output is achieved for the one layer and 20 neurons case. The result for AR15 material (Table 1) without noise is presented in Fig. 6 and with noise in Fig. 7. The measurements made for Arlon© CuClad 250GX-0620 55 11 are introduced to the trained network and the output obtained is presented in Fig. 8. Simulations show that GA and PSO have better performance than ANN except for the case of AR15 with noise.

4 Conclusions and Future Work

The application of bio-inspired techniques to dielectric constant estimation via free-space measurements has been presented. Results obtained are promising and demonstrate the validity of this approach. GA and PSO show better performance for actual measurements if the error is not high. On the other hand, ANN present better behavior in presence of high noise. Future work must include the training of networks with a set of actual measurements, and the experimentation with other architectures/topologies.

References

1. K.-P. Schmitt and E. Wölfle, '*Non-Cooperative Target Identification by Radar; State of the Art and Future*'. Proc. RTO Meeting on Non-Cooperative Air Target Identification Using Radar, Mannheim, Germany. (1998)
2. R. Miller, D. Shephard, '*Aspects of NCTR for Near-Future Radar*'. Proc. of SET-080 Target Identification and Recognition Using RF Systems, Oslo, Norway. (2004)
3. L. Du, H. Liu, Z. Bao and J. Zhang, '*A two-distribution compounded statistical model for radar HRRP target recognition*'. IEEE Transactions on Signal Processing, 54(6 I), 2226-2238. (2006)
4. I.Montiel, D. Poyatos, I. González, D. Escot, C. García, E. Diego, '*FASCRO Code and the Synthetic Database Generation Problem*'. Proc. of SET-080 Target Identification and Recognition Using RF Systems, Oslo, Norway. (2004)
5. A.W. Rihaczek, S.J. Hershkowitz, '*Theory and Practice of Radar Target Identification*'. Artech House Boston. (2000)
6. D.K. Misra, '*On the measurement of the complex permittivity of materials by an open-ended coaxial probe*'. IEEE Microwave Guided Wave Letter 5, 161163. (1995)
7. B. Meng, J. Booske, and R. Cooper, '*Extended cavity perturbation technique to determine the complex permittivity of dielectric materials*'. IEEE Trans Microwave Theory Tech MTT-43, 26332636. (1995)
8. J. Musil, F. Zacek, '*Microwave Measurements of Complex Permittivity by Free-Space Methods and Their Applications*'. Elsevier New York. (1986)
9. I. Montiel, '*INTA'S Free Space NRL arch System and Calibration for Absorber Material Characterization*'. AMTA 17th Meeting and Symposium, 323-328. Williamsburg, VA. (1995)
10. D.S. Weile, E. Michielssen, '*Genetic algorithm optimization applied to electromagnetics: A review*'. IEEE Transactions on Antennas and Propagation, 45 (3), pp. 343-353. (1997)
11. J. Robinson and Y. Rahmat-Samii, '*Particle Swarm Optimization in Electromagnetics*'. IEEE Transactions on Antennas and Propagation, vol. 52, No. 2. (2004)
12. D.W. Boeringer, D.H. Werner, '*Particle swarm optimization versus genetic algorithms for phased array synthesis*'. IEEE Transactions on Antennas and Propagation, 52 (3), pp. 771-779. (2004)
13. C. Christodoulou, M. Georgiopoulos, '*Applications of Neural Networks in Electromagnetics*'. Artech House Boston. (2001)

Context Information for Human Behavior Analysis and Prediction*

J. Calvo[1], M.A. Patricio[2], C. Cuvillo[1], and L. Usero[3]

[1] Dpto. de Organización y Estructura de la información, Universidad Politécnica de Madrid, Spain
[2] Grupo de Inteligencia Artificial Aplicada. Dpto. de Informática. Universidad Carlos III de Madrid, Spain
[3] Dpto. Ciencias de la Computación. Universidad de Alcalá, Spain
{ccuvillo,jcalvo}@eui.upm.es, mpatrici@inf.uc3m.es, luis.usero@uah.es

Abstract. This work is placed in the context of computer vision and ubiquitous multimedia access. It deals with the development of an automated system for human behavior analysis and prediction using context features as a representative descriptor of human posture. In our proposed method, an action is composed of a series of features over time. Therefore, time sequential images expressing human action are transformed into a feature vector sequence. Then the feature is transformed into symbol sequence. For that purpose, we design a posture codebook, which contains representative features of each action type and define distances to measure similarity between feature vectors. The system is also able to predict next performed motion. This prediction helps to evaluate and choose current action to show.

1 Introduction

In the last decade, we have witnessed a more user-centered implementation of computer science research. Ambient Intelligence (AmI) is a new paradigm that promotes the advancement of science and technology to build smart environments. AmI proponents advocate an invisible technological support layer of information processing to improve the quality of life in public and private spaces [1]. AmI puts forward the criteria for the design of intelligent environments, with ears and eyes [2]. AmI can monitor the user and can create a safety net around them, making their surroundings more secure and pleasant to live and inhabit. In AmI application, machines are able to understand information of their environments. Computer vision researches deal with this kind of problems, dedicated to interpret image sequences [3]

Last investigations in computer vision, since fixed images until recorded video sequences, has mainly concentrated in research on the evaluating of behavior recognition. Experiments in computer science research are concentrated on the

* Funded by project IMSERSO-AUTOPIA.

J. Mira and J.R. Álvarez (Eds.): IWINAC 2007, Part II, LNCS 4528, pp. 241–250, 2007.

development of new methods in analysis of data. Main tasks consist in carrying out extraction and processing as much information as possible about objects and humans in the scene [4]. Regards to behavior understanding, several methods for detection in specific domains can be found. Our approach focus in identification and classification human actions. This approach belongs to motion analysis, specifically, activity classification and motion detection.

Our paper presents an approach which is included into Indirect Model Use [5] from a blob (surrounding box of a mobile object) representation in two dimensions, 2D. We reference next some of the last related works that are carrying out nowadays, using similar techniques.

1.1 Related Works

Blob representation is normally described by some of the figure-ground segmentation approaches. The object or the subject is represented as a blob or number of blobs each having similar features. The similarities can be coherent flow [6], similar colours [7], or both [8]. The main philosophy of grouping information according to similarities in inspired by research into the human visual system by the Gestalt school in the 1930s [9].

Regards to pose estimation, our approach uses an indirect model. Example-based approaches use a database that describe poses in both image space and pose space. Our work is framed into look-up table, in our case this query information means structured data situated in codebooks, we will explain it in next sections. The methods in this class use an *a priori* model when estimating the pose of the subject. They use the model as a reference (similar to [10]) or query tables where relevant information may be obtained to drive the representation of extracted data. Mori and Malik [11] extract external and internal contours of an object. Shape contexts are employed to encode the edges. In an estimation step, the stored exemplars are deformed to match the image observation. In this deformation, the location of the hand-labelled 2D locations of joints also changes. The most likely 2D joint estimate is found by enforcing 2D image distance consistency between body parts. Shape deformation is also used by Sullivan and Carlsson [12]. To improve the robustness of the point transferral, the spatial relationship of the body points and color information is exploited. Loy et al. [13] perform interpolation between key frame poses based on [14] and additional smoothing constraints. Manual intervention is necessary in certain cases.

Referring to dimensionality, 2D models are the most suitable for our representation. These are appropriate for motion parallel to the image plane and they are sometimes used for motion analysis like in our approach. Ju et al. [6], Howe et al. [15] used a so-called Cardboard model in which the limbs are modeled as planar patches. Each segment has 7 parameters that allow it to rotate and scale according to the 3D motion. Navaratnam et al. [16] take a similar approach but model some parameters implicitly. In [17], an extra patch width parameter was added to account for scaling during in-plane motion. In [18], [19], [20], the human body is described by a 2D scaled prismatic model [21].

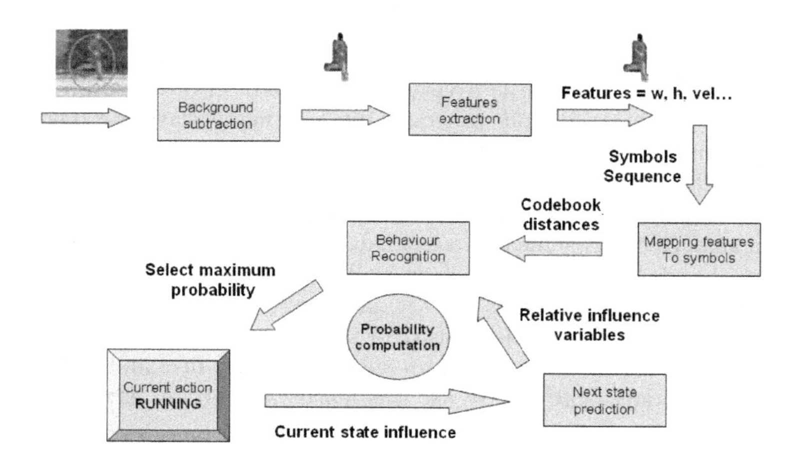

Fig. 1. General Scheme

2 System Overview

The proposed system is depicted in Figure 1. The system architecture is divided in four main parts, feature extraction, mapping features to codebooks, current state behavior recognition and next state behavior prediction. These parts are widely explained along next section.

For **feature extraction** we use background subtraction and threshold the difference between current frame and background image to segment the foreground object. After the foreground segmentation, we extract the blob features using techniques that will be described after this section. The extracted features are used for latter action recognition.

After feature extraction, a Symbols Sequence Vector is used to store the **mapping** between the codebooks of each behavior, (e.g. walking, running) and a mobile object. Each behavior to analyze is represented by a codebook. A virtual window runs along the Symbols Sequence, it is able to gets last states.

State **behavior recognition** is carried out by means of matches. These matches belong to one kind of motion, so that, we calculate the number of matches belonging to each behavior using distance measure.

Using some probability computation, we add the necessary information about **next state prediction**. Prediction has a feedback with the chosen current state and so on (see Figure 1).

2.1 Feature Extraction

Human action is composed of a series of postures over time. Once an mobile object is extracted from the background, it is represented by its boundary shape. A filtering mechanism to predict each movement of the detected object is a common tracking method. The filter most commonly used is the Kalman filter [22].

A good feature extraction is a very essential part in order to be successful. Features can be obtained from processed blob in many ways, we have selected simple information like blob width and height, and register blob position changes by means of *Kalman filter*.

The *Kalman filter* is a set of mathematical equations that provides an efficient computational (recursive) means to estimate the state of a process, in a way that minimizes the mean of the squared error. The filter is very powerful in several aspects: it supports estimations of past, present, and even future states, and it can do so even when the precise nature of the modeled system is unknown.

These equations provides us features such as velocity in x and y axes, height, width and blob identification. We use this features to obtain data over time.

Input: Blob representation

Output: A set of blob features

1. Calculating a global velocity, which is able to give us information about blob position, regards x and y axes at the same time. We use this velocity to estimate the variation of the blob position with regards pixels over frame.

$$BlobVelocity = \sqrt{v_x^2 + v_y^2}$$

   where v_x and v_y are the velocity component in axis x and y obtained by Kalman filter. We store this changes of position inside an static array that can preserve its value along the execution algorithm.
2. Obtaining a difference between height and width. It will be a very useful data, as we will see later, in order to measure human motion changes.

$$difHW = BlobHeight - BlobWidth$$

   We store again the relation between height and width during the whole execution.
3. Storing blob area from Kalman data. This number is given beyond the post processing blob tracking over Kalman equations.
4. Identifying blobs. Each blob registered by tracking will have a number of ID.

2.2 Mapping Features to Symbol

To apply our probability model and next state prediction to time-sequential video, the extracted features must be transformed into symbol sequence for latter action recognition. This is accomplished by a technique called: *Symbols Sequence*.

Symbols Sequence Vector. In order to explain symbols sequence vector, we must introduce the concept of *Codebook*. A Codebook is a number of symbols. Each symbol matches with a representative posture feature inside a human action. We have one Codebook per human action (e.g. codebook for walking).

C_{jn} *where* $j \in$ *Human Action* $\forall\, n \in$ *Number Of Symbols Set.*
(*e.g.* C_{r18} *would be symbol* 18 *inside running codebook.*)

These numbers can vary depending on the target of the application (e.g. security will add a new codebook for *figthing*). We can also extend the number of symbols representing each action for improving the level of accuracy.

Each symbol in each codebook has some associated properties. When the extracted blob is processed, we obtain the features mentioned before. This features represent the action that the human is carrying out. So that, we compare the symbols of each codebook with the extracted features by minimal distance. It is the moment when we introduce the selected symbols in the Symbols Sequence Vector (SSV). The SSV has the same size number as the number of frames in the video sequence.

2.3 Behavior Recognition

The action recognition module involves two submodules: recognition and prediction. These two phases are working together and the selected data are moving from one to another submodule. There exists a kind of feedback, we can not know the behavior without the help of prediction, and of course, we can not predict without the knowledge of the current state of behavior.

Once we have a number of symbols in the SSV, the system begins to scan this vector through a systematic procedure called sliding Window. This Window has a variable *size W*, which depends, once again, on the application target. This Window allows us to know the current symbol that matches with the current state of behavior and we can also use the symbols that matches with the last *n states* of behavior. The system gets these last and current symbols and it compares them with the declared codebooks. When a symbol matches with one of the symbols inside a codebook, we add a new value to a vector associated with each codebook. At the end of this process we check how many values have been added in these vectors. In order to know the probability of a current state, the system multiply these added values by a number. This number is obtained dividing the maximum probability of an event by the size of the Window, that is, the number of states that we want to observe. Finally, the event with the greater probability is the selected state.

It is very important to bear in mind that this overall process happens very fast. If we choose, for example, to observe the last five frames in order to determine the current state of behavior, we are watching no more than half a second in the video sequence.

2.4 Next State Prediction

The following is a more complex situation. What we have mentioned above could be used for classifying single actions. A man performs a series of actions, and we recognize which action he is performing now. One may want to recognize the action by classification of the posture at current time T. However, there is a problem. By observation we can classify postures into two classes, including key postures and transitional postures. Key postures uniquely belong to one action so that people can recognize the action from a single key posture. Transitional postures are interim between two actions, and even human cannot recognize the action from a single transitional posture. Therefore, human action can not recognized from posture of a single frame due to transitional postures. So, we refer a period of posture history to find the action human is performing. We need to use the same sliding window as before for real-time action recognition. At current time T, symbol subsequence between T-W and T, which is a period of posture history, is used to recognize the current action by computing the maximal likelihood. As we have seen before, each codebook is composed by symbols. These symbols are ordered over the Real Line, (e.g. symbols of the walking codebook could be 13, 14, 15, 16). With the value of these symbols we can make an arithmetic mean (e.g. in last example it would be 14.5). Therefore, this arithmetic mean can be made over any set of symbols that belong to codebooks.

In fact, this is the case of the Window. Depending of its size W, we can also get an arithmetic mean. Thus, we can compare the arithmetic mean that belongs to the codebook, that represents the current posture, with the arithmetic mean of symbols inside the sliding Window scheme. By means of measuring distances between the arithmetic means of the possible future postures, we assign a relative influence to two variables, defined as *alfa* (α) *and beta* (β).

If the distance between the arithmetic mean of the codebook symbols, that represent current posture, is smaller than its own arithmetic mean, we can guess that the posture is changing to another. Therefore, the system assigns a big relative influence to α, otherwise, big relative influence goes to β. It is obvious that we can find several cases. So that, α and β values are continuously changing. In order to predict the next state, we use these variables together with current posture number and a predicted posture one. These numbers are very easy to obtain. They came from the last iteration, it is very simple, each state has a number. The result of the next equation is also a number that is next to one of those state numbers mentioned before, therefore, the result would be the next state. It is well known that the predicted posture can be the same as the current one if the human is not performing any change.

$$NextState = (\alpha * CurrentStateNumber) + (\beta * PredictedStateNumber)$$

In one word, the system is selecting two positions per frame, current and predicted position.

3 Experimental Results

In order to evaluate the quality and the performance of our work, we have implemented a system that is able to identify and recognize a series of actions. Recognition over series of actions is carrying out in a real-time framework. The proposed system is able to show the current action that the human is performing and the predicted behavior. Predicted behavior has a relative influence in current action as we have explained before.

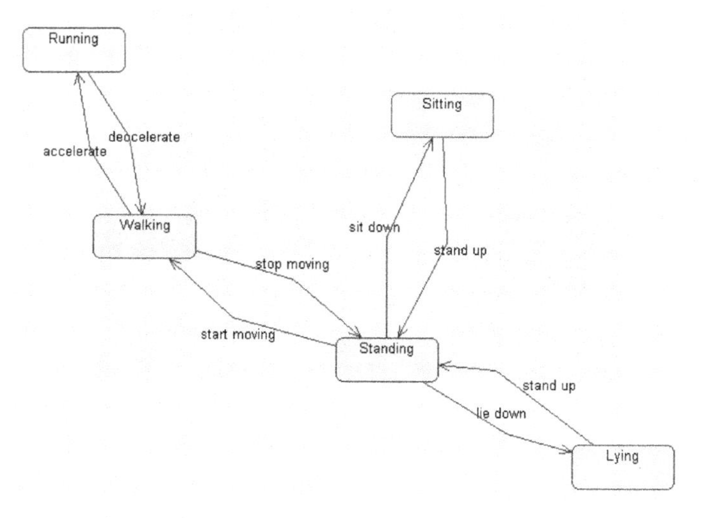

Fig. 2. States Diagram used in experiments

The video content was captured by a digital camera. The number of frames is chosen experimentally. Too short sequences do not provide enough time for refreshing the necessary data, on the other side, too long sequences are quite difficult to manage. Video sequences have been recorded outside. We remembered not to carry out many assumptions, therefore we consider that the system is quite solid. Recording outside is even more difficult but more realistic, we do not believe in prepared scenarios that bear in mind a big quantity of assumptions, (see Figure 4).

In our sequence of experiments, we have tried to keep one subject performing the actions. Five Codebooks have been declared, one codebook per action. Each codebook is composed by four symbols, (see Figure 3).We can observe the different actions considered in the states diagram, (see Figure 2).Transitions in the diagrams represent the level of freedom for human behavior.

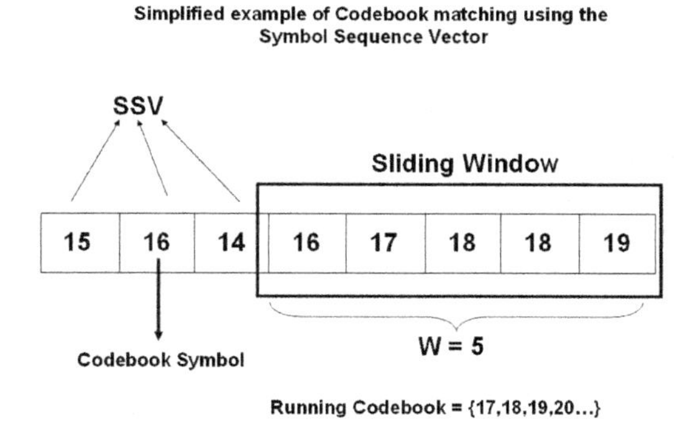

Fig. 3. Simplified example of matching

Fig. 4. Example of recorded frame

%	Run	Walk	Stand	Sit	Lie down
Run	89	2			
Walk	11	98	4		
Stand			91	10	
Sit			5	88	8
Lie down				2	92

Fig. 5. Confusion matrix for recognizing a series of action

Observing Confusion Matrix, (see Figure 5), it is easy to appreciate results with more than 90% of success. These results belong to the proposed experiment mentioned above. Percentage refers to the total number of recorded frames. Each frame has its ground truth, so that, we have measure how many frames were correctly recognized. The left side is the ground truth of action types, and the upper side is the recognition action types. The number over the diagonal is the percentage of each action which is correctly classified. The percentages which are out of the diagonal are errors related to action recognition, but we can know which kind of action is identified by the system. Observing the table we can appreciate that most of the postures sequences were correctly classified. A great recognition rate around 90% was achieved by the proposed method. Confusions happen when the human is not moving, depending on the accurate of tracking, centroid is shaking, so that, system interprets it like a movement. The unknown period is the time during which human performs actions that are not defined, transitions between postures.

4 Conclusions and Future Work

We have showed an interesting and solid system for human action recognition based on the information of the extracted features. These symbols are transformed into numbers when they are mapped into a structure called Symbols Sequence Vector. Action recognition is achieved by distances between these extracted features and codebooks, one codebook per human action. The recognition accuracy could still be improved whether tracking system were optimized. The system is also able to predict the next state, helping with some relative influence to show the current action performed. The system is able to identify a series of actions in an outside background that can be changeable. Not many assumptions and constraints are required.

The system could be improved in some ways. We can transform that distance method into a fuzzy logic structure. Therefore, the future of our approach is training the system. The system could also be scaled up, that is, we can add more and more kind of actions, depending on the target of the system. It would be easy to achieve some experiments, i.e security applications, we only should add new states representing the chosen field with the suitable codebooks like *fighting, snooping, stealing*, etc.

References

1. Nigel Shadbolt. Ambient intelligence. *IEEE Intelligent Systems*, 18(4):2–3, 2003.
2. Paolo Remagnino, Gian Luca Foresti, and Tim Ellis. *Ambient Intelligence: A Novel Paradigm*. Springer-Verlag Telos, 2004.
3. F. Castanedo, M. A. Patricio, J. García, and J. M. Molina. Extending surveillance systems capabilities using bdi cooperative sensor agents. In *VSSN '06: Proceedings of the 4th ACM international workshop on Video surveillance and sensor networks*, pages 131–138, New York, NY, USA, 2006. ACM Press.
4. M. A. Patricio, J. Carbó, O. Pérez, J. García, and J. M. Molina. Multi-agent framework in visual sensor networks. *EURASIP Journal on Advances in Signal Processing*, 2007:Article ID 98639, 21 pages, 2007. doi:10.1155/2007/98639.
5. Thomas B. Moeslund and Erik Granum. A survey of computer vision-based human motion capture. *Computer Vision and Image Understanding: CVIU*, 81(3):231–268, 2001.
6. S. X. Ju, M. J. Black, and Y. Yacoob. Cardboard people: A parameterized model of articulated image motion. In *Proceedings of the International Conference on Automatic Face and Gesture Recognition (FGR96)*, page 3844, Dept. of Comput. Sci., Toronto Univ., Ont., 1996.
7. B. Heisele and C. Wohler. Motion-based recognition of pedestrians. In *In Intenational Conference of Pattern Recognition*, Brisbane, Qld., Australia, 1998.
8. Christoph Bregler. Learning and recognizing human dynamics in video sequences. In *Computer Society Conference on Computer Vision and Pattern Recognition (CVPR'97)*, U.C Berkeley, 1998.
9. K. Koffka. *Principle of Gestalt Psychology*. Harcourt Brace, 1935.
10. Óscar Pérez, Massimo Piccardi, Miguel Ángel Patricio, Jesús García, and José M. Molina. Comparison between genetic algorithms and the baum-welch algorithm in learning hmms for human activity classification. In *EvoWorkshops*, Lecture Notes in Computer Science. Springer, 2007.
11. G. Mori and J. Malik. Recovering 3d human body configurations using shape contexts. In *IEEE Transactions on Pattern Analysis and Machine Intelligence (PAMI)*, page 10521062, Div. of Comput. Sci., California Univ., Berkeley, CA, USA, 2006.

12. D. Liebowitz and Stefan Carlsson. Uncalibrated motion capture exploiting articulated structure constraints. In *International Journal of Computer Vision 51*, page 171187, KTH, SE-100 44 Stockholm, Sweden, 2003.
13. G. Loy, M. Eriksson, J. Sullivan, and S. Carlsson. Monocular 3d reconstruction of human motion in long action sequences. In *Proceedings of the European Conference on Computer Vision (ECCV04)*, page 442455, Prague, Czech Republic, 2004.
14. C. J. Taylor. Reconstruction of articulated objects from point correspondences in a single uncalibrated image. In *Computer Vision and Image Understanding CVIU)*, page 349363, Dept. of Comput. & Inf. Sci., Pennsylvania Univ., Philadelphia, PA, 2000.
15. N. R. Howe, M. E. Leventon, and W. T. Freeman. Bayesian reconstruction of 3d human motion from single-camera video. In *Advances in Neural Information Processing Systems (NIPS)*, page 820826, Denver, CO, 2000.
16. R. Navaratnam, A. Thayananthan, P. Torr, and R. Cipolla. Hierarchical part-based human body pose estimation. In *Proceedings of the British Machine Vision Conference (BMVC05)*, Oxford, United Kingdom, 2005.
17. Y. Huang and T. S. Huang. Model-based human body tracking. In *Proceedings of the International Conference on Pattern Recognition (ICPR02)*, page 552555, Quebec, Canada, 2003.
18. T.-J. Cham and J. M. Rehg. A multiple hypothesis approach to figure tracking. In *Proceedings of the Conference on Computer Vision and Pattern Recognition (CVPR99)*, page 239245, Ft. Collins, CO, 1999.
19. A. Agarwal and B. Triggs. Tracking articulated motion using a mixture of autoregressive models. In *Proceedings of the European Conference on Computer Vision (ECCV04) - volume 3, no. 3024 in Lecture Notes in Computer Science*, page 5465, Prague, Czech Republic, 2004.
20. T. J. Roberts, S. J. McKenna, and I. W. Ricketts. Human pose estimation using learnt probabilistic region similarities and partial configurations. In *Proceedings of the European Conference on Computer Vision (ECCV04) - volume 4, no. 3024 in Lecture Notes in Computer Science*, page 291303, Prague, Czech Republic, 2004.
21. J. M. Rehg D. D. Morris. Singularity analysis for articulated object tracking. In *Proceedings of the Conference on Computer Vision and Pattern Recognition (CVPR98)*, page 289297, Santa Barbara, CA, 1998.
22. Markus Kohler. Using the kalman filter to track human interactive motion - modelling and initialization of the kalman filter for translational motion. Technical Report 629/1997, Fachbereich Informatik, Universität Dortmund, 44221 Dortmund, Germany, 1997.

Road Sign Analysis Using Multisensory Data

R.J. López-Sastre, S. Lafuente-Arroyo, P. Gil-Jiménez, P. Siegmann, and S. Maldonado-Bascón

University of Alcalá, Department of Signal Theory and Communications
Polytechnic School, A-2 Km. 33,600 - 28805 - Alcalá de Henares - Madrid, Spain
{robertoj.lopez, sergio.lafuente, pedro.gil, philip.siegmann, saturnino.maldonado}@uah.es

Abstract. This paper deals with the problem of estimating the following road sign parameters: height, dimensions, visibility distance and partial occlusions. This work belongs to a framework whose main applications involve road sign maintenance, driver assistance, and inventory systems. From this paper we suggest a multisensory system composed from two cameras, a GPS receiver, and a distance measurement device, all of them installed in a car. The process consists of several steps which include road sign detection, recognition and tracking , and road signs parameters estimation. From some trigonometric properties, and a camera model, the information provided by the tracking subsystem and the distance measurement sensors, we estimate the road signs parameters. Results show that the described calculation methodology offers a correct estimation for all types of traffic signs.

1 Introduction

Several works have recently focused on traffic sign detection and recognition [1], [2], [3], [4], [5] and [6]. Specifically, in [7] it is described the framework that we use in this research. In this paper we handle the task of automatically estimating height, dimensions, visibility distance and partial occlusions of road signs. The aim of this work is to get an efficient framework which could be used in a inventory system, which could provide all of these estimated parameters for each sign, and not only the type. From this work, we present a complete development based on trigonometric relationships and a camera model to estimate the listed before parameters of road signs.

This paper is organized as follows. Section II shows a global vision of our system, and in section III we describe all the implemented algorithms to achieve the traffic sign parameters estimation. Sections IV and V report the obtained results and present the conclusions of this work, respectively.

2 System Overview

For this work we have built a complete inventory system as it is shown in Fig. 1, which has the following subsystems:

J. Mira and J.R. Álvarez (Eds.): IWINAC 2007, Part II, LNCS 4528, pp. 251–260, 2007.

- **Capture subsystem.** We have mounted two firewire cameras on a car. The first one is entrusted to capture the road signs with a framerate of 15 fps (frames per second), and with the second one we want to capture the milestones of the route. This subsystem synchronizes the capture of these two cameras, and it saves the captured images in the hard disk of a laptop.
- A **Positioning subsystem** which is composed of a GPS (Global Positioning System) receiver and a Distance Measurement Device (DMD). The aim of this block is to acquire the GPS position of the vehicle, each second, and to measure the covered distance by the car. This subsystem sends all of these measured data to the Traffic sign analysis subsystem.
- **Traffic sign analysis subsystem**, which will develop two task. Firstly, the systems tries to detect, recognize and track the traffic signs in the images, and in a second step, the system performs the estimation of parameters.

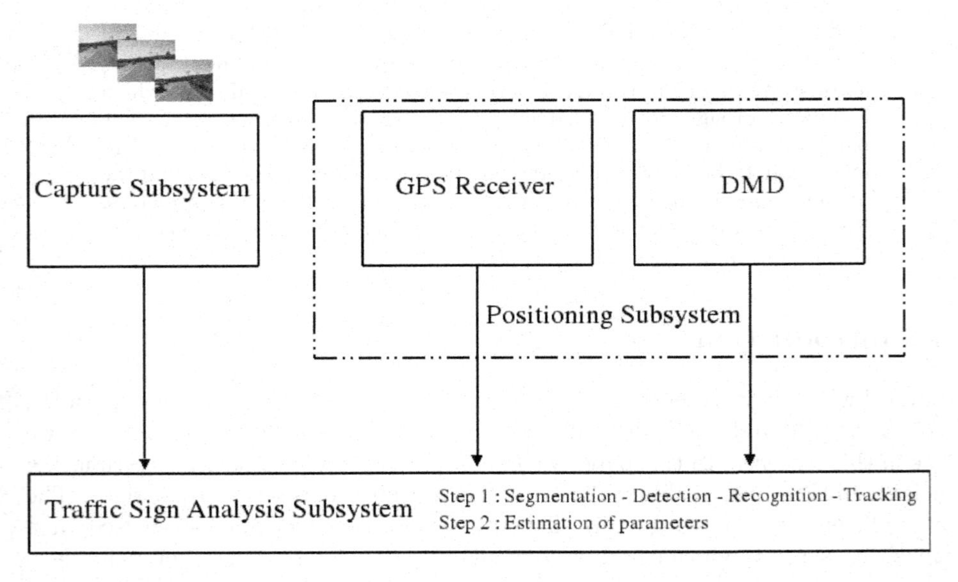

Fig. 1. Diagram with all subsystems which are part of the complete system

In Fig. 2 we show a complete flow diagram with all the tasks that this system realizes. In the last step it is where we implement the parameters estimation process which is the main focus of this paper. It is important to note that this estimation step is possible only after the system has completed the identification of a road sign, which implies four stages: segmentation, detection, recognition and tracking. Specifically, in [7] the author describes all of these steps.

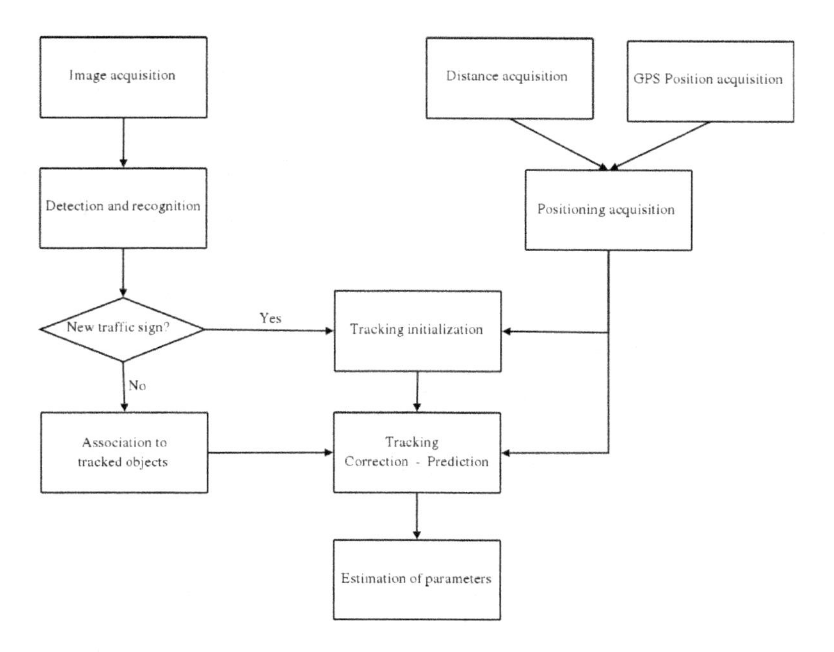

Fig. 2. System flow diagram

3 Traffic Sign Parameters Estimation

There are at least 350 of different road sign types. As we have described in the previous section, our system allows to detect and identify road signs in a sequence of video. This kind of systems require detailed knowledge of the sign, with the additional purpose of reducing false alarms and to make an inventory not only with the type of the found road sign during the route, but also with some parameters of each sign: height, dimensions, visibility distance, partial occlusion, GPS position, distance of separation between them. For an inventory system, these kind of data could be used to verify if the dimensions are standard, if two consecutive road signs are too closer, or if it exists any kind of partial occlusion which complicates that a driver can perceive the signal.

To obtain these estimations of parameters we have considered a pin-hole camera model, as shows Fig. 3, where they are known the following parameters: the focal length (f), size of each unit cell (u_{ch} and u_{cw}, which are the height and width of a cell, respectively) in the Charge-Coupled Device (CCD) of the camera, the height of the camera (h_c), the height of the middle point of a captured image (h_m), and the distance between the camera plane and the middle point plane of a captured image d_m. Then, before starting to record a route, we have to complete a calibration step, which consists in to measure h_c, h_m and d_m. We develop this step on a known area, where we can measure these distances without difficulties.

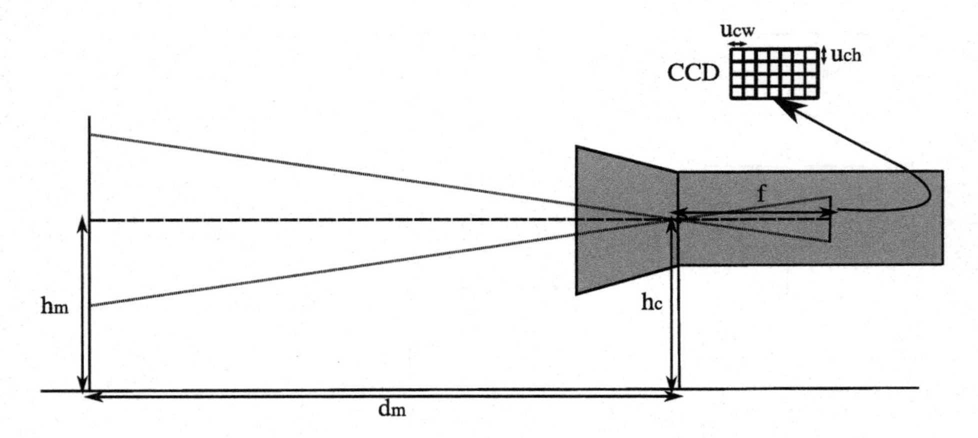

Fig. 3. Calibration of the system

3.1 Height Estimation

Our first objective is to estimate the height of the sign. Ideally, we could consider that the plane where is the road sign, and the plane where is projected the image in the camera, are parallel, but it does not happen in a real situation, because the sign or the camera, or both, can be inclined. We have to take into account this consideration to design a model which lets us to estimate the height of an object. Anyway, we must to define the model where camera and sign plane are parallel, and then, we will extend this explanation to the real situation.

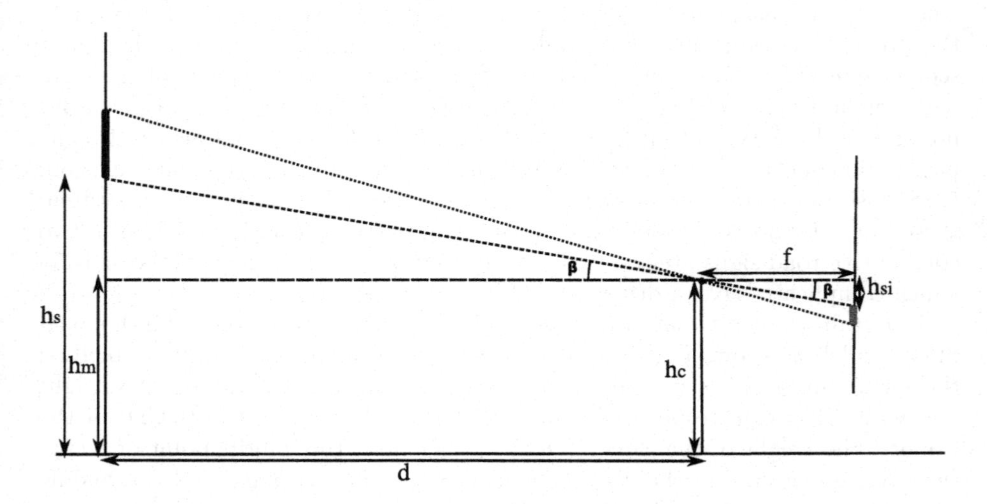

Fig. 4. Ideal situation where camera and road sign plane are parallel

Figure 4 shows the created model to estimate the height of a road sign. After the detection and recognition process, we have the position of the identified road sign in the image $h_{si} = h_{sp}u_{ch}$, where h_{sp} is the height in pixels and u_{ch} is the height in mm of each pixel. The Positioning subsystem gives the covered distance between two images where the recognition subsystem had identified the same road sign. Then, knowing these data and that $h_c = h_m$ we can establish that

$$\tan(\beta) = \frac{h_{si}}{f} , \tag{1}$$

$$\tan(\beta) = \frac{h_s - h_c}{d + x} , \tag{2}$$

where x is the covered distance between two consecutive captured images where the signal had been recognized. For the second captured image we can define

$$\tan(\beta') = \frac{h'_{si}}{f} , \tag{3}$$

$$\tan(\beta') = \frac{h_s - h_c}{d} . \tag{4}$$

From both pair of equations we can write

$$\frac{h_{si}}{f} = \frac{h_s - h_c}{d + x} , \tag{5}$$

$$\frac{h'_{si}}{f} = \frac{h_s - h_c}{d} . \tag{6}$$

These last two equations form the following system of equations

$$h_s - \frac{h'_{si}}{f} d = h_c , \tag{7}$$

$$h_s - \frac{h_{si}}{f} d = h_c + \frac{h_{si}}{f} x , \tag{8}$$

where the unknowns are h_s and d.

Our objective is to obtain the height of the road sign h_s which is

$$h_s = h_c + \frac{h'_{si} h_{si} x}{f\left(h'_{si} - h_{si}\right)} . \tag{9}$$

We only have considered the situation in which the road sign is over the middle point of the captured image. Of course, we have two possible situations more. The traffic sign can be at the same height that the middle point, then $h_s = h_m$. The last situation is that $h_s < h_m$, and then we can write

$$\tan(\beta) = \frac{h_{si}}{f} , \tag{10}$$

$$\tan(\beta) = \frac{h_c - h_s}{d + x} , \tag{11}$$

$$\tan(\beta') = \frac{h'_{si}}{f} \quad , \tag{12}$$

$$\tan(\beta') = \frac{h_c - h_s}{d} \quad . \tag{13}$$

Now h_s is defined as follows

$$h_s = h_c - \frac{h'_{si} h_{si} x}{f\left(h'_{si} - h_{si}\right)} \quad . \tag{14}$$

In a real approach we must consider that the road sign and the camera plane are not parallel. Figure 5 shows this situation. Now, the angle $\alpha \neq 0$, and in a first stage of calibration we must measure this angle by

$$\alpha = \arctan\left(\frac{h_m - h_c}{d_m}\right) \quad . \tag{15}$$

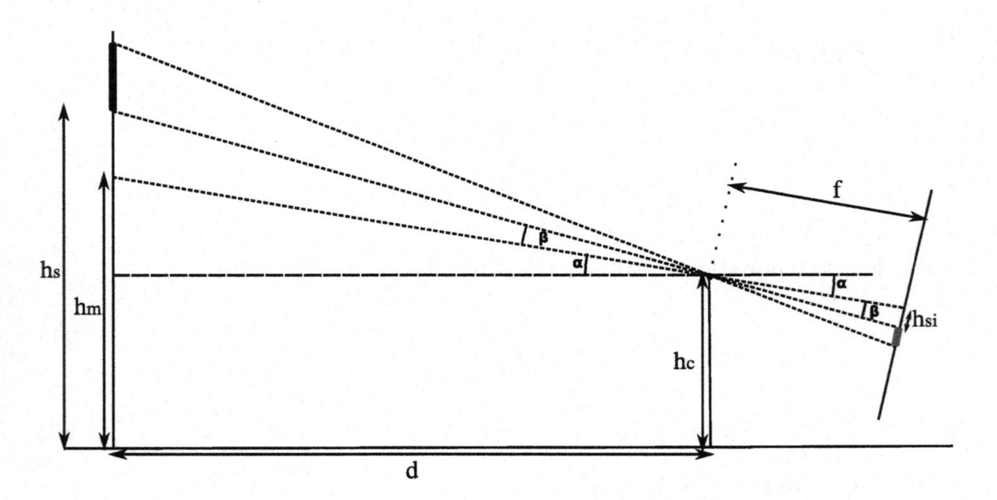

Fig. 5. Real situation where camera and road sign plane are not parallel

From this point, with $\alpha > 0$, we can define

$$\beta = \arctan\left(\frac{h_{si}}{f}\right) \quad , \tag{16}$$

$$\tan(\alpha + \beta) = \frac{h_s - h_c}{d + x} \quad . \tag{17}$$

For the next captured image we have

$$\beta' = \arctan\left(\frac{h'_{si}}{f}\right) \quad , \tag{18}$$

$$\tan(\alpha + \beta') = \frac{h_s - h_c}{d} \ . \tag{19}$$

Then, we have the following system of equations

$$h_s - \tan(\alpha + \beta)\, d = h_c + \tan(\alpha + \beta)\, x \ , \tag{20}$$

$$h_s - \tan(\alpha + \beta')\, d = h_c \ . \tag{21}$$

From this point, following the same manner of reasoning that in the ideal situation, we will have that h_s must be estimated as follows

$$h_s = h_c - \frac{x \tan(\alpha + \beta) \tan(\alpha + \beta')}{\tan(\alpha + \beta) - \tan(\alpha + \beta')} \ , \tag{22}$$

Also, if we consider the situation where $h_s < h_m$, then h_s will be estimated as

$$h_s = h_c - \frac{x \tan(\alpha - \beta) \tan(\alpha - \beta')}{\tan(\alpha - \beta) - \tan(\alpha - \beta')} \ . \tag{23}$$

This mathematical study, about the trigonometric relationships between the real world and the projected image in a camera, lets us to estimate the height of a sign successfully.

3.2 Dimensions Estimation

To estimate the dimensions of a road sign is a task which is related with the type of the sign. We must know, before the dimensions estimation step, the answer to the following question: what kind of traffic sign do we have recognized?. With this information, we only have to apply the same algorithm that we have described in the previous section, but not only for the botton point of the sign. The detection and recognition subsystem gives us the vertexes of the rectangle where the road sign is confined. For a correct dimensions estimation, we have to estimate the height of the top and botton corner of this rectangle. Then, we can compute all the dimensions for every type of road sign.

3.3 Visibility Distance Estimation

We define the visibility distance as the distance where the road sign has been detected for the first time by our system. This parameter is crucial for and inventory system which wants to measure the level of safety of a road. To estimate this parameter, we have to take up again the approach to estimate the height of a road sign. Figure 4 shows the model for a ideal situation where road sign and camera plane are parallel. Equations (7) and (8) compose a systems of equations, where the unknowns are d and h_s. We can work out the value of d, which is the visibility distance that we are searching.

$$d = \frac{h_{si} x}{h'_{si} - h_{si}} \ . \tag{24}$$

In a real situation we have to use the model presented in Fig. 5. Then, we can estimate the visibility distance as

$$d = \frac{\tan(\alpha + \beta)\, x}{\tan(\alpha + \beta') - \tan(\alpha + \beta)} \quad \text{if } h_s > h_m \tag{25}$$

$$d = \frac{\tan(\alpha - \beta)\, x}{\tan(\alpha - \beta') - \tan(\alpha - \beta)} \quad \text{if } h_s < h_m \tag{26}$$

3.4 Partial Occlusion Estimation

Many times, traffic signs appear occluded by other objects like trees, vehicles, other road signs, etc. This traffic signs analysis system gives an estimation of partial occlusions, and to have this information of each sign results crucial for a complete inventory system which measures the level of maintenance of a road, including the safety. To estimate this parameter, we use the information provided by the tracking subsystem: if the same road sign is detected in non-consecutive frames, we can determine that a partial occlusion was happen.

4 Results

The current data of several traffic signs have been obtained for a sequence in a two ways road. For testing our approach we have measured the height and dimensions of some road signs, which are presented in Fig. 6.

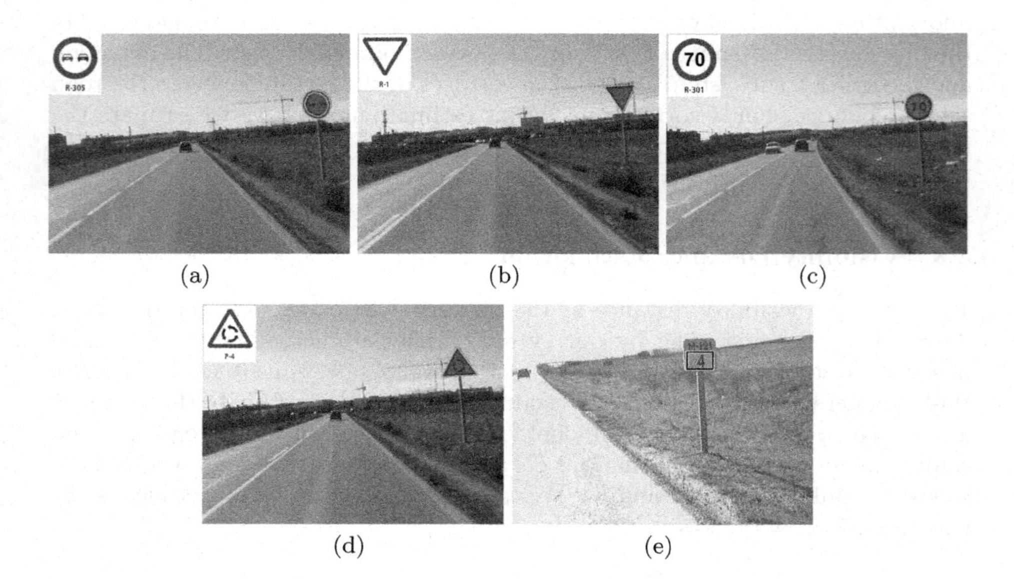

Fig. 6. Some detected and recognized road signs

Table 1. Parameters estimation for some road signs

Road Sign	Real Height (m)	Estimated Height (m)	Error (%)
a)	1.95	1.90	2.5
b)	2.16	2.13	1.38
c)	1.84	1.81	1.6
d)	1.92	1.87	2.6

Our test sequences of images have been recorded with two firewire cameras fixed onto the hood and the roof of the car. The captured images have a resolution of 640×480, and in a calibration step we have measured h_c, h_m, f, d_m and u_{ch}. Figure 6 shows in image e) a milestone detected by our system, and Table 1 shows the estimated height for each type of sign and the error commited.

We can observe that the error obtained is not too high, and the parameter estimated is closer to the real value. The worse the measurement in the calibration step, the larger the error in the estimation. The most important cause of error in the estimation process is the measurement of the variable x, which indicates the covered distance between two captured images. The Positioning subsystem refreshes this variable each 1.6 meters, because this is the resolution of our DMD, then we do not have the exact value of this parameter. The presented mathematical procedure works well when all the variables are measured correctly, and the error obtained in this situation is about 0.8%. But the error in this estimation can be caused by some other factors: lens distortion, an incorrect detection of the sign in the image. The dimensions of each sign are estimated following the same process, but for two point of the signal, as we have described in the previous sections.

Table 2 shows some estimated visibility distances.

Table 2. Visibility Distance Estimation

Road Sign	Visibility Distance (m)
a)	41.2
b)	37.6
c)	42.7
d)	21.6

5 Conclusions

This paper describes a complete framework to estimate height, dimensions, partial occlusions and visibility distance of traffic signs. The integration of the recognition, tracking and positioning subsystems, allows to estimate these parameters automatically, following the mathematical approach presented from this work.

Future lines of work will focus on estimating more attributes of each sign, such as deformation level, distance from the sign to the margin of the road, level of slope. On the other hand, we will work in a more complete system which will have calibrated cameras and a Positioning subsystems which improves the estimation of covered distance. Another future objective will be to complete the system with an stereo vision framework, in order to improve these estimations of parameters.

Acknowledgment

This work was supported be the project of the Ministerio de Educación y Ciencia of Spain number TEC2004/03511/TCM.

References

1. Maldonado, S., Lafuente, S., Gil, P., Gómez, H., López, F.: Road-sign detection and recognition based on support vector machines. IEEE Trans. on Intelligent Transportation Systems (2006)
2. de la Escalera, A., Armingol, J.M., Pastor, J.M., Rodríguez, F.J.: Visual sign information extraction and identification by deformable models for intelligent vehicles. IEEE Trans. on Intelligent Transportation Systems **15** (2004) 57–68
3. Fang, C., Chen, S.: Road sign detection and tracking. IEEE Trans. on Vehicular Technology **52** (2003) 1329–1341
4. Lafuente Arroyo, S., García Díaz, P., Acevedo Rodríguez, F., Gil Jiménez, P., Maldonado Bascón, S.: Traffic signs classification invariant to rotations using support vector machines. In: Proceedings of Advabced Concepts for Intelligent Vision Systems, Brussels, Belgium (2004)
5. de la Escalera, A., Moreno, L.E., Salichs, M.A., Armingol, J.M.: Road traffic sign detection and classification. IEEE Trans. on Industrial electronics **44** (1997) 848–859
6. Aoyagui, Y., Asakura: A study on traffic sign recognition in scene image using genetic alhorithms and neural networks. In: Proceedings IEEE Int. Conf. Industrial Electronics, Control and Instrumentation. Volume 3., Taipei, Taiwan (1996) 1838–1843
7. Lafuente-Arroyo, S., Maldonado-Bascón, S., Gil-Jiménez, P., Gómez-Moremo, H., López-Ferreras, F.: Road sign tracking with a predictive filter solution. In: Proc. of IEEE IECON, Paris, France (2006) 3314–3319

Video Tracking Association Problem Using Estimation of Distribution Algorithms in Complex Scenes*

Miguel A. Patricio, J. García, A. Berlanga, and José M. Molina

Applied Artificial Intelligence Group (GIAA)
Universidad Carlos III de Madrid
Avda. Universidad Carlos III 22, 28270-Colmenarejo, Spain
{mpatrici,jgherrer}@inf.uc3m.es, {molina,aberlan}@ia.uc3m.es

Abstract. In this work an efficient and robust technique of data association will be developed as a search in the hypotheses space defined by the possible association between detections and tracks. The full data association problem in visual tracking is formulated as a hypotheses search with a heuristic evaluation function to take into account structural and specific information such as distance, shape, colour, etc. This heuristic should represent the real problem so that its optimization leads to the solution of each situation. In order to guarantee performance in real time, the search process will have assigned a bounded amount of time to give the solution. The number of evaluations is restricted to accomplish this bound. The use of Estimation Distribution Algorithms (EDA) allows the application of an Evolutionary Computation technique to search in the hypothesis space. The performance of alternative algorithms used to provide the solution with this time constraint will be compared considering complex situations.

1 Introduction

Video surveillance systems are one of the real-world applications of computer vision which have received more attention in the recent years. These systems are intended to continuous visual monitoring indoor/outdoor areas such as protected buildings, commercial areas, public transportation, parking, ports, etc. [1]. An expected feature of these systems is the capability to track and maintain identity (ID) of all detected objects over time. Therefore, a key aspect to have a visual surveillance [2] system is robust movement segmentation, which has been the objective of many research works. Although plenty of techniques have been applied for video segmentation, it is still a difficult and not resolved problem in the general case and under complex situations. The basic aspects to address are: extract moving objects from the background and precise separation of individual objects when their images get close [3]. This process follows the typical steps of

* Funded by project CAM 15 MADRINET S-0505/TIC/0255, IMSERSO AUTOPIA, CICYT TSI2005-07344-C02-02.

J. Mira and J.R. Álvarez (Eds.): IWINAC 2007, Part II, LNCS 4528, pp. 261–270, 2007.

a Multi-target tracking application: (1) extracting detections, (2) gating detections with existing tracks, (3) associating detections to tracks and (4) filtering the tracks with new detections. Target splitting and merging distinguish video data processing with respect to other sensor data sources, making the data association (or correspondence) task demand for powerful and specific techniques [4].

In this work, we propose a hybrid approach that explores data association hypotheses (even considering the multi-scan case) and makes use of object structural information in the images. The combinatorial search is addressed with the use of efficient algorithms coming from Evolutionary Computation. So, it is searched a solution exploring the optimal decisions, but limiting computation load in order to process in real time the sequence of frames (25 fps). Genetic Algorithms (Gas) have been previously applied in the data association problem in radar context by Angus et al [5] within a single scan, and by Hillis [6] to deal the multi-scan data association problem. The application to visual tracking of genetic algorithms is addressed in this work. It implies the adaptation to the visual problem (split and merge effects, evaluation heuristics) and a very fast capability to process the video flow in real time. In order to achieve the second objective, the performance of a family of very efficient EC algorithms will be analyzed when applied to this field. This new approach is called Estimation of Distribution Algorithm (EDA) [7]. In EDAs, the problem-specific interactions among the variables of individuals are taken into account. The interrelations are expressed explicitly through the joint probability distribution associated with the individuals of variables selected at each generation. Then, sampling this probability distribution generates offspring. The estimation of the joint probability distribution associated to the selected individuals is the most complex task, and the diverse proposals in this field differ in the approach selected in this issue. For instance, UMDA [8] assumes linear problems of independent variables; PBIL [9] uses vector probabilities instead of population, also with independent variables, MIMIC [10] searches in each generation permutations of variables to find the probability distribution using Kullback-Leibler distance (two-order dependencies); EBNA [12] uses Bayesian Network for the factorization of the joint probability distribution; cGA [13] represents the population as a probability distribution over the set of solutions, etc.

2 Association Problem in Video Tracking Systems

The problem of segmenting interesting objects from a video sequence can be dealt as a typical multi-target tracking application [14]. A processing node receives the stream of frames over time, each one containing the detection of interesting objects (or groups of closely spaced objects) together with false alarms [15]. The objective is to successfully segment objects from image background and automatically initiate, maintain and terminate tracks. This tracking process must achieve enough reliability about the tracks provided, keeping a unique track per real object and minimizing the losses and switches of tracks even under complex conditions. Objects should be tracked without interruption even in the case that the low-level detection algorithms fail to segment them as a single foreground regions (blobs). This occurs when other region occludes the object (a fixed object in the scene or other moving object), the object image is split into pieces or even the images from different objects get merged because of close projection on the camera plane. An important problem in the case of merged region is to correctly recover the original trajectories when the objects "reappear" after the time interval of interaction [16]. The performance of multi target tracking (MTT) system critically depends on two strictly coupled subtasks: the data association method used for assigning observations to tracks and the model selected to estimate the movement of an object. Target-tracking community has usually formulated the total process as an optimal state estimation and a data association problem, with a Bayesian approach to recursively filter observations coming from sensors. Only in the case that a single target appears, with no false alarms, there is no association problem and optimal Bayesian filters can be applied, such as Kalman filter under ideal conditions, or suboptimal Bayesian filters such as Multiple Models (IMM)[17] for realistic manoeuvring situations and Particle Filter (PF) [18] in non-Gaussian conditions. In a general case, the tracking system should handle complex motions and interactions such as passing, occlusions and stopping. Each real object may generate multiple observations and the problem of searching for an optimal or near-optimal hypothesis requires a deeper analysis. A number of statistical data association techniques have been developed. The problem can be also extended across multiple scans (frames) to search the best hypothesis for the associations, using a tree of open hypotheses to delay assignment decisions until more evidence is available. Multi-scan, multiple-target algorithms commonly used include multi-hypothesis trackers (MHTs) and Joint Probabilistic Data Association (JPDA) [19]. To reduce the computational complexity of JPDA and MHT, probabilistic multiple hypotheses tracking (PMHT) algorithm [20]. Recently, data association algorithms have received wider attention by the computer vision community. For instance, the JPDA filter has been applied to vision 3-D reconstruction [21]. However, the direct formulation of these algorithms (JPDA and MHT) presents important limitations. The first one is that they assume that a target can generate at most one measurement per scan and, conversely, a measurement could have originated from at most one target. For instance, Cox [22] proposes an implementation of MHT algorithm with visual sensors, but objects are simplified to points, no considering the problem of data. The full problem of data association in the visual context is a correspondence of multiple blobs to multiple tracks. It is needed to remove the one-to-one constraint and extend the algorithms to consider multiply fragmented or merged blobs updating each track. Besides, the evaluation function, such as a distance considering only residuals between centroids disregards structural information about objects. So it is necessary to build an extended function to evaluate the assignment hypotheses considering structural information too. In the second place, the required computational burden may be too heavy with large numbers of targets and observations.

The problem of split and merging has been considered from different points of view. Some proposals do not consider explicit hypotheses and use lower-level image information to address the problem. For instance, w4 system for tracking people [23], is based on low-level correlation operators to resolve occlusions and people-group tracking. Kim et al [24] propose object-tracking through a "natural correspondence" derived from the use of labels retained from a frame to the next one, created by a distributed genetic algorithm for spatial segmentation. The segmentation result from previous frame is evolved to next one, so labels are propagated from one site to the next by selection. The authors of this work [4] have previously proposed a method using a fuzzy system which evaluates heuristics computed from overlapping blobs and tracks. It computes "confidence levels" that are used to weight each gated blob's contribution to update the target track and its associated shape. These proposals build a sub-optimal solution avoiding the combinatorial complexity of enumerating the possible data association alternatives. Regarding the explicit evaluation of hypotheses, Genovesio et al [15] propose the use of synthetic virtual measurements for use in a JPDA filter. This extension allows this filter evaluate all feasible associations considering the mentioned effects of merge and split on images, although no scheme is proposed to deal the combinatorial optimization problem appearing when the number of potential possibilities grows at a combinatorial rate. Kumar et al [25] search an efficient computational solution decomposing the matching problem into several sub-problems in order to apply Dynamic Programming for data association.

3 Application of EDA's to Video Association Problem

The association problem has been defined as a search over possible blob assignments. This problem could be defined as minimizing an heuristic function to evaluate blob assignments by an efficient algorithm (Estimation of Distribution Algorithm). The heuristic function takes a Bayesian approach to model the errors in observations. The formulation of data association as a minimization problem solved by a genetic technique is not a handicap with respect to the required operation in real time. A worst-case number of operations can be fixed and bound the time consumed by the algorithm, if we restrict the maximum number of evaluations. Then, given a certain population size, the algorithm will run a number of generations limited by this bound on the number of evaluations. The most important aspect is that the EDA should converge to acceptable solutions with these conditions of limited population size and number of generations.

3.1 Heuristic to Evaluate Assignments

The description of the heuristic of the search, that determines the quality of the solutions and guides the search towards the optimal one, is shown in this section. An extended distance is used as evaluation function for groups of detected blobs assigned to tracks according to matrix A (A represents each hypothesis to be evaluated). The heuristic is aimed at providing a measure of probability density of assigned observations to tracks. This likelihood function considers several types of terms and their probabilistic characterization: the separation between tracks and centroids of groups of blobs, the "similarity" between track-smoothed target attributes and those extracted from blob groups, and the events related with erasing existing tracks and creating new ones. As mentioned in the introduction, the final objective is to achieve a good trade-off between capability to re-connect image regions, keeping a single track per target, and avoid at the same time the miss-assignment of blobs coming from different objects or from extraneous sources.

The extended distance allows the evaluation of a certain hypothesis for grouping blobs in sets and assigning them to tracks. The term considering centroid residual, typically used in other approaches, is enriched with terms for attributes to take into account the available structural characteristics of targets which can be extracted from data. There are also terms considering that hypotheses may label some blobs as false alarms or may leave confirmed tracks with no updating blobs:

$$\begin{aligned} \log(P(\mathbf{b}[k]|\mathbf{A}[k], \hat{x}^{1,\dots,M}[k-1])) &= \log \prod_{j-th,track} \mathrm{D}_{\text{Group-Track(j)}} = \\ &= \sum_{j-th,track} \log \mathrm{D}_{\text{Group-Track(j)}} = \\ &= \sum_{j-th,track} \mathrm{d}_{\text{Group-Track(j)}} \end{aligned} \quad (1)$$

If we denote the blobs assigned to j-th track as Group(i)-Track(j)=$\{b_i[k]|\mathbf{A}_{ij}[k]=1\}$

$$\mathbf{d}_{ij}=\mathbf{d}_{Group(i)-Track(j)}= \mathbf{d}_{Centroid(i,j)}+\mathbf{d}_{Attributes(i,j)}+\mathbf{d}_{PD(i,j)}+\mathbf{d}_{PFA(i)} \quad (2)$$

where sub-indices i, j refer to the i-th group of blobs and j-th track:

- $\mathbf{d}_{Centroid(i,j)}$: it is the normalised residual between j-th track prediction and centroid of the assigned group of blobs under i-th hypothesis.
- $\mathbf{d}_{Attributes(i,j)}$: it is the normalised residual between track attributes and those extracted from the group. Its value is given, assuming Gaussian distribution and attribute independence.
- $\mathbf{d}_{PD(i,j)}$: assesses the cost of no updating a confirmed track for those hypotheses in which no blob is assigned to j-th track. It considers the probability of updating each track.
- $\mathbf{d}_{PFA(i)}$: assesses the cost of labelling a blob as a false alarm, also assuming a certain probability of false alarm, PFA.

3.2 Encoding and Efficient Search with EDA Algorithms

The association consists of finding the appropriate values for assignment matrix A, where element A(i, j) is 1 if blob i is assigned to object j and 0 in the opposite case. In order to be able to use the techniques of evolutionary algorithms, the matrix A is codified in a string of bits, being the size of matrix **A** NxM, with N the number of extracted blobs and M the number of objects in the scene. A first possibility for problem encoding was tried with a string of integer numbers representing the possible M objects to be assigned for each blob, including the "null" track 0, as shown in Figure 1.

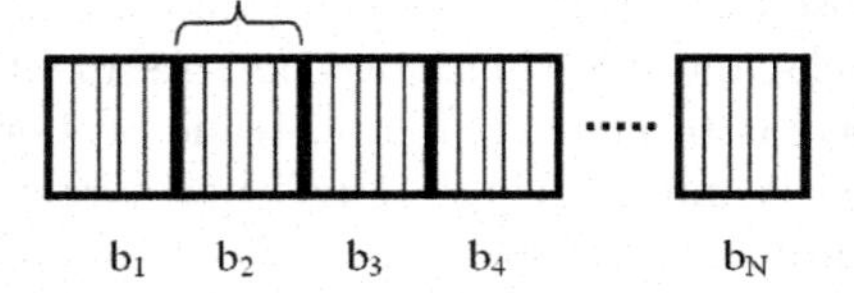

Fig. 1. Simple encoding for blob assignment

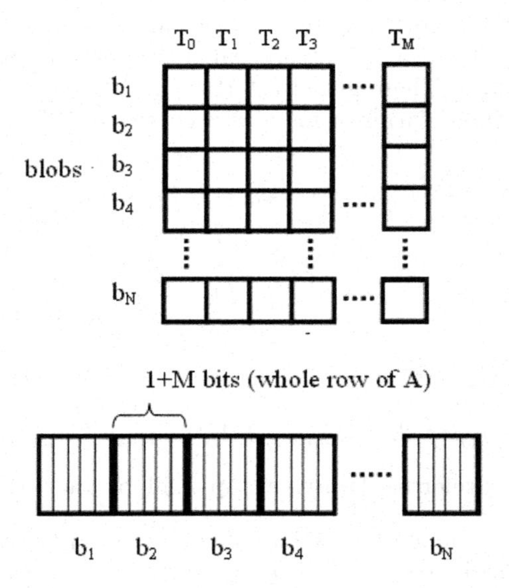

Fig. 2. Direct encoding for whole A matrix

This encoding requires strings of $N\log_2(1+M)$ bits and has the problem of constraining search to solutions in which each blob can belong to one object at most. This could be a problem in situations where images from different objects get overlapped and may leave some tracks unassigned and lost. Then, a direct encoding of A matrix was used for general solutions, where the positions in the string represent the assignations of blobs to tracks. With this codification, where individuals need $N(1+M)$ bits, a blob can be assigned to several objects, see Figure 2.

Finally, in order to allow and effective search, the initial individuals are not randomly generated but they are fixed to solutions in which each blob is assigned to the closest object. So, the search is performed over combinations starting from this solution in order to optimize the heuristic after changing any of this initial configuration. Besides, for the case of EDA algorithms, the vector probabilities are constrained to be zero for the case of very far pairs and those blobs which fall in spatial gates of more than one track have a non-zero change probability.

4 Evaluation

In this section we present the comparison between EDA algorithms presented in this work and a high performance tracking algorithm based on a combination of Particle Filter and Mean Shift algorithms [26]. The surveillance system and the tested algorithms were implemented in C++ under Microsoft Visual Studio, using the "visual surveillance" algorithms incorporated in the Open Source Computer Vision Library (OpenCV). The system was tested on a AMD Athlon^{TM} 64 Processor 3200+ with 2.01 GHz and 1 Gb of RAM. The OpenCV "visual surveillance" algorithms use a pipeline structure. The input data for the pipeline are image of current frame and the output data are the information about the tracks position and size. The "FG/BG Detection" module performs foreground/background segmentation for each pixel; the "Blob Entering Detection" module uses the result of "FG/BG Detection" module to detect new blob object entered to a scene on each frame; and the "Blob Tracking" module initialized by "Blob Entering Detection" results and tracks each new entered blob. This pipeline structure allows us to exchange different algorithms for "Blob Tracking" module, and to maintain the same execution conditions, i.e. using the same "FG/BG Detection" and "Blob Entering Dectection" modules. We have used five "Blob Tracking" modules (see Table 1) and an implementation of the Particle Filtering algorithm (MSPF) from OpenCV library. We have fixed the same "FG/BG Detection" module for every test that we carried out. The selected module was the OpenCV implementation of the adaptive background mixture models for real-time tracking.

The dataset used throughout this paper came from a DV camcorder and show Maritime scenes (named BOAT in the rest of paper). The videos were recorded in an outdoor scenario using a DV videorecorder. The videos have a high quality with a resolution of 720x480 pixels with 15 fps. The videos feature several boats in an outdoor environment lit by the sun. The videos are very interesting due to the complex segmentation of maritime scenes. The sea has continuous movement, which contributes to the creation of a great amount of noisy blobs.

Before comparing the obtained results of algorithms used for visual tracking accomplishment, the first step is to determine the metric that allows making the comparisons of the algorithms behaviour. A specific quality measurement of the association has not been used and the global behaviour has been observed. In the analysis, a special emphasis in the speed of algorithms convergence is

Table 1. Algorithms applied to the three datasets

Algorithm	*Description*
CGA	Compact Genetic Algorithm
UMDA	Univariate Marginal Distribution Algorithm
PBIL	Population-Based Incremental Learning
MSPF	Particle Filtering based on Mean-Shift weight
GA	Genetic Algorithm

Fig. 3. A scene of BOAT dataset

Table 2. Measures of quality of the algorithms applied to BOAT

	mean TPF (ideal=3)	**sd TPF**	**FPS**	**mean TPT (millisecond)**	**sd TPT**
CGA	2.98	0.07	4.29	2.22	0.08
UMDA	2.93	0.17	4.26	4.16	0.75
PBIL	2.98	0.07	4.04	9.14	1.93
MSPF	2.98	0.07	2.33	67.93	0.78
GA	2.93	0.17	2.21	79.37	18.23

made in order to evaluate the application of the proposal development in real-time video tracking. The measures taken into account are: Tracks per Frame, Frames per Second and Time per Track. Table 2 shows the values of the quality parameters obtained for BOAT scenario.

The results obtained for sequence BOAT show clearly the advantage of the EDAS on GA and particle filter. Observing the referring columns TPF, the quality of the tracking is the same for all the algorithms, very close to the ideal value. UMDA and GA are slightly less stable compared to the rest, but the difference is despicable. The speed that the EDAS solve the combinatorial problem of the association of blobs to tracks is quite superior (50%) to GA and particles filter, this result is also observed in the histograms of the number of evaluations necessary to obtain the solution, the number of evaluations necessary to obtain the same quality is several orders of magnitude greater.

5 Conclusions

In this paper, a hybrid approach exploring data association hypotheses and using of object structural information in the images is developed. The association problem is defined as a search problem, but limiting computation load in order to process in real time the sequence of frames (25 fps). Evolutionary Computation techniques have been applied to this search problem, in particular Estimation Distribution Algorithms (EDA) show an efficient computational behaviour to real time problems.

References

1. James M. Ferryman, Stephen J. Maybank, and Anthony D. Worrall. Visual surveillance for moving vehicles. *Int. J. Comput. Vision*, 37(2):187–197, 2000.
2. M. A. Patricio, J. Carbó, O. Pérez, J. García, and J. M. Molina. Multi-agent framework in visual sensor networks. *EURASIP Journal on Advances in Signal Processing*, 2007:Article ID 98639, 21 pages, 2007. doi:10.1155/2007/98639.
3. Óscar Pérez, Miguel Ángel Patricio Guisado, Jesús García, and José M. Molina. Improving the segmentation stage of a pedestrian tracking video-based system by means of evolution strategies. In *EvoWorkshops*, volume 3907 of *Lecture Notes in Computer Science*, pages 438–449. Springer, 2006.
4. J. Garcia, J. M. Molina, J. A. Besada, and J. I. Portillo. A multitarget tracking video system based on fuzzy and neuro-fuzzy techniques. *EURASIP Journal on Applied Signal Processing*, 14:2341–2358, 2005.
5. J. Angus, H. Zhou, C. Bea, L. Becket-Lemus, J. Klose, and S. Tubbs. Genetic algorithms in passive tracking. Technical report, Claremont Graduate School, Math Clinic Report, 1993.
6. D. B. Hillis. Using a genetic algorithm for multi-hypothesis tracking. In *ICTAI '97: Proceedings of the 9th International Conference on Tools with Artificial Intelligence*, page 112, Washington, DC, USA, 1997. IEEE Computer Society.
7. Pedro Larranaga and Jose A. Lozano. *Estimation of Distribution Algorithms: A New Tool for Evolutionary Computation.* Kluwer Academic Publishers, Norwell, MA, USA, 2001.
8. Heinz Mühlenbein. The equation for response to selection and its use for prediction. *Evolutionary Computation*, 5(3):303–346, 1997.
9. Shumeet Baluja. Population-based incremental learning: A method for integrating genetic search based function optimization and competitive learning,. Technical Report CMU-CS-94-163, Pittsburgh, PA, 1994.
10. Jeremy S. de Bonet, Charles L. Isbell, Jr., and Paul Viola. MIMIC: Finding optima by estimating probability densities. In Michael C. Mozer, Michael I. Jordan, and Thomas Petsche, editors, *Advances in Neural Information Processing Systems*, volume 9, page 424. The MIT Press, 1997.
11. Heinz Mühlenbein and Thilo Mahnig. The factorized distribution algorithm for additively decompressed functions. In *1999 Congress on Evolutionary Computation*, pages 752–759, Piscataway, NJ, 1999. IEEE Service Center.
12. Martin Pelikan, David E. Goldberg, and Erick Cantú-Paz. Linkage problem, distribution estimation, and Bayesian networks. Technical Report 98013, Urbana, IL, 1998.
13. G. R. Harik, F. G. Lobo, and D. E. Goldberg. The compact genetic algorithm. *IEEE Transactions on Evolutionary Computation*, 3(4):287, November 1999.
14. F. Castanedo, M. A. Patricio, J. García, and J. M. Molina. Extending surveillance systems capabilities using bdi cooperative sensor agents. In *VSSN '06: Proceedings of the 4th ACM international workshop on Video surveillance and sensor networks*, pages 131–138, New York, NY, USA, 2006. ACM Press.
15. Auguste Genovesio and Jean-Christophe Olivo-Marin. Split and merge data association filter for dense multi-target tracking. In *Proceedings of the Pattern Recognition, 17th International Conference on*, volume 4, pages 677–680, Washington, DC, USA, 2004. IEEE Computer Society.
16. Hai Tao, Harpreet S. Sawhney, and Rakesh Kumar. Object tracking with bayesian estimation of dynamic layer representations. *IEEE Trans. Pattern Anal. Mach. Intell.*, 24(1):75–89, 2002.

17. M. Yeddanapudi, Y. Bar-Shalom, and K. Pattipati. Imm estimation for multitarget-multisensor air traffic surveillance. In *Proceedings of the IEEE*, volume 85, pages 80–96, 1997.
18. Xinyu Xu and Baoxin Li. Rao-blackwellised particle filter for tracking with application in visual surveillance. In *Visual Surveillance and Performance Evaluation of Tracking and Surveillance, 2005. 2nd Joint*, pages 17–24, 2005.
19. Samuel S. Blackman and Robert Popoli. *Design and Analysis of Modern Tracking Systems*. Artech House, INC., 1999.
20. Y. Ruan and P. Willett. Multiple model pmht and its application to the benchmark radar tracking problem. *IEEE Transactions on Aerospace and Electronic Systems*, 40(4):1337–1350, October 2004.
21. Y.L. Chang and J.K. Aggarwal. Neural network optimization for multi-target multi-sensor passive tracking. In *Proc. IEEE Workshop on Visual Motion*, pages 268–273, 1991.
22. Ingemar J. Cox and Sunita L. Hingorani. An efficient implementation of reid's multiple hypothesis tracking algorithm and its evaluation for the purpose of visual tracking. *IEEE Trans. Pattern Anal. Mach. Intell.*, 18(2):138–150, 1996.
23. Ismail Haritaoglu, Davis Harwood, and Larry S. David. W4: Real-time surveillance of people and their activities. *IEEE Trans. Pattern Anal. Mach. Intell.*, 22(8):809–830, 2000.
24. Eun Yi Kim and Se Hyun Park. Automatic video segmentation using genetic algorithms. *Pattern Recogn. Lett.*, 27(11):1252–1265, 2006.
25. Pankaj Kumar, Surendra Ranganath, Kuntal Sengupta, and Weimin Huang. Cooperative multi-target tracking and classification. In *ECCV 2004, 8th European Conference on Computer Vision*, volume 3021 of *Lecture Notes in Computer Science*, pages 376–389. Springer, 2004.
26. Katja Nummiaro, Esther Koller-Meier, and Luc J. Van Gool. An adaptive color-based particle filter. *Image Vision Comput.*, 21(1):99–110, 2003.

Context Information for Understanding Forest Fire Using Evolutionary Computation

L. Usero[3,4], A. Arroyo[2], and J. Calvo[1]

[1] Dpto. de Organización y Estructura de la información,
Universidad Politécnica de Madrid, Spain
[2] Dpto. de Sistemas Inteligentes Aplicados
Universidad Politécnica de Madrid, Spain
[3] Dpto. Ciencias de la Computación
Universidad de Alcalá, Spain
[4] Center for Spatial Technologies and Remote Sensing, U. California. One Shields Ave. 95616-8617 Davis, CA. USA
aarroyo@eui.upm.es, luis.usero@uah.es, jcalvo@tdi.eui.upm.es

Abstract. One of the major forces for understanding forest fire risk and behavior is the fire fuel. Fire risk and behavior depend on the fuel properties such as moisture content. Context information on vegetation water content is vital for understanding the processes involved in initiation and propagation of forest fires. In that sense, a novel method was tested to estimate vegetation canopy water content (CWC) from simulated MODIS satellite data. An inversion of a radiative transfer model called Forest Light Interaction-Model (FLIM) from performed using evolutionary computation. CWC is critical, among other applications, in wildfire risk assessment since a decrease in CWC causes higher probability to have wildfire occurrence. Simulations were carried out with the FLIM model for a wide range of forest canopy characteristics and CWC values. A 50 subsample of the simulations was used for the training process and 50 for the validation providing a RMSE=0.74 and r2=0.62. Further research is needed to apply this method on real MODIS images.

Keywords: Genetic Programing, Vegetation Water Content, Forest Fire Understanding.

1 Introduction

Detecting the water content (Cw) is useful to monitor vegetation stress even forest fire. Context information gathered by remote sensing is vital to understand the forest fire risk and behavior. So it is significant to use of remote sensing to measure spectral properties of leaves can provide an indirect structural canopy variables estimation in order to obtain a comprehensive spatial and temporal distribution.

Vegetation canopy water content (CWC) is the weight of the water per leaf area unit and per ground area unit. CWC retrieval results critical for several environmental applications including wildfire risk [2]. Fires front advances when

J. Mira and J.R. Álvarez (Eds.): IWINAC 2007, Part II, LNCS 4528, pp. 271–276, 2007.

the CWC is dried out. Empirical methods have been commonly applied to derive CWC from satellite data based on the response to changes in reflectance in the near infrared and shortwave infrared part of the spectrum [7]. They work well for a specific location, but they need to be calibrated from site to site. Radiative transfer models model the response in reflectance of the vegetation canopy, to account for a wide range of biophysical conditions [3]. Therefore, these models can be applied to derive CWC for different sites of diverse ecosystem conditions. The simplest radiative transfer models assume that the vegetation forms a continuous canopy of leaf layers. More sophisticated models consider the effect of the tree shadows in the reflectance response, assuming trees are homogeneously distributed and equal in size. Further complete models take into account heterogeneous canopies with tree of different sizes and even understorey layer [6]. Simpler models make more assumptions, so they could be far from reality, but they are easier to parameterize, with the less number of input variables.

One limitation of the radiative transfer models is that inversion to derive CWC is computationally very expensive. In order to reduce this limitation neural networks and genetic algorithms have been tested [5]. Forest Light Interaction-Model (FLIM) assumes a homogeneous forest canopy, accounting for the tree shadows [4]. This paper uses FLIM to generate CWC from simulated MODIS satellite data. The model was selected since it is fairly complicated, but simpler to parameterize than models that account for the heterogeneity in the tree distribution. Evolutionary computation was applied to test the sensibility of several vegetation indexes to obtain CWC and to provide a robust model to predict this variable from the reflectance response.

In the last years, several new intelligent approaches emerge to obtain the content and spatial distribution of vegetation biochemical information over local to regional and eventually global scales through remote sensing data. These new approaches are related to soft computing techniques close to the computational intelligence. In [8], EWT and DM on dry samples estimations with neural nets were as good as other methods tested on the same dataset, such as inversion of radiative transfer models. DM estimations on fresh samples using ANN ($r2=0.86$) improved significantly the results using inversion of radiative transfer models ($r2=0.38$). Applications of the genetic algorithms (GA) to a variety of optimization problems in remote sensing have been successfully demonstrated [9]. In [9] estimated LAI by integrating a canopy RT model and the GA optimization technique. This method was used to retrieve LAI from field measured reflectance as well as from atmospherically corrected Landsat ETM+ data. Four different ETM+ band combinations were tested to evaluate their effectiveness. The impacts of using the number of the genes were also examined.

The major aim of this work is to assess the accuracy of estimating LAI information by means of evolutionary computation. Following section, we depict our evolutionary computational method. Finally, the authors present several successfully experimentations and conclusions.

2 Experimental Results

The objective of this experiment is to find an index that is able to correlate as far as possible with CwLAI value. This index will be formed by a data combination obtained beyond the first seven bands of the MODIS sensor.

Indexes those are normally used in remote sensing (e.g. NDVI) are not useful for this purpose. These indexes are lack of correlation with the searched CwLAI values. (All of them have coefficient determination inferior than 0.1).

Table 1.

Variations in Cross and Reproduction Operators	
Population size	1000 characters
Likelihood cross	0.9
Likelihood reproduction	0.1
Kind of Selection	tournament
Tournament size	10
Elitism	among 1 and 5 characters
Final Nodes Seven bands	MODIS (M1 .. M7)

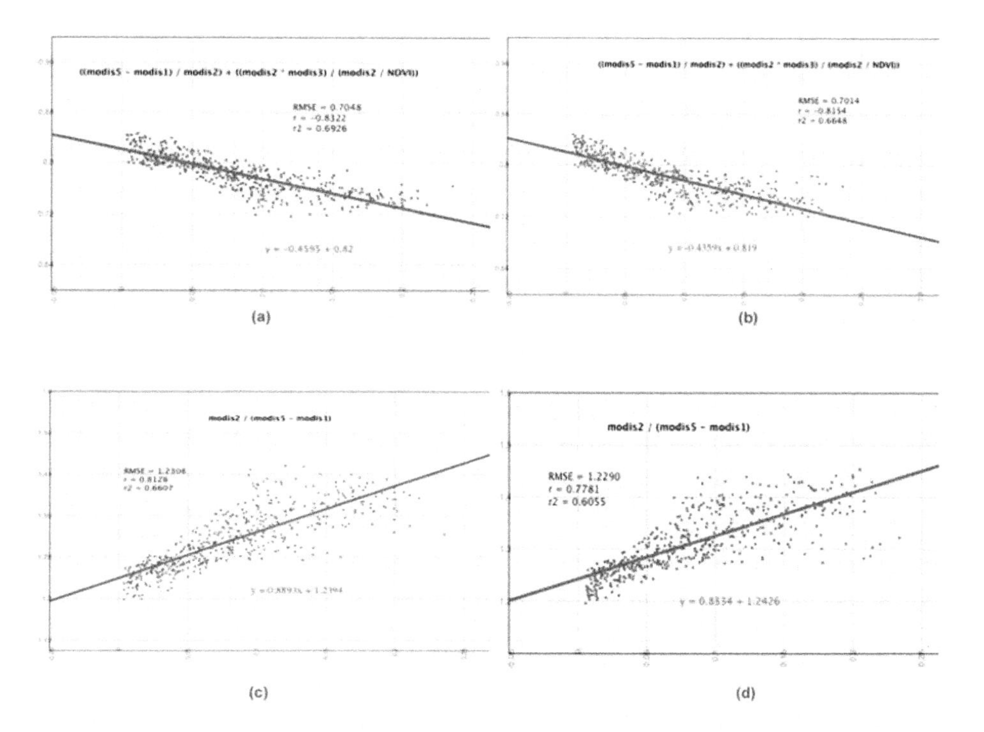

Fig. 1. Correlation between CwLAI values and the ones which have been obtained with the index that is pointed out in the graph

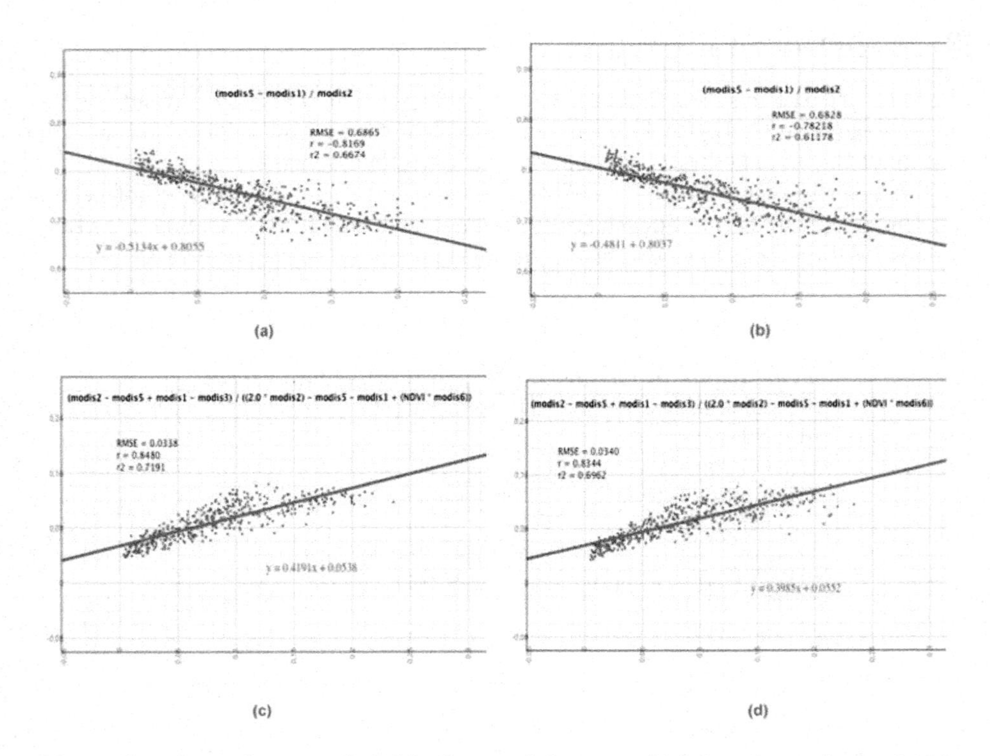

Fig. 2. Correlation between CwLAI values and the ones which have been obtained with the index that is pointed out in the graph

In order to find this index inside the set of possible ones those are formed by means of data combination we have decided to use Evolutive Computation Techniques, and concretely, Genetic Algorithm (GA). Data combinations are obtained from MODIS sensor together with allowed operations for creating the index.

In order to achieve the test with Genetic Algorithm, we have used a system of investigation in Evolutive Computation based in Java (EJC) [10] developed by Evolutionary Computation Laboratory (ECLab) George Mason University.

For the whole test system we have used a set of 1000 samples obtained by means of FLIM model. From this set, 500 samples have been spent in training phase, the other 500 ones have been used for evaluating obtained solutions in training phase (see Table 1).

Tests have also been achieved bearing in mind typical indexes in remote sensing like final nodes. Modifications in aptitude function have been carried out in order to incorporate RMSE values in optimization process.

From Figure 1 to Figure 3, we can observe indexes that present the greatest correlation among all the possibilities in the different executions from the chosen evolutive schema are depicted. For each one of these indexes, a tested graph is showed. Testing has been achieved with training data set and test data set. We can observe RMSE values, degree Pearson correlation (r) and coefficient

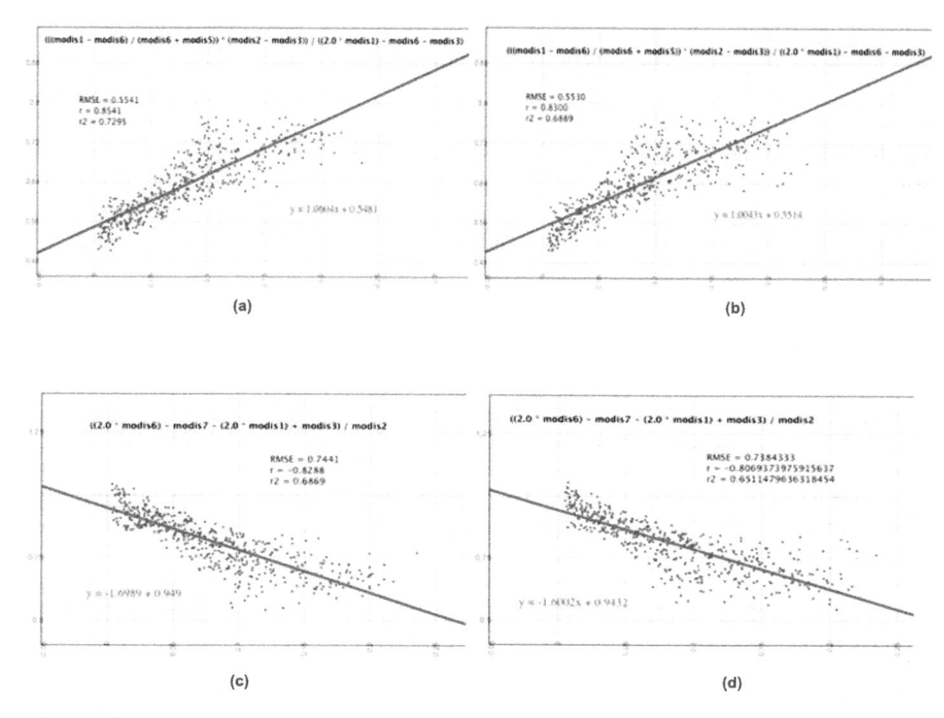

Fig. 3. Correlation between CwLAI values and the ones which have been obtained with the index that is pointed out in the graph

determination (r2). Each figure is divided into four graphics. Each row represents the same expression but one is using a training data set and the other is working with a test data set.

Training data sets are, obviously, offering better results than test data set. Training set spend a certain period of time in learning how to improve the solution.

We can appreciate twelve different results for the proposed index. Six with training data set (right side in the figures) and six with test data set (left side in the figures) as we have explained before. Ideally, the best result would be the closest to 1, despite of the obtained indexes are close to 0.7 (acceptable correlation degree), it is very important to remark that we have tried not to execute complicated expressions in the genetic algorithm in order to obtain useful indexes. So that, results are easy to manage by a real teledetection system.

3 Conclusions and Future Work

We have shown how Genetic Algorithm improves the estimation of vegetation water content. This context information is vital to assess forest fire risk and behavior. Authors plan to enhance this evolutionary technique incorporating some new features into the evolutive schema.

Author are immersed in new experimentation, where they use a multi population schema where characters have the possibility of migrating from the current population, promoting diversity of the characters in different populations. Authors are working as well, in a new way of representing characters that were able to implement different operators of symbiotic variation, so that, we can make the most of this multi population schema.

References

1. Gao, B. -C., and Goetz, A. F. H. (1995). Retrieval of equivalent water thickness and information related to biochemical components of vegetation canopies from AVIRIS data. Remote Sensing of Environment, 52(3), 155?162.
2. Chuvieco, E., Cocero, D., Riaño, D., Martin, P., Martnez-Vega, J., de la Riva, J., et al. (2004). Combining NDVI and Surface Temperature for the estimation of live fuels moisture content in forest fire danger rating. Remote Sensing of Environment, 92(3), 322?331.
3. Goel, N. S. (1988). Models of vegetation canopy reflectance and their use in estimation of biophysical parameters from reflectance data. Remote Sensing Reviews, 4, 1 - 212.
4. Rosema, A., Verhoef, W., Noorbergen, H., and Borgesius, J. J. (1995). A new forest light interaction model in support of forest monitoring. Remote Sensing of Environment, 42, 23- 41.
5. Xiao, X., Boles, S., Liu, J., Zhuang, D., Frolking, S., Li, C., et al. (2005). Mapping paddy rice agriculture in southern China using multi-temporal MODIS images. Remote Sensing of Environment, 95(4), 480?492.
6. Ceccato, P., N. Gobron, S. Flasse, B. Pinty and S. Tarantola 2002, Designing a spectral index to estimate vegetation water content from remote sensing data: Part 1 - Theoretical approach. Remote Sensing of Environment. 82(2-3): 188-197.
7. Riaño, D., P. Vaughan, E. Chuvieco, P.J. Zarco-Tejada and S.L. Ustin in press, Estimation of fuel moisture content by inversion of radiative transfer models to simulate equivalent water thickness and dry matter content. Analysis at leaf and canopy level. IEEE Transactions on Geoscience and Remote Sensing
8. David Riao, Susan L. Ustin, Luis Usero Estimation of fuel moisture content using neural networks. IWINACC 2005
9. Fang, H., Liang, S., and Kuusk, A. (2003). Retrieving leaf area index using a genetic algorithm with a canopy radiative transfer model. Remote Sensing of Environment, 85, 257-270.
10. Evolutionary Computation Laboratory (ECLab). Evolutive Computation based in Java (EJC). George Mason University [http://cs.gmu.edu/ eclab/projects/ecj/]

Feed-Forward Learning: Fast Reinforcement Learning of Controllers*

Marek Musial and Frank Lemke

Real-Time Systems and Robotics (PDV)
Technische Universität Berlin

Abstract. Reinforcement Learning (RL) approaches are, very often, rendered useless by the statistics of the required sampling process. This paper shows how *very fast* RL is essentially made possible by abandoning the state feedback during training episodes. The resulting new method, *feed-forward learning* (FF learning), employs a return estimator for pairs of a state and a *feed-forward policy's* parameter vector. FF learning is particularly suitable for the learning of controllers, e.g. for robotics applications, and yields learning rates unprecedented in the RL context.

This paper introduces the method formally and proves a lower bound on its performance. Practical results are provided from applying FF learning to several scenarios based on the collision avoidance behavior of a mobile robot.

1 Introduction

Reinforcement learning (*RL*) is a well-known approach that *must* be applied whenever some agent is to learn a behavior and the agent's information is limited to state observations and a reward signal. The general RL background has been elaborately presented e.g. in the excellent book [1] by Sutton and Barto. Numerous algorithms and algorithm variants have been devised within the RL framework, with temporal difference (TD) learning [2] and Q-learning [3] among the best-known thereof. Especially for model-free policy optimization, it is required for existing formal convergence properties [4] that *all state-action pairs [...] be visited an infinite number of times in the limit of an infinite number of episodes* [1]. Speedups have been reported, but mainly result from better utilization of the acquired sampling data, e.g. [5], but not from a simplified sampling process. Moreover, when the action resolution is increased to approach continuous-time modeling, the statistical relevance of the "free" initial action in the state-action-pairs with regard to the expected return vanishes. This leads to the fact that for practical application in non-toy cases, the "infinite number" mentioned in the theorems also constitutes a far-too-good hint at the practically achievable convergence rate.

In order to particularly address the properties of control learning applications, this paper suggests a completely new RL approach. Instead of estimating

* This work has been conducted within the *NeuRoBot* project, funded by German Research Foundation (DFG).

J. Mira and J.R. Álvarez (Eds.): IWINAC 2007, Part II, LNCS 4528, pp. 277–286, 2007.

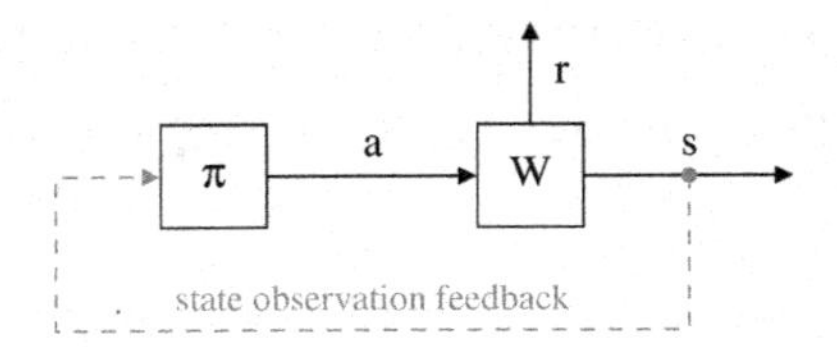

Fig. 1. System diagram of the RL training process: FF learning abandons the state observation feedback

returns for state-action-pairs, as done by classical model-free RL methods like Q-learning, the new method involves a return estimator for *state-policy pairs*. Here, the policies are *deterministic* and *state-independent*, actually parameterized functions over time. The sampling process is completely independent of the policy *currently favored by the agent*, so that all acquired return samples are final and do not require any updates (*backups*) in the course of policy iteration. This results in a dramatic simplification of the sampling statistics, and in turn leads to *very fast* reinforcement learning.

Figure 1 depicts the training process of RL: Some policy π issues actions a to effect the environment (world) W, which in turn yields state observations s and a reward signal r, the use of which is not detailed in this diagram. While traditional RL approaches do use state feedback, depicted through the dashed arrow, to govern the sampling process during training episodes, the new approach eliminates this feedback completely. It is therefore termed *feed-forward learning* (FF learning).

The remainder of this paper is organized as follows: The subsequent section 2 explains the FF learning method formally, section 3 addresses formal properties such as convergence issues and performance bounds, and section 4 gives insight into some relevant implementation details. Result examples from an collision avoidance application for a ground-based mobile robot are presented in section 5 in order to emphasize the applicability and performance of FF learning. Section 6 concludes the paper with an outline of the merits and drawbacks of FF learning.

2 FF Learning Method

This section describes the FF learning algorithm formally. After introducing the basic entities, the description distinguishes between the training phase and the application phase.

For the formal treatment, continuous state and action spaces and continuous time are assumed. However, quantization of some or all of these sets is possible and may be required in an implementation. The learning task is characterized by a set of (observable) *states* $s \in S \subseteq \mathbb{R}^K$ and a set of permissible *actions* $a \in A \subseteq \mathbb{R}^N$. The environment W evolves according to some (unobservable) internal state and the applied action signal $a(t)$, while outputting observable state information $s(t) \in S$ and a *reward* signal $r(t) \in \mathbb{R}$. Furthermore, let $s(t), r(t)$ be piecewise continuous for continuous $a(t)$. There is no assumption whether the process underlying W is a Markov process, or whether it is deterministic.

For return calculation, an infinite horizon, discounted reward problem is generally assumed. However, finite episodes and undiscounted reward can easily be introduced as special cases. An *episode* starts always at $t = 0$, and the *return* $R \in \mathbb{R}$ assigned to time t_0 is defined as

$$R(t_0) = \int_{t=t_0}^{\infty} r(t) \cdot e^{-\gamma(t-t_0)} \, dt \tag{1}$$

with discount rate $\gamma \geq 0$ (with $\gamma = 0$ modeling undiscounted reward). The end of a single episode, either induced by the environment or arbitrarily by the training algorithm, is denoted by $t = T$.

Now, FF learning introduces a set of *feed-forward policies*

$$\pi_p : \mathbb{R}_0^+ \to A \tag{2}$$

defining action trajectories $a(t) = \pi_p(t)$ in dependence of a D-dimensional parameter vector $p \in P \subseteq \mathbb{R}^D$. It is further postulated that the feed-forward policies are parameterized such that the postfix of any of them can be regarded as a new episode start by selecting a suitable parameter vector, i.e. formally:

$$\forall p \in P, \ t_0 \geq 0 \quad \exists p' \in P \quad \forall t \geq 0 \ : \ \pi_{p'}(t) = \pi_p(t_0 + t) \tag{3}$$

2.1 Training Phase

All episodes in the *training phase* are executed, without state feedback (see section 1 and Figure 1), by selecting a single parameter vector $p \in P$ of a feed-forward policy. This fully determines the agent's action signal as:

$$a(t) = \pi_p(t) \qquad t \in [0; T] \tag{4}$$

p is selected at the start of the training episode, which may be done randomly or in some way involving the current "application phase" policy (see 2.2), perhaps *ϵ-greedy* as known from classical exploration regimes. But the crucial fact is that p is kept constant throughout every training episode.

The training process on the whole consists in sampling triples

$$(s, p, R) \in S \times P \times \mathbb{R} \tag{5}$$

in order to train a function approximator to estimate a return $\bar{R}(s, p) \in \mathbb{R}$ from a pair of state s and policy parameters p. This is trivially possible for $t_0 = 0$, using $s(0)$, p, and $R(0)$. But due to (3), any number of return triples may be extracted from a single training episode, by using $s(t_0)$, p', and $R(t_0)$ for any instantiation of the variables according to (3). In other words, any episode postfix can be considered and evaluated just as a full episode on its own.

Therefore, the following two facts are met:

1. Samples are obtained as fast as in traditional "backup" algorithms, like Q-learning or TD learning, e.g. one sample per simulation step.

2. All samples obtained are final, i.e. do not depend on the "application policy", and therefore, need not be corrected at any later time. Traditional "backup" algorithms, in contrast to that, must continuously adapt their return estimates together with their policy evolution, which leads to a twofold, convoluted convergence process.

The implementation of the return estimator is irrelevant for the FF learning algorithm as such. Any sort of neural network can be employed, and table-based discretization is also basically possible.

2.2 Application Phase

In the application phase, it is neither desirable nor possible to limit the agent's behavior to one of the feed-forward policies π_p. Instead, the agent uses the *initial* action of the *most promising* feed-forward policy, at any time step. This gives the "application phase" policy $\pi^* : S \rightarrow A$ as:

$$\pi^*(s) = \pi_{\arg\max_{p \in P} \bar{R}(s,p)}(0) \tag{6}$$

Here, assume the *arg max* operator works deterministically in a way that keeps π^* piecewise continuous over time, e.g. by always yielding the geometric center of subspaces with equal maximum return.

The implementation of the *arg max* operator needs to be addressed by an actual implementation of FF learning. Clearly, determining this maximum will usually be approximate and may still involve substantial computation cost.

The application phase *does* employ state observation feedback, according to Figure 1.

3 Formal Properties

This section examines the formal performance properties of FF learning. As a part of this, the characteristics of the environment and of the feed-forward policies must be investigated, in relationship to one another.

First of all, it is immediately clear that (6) does *not* denote an optimal policy with respect to (1) with any generality. This simply results from the fact that the look-ahead conducted in action selection according to (6), even in case of a perfect estimator $\bar{R}$, is limited to the family of feed-forward policies used in training and will thus ignore non-representable action trajectories with potentially higher return.

On the other hand, the positive statement is conveyed in the following theorem:

Theorem 1. *Assuming that the process underlying W is Markov*[1] *and provided that the estimator $\bar{R}(s,p)$ in (6) is perfect (i.e. equals the expected return of the*

---

[1] For the formal proof, the Markov property is required, which is common for the formal examination of RL algorithms. This should *not* be misconceived as restricting the application of FF learning to Markov processes.

feed-forward policy for any p applied from state s onward), then $\bar{R}$ delivers a lower bound on the return of π^*:

$$E\{R(t) \mid s(t), \pi^*\} \geq \max_{p \in P} \bar{R}(s(t), p) \tag{7}$$

Proof Sketch: Starting with a policy that permanently applies π_p for the *arg-max* p from the right-hand side of (7) and delivers an expected return of $\bar{R}(s(t), p)$, consider one adds a new point τ in the episode time at which p is replaced by the "new" *arg max* according to (6), say p'. Denote the "old" expected return starting from τ with $E\{R(\tau)\}$ and the "new" one after adding the additional parameter lookup at τ with $E\{R'(\tau)\}$. Now, it is

$$E\{R'(\tau)\} \geq E\{R(\tau)\} \tag{8}$$

as π^* picks, for any state $s(\tau)$, the p' with the highest expected return, always considering to keep the "old" feed-forward policy (described by p at the starting time t) because of (3). By applying this argument repeatedly for e.g. $\tau, 2\tau, 3\tau, \ldots$ and using an arbitrarily small τ, the expected return of the so constructed policy converges to the left-hand side of (7) without ever decreasing. □

While Theorem 1 does not provide a particular good quantitative estimate of the return to be expected from π^*, looking at the best-possible performance of the feed-forward policies covered by π_p is usually quite easy and intuitive. By all means, Theorem 1 will be the primary reference when considering how many parameters (degrees of freedom, DOF) D need to be used in the π_p family of feed-forward policies in order to meet the requirements of any specific application. In the limit of infinitely many DOF in π_p, the right-hand side of (7) converges to the optimal value, which then makes π^* an optimal policy according to Theorem 1.

4 Implementation Details

This section explains some details of the example implementation of FF learning that has been used to generate the results presented in section 5. These details are *not* inherent to FF learning, but could be addressed in different ways as well.

4.1 $\bar{R}$ Approximator

As the approximator of the $\bar{R}$ estimate, a standard multilayer perceptron (MLP) with backpropagation training has been used. For its implementation, the *Fast Artificial Neural Network* library (FANN, [6]) has been employed. This library is available in source and provides the standard network structures and training algorithms with very high computation performance. During training, the samples according to (5) are collected in a single training set, which is occasionally used for training the estimator network. In training, a fraction (30%) of this set is randomly selected as a test set for cross-validation and early stopping to avoid overfitting. The network uses two hidden layers with sigmoid activation functions and a single linear output unit for the estimated return. As the actual training algorithm, FANN's *iRPROP* has been used (see [6] for details).

4.2 *arg max* Operator

The implementation of the *arg max* operator in (6), especially for increasing dimensionality of p, is a central issue of FF learning. Due to the potential non-continuity of the *arg max* operator, it does not make sense to try to train an estimator to directly obtain the most promising p vector. Instead, an on-line search has been implemented, in the form of a two step scheme:

1. A global maximum search is performed, starting at the best p vector from the last time step, by checking a number of fixed displacements along all axes. The displacements used, for $P = [-1; +1]^D$, are $\pm 0.01, \pm 0.02, \pm 0.04, \pm 0.1, \pm 0.2, \pm 0.3, \pm 0.5, \pm 1, \pm 1.5$.
2. The best p vector from step 1 is used as the starting point of a line search in the direction of the estimate's gradient, which can be determined based on the network's δ-terms (see [6,7])[2].

While step 1 is intended to allow switching between different local maxima of the return estimate toward the global optimum, step 2 yields a good approximation of the true local (and hopefully, global) optimum at low computation cost.

Clearly, this is a heuristic approach that in no way guarantees to determine the true global maximum. On the other hand, with a small time step size, it is practically sufficient to reach the "correct" local optimum after (or within) a couple of time steps. The application results obtained so far indicate that this heuristic optimum search does perform well enough in practice.

4.3 Sampling Control

Basically, the samples according to (5) need *not* be taken IID according to the probability distribution related to any specific intermediate policy, because the feed-forward policy π_p fully defines the statistics of their emergence. Nevertheless, the least-square-error approximation performed by the approximator's learning process *does* depend on the distribution of the samples according to (5). The sampling process conducted by FF learning, based on the family of feed-forward policies, accounts for rather low mean returns as compared to the final π^* policy. Therefore, it is highly reasonable to exert a special *sampling control* regime to minimize the effect of this low mean on the training of the approximator network.

For this purpose, two special rules have been employed in the sample implementation of FF learning:

1. For any new (s, p) pair to be inserted into the training set, its minimum Euclidean distance $\delta_{\min}$ from any element already in the training is computed. The pair is discarded if it is too close (similar) to some prior one (in the examples: if $\delta_{\min} < 0.2$).

[2] For this purpose, the FANN library has been slightly modified to allow direct access to the δ-terms of the network.

2. The range of possible returns of feed-forward policies is divided into a number B of equally sized bins, each of which does not receive more than $1/B$ of the total maximum number of training samples. That is to stop sampling for frequent low returns earlier than for rare better returns (in the examples: $B = 5$ with a maximum of 20000 samples per bin).

Both measures together aim at enforcing the best-possible approximation of an equal distribution of samples over $S \times P$. This reduces crosstalk from bad to good returns in the approximator's training.

5 Result Examples

This section presents a group of sample applications with varying degree of freedom (DOF) and the learning results obtained therein. All examples pertain to a simulated ground-based mobile robot the movement of which is determined by a translation speed v and a rotation rate ω that are in principle independent. That is, the robot moves, unless $\omega = 0$ or $v = 0$, along a circular arc with radius v/ω. Physically, robots with two independently driven wheels plus one or more support wheels exhibit this kind of kinematics. In the experiments, v is sometimes kept constant, leaving only $N = 1$ DOF in the action $a = \omega$, and sometimes controlled by the policy, which results in the $N = 2$ DOF action $a = (v, \omega)$. In the former case, $v = 50\text{mm/s}$, $-70°/\text{s} \leq \omega \leq 70°/\text{s}$ have been used; in the latter one, $0 \leq v \leq 50\text{mm/s}$, $-30°/\text{s} \leq \omega \leq 30°/\text{s}$.

The (non-Markovian) state input to the return estimator consists of 21 simulated distance sensors, each of which covers a single view sector with an opening angle between 4° and 16°, resulting in a total view angle of 164°. Figure 2 depicts the arrangement of the view sectors. Every component s_i of the state vector $s \in S = [0; 1]^{21}$ indicates the distance r_i between the robot's center and the closest obstacle point in the corresponding view sector relative to the robot's radius $R = 32\text{mm}$ via $s_i = R/r_i$. Using r_i as the denominator ensures higher resolution in the more relevant region of small distances and provides a convenient range for the s_i.

The continuous reward signal $r(t)$ is meant to encourage distance coverage compared to rapid turning in small circles:

$$r(t) = -0.003\frac{\text{s}}{\text{deg}} \cdot \|\omega(t)\| + 0.015\frac{\text{s}}{\text{mm}} \cdot \frac{\|x(T) - x(t)\|}{T - t} \tag{9}$$

The second summand denotes the mean "topological" velocity over the rest of the episode and $x(t)$ the robot's position in the world. A collision with a wall or obstacle terminates the episode and is penalized with an immediate reward impulse of $\int r(t)dt = -0.75$. The discount rate has been set at $\gamma = 0.1\text{s}^{-1}$.

The experiments use either *constant* action feed-forward policies for each action component, $a_i(t) = A$, with $M = 1$ and parameter $p = A$, or *exponentially* varying feed-forward policies for each action component, $a_i(t) = A + B \cdot e^{-kt}$, with $M = 2$ and parameters $P = (A, B)$. In the latter case, $k = 0.3s^{-1}$ has been used. Thus, the total DOF of the feed-forward policy π_p according to (2) is always $D = N \cdot M$.

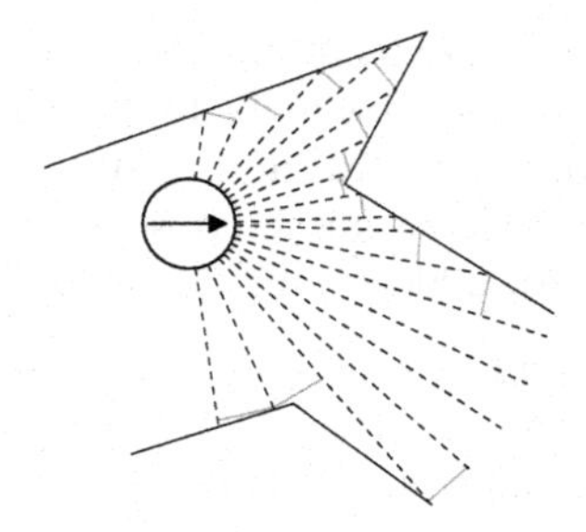

Fig. 2. Obstacle sensor model, comprised of 21 virtual view sectors

In all experiments, *training episodes* have been performed according to section 2.1 and *test episodes* according to section 2.2. The latter ones only serve for monitoring the training progress. After each group of 90 training and 10 test episodes, the return estimator is newly trained from the full training set (see 4.1). Training episodes are terminated after 50s of simulation, test episodes after 500s, unless a collision has occurred earlier.

5.1 "Track" Example

The first examined environment is roughly similar to a race track and shown in Figure 3. The track width is about 125mm. The grey line in Figure 3 depicts the first "good" (i.e. collision-free) test episode for the most primitive $N = M = 1$ case, which occurred after no more than 90 training episodes and involved only 528 training samples according to (5), with only 70% of them actually used as the training set. Please note that the inferred controller is mostly oscillation-free, which is extraordinary after so few trials, and that the plotted trajectory

Fig. 3. Track example with $N = 1$, $M = 1$. Collision-free navigation without oscillation after only 90 training episodes.

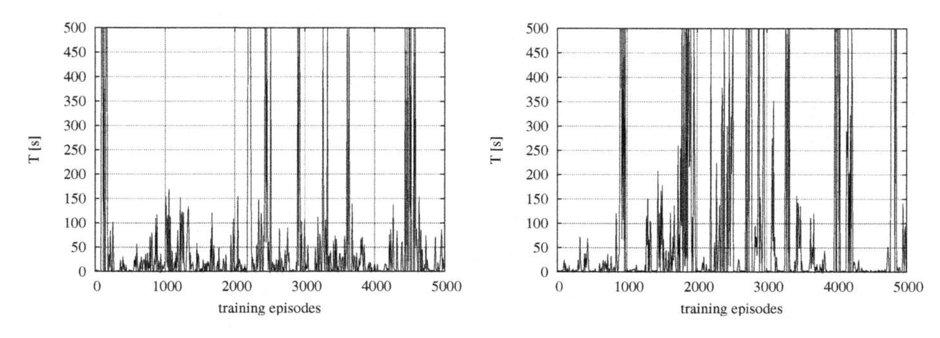

Fig. 4. Learning progress for track example, $N = M = 1$ (left) and $N = 1, M = 2$ (right)

actually comprises 3 "laps" around the track with very high repeat accuracy. With exponential FF policy ($N = 1, M = 2$), the first successful round-track test requires 900 training episodes; with additional v-control ($N = M = 2$) this worsens to 3240 training episodes. Thus, adding more DOF tends to increase learning time, probably because the additional DOF are redundant in this case.

Figure 4 shows the test episode durations over the number of training episodes for the two $N = 1$ cases. Clearly, FF learning does not show a smooth, exponential-decay-type convergence process. This is mainly due to the potentially non-continuous nature of the *arg max* operator in (6), which is very sensitive to "noise" in the estimator training . On the other hand, this coincides with an extremely steep initial learning curve. In practice, one should appoint some additional global criterion for selecting the "desired" agent (e.g. the fraction of "good" test episodes in this case).

5.2 "Corridor" Example

This experiment addresses the learning of non-scalar actions ($N = 2$). The corridor shown in Figure 5 strictly requires reduced speed for turning. With exponential FF policy ($M = 2$), a close-to-optimal behavior occurs for the first time after 1620 training episodes. With constant-action FF policy ($M = 1$), the utility of reduced-speed turns cannot be predicted by the learning algorithm, and FF learning does not effectively succeed before 54810 episodes of training – a success that can well be labeled random. Thus, FF learning does benefit from increased DOF in the FF-policy if the task is not suitably addressed otherwise.

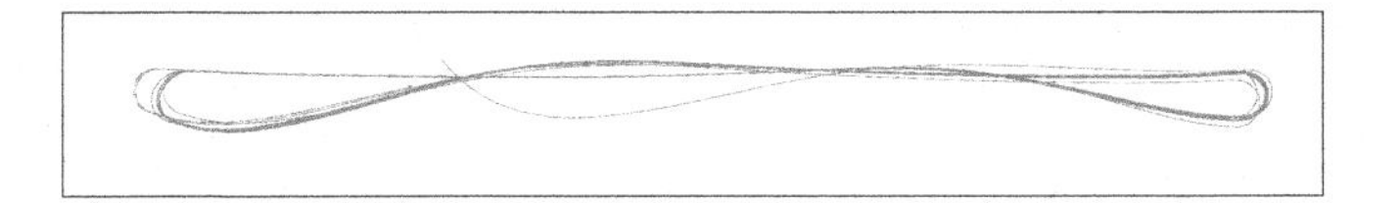

Fig. 5. Corridor example with $N = 2$, $M = 2$. Collision-free test run evolved after 1620 training episodes; turning speed is below 14.4 mm/s, average speed is 38.6 mm/s.

5.3 Results Summary

Table 1 summarizes the learning results for all examined combinations of environment and DOF. For all examples with velocity control, the average speed is shown, indicating that the reward for fast traveling is well honored in all cases.

Table 1. Summary of learning results

World	N	M	episodes to 1st good	best average speed [mm/s]	episodes to best speed
Track	1	1	90	–	–
Track	1	2	900	–	–
Track	2	2	3240	44.0	4230
Corridor	2	1	40410	41.5	54810
Corridor	2	2	1620	40.2	4050

6 Conclusion

FF learning has been shown capable of solving near-real-world control learning problems with very low learning cost. It may even to suitable for tasks with no environment simulation available. Its main drawback is the convolution between the FF policies chosen in training and the characteristics of the environment and task. However, Theorem 1 provides a suitable guideline for selecting appropriate FF policies.

"Fast" biological control learning is sometimes based on an RL situation as well, i.e. solely on exploration and some kind of reward. Thus, FF learning can be considered to close the performance gap between biological and artificial RL to some extent, although there is no methodological foundation from biology involved.

References

1. Sutton, R.S., Barto, A.G.: Reinforcement Learning: An Introduction. MIT Press, Cambridge, MA (1998)
2. Sutton, R.S.: Learning to predict by the methods of temporal differences. Machine Learning **3** (1988) 9–44
3. Watkins, C.J.C.H.: Learning from Delayed Rewards. PhD thesis, King's College, University of Cambridge, Cambridge, UK (1989)
4. Bertsekas, D.P.: Dynamic Programming and Optimal Control. second edn. Volume two. Athena Scientific, Belmont, MA (2001)
5. Boyan, J.A.: Least-squares temporal difference learning. In: Proc. 16th International Conf. on Machine Learning, San Francisco, CA, Morgan Kaufmann (1999) 49–56
6. Nissen, S., Nemerson, E.: Fast Artificial Neural Network Library, Version 1.2.0 Reference Manual. http://fann.sourceforge.net (2004)
7. Hertz, J., Krogh, A., Palmer, R.G.: An Introduction to the Theory of Neural Computation. Lecture Notes Volume I. Addison Wesley (1991)

An Adaptive Michigan Approach PSO for Nearest Prototype Classification

Alejandro Cervantes, Inés Galván, and Pedro Isasi

Department of Computer Science, University Carlos III de Madrid
Avda. Universidad, 30. 28911 Leganés, Madrid, Spain
alejandro.cervantes@uc3m.es, inesmaria.galvan@uc3m.es, pedro.isasi@uc3m.es

Abstract. Nearest Prototype methods can be quite successful on many pattern classification problems. In these methods, a collection of prototypes has to be found that accurately represents the input patterns. The classifier then assigns classes based on the nearest prototype in this collection. In this paper we develop a new algorithm (called AMPSO), based on the Particle Swarm Optimization (PSO) algorithm, that can be used to find those prototypes. Each particle in a swarm represents a single prototype in the solution; the swarm evolves using modified PSO equations with both particle competition and cooperation. Experimentation includes an artificial problem and six common application problems from the UCI data sets. The results show that the AMPSO algorithm is able to find solutions with a reduced number of prototypes that classify data with comparable or better accuracy than the 1-NN classifier. The algorithm can also be compared or improves the results of many classical algorithms in each of those problems; and the results show that AMPSO also performs significantly better than any tested algorithm in one of the problems.

Keywords: Classification, Data Mining, Nearest Neighbor, Particle Swarm, Swarm Intelligence.

1 Introduction

This paper introduces a new technique to perform prototype selection to be applied in Nearest Neighbor classification. Nearest Neighbor (NN) is a lazy learning method where the class assigned to a pattern is the class of the nearest pattern known to the system, measured in terms of a distance defined on the feature (attribute) space where patterns are represented as points.

To perform NN classification, data is split in a training and a test set. Each pattern in the test set is classified with the class of the nearest pattern in the training set. This means that each pattern in the training set defines a region of the feature space where it determines the expected class of the test patterns. Those regions are called Voronoi regions. When Euclidean distance is used, the Voronoi regions are delimited by linear borders. This may or may not provide an optimum solution, as the actual solution of the problem may require non-linearly delimited Voronoi regions, which may be accomplished using the k-NN algorithm or non-Euclidean distances.

J. Mira and J.R. Álvarez (Eds.): IWINAC 2007, Part II, LNCS 4528, pp. 287–296, 2007.

Both computational reasons and presence of noise in data lead to the development of techniques that reduce the number of patterns evaluated without increasing the classification error. These methods calculate a reduced set of prototypes, that are the only ones used for classification. In instance selection methods [1] prototypes are a subset of the pattern set, but this is not true for all the prototype selection methods [2], where prototypes may be selected in any position in the attribute space.

Our method selects those prototypes by first allocating a population of solutions that represent the prototypes and then moving them in the attribute space in a way inspired by the Particle Swarm Optimization (PSO) algorithm [3].

The PSO algorithm is a biologically-inspired algorithm motivated by a social analogy. The algorithm is based on a set of potential solutions which evolves to find the global optimum of a real-valued function (fitness function) defined in a given space (search space). Particles represent the complete solution to the problem and move in the search space using both local information (the particle memory) and neighbor information (the knowledge of neighbor particles). In the standard approach of PSO, a potential solution is encoded in each particle and the swarm finally converges to a single solution.

However, our method, called AMPSO (Adaptive Michigan PSO) has important differences as it uses a Michigan Approach; this term is borrowed from the area of genetic classifier systems ([4][5]). In this approach a member of the population does not encode the whole solution to the problem, but only part of the solution. To implement this behavior, movement and neighborhood rules of the standard PSO are changed. A related approach was used in [6], where a Michigan binary version of PSO was able to discover a set of induction rules for discrete classification problems.

The advantages of the Michigan approach versus the conventional PSO approach are: a) scalability and computational cost, as particles have much lower dimension; and b) flexibility and reduced assumptions, as the number of prototypes in the solution is not fixed.

Moreover, our algorithm does not use a fixed population of particles; given certain conditions, we allow particles to reproduce to adapt to some situations, and also we destroy useless particles to reduce the computational cost and the complexity (number of final prototypes) of the solution.

This paper is organized as follows: section 2 shows how the solution to the classification problem is encoded in the particles of the swarm; section 3 details the proposed AMPSO algorithm; section 4 describes the experimental setting and results of experimentation; finally, section 5 discusses our conclusions and future work related to the present study.

2 Encoding Prototypes in Particles

The PSO algorithm uses a population of particles each encoding a complete solution to an optimization problem. The position of each particle in the search space changes depending on the particle's fitness and the fitness of its neighbors.

In the Michigan-approach PSO we propose, each particle represents a potential prototype to be used to classify the patterns using the nearest neighbor rule. The particle position is interpreted as the position of a single prototype. Each particle has also a class; this class does not evolve following the PSO rules, but remains fixed for each particle since its creation.

The dimension of the particles is equal to the number of attributes of the problem. The attributes of the problem are transformed in continuous values in the $[0, 1]$ range; this means that maximum and minimum values for the particles' positions are fixed in all the dimensions.

A particle classifies a pattern when it is the closer particle (in terms of Euclidean distance) to that pattern.

3 The Adaptive Michigan PSO Algorithm

3.1 Algorithm Pseudo Code and Movement

Our algorithm is based in the PSO algorithm but performs some extra calculations and has an extra cleaning phase. The overall procedure follows. Our additions are explained in the following sections.

1. Load training patterns
2. Initialize swarm; dimension of particles equals number of attributes.
3. Insert N particles of each class in the training patterns.
4. Until max. number of iterations reached or success rate is 100%:
   (a) Check for particle reproduction and deletion. (see 3.6)
   (b) Calculate which particles are in the competing and non-competing sets of particles for every class (see 3.4).
   (c) For each particle,
      i. Calculate Local Fitness. (see 3.3)
      ii. Calculate Social Adaptability Factor. (see 3.5)
      iii. Find the closest particle in the non-competing set for the particle class (attraction center).(see 3.4)
      iv. Find the closest particle in the competing set for the particle class (repulsion center).(see 3.4)
      v. Calculate the particle's next position. (see Eq. 1 and Eq. 2 in 3.2)
   (d) Move the particles
   (e) Assign classes to the patterns in the training set using the nearest particle.
   (f) Evaluate the swarm classification success
   (g) If the swarm gives the best success so far, record the particles' current positions as "current best swarm".
5. Delete, from the best swarm found so far, the particles that can be removed without a reduction in the classification success value.
6. Evaluate the swarm classification success over the validation set and report result.

In step 5 of the previous procedure a reduction algorithm is applied after the swarm reaches its maximum number of iterations. Its purpose is to delete unused particles from the solution. Particles are removed one at a time, starting with the one with the worst local fitness value, only if this action does not reduce the swarm classification success rating over the training set. The "clean" solution is then evaluated using the validation set.

3.2 Movement Equations

In order to take into account the concepts previously described, the equation that determines the velocity at each iteration becomes the following (1).

$$v_{id}^{t+1} = \chi(w \cdot v_{id}^t + c_1 \cdot \psi_1 \cdot (p_{id}^t - x_{id}^t) +$$

$$c_2 \cdot \psi_2 \cdot sign(a_{id}^t - x_{id}^t) \cdot Sf_i + c_3 \cdot \psi_3 \cdot sign(x_{id}^t - r_{id}^t) \cdot Sf_i) . \quad (1)$$

where the meanings of symbols are: v_{id}^t, component in dimension d of the ith particle velocity in iteration t; x_{id}^t, same for the particle position; c_1 ,c_2, c_3, constant weight factors; p_i, best position achieved so far by particle i; ψ_1 ,ψ_2, ψ_3, random factors in the [0,1] interval; w, inertia weight; χ, constriction factor; a_i, attraction center for particle i ; r_i, repulsion center for particle i ; Sf_i, Social Adaptability Factor ;

This expression allows the particle velocity to be updated depending on four different influences: the current velocity of the particle is retained from iteration to iteration (inertia term, $w \cdot v_{id}^t$); the particle's memory (individual term, $c_1 \cdot \psi_1 \cdot (p_{id}^t - x_{id}^t)$); the current position of a particle to which the particle is attracted (attractive term, $c_2 \cdot \psi_2 \cdot sign(a_{id}^t - x_{id}^t) \cdot Sf_i$); and the current position of a particle from which the particle is repelled (repulsive term, $c_3 \cdot \psi_3 \cdot sign(x_{id}^t - r_{id}^t) \cdot Sf_i$). Also, if a_i or r_i does not't exist the respective term (attractive term or repulsive term) is ignored.

After velocity update, the particle position is calculated using (2).

$$x_{id}^{t+1} = x_{id}^t + v_{id}^{t+1} . \quad (2)$$

3.3 Local Fitness Function

In our approach, each particle has a Local Fitness value that measures its performance independently of the other particles. For this purpose, we use (3).

$$\text{Local Fitness} = \begin{cases} 0 & \text{if } \{g\} = \{b\} = \emptyset \\ \frac{G_f}{Total} + 2.0 & \text{if } \{b\} = \emptyset \\ \frac{G_f - B_f}{G_f + B_f} + 1.0 & \text{otherwise} \end{cases} , \text{ where } \quad G_f = \sum_{\{g\}} \frac{1}{d_{g,i}+1.0} \quad B_f = \sum_{\{b\}} \frac{1}{d_{b,i}+1.0} . \quad (3)$$

where $\{g\}$ is the set of patterns of the same class classified by the particle; $\{b\}$ is the set of patterns of different class classified by the particle; $d_{g,i}, d_{b,i}$ are the Euclidean distances between the patterns and the particle; and $Total$ is the number of patterns in the training set.

This fitness function gives higher values (greater than +2.0) to the particles that only classify patterns whose class matches the class of the particle, and assigns values in the range [0.0, +2.0] to particles that classify some patterns of a wrong class.

In the lowest range, the particles only take into account local information (the proportion of good to bad classifications made by itself). In the highest range, the particle fitness uses some global information (the total number of patterns to be classified), to be able to rank the fitness of particles with a 100% accuracy (particles for which $\{b\} = \emptyset$).

By including in the formula the distance between patterns and particles we give higher fitness to particles close to the patterns they classify correctly, so they can move closer to the area where the patterns are located. This tendency may be compensated by the repulsion term.

3.4 Neighborhood for the Michigan PSO

Our algorithm uses a dynamic neighborhood, which means that on each iteration, movement of each particle may be influenced by different particles in the swarm:

- For each particle of class C_i, non-competing particles are all the particles of classes $C_j \neq C_i$ that currently classify at least one pattern of class C_i.
- For each particle of class C_i, competing particles are all the particles of class C_i that currently classify at least one pattern of that class (C_i).

When the movement for each particle is calculated, that particle is both:

1. Attracted by the closest (in terms of Euclidean distance) non-competing particle in the swarm, which becomes the "attraction center" (a_i) for the movement. In this way, non-competing particles guide the search for patterns of a different class.
2. Repelled by the closest competing particle in the swarm, which becomes the "repulsion center" (r_i) for the movement. In this way, competing particles retain diversity and push each other to find new patterns of their class in different areas of the search space.

Other authors have already used the idea of repulsion in PSO in different ways. For instance, in [7] and [8], repulsion is used to increase population diversity in the standard PSO and to allow the swarm to dynamically adapt to change. In [9] a repulsive force is introduced to achieve particle diversification in a binary PSO.

3.5 Social Adaptability Factor

The social part of the algorithm (influence from neighbors) determines that particles are constantly moving towards their non-competing neighbor and far from their competing neighbor. However, particles that are already located in the proximity of a good position for a prototype should rather try to improve their position and should possibly avoid the influence of neighbors.

To implement this effect we have generalized the influence of fitness in the sociality terms by introducing a new term in the PSO equations, called "Social Adaptability Factor" (S_f), that depends inversely on the "Best Local Fitness" of the particle. In particular, we have chosen plainly the expression in (4); value for the exponent s_{exp} was set to 1.0 after some experimentation.

$$Sf_i = 1/(\text{Best Local Fitness}_i + 1.0)^{s_{exp}} \quad (4)$$

3.6 Reproduction and Deletion of Particles

Though the swarm starts with a fixed number of particles, we introduce two features that adjust the number of particles and their classes to the problem's requirements. The rules for reproduction and deletion of particles are:

- Particles that classify a set of patterns of several classes have a probability to give birth to one particle for each of classes in that set. The chance of "reproduction" is inversely proportional to the particle fitness (so the worst particles are more likely to reproduce) and proportional to a parameter (P_r). The particles are placed in the position of the "parent" particle and their velocities are randomized.
- Particles that do not classify any pattern have a chance to be deleted that grows linearly from 0 to the value of a parameter (P_d) in the last iteration.

4 Experimentation

4.1 Global Swarm Evaluation

The local fitness function defined above, is used for particles' movement. However, to evaluate the goodness of the swarm as a classifier system, we use the classification success rate (5).

$$\text{Swarm Evaluation} = \frac{Good\ classifications}{Total\ patterns} \cdot 100 \quad (5)$$

Given the NN criterion for classification, unclassified patterns are not possible, as every pattern is assigned the class of the nearest prototype.

The system stores the best swarm evaluation obtained when performing classification on the training set. This function is also used to evaluate the final best swarm success rate over the validation set.

4.2 Problems' Description

We perform experimentation on the problems summarized in Table 1. The first (diagonal) is an artificial bi-dimensional problem that is very simple for lineal classifiers, and it is used to understand the new algorithm's properties. We generate 500 random training patterns and 1500 validation patterns with coordinates in the $[0, 1]$ range. Patterns where $x > y$ are assigned class 1, and the rest are assigned class 0.

The rest are well-known real problems taken from the UCI collection, used for comparison with other classification algorithms. All the problems have real-valued attributes, so no transformation was done on data besides normalization. Success rates from several classification algorithms over some of these problems can be found in [2].

Table 1. Problems used in the experiments

Name	Instances	Attrbs.	Classes	Class Distribution	Validation
Diagonal	2000	2	2	1000 / 1000	Train & Test
Diabetes	768	8	2	500 / 268	10-fold CV
Bupa	345	6	2	200 / 145	10-fold CV
Glass	214	9	6	70 / 76 / 17 / 13 / 9 / 29	10-fold CV
Thyroid (new)	215	5	3	150 / 35 / 30	10-fold CV
Wisconsin	699	6	2	458 / 241	10-fold CV

4.3 Parameter Selection

The values of the swarm parameters were selected after some preliminary experimentation.

In all cases we found that it was better to use a small value for the inertia coefficient ($w = 0.1$); the number of iterations was set to 300 after checking that number was roughly equal to double the average iteration in which the best result was achieved.

For the PSO coefficients, we used low values for the parameters in the diagonal problem ($c_1 = 0.5$, $c_2 = 0.15$, $c_3 = 0.05$), which means that movement is performed in small steps. In the other problems we used the same set of values ($c_1 = c_2 = 1.0$, $c_3 = 0.5$). Velocity was clamped to the interval $[-1.0, +1.0]$ and $\chi = 0.1$ in all the experiments.

To test reproduction and deletion of particles, we performed three series of experiments, the first with no reproduction nor deletion (Series 1, $P_r = P_d = 0.0$), the second with only reproduction (Series 2, $P_r = 0.1$, $P_r = 0.0$) and the third with reproduction and deletion(Series 3, $P_r = P_d = 0.1$).

In each one of the series we performed 100 runs for the Diagonal problem, and 10 runs for the problems that use 10-fold cross validation, which also gives a total of 100 runs over each. In all the experiments, we used 10 particles per class for the initial swarm.

4.4 Experimental Results

The results of AMPSO on all the data sets are shown in Table 2; in this table we compare our results with the results of our own tests with other algorithms,

Table 2. Experiment results (success rate) for the UCI data sets; the three AMPSO series compared to several commonly-used classifiers

Domain	AMPSO Series 1	AMPSO Series 2	AMPSO Series 3	IBK (K=1)	IBK (K=3)	J48	PART	Naive Bayes	SMO
Diagonal	94.26	95.71	95.56	97.93	97.64	95.22	95.50	96.30	97.76
Diabetes	74.14	74.45	74.25	70.44	73.88	74.48	74.18	75.69	76.63
Bupa	63.29	64.51	65.05	62.90	63.16	66.01	63.08	55.63	57.95
Glass	82.04	84.46	83.17	71.99	70.58	72.86	73.79	47.25	57.10
Thyroid	94.03	93.81	94.88	97.21	93.46	92.06	93.92	96.75	89.74
Wisconsin	96.46	96.66	96.48	95.14	96.42	94.70	95.14	95.99	96.85

using WEKA with the default parameters for each of the algorithms. In the table, IBK(k=1) means plain NN, and IBK(k=3), 3-NN classification.

The results are very similar though both Series 2 and 3 are slightly better than the AMPSO algorithm with no reproduction nor deletion (Series 1) in all the problems but the Thyroid problem.

The success rate in terms of accuracy for AMPSO is comparable or clearly better than the result of IBK with K=1 in all but the Diagonal and the Thyroid problems. In the rest of the problems, the patterns are correctly represented with the particles in the solution swarm.

For the Diagonal problem, the reason is that the learning bias of the algorithm is not favoring the accuracy of the solution. A very good solution can be found just placing one prototype of each class, located at the central position of each cluster; the solutions found by our method are worse than that trivial solution due to the fact that AMPSO explicitly searches near the classes boundary (the diagonal) where a solution is more difficult to find.

In the thyroid problem our hypothesis is that NN classification requires more prototypes than the particles generated by the algorithm. It may be difficult to beat the performance of 1-NN in this problem as it seems that it significantly improves any other method. Reduction of the prototype number is more likely to decrease accuracy in this case.

The algorithm significantly improved the accuracy of all the rest in the Glass problem. Good performance on this problem is possible due to the fact that non-euclidean distances seem not to improve the results of Euclidean distance in this case (see [2]). Our algorithm seems to significantly improve the results of all the other classifiers, being to our knowledge the best result over this data set.

In Table 3, we compare the results of the three series, to show the impact of the introduction of particle reproduction and removal over the algorithm's performance and the complexity of the solutions found, measured in terms of the average number of prototypes in the solution.

The results in Table 3 show that Series 2 produces a greater number of prototypes, that is, solutions are of greater complexity. Particle deletion reduces this number of prototypes in the final solution as expected. Those results also show that series 2 and 3 have greater computational costs (average number of iterations).

Table 3. Comparison of the three series of experiments

	Series	Diagonal	Diabetes	Bupa	Glass	Thyroid	Wisconsin
Prototypes	1	13.37	10.47	10.89	10.30	11.37	11.12
	2	14.48	13.52	12.61	12.76	14.67	19.32
	3	13.38	13.00	12.01	11.73	13.80	19.00
Evaluations	1	3222	3114	2918	10620	5235	3198
	2	3185	4116	3550	14236	7114	5959
	3	2909	4224	3908	12760	6737	8762

5 Conclusions

The purpose of this paper is to develop an effective algorithm for prototype creation, to be applied to classification problems. Our algorithm is based on a new approach of Particle Swarm Optimization (PSO); it determines a set of prototypes that represent the training patterns and can be used as a classifier using the nearest neighbor (1-NN) rule.

In the standard PSO, a straightforward encoding of a set of prototypes in each particle would produce a search space of high dimension that might hinder the swarm success. For this reason, we propose a Michigan Approach PSO (AMPSO), in which each particle represents a single prototype. In AMPSO the population of particles also changes over time to adapt to the problem.

The AMPSO algorithm introduces a local fitness function to guide the particles' movement and dynamic neighborhoods that are calculated on each iteration to ensure particles are influenced only by others that either compete or cooperate to classify part of the patterns.

We have tested the algorithm in an artificial problem and six well-known benchmark problems and we have found that the results are always close to or better than the standard nearest neighbor classifier (1-NN) with the limited number of prototypes that compose the solutions found by the algorithm. This proves that PSO can be used to produce a small representative set of prototypes.

When the results are compared to other classifiers, AMPSO can produce competitive results in all the problems, specially when used in problems where 1-NN classifier does not perform very well. Finally, AMPSO outperforms significantly all the algorithms on the Glass Identification data set, where it achieves more that 10% improvement on average.

The introduction of particle reproduction and deletion allows the number of particles in the swarm to grow to produce solutions of greater complexity but the extra cost in terms of computational cost has to be evaluated.

It is clear than further work could improve the algorithm performance if it makes AMPSO able to adaptively tune important parameters (such as the reproduction and deletion rate) to the problem. Also, any technique that may improve Nearest Neighbor classifiers' performance could be applied to AMPSO.

Acknowledgments

This article has been financed by the Spanish founded research MEC project OPLINK::UC3M, Ref: TIN2005-08818-C04-02 and CAM project UC3M-TEC-05-029.

References

1. Henry Brighton and Chris Mellish. Advances in instance selection for instance-based learning algorithms. *Data mining and knowledge discovery*, 6(2):153–172, 2002.
2. Fernando Fernández and Pedro Isasi. Evolutionary design of nearest prototype classifiers. *Journal of Heuristics*, 10(4):431–454, 2004.
3. J. Kennedy, R.C. Eberhart, and Y. Shi. *Swarm intelligence.* Morgan Kaufmann Publishers, San Francisco, 2001.
4. J.H. Holland. Adaptation. *Progress in theoretical biology*, pages 263–293, 1976.
5. Steward W. Wilson. Classifier fitness based on accuracy. *Evolutionary Computation*, 3(2):149–175, 1995.
6. Alejandro Cervantes, Pedro Isasi, and Inés Galván. A comparison between the pittsburgh and michigan approaches for the binary pso algorithm. In *Proceedings of the 2005 IEEE Congress on Evolucionary Computation, CEC 2005*, pages 290–297, 2005.
7. T. M. Blackwell and Peter J. Bentley. Don't push me! collision-avoiding swarms. In *Proceedings of the IEEE Congress on Evolutionary Computation (CEC)*, pages 1691–1696, 2002.
8. T. M. Blackwell and Peter J. Bentley. Dynamic search with charged swarms. In *Proceedings of the Genetic and Evolutionary Computation Conference 2002 (GECCO)*, pages 19–26, 2002.
9. Alejandro Cervantes, Pedro Isasi, and Inés Galván. Binary particle swarm optimization in classification. *Neural Network World*, 15(3):229–241, 2005.

Behavioural Modeling by Clustering Based on Utility Measures

Philip Hoelgaard[1], Ángel Valle[1,2], and Fernando Corbacho[1,2]

[1] Cognodata Consulting, C/ Caracas 23, 28010 Madrid, Spain
[2] Escuela Politécnica Superior, Universidad Autónoma de Madrid, 28049 Madrid, Spain

Abstract. This paper presents a new framework for behavioural modelling that allows to unravel the key drivers that direct specific cognitive behaviours. In order to do so, a novel framework for clustering based on utility measures is presented that allows to understand the different behaviours that different groups of people may have, and allows the creation of profiles that are relevant with respect to the utility measure. The proposed method is not contrary to other clustering methods but rather builds on the functionality of 'basic' clustering algorithms. A common aim of clustering consists of partitioning a set of patterns into different subsets of patterns which have homogeneous characteristics. In this paper we suggest a more ambitious goal that additionally tries to maximize a utility measure. The paper also describes the results obtained when the method is used to analyze human behaviour in the area of customer intelligence. Specifically, the paper analyzes human behaviour with respect to different socio-demographic and economic indicators and allows to uncover the underlying characteristics that may explain the observed cognitive behaviour.

1 Introduction

Computational modeling of cognitive tasks deals with how cognition can be explained by different processes (perception, learning, language, etc.) as well as the underlying mechanisms that give rise to the observed cognitive behaviour. More specifically, behavioural modeling, among other aims, attempts to understand the key drivers that direct behaviour. This paper provides an analysis of human behaviour in the area of customer intelligence. Specifically, this paper analyzes human behaviour with respect to different socio-demographic and economic indicators and allows to uncover the underlying characteristics that may explain the observed behaviour.

We provide a novel framework (C.U.M) of clustering based on Utility Measures that allow to unravel the different behaviours that different groups of people may have, and allow the creation of profiles that are relevant with respect to the Utility Measure. Next we succinctly describe the proposed method and place it in context with regard to existing clustering algorithms. Notice that the proposed method is not contrary to other clustering methods but rather builds

J. Mira and J.R. Álvarez (Eds.): IWINAC 2007, Part II, LNCS 4528, pp. 297–306, 2007.

on the functionality of 'basic' clustering algorithms. A common aim of clustering consists of partitioning a set of patterns into different subsets of patterns which have homogeneous characteristics *(Rui Xu, [6])*. In this paper we suggest a more specific goal that additionally incorporates the concept of the utility of a particular clustering. The number of possible clusterings B_N (Bell Number) upon a set is exponential in its size, N. E.g. $B_{100} >> 1e100$. Thus, it is unfeasible to generate all possible different clusterings; let alone for an analyst to select one for a particular problem. In the C.U.M. framework many different possible clusterings are generated and a Utility Measure applied to achieve an ordering of these and facilitate the selection of one or more useful clusterings.

Thus, the overall objective that we propose is two-fold, on one hand to find homogeneous clusters with respect to the input attributes and, on the other hand, find clusterings that maximize a Utility Measure.

In the empirical risk minimization framework (ERM) *(Vapnik, [1])* clustering corresponds to learning the following mapping:

$$f(x, w) : x \to c(x) \tag{1}$$

where x is the input pattern, w is the parameter set and $c(x)$ denotes the cluster center coordinates for a pre-specified number of clusters. A common loss function is the squared error distortion

$$L(f(x, w)) = (x - f(x, w)) \circ (x - f(x, w)) \tag{2}$$

where $\circ$ denotes the usual inner product. Minimizing the risk functional

$$R(w) = \int (x - f(x, w)) \circ (x - f(x, w))p(x)dx \tag{3}$$

gives an optimal vector quantization based on the observed data.

Our framework also attempts to learn this mapping but subject to an additional constraints, namely to maximize a Utility Measure which will be presented in section 2.1.

This type of utility-based clustering has proven to be useful in several business intelligence applications such as customer segmentation where there are specific Utility Measures to evaluate the usefulness of the resulting clusterings. We claim that this framework has a very wide applicability. In the following sections we present the framework and all the components, we then explain the differences with respect to the induction of decision trees when a target is present.

The results section presents a customer segmentation based on the Utility Measure which has been applied to different business intelligence applications. We close with conclusions on the applicability and generalization of the proposed method.

2 Framework for Clustering Based on Utility Measures

In the following we shall review a methodology that applies a Utility Measure upon a set of clusterings to identify a clustering that is useful for describing the relationship between input attributes and some exogeneous information.

The exogenous information is not part of the patterns used to perform the clustering - it represents associated information wrt. which the patterns should be characterized. As an example, consider a pattern consisting of the demographic attributes Age, Sex and Marital Status: a possible exogenous variable is Income and one might be interested in discovering how the Income varies over the three-dimensional attribute space Age, Sex, Marital Status.

The concrete Utility Measure proposed here reflects the behaviour of the exogenous information in each cluster of a clustering.

2.1 General Procedure of Generation, Measurement and Analisis of Clusterings

Our aim is to obtain clusterings for which

i) Each cluster is homogeneous wrt. the attributes
ii) The clusters are well separated
iii) Within each cluster, the exogeneous variable has a narrow distribution
iv) Between clusters, there is a high degree of variability of the exogeneous variable

Note that the two first requirements *(i)* and *(ii)* belong to the classical framework of clustering *(eq. 3)* while the last two relate specifically to the exogeneous variable and the concept of Utility. For the generation of each particular clustering, only the two first requirements are considered, the exogeneous variable is ignored. The procedure for finding one or more useful canditate clusterings is as follows:

1) A large number of clusterings are generated with respect to the attributes. To obtain a diverse set of clusterings, each clustering is generated using different initial conditions[1] *and a different subset of the available attributes.*[2]

2) A measure of utility of a clustering is obtained based on (iii)-(iv) and applied to the resulting set of clusterings to provide an ordering of these.

3) According to the ordering, a small number (e.g. 5-10) of clusterings are picked out and represented graphically. A human analisis of these clusterings is applied to obtain satisfactory descriptions of the relationships between attributes and exogeneous information.

In the following we will put forward the individual measures corresponding to *(iii)-(iv)* and the Utility measure described in step *(2)*.

[1] E.g. randomly chosen initial cluster centres from a uniform distribution on the attribute space.

[2] The subset of attributes should always be of the same size and the attributes normalized (e.g. to have the same standard deviation). These requirements permit that the degree of homogeneity of different clusterings may be compared directly. Note also that if the attributes are not normalized, the clusterings should be constructed using a non-euclidean distance measure (e.g. Mahalanobis).

2.2 The Utility Measure

We propose a measure on the exogenous information that adress *(iii)* and *(iv)* from section *2.1*:

iii) Within each cluster, the exogenous variable has a narrow distribution.
iv) Between clusters, there is a high degree of variability of the exogenous variable.

In the following, let D measure the centredness of *(iii)* and T the variability of *(iv)*. Let V denote the exogenous variable, C a particular clustering, and $E_{C_i}(V)$ and $\sigma_{C_i}(V)$ the mean and standard deviation of V in cluster C_i, respectively. With this we may formulate the following:

$$D(C,V) = E\left(\bigcup_i \sigma_{C_i}(V)\right) \tag{4}$$

$$T(C,V) = \sigma\left(\bigcup_i E_{C_i}(V))\right) \tag{5}$$

We define the Utility Measure, U, for a clustering simply as

$$U(C,V) = D(C,V) + T(C,V) = E\left(\bigcup_i \sigma_{C_i}(V)\right) + \sigma\left(\bigcup_i E_{C_i}(V)\right) \tag{6}$$

We remind that the purpose of the Utility Measure is to identify clusterings that comply with *(iii)* and *(iv)*. In practice the present definition serves this purpose well but it should be noted that there are several possible variations. Two examples: 1) the terms D and T may be pondered differently and 2) to focus only on a subset of clusters that comply with *(iii)*, $E\left(\bigcup_i \sigma_i(V)\right)$ may be substituted by $E_n\left(\bigcup_i \sigma_i(V)\right)$, where E_n denotes that the n highest spreads are excluded from the mean.

2.3 Behavioural Analisis of High-Utility Clusterings

For the human analisis of the behavioural aspects of a specific clustering, we consider a graphical representation of each of its clusters. Please refer to the example in section *4*: here are shown diagrams for 2 clusters, A and B (they belong to a particular clustering with 6 clusters). A diagram consists of two parts:

- **left part:** Distributions of attributes, in cluster (solid) and globally (line).
- **right part:** Distribution of exogenous variable, in cluster (solid) and globally (line).

The purpose of the diagrams is to profile the attributes and the exogeneous variable in the clusters and to relate the behaviour of the latter to the former. To discover the relationship between the attributes and the exogenous variable we proceed as follows:

a) *For each of the clusters, mark the attributes whose distributions differ the most from the corresponding global distributions. In the example in section 4, clusters A and B, we have marked these attributes with a lightning symbol.*
b) *Describe the main characteristics of the distribution of the exogeneous variable.*
c) *For each cluster form a hypothesis for the relationship between the attributes and the exogenous variable.*
d) *Check to see if the hypothesis should involve more variables or if some are unnecessary. Mark or unmark these and return to step (c). Observe that the description of the distribution of an attribute should be in the context of the current hypothesis. Specifically, the global distribution should no longer be the main reference.*
e) *For each cluster, sintesize the hypothesis into a short description.*

Note that all the hypothesis should make sense in terms of the fundamental understanding of the problem, it's intended use and actionability.

3 Comparison with Decision Trees

It is important to emphasize that in the present framework, the clusterings themselves do not depend on the exogenous variable: the Utility Measure is applied a posteriori on the generated clusterings and only affects *which* clusterings are selected for the final human evaluation. If the Utility Measure were used to direct the generation of the clusterings, the vast majority would violate the fundamental and classical requirements *(i)* and *(ii)* in section *2.1*: if the exogenous variable, V, were included for the generation of the clustering, this would imply using a distance measure on the combined space $\{A \times V\}$. But, depending on the two points selected, a distance measured in the combined space $\{A \times V\}$ may be either greater or smaller than that measured in the attribute space $\{A\}$. Therefore, a cluster of points that are 'close' to each other in $\{A \times V\}$ are not necessarily 'close' in $\{A\}$.

Decision trees constitute one of the most popular methods for extracting knowledge from a given set of patterns (Breiman, [4]). A decision tree performs induction on the attributes to discriminate the exogenous variable. The result is a set of leaf nodes that each define a segment in attribute space, $\{A\}$. The segments may suffer from three drawbacks:

1) *The definition of a segment is a direct consequence of the exogeneous variable and no criteria of homogeneity of the attributes is applied.*
2) *A segment is a hypercube: each lateral bounds the value of one of the attributes. This way of defining a segment of patterns is often too rough for real world problems.*
3) *The hypercube is normally open - not all the attributes are used to define the node. This means that the distributions of the remaining attributes may have very wide spreads.*

Recent methods address *(1)* by using decision trees to partition the attribute space without using the exogeneous information *(e.g. Basak & Krishnapuram, 2005)*. The branching decision at each node of the tree is taken based on the clustering tendency of the data available at the node. The final decision tree provides a hierarchical clustering of the input data.

But, as described in the introduction, similarly to other clustering methods, this method lacks the Utility Measure of C.U.M that allows an ordering to be applied to a large set of clusterings (from the vast space of all possible clusterings) and a relevant clustering to be chosen from this.

C.U.M results in clusters that are fuzzy clouds of points. This is in contrast to the sharp boundaries defined by decision trees (described in *(2)* and *(3)* above). For a cluster, the most important aspects of the distributions of its attributes are synthesized into a natural language description. A natural language description is satisfactory for natural phenomena that possess a high degree of intrinsical fuzziness.

4 Results

We present a problem of behavioural modeling in the context of an automobile insurance company. In this case, the endogenous information for C.U.M are the attributes that characterize the insured: Age, Sex, Marital Status and his situation in the company: Policy Age, Automobile Price, Number of Installments, or his Expected Policy Lifespan. The exogenous information of the problem is the 'Annual Policy Profit' generated by the client.

We shall analize a clustering that has a high value of the Utility Measure. Figure 1 shows a representation of this clustering. Each circle corresponds to a cluster. The abscissa represents the value of the exogenous variable *Annual Policy Profit* in a cluster. The ordinate represents the size of a cluster. We note

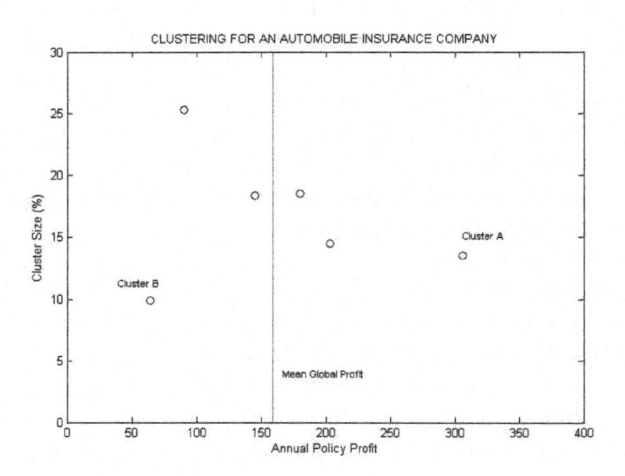

Fig. 1.

that the clusters are well separated wrt. the exogenous variable. This is useful because it reflects that the behaviour is differentiated across clusters. In this example the clustering contains 6 clusters. Below, we discuss 2 of these: One with a high *Annual Policy Profit* and another with low profit.

4.1 Analysis of Cluster A

Attribute	Description	Global Mean
Annual policy profit	Higher policy profit (mean 306 Euros)	159 Euros
Client age	Younger clients (mean 34 years)	46 years
Sex	Male clients (99% male)	76%
Expected policy lifespan	Shorter lifespans (mean 3.8 years)	4.5 years

The above table is the result of the first step, *(a)*, in the behavioural analysis described in section *2.3*. The next steps iterate *(b)-(c)* from section *2.3* to obtain a hypothesis for the relationship between the atributes and the exogenous variable.

Young male clients in their globality constitute a high risk group and therefore pay a high premium. However, the present cluster identifies a group of younger males whose risk of accident is slightly lower than that of the global population and (not shown) considerably lower than expected for all males in the same age-group. Likewise, it is observed that the cluster contains a higher rate of married clients than expected for the age-group.

Thus, our hypothesis is that in this cluster, the Annual Policy Profit is high because these clients pay a high premium compared to their actual risk (medium expenses from accidents) and that the circumstance of being married implies a more mature behaviour (less risk).

Observe that we have only presented the final hypothesis. In the iteration process, the attribute *Expected Policy Lifespan* has been eliminated and the attribute *Marital Status* added. The distribution of *Expected Policy Lifespan* can be considered a consequence (shorter policy lifespan) of the distribution of *Client Age*: young people are generally more flexible and less afraid of accepting a 'better offer'. On the other hand, *Marital Status* is interesting *because* its distribution is similar to the global: this is not generally so in this age-group.

4.2 Analysis of Cluster B

Now, the steps *(b)-(c)* from section *2.3* are iterated to obtain a hypothesis for the relationship between the atributes and the exogenous variables:

Older clients with high seniority are normally considered 'priority clients' and are given high bonuses on their premium. In this cluster almost all the clients

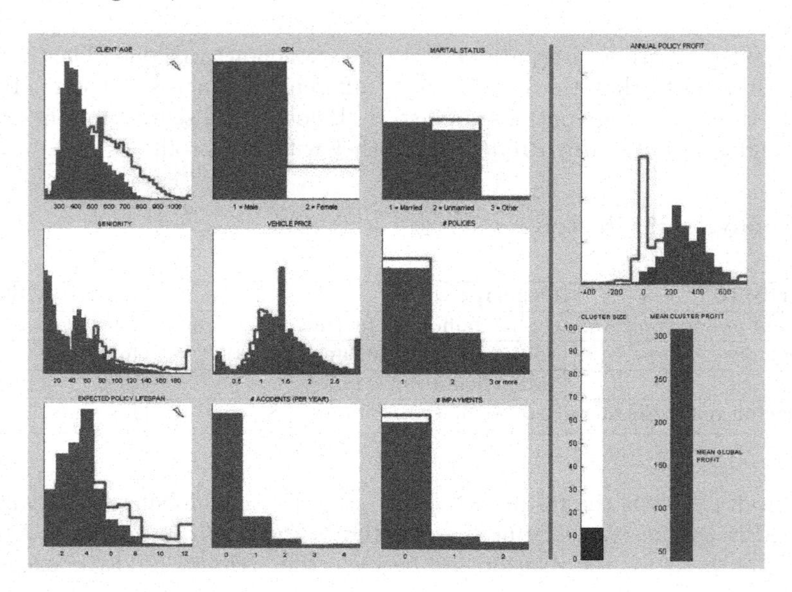

Fig. 2. Cluster A

are married and predominantly have more than one policy (automobile). This situation normally implies that a family member with a higher risk (e.g. a son) may be using the automobile. In fact, it is observed that the risk of accident (*Number of Accidents per Year*) is a little higher than the global: it would be expected to be much lower for these experienced drivers.

Attribute	**Description**	**Global Mean**
Annual policy profit	Lower policy profit (mean 64 Euros)	159 Euros
Client age	Older clients (mean 57 years)	46 years
Sex	Male clients (92% male)	76% male
Marital Status	Married clients (96% married)	47% married
Seniority	High seniority (mean 12 years)	4.9 years
Number of policies	More policies (mean 1.9 policies	1.44 policies
Expected policy lifespan	Shorter lifespans (mean 7.8 years)	4.5 years

Thus, our hypothesis is that in this cluster, the 'Annual Policy Profit' is low because these clients pay a low premium compared to their actual risk (medium expenses from accidents) due to the circumstance that the vehicle is being used by other familiy members.

Regarding the fact that in the cluster, 92% are male clients (76% in global population): Our hypothesis of having identified older family heads is consistent with the fact that traditionally these are male.

And finally (step *(d)*), we arrive at the following general description of clusters A and B:

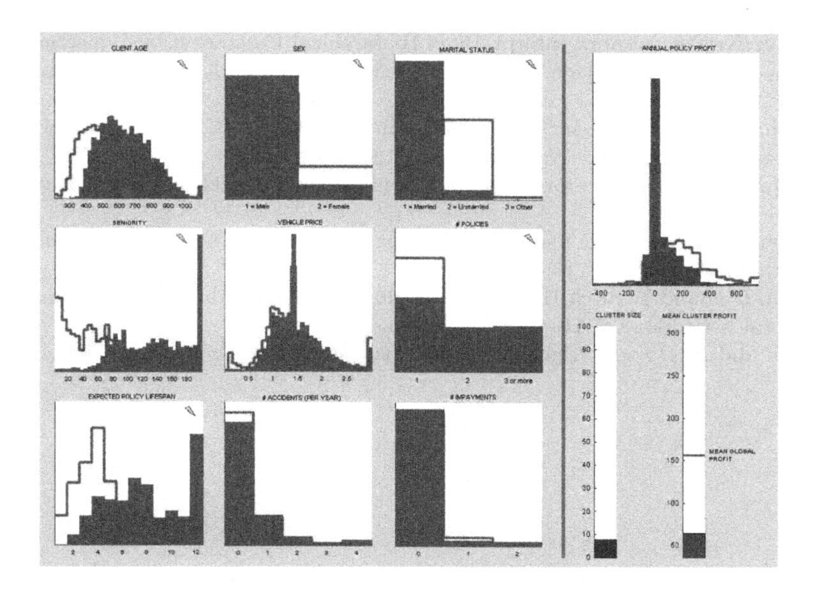

Fig. 3. Cluster B

Cluster A: Younger male clients; more married and with less risk than expected for their age group and as a consequence, high Annual Policy Profit.
Cluster B: Priority clients with family use of vehicle and higher than expected risk - resulting in a low Annual Policy Profit.

We conclude that the hypothesis found for clusters A and B explain the observed behaviour in a satisfactory manner.

5 Conclusions

We would like to claim that clustering based on utility measures is a general and powerful framework for cognitive behavioural modelling. This paper provides an example in the area of customer intelligence. Many other examples in customer intelligence have been developed by Cognodata Consulting using this framework. The general applicability of the method only depends on the selection of an exogenous variable to be introduced in a specific Utility Measure which, we claim, can be derived in many different domains of application.

References

1. Vladimir N. Vapnik.: An Overview of Statistical Learning IEEE transactions on Neural Networks, vol.**10**, 1999
2. R. Duda, P. Hart, and D. Stork.: Pattern Classification Second ed. New York: Wiley 2001

3. Basak & Krishnapuram.: Interpretable Hierarchical Clustering by Constructing an Unsupervised Decision Tree. IEEE Transactions on knowledge and data engineering, vol **17**, 2005
4. L. Breiman, J.H. Friedman, R.A. Olshen, and C.J. Stone.: Classification and regression trees, Belmont, CA: Wadsworth Int. 1984
5. J.R. Quinlan.: Programs for Machine Learning. San Francisco: Morgan Kaufmann, 1993
6. J.R. Quinlan.: Improved Use of Continuous Attributes in C4.5. J. Artitifical Intelligence, vol. **4**, pp.77-90, 1996
7. Rui Xu, Domand Wunsch II.: Survey of Clustering Algorithms. IEEE Transactions on neural networks, vol **16**, N°3 May 2005.
8. A. Jain and R. Dubes.: Algorithms for Clustering Data. Englewood Cliffs, 1998

Two-Stage Ant Colony Optimization for Solving the Traveling Salesman Problem

Amilkar Puris[1], Rafael Bello[1], Yailen Martínez[1], and Ann Nowe[2]

[1] Department of Computer Science, Central University of Las Villas, Cuba
{ayudier, rbellop, yailenm}@uclv.edu.cu
[2] CoMo Lab, Department of Computer Science, Vrije Universiteit Brussel, Belgium
ann.nowe@vub.ac.be

Abstract. In this paper, a multilevel approach of Ant Colony Optimization to solve the Traveling Salesman Problem is introduced. The basic idea is to split the heuristic search performed by ants into two stages; in this case we use both the Ant System and Ant Colony System algorithms. Also, the effect of using local search was analyzed. We have studied the performance of this new algorithm for several Traveling Salesman Problem instances. Experimental results obtained conclude that the Two-Stage approach significantly improves the Ant System and Ant Colony System in terms of the computation time needed.

1 Introduction

Ant Colony Optimization (ACO) is a metaheuristic used to guide other heuristics in order to obtain better solutions than those generated by local optimization methods. In ACO a colony of artificial ants cooperates to look for good solutions to discrete problems. Artificial ants are simple agents that incrementally build a solution by adding components to a partial solution under construction. This computational model was introduced in 1991 by M. Dorigo and co-workers [12] and [13]. Information about this metaheuristic can be found in [1], [3] and [5].

Ant System (AS) is the first ACO algorithm; it was introduced using the Traveling Salesman Problem (TSP) [2] and [4]. In TSP, we have a set of N fully connected cities $\{c_1,...,c_n\}$by arcs (i,j); each arc is assigned a weight d_{ij} which represents the distance between cities i and j, the goal is to find the shortest possible tour visiting each city once before returning to initial city. When ACO is used to solve this problem, pheromone trails (τ_{ij}) are associated to arcs which denote the desirability of visiting city j directly from city i. Also, the function $\tau_{ij}= 1/d_{ij}$ indicates the heuristic desirability of going from i to j. Initially, ants are randomly associated to cities. In the successive steps ant k applies a random proportional rule to decide which city to visit next according to 1:

$$p_{ij}^k = \frac{(\tau_{ij})^{\alpha} * (\eta_{ij})^{\beta}}{\sum\limits_{l \in N_i^k} (\tau_{il})^{\alpha} * (\eta_{il})^{\beta}} \; if \; j \in N_i^k \; (neighborhood \; of \; ant \; k) \qquad (1)$$

J. Mira and J.R. Álvarez (Eds.): IWINAC 2007, Part II, LNCS 4528, pp. 307–316, 2007.

where α and β are two parameters to point out the relative importance of the pheromone trail and the heuristic information, respectively. After all ants have built their tours the values τ_{ij} are updated in two stages. First, τ_{ij} values are decreased by evaporation, $\tau_{ij} = (1 - \delta) * \tau_{ij}$, using the parameter δ, where $0 \prec \delta \prec 1$. This to avoid unlimited accumulation of pheromone. Secondly, all ants increase de value of τ_{ij} on the arcs they have crossed in their tours, $\tau_{ij} = \tau_{ij} + Inc_{ij}$, where Inc_{ij} is the amount of pheromone deposited by all ants which included the arc(i, j) in their tour. Usually, the amount of pheromone deposited by ant k is equal to $1/C_k$, where C_k is the length of the tour of ant k.

Some direct successor algorithms of Ant Systems are: Elitist AS, Rank-based AS and MAX-MIN AS. A more different ACO algorithm is Ant Colony System (ACS). ACS uses the following pseudorandom proportional rule to select the next city j from city i.

$$j = \begin{cases} \arg\max_{l \in N_i^k} \left\{ \tau_{ij} * (\eta_{il})^{\beta} \right\} & if\ q \preceq q_0 \\ \text{random } selection\ according\ to\ (1) & otherwise \end{cases} \quad (2)$$

where q is a random variable uniformly distributed in $[0, 1]$, q_0 which is a parameter taken in the interval $[0, 1]$, controls the amount of exploration, and $\alpha = 1$ in the random selection (expression 1). In ACS, ants have a local pheromone trail update ($\tau_{ij} = (1 - \xi) * \tau_{ij} + \xi * \tau_{ij}(0)$) applied after crossing an arc (i, j), where $\tau_{ij}(0)$ represents the initial value for the pheromone, and a global pheromone trail update ($\tau_{ij} = (1 - \delta) * \tau_{ij} + \delta * Inc_{ij}$) executed only by the best-so-far ant.

In this paper, we propose a new approach to ACO in which the search process developed by ants is splitted into two stages. We have studied the performance of this proposal using the AS and ACS algorithms. In the following, we analyze some related works and introduce the new algorithm. After that, the performance of it is studied in the case of the Travelling Salesman Problem (TSP). At last the results are shown.

2 Related Works

The problem of finding low cost tours in reasonable time rather than solving the problem to optimality in the case of the TSP using a multilevel approach was addressed in [8]. The multilevel idea was first proposed by Bernard and Simon [9] as a method in speeding-up the recursive spectral bisection algorithm partitioning unstructured problems. It has been recognized that an effective way of accelerating search algorithms is to use multilevel techniques. The approach presented in [8] progressively coarsens the TSP, initializes a tour and then employs either the Lin-Kernighan or the Chained Lin-Kernighan algorithms to refine the solution on each coarsened problem; the resulting multilevel algorithm is shown to considerably enhance the quality of tours.

On the other hand, authors in [10] present a multilevel approach to ACO, for solving the mesh partitioning problem in the finite element methods. The multilevel ant colony algorithm performs very well and it is shown to be superior to several classical mesh partitioning methods. Their studies show that ACO was successful to solve the graph-partitioning problem in the case of graphs of smaller size as a result of this, they enhanced the basic ACO with a multilevel technique. That is, a set of the largest independent subgraph is created from the original graph, these are optimized using ACO, and then the optimized partition is expanded.

In the model proposed in this paper we use a multilevel approach to solve the TSP motivated by the fact that it was shown that a multilevel strategy is beneficial in order solve the TSP [14]. In this case, we introduce the multilevel approach to ACO. However, this model differs from the previous approach [10] in some aspects: (i) the search process developed by the ants is divided in two stages instead of on a grouping of cities; (ii) the search process developed by ants is partitioned into two more simple search processes instead of partitioning the problem into more simple subproblems; (iii) the partial solution obtained in the first level is used as initial state for the search of the ants in the second level.

3 Two-Stage Ant Colony Optimization (TS-ACO)

The Two-Stage Ant Colony Optimization meta-heuristics (TS-ACO) proposed in this investigation is based on the following idea: to divide the search process made by the ants into two stages, so that, in the first stage preliminary results are reached (partial solutions) that serve as an initial state for the search made by the ants in the second stage. In the case of TSP, this means that tours containing a subset of cities are generated in the first stage, in the second stage, these routes will serve like an initial state for the ants.

The determination of the initial state represents a significant issue in heuristic search. It is well known that the initial state has an important effect on the search process. The aim is to be able to approach the initial state to the goal state. Of course, it is necessary to consider an adequate balance between the computational cost obtained in the initial state and the total cost; in other words, the sum of the cost of approaching the initial state to the goal state plus the cost of finding the solution to "improved" initial state should not be greater than the cost of doing this without any approximation to the initial state.

More formally, the purpose is the following. Let E_i be an initial state randomly generated, or obtaining it by any other method without a significant computational cost, E_i^* is a mid state generated by some method M that approaches it to the goal state, $C_M(E_i^*)$ denotes the cost of obtaining the state E_i^* from E_i using the method M, and $CC_{HSA}(x)$ is the computational cost to find a solution from state x using a Heuristic Search Algorithm (HSA). Then, the objective is that $C_M(E_i^*) + CC_{HSA}(E_i^*) \prec CC_{HSA}(E_i)$.

3.1 Two-Stage Ant System (TS-AS) and Two-Stage Ant Colony System (TS-ACS)

Both AS and ACS algorithms have been selected from ACO meta-heuristics to prove the approach described above. So, in both the TS-AS and TS-ACS algorithms proposed herein, the procedure to generate E_i^* and the HSA are both the AS and ACS algorithms, so the objective is $C_{ACS}(E_i^*)+CC_{ACS}(E_i^*) \prec CC_{ACS}(E_i)$ or $C_{AS}(E_i^*)+CC_{AS}(E_i^*) \prec CC_{AS}(E_i)$. As AS and ACS are used in both stages, the difference between the 2 stages is obtained by giving different values to some parameters of the model in each stage.

A ratio (r) is introduced in order to establish the relative setting of the values of the algorithm's parameters in both stages; the ratio indicates which proportion of the complete search is given to the first stage. For instance, if r=0.3, means that the first stage will cover 30% of the search process and the second stage the remaining 70% (see an example of the application of this ratio in next section).

The setting of the ratio r has a high influence on the overall performance of the algorithm. A high value of r, say almost 1, causes the state E_i^* to be closer to the goal state, by doing so the value of $C_{ACS}(E_i^*)$ may increase and the value of $CC_{ACS}(E_i^*)$ will decrease. But, in addition to this balance between the costs of $C_{ACS}(E_i^*)$ and $CC_{ACS}(E_i^*)$, we have the problem about how much the space search is explored; while more greater is the rate r, the search in the second stage decreases for several reasons: (I) there are less ants working, (II) the amount of cycles decreases, and (III) although the quantity of possible initial states for the second stage must grow when r grows, that amount is already limited by the result of the previous stage.

Therefore, a key point is to study what value of rate r is the best in order to obtain the best balance between the searches in both stages. This value must allow:

- To minimize the value of $C_{ACS}(E_i^*)+CC_{ACS}(E_i^*)$ or $C_{AS}(E_i^*)+CC_{AS}(E_i^*)$
- To allow an exploration of the search space that guarantees to find good solutions

4 TS-AS and TS-ACS in the Travelling Salesman Problem

When applying the algorithms AS and ACS to the TSP, the ants begin the search starting from random initial states; that is, in each cycle an ant begins its tour in a randomly selected city, and chooses the next city to visit using rule 2. In the beginning no pheromone information is available to guide the search, only the heuristic information is present.

On the contrary, the TS-AS and TS-ACS construct partial tours (they do not include all the cities) in the first stage; this information serves as initial state for the ants in the second stage of the search process. In other words, instead of restarting the search from scratch every cycle, ants use the partial tours built in the first stage as the starting point in the second stage.

In TSP the parameters whose values are set depending on the ratio r are: the quantity of ants (m) to use in each stage, the number of cycles to execute in each stage (nc), and the amount of cities (cc) that must be included in the tour in each stage. In the experiments reported below we compare the conventional single stage ACS with the proposed Two Stages ACS; and AS with TS-AS.

The setting of the parameters is done in the following way. Suppose a setting of the parameters for both conventional AS and ACS algorithms as follows: $m = 100$, $nc = 100$, and $cc = 30$, and a setting of the ratio as $r = 0.3$, then the values of these parameters for both TS-AS and TS-ACS were set as follows $m_1 = 100 * 0.3 = 30$, $nc_1 = 100 * 0.3 = 30$ and $cc_1 = 30 * 0.3 = 9$ for the first stage; and $m_2 = 100 * 0.7 = 70$, $nc_2 = 100 * 0.7 = 70$ and $cc_2 = 30$. Meaning that 30 ants execute the AS or ACS algorithms during 30 cycles starting from random cities and constructing tours of 9 cities. In the second stage, 70 ants will execute AS and ACS algorithms during 70 cycles forming tours of 30 cities.

This means that in the first stage 30% of the ants, search for solutions which is reduced in size (tours including only 30% of the cities), and this in 30% of the total number of cycles. In the second stage the remaing 70% of the ants are used, they get 70% of the total number of the cycles, in order to find solutions to the complete problem. When the first stage is finished, we select a subset of solutions (denoted by EI) containing a quantity (cs) of the best solutions (tours with shortest distance) found in the first stage.

A refinement algorithm based on a local search strategy is introduced to improve the results of the first stage which explores small regions of the solution space. This means, the partial solutions in the set EI resulting from stage 1 are improved using the well known 2-opt procedure [11]. The TSP 2-exchange neighborhood of a candidate solution s consist of the set of all solutions s^* that can be obtained from s by exchanging two pairs of arcs in any possible way [5].

This improved EI subset, provides the initial states for the second stage. This means that in each cycle of the second stage each ant chooses, in a random way, an element of EI (a partial tour) and positions itself in a randomly selected city belonging to the tour. After that, the ant will add other nc_2 cities to this tour using the AS and ACS algorithms. The initial values of the pheromones in the second stage are those attained at the end of the first stage.

The TS-ACS-TSP algorithm is given below:

Given the parameters (beta, rho, epsilon, cc, ratio r, number of solutions in EI (cs))

Algorithm 1. *P_0: Define the quantity of ants (m) either like an input data or by using some method depending on the number of cities.*

P_1: Stage 1

$P_{1.1}$: Calculate the parameters for the first stage:

$m_1 = r * m$

$nc_1 = r * nc$

$cc_1 = r * cc$

$P_{1.2}$: Apply the ACO algorithm, which in the first stage develops nc_1 cycles

$P_{1.3}$: Set of tours $\leftarrow$ Tours generated by ACO algorithm in the first stage

P_2: Stage 2

$P_{2.1}$: Calculate the parameters for the second stage:

$m_2 = m - m_1$

$nc_2 = nc - nc_1$

$cc_2 = cc$
$EI \leftarrow$ Selecting the best cs solutions from Set of tours
$P_{2.2}$: Apply a local search to improve partial solutions in EI set
$P_{2.3}$: Apply the ACO algorithm, which in the second stage develops nc_2 cycles, using the elements of EI like initial states for the ants in the second stage

5 Experimental Results

A comparative study of AS and TS-AS, and so for the ACS and TS-ACS algorithms in the TSP was done by using public available (or benchmark) data base for this problem [7].

For the experiments regarding AS and TS-AS we use the following parameter settings: $\delta = 0.1$, $\xi = 0.1$, $\alpha = 2$, $\beta = 3$, the number of ants (m) is equal to the number of cities of the TSP instances, and a number of cycles $nc = 100$.

The column "Best Solution" indicates the best solution reported back for that problem in the corresponding database, and the column "Best solution with AS" contains the best solution found by our implementation of AS algorithm. Besides, we developed experiments to determine an adequate value for the ratio. We run the TS-AS-TSP algorithm using several values for the ratio $r = \{0.2, 0.25, 0.3, 0.4, 0.5\}$. Table 1 shows the results averaged over 6 runs. The quality of the solution and the time cost were measured.

Table 1. Best solutions obtained using AS algorithm

DataSet	BS	BS with AS	Time (ms)	BS with TS-AS	Time (ms)
bays29.tsp	2020	2077	330	2042(0.2)	173
berlin52.tsp	7542	7601	2193	7594(0.2)	1182
st70.tsp	675	752	6231	750(0.25)	2587
rd100.tsp	7910	8603	20194	8551(0.2)	11032
ch150.tsp	6528	6696	83995	6616(0.3)	32634
kroA200.tsp	29368	31662	240559	29727(0.2)	132603
a280.tsp	2579	2687	835179	2590(0.2)	472434
lin318.tsp	42029	43651	1332324	42906(0.2)	752929
pcb442.tsp	50778	53469	4520998	52545(0.3)	1880513
rat783.tsp	8806	8950	5081400	8850(0.25)	2460528

For the experiments regarding ACS and TS-ACS we use the following parameter setting: $\alpha = 0.1$, $\xi = 0.1$, a quantity of ants $m = 10$ and a number of cycles $nc = 1000$. Firstly, we studied the values {1, 3, 5} and {0.3, 0.6, 0.9} for β and q_0 respectively. The values $\beta = 5$ and $q_0 = 0.6$ were selected to develop the next experiments which results are shown in Table 2. The column "Best solution" indicates the best solution reported back for that problem in the corresponding database, and the column "Best solution with ACS" contains the best solution found by our implementation of the ACS algorithm. We ran the TS-ACS-TSP

Table 2. Best solutions obtained using ACS algorithm

Database	BS	BS with ACS	Time (ms)	BS with TS-ACS	Time(ms)
bays29.tsp	2020	2033	1026	2022(0.2)	688
berlin52.tsp	7542	7590	3801	7550(0.2)	1980
st70.tsp	675	722	7632	738(0.25)	5344
rd100.tsp	7910	8386	18140	8351(0.2)	11032
ch150.tsp	6528	6604	54001	6590(0.25)	21391
kroA200.tsp	29368	31362	112982	29658(0.2)	64829
tsp225.tsp	3919	4285	148351	4295(0.25)	74109
a280.tsp	2579	2604	271179	2593(0.2)	166810
lin318.tsp	42029	43500	398572	43028(0.3)	233437
pcb442.tsp	50778	53465	973212	52303(0.25)	468578
rat783.tsp	8806	8895	5081400	8834(0.3)	2460528

algorithm using several values for the ratio, $r = \{0.2, 0.25, 0.3, 0.4, 0.5\}$ to determine an adequate value for this parameter.

The statistical analysis performed in order to compare the solution values for the mentioned algorithms using Monte Carlo Significance of Friedman's test=0.000 is displayed in Figure 1 a), whereas b) provides the comparison among the algorithms with respect to the time needed to get the solution. Mean Ranks with a common letter denote non-significant difference acoording to Wilcoxon's test, proving that there are not significatives differences between the ACS and TS-ACS in solutions values, and showing an important difference between them in the time cost.

	Mean Ranks
Optimum	1.00^{a}
TS-ACS(r=0.2)	2.6363^{b}
ACS	3.1818^{b}
TS-ACS(r=0.25)	3.5454^{bc}
TS-ACS(r=0.3)	4.6363^{c}

a) Statistical analysis for solutions values

	Mean Ranks
TS-ACS(r=0.3)	1.18^{a}
TS-ACS(r=0.25)	2.18^{b}
TS-ACS(r=0.2)	2.82^{b}
ACS	3.82^{c}

b) Statistical analysis for time cost

Fig. 1. Statistical Analysis

The effect of introducing local search in the ACS and TS-ACS algorithms is presented in Tables 3 and 4. The results are averaged over10 runs. The quality of the solution and the time cost were measured.

The comparison with respect to the quality of the solution is presented in Table 3. The column "Best solution with ACS plus 2-opt" contains the best solution found by the ACS algorithm plus a local search developed with 2-opt procedure; the column "Best solution with TS-ACS" contains the best solution found by TS-ACS algorithm and the corresponding ratio with which it was found; the column "Best solution with TS-ACS plus 2-opt after first stage" contains the best solution found by TS-ACS algorithm plus a local search developed with 2-opt procedure applied to the solutions obtained after stage 1, and the column "Best solution with TS-ACS plus 2-opt after both stages" contains the best solution found by TS-ACS algorithm plus a local search developed with 2-opt procedure applied to the solutions obtained after the two stage. A similar comparison is presented in Table 4 respect to the time cost (measured in seconds).

Table 3. A comparison between ACS and TS-ACS in TSP (quality solution)

Database	BS	ACS-2opt	TS-ACS-2opt after 1st stage	TS-ACS-2opt after both stages
bays29.tsp	2020	2028	2045 (0.2)	2020 (0.2)
berlin52.tsp	7542	7564	7563 (0.25)	7542 (0.25)
st70.tsp	675	719	730 (0.25)	722 (0.25)
rd100.tsp	7910	8345	8358 (0.3)	8159 (0.2)
ch150.tsp	6528	6573	6696 (0.3)	6570 (0.2)
kroA200.tsp	29368	31298	31889 (0.2)	29470(0.25)
tsp225.tsp	3919	4115	4204 (0.25)	4102 (0.2)
a280.tsp	2579	2604	2644 (0.25)	2618 (0.25)
lin318.tsp	42029	43006	43503 (0.3)	42086 (0.2)
pcb442.tsp	50778	52861	53369 (0.3)	51790 (0.2)
rat783.tsp	8806	8855	8904(0.25)	8815(0.3)

Table 4. A comparison between ACS and TS-ACS in TSP (time cost)

Database	ACS-2opt	TS-ACS-2opt after 1st stage	TS-ACS-2opt after both stages
bays29.tsp	1231	606	651
berlin52.tsp	3937	1719	1734
st70.tsp	7851	3609	3687
rd100.tsp	18721	12492	14202
ch150.tsp	57803	21984	31874
kroA200.tsp	123852	66781	74873
tsp225.tsp	155151	76120	78677
a280.tsp	295464	135987	146898
lin318.tsp	430637	172492	190220
pcb442.tsp	1040741	425612	478066
rat783.tsp	5315734	2490555	2983343

These experimental results can be summarized in the following way:

- The Two-Stage approach has a relevant impact in both AS and ACS algorithms because the time cost is reduced in a 40-60 percent, obtaining similar solutions values.
- A value of rate r between 0.2 and 0.3 produces the best results.

- Greater values for the ratio r decrease the time cost but also the solution quality (because the search space is not sufficiently covered).
- The introduction of the local search developed with the 2-opt procedure increases the quality of the solution in ACS and TS-ACS algorithms.
- The quality of the solutions are similar, but usually the best solutions are obtained by using TS-ACS plus 2-opt after both stages.

6 Conclusions

We have presented an improvement of the Ant Colony Optimization based in a multilevel approach. It consists on splitting the search process developed by ants into two stages. The study was developed using the Ant System and Ant Colony System algorithms. In this approach the values of some parameters (number of ants, quantity of cycles, number of cities included in each stage, etc.) are assigned a different value in each stage according to a ratio which indicates what proportion of the complete search corresponds to each stage.

We studied the performance using different ratio values in the Traveling Salesman Problem. The best results were obtained when this value is about 0.3. Moreover, the positive effects of introducing local search is studied and showed in this paper.

This new approach to ACO produces an important reduction of the computation time cost, yet preserving the solution quality and can be successfully applied to any algorithm of the ACO metaheuristic.

References

1. Dorigo, M. et al. The Ant System: Optimization by a colony of cooperating agents. IEEE Transactions on Systems, Man and Cybernetics-Part B, vol. 26, no. 1, pp. 1-13. 1996.
2. Dorigo, M. and Gambardella, L.M.. Ant colonies for the traveling salesman problem. BioSystems no. 43, pp. 73-81, 1997.
3. Dorigo, M. et al. Ant algorithms for Discrete optimization. Artificial Life 5(2), pp. 137-172. 1999.
4. Dorigo, M. and Stützle, T.. ACO Algorithms for the Traveling Salesman Problem, 1999, Evolutionary Algorithms in Engineering and Computer Science: Recent Advances in Genetic Algorithms, Evolution Strategies, Evolutionary Programming, Genetic Programming and Industrial Applications, John Wiley & Sons., EEUU.
5. Dorigo, M. and Stutzle, T.. Ant Colony Optimization. MIT Press. 2004.
6. Stefanowski, J.. An experimental evaluation of improving rule based classifiers with two approaches that change representations of learning examples. Engineering Applications of Artificial Intelligence 17, pp. 439-445. 2004.
7. The TSPLIB Symmetric Traveling Salesman Problem Instances, http://elib.zib.de/pub/Packages/mp-testdata/tsp/tsplib/index.html
8. Walshaw, C.. A multilevel approach to the traveling salesman problem. Operations Research 50(5), 862-877, 2002.

9. Bernard, S.T. and Simon, H.D.. A fast multilevel implementation of recursive spectral bisection for partitioning unstructured problems. Concurrency: Practice and Experience 6 (2), pp. 101-117. 1994.
10. Korosec, P. et al.. Solving the mesh-partitioning problem with an ant-colony algorithm. Parallel Computing 30, pp. 785-801. 2004.
11. Michalewicz, Z. and Fogel, D.B.. How to solve it: modern heuristics. Springer-Verlag. 2004.
12. Dorigo, M., Maniezzo, V., Colorni, A.: Positive feedback as a search strategy. Technical Report 91-016, Dipartimento di Elettronica, Politecnico di Milano, Milan, Italy (1991).
13. Dorigo, M.: Ottimizzazione, apprendimento automatico, ed algoritmi basati su metafora naturale. PhD thesis, Dipartimento di Elettronica, Politecnico di Milano, Milan, Italy (1992).
14. Walshaw, C. et al.. The multilevel paradigm: a generic Meta-heuristic for Combinatorial Optimization Problems?. Technical Report 00/IM/63, Univ. Greenwich, London SE10 9LS, UK. 2000.

Solving Dial-a-Ride Problems with a Low-Level Hybridization of Ants and Constraint Programming

Broderick Crawford[1,2], Carlos Castro[2], and Eric Monfroy[2,3]

[1] Pontificia Universidad Católica de Valparaíso, Chile
FirstName.Name@ucv.cl
[2] Universidad Técnica Federico Santa María, Valparaíso, Chile
FirstName.Name@inf.utfsm.cl
[3] LINA, Université de Nantes, France
FirstName.Name@univ-nantes.fr

Abstract. This paper is about Set Partitioning formulation and resolution for a particular case of VRP, the Dial-a-ride Problem. Set Partitioning has demonstrated to be useful modeling this problem and others very visible and economically significant problems. But the main disadvantage of this model is the need to explicitly generate a large set of possibilities to obtain good solutions. Additionally, in many cases a prohibitive time is needed to find the exact solution. Nowadays, many efficient metaheuristic methods have been developed to make possible a good solution in a reasonable amount of time. In this work we try to solve it with Low-level Hybridizations of Ant Colony Optimization and Constraint Programming techniques helping the construction phase of the ants. Computational results solving some benchmark instances are presented showing the advantages of using this kind of hybridization.

Keywords: Ant Colony Optimization, Constraint Programming, Hybrid Algorithm, Dial-a-ride Problem, Set Partitioning.

1 Introduction

In the last five decades, it is possible to see an always increasing number of vehicles. Although new streets and highways are constructed, bottlenecks and congestion are not isolated situations and they seem to be intrinsic part of the transportation activity. The true is that the saturation of the transportation infrastructures is a global phenomenon. The paradox is that while nowadays people needs to move more quickly within the city centre and from peripheral areas, our overcrowded streets give us to longer delays, more frequent accidents and higher pollution emissions. Furthermore, the possibility of increasing the capacity of the roadway system is limited, due to several factors: lack of large spaces in urban areas, considerable environmental impact and high cost of construction. Then, the use of Optimization Techniques and Information Technologies represents a good alternative to face the Transportation on demand (TOD) problem. TOD

J. Mira and J.R. Álvarez (Eds.): IWINAC 2007, Part II, LNCS 4528, pp. 317–327, 2007.

is concerned with the transportation of passengers or goods between specific origins and destinations at the request of users. Common examples are dial-a-ride transportation services for the disabled and the elderly, urban courier services, aircraft sharing, and emergency vehicle dispatching. The growing emphasis on electronic commerce, cycle-time compression and just-in-time deliveries has increased the need for demand-responsive freight transportation systems.

The research on Operations Research (OR) and Artificial Intelligence (AI) has deserved an increasing interest in this context, providing a varied set of models and techniques to solve this kind of problems. For example, many real life problems can be formulated as Set Partitioning Problems (SPP). Although the best known application of the SPP is Airline Crew Scheduling [4,21], several other applications exist, including Vehicle Routing Problems [2,28,5] and Query Processing [25].

The main disadvantage of SPP-based models is the need to explicitly generate a large set of possibilities to obtain good solutions. Additionally, in many cases a prohibitive time is needed to find the exact solution. Today, many efficient metaheuristic methods have been developed to make possible a good solution in a reasonable amount of time for different problems. In relation with VRP its first SPP formulation for Vehicle Routing Problems (VRP) was proposed in 1964 [5] and recent contributions are including the use of metaheuristic to solve it [12,37,27]. Furthermore, Set Partitioning Problems occur as subproblems in various combinatorial optimization problems. In Airline Scheduling, a subtask called Crew Scheduling, takes as input data a set of crew pairings, where the selection of crew pairings which cause minimal costs and ensure that each flight is covered exactly once, can be modelled as a set partitioning problem [4]. In [9] solving a particular case of VRP, the Dial-a-ride Problem (DARP), also uses a SPP decomposition approach. DARP consists of designing vehicle routes and schedules for n users who specify pick-up and drop-off requests between origins and destinations. The aim is to plan a set of m minimum cost vehicle routes capable of accommodating as many users as possible, under a set of constraints. The most common example arises in door-to-door transportation for elderly or disabled people. In the past few years several new and powerful algorithms have been developed to solve DARP. The best exact solution methodologies are based on branch-and-cut and on branch-and-cut-and-price. Nowadays, new solvers employ a variety of techniques including metaheuristics like tabu search, simulated annealing, variable neighbourhood search, and large neighbourhood search. In [12] there is a review of the scientific literature on the DARP.

Because, the SPP formulation have demonstrated to be useful modeling VRP problems (or their phases), it is our interest to solve it with novelty techniques. In this work, we solve some test instances of SPP with Ant Colony Optimization (ACO) algorithms and some hybridizations of ACO with Constraint Programming (CP) techniques like Forward Checking. There exist some problems for which the effectiveness of ACO is limited, among them the strongly constrained problems. Those are problems for which neighbourhoods contain few solutions, or none at all, and local search has a very limited use. Probably, the most significant of those problems is the SPP and a direct implementation of the basic ACO framework is unable of obtaining feasible solutions for many SPP standard tested instances [33]. The best performing metaheuristic for SPP is a genetic algorithm due to Chu and Beasley [11,7]. There already exists some first approaches applying ACO to the SCP. In [1,30] ACO has only been used as a construction algorithm and the approach has only been tested on some small SCP instances. More recent works [26,31,24] apply Ant Systems to the SCP and related problems using techniques to remove redundant columns and local search to improve solutions. Taking into account these results, it seems that the incomplete approach of Ant Systems could be considered as a good alternative to solve these problems when complete techniques are not able to get the optimal solution in a reasonable time.

In this paper, we propose the addition of a Constraint Programming mechanism in the construction phase of ACO thus only feasible partial solutions are generated. The CP mechanism allows the incorporation of information about the instantiation of variables after the current decision. This idea differs from the one proposed by [35] and [23], those authors propose a lookahead function evaluating the pheromone in the Shortest Common Supersequence Problem and estimating the quality of a partial solution of a Industrial Scheduling Problem, respectively. So, there are two main motivations for our work. On one hand, we try to improve the performance of ACO algorithms when dealing with industrial problems. On the other hand, we are interested in the development of robust algorithms integrating complete as well as incomplete techniques, because we are convinced that this is a very promising approach to deal with hard combinatorial problems.

This paper is organised as follows: Section 2 is dedicated to the presentation of the problem and its mathematical model. In Section 3, we describe the applicability of the ACO algorithms for solving SPP. In Section 4, we present the basic concepts to adding Constraint Programming techniques to the two basic ACO algorithms: AS and ACS. In Section 5, we present results when adding Constraint Programming techniques to the two basic ACO algorithms to solve benchmarks available in the OR-Library of Beasley [6]. Finally, in Section 6 we conclude the paper and give some perspectives for future research.

2 Problem Description

This paper is about Set Partitioning Problem resolution for Vehicle Scheduling in Dial-a-ride Systems. The main objective of the DARP is to minimize operation costs for renting pieces of work from the service providers. Customer satisfaction is another important goal, it is treated by means of the time windows. The solution approach considered in this work, starting from a network formulation of the DARP, decomposes the problem into a Clustering and a Chaining phase, solving both phases like SPP. In [9] it is suggested the following two step decomposition approach to the DARP:

- Clustering Step: Construct a set of feasible clusters.
- Chaining Step: Chain clusters to a set of tours that constitute a feasible schedule.

The decomposition is based on the concept of a cluster. A cluster is a segment of a vehicle tour satisfying the local constraints: Pairing precedence, time windows, no stop, and capacity. Clusters are useful for vehicle scheduling because they can serve as the building blocks of vehicle tours. Then, we can chain clusters to feasible tours just as we constructed clusters from the individual requests. As the clusters already satisfy the local constraints, the chaining can concentrate on the remaining global constraints.

Clustering and Chaining SPP are of identical structure. The objective of the clustering step is to construct a set of clusters that can be chained to an optimal solution of the DARP. Then DARP results in the following optimization problem over clusters: *Given the customer requests, find a set of clusters such that each request is contained in exactly one cluster and the sum of the cluster objectives is minimal.*

The formulation of the clustering step aims at inputs for the chaining phase and can be formulated as a Set Partitioning Problem. SPP is the NP-complete problem of partitioning a given set into mutually independent subsets while minimizing a cost function defined as the sum of the costs associated to each of the eligible subsets. In the SPP matrix formulation we are given a $m \times n$ matrix $A = (a_{ij})$ in which all the matrix elements are either zero or one. Additionally, each column is given a non-negative cost c_j. We say that a column j can cover a row i if $a_{ij} = 1$. Let J denotes the set of the columns and x_j a binary variable which is one if column j is chosen and zero otherwise. The SPP can be defined formally as follows:

$$Minimize \qquad f(x) = \sum_{j=1}^{n} c_j \times x_j \tag{1}$$

$$Subject\ to \qquad \sum_{j=1}^{n} a_{ij} \times x_j = 1; \qquad \forall i = 1, \ldots, m \tag{2}$$

These constraints enforce that each row is covered by exactly one column. In this formulation, each row represents a customer request that must be contained in exactly one cluster. The columns represent clusters. c_j is the vector of cluster objectives. Having decided for a set of clusters we can treat the chaining step in exactly the same way as we just did with the clustering step.

Approximating the objective value of the DARP as a sum of objectives of individual tours, the DARP for fixed clusters simplifies to the following optimization problem over tours: *Given a clustering, find a set of vehicle tours such that each cluster is contained in exactly one tour and the sum of the tour objectives is minimal.*

Natural objectives associated to tours are operation costs for vehicles and/or customer satisfaction criteria like accumulated waiting time. It can also be modelled as a set partitioning problem too. Other DARP solution approaches, similar to the found in [9], using the concepts of Clustering and Chaining have been discussed in different publications [16,14].

3 Ant Colony Optimization for Set Partitioning Problems

In this section, we briefly present ACO algorithms and give a description of their use to solve SPP. More details about ACO algorithms can be found in [18,19].

The basic idea of ACO algorithms comes from the capability of real ants to find shortest paths between the nest and food source. From a Combinatorial Optimization point of view, the ants are looking for *good solutions*. Real ants cooperate in their search for food by depositing pheromone on the ground. An artificial ant colony simulates this behavior implementing artificial ants as parallel processes whose role is to build solutions using a randomized constructive search driven by pheromone trails and heuristic information of the problem. An important topic in ACO is the adaptation of the pheromone trails during algorithm execution to take into account the cumulated search experience: reinforcing the pheromone associated with good solutions and considering the *evaporation* of the pheromone on the components over time in order to avoid premature convergence. ACO can be applied in a very straightforward way to SPP. The columns are chosen as the solution components and have associated a cost and a pheromone trail [20]. Each column can be visited by an ant only once and then a final solution has to cover all rows. A walk of an ant over the graph representation corresponds to the iterative addition of columns to the partial solution obtained so far. Each ant starts with an empty solution and adds columns until a cover is completed. A pheromone trail τ_j and a heuristic information η_j are associated to each eligible column j. A column to be added is chosen with a probability that depends of pheromone trail and the heuristic information. The most common form of the ACO decision policy (*Transition Rule Probability*) when ants work with components is:

$$p_j^k(t) = \frac{\tau_j * \eta_j^\beta}{\sum_{l \notin S^k} \tau_l [\eta_l]^\beta} \quad \text{if } j \notin S^k \tag{3}$$

where S^k is the partial solution of the ant k. The β parameter controls how important is η in the probabilistic decision [20,31].

Pheromone trail τ_j. One of the most crucial design decisions to be made in ACO algorithms is the modelling of the set of pheromones. In the original ACO implementation for TSP the choice was to put a pheromone value on every link between a pair of cities, but for other combinatorial problems often can be assigned pheromone values to the decision variables (first order pheromone values) [20]. In this work the pheromone trail is put on the problems component (each eligible column j) instead of the problems connections. And setting a good pheromone quantity is not a trivial task either. The quantity of pheromone trail laid on columns is based on the idea: *the more pheromone trail on a particular item, the more profitable that item is* [30]. Then, the pheromone deposited in each component will be in relation to its frequency in the ants solutions. In this work we divided this frequency by the number of ants obtaining better results.

Heuristic information η_j. In this paper we use a dynamic heuristic information that depends on the partial solution of an ant. It can be defined as $\eta_j = \frac{e_j}{c_j}$, where e_j is the so called cover value, that is, the number of additional rows covered when adding column j to the current partial solution, and c_j is the cost of column j. In other words, the heuristic information measures the unit cost of covering one additional row. An ant ends the solution construction when all rows are covered.

In this work, we use two instances of ACO: Ant System (AS) and Ant Colony System (ACS) algorithms, the original and the most famous algorithms in the ACO family [20]. ACS improves the search of AS using: a different transition rule in the constructive phase, exploiting the heuristic information in a more rude form, using a list of candidates to future labelling and using a different treatment of pheromone. ACS has demonstrated better performance than AS in a wide range of problems [19].

Trying to solve larger instances of SPP with the original AS or ACS implementation derives in a lot of unfeasible labelling of variables, and the ants can not obtain complete solutions. In this paper we explore the addition of a lookahead mechanism in the construction phase of ACO thus only feasible solutions are generated. A direct implementation of the basic ACO framework is incapable of obtaining feasible solution for many SPP instances [13]. Each ant starts with an empty solution and adds columns until a cover is completed. But to determine if a column actually belongs or not to the partial solution ($j \notin S^k$) is not good enough. The traditional ACO decision policy, Equation 3, does not work for SPP because the ants, in this traditional selection process of the next columns, ignore the information of the problem constraints when a variable is instantiated. And in the worst case, in the iterative steps is possible to assign values to some variable that will make impossible to obtain complete solutions. To improve it, the procedure used is similar to the Constraint Propagation technique from Constraint Programming [8,3].

4 ACO with Constraint Programming

Particularly promising possibilities of combining Metaheuristics with Constraint Programming techniques were pointed out in [36,22]. The hibridization between different types of Metaheuristics fall into the Low-level hybrid category and it can also be extended towards hybrids of Metaheuristics with OR and CP techniques. Recently, some efforts have been done in order to integrate Constraint Programming techniques to ACO algorithms [34,10,22]. An hybridization of ACO and CP can be approached from two directions: we can either take ACO or CP as the base algorithm and try to embed the respective other method into it. A form to integrate CP into ACO is to let it reduce the possible candidates among the not yet instantiated variables participating in the same constraints that the actual variable. A different approach would be to embed ACO within CP. The point at which ACO can interact with CP is during the labelling phase, using ACO to learn a value ordering that is more likely to produce good solutions.

In this work, ACO use CP in the variable selection (when adding columns to partial solution). The CP algorithm used in this paper is Forward Checking with Backtracking [15]. It performs Arc Consistency between pairs of a not yet instantiated variable and an instantiated variable, i.e., when a value is assigned to the current variable, any value in the domain of a future variable which conflicts with this assignment is removed from the domain.

The Forward Checking procedure, taking into account the constraints network topology (i.e. wich sets of variables are linked by a constraint and wich are not), guarantees that at each step of the search, all constraints between already assigned variables and not yet assigned variables are arc consistent. Then, adding Forward Checking to ACO for SPP means that columns are chosen if they do not produce any conflict with the next column to be chosen. In other words, the Forward Checking search procedure guarantees that at each step of the search, all the constraints between already assigned variables and not yet assigned variables are arc consistency. This reduces the search tree and the overall amount of computational work done. But it should be noted that in comparison with pure ACO algorithm, Forward Checking does additional work when each assignment is intended to be added to the current partial solution. Arc consistency enforcing always increases the information available on each variable labelling. Figure 1 describes the hybrid ACO+CP algorithm to solve SPP.

```
1   Procedure ACO+CP_for_SPP
2   Begin
3    InitParameters();
4    While (remain iterations) do
5       For k := 1 to nants do
6          While (solution is not completed) and TabuList <> J do
7             Choose next Column j with Transition Rule Probability
8             For each Row i covered by j do                 /* constraints with j     */
9               feasible(i):= Posting(j);                    /* Constraint Propagation */
10            EndFor
11            If feasible(i) for all i then AddColumnToSolution(j)
12                                     else Backtracking(j); /* set j uninstantiated   */
13            AddColumnToTabuList(j);
14         EndWhile
15      EndFor
16     UpdateOptimum();
17     UpdatePheromone();
18   EndWhile
19   Return best_solution_founded
20  End.
```

Fig. 1. ACO+CP algorithm for SPP

5 Experiments and Results

Table 1 presents the results when adding Forward Checking to the basic ACO algorithms for solving test instances taken from the Orlib [6]. It compares performance with IP optimal, Genetic Algorithm of Chu and Beasley [11], Genetic Algorithm of Levine [32] and the most recent algorithm by Kotecha et al. [29].

The first five columns of Table 1 present the problem code, the number of rows (constraints), the number of columns (decision variables), the best known cost value for each instance, and the density (percentage of ones in the constraint matrix) respectively. The next three columns present the results obtained by better performing metaheuristics with respect to SPP. And the last four columns present the cost obtained when applying Ant Algorithms, AS and ACS, and combining them with Forward Checking. An entry of "X" in the table means no feasible solution was found. The algorithms have been run with the following parameters settings: influence of pheromone (alpha)=1.0, influence of heuristic information (beta)=0.5 and evaporation rate (rho)=0.4 as suggested in [30,31,20]. The number of ants has been set to 120 and the maximum number of iterations to 160, so that the number of generated candidate solutions is limited to 19.200. For ACS the list size was 500 and Qo=0.5. Algorithms were implemented using ANSI C, GCC 3.3.6, under Microsoft Windows XP Professional version 2002.

Table 1. ACO with Forward Checking

Problem	Rows	Columns	Optimum	Density	Beasley	Levine	Kotecha	AS	ACS	AS+FC	ACS+FC
sppnw06	50	6774	7810	18.17	7810	-	-	9200	9788	8160	8038
sppnw08	24	434	35894	22.39	35894	37078	36068	X	X	35894	36682
sppnw09	40	3103	67760	16.20	67760	-	-	70462	X	70222	69332
sppnw10	24	853	68271	21.18	68271	X	68271	X	X	X	X
sppnw12	27	626	14118	20.00	14118	15110	14474	15406	16060	14466	14252
sppnw15	31	467	67743	19.55	67743	-	-	67755	67746	67743	67743
sppnw19	40	2879	10898	21.88	10898	11060	11944	11678	12350	11060	11858
sppnw23	19	711	12534	24.80	12534	12534	12534	14304	14604	13932	12880
sppnw26	23	771	6796	23.77	6796	6796	6804	6976	6956	6880	6880
sppnw32	19	294	14877	24.29	14877	14877	14877	14877	14886	14877	14877
sppnw34	20	899	10488	28.06	10488	10488	10488	13341	11289	10713	10797
sppnw39	25	677	10080	26.55	10080	10080	10080	11670	10758	11322	10545
sppnw41	17	197	11307	22.10	11307	11307	11307	11307	11307	11307	11307

The effectiveness of Constraint Programming is showed to solve SPP, because the SPP is so strongly constrained the stochastic behaviour of ACO can be improved with lookahead techniques in the construction phase, so that almost only feasible partial solutions are induced. In the original ACO implementation the SPP solving derives in a lot of unfeasible labelling of variables, and the ants can not complete solutions.

6 Conclusions and Future Directions

Our main contribution is the study of the combination of Constraint Programming and Ant Colony Optimization in a Low-level hibridization solving benchmarks of the Set Partitioning Problem. The main conclusion from this work is that we can improve ACO with CP. Computational results also indicated that our hybridization is capable of generating optimal or near optimal solutions for many problems.

The concept of Arc Consistency plays an essential role in Constraint Programming as a problem simplification operation and as a tree pruning technique during search through the detection of local inconsistencies among the uninstantiated variables. We have shown that it is possible to add Arc Consistency to any ACO algorithms and the computational results confirm that the performance of ACO can be improved with this type of hybridisation. Anyway, a complexity analysis should be done in order to evaluate the cost we are adding with this kind of integration. We strongly believe that this kind of integration between complete and incomplete techniques should be studied deeply. Because, in many cases a prohibitive time is needed to find the exact solution of SPP modelling Transportation on Demand problems (or a phase of them). Then, efficient metaheuristic methods can be very useful making good solutions in a reasonable amount of time.

We are working in the development of a computational application that can integrate the resolution for both Clustering and Chaining in order to provide a whole solver for the problem.

References

1. D. Alexandrov and Y. Kochetov. Behavior of the ant colony algorithm for the set covering problem. In *Proc. of Symp. Operations Research*, pages 255–260. Springer Verlag, 2000.
2. G. B. Alvarenga and G. R. Mateus. A two-phase genetic and set partitioning approach for the vehicle routing problem with time windows. In *HIS '04: Proceedings of the Fourth International Conference on Hybrid Intelligent Systems (HIS'04)*, pages 428–433, Washington, DC, USA, 2004. IEEE Computer Society.
3. K. R. Apt. *Principles of Constraint Programming.* Cambridge University Press, 2003.
4. E. Balas and M. Padberg. Set partitioning: A survey. *SIAM Review*, 18:710–760, 1976.
5. M. L. Balinski and R. E. Quandt. On an integer program for a delivery problem. *Operations Research*, 12(2):300–304, 1964.
6. J. E. Beasley. Or-library:distributing test problem by electronic mail. *Journal of Operational Research Society*, 41(11):1069–1072, 1990.
7. J. E. Beasley and P. C. Chu. A genetic algorithm for the set covering problem. *European Journal of Operational Research*, 94(2):392–404, 1996.
8. C. Bessiere. Constraint propagation. Technical Report 06020, LIRMM, March 2006. also as Chapter 3 of the Handbook of Constraint Programming, F. Rossi, P. van Beek and T. Walsh eds. Elsevier 2006.
9. R. Borndorfer, M. Grotschel, F. Klostermeier, and C. Kuttner. Telebus berlin: Vehicle scheduling in a dial-a-ride system. Technical Report SC 9723, Konrad-Zuse-Zentrum fur Informationstechnik, 1997.
10. C. Castro, M. Moossen, and M. C. Riff. A cooperative framework based on local search and constraint programming for solving discrete global optimisation. In *Advances in Artificial Intelligence: 17th Brazilian Symposium on Artificial Intelligence, SBIA 2004*, volume 3171 of *Lecture Notes in Artificial Intelligence*, pages 93–102, Sao Luis, Brazil, October 2004. Springer.

11. P. C. Chu and J. E. Beasley. Constraint handling in genetic algorithms: the set partitoning problem. *Journal of Heuristics*, 4:323–357, 1998.
12. J.-F. Cordeau and G. Laporte. The dial-a-ride problem (darp): Variants, modeling issues and algorithms. *Journal 4OR: A Quarterly Journal of Operations Research*, 1(2):89–101, 2003.
13. B. Crawford and C. Castro. Integrating lookahead and post processing procedures with aco for solving set partitioning and covering problems. In L. Rutkowski, R. Tadeusiewicz, L. A. Zadeh, and J. Zurada, editors, *ICAISC*, volume 4029 of *Lecture Notes in Computer Science*, pages 1082–1090. Springer, 2006.
14. F. Cullen, J. Jarvis, and D. Ratliff. Set partitioning based heuristics for interactive routing. *Networks*, 11:125–144, 1981.
15. R. Dechter and D. Frost. Backjump-based backtracking for constraint satisfaction problems. *Artificial Intelligence*, 136:147–188, 2002.
16. M. Desrochers, J. Desrosiers, and M. Solomon. A new optimization algorithm for the vehicle routing problem with time windows. *Oper. Res.*, 40(2):342–354, 1992.
17. M. Dorigo, M. Birattari, C. Blum, L. M. Gambardella, F. Mondada, and T. Stützle, editors. *Ant Colony Optimization and Swarm Intelligence, 4th International Workshop, ANTS 2004, Brussels, Belgium, September 5 - 8, 2004, Proceedings*, volume 3172 of *Lecture Notes in Computer Science*. Springer, 2004.
18. M. Dorigo, G. D. Caro, and L. M. Gambardella. Ant algorithms for discrete optimization. *Artificial Life*, 5:137–172, 1999.
19. M. Dorigo and L. M. Gambardella. Ant colony system: A cooperative learning approach to the traveling salesman problem. *IEEE Transactions on Evolutionary Computation*, 1(1):53–66, 1997.
20. M. Dorigo and T. Stutzle. *Ant Colony Optimization.* MIT Press, USA, 2004.
21. A. Feo, G. Mauricio, and A. Resende. A probabilistic heuristic for a computationally difficult set covering problem. *OR Letters*, 8:67–71, 1989.
22. F. Focacci, F. Laburthe, and A. Lodi. Local search and constraint programming. In *Handbook of metaheuristics.* Kluwer, 2002.
23. C. Gagne, M. Gravel, and W. Price. A look-ahead addition to the ant colony optimization metaheuristic and its application to an industrial scheduling problem. In J. S. et al., editor, *Proceedings of the fourth Metaheuristics International Conference MIC'01*, pages 79–84, July 2001.
24. X. Gandibleux, X. Delorme, and V. T'Kindt. An ant colony optimisation algorithm for the set packing problem. In Dorigo et al. [17], pages 49–60.
25. R. D. Gopal and R. Ramesh. The query clustering problem: A set partitioning approach. *IEEE Trans. Knowl. Data Eng.*, 7(6):885–899, 1995.
26. R. Hadji, M. Rahoual, E. Talbi, and V. Bachelet. Ant colonies for the set covering problem. In M. D. et al., editor, *ANTS 2000*, pages 63–66, 2000.
27. R. M. Jorgensen and L. Jesper. Solving the dial-a-ride problem using genetic algorithms. *Journal of the Operational Research Society*, 2007. Forthcoming.
28. J. P. Kelly and J. Xu. A set-partitioning-based heuristic for the vehicle routing problem. *INFORMS J. on Computing*, 11(2):161–172, 1999.
29. K. Kotecha, G. Sanghani, and N. Gambhava. Genetic algorithm for airline crew scheduling problem using cost-based uniform crossover. In *Second Asian Applied Computing Conference, AACC 2004*, volume 3285 of *Lecture Notes in Artificial Intelligence*, pages 84–91, Kathmandu, Nepal, October 2004. Springer.
30. G. Leguizamón and Z. Michalewicz. A new version of ant system for subset problems. In *Congress on Evolutionary Computation, CEC'99*, pages 1459–1464, Piscataway, NJ, USA, 1999. IEEE Press.

31. L. Lessing, I. Dumitrescu, and T. Stützle. A comparison between aco algorithms for the set covering problem. In Dorigo et al. [17], pages 1–12.
32. D. Levine. A parallel genetic algorithm for the set partitioning problem. Technical Report ANL-94/23 Argonne National Laboratory, May 1994. Available at http://citeseer.ist.psu.edu/levine94parallel.html.
33. V. Maniezzo and M. Milandri. An ant-based framework for very strongly constrained problems. In M. Dorigo, G. D. Caro, and M. Sampels, editors, *Ant Algorithms*, volume 2463 of *Lecture Notes in Computer Science*, pages 222–227. Springer, 2002.
34. B. Meyer and A. Ernst. Integrating aco and constraint propagation. In Dorigo et al. [17], pages 166–177.
35. R. Michel and M. Middendorf. An island model based ant system with lookahead for the shortest supersequence problem. In *Lecture notes in Computer Science, Springer Verlag*, volume 1498, pages 692–701, 1998.
36. G. R. Raidl. A unified view on hybrid metaheuristics. In F. Almeida, M. J. B. Aguilera, C. Blum, J. M. Moreno-Vega, M. P. Pérez, A. Roli, and M. Sampels, editors, *Hybrid Metaheuristics*, volume 4030 of *Lecture Notes in Computer Science*, pages 1–12. Springer, 2006.
37. B. Rekiek, A. Delchambre, and H. A. Saleh. Handicapped person transportation: An application of the grouping genetic algorithm. *Engineering Applications of Artificial Intelligence*, 19(5):511–520, 2006.

Profitability Comparison Between Gas Turbines and Gas Engine in Biomass-Based Power Plants Using Binary Particle Swarm Optimization

P. Reche López[1], M. Gómez González[2], N. Ruiz Reyes[1], and F. Jurado[2]

[1] Telecommunication Engineering Department, University of Jaén
Polytechnic School, C/ Alfonso X el Sabio 28, 23700 Linares, Jaén, Spain
[2] Electrical Engineering Department, University of Jaén
Polytechnic School, C/ Alfonso X el Sabio 28, 23700 Linares, Jaén, Spain

Abstract. This paper employs a binary discrete version of the classical Particle Swarm Optimization to compare the maximum net present value achieved by a gas turbines biomass plant and a gas engine biomass plant. The proposed algorithm determines the optimal location for biomass turbines plant and biomass gas engine plant in order to choose the most profitable between them. Forest residues are converted into biogas . The fitness function for the binary optimization algorithm is the net present value. The problem constraints are: the generation system must be located inside the supply area, and its maximum electric power is 5 MW. Computer simulations have been performed using 20 particles in the swarm and 50 iterations for each kind of power plant. Simulation results indicate that Particle Swarm Optimization is a useful tool to choose successful among different types of biomass plant technologies. In addition, the comparison is made with reduced computation time (more than 800 times lower than that required for exhaustive search).

1 Introduction

Gumz (1950) is the earliest reference found describing the concept of combining a pressurized gasifier with a gas turbine engine, although Gumz himself references an earlier work proposing this concept. He also states that the combination could certainly benefit from future development of pressurized hot gas cleaning to avoid excessive turbine blade wear. Gumz was speaking of coal-fueled plants but the concept is similar when using biomass as fuel. [1].

Stationary engines are rated by the amount of power that can be continuously delivered at the coupling. Speed ratings are based upon mechanical stresses and the ability of the piston and the piston rings to receive adequate lubrication and to seal combustion gases. Like-model engines operating at different installations may experience varying consumption rates due to variations in operations conditions, purification standards, product quality, and in-service-hours [2].

The gasifier heats with limited oxygen supply the forest residues, the final result is a very clean-burning gas fuel suitable for direct use in gas turbines or gas engine.

J. Mira and J.R. Álvarez (Eds.): IWINAC 2007, Part II, LNCS 4528, pp. 328–336, 2007.

Choosing between biomass gas turbines or biomass gas engine is a computationally heavy task, if the electric power generated by the plant is about 5 MW [3]. When a realistic problem formulation is to be solved, most analytical, numerical programming or heuristic methods are unable to work well. In recent years, Artificial Intelligence (AI)-based methods, such as Genetic Algorithms (GAs), have been applied to similar problems with promising results [4]. Meanwhile, some new AI-based methods are introduced and developed. Although these AI-based methods do not always guarantee the globally optimal solution, they provide suboptimal (near globally optimal) solutions in a short CPU time. This paper employs a modern AI-based method, Particle Swarm Optimization (PSO) [5][6][7], to solve the problem of determining the most profitable technology (gas turbine o gas engine), after determining optimal location for biomass plant supplied with forest residues. In this work, the fitness function for the PSO algorithm is the net present value (eq. (10)) .

PSO is a nature-inspired evolutionary stochastic algorithm developed by James Kennedy and Russel Eberhart in 1995 [5]. This technique, motivated by social behavior of organisms such as bird flocking and fish schooling, has been shown to be effective in optimizing multidimensional problems. PSO, as an optimization tool, provides a population-based search procedure in which individuals, called particles, change their positions (states) with time. In a PSO system, particles fly around in a multidimensional search space. During flight, each particle adjusts its position according to its own experience, and the experience of neighboring particles, making use of the best position encountered by itself and its neighbors. The main advantages of PSO are: it is very easy to implement and there are few parameters to adjust.

2 Particle Swarm Optimization

2.1 The Classical Approach

The classical PSO algorithm is initialized with a swarm of particles randomly placed on the search space. In the $(t+1)$-th iteration, the position of i-th particle is update adding to its previous position the new velocity vector, according to the following equation:

$$\boldsymbol{x}_i^{t+1} = \boldsymbol{x}_i^t + \boldsymbol{v}_i^{t+1}, \quad i = 1, ..., P \tag{1}$$

where $\boldsymbol{x}_i^t$ denotes the position vector of the i-th particle at the t-th iteration, $\boldsymbol{v}_i^t$ represents the velocity vector at the t-th iteration, both $\boldsymbol{x}_i^t$ and $\boldsymbol{v}_i^t$ are N-dimensional vectors, N being the number of variables of the function to be optimized. P is the number of particles in the swarm.

The velocity vector is updated according to the following equation:

$$\boldsymbol{v}_i^{t+1} = \omega \cdot \boldsymbol{v}_i^t + c_1 \cdot rand_1 \cdot (\boldsymbol{p}_{i,best}^t - \boldsymbol{x}_i^t) + c_2 \cdot rand_2 \cdot (\boldsymbol{g}_{best}^t - \boldsymbol{x}_i^t) \tag{2}$$

where $\boldsymbol{p}_{i,best}^t$ is the best solution achieved for the i-th particle at the t-th iteration and $\boldsymbol{g}_{best}^t$ is the best position found for all particles in the swarm at the

t-th iteration. c_1 and c_2 are positive real numbers, called learning factors or acceleration constants, that are used to weight the particle individual knowledge and the swarm social knowledge, respectively. $rand_1$ and $rand_2$ are real random numbers uniformly distributed between 0 and 1, that make stochastic changes in the particle trajectory. Finally, ω is the inertia weight factor and represents the weighting of a particles previous velocity.

From equation (2), we can find that the current flying velocity of a particle comprises three terms. The first term is the particles previous velocity revealing that a PSO system has memory. The second term and the third term represent a cognition-only model and a social-only model, respectively.

2.2 The Proposed Binary Approach

In spite of usual formulation for PSO uses real-number coding, we have applied in this work a discrete PSO algorithm which uses binary-number coding. In the proposed binary PSO algorithm, $\boldsymbol{x}_i^t$ and $\boldsymbol{v}_i^t$ are N-length binary vectors. Equation (1) is applied using the exclusive-or ('XOR') operator instead of real adding:

$$\boldsymbol{x}_i^{t+1} = \boldsymbol{x}_i^t \oplus \boldsymbol{v}_i^{t+1}, \quad i = 1, ..., P \tag{3}$$

Here, the velocity vector can be interpreted as a change vector. Thus, if $v_i^t[j]$='1', then $x_i^{t+1}[j] = \bar{x}_i^t[j]$, $\bar{x}_i^t[j]$ being the logical negation of $x_i^t[j]$. However, if $v_i^t[j]$='0', then $x_i^{t+1}[j] = x_i^t[j]$ (no change happens).

The velocity vector (change vector) is updated by applying the following equation:

$$\boldsymbol{v}_i^{t+1} = \bar{\boldsymbol{\omega}}_i^t + \boldsymbol{\omega}_i^t \cdot \left(\boldsymbol{c}_{i,1} \cdot (\boldsymbol{p}_{i,best}^t \oplus \boldsymbol{x}_i^t) + \boldsymbol{c}_{i,2} \cdot (\boldsymbol{g}_{best}^t \oplus \boldsymbol{x}_i^t)\right) \tag{4}$$

where the inertia vector $\boldsymbol{\omega}_i^t$ is a random N-length binary vector and $\bar{\omega}_i^t$ its logical negation. $\boldsymbol{c}_{i,1}$ and $\boldsymbol{c}_{i,2}$ are also random N-length binary vectors. In equation 4, symbol + represents the logical OR operator and symbol $\cdot$ represents the logical AND operator.

In our approach, parameter p_i has been defined. It represents a probability which decreases with the number of iterations. Here, parameter p_i is applied to generate inertia vector $\boldsymbol{\omega}_i^t$ as follows: $\omega_i^t[j]$='0' with p_i probability, in such a way that at the initial iterations (high p_i values) the algorithm explores the search space and at the last iterations (low p_i values) the algorithm exploit the search space. The idea is to allow particle swarm to perform a random exploration over the space search at the initial iterations. Later, when the particle swarm has acquired enough knowledge about the problem, its movement is conducted by the best solution $\boldsymbol{p}_{i,best}^t$ and the best position $\boldsymbol{g}_{best}^t$ at the t-th iteration, in order to reach a suboptimal solution with reduced computational cost.

3 Problem Description

The problem to be solved consists on comparing the use of a gas turbines or a gas engine in biomass-based power generation systems. The size of the generation system depends on: 1) biomass quantity that can be collected, 2) selection of parcels where to collect the biomass. Location of power plant (parcel p) mainly depends on the characteristics of the considered parcels. In this work, K parcels of constant area have been regarded, all of them characterized by a predominant biomass type (forest residues in this work). These parcels also present other relevant characteristics, such as accessibility [8].

The values of the variables involved in the problem are obtained from databases or Geographic Information Systems (GIS). These are the following:

- S_i: Area of parcel i (km^2).
- U_i: Usability coefficient of parcel i. It is applied to take into account only the usable surface.
- D_i: Net density of dry biomass yield from parcel i ($ton/(km^2 \cdot yr)$).
- LHV_i: Lower heat value of biomass in parcel i (MWh/ton).
- L_p: Length of the electric line that connects the power plant to the grid (Km).
- $dist(p, i)$: Distance between parcel i and the power plant, which is located in parcel p(km).
- C_{cu_i}: Biomass collection unit cost in parcel i (Euro/ton).

Therefore, given the total mean efficiency of the electric generation system, eff, the electricity produced, E_g (MWh/yr), is equal to:

$$E_g = eff \cdot \sum_{i=1}^{K} (S_i \cdot U_i \cdot D_i \cdot LHV_i) \tag{5}$$

Assuming a plant running time of T(h/yr), the electric power, P_e(MW) is:

$$Pe = \frac{E_g}{T} \tag{6}$$

4 Objective Function: Net Present Value

The objective function takes into consideration costs and benefits. Specifically, initial investment and collection, transportation, maintenance and operation costs are considered, together with benefits from the sale of electrical energy. Therefore, the net present value is chosen as the objective function.

In this section some interesting parameters to evaluate the net present value of the project are reviewed. The initial investment, the present value of cash inflows (benefits) and cash outflows (costs) are studied and adapted to the particularities of this work.

4.1 Initial Investment

The initial investment (INV) for the design, construction of the generation plant and required equipment is expressed as:

$$INV = INV_f + I_s \cdot P_e + C_L \cdot L_p \tag{7}$$

where INV_f is the fixed investment (Euro), I_s is the specific investment (Euro/MW) and C_L the electric line cost (Euro/km).

4.2 Cash Inflows

The present value of cash inflows (PV_{IN}) is obtained from the sold electric energy during the useful lifetime, V_u. It can be written as:

$$PV_{IN} = p_g \cdot E_g \cdot \frac{K_g \cdot (1 - K_g^{V_u})}{1 - K_g} \tag{8}$$

where p_g is the selling price of the electric energy injected to the network (Euro/MWh), E_g the sold and produced electric energy (MWh/yr) and $K_g = \frac{1+r_g}{1+d}$, r_g being the annual increase rate of the sold energy price and d the nominal discount rate.

4.3 Cash Outflows

The present value of cash outflows (PV_{OUT}) is the sum of the following costs during the useful lifetime of the plant: annual collection cost, C_c, annual transport cost, C_t and annual maintenance and operation costs, C_{mo}.

The annual cost of biomass collection is $C_c = \sum_{i=1}^{K}(C_{cu_i} \cdot U_i \cdot S_i \cdot D_i)$.

The annual cost of biomass transport is $C_t = \sum_{i=1}^{K}(C_{tu_i} \cdot S_i \cdot D_i \cdot dis(p,i))$, where C_{tu_i} is the biomass transport unit cost in parcel i (Euro/(ton · km)).

The annual maintenance and operation costs are $C_{mo} = m \cdot E_g$, m being average maintenance costs (Euro/MWh) and E_g the produced electric energy (MWh/yr).

Finally, the present value of cash outflows is:

$$PV_{OUT} = C_c \cdot \frac{K_c \cdot (1 - K_c^{V_u})}{1 - K_c} + C_t \cdot \frac{K_t \cdot (1 - K_t^{V_u})}{1 - K_t} + C_{mo} \cdot \frac{K_{mo} \cdot (1 - K_{mo}^{V_u})}{1 - K_{mo}} \tag{9}$$

where $K_c = \frac{1+r_c}{1+d}$, $K_t = \frac{1+r_t}{1+d}$ and $K_{mo} = \frac{1+r_{mo}}{1+d}$, r_c being the annual increase rate of C_c, r_t the annual increase rate of C_t and r_{mo} the annual increase rate of C_{mo}.

4.4 Net Present Value

The net present value (NPV) is defined as follows:

$$NPV = PV_{IN} - PV_{OUT} - INV \tag{10}$$

An investment is profitable when $NPV > 0$.

5 Experimental Results

The region considered to apply the proposed method consists of $128 \times 128 = 16384$ parcels of constant surface, $S_i = 2$ km^2. The region is covered by natural forest vegetation. Therefore, forest residues constitute the biomass type. The available information for each parcel comprises S_i, U_i, D_i, LHV_i, L_p, $dist(p,i)$ and C_{cu_i}. Other parameter values are shown in table 1:

Table 1. Standard values for parameters

Parameter	Value	Parameter	Value
$C_{tu_i}(Euro/(Ton \cdot km))$	0.3	$C_L(Euro/km)$	$3 \cdot 10^4$
$T(h/yr)$	7500	$INV_f(Euro)$	$1.5 \cdot 10^6$
$p_g(Euro/MWh)$	100	d	0.08
r_g	0.04	r_c	0.06
r_i	0.08	r_{mo}	0.04

Parameters which are characteristics of the type of unit generation are listed in table 2. The gas turbine unit generation requires a higher specific investment than gas engine, but gas engine maintenance costs are twice higher than gas turbine maintenance costs and less useful lifetime:

Table 2. Specific values for unit generation

Gas turbine	Value	Gas engine	Value
$m(Euro/MWh)$	0.05	$m(Euro/MWh)$	0.6
eff	0.3	eff	0.2
$I_s(Euro/MW)$	$1.2 \cdot 10^6$	$I_s(Euro/MW)$	$0.2 \cdot 10^6$
$V_u(yr)$	15	$V_u(yr)$	10

Figure 1 presents the theoretical biomass potential, which is defined from the net density of dry biomass that can be obtained at any parcel, D_i (ton/(Km$^2 \cdot$ yr)), and provides a measure of the primary biomass resource. Location of the electric line inside the considered region is also shown in figure 1.

Figure 2 shows the available biomass potential. It has been created taking the following parameters into account: D_i(ton/(Km$^2 \cdot$ yr)), U_i, S_i(Km^2) and LHV_i(MWh/ton). By multiplying the four variables for all the parcels that comprise the entire region, it results the available biomass potential, expressed in (MWh/yr), as depicted in figure 2.

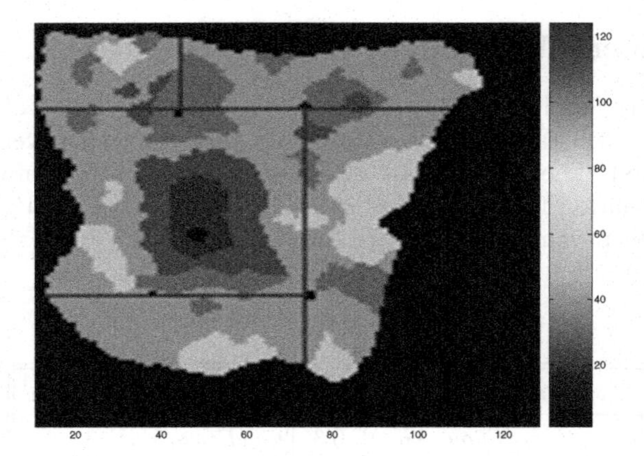

Fig. 1. Theoretical biomass potential (ton/(Km^2 · yr))

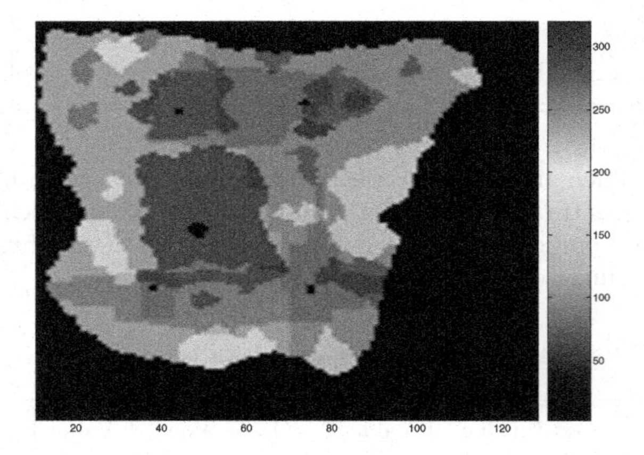

Fig. 2. Available biomass potential (MWh/yr))

Table 3. Output values

Gas turbine	Value	Gas Engine	Value
$NPV(KEuro)$	16586.23	$NPV(KEuro)$	9117.04
$P_e(MW)$	4.75	$P_e(MW)$	4.98
Supply area	880.0	Supply area	1584.0

Simulation data are: $P = 20$, $N = 20$ and 65 iterations. The constraints for simulation are: 1) The electric power generated by the plant is limited to 5 MW; 2) The generation system must be located inside the supply area.

The proposed PSO algorithm provides the output values in table 3. Gas turbine has been shown more profitable (higher net present value) than gas engine.

Figure 3 shows the optimal location and supply area for the gas turbine plant and for the gas motor plant in a typical realization. Note that the optimal location is different in each case.

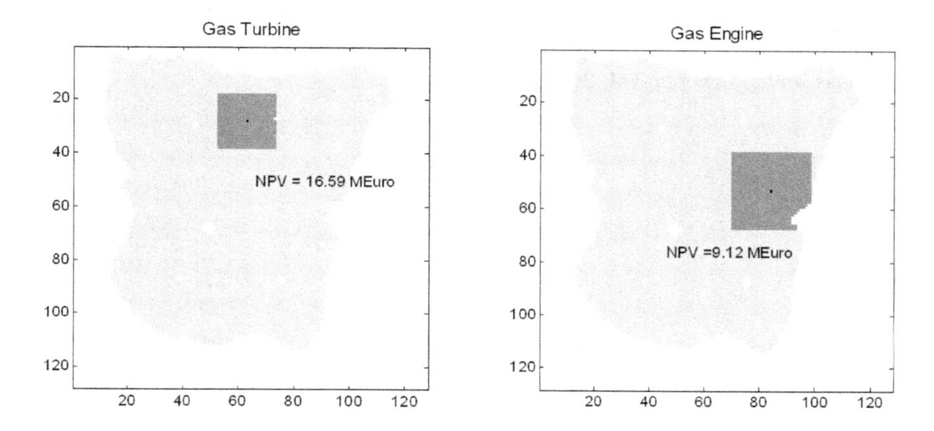

Fig. 3. Optimal location and supply area for the biomass plant

6 Conclusions

This paper has presented an AI-based method to determine the optimal supply area and location for an electric generation system based on biomass. The proposed AI-based method is a discrete binary version of the PSO algorithm, which makes use of the profitability index as objective function. The proposed approach have been assessed using a region composed of 16384 parcels, all of them with the same area (S_i=2 km^2). In the region under study, gas turbine has been shown more profitable than gas motor, and the net present value of the gas turbine-based project has achieved 16.58 MEuro. Computer simulations have shown the good performance of the proposed method. Convergence is reached in few iterations (about 25) and computational cost more than 800 times lower than that required for exhaustive comparison.

References

1. Gumz, W."Gas Producers and Blast Furnaces", *Wiley*, New York, pp. 166-167, 1950.
2. Logan. E. Jr."Handbook of Turbomachinery", *Marcel Dekker*, New York, 1994.
3. Jurado F., Ortega M., Cano A., Carpio J. "Biomass gasification, gas turbine, and diesel engine", *Energy Sources*, 23 (10): 897-905, 2001.
4. Boone, G. and Chiang, H.D. "Optimal capacitor placement in distribution systems by genetic algorithm", *Electr. Power Energy Syst.*, vol. 15, pp. 155162, 1993.

5. Kennedy, J. and Eberhart, R. "Particle Swarm Optimization", *Proc. IEEE Int. Conf. on Neural Networks*, pp. 1942-1948, 1995.
6. Kennedy, J. "The particle swarm: Social adaptation of knowledge", *Proc. IEEE Int. Conf. Evol. Comput.*, Indianapolis, IN, pp. 303308, 1997.
7. Eberhart, R. and Kennedy, J. "A new optimizer using particle swarm theory", *Proceedings of the Sixth International Symposium on Micro Machine and Human Science*, pp. 3943, 1995.
8. Freppaz, D., Minciardi, R., Robba, M., Rovatti, M., Sacile, R. and Taramasso, A. "Optimizing forest biomass exploitation for energy supply at a regional level", *Biomass and Bioenergy*, vol. 26, pp. 15-25, 2004.

Combining the Best of the Two Worlds: Inheritance Versus Experience*

Evolutionary Knowledge-Based Control and Q-Learning to Solve Autonomous Robots Motion Control Problems

Darío Maravall[1], Javier de Lope[2], and José Antonio Martín H.[3]

[1] Department of Artificial Intelligence
Universidad Politécnica de Madrid
dmaravall@fi.upm.es

[2] Department of Applied Intelligent Systems
Universidad Politécnica de Madrid
javier.delope@upm.es

[3] Departamento de Sistemas Informáticos y Computación
Universidad Complutense de Madrid
jamartinh@fdi.ucm.es

Abstract. In this paper a hybrid approach to the autonomous navigation of robots in cluttered environment with unknown obstacles is introduced. It is shown the efficiency of the hybrid solution by combining the optimization power of evolutionary algorithms and at the same time the efficiency of the Reinforcement Learning in real-time and on-line situations. Experimental results concerning the navigation of a L-shaped robot in a cluttered environment with unknown obstacles in which appear real-time and on-line constraints well-suited to RL algorithms and extremely high dimension of the state space usually unpractical for RL algorithms but at the same time well-suited to evolutionary algorithms, are also presented. The experimental results confirm the validity of the hybrid approach to solve hard real-time, on-line and high dimensional robot motion control problems.

1 Introduction: Bio-inspired Approaches to Solve Autonomous Robot Motion Control Problems Based on the Sensory-Motor Coordination Paradigm

As the complexity and sophistication of robotics systems and applications have steadily increased in the last years, novel methods and approaches in the field have consequently appeared; in particular, bio-inspired and artificial intelligence-based ones. Also, the higher demands on the tasks to be performed by the current robots have made necessary the introduction of new and more complex sensor

* This work has been partially funded by the Spanish Ministry of Science and Technology, project: DPI2006-15346-C03-02.

J. Mira and J.R. Álvarez (Eds.): IWINAC 2007, Part II, LNCS 4528, pp. 337–346, 2007.

subsystems, and, the same time, new and more sophisticated actuators and mechanical subsystems. Therefore, as a consequence, the design of the new robot controllers basically revolve around the central issue of coordinating perception (i.e. the information from the environment's states gathered by the robot's sensors) and action (i.e. the response decided by the robot's controller through the robot's actuators). Such robot's actions are aimed at the performance of some predefined tasks or objectives injected by the designer, so that the main objective of the controller is to drive the pair robot-environment from any arbitrary initial state to a desired, optimum final state (see Section 3 below for a formal and general statement of this control problem).

2 Guidelines for the Design of an Optimum (Almost Universal) Controller for Autonomous Robots by the Combination of Evolutionary Algorithms and Reinforcement Learning

As commented above, the current robotics systems require controllers able to solve complex problems under very uncertain and dynamic environmental situations. The well-known Reinforcement Learning (RL) paradigm is probably the approach best suited to the implementation of these controllers as it is based on the idea of choosing control actions that drive the system from an arbitrary initial state to a final desired state by means of applying the optimum actions available to the robot at each instant of time. RL is particularly attractive and efficient in the very common and hard situation in which the designer does not have all the information necessary a priori for designing the optimum control sequence of actions, so that the controller has to learn in real time the optimum decisions sequence through the evaluation of the rewards and punishments received by the controller from the environment [1]. Furthermore, from a theoretical point of view, the RL approach seamlessly fits the usual modeling of the robot-environment interaction (i.e. a finite state automaton) as a Markov Decision Process (MDP) so that well-known control techniques for MDPs like Dynamic Programming may be applied to the design of the robot controller. However, the drawback associated to the RL paradigm is related to the curse of dimensionality that appears in the usual case of complex real-life environments of a very high dimension of the state variable. In order to avoid this serious problem we propose in this paper to initialize a typical Q-learning algorithm with a look-up table of situations-actions by means of an evolutionary algorithm so that the Q-learning algorithm departs its exploration of the environment states from the optimum knowledge base provided by the evolutionary algorithm and, thus speeding up the real-time search of the reinforcement learning-based controllers as explained in the sequel.

The first step in our proposed method is (1) the identification and (2) the subsequent granulation of the state variables associated to the pair robot-environment which is highly application dependant.

As for the granulation of state variables: it can be performed by applying either the fuzzy set concept in which case we should obtain a Fuzzy Knowledge Rule Base (FKRB) or a Boolean sets-based granulation in which case we should obtain instead a Boolean Knowledge Rule Base (BKRB). As for the advantages and disadvantages of both types of knowledge rules (i.e. fuzzy or Boolean) the reader can consult a large technical literature. In particular, we believe that it is much more constructive to take an hybrid and all-embracing perspective as both fuzzy and Boolean sets concepts can be considered as similar and complementary rather than opposing under the Computing with Words (CW) paradigms [2].

Once the variables have been properly granulated, the next step is obtaining the knowledge base of production or control rules of the type "*if* situation *then* action", that is, the output of this step is the establishment of the general, almost universal associative matrix or look-up table of situations/actions associated to the specific problem or application. In the most common cases of high number of possible control rules the search of the optimum knowledge base of rules can only be undertaken by efficient, parallel population-based search methods and techniques like evolutionary algorithms.

RL is the conventional and well-established on-line approach to solve hard robot motion control problems but this real time solution may have some problems for real life cases in which appears a combinatorial explosion of the state variables space. However, parallel population-based techniques like evolutionary algorithms are well-known for its robustness and efficiency for solving this kind of highly combinatorial explosion problems; thus, to illustrate this issue, we have chosen a concrete hard robot motion control problem in which the number of state variables is extremely high making the exploration nature of the RL algorithm unpractical and for that reason it is necessary the application of evolutionary algorithms for the exploration phase of the optimum solution. Summarizing, in the sequel we propose a very hard control motion problem and we show the difficulties of a standard Q-learning algorithm to obtaining the knowledge rule base and we also show that a genetic algorithm is able to converge to the optimum knowledge rule base so that we can inject this genetic-based knowledge rule base as the initialization of the standard Q-learning algorithm. Summarizing, the proposed method consist of the following stages: (1) firstly as usual the selection and subsequent granulation of the state variables involved which is a designer-depended task is undertaken; (2) secondly, we proceed to the obtaining of the knowledge rule base by means of a genetic algorithm; (3) finally, once the obtained optimum knowledge rule base provided by the genetic algorithm we can proceed with the third step: in which the standard Q-learning algorithm starts its own exploration of the environment in order to build its Q-table and, at the same time, it exploits the knowledge provided by the genetic algorithm. As a conclusion the combination of the genetic algorithm's power in the search of the optimal solution and the real-time nature of the Q-learning is the best approach for solving the proposed hard robot motion control problem. These ideas are shown in the Fig. 1

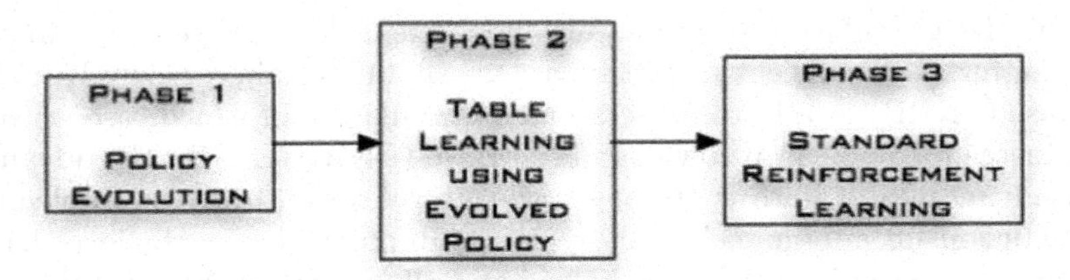

Fig. 1. Phases in which is divided the proposed method

3 A Hard Motion Robot Control Problem

To illustrate the above discussion we have chosen an interesting autonomous robot motion control problem: a two-link L-shaped robot moving in a cluttered environment with polygonal obstacles (Fig. 2). The two-link robot has several degrees-of-freedom. The first one is the linear movement along the XY Cartesian axes of the robot's middle joint. Another rotational movement of the second link is also considered (ϕ) in order to allow controlling the robot's orientation in the plane. Finally two additional independent rotational movements for both links are also permitted (θ_1 and θ_2).

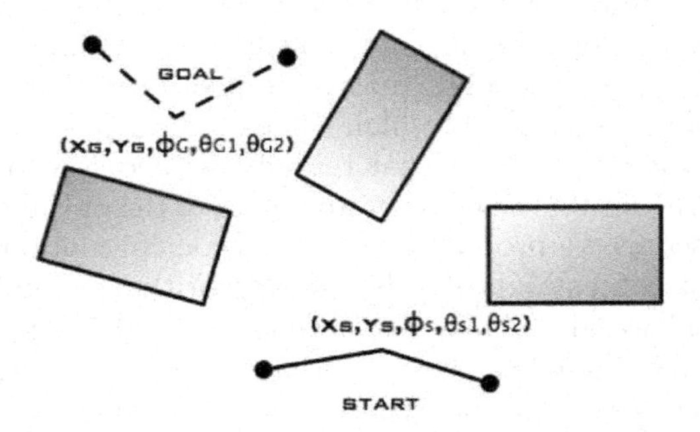

Fig. 2. L-shape multi-link robot used for experimentation

4 Evolving the Table of Situations-Actions

By following the general method explained in Section 2 our first step for obtaining the final knowledge rule base is the selection and subsequent granulation of the state variables of the particular application considered in this paper.

We have defined the table in such a way that we are able to effiently and completely describe the robot state in a simulated environment with obstacles. To completely describe *all* the possible states of the robot-environment pair (or, in another words, *all* the possible situations of the robot in the environment) we have distinguished two different groups or classes of state variables:

(a) those concerning the primary robot task of reaching a desired goal position, and (b) those variables concerning the additional robot task of collision avoidance. Therefore, we have used three different state variables: ε_d, the error between the current robot position and target position, ε_ϕ, the error between the current robot orientation and the target orientation, and ρ_o, the distance to the nearest obstacle.

For each state variable we have defined several linguistic granules in order to establish partitions in the values determined by the sensors. We have defined the bounds of these zones as thresholds although they could be defined as a fuzzy-like way; i.e. we have chosen for simplicity in our controller a Boolean partition of the linguistic variables involved. The thresholds are assigned in designing time but they could be evolved in the evolutionary process.

For the position and orientation errors, ε_d and ε_ϕ respectively, we have defined three thresholds. The first determines positions and orientation very near to the goal, where the distance is practically zero (Z). The other two thresholds define small (S) and big (B) distances.

For the distance to nearest obstacle variable, ρ_o, we have defined four thresholds for the next situations: when the distance is very small and the robot could collide (VS), when the distance is small (S), when the distance is big (B) and, finally, for the cases in which the obstacle is very far (VB).

Thus, the table includes 36 possible states which characterize all the possible situations in which the robot can be in the world.

We have also determined the actions that could be executed by the robot. We have discretized these actions in order to generate rules of the type of "*if* situation *then* action", as we have previously commented. The actions are the 8 different linear movements that the robot can make related to its own coordinate frame, i.e. move forward, move backward, move left, move right, and the four diagonal movements. Also, we have defined 2 turn movements, left and right. Thus, we have defined 10 movements in total. Then, the total number of control rules is as high as 10^{36}.

The evolutionary algorithm designed takes every table as the genotype of an individual (i.e. the Michigan approach to learning classifiers). It is only based on the mutation operator although we have checked and compared the results to a classical genetic algorithm. We propose its use due to the crossover operator could not be clearly identified for the proposed problem.

Each action of each individual is mutated in each generation in such a way that we can obtain the same action than the previous generation (really, no mutation case) or the two possible adjacent actions defined in an internal table. This internal table is fixed and "circular", we mean, when one of the bounds are encountered, the other bound is used as the value for the mutation.

Also we have implemented a "replacing policy" in which the n worst individuals are eliminated and replaced by new randomly generated ones. Thus, we avoid that the populations keep trapped by local minima as well as maintain the better solutions between two successive generations.

5 The Transition from Innate Behavior to Knowledge-Based Behavior by Means of On-Line Experience

In MDP terminology a stationary deterministic policy π_d is a policy that commits to a single action choice per state, that is, a mapping $\pi_d : S \rightarrow A$ from states to actions, in this case, $\pi_d(s)$ indicates the action that the agent takes in state s. Hence the table generated in the previous section by means of evolutionary techniques is just a stationary deterministic policy π_d which represent the *innate behavior* of the robot controller. The goal of this development stage is to produce a robot controller that based initially in its innate-behavior experiments a transition to a knowledge-based behavior by means of on-line experience.

Maybe the most important and relevant paradigm of knowledge-based behavior by means of on-line experience is the RL paradigm. One of the most important breakthroughs in RL was the development of an off-policy algorithm called Q-learning [3]. In a simple form its learning rule is:

$$Q(s,a) = Q(s,a) + \alpha[r + \gamma \max_a Q(s_{t+1}, *) - Q(s,a)] \tag{1}$$

Note that this is a classical update rule which adaptively approximates the average μ of a set $x = (x_1 ... x_\infty)$ of observations:

$$\mu = \mu + \alpha[x_t - \mu], \tag{2}$$

thus replacing the factor x_t in (2) with the current reward r plus a fraction $\gamma \in [0, 1]$ of the best possible approximation to the maximum expected future reward $\max_a Q(s_{t+1}, *)$:

$$\mu = \mu + \alpha[r + \gamma \max_a Q(s_{t+1}, *) - \mu], \tag{3}$$

we are indeed approximating the average of the maximum future reward when taking the action a at the state s.

The development of an off-policy RL algorithm has very important consequences from the theoretical point of view indeed based on this fact in [4] the authors proved the convergence of Q-learning to an optimal control policy with probability **1** under certain strong assumptions see [1, p.148].

We will use this special feature of Q-learning (off-policy) as the main theoretical fundament of our transitional method from innate behavior to experience-based behavior.

We must note the fact that being Q-learning an off-policy algorithm we can separate the policy used to select actions from the policy that the experience-based behavior controller is learning. Thus, we can use the stationary deterministic policy π_d as the initial behavioral policy while learning in parallel a non-deterministic policy π which will be based on the information provided by the Q-table.

Of this form, we can take effective advantage of the knowledge inherited from the evolutive stage. Basically, the method consist of behaving with innate capabilities until enough experience has been gained from the interaction with the environment and the Q-table has converged to the optimal values that represents the greedy policy π_d.

Once the Q-learning controller has been trained, the policy π_d can be inhibited letting the controller behave with the new policy π. Hence, the final controller will be an adaptive on-line experience-based behavior controller, that is:

1. It will remain adaptive in the sense that it can adapt to changes in the environment.
2. It is rational in the sense that it has an action selection mechanism over different choices based on cause and effect relations stored in the form of a Q-table.
3. It will perform on-line, that is, can learn continuously by direct real-time interaction with the environment.

6 Experimental Results

We have divided this section in two parts, one concerning a "simple problem" (i.e. a robot motion control problem in which the dimension of the state space is not extremely high) and another one a "complex problem" (i.e. in which the dimension of the state variables suffers a combinatorial explosion).

6.1 The Simple Case: Moving a Stick in a Cluttered Environment with Unknown Obstacles

The state variables associated to this problem have been described in the Section 4 (in particular remember the variables ε_d, ε_ϕ and ρ_o, that define the state of the robot relative to the goal position and also the variables that define the state of the robot relative to the obstacle situation and their subsequent proposed granulation).

Pure Q-Learning is Able to find the Optimum Solution. We mean by pure Q-learning the standard Q-learning algorithm in which the initialization of the Q-table is arbitrary, usually to zero. Fig. 3 shows the paths followed by the robot in several environments with randomly selected goal positions.

Genetic Algorithm is also Able to find the Optimum Solution. As a general conclusion regarding the performance of RL and evolutionary paradigms in the solution of the proposed robot motion control problem we have not found any remarkable distinction among the solutions obtained by the two methods. Fig 4 show examples of the solutions obtained independently by both methods in which we can see that the stick is always successfully translated for an initial position to a goal position without colliding with the obstacles.

6.2 The Complex Case: Controlling a Two-Link L-Shaped Robot

We have already described this autonomous robot motion control problem (see Section 3 for a detailed description of this problem). As for the granulation of

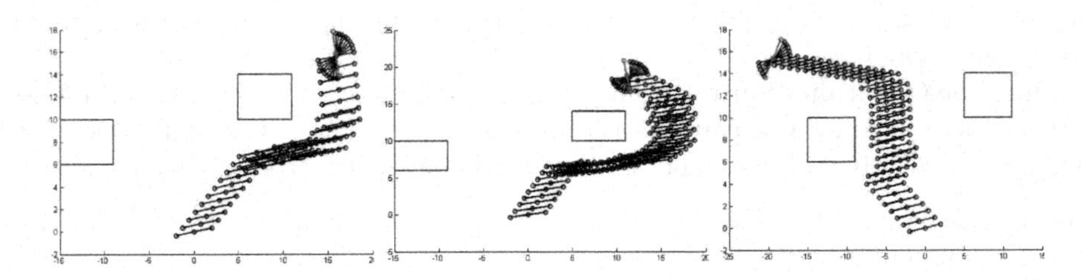

Fig. 3. Several paths followed by the robot in simulated cluttered environments

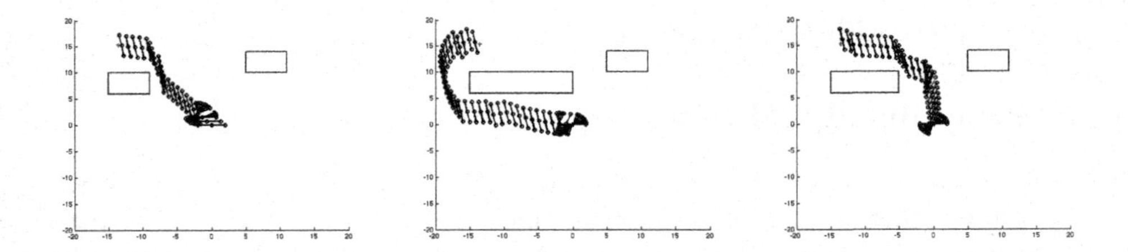

Fig. 4. Trajectories of the evolutionary controller. As we can see there is no remarkable difference between both approaches for the proposed task.

the state variables of this problem we have defined three linguistic terms for each variable: zero or near zero, small and big. So that with the granulation stick itself we have a total number of states as high as $3 \times 3 \times 4 \times 3 \times 3 = 324$, a considerable complexity as compared to the $3 \times 3 \times 4 = 36$ of the simple case.

Beside this remarkable increase in the state space we also remind that the actions available to the robot are even higher, two more movements per link amounting to 14 total number of actions; therefore the total number of control rules is 14^{324}, making the search of the optimum set of control rules a truly hard problem.

Pure Q-Learning Approach. Pure Q-learning is unable to converge to the optimum solution. As expected, the powerful real time Q-learning algorithm has troubles in the exploration phase of very highly dimensional spaces as it is shown in Fig. 5.

Evolutionary Approach. The genetic algorithm is able to find the optimum solution. We have found the excellent quality of genetic algorithm in the exploration task of very highly dimensional spaces as the one considered in our paper (see Fig. 6 in which individual of generation 500 translates the robot to the goal position although good controllers can be obtained in the generation 150).

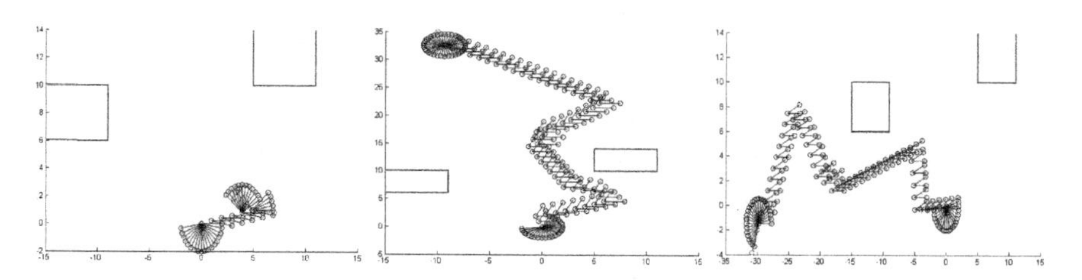

Fig. 5. Trajectories of the evolved controller (500 generations)

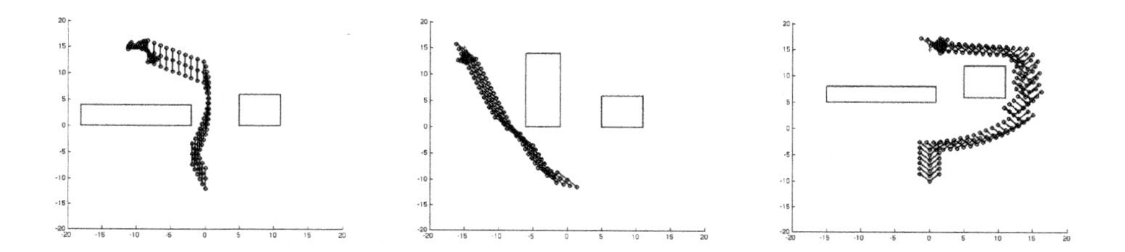

Fig. 6. Trajectories of the evolved controller (500 generations)

7 Conclusions

In this paper the advantages and disadvantages of two well-established approaches like Reinforcement Learning (RL) and Evolutionary Algorithms (EA) are discussed as regarding their respectives performances in the solutions of a particular hard robot motion control problem. RL presents very attractive features concerning real-time applications, on-line applications, however, RL sometimes presents difficulties when the dimension of the state variables suffers a combinatorial explosion. On the other hand EA are extremely powerful off-line optimizers, in particular, for extremely high dimension spaces. However, at the same time they do not seem as well suited to real-time, on-line environments as RL algorithms are. For such reasons in this paper we have proposed the combination of both paradigms for the solution on real-time, extremely high dimensional robot motion control problems. We have shown the efficiency of hybrid approach for autonomous navigation of a L-shaped two links robot in a cluttered environment with unknown obstacles.

An extremely important and vital step in the design of the intelligent autonomous robot controller is the selection and the subsequent granulation of the state variables involved in the particular application. Concerning the granulation two approaches are possible: fuzzy or Boolean granulation. In this paper we have chosen a Boolean granulation and we expect in the future to try a fuzzy granulation in order to test the power of the fuzzy logic controller as universal approximators [5,6].

Concerning the evolutionary algorithm we have chosen the coding in an individual the complete set of knowledge rules (i.e. the Michigan approach to learning classifiers). Another possibility is the coding of each individual as a single knowledge rule (i.e. the Michigan approach [7]) in which case there appears a hard problem of rules competition and coordination. Therefore, in the future we plan to explore this issue concerning two different approaches to the learning classifiers paradigm, i.e. Pittsburgh *versus* Michigani, for the solution of real-time, on-line high dimensional robot motion control problems.

References

1. Sutton, R.S., Barto, A.G. (1998) Reinforcement Learning: An Introduction. MIT Press, Cambridge, MA
2. Wang, P.P (2001) Computing with Words. Wiley Interscience, New York
3. Watkins, C.J. (1989) Models of Delayed Reinforcement Learning, PhD Thesis Dissertation, Psychology Department, Cambridge University, Cambridge, UK
4. Watkins, C.J., Dayan, P. (1992) Technical note Q-learning. Machine Learning, **8**:279
5. Mendel, J., Wang, L. (1992) Generating Fuzzy Rules by Learning Through Examples, IEEE Trans. on Systems, Man and Cybernetics, **22**:1414-1427
6. Pedrycz, W. (1993) Fuzzy Control and Fuzzy Systems (2nd edition). Addison Wesley Longman, Menlo Park, CA
7. Holland, J.H. (1986) Escaping brittleness: The possibilities of general purpose learning algorithms applied in parallel rule-based systems. In R.S. Michaiski, J.G. Carbonell, & T.M. Mitchell (Eds.), Machine Learning II, pp. 593-623

Towards the Automatic Learning of Reflex Modulation for Mobile Robot Navigation

C. Galindo, J.A. Fernández-Madrigal, and J. González

University of Málaga, Campus Teatinos, 29071 Málaga, Spain
{cipriano,jafma,jgonzalez}@ctima.uma.es
http://www.isa.uma.es

Abstract. Reflexes are meant to provide animals with automatic responses for a better adaptation to their niches. In particular, humans have the capability to voluntarily modify these responses in certain situations to attain specific goals. The ability of using past experiences to tune automatic responses (reflexes) has contributed to a better adaptation to our environments and thus, the question arises of applying this to machines. In the robotic arena, imitating animal reflexes has been largely explored through fixed stimuli-behavior schemas included in reactive or hybrid architectures. In this paper we consider the less explored direction of permitting a mobile robot to modify its reflexes according to its experience, i.e. ignoring the reflex of stopping when approaching an obstacle if the robot goal is close. We explore reinforcement learning as a mechanism to automatically learn when and how modulate reflexes over the robot operational life. Advantages of our mechanism are illustrated in simulations.

1 Introduction

Biologic inspiration has been usually considered as a good starting point to address challenges in robotics. The robotic community has paid special attention to the way in which simple animals survive in their niches, performing efficiently actions that we pursuit in robotic applications, i.e. reliable navigation, self-localization, automatic battery recharge (feeding), etc.

In general, the animal's survival depends on the ability for self-adaptation to the environment through some type of interaction, which generally is implemented as automatic, wired responses to stimuli. This evidence has served as an inspiration to implement robotic architectures [1], in which the robot is endowed with a set of simple behaviors, like collision avoidance, that are automatically triggered when some external circumstance occurs.

However, this type of simple adaptation to the environment is not sufficiently appropriate for cognitive animals, that is, those that can deliberate on their actions. Cognitive animals, including humans, do not rely only on automatic responses (*reflexes*) but also on *voluntary* actions. Although in neural science the border between reflex and voluntary actions is not well defined [3] we understand here as voluntary actions those that require a certain planning or decision

J. Mira and J.R. Álvarez (Eds.): IWINAC 2007, Part II, LNCS 4528, pp. 347–356, 2007.

processing in contrast to reflexes which are pre-defined. Cognitive animals make use of both types of actions, providing them with a better adaptation to the environment. But, what happens when a voluntary action contradicts a reflex?

Reflexes, defined as involuntary actions, prevent animals from potentially harmful situations. For example, when something comes into contact with our eyes, the eyelid reflex quickly closes it. But, as stated in neuroscience, humans voluntarily modify some automatic responses arising from reflexes. Such a voluntary reflex modification can be classed into two groups: *modulation* and *suppression*. Modulation implies a modification (increase or decrease) of the strength of the reflex action, while the suppression of a reflex causes its cancellation. Some works that support this can be found in [4], which provides evidences of the modulation of the post-auricular reflex that alerts us of a startle acoustic stimulus. There are also evidences that we can suppress completely some reflexes, like the "visual grasp reflex", that makes us to focus on new visual stimuli appeared in the peripheral visual field through a quick eye movement called *saccade*. It has been proved that humans can suppress this reflex; even override it by voluntary focusing on the opposite part of the visual field where the stimulus appears [6]. Other examples of voluntary reflex modification can be found in [5,7].

In any case, the reflex modification is strongly tight to the environmental context, the individual, and his/her goals. This paper, inspired on the human capability to modify the response to certain reflexes, is a first step towards the development of mechanisms to enable a mobile robot to modulate reflexes in order to better achieve a particular goal. Our approach relies on reinforcement learning techniques [8] to permit a robot to gain knowledge about when and how reflexes should be modulated given a particular configuration, i.e. robot state, action being performed, current goal, etc. In particular, results from our simulations demonstrate the improvement in the performance of mobile robot navigation when it is allowed to modulate collision reflexes, in comparison to its performance considering unmodulated reflexes.

The structure of the paper is as follows. Section 2 gives a brief overview of the human nervous system on which our approach is inspired. Section 3 describes the proposed mechanism for the automatic learning of reflex modulation. Section 4 details the learning process in simulations. The resultant policy is tested and compared with two hand-made policies in section 5 through simulated experiments. Finally some conclusions and future work are outlined.

2 Reflexes and Voluntary Actions in the Human Nervous System and in a Robotic Architecture

Our work is inspired on the way in which reflexes and voluntary signals are treated by the human brain. Roughly speaking, the human nervous system is divided into the *Central Nervous System* (CNS), the *Peripheral Nervous System* (PNS), and the *Automatic Nervous System* (ANS). The CNS (composed of the brain and the spinal cord) sends voluntary impulses to every part of the body by means of the PNS. Drawing a comparison between the human nervous system

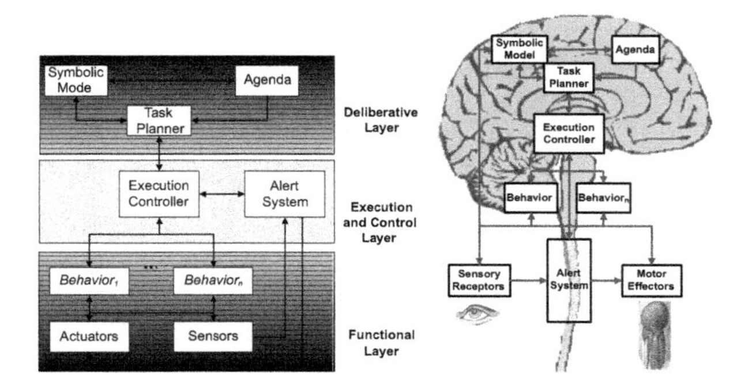

Fig. 1. A scheme of a hybrid robotic architecture with reflex support [2] and its analogy with the human nervous system. Left) The deliberative layer manages symbolic information to produce voluntary actions. The middle one acts as an interface to the functional layer, which manages the physicals of the robot (motors, sensors, etc.) and also implements wired behaviors. Right) Components of the deliberative layer would correspond to the more evolutioned area of the brain, while the executive and functional layers cover the inner parts of the brain (cerebellum), the spinal cord, and the sensorimotor organs.

and a typical hybrid robotic architecture (see fig. 1), the CNS would correspond to the deliberative layer, while the PNS corresponds to the intermediate and functional layers [9]. Thus, in the same manner that a voluntary action to move a leg is transmitted from the CNS to the proper neuron motor, in the mobile robot comparison, a movement command is transmitted from the deliberative layer of the architecture to the correspondent navigational algorithm. The flow of information is bidirectional in both cases: information gathered by a sensory neuron (robot sensor) is communicated to the CNS (deliberative layer).

Our nervous system has evolved to properly adjust motor responses when certain stimuli are perceived. Thus, touching a hot stove results in a rapid, automatic, and forceful withdrawal of the arm, that is, a reflex act. In most cases, the signal that causes the reflex is managed at the spinal cord, though it is also communicated to the brain. For example, when we step on a tack (see fig. 2), the sensorial neurons of our foot detect such a stimulus and transmit a signal to the spinal cord. There, a group of neurons reacts activating the correspondent motor neurons that cause a quick movement of the leg to pull our foot away.

From the robotic perspective, reflexes have also been considered to avoid critical situations, like collisions. For example, proximity sensors can be employed to detect obstacles and force a (reactive) navigational algorithm to head a free-obstacle direction. In these cases, reflexes are managed at a low level of the robotic architecture and thus, only consider local information to take a decision. Nevertheless, as commented, reflexes are also communicated to the human brain through a neural connection ((4) in fig. 2) which enable the brain to inhibit or modulate the neuron that produces the motor responses.

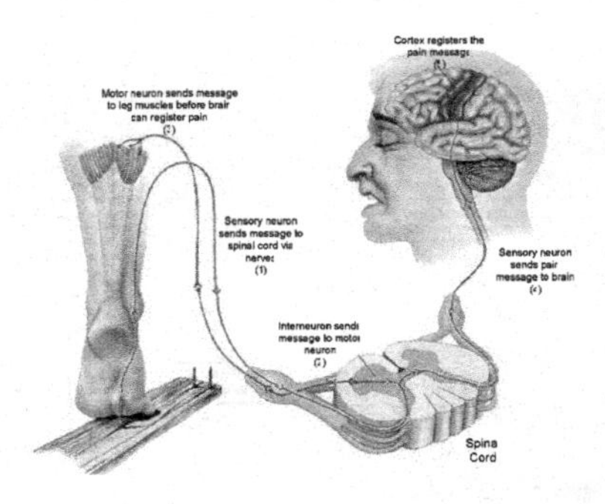

Fig. 2. Example of a reflex act. Sequence (numbered in brackets) of neurons activation during a reflex. Elements involved in (1)-(3) make up the corresponding reflex arc. Notice that though the spinal cord sends the motor response to the motor neuron, there is also a connection to the brain. Such a connection communicates the pain signal to the cognitive centers and can be utilized to inhibit the reflex response. (Image taken from [10]).

In general, the purpose of modulating reflexes is to achieve a long-term objective that could contradict a reflex response. For instance, in spite of the reflex action triggered when we touch a hot surface, we can voluntarily inhibit it if we are aware that the reflex of taking away our hands will knock our food over. Our aim, thus, is to provide a mobile robot with a similar capability to modulate/suppress reflexes based on its current context and experience. In our approach, this ability is provided by a reinforcement learning algorithm [8], which permits the *Execution Controller* of a typical hybrid architecture to learn from the robot experience when and how to modulate reflexes.

3 Automatic Learning of Reflex Modulation/Suppression

This section details the proposed automatic learning process for modulating robot reflexes. First, a brief introduction to Reinforcement Learning (RL) through the Q-learning approach is given. Then, we particularize this for learning when and how to modify reflexes for mobile robots.

3.1 Reinforcement Learning

Our mechanism for learning reflex modulation is based on Reinforcement Learning (RL) [8]. In short, RL is a machine learning paradigm based on the experience of an agent who knows that being in a certain state s and executing an action a reaches a new state, let's say s', and gets a reinforcement signal or reward r. These tuples, (s,a,s',r), obtained by the agent experience, are used for finding a policy (a sequence of actions) that maximizes some optimality function based on the rewards. RL assumes that the environment where the agent performs is a non-deterministic one. This means that taking the same action in the same state on different occasions may yield different next states and/or rewards.

In our work we have chosen the well-known Q-learning approach [8] to implement RL since it provides policies that tend to the optimal ones in spite of not having a model of the state transition function. For learning such near-optimal policies, the Q-learning process recursively computes a matrix Q of (number of states) x (number of actions) as follows:

$$Q(s,a) = Q(s,a) + \alpha(r + \gamma \max_{a'} Q(s',a') - Q(s,a)) \quad (1)$$

where α is the learning rate constant, r is the reward obtained when executing action a at the state s, γ is the discount factor (which represents the importance of future rewards), and $Q(s',a')$ refers to the Q-values of the next state s' for any action a'. Matrix Q represents the best policy under the agent experience, and therefore, the best known algorithm to follow when the agent is confronted again with the same world state: for a given state, the action for which Q is maximum becomes the best decision.

3.2 Reinforcement Learning for Reflex Modulation/Suppression

Now we delve into the application of Q-learning for modulating robot reflexes, illustrating it with the problem of learning a good policy for modulating the instinctive reflex of reducing the robot velocity when an unexpected obstacle appears. The resultant policy Q will determine the robot action given a particular robot state, which will include, among other variables, the strength of the current reflex signal and the distance to the obstacle (as commented further on). A similar study can be developed for other reflexes like the one of changing the robot heading when facing an obstacle or the stop reflex because of an imminent battery failure.

In our approach, we consider that the collision reflex will only have influence on the robot velocity: its velocity will be inversely proportional to the strength of the resultant reflex after its modulation. Thus, the robot could take advantage of using a proper reflex modulation policy to arrive at the destination quicker than through a non reflex-modulated navigation.

The learning process (see section 4 has been carried out in simulations since Q-learning usually needs long experiments to converge to a good policy, which would be impractical for a real robot platform. Moreover, for simplicity, we will only deal with the problem of finding a good policy for modulating/suppressing that single reflex, though the proposed scheme can be adapted to consider more complicate situations in which more than one reflex act are considered. In that case, all the possible modulations on the reflexes should be included in the same Q-learning implementation, since one can affect to another.

We define our Q-learning problem as follows:

- *States.* States are combinations of external stimuli and/or internal information that summarize the particular conditions of the robot workspace at any time. In our approach, states are 4-tuples $(d, rs, rm, actime)$, where d is the distance to the goal, rs represents the current reflex strength (inversely proportional to the distance to the closest obstacle), rm the current modulation applied to the reflex, and *actime* indicates the time interval during which the robot has been continuously modulating (or suppressing) the reflex. This last factor is considered for avoiding the continuous effort of modulation (accelerations that could damage or wear down the robot mechanisms).

- *Actions.* The set of possible actions considered for managing reflexes, which for our purposes have been set as: suppress the reflex, modulate it to 50% of its strength, modulate it to 25% of its strength, or leaving it unaltered. Remember that this modulation action along with the reflex strength will produce the robot velocity, so our actions modify the robot performance during navigation.
- *State Transitions.* The transition function produces the next state of the world given the current one and the action carried out. For simulating our setup, we can define the state transition functions (one per each component of states) as follows:
  - Transition of distance (d). It depends on the robot velocity during the execution of each learning step: the robot will navigate at constant speed defined by its modulation of reflexes.
  - Transition of reflex strength (rs). It is calculated based on the distance to the closest obstacle with respect to the forward direction of the robot. For this aim, we have considered a proximity sensor that yields the distance to obstacles located in the trajectory of the robot.
  - Transition of reflex modulation (rm). This transition is straightforward since the rm component of the new state is directly the selected Q action.
  - Transition of the accumulated modulation time ($actime$). The accumulated modulation time is set to zero when the selected action is "leave the reflex unaltered", otherwise, it is increased at every step of the learning algorithm.
- Reward. The reward function guides the learning process by representing the goodness of the robot actions given a particular state. We consider the following factors: 1) time consumed to arrive at the destination (that is, to achieve the voluntary goal), 2) accumulated time during which reflexes have being modulated, and 3) number of collisions with obstacles due to the modulation of reflexes. The reward function will favour those states in which factors 1), 2) and 3) are minimum.

In the next section, our learning process of a good policy is detailed in depth. It will be experimentally compared to intuitive hand-made policies (in section 5), demonstrating the suitability of the proposed mechanism.

4 The Learning Process

The learning process of the Q matrix has been carried out in simulations considering an open squared scenario of 10x10m (see fig. 3) where the robot is successively commanded to navigate between random locations following a straight

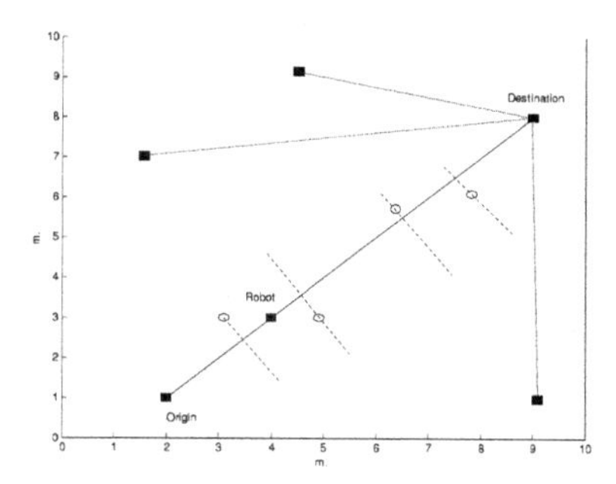

Fig. 3. Scenario for the learning process. Within this setup, the robot was commanded to go to a destination (at coordinates (9,8) from 100 random locations (some are marked with a filled rectangle). Each navigation involves 1000 learning steps. Random dynamic obstacles are programmed to cross the robot trajectory at different speeds.

line (the average length of the trajectories is around 10 m.). Dynamic obstacles cross the robot trajectory at random, triggering collision reflexes.

Within this scenario the robot has to adapt its maximum speed (30 cm/s.) according to the sensed reflexes (decreasing it if a reflex is triggered and not completely modulated yet). The objective is to arrive at the destination as soon as possible, and without collisions, even in presence of obstacles that force the robot to slow or stop. In the equation (1), the *learning rate*, α, is set initially to 1 and decreased over time[1] , and the *discount learning factor* (γ) to 0.9.

Since Q-learning can deal only with discrete state and action spaces, the parameters that represent the robot state $(d, rs, rm, actime)$ are discretized in the set {1,2,3,4}. Modulation actions are also discretized to 1 (no modulation), 2 (modulate at 25%), 3 (modulate at 50 %), and 4 (suppress the reflex). Discretized state values, denoted with the superscript d, and their corresponding partitions, are shown in fig. 4.

	1	2	3	4
Distance to the goal (d^d)	very close (less than 1 m.)	close (less than 3 m.)	near ([3-6] m.)	far (more than 6 m.)
Reflex strength (rs^d)	non-reflex (obstacle farther than 2m.)	weak (obstacle nearer than 2 m.)	normal (obstacle nearer than 1 m.)	extreme (obstacle nearer than 20 cm.)
Reflex modulation (rm^d)	Suppressed	25% modulated	50% modulated	No-Modulated
Accumulated modulation time ($actime^d$)	none	short (less than 30 sec.)	normal (less than 1 min.)	maximum (more than 1.5 minutes

Fig. 4. State parameters discretization. We consider three levels for reflexes: weak, normal and extreme.

[1] We have chosen $\alpha(i) = 1/i^c$, where i is the current learning step of the Q-learning algorithm and c is a constant that we have calculated for $\alpha = 0.05$ in the last iteration. This function is demonstrated to produce a good convergence rate [11].

The reward function is computed from these discrete values and normalized within [0,1] as:

$$R = \begin{cases} \eta_1 * R_1 + \eta_2 * R_2 & \text{if the robot is not stopped to avoid a collision} \\ 0 & \text{otherwise} \end{cases} \quad (2)$$

where η_1 and η_2 are weighting constants such that $\eta_1 + \eta_2$=1, R_1 rewards the velocity of the robot, and R_2 rewards not to modulate when we are near the goal and not to accumulate modulation actions for long times. More precisely, R_1 is computed as:

$$R_1 = 1 - \frac{rs^d}{4} * \frac{rm^d}{4} \quad (3)$$

and R_2 is computed as:

$$R_2 = \begin{cases} 1 - \mathrm{rm}^d/4 & \text{if } \mathrm{d}^d = (close \text{ or } very\ close)\ \&\ \mathrm{rs}^d = reflex \\ 0 & \text{if } \mathrm{d}^d = (close \text{ or } very\ close)\ \&\ \mathrm{rs}^d = non - reflex \\ 1 - \mathrm{actime}^d/4 & \text{if } \mathrm{d}^d = (near \text{ or } far) \end{cases} \quad (4)$$

We have set the parameters to $\eta_1 = 0.85$ and $\eta_2 = 0.15$, to weigh up policies in which the robot obtains a certain advantage from reflex modulation in terms of the time (or equivalently, speed) to reach the goal.

5 Simulated Experiments

In this section we show some simulated experiments in which the policy learnt in section 4 has been used for modulating the collision reflex of a mobile robot. We have simulated a mobile robot navigating within the scenario depicted in fig. 5 by tracking five trajectories previously defined (similar to the one shown in the figure). In these experiments, dynamic obstacles may intercept the robot trajectory randomly with a probability similar to the one employed when learning the matrix Q, which is assumed to be characteristic of the environment. The robot response to obstacles consists of regulating its speed according to the learnt policy. In our experiments non modulating extreme reflexes is punished with an emergency stop, that simulates a safety mechanism to avoid physical damages, i.e. a bumper that makes the robot to stop. Next, we compare the results of the learnt policy with two hand-made policies:

- An intuitive one that decides to suppress reflexes when the robot is close to the destination, to modulate them at 50% when it is near, and not to modulate otherwise.
- A no-modulation: a policy consisting of never modulate reflexes.

The results of 100 navigational experiments are shown in fig. 6. Note that the no-modulation policy has a high reward (though less than Q) since it never inhibits reflexes, and thus non emergency stops are needed to avoid collisions, but at the expense of navigating slowly, as shown in b). Also note how the best results correspond to the Q-learning algorithm in all cases.

In fig. 6(a) the reward of the Q-algorithm is the highest one, while the reward of the intuitive policy is higher than the one that never modulates. Fig. 6(b) shows

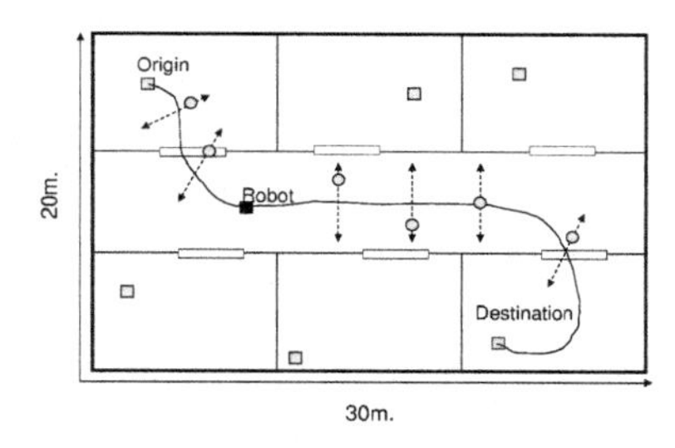

Fig. 5. Simulated environment for testing the reflex modulation based on a previously learnt policy. In this scenario, the robot is instructed to navigate between given locations (one in each room) by tracking trajectories previously defined, while considering dynamic obstacles that may appear at random (circles in the figure).

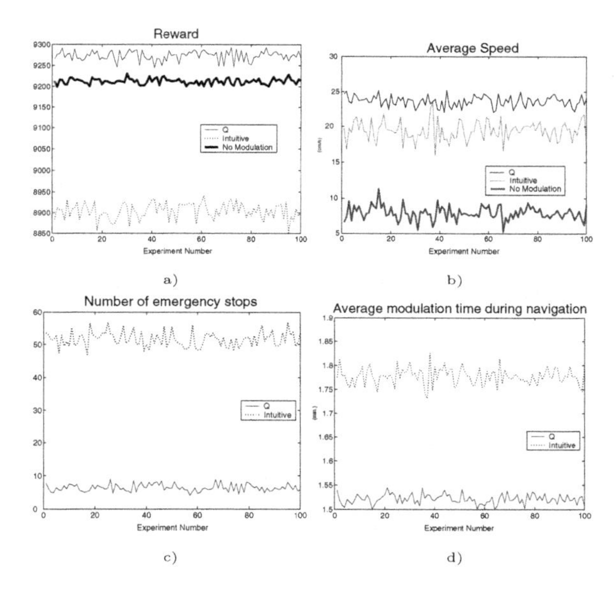

Fig. 6. Comparison of the Q-learnt policy and 2 intuitive hand-coded policies in 100 experiments. a)Obtained reward for the 3 policies. b)Average robot speed. Note that the learnt policy permits the robot to navigate quickly with a reduced number of emergency stops (c). d)Modulation effort measured as the average time of modulation. No-modulation is not shown in c-d since these charts are not applicable to this policy.

that the Q policy enables the robot to achieve the goal quicker than the others. This is due to its capability to smartly modulate reflexes and therefore to enable the robot to navigate closely to the maximum speed (30 cm/s) as frequently as possible, but not at the expense of producing more emergency stops, as shown in fig. 6(c). The modulation effort measured as the time in which a modulation action has been considered is depicted in fig. 6(d). Note that both Q and Intuitive policies have a similar modulation effort to achieve the goal (obviously the no-modulation policy does not appear in the figure since it never modulates).

6 Conclusions and Future Work

In this paper we have taken the first steps to explore the utility of modulation/suppression of reflexes in robotics. Though most of the works in literature rely on wired reflexes to protect robots from dangerous situations, in some cases it should be convenient that deliberative processes could modulate or even suppress them in order to achieve a desirable goal. We have implemented an automatic learning process, based on Reinforcement Learning, that guesses from experience the best way to modulate a reflex. In our approach the learning process is carried out in simulations and the resultant policy is applied in a simulated robotic scenario. In the future we will research on the development of an on-line learning process and its application to real robots.

References

1. Brooks R.A.: A Robust Layered Control System for a Mobile Robot. IEEE J. on Robotics and Automation, **2** 1986 14–23.
2. Galindo C., Gonzalez J., Fernandez-Madrigal J.A. A Control Architecture for Human-Robot Integration: Application to a Robotic Wheelchair. IEEE Trans. on Systems, Man, and Cyb.–Part B **36** 2006 1053–1068.
3. Prochazka A., Clarac F., Loeb G., Rothwell J., Wolpaw J.: What do reflex and voluntary mean?. Exp Brain Res, **130** 2000 417–432.
4. Benning, S. D., Patrick, C. J., and Lang, A. R.: Emotional modulation of the post-auricular reflex. Psychophysiology **41** 2004 426–432.
5. Cullen K., Chen-Huang C. and McCrea R.: Firing Behavior of Brainstem Neurons during Voluntary Cancellation of the Horizontal V-O Reflex. J. Neurophysiology **70** 1993 844–856.
6. Everling S, Dorris MC, and Munoz DP.: Reflex suppression in the anti-saccade task is dependent on prestimulus neural processes. J. of Neurophysiology **80** 1998.
7. Jefferson E. Roy and Kathleen E. Cullen: Vestibuloocular Reflex Signal Modulation during Voluntary and Passive Head Movements. J. of Neurophysiology **87** 2002.
8. Kaelbling L.P., Littman M.L., Moore A.W. Reinforcement Learning: A Survey. Journal of Artificial Intelligence Research **4** 1996 237–277.
9. Murphy R.B. Introduction to AI Robotics. The MIT Press, Cambridge, 2000.
10. http://www.sruweb.com/~walsh/spinal_reflex.jpg, courtesy Prof. A.A. Walsh.
11. Even-Dar E. and Mansour Y.: Learning Rates for Q-learning, Journal of Machine Learning Research, **5** 2003 1–25.

Evolving Robot Behaviour at Micro (Molecular) and Macro (Molar) Action Level

Michela Ponticorvo[1] and Orazio Miglino[1,2]

[1] Department of Relational Sciences "G.Iacono", University of Naples "Federico II", Naples, Italy
[2] Institute of Cognitive Sciences and Technologies, National Research Council, Rome, Italy

Abstract. We investigate how it is possible to shape robot behaviour adopting a molecular or molar point of view. These two ways to approach the issue are inspired by Learning Psychology, whose famous representatives suggest different ways of intervening on animal behaviour.

Starting from this inspiration, we apply these two solutions to Evolutionary Robotics' models. Two populations of simulated robots, controlled by Artificial Neural Networks are evolved using Genetic Algorithms to wander in a rectangular enclosure. The first population is selected by measuring the wandering behaviour at micro-actions level, the second one is evaluated by considering the macro-actions level. Some robots are evolved with a molecular fitness function, while some others with a molar fitness function. At the end of the evolutionary process, we evaluate both populations of robots on behavioral, evolutionary and latent-learning parameters.

Choosing what kind of behaviour measurement must be employed in an evolutionary run depends on several factors, but we underline that a choice that is based on self-organization, emergence and autonomous behaviour principles, the basis Evolutionary Robotics lies on, is perfectly in line with a molar fitness function.

1 Introduction

Designing mobile robot's behaviour is far from trivial: designing them by hand from scratch is a very difficult task for humans, while designing them automatically doesn't assure to scale up to complicated tasks. A lot of energy in this process is devoted to behaviour description. We usually describe the behaviour we wish to model in qualitative terms that are then translated in the equation required for automatic evaluation. But, in doing this, we can assume various points of view, different analysis levels. For example, if we want a robot to play soccer, to follow the shortcut up to a target or just to wander in an environment without sticking into obstacles, we can concentrate on constraints on a micro-level (the sensors or motors state) or on global behaviours (score goals or explore an arena). This latter frame of description is the one assumed, for instance, by Behavior Based Robotics [1], an approach in which the desired behaviour is divided into a set of simpler behaviours. With this approach, however, the designer

J. Mira and J.R. Álvarez (Eds.): IWINAC 2007, Part II, LNCS 4528, pp. 357–366, 2007.

must decide which are these simpler behaviours. Also in Robot Shaping [3], a technique that deals with designing and building learning autonomous robots in which the human designer proposes how a task should be carried out and how it should be decomposed into sub-tasks that are then implemented at the robot level and reinforced, the focus is on behaviour level.

The decision about behaviour description to be kept for robot behaviour is not different from what must be defined in the supervision of learning in natural organisms. In fact we talk about *shaping*, a term coming from experimental psychology [19] where it describes a particular technique to train animals. In other words, Robot Shaping, Behavior Based Robotics and, as we will see, also other techniques for robot behaviour design, must face the same problem that encountered Behaviorist Psychology and encounters Learning Psychology, that is to choose the right analysis level that allows to evaluate the efficiency of training procedures on animals and human beings. What does an experimenter (but also a teacher or a breeder) must concentrate on to improve learning? This is a question many psychologists have tried to answer, proposing different interpretations with their experimental or clinical work. Hill[7]in his work on Learning Psychology has distinguished these potential solutions into two families: molecular vs. molar. What does this means?

Molecular and molar are two words derived from chemistry. The first one refers to molecules, "the smallest unit into which a substance can be divided without changing its chemical nature" (Oxford Wordpower Dictionary), while the second one refers to mole, the number of grams of a certain molecule that is equal to the number that indicates its molecular weight. More generically we can consider a mole a set of various molecules of the same kind. In Learning Psychology these terms are adopted to indicate theories that consider the smallest components of a behaviour such as a muscle that moves (molecular) or the behaviour as a whole(molar), for example going out of a maze.

Some psychologists, the molecular approach supporters, believed that the way to follow was to concentrate on micro-behaviours that formed animal's performance (Pavlov [16] for aspects related to micro-actions stabilization and Guthrie [4,5] for the importance to focus on micro-actions to understand behaviour). On the contrary the molar approach (cfr.Tolman[17]) preferred to reinforce macro-behaviours that lead to a satisfying final outcome, for example to localize a target area in a complex labyrinth.

These two approaches are indeed complementary: we can indeed imagine "molar" and "molecular" not as dichotomous entities, but as the extremes of a continuum for behaviour definition that may be applied to Robotics and robots'design too, as it is represented in figure 1 for a wandering behaviour. This distinction is echoed in Evolutionary Robotics, the technique we have used in the experiments described in this paper.

Evolutionary Robotics [15,6] is a discipline belonging to Artificial Life [8] whose goal is to obtain artificial agents, both physical or simulated. This methodology is inspired by darwinian selection, according tho which only fittest organisms can reproduce. To run an artificial evolutionary process according to

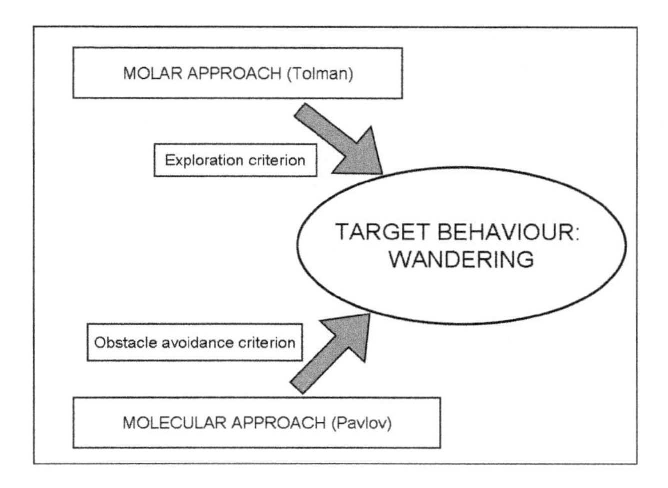

Fig. 1. Two ways to produce a wandering behaviour: we can described it at the lower level of a set of micro-actions (measures of sensors and motors activation or internal states) in the molecular approach or at the level of macro-actions in the molar approach

Evolutionary Robotics' dictates it is necessary to define a criterion that addresses the entire evolution. A typical experiment in Evolutionary Robotics can be described like this: an initial population of robots whose features are defined in an artificial genotype, is tested and evaluated according to a criterion, usually referred to as the fitness function. For example we may define in the genotype the connections' weights of an Artificial Neural Network, the robot controller that determines the behaviour. The robots that obtain the best scores are allowed to reproduce: their genotypes, opportunely mutated or crossed, will constitute the genotypes of the second generation and will thus determine the second generation phenotypes and behaviours. The testing-evaluating-reproducing loop is iterated for a certain number of generations or until at least some robots display the desired behaviour or solve the predefined task. It is clear from this brief description the fundamental role played by the fitness function in the evolutionary process, as this function is used to evaluate the performance of robots and to select the ones that will reproduce. In fact robots are evaluated about their ability according to the criterion defined by the experimenter, measured by the fitness function. The probability that a robot reproduces depends on the score obtained in respect to this function. Consequently fitness formula's design is fundamental in every Evolutionary Robotics'experiment.

This aspect has been underlined since the seminal work by Nolfi and Floreano [15] who proposed a framework for describing various fitness functions: the Fitness Space. This fitness space is a three-dimensional space with axes representing three continuous dimensions that are relevant for fitness functions. A fitness function can then be imagined as a point in this space.

The first dimension goes from "functional" to "behavioral". A functional fitness function focuses on functioning modes of the controller, while a behavioral one evaluates its behavioral outcome.

The second dimension "explicit vs.implicit" considers the amount of constraints that are taken into account in the fitness function. An explicit fitness function considers many precise components, while an implicit one has just few constraints or not at all.

The latter dimension has "external" and "internal" as extremes, indicating if the variables in the fitness function are accessible to the evolving robot computation. In an internal fitness function variables are calculated on sensors' activation of the robots, while in an external one on information available only for the experimenter.

Choosing a fitness function is not a trivial decision that strictly depends on the purpose of the evolutionary process and will address the whole bulk of results.

Also in this case, it is necessary to find the right mix between molar and molecular. For example, when Nolfi and Floreano [15] want to obtain an obstacle avoidance behaviour, they use a molecular fitness function that rewards particular micro-actions(we will describe it in detail in the next section). If, on the contrary, the task is to localize an area inside an arena, it's not compulsory to consider micro-actions and it can be more useful to use a molar function [9]. With more complex tasks, integration of the two approaches may be a right choice [12]. Another example of differently conceived fitness functions can be found in co-evolution experiments. To study prey-predator dynamics Nolfi and Floreano [14] use a molar fitness function that assigns simply 1 point for the predator and 0 for the prey if the predator is able to catch the prey and 0 for the predator and 1 for the prey if it escapes the predator. On the contrary Cliff and Miller [2] used in their experiments on co-evolution a more complex fitness function that includes more constraints that address evolution, for example predators are also scored for their ability to approach the prey.

So it seems to us very fruitful to analyze the possible outcome of differently conceived fitness functions as it might shed light on how to guide learning processes. In this paper this is what try to do, that is to compare two differently characterized fitness functions (molecular vs. molar) on the same behaviour to verify what happens at behavioral, learning or evolutive level.

2 Wandering in a Closed Arena Without Sticking into Obstacles: Two Possible Fitness Functions

The task we have tested the fitness functions on is a simple wandering task. We just want the robot to move in a rectangular enclosure without bumping into walls and into cylindric obstacle that are inside the arena.

In order to compare two differently conceived fitness functions we analyze the molecular fitness function used by Nolfi and Floreano [15] to obtain an obstacle avoidance behaviour and a molar one proposed by Walker and Miglino [20]. The molecular fitness function is composed by three distinct components that reward three variables: the average of robot wheels' speed, the wheels'differential and the activation value of the most active infrared sensor. These three components encourage motion, straight movement and obstacle avoidance.

For the molar fitness function we divide the rectangular arena into 50 cells (10*5) and we give the robot a reward that corresponds to the number of cells visited for the first time.

3 Method

In our experiments, we run two different evolutionary process to obtain a wandering behaviour. In the first one the population of robots is selected with the molecular fitness function by Nolfi and Floreano [15] described in the previous section. In the second one the robots are selected using the molar fitness function. We use the EvoRobot simulator [13] to run the evolutionary process on the software robots.

3.1 Robots

Each robot consists of a physically accurate simulation of a round robot, with a diameter of 5.5 cm. Each robot is equipped with 8 infrared proximity sensors (capable of detecting objects within 3 cm of the sensor). The robots move using 2 wheels (one on each side of the robot) powered by separate, independently controlled motors. The control system is an Artificial Neural Network: a perceptron whose input layer is formed by 8 input units that codify the activation of Infrared Sensors that receive stimulation from obstacles up to 5cm. In the output layer there are 2 output neurons, totally connected to all input units, that control wheels.

3.2 Training Environment

In our experiments, we use this training environment: a rectangular arena (500*1000 mm.) in which there are 5 obstacle in randomly chosen position. These obstacles are cylinders with 27.5 mm. radius.

3.3 Training Procedure

The robots are trained using a Genetic Algorithm [10].

At the beginning of each experiment, we create 100 simulated robots with random connection weights. We then test each robot's ability to wander inside the arena. The robot is positioned in a random location and allowed to move around for 100 computation cycles (1ms per cycle). The robots are rewarded according to the two fitness functions described before: a molar and a molecular one. At the end of this procedure, the 80 robots with the lowest scores are eliminated (truncation selection). The remaining 20 robots are then cloned (asexual reproduction). Each "parent" produces five "offspring". During cloning, 25 per cent of neural connections are incremented by random values uniformly distributed in the interval [-1, +1]. The testing/selection/cloning cycle is iterated for 100 "generations". For each population the simulation is repeated twenty times with the same parameters and randomly generated initial connecting patterns.

4 Results

4.1 Evolutionary Patterns

In figure 2 and 3 the evolutionary trends for the two populations are represented.

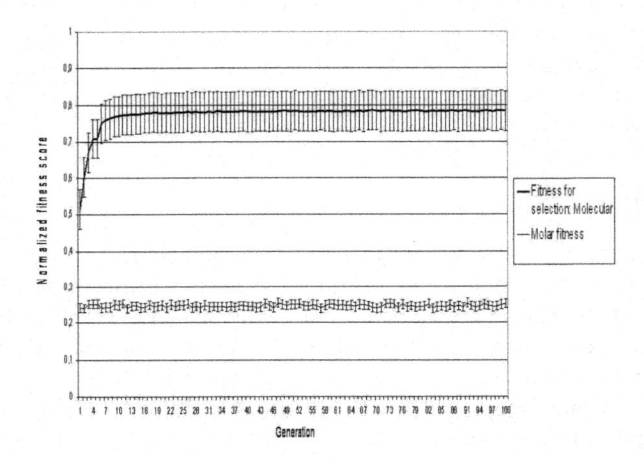

Fig. 2. Evolutionary trend for robots selected by the molecular fitness function. On x axis there are generations, on y axis fitness scores. The thin line indicates the score gained on the molar fitness function, that is not considered for selection.

The scores regard the average of the best robots of each seed along generations. As it can be seen standard deviation is lower for the molar fitness function. In the figures above the scores on the fitness function not used for evolution are represented, as these data are relevant for another analysis we will describe in the following section.

4.2 Training Speed

We take then into account an evolutionary parameter that is the time lapse from initial generation up to the peak of fitness. This is a measure of evolutionary process speed, a parameter that is crucial for Evolutionary Robotics, as robot evaluation may require a long time. It is also an indirect measure of how "easy" a task is for the evolving robot. If less time is required to reach the maximum in terms of fitness scores, this means that the task is easier to accomplish for the system formed by the robot and the environment that constraints its action.

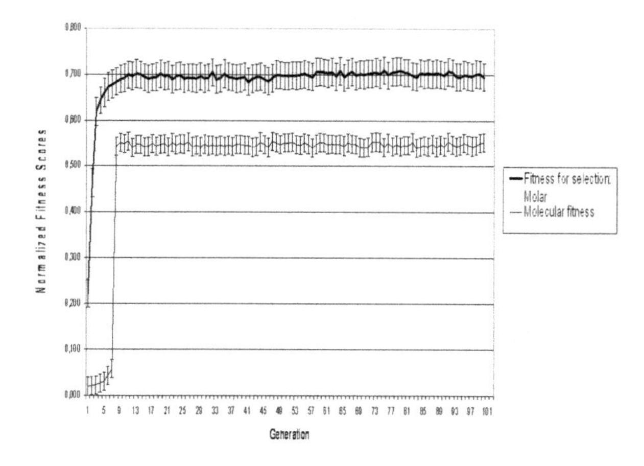

Fig. 3. Evolutionary trend for robots selected by the molar fitness function. On x axis there are generations, on y axis fitness scores.The thin line indicates the score gained on the molecular fitness function, that is not considered for selection.

We thus compare the emergence time of fitness peaks in the two populations of robots. We observe that for the robots evolved with the molar fitness function the best solution appears after less generations than for the molecular fitness function. This difference is statistically significant.

$$t(38) = 6,41; p = 0,00.$$

4.3 Fitness Functions Comparison

We compare the two fitness function on what we call latent learning. This is a concept close to the one of generalization, but it is more focused on generalization lying inside a certain task. This means that we wonder if, during the evolutionary process, robots learn something else about the task that is not explicitly rewarded. In this case we would like to know if, while evolved for wandering with a molar function they maximize also the constraints of the molecular one and vice-versa. From figures 2 and 3 we can infer that robots evolved with the molar fitness function gain good scores in the molecular one, but robots evolved with the molecular fitness function score bad on the molar one. So we run another test on this issue: what happens if we evaluate the best 20 robots evolved with the molar fitness function using the molecular one? The fitness scores, even if less high in average than the corresponding values obtained with molecular fitness function, are not significantly different.

$$t(38) = 0,95; p = 0,34.$$

What happens if, on the contrary, we go the other way around, testing the best 20 robots evolved with molecular fitness function using the molar one? The results obtained by these robots are significantly lower than the ones obtained with the molar function.

$$t(38) = 9,92; p = 0,00.$$

This means that the molar fitness function, even if not explicitly designed to maximize the three components that form the molecular one, "contains" nevertheless these variables implicitly and addresses the evolutionary process in order to maximize them. On the contrary, the molecular fitness function does not allow the emergence of the exploratory behaviour rewarded by the molar function. This means that the molar fitness function "includes" the molecular one in terms of latent learning but not vice-versa.

4.4 Behavioral Analysis

What happens at the behavioral level? Figures 4 and 5 show the trajectories by two very efficient robots from the populations evolved with molecular and molar fitness functions.

As it can be seen from the figures above, the behaviours displayed by robots evolved with different fitness functions are quite different. The robots evolved with the molecular fitness function proceed straight until they reach a wall or an obstacle, at this point they turn left or right, avoid the obstacle and continue their run.

On the other side, robots that have been evolved with the molar fitness function proceed fast and avoid obstacle but, in many cases they do not go straight. In fact between these robots we can find, together with some agents that displace straightly, many robots that follow curved trajectories, drawing curves

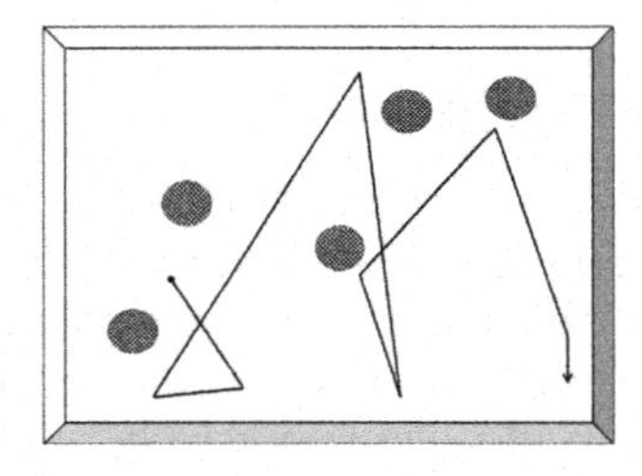

Fig. 4. Behaviour displayed by an individual evolved with the molecular fitness function

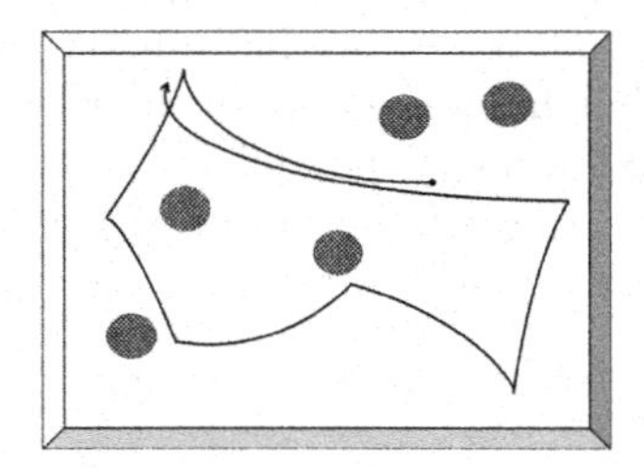

Fig. 5. Behaviour displayed by an individual evolved with the molar fitness function

with different radius and avoiding obstacles when encountered. The behaviours that emerge with the molar fitness function are much more varied.

5 Conclusions

The data exposed above suggest that a wandering behaviour can emerge adopting both a molar and a molecular point of view. Why should we prefer one or the other? As happens in natural organisms' learning supervision, it depends on what we want to achieve. The molecular function can be the right choice in some cases and the molar function in some others.

The presented results lead us to sustain, for the simple task we analyzed, the molar one. Why? It allows the system to build its own solution freely. Regardless of what an experimenter thinks a solution should be like in details, a global solution emerge by itself. This is possible because the system formed by the robot and the environment can self-organize, thus exploring ways to solution that may be not considered *a priori*. What does this emergent solution look like? First of all it includes a wider set of strategies.

To explore an environment while avoiding obstacles we could believe that the best is to go straight, but following curved trajectories may be a good way to solve the same task. This kind of solution, that does not emerge in the case of the molecular fitness function, is equally efficient and enriches the whole of solutions between which evolution can look for. This is surely an advantage in an evolutionary perspective. With this kind of function, robots are favored to establish a useful relation with the environment exploiting a very precise coordination between input and output, thus adapting to external constraints.

Good hints can be found at the evolutionary level too: the evolutionary process appears to be faster with the molar function as the fitness peak is reached in fewer generations. Moreover the standard deviation in fitness scores is low, thus indicating that a great amount of robots is able to obtain high fitness values. Another positive aspects is about latent learning. The molar function permits to improve also the scores on the molecular constraints, even if not explicitly considered. In other words, it shows a good latent learning. The last, but not least, reason has been already partly discussed and is about the preference for a self-organizing solution. Letting the system the possibility to self-organize, unexpected and more efficient solutions can emerge, also suggesting new ways of approaching a certain problem, an important issue in scaling to more complex behaviours.

In fact, in building an intelligent robot a mechanical creature which can function autonomously" (pag.3), that is one of to purpose of Artificial Intelligence, the science of making machines act intelligently(pag.15) [11], we must keep in mind that autonomy is fundamental, also in the sense that the robot must operate without human-intervention, supervision or instruction and adapt to changes in the environment and itself. If this is the goal, also the means to reach it should be informed to the maximum possible autonomy, what we can do choosing a molar fitness function in Evolutionary Robotics rather than a molar one.

In closing we would like to underline that, even if these results do not apply directly to natural organisms' learning supervision, they nonetheless supply interesting insights on this controversial issue.

References

1. Brooks, R.A.: Intelligence without representation. Artificial Intelligence (47), 139-159 (1991)
2. Cliff, D., Miller, G.F.: Tracking the read queen: Measurement of adaptive progress in coevolutionary simulations. In F. Moran, A. Moreno, J.J. Merelo, P. Chacon (eds.): Advances in Artificial Life: Proceedings of the Third European Conference on Artificial Life, Berlin:Springer Verlag (1995)
3. Dorigo, M., Colombetti, M.: Robot Shaping. An Experiment in Behavior Engineering. Intelligent Robotics and Autonomous Agents series, vol. 2. MIT Press (1998)
4. Guthrie, E.R.: The psychology of learning. Gloucester, MA: Simith (1960)
5. Guthrie, E.R., Horton, G.P.: Cats in a puzzle box. New York: Rinehart (1946)
6. Harvey, I., Husbands, P., Cliff, D., Thompson A., Jakobi N. Evolutionary robotics: The Sussex approach. Robotics and Autonomous Systems, 20:205-224(1997)
7. Hill, W.F.: Learning: A survey of psychological interpretations. Paperback (1973)
8. Langton, C.G.: Artificial Life: An Overview.Cambridge, MA: The M. I. T. Press/A Bradford Book (1995)
9. Miglino, O., Lund, H.H.: Do rats need euclidean cognitive maps of the environmental shape? Cognitive Processing, 1-9(2001)
10. Mitchell, M.: An Introduction to Genetic Algorithms. Cambridge, MA: MIT Press(1996)
11. Murphy, R. R.: Introduction to AI robotics. Cambridge, MA: MIT Press/Bradford (2000)
12. Nolfi, S.: Evolving non-trivial behaviors on real robots: a garbage collecting robot. Robotics and Autonomous System, 22: 187-198(1997)
13. Nolfi, S. (2000). Evorobot 1.1 User Manual. Technical Report, Institute of Psychology, Rome
14. Nolfi, S., Floreano, D.: How co-evolution can enhance the adaptive power of artificial evolution: Implications for evolutionary robotics. In P. Husbands and J.-A. Meyer (Eds.), Proceedings of the First European Workshop on Evolutionary Robotics, Berlin: Springer Verlag, pp.22-38 (1998)
15. Nolfi, S.,Floreano, D.: Evolutionary Robotics: The Biology, Intelligence and Technology of Self-Organizing Machines. Cambridge, MA: MIT Press/Bradford (2000)
16. Pavlov, I.P.: Conditioned reflexes. Oxford University Press (1927)
17. Tolman, E.C.: Purposive behavior in animals and men. New York: Appleton-Century-Crofts (1932)
18. Sharkey, N.E., Heemskerk, J.: The neural mind and the robot, In A. J. Browne (Ed.), Neural Network Perspectives on Cognition and Adaptive Robotics, Bristol U.K. :IOP press (1997)
19. Skinner, B.F.: The behavior of organisms: An experimental analysis, New York: Appleton-Century-Crofts (1938)
20. Walker, R., Miglino, O.: Simulating exploratory behavior in evolving Artificial Neural Networks. Proceedings of Gecco1999, Morgan Kauffman (1999)

Discretization of ISO-Learning and ICO-Learning to Be Included into Reactive Neural Networks for a Robotics Simulator

José M. Cuadra Troncoso, José R. Álvarez Sánchez, and Félix de la Paz López

Departamento de Inteligencia Artificial, UNED, Spain
{jmcuadra,jras,delapaz}@dia.uned.es

Abstract. Isotropic Sequence Order learning (ISO-learning) and Input Correlation Only learning (ICO-learning) are unsupervised neural algorithms to learn temporal differences. The use of devices implementing this algorithms by simulation in reactive neural networks is proposed. We have applied several modifications to original rules: weights sign restriction, to adequate ISO-learning and ICO-learning devices outputs to the usually predefined kinds of connections (excitatory/inhibitory) used in neural networks, and decay term inclusion for weights stabilization. Original experiments with these algorithms are replicated as accurate as possible with a simulated robot and a discretization of the algorithms. Results are similar to those obtained in original experiments with analogue devices.

1 Introduction

Valentino Braitenberg explored in [2] the possibilities of modeling increasingly complex behaviors by means of connecting threshold devices, through connections with temporal sequences association learning. This kind of learning is related to psychological theory on classical conditioning, and it has been studied in AI from several points of view, a survey can be found in [8]. In order to simulate classical conditioning Porr and Wörgötter devised an unsupervised, differential Hebbian rule named Isotropic Sequence Order learning (ISO-learning) [9,8,6,7,5], later they presented a modification named Input Correlation Only learning (ICO-learning) [10]. In this paper we will describe briefly both methods and propose modifications to include them into reactive neural networks for a robotic simulator. These modifications are: discretization of learning rules, for a simpler software implementation, decay term inclusion, for weights stabilization, and weights sign restriction, to force output kind (excitatory/inhibitory). Paper finalizes describing experiments showing the effects of these modifications.

2 ISO-Learning and ICO-Learning

The unconditioned stimulus (US) appearance triggers organism unconditioned response (UR), this is a reflex. If some stimulus, let's call it conditioned stimulus

J. Mira and J.R. Álvarez (Eds.): IWINAC 2007, Part II, LNCS 4528, pp. 367–378, 2007.

(CS), starts to be perceived before than US perception, organism is able to learn that CS is followed by US and CS perception can elicit UR, preventing, possibly, subsequent US appearance. This is the classical conditioning theory, in order to imitate it Porr and Wörgötter [9] have designed a neural circuit, similar to those shown in figure 1, that operates in continuous time domain. US signal arrives to only one input, whereas CS signal does to several ones, N, in order to learn long temporal intervals. Let's call x_U and x_C to US and CS input signals. Each input signal is band-pass filtered producing a filtered input u_k, in experiments they use resonators, second order IIR peak filters showing resonance. Filtered inputs, from now on we will name them only inputs, are linearly combined to yield circuit output, v,

$$v = \rho_0 u_0 + \sum_{k=1}^{N} \rho_k u_k ,$$

subindex $k = 0$ states US and ρ_k are weights, in all robot experiments they keep ρ_0 constant (mostly $\rho_0 = 1$), like an innate reflex. Resonator corresponding to US has to reach its maximun response quickly, in order to trigger reflex response as soon as possible.

Weights variation, learning rule, is expressed as

$$\frac{d}{dt}\rho_k = \mu u_k \frac{d}{dt} v ,$$

being μ the learning rate, $\mu \ll 1$. Learning rule depends on correlation between inputs and output temporal derivative, thus it is a differential Hebbian one. Due to its Hebbian nature ISO-learning exhibits convergence problems [3]. Theoretically under closed-loop condition, when learning is achieved, and thus US does not appear $x_U = 0$, weights change stabilizes if $\mu \to 0$, but in practical situations μ is a positive constant and weights diverge, as experiments show. In 2006, five years after first paper on ISO-learning, the same authors propounded a solution for convergence problem: now learning rule, ICO-learning, depends only on correlation between inputs.

$$\frac{d}{dt}\rho_k = \mu u_k \frac{d}{dt} u_0 \qquad k > 0.$$

Thus as soon as x_U becomes 0 weights stabilize even with relatively high learning rates, making possible learning with few training sessions or with only one, one-shot learning.

The main part of the research for present work was done before ICO-learning introduction, but modifications proposed bellow are also applicable to this algorithm and tests using ICO-learning are included in experiments section.

3 Discrete ISO-Learning and ICO-Learning

3.1 Circuit Discretization

Although Porr and Wörgötter used simulators for some initial experiments but they made mostly analogue circuits for real robots experiments. Usually only

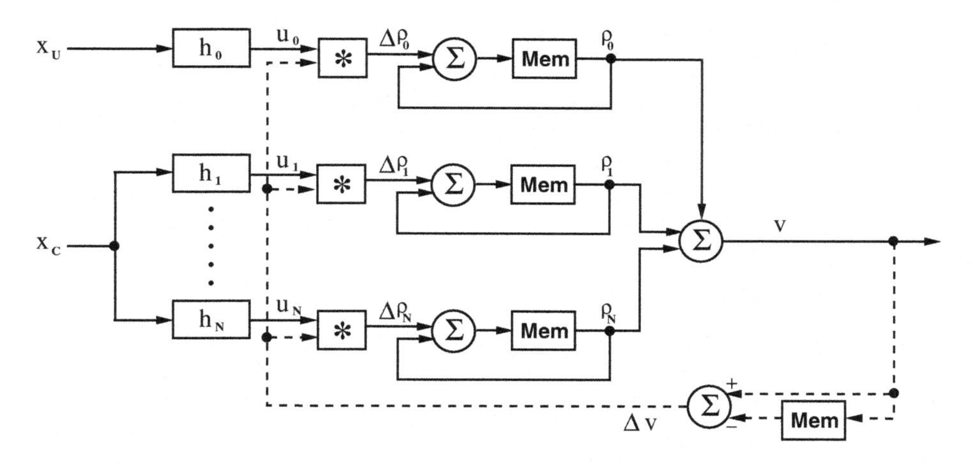

Fig. 1. ISO-learning neural circuit. Unfiltered signals, x_U and x_C, arrive to filters with transfer function h_k yielding filtered signals u_k. Weights are represented by ρ_k and v is the output. Symbol "$*$" means product and rectangles labeled with "Mem" are memory latches. Dashed lines show circuit feedback. Weight associated to US, ρ_0, is usually kept constant.

software is used to control robots. Robotic software operates mainly in discrete time domain, then a simulated discrete version of ISO-learning would be useful. Our circuit version is shown in figure 1. In order to formulate this discrete version we have to discretize resonators and output derivative. Derivative discretization is done straightforwardly replacing it by output first difference$\triangle v$, so our learning rule is

$$\triangle \rho_k = \mu u_k \triangle v \,.$$

Following Porr and Wörgötter [9,6] we are interested in filters time response in order to discretize resonators. The continuous impulse response is

$$h(t) = \frac{1}{b} \sin(bt) e^{-at} \,,$$

a is a damping coefficient and b is related to oscillation period. The corresponding discrete impulse response, see [1], is

$$h[n] = r^n \frac{\sin((n+1)\theta)}{\sin\theta} \,.$$

Making equal continuous time, t, and discrete time, nT, being T filters sampling period, we get $r = e^{-aT}$ and $\theta = bT$. We do not take constants in account because we will use normalized filters output to have an impulse response with first maximum value of 1, in order to get similar contributions from every resonator when weights are equal. Discrete filter response is then expressed, before normalization, as:

$$u[n] = x[n] + 2r\cos\theta \cdot u[n-1] + r^2 \cdot u[n-2] \,.$$

It is usual in filters literature to define filters by their frequency f and their quality Q, being the relations among them and a and b: $a = \frac{\pi f}{Q}$, $b = \sqrt{(2\pi f)^2 - a^2}$.

However, although impulse response, after normalization, is almost equal for both cases, responses to other arbitrary signals would be stronger in the discrete case. For example, step responses would be thousand times stronger for values we were dealing with. So, in order to get tractable outputs, we used in our experiments damping factors 50 times bigger to compensate discretization errors.

3.2 Weights Stabilization

In order to stabilize weights there are several possibilities [4]. The common case is to introduce in the learning rule a subtractive decay term being proportional to weight, but if US does not appear for a long time weights decay almost to 0, and thus learning could be forgotten. In this case if a robot run across a really wide area free of obstacles, it could forget a previously learned obstacles avoidance behavior. So we are going to use a multiplicative non-linear decay term leading to a learning rule in the form:

$$\triangle \rho_k = \mu e^{-c|\rho_k|} u_k \triangle v \ ,$$

with $c \gg 0$. If we interpret

$$\mu' = \mu e^{-c|\rho_k|}$$

as the actual learning rate, we get $\mu' \to 0$ if ρ_k absolute value increase and then we will be nearer from theoretical convergence conditions, see section 2.

3.3 ISO-Learning and ICO-Learning Inclusion in a Reflex Neural Network

ISO-learning and ICO-learning weights sign evolution depends on output effect, in addition output effect tries to imitate unconditioned response, so actually weights sign evolution relies on this reflex response. If response is excitatory signs will become mainly positive and they will do mainly negative for an inhibitory response. When learning is achieved, US appearance stops and then main weights sign will determine circuit output sign.

Let's call, from now on, ISO-learning and ICO-learning circuits as ISO-device and ICO-device respectively, in order to consider them as neurons classes to be inserted in a reflex neural network made of threshold devices. We will also talk only on ISO-devices but ICO prefixes are interchangeable.

Porr and Wörgötter connected robot sensors directly to ISO-device inputs and ISO-device output directly to robot motors, thus output sign has a direct connection, so that for example, positive output could mean "go forward" and negative output could mean "go backward". But when we insert an ISO-device in a neural network we want sign of output being determined by defined synapse kind (excitatory/inhibitory), thus ISO-device output sign could be misinterpreted. To avoid this problem we want all weights have the same sign, in spite of their natural trend. So weights are restricted to be non-negative, thus learning rule is modified as: "if, after applying increments, a particular weight become negative then set it to 0".

4 Experiments

In order to assess performance of discretization modifications, we tried to simulate experiments by software with similar conditions to those done by Porr and Wörgötter with analogue devices and a real robot [9,10]. Given that there still not exist quantitative methods to evaluate ISO-learning or ICO-learning performance, we have to evaluate robot behavior mainly from an external observer point of view, using subjective criterions such as "smoothness" of trajectories, distance to objects in apparent collisions and ellapsed time to achieve this behavior, etc. In this section we will describe original experiments and our simulations. Simulations were performed using a software simulator in development by our team.

4.1 Original Experiments

Porr and Wörgötter used a simple (Rug Warrior) robot with two front infrared sensors and three collision sensors, two front and one rear, rear sensor was not used in these experiments. Their signals arrived to a D/A converter and were filtered by two filter sets, one to control translational velocity and the other to control turn velocity. Collision signal, US, response is a retraction movement consisted of a backward movement with some turn towards collision side. Robot had also two front infrared range sensors, their signals, CS, do not fire any response at the beginning of training, sensors had a range of 15 cm. Sensors and motors were connected to filter sets as described in section 2. When a collision is going to happen CS signals start to arrive to ISO-devices and when collision really happens US signal arrives to them. ISO-devices learn to relate range sensors signal arrival with later bumpers signal arrival and subsequent response. When learning is achieved weights corresponding to CS are big enough to elicit an early conditioned response avoiding collision.

Weight corresponding to US is constantly set to 1, weights corresponding to CS are initially set to 0. Resonators frequencies and qualities were designed to cover temporal intervals between 50 and 500 milliseconds. Ten resonators were used for CS in each ISO-device. In absence of stimuli, robot was programmed to run at a constant speed of 45 cm/s. Robot learned to avoid collision in about 100 seconds.

4.2 Simulated Experiments Description

Robot. We use a simulated Khepera robot. Real basic Khepera is equipped with six front infrared sensors and two rear ones. Infrared sensors have a range of 10 cm and their sampling period is 20 ms. We added two simulated front bumpers and a rear one, see LB, RB and BB in figure 2.a, this last bumper did not take part in original experiments, but we added them in order to use rear proximity sensors in experiments. Then we set cruising speed to 30 cm/s in order to have speeds proportional to sensors range, so we have to cover similar duration intervals to those Porr and Wörgötter had to cope with. Khepera robot

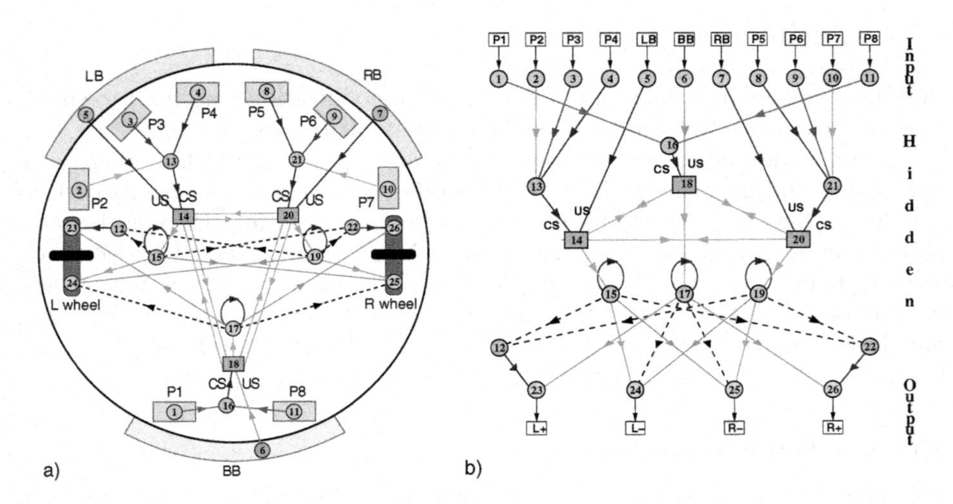

Fig. 2. Simulated robot and neural network diagram. P1 to P8 are infrared proximity sensors. LB, BB and RB are bumpers. L+, L-, R+ and R- represent left/right motor forwards and backwards. Numbered circles are threshold devices and numbered rectangles are ISO-devices. Continuous lines are excitatory connections and dashed lines are inhibitory connections, gray levels of lines show weights strength (darker means stronger). See detailed description in the text. a) Robot scheme with net inside. b) Net only scheme layered view.

has two independent driving wheels. In simulated Khepera robot velocities command were speed orders to both wheels, however, in Porr and Wörgötter robot velocities command were translation and rotation speeds orders.

Neural Net. Neural network is depicted in figure 2.b. Input neuron, 5, connected to left bumper, LB, sends its output, carrying US signal, to ISO-device number 14. Input neurons 2, 3 and 4 connected to left proximity sensors send their outputs to a unique hidden neuron, number 13, for design simplicity. This neuron sums infrared perception from left side and sends its output, carrying CS signal, to ISO-device. ISO-device sends its output to a hidden neuron, number 15, that has an auto-connection with weight less than 1, in order to get damped repetition of its output. We introduced command neurons, 12 and 22, in order to be used by the program user to send velocity commands, in this experiment their outputs were fixed to get a forward speed of 30 cm/s in each wheel, in abscense of stimuli. Neuron 15 output is sent to left output neurons 24 and 25 and to command neurons, numbers 12 and 22, in such a way they yield the retraction movement described in section 4.1. Right and rear sensory areas of the robot are connected in a similar fashion. ISO-devices 14 (left), 18 (right) and 20 (rear) are identical.

During preliminary experiments we observed that if, by chance, at the beginning of experiment most of the collisions happened at the same sensory area, ISO-device belonging to that area achieved learning but the others almost did not start it, of course. So we devised a new kind of connection inspired by fibers

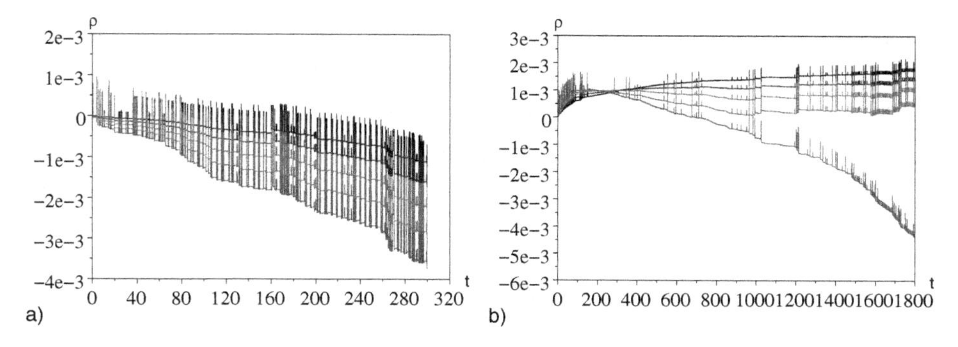

Fig. 3. ISO-device 20: Some CS resonators weights evolution. High peaks are collision processed by this ISO-device, small peaks are collisions coming from the other ISO-devices via learning links. Weights are sampled every 0.02 s. a) shows results of a experiment without sign restriction or decay term inclusion 300 s. long with $\mu = 0.00008$. b) show results of a experiment also without sign restriction but with decay term inclusion 1800 s. long with $\mu = 0.0001$ and $c = 100$.

in *corpus callosum*. These connections (learning links) between ISO-devices, 14, 18 and 20, map resonators of their output ISO-devices onto resonators of their input ISO-devices. This kind of connection increase the weight of a resonator of its input ISO-devices proportionally to weight increment of the equivalent resonator of its output ISO-device.

Resonator weight for US was fixed to 1 and weights for CS was set initially to 0, as in original experiments.

4.3 Simulated Experiments Results

Leaving aside discretization, we have presented two main modifications to Porr and Wörgötter algorithm: weights sign restriction and decay term inclusion. We are going to present our results in cases depending on the combinations of inclusion or not of both modifications. As we will show below, both modifications were necessary to obtain the desaired learning behavior. Finally we comment experiments with both modifications applied to ICO-learning instead of ISO-learning.

Experiments Without Signs Restriction or Decay Term Inclusion. When trying the discrete version of the original algorithm it is always observed a evolution of every resonator weight to negative values, leading to negative output and thus a misinterpretation of this output happens, as commented at the end of section 3.3. This negative output prevents collisions avoiding. Figure 3.a shows the evolution to negative values of weights and where the peaks correspond to the collisions. ISO-device negative outputs also inhibit partially reflex US response, forcing robot to walk close to the walls (see figure) 4.b. Learning does no take place and thus no behavior enhancements happen during experiment, as figures 4.a and 4.b show.

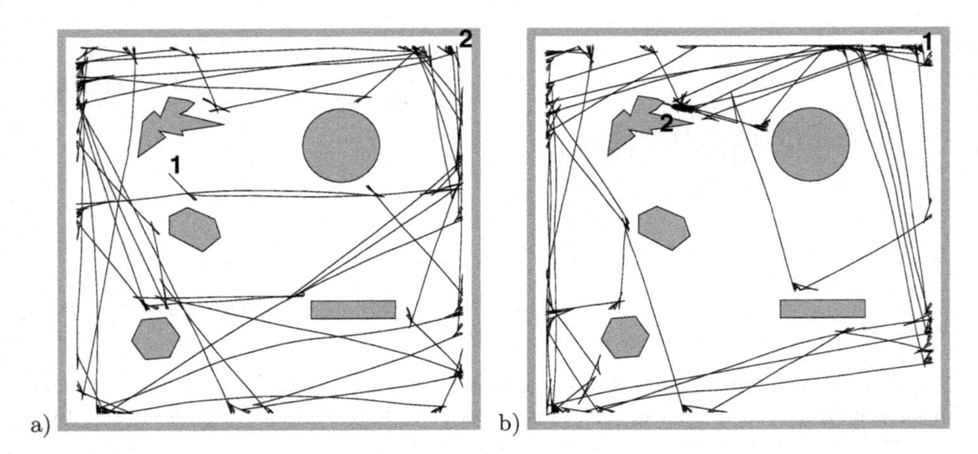

Fig. 4. Robot paths during experiments without signs restriction or decay term inclusion. Robot starts at point 1 and it ends at point 2. a) from $t = 0$ s. to $t = 150$ s. b) from $t = 150$ s. to $t = 300$ s.

Experiments with Signs Restriction but Without a Decay Term. In this case, weights become always null, as it was expected, with a continuous decreasing and a non-negative restriction.

Experiments Without Signs Restriction but Including a Decay Term. Figure 3.b shows a half an hour long experiment. During approximately first 150 s. a lot of collisions and a quick global increment of weights are observed. At this point learning is achieved and collisions appears only occasionally, but after $t = 700$ s., when one of the weights become negative, collisions begin to appear more frequently. After $t = 1200$ s. the negative weight effect exceed all other weights effects, yielding negative output and robot collides continuously. Weights evolution in this experiment before weights become negative are similar to those described in detail in next experiment.

Experiments with Signs Restriction and Including a Decay Term. Figures 5.a and 5.b show first stages of an hour and a half long experiment, in order to clearly observe initial weights evolution and robot behavior. Figures 5.c and 5.d show the rest of experiment, in order to observe if weights stabilization is achieved. In a similar way as in previous experiment case, at the beginning a lot of collisions and a quick increment of weights are observed, but in this case learning is achieved at 90 s. approximately. At this moment a weight corresponding to resonator with biggest oscillation period starts to decrease. Some of other weights follow it later starting their decrement in descending order of periods. Weights decrement does not reach to weights corresponding to resonators with small oscillation period.

Robot behavior at the beginning of experiment is shown in figure 6.a. Initially collisions cause a lot of strong retraction movements, shown as big curved triangles in path. But when learning is achieved these retraction movements turn to

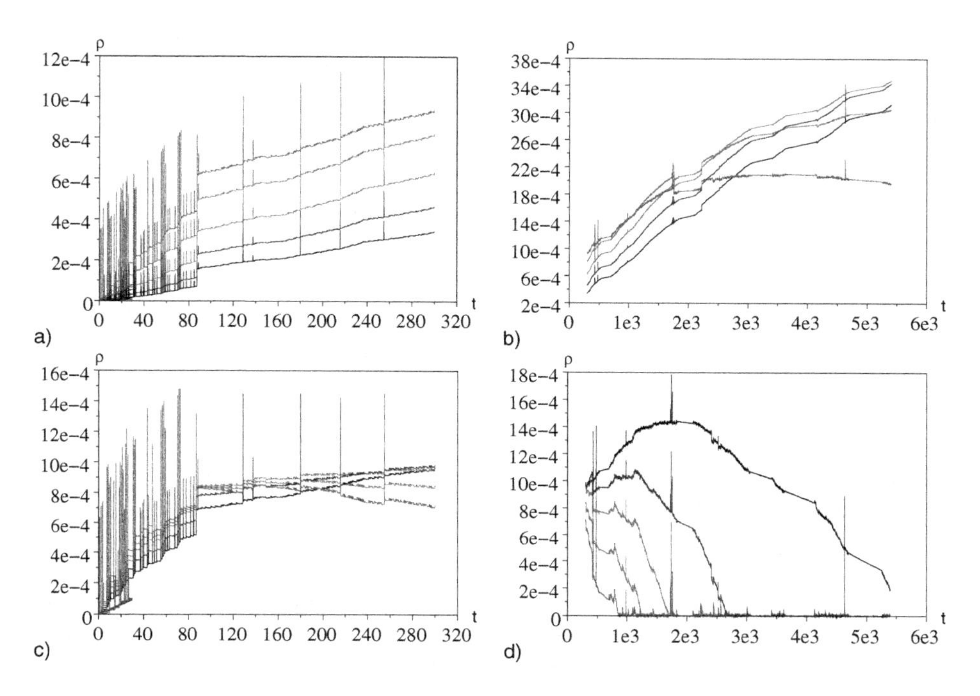

Fig. 5. ISO-device 20: CS resonators weights evolution with sign restriction and decay term inclusion, $\mu = 0.0001$ and $c = 100$. High peaks are collision processed by this ISO-device, small peaks are collisions coming from the other ISO-devices via learning links. a) and b) show weights of resonators with small oscillation periods, a) from $t = 0$ to $t = 300$ s., weights are sampled every 0.02 s., b) from $t = 300$ to $t = 5400$ s., weights are sampled every second. c) and d) show weights of resonators with high oscillation periods, c) from $t = 0$ to $t = 300$ s., weights are sampled every 0.02 s., d) from $t = 300$ to $t = 5400$ s., weights are sampled every second.

be softly curved segments, see figure 6.b. This is due CS response is earlier than US one, but weaker.

Differences with previous experiment appears when the weight of the biggest oscillation period resonator becomes 0, given that weight sign is restricted to be non-negative. Learning is not forgotten, we can see only occasional collisions in 5.b and in 5.d. At the end of experiment conditioned response is only due to resonators with small oscillation periods, that is, those eliciting an earlier response. In original experiments weights sign become mainly positive if response is excitatory and they do mainly negative if response is inhibitory, and they do not become 0 normally. This situation is different from those observed in our experiments. So our modifications tend to elicit earlier responses than original algorithm does.

Robot behavior at the very end of experiment ($t = 5400$ s) is similar to those shown in figure 6.b, but movements are even softer, earlier responses make possible a more efficient navigation.

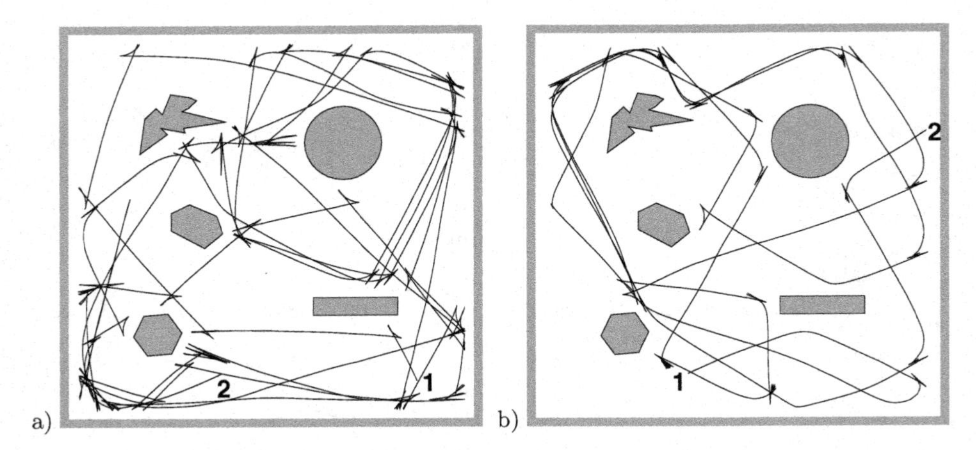

Fig. 6. Robot paths during experiments with signs restriction and decay term inclusion.. Robot starts at point 1 and it ends at point 2. a) from $t = 0$ s. to $t = 150$ s. b) from $t = 150$ s. to $t = 300$ s.

ICO-Learning with Signs Restriction and Including a Decay Term. In this case, we found weights evolution similar to original experiments [10]. Weights absolutely stabilize as soon as learning is achieved even with high learning rates, as in original experiments. Higher weights correspond to bigger oscillation periods, as in previous experiment case. Using, for example, $\mu = 0.001$ and $c = 100$, one-shot learning can be achieved.

4.4 Experiments Discussion

We have observed that decay term inclusion enables initial weights evolution. This is an interesting effect of a term introduced to have only appreciable effect at advanced stages of weights evolution. At initial stages of learning, weights grow in a similar way to those observed in original experiments [9], except maybe weights sign. But at advanced stages weights evolution differs from original ISO-learning experiments. In ICO-learning case, as weights evolution stops when learning is achieved, we found same results as in original experiments. In our ISO-learning experiments weights corresponding to resonators that elicit a later response, start to decrease and they become 0. Weights corresponding to resonators that elicit an earlier response, continue growing slowly. So output is more or less stabilized, although stronger decay term could be used if need it. This tendency to elicit earlier responses as learning progresses, is not clearly observed in original experiments. Earlier responses to stimuli generally yield more efficient behaviors and could be understood as an advantage for organisms survival.

5 Conclusions and Future Work

We have presented several adaptations to ISO-learning and ICO-learning algorithms. First transformation has been full discretization of learning rule, so we can avoid to simulate derivative and continuous filters (resonators), thus improving computational speed and making easy the inclusion of these algorithms in robotics software. Second modification was to restrict weights sign, in order to include ISO-learning and ICO-learning devices into structures more complex than those Porr and Wörgötter used in their experiments. In this work we inserted ISO-devices and ICO-devices into a reactive neural net, where kind of the connections (excitatory/inhibitory) was previously defined, so we had to control weights sign. Third modification was the introduction of a decay term. ISO-learning resonators weight suffer stability problems due to ISO-learning Hebbian nature. We discarded to introduce the usual subtractive decay term, because it leads to forget learning after learning ends. Instead of it, we have employed a multiplicative term that could be interpreted as a reducing factor for learning rate, when weight magnitude increases.

Experiments were performed in a simulated Khepera robot, equipped with eight infrared proximity sensors and three collision sensors, a more complex sensory system than those used in original, continuous rule experiments. Our experiments show that a full discretization of original learning rules works appropriately, when resonators damping factors are adjusted to compensate discretization errors. Resonators weights signs can be restricted, in order to have no contradiction between device output and the predefined kind of its output synapses. However, decay term inclusion is needed to enable initial weights evolution, when their signs are restricted.

In our experiments, weights behavior after learning is achieved tends to elicit earlier responses, this fact makes a difference with respect to original experiments. Given that to obtain earlier responses could be an important issue in quite a lot situations, a discussion between ISO-learning and ICO-learning comparision of performances may be opened. ICO-learning is preferred to ISO-learning due to its, possibly fast, convergence [10]. But when weights evolution stops, ICO-learning ability to elicit earlier responses is lost and maybe modified ISO-learning could be preferable.

In order to evaluate our modifications in real situations, we are adapting these experiments, and creating new ones more advanced, for a real Pioneer 3 AT robot. In a more theoretical field we are trying to simplify filters, given that any type of band-pass filter [9,10] could be used within the rules we have studied.

Acknowledgement

This work was partially supported by TIN2004-07661-C02-01 project from the Spanish Ministerio de Ciencia y Tecnología (MCyT).

References

1. Oppenheim Alan V., Willsky Alan s., and Nawab S. Hamid. *Signals and systems.* Prentice Hall, second edition, August 1996.
2. Valentino Braitenberg. *Vehicles: experiments in synthetic psychology.* MIT, 1986.

3. Wulfram Gerstner and Werner M. Kistler. Mathematical formulations of Hebbian learning. *Biological Cybernetics*, 87:404–415, 2002.
4. Richard Kempter, Wulfram Gerstner, and L. Leo van Hammen. Intrinsic stabilization of output rates by spike-based Hebbian learning. *Neural Computation*, 13:2709–2741, 2001.
5. B. Porr, C. von Ferber, and F. Wörgötter. ISO-learning aproximates a solution to the inverse-controller problem in an unsupervised behavioural paradigm. *Neural Computation*, (15):865–884, 2003.
6. B. Porr and F. Wörgötter. Isotropic sequence order learning. *Neural Computation*, (15):831–864, 2003.
7. B. Porr and F. Wörgötter. Isotropic sequence order learning in a closed-loop behavioural system. *Roy. Soc. Phil. Trans. Mathematical, Physical & Engineering Sciences*, 361(1811):2225–2244, 2003.
8. Bernd Porr. *Sequence-Learning in a Self-Referential Closed-Loop Behavioural System.* PhD thesis, Stirling University, May 2003.
9. Bernd Porr and Florentin Wörgötter. Temporal Hebbian learning in rate-coded neural networks: A theoretical approach towards classical conditioning. *Lecture Notes in Computer Science*, 2130:1115–1120, 2001.
10. Bernd Porr and Florentin Wörgötter. Strongly improved stability and faster convergence of temporal sequence learning by utilising input correlations only. *Neural Computation*, 18(6):1380–1412, 2006.

Towards Automatic Camera Calibration Under Changing Lighting Conditions Through Fuzzy Rules

M. Valdés-Vela, D. Herrero-Pérez, and H. Martínez-Barberá

Dept. Information and Communication Engineering,
University of Murcia, 30100 Murcia, Spain
mvaldes@dif.um.es, dherrero@dif.um.es, humberto@um.es

Abstract. The context of this paper is the auto calibration of a CCD low cost camera of a robotic pet. The underlined idea of the auto calibration is to imitate the human eye capabilities, which is able to accommodates changing lighting conditions, and only when all functionalities works properly, light is converted to impulses to the brain where the image is sensed. In order to choose the more appropriated camera's parameters, a fuzzy rules model has been generated following a neuro-fuzzy approach. This model classifies images into five classes: from very dark, to very light. This is the first step to the generation of a subsequent fuzzy controller able to change the camera setting in order to improve the image received from an environment with changing lighting conditions.

Keywords: Fuzzy Rules, Neuro-Fuzzy Systems, Bio-inspired solutions to engineering.

1 Introduction

The parts of a camera work in a manner similar to the human eye, but the human eye is remarkable. It is able to accommodate to changing lighting conditions, and when all eye functionalities works properly, light is converted to impulses to the brain where the image is sensed. This work aims to classify the images using their lighting conditions, and then it accommodates the camera parameters given the environment conditions. In order to classify the images, a fuzzy rules model is used. Fuzzy rules describe qualitative relationships among the variables in the system being modeled. On one hand, they can be regarded as flexible mathematical structures able to perform nonlinear mappings between input and output data. On the other, they look like natural language statements and, therefore, they are suitable for interpretation. Besides, fuzzy rules models can be used to achieve classification tasks given that a fuzzy classifier is only a fuzzy rules model with crisp and discrete outputs.

Fuzzy rules models obtained from expert knowledge may show poor performance. This can be sorted out applying techniques to build fuzzy models from numerical input-output data, the so-called Data Driven Fuzzy Modeling

J. Mira and J.R. Álvarez (Eds.): IWINAC 2007, Part II, LNCS 4528, pp. 379–388, 2007.

(DDFM) approach [1,13]. However, they pay little attention to the transparency of the resulting rules what is their main drawback. However, in the last years several approaches to achieve a trade-off between performance and transparency have arisen [12].

In this job we generate a TSK model [11] to classify images according to their iluminancy characteristics. TSK rules will be introduced later. The DDFM process is performed starting from a set of images that have been taken during two weeks (about 5000 images), in different times, weather conditions and with different camera configurations. Then, they are manually labeled (from *VERY DARK* to *VERY LIGHT*) in order to obtain a carpus of annotated images which is the input to a DDFM technique. DDFM is the set of techniques to build fuzzy models from numerical input-output data. Concretely , it has been applied a clustering technique [2] followed by an Adaptive Neuro-Fuzzy System (ANFIS) [5] to build the model. Afterward, this classifier will used as the base of a fuzzy controller able to decide which the proper camera's parameters are, according to the given lighting conditions. Nevertheless, the current job must be regarded as a first approximation to the application of fuzzy models to the current problem. Further researches will lead us to the obtention of a fuzzy controller for the auto calibrating of the camera taking into account the lighting conditions. In this way, given an image classified as non-correct, the controller will select the more suitable setting of the camera parameters.

The robot used is the *Sony AIBO (Artificial Intelligence roBOt)*, which is one of several types of robotic pets designed and manufactured by Sony. The AIBO has been used as an inexpensive platform for artificial intelligence research, because it integrates a computer, vision system and articulators in a package cheaper than conventional research robots. The vision system consists on a low cost CCD camera, which is used by the robot as main exteroceptive system. This robot is the standard platform used in the Sony Four Legged League, which is one of the official leagues of the Robocup [8]. Thus, the camera calibration is performed in the official field of this robotics soccer league.

This standard field is colour coded: there are four uniquely coloured landmarks, two goal nets of different colour, the ball is orange, and the robots wear coloured uniforms. Currently, the dimensions of the field are $6x4$ meters. The lighting conditions are changing because the field is in a warehouse with several big windows at several meters high. Therefore, the lighting conditions are very affected by the weather, including clouds, sunrise and sunset. In addition, there are four spot lights around the field.

The low cost camera of the AIBO only allows set three parameters; white balance, gain and shutter speed. White balance is the process of removing unrealistic colours, which are due to the "colour temperature" of the light source, i.e., relative warmth or coolness of white light. This "colour temperature" is usually measured in Kelvins. Human eyes are very good at judging what is white under different light sources; however information from digital images often has great difficulty to make a decision. The white balance possibilities are only three for this camera: indoor (2856k), fluorescent (5000k) and outdoor (6500k) mode. The

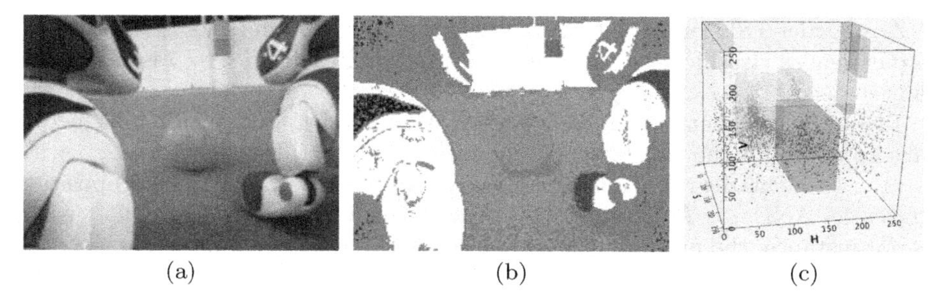

(a) (b) (c)

Fig. 1. Vision system for detecting the colours with the low cost camera: (a) raw image, (b) colour segmented and (c) colour calibration for eight colours

gain of the CCD cameras is the value of the amplifier of the charge-coupled device (CCD), which provides the value of a pixel of the image. This gain provides more light to the pixels, although the use of gain raises noise in the resulting image. We only have three possible configurations for the camera's gain: low (-6dB), mid (0dB) and high (+6dB). Finally, the shutter speed is the length of time the shutter remains opened, i.e., the time the CCD is exposed to the light. The longer the shutter remains open, the greater amount of light is allowed into the CCD. The shutter speed possibilities are also three, which are expressed as a fraction of second: slow (1/50sec), mid (1/100sec) and fast (1/200sec). Note the relationship between the gain and the shutter speed.

The problem consists on the classification the resulting images using a set of camera's parameters. If these parameters are correct, the vision system is able to detect the different colour of the standard environment where is operating. Fig. 1 shows an example of a correct image and the result of the colour segmentation. If the image has right light properties, the segmentation based on colours is usually correct.

The document is structured as follows: section 2 is devoted to the description of the sort of fuzzy rules model used in this job. Section 3 explains the DDFM process carried out to generate our fuzzy classifier. Section 4 is devoted to show some measure to quantify the classifier performance and finally, conclusions and further works are described in section 5.

2 Fuzzy Rules Model

In this paper we are concerned with zero-order TSK rules [11,10]:

$$
\begin{array}{l}
\text{IF } x_1 \text{ is } A_{11} \text{ AND} \ldots \text{ AND } x_p \text{ is } A_{1p} \text{ THEN } y = a_{10} \\
\text{IF } x_1 \text{ is } A_{21} \text{ AND} \ldots \text{ AND } x_p \text{ is } A_{2p} \text{ THEN } y = a_{20} \\
\ldots \\
\text{IF } x_1 \text{ is } A_{r1} \text{ AND} \ldots \text{ AND } x_p \text{ is } A_{rp} \text{ THEN } y = a_{r0}
\end{array}
$$

where r is the number of rules, p is the number of input variables, $x_1, x_2, \ldots, x_p$ are de input variables and y is the output one. $A_{i1}, A_{i2}, \ldots, A_{ip}$ are the antecedent fuzzy sets of i-th rule.

Zero-Order TSK rules are more suitable for classification tasks. As we will see in our images classification problem a natural number from 1 to 5 is associated to each class.

To obtain the output inferred with a TSK model, first of all it is necessary to compute the *degree of fulfillment* of the ith rule, denoted as τ_i, with $i = 1, \ldots, r$. This degree of fulfillment is given by the true value of the proposition "x_1 is A_{i1} AND ... AND x_p is A_{ip}", whose meaning is given through a t-norm as, for instance, the product operator. Therefore, given the input $x_1, \ldots, x_p$, the degree of fulfillment of the i-th rule is computed as:

$$\tau_i = A_{i1}(x_1) \cdot A_{i2}(x_2) \cdot A_{ip}(x_p)$$

Finally, the FIS output y is obtained adding the partial outputs of each rule, weighted with their respective degrees of fulfillment through the equation:

$$y = \frac{\sum_{i=1}^{r} \tau_i y_i}{\sum_{i=1}^{r} \tau_i}$$

where y_i is the output of the i-th rule $y_i = a_{i0}$.

Next, we describe the DDFM process. After the data adquisition and preprocessing, the input variables are selected, then the input-output space is divided by clustering to obtain a rough set of fuzzy rules, and finally, the parameters of the fuzzy sets involved in the rules are optimised with a neuro-fuzzy approach.

3 Generating the Images Classifier: The DDFM Process

In general, after the data (images) adquisition and preprocessing, four main stages compose a DDFM process: first, the relevant input variables must be selected. Then the number of rules must be detected (*rules number identification*). After that, a rough approximation to the set of rules is obtained (*rules generation*). Afterward the fuzzy sets involved in the rules are adjusted in the *parameter optimization* stage to better mimic the system behaviour.

Concretely, our strategy, the selection of variables is performed manually according to the visual distribution of the available images. Afterwards *a subsequent clustering method* is applied to the training set of data. Every cluster can be regarded as a rough fuzzy rule. So, if they are projected into every axis of the space under consideration initial fuzzy sets of the fuzzy rules are obtained. So, these point-wise fuzzy sets will be tuned using the adaptive capabilities of a *neuro-fuzzy architecture.*

3.1 Data Acquisition

The data acquisition stage consists on collecting an ample amount of images from representative locations where the robotic pet has to operate. The environment is a standard soccer field of the Robocup [8]; in particular, the official field of

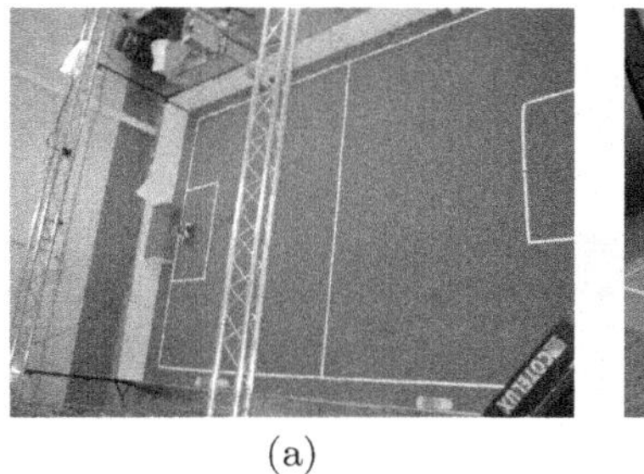
(a)

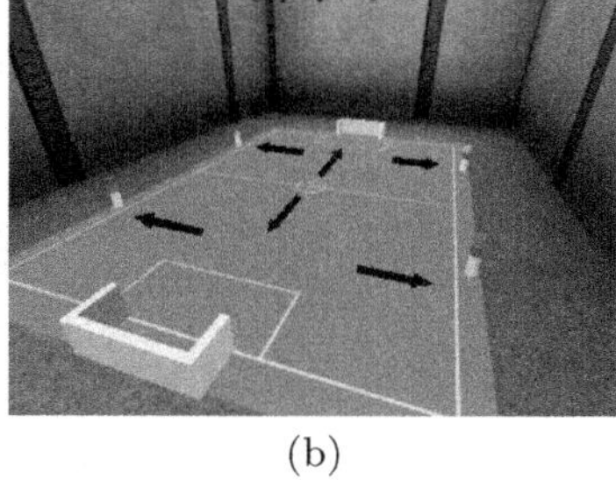
(b)

Fig. 2. Experimental setup: (a) soccer field where the data are collected and (b) the six locations from the pictures are taken

the Four Legged Robot League, where the robotic pet is the platform used for the robotic competition (see figure 2(a)).

The standard field is colour coded: there are four uniquely coloured landmarks, two goal nets of different colour, the ball is orange, and the robots wear coloured uniforms. Currently, the dimensions of the field are 6×4 meters. The lighting conditions are changing because the field is in a warehouse with several big windows at several meters high. Therefore, the lighting conditions are very affected by the weather, including clouds, sunrise and sunset. In addition, there are four spot lights around the field.

The data are collected from six specific locations of the field during two weeks. In particular, we store the number of spot lights switched on, the robot's location and the camera's parameters for each image taken. Thus, we take forty five images from each location, corresponding to the twenty seven possible camera's parameters by the five possible spot lights conditions, i.e., one, two, three, four or no-one switched on.

The images are manually labeled in five classes: VERY DARK, DARK, CORRECT, LIGHT and VERY LIGHT. Given that fuzzy rules consequents are numerical values we assign a number from 1 to 5 to each one of the classes, from VERY DARK to VERY LIGHT.

Once we have the preprocessed data, a part is reserved for learning (70%) and the other for validate the model (30%).

3.2 Selection of Input Variables

The selection of input variables stage consists on finding out the relevant inputs that affect the output. Concretely, in our images classification problem we must decide what statistical descriptive measures of an image have a better classification power. The characteristics selected will be the inputs to our fuzzy rules model. Of course, the output variable will be the class.

In our case, variable selection is performed by hand. Given that centered broad histograms correspond to proper images. Logically, if the image is dark the histogram become to low values and viceversa, when the image is light the

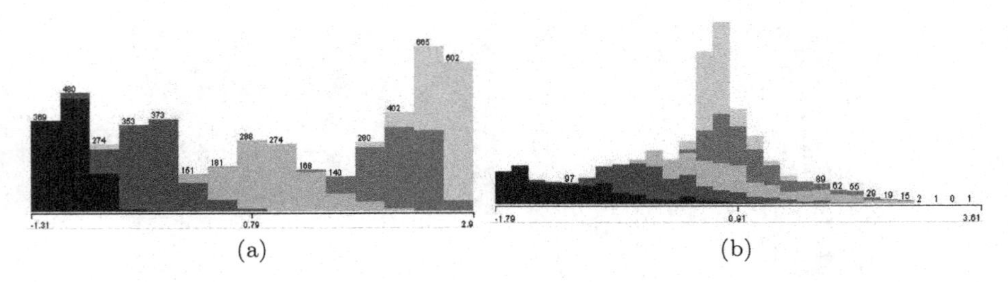

(a) (b)

Fig. 3. Histograms of attribute (a) mean and (b) standard deviation to the class

histogram become to high values. In the saturation state, both darkness and lightness, the histogram is thinner than in proper images [7].

Mean and standard deviation give the above mentioned information about an image histogram. The means and deviation histograms against the class are shown in figures 3(a) and 3(b) where different tones correspond to different classes, it can be observed that both the mean and stantdard deviation has a good splitting capability although standard deviation overlaps light and very light images. However, missclassification of light images into very light ones and vice versa is not a great problem since light and very light snapshots become correct applying similar camera configurations. On the other side, very light images are very uncommon as is reflected in the sample.

3.3 Generation of the Initial Fuzzy Rules

The Fuzzy C-Means Clustering Method (FCM) [2] is one of the most used in order to optimize an initial partition of the data. Given a set of examples, the algorithm must be provided with a previously established number c, a set of c centroids and an initial fuzzy c-partition of the data that is to say, each data belongs to every cluster with a degree in the range $[0, 1]$. Then, this algorithm optimices the c-partition in such a way that certain *cost function* J is minimiced. This function is the weighted within groups sum of squared-errors.

At each step, three operation are sucessively carried out until the centroids are stable with respect to a given tolerance (ϵ): (1) computation of the centroids (assuming that the membership degrees composing the partition are constants numbers); (2) calculus of the distances from each data to every centroid (the more often used measure of distance is the euclidean, it induces spherical clusters); (3) update the membership degree of each data point to every cluster (asumming that the centroids are constants numbers).

As it has been mentioned, FCM needs the number c of clusters as initial parameter. Then, in order to select a good value for c the FCM is lauch with different c values and the clusters model with the best performance is selected. Finally, c is set to 16.

The resulting set of clusters considered as a first approximation to the target fuzzy rules model. These clusters centroids are the initial parameters for the fuzzy sets involved in the fuzzy rules model. The next step is the tuning of those parameters. This stage is described in the next section.

3.4 Neuro-based Fuzzy Rules Tuning

In order to optimise the fuzzy sets and consequent parameters, we use ANFIS [5] that is the acronym of *Adaptive-Network-based Fuzzy Inference Systems.* ANFIS is functionally equivalent to a TSK FIS. It is compound of five layers, from 0 to 4. The nodes in layer 1 and layer 4 correspond to the antecedent and consequent parameters of the TSK FIS respectively. We have supposed fuzzy sets with gaussian membership functions. Therefore, the mean of the fuzzy sets in the antecedents layer are initialized with the centroids obtained in the previous phase of DDFM and then the training is carried out during several *epoches* (iterations). Every epoch of the training is compound of a *forward* and a *backward pass.* ANFIS uses the Least-Square method to tune the layer 4 parameters and hence to identify the rules consequents (forward pass). Afterwards, errors for every training data are calculated and, in the backward pass, error signals are propagated and the rule antecedents are modified through backpropagation. Its learning mechanism is called *hybrid learning* since it combines to different learning mechanisms.

4 Results

Final and intermediate fuzzy rules models were validated using 30% data. In table 1 training and validation errors for every model are deployed, where RMSE stands for Root Means Squared Error (RMSE) and is computed by the equation:

$$RMSE = \sqrt{\frac{\sum_{i=1}^{n}(y_i^t - \hat{y}_i^t)^2}{n}}$$

being y_i^t and $\hat{y}_i^t$ the inferred and real outputs for the i-th datum and being n the number of data. The first row refers to the errors of a classification model based on the clusters generated by FCM. The second row corresponds to the rough fuzzy rules model resulting from the projection of the fuzzy clusters into each axis. The third one shows the errors of the TSK obtained after trainning ANFIS during 150 epoches. It must be remarked that training and validation errors were comparable, indicating the absence of overtraining.

Fig.4 shows evolution of the RMSE, which is stabilized at 0.36.

The fuzzy classifier output will be the input to a subsequent fuzzy controller for the automatic calibration of a robot camera. This controller will be obtained by means of the application of the Fuzzy Actor-Critic Reinforcement Learning

Table 1. RMSE of the different fuzzy models

model	RMSE Tr.	RMSE Va.
after FCM	0.50	0.52
after proj. TSK	0.49	0.51
after ANFIS	0.36	0.36

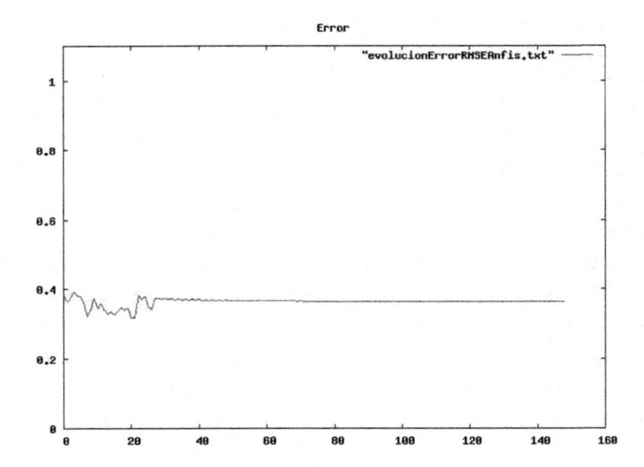

Fig. 4. Evolution of the root mean square error (RMSE)

(FACL) [6]. Reinforcement learning is a trial-error based method in which the robot learns through the interaction with its environment. Its actions have direct consequences in the form of punishments and rewards but the correct actions are not provided. The robot has to discover by itself what actions leads to a better overall performance. In fact when actions are triggered in presence of certain input signals, that improve performance, they become associated with these signals [6].

In the problem under consideration, the robot will perceive an image through the fuzzy classifier. If the image is classified as non-correct it will change the camera settings according to a preestablished policy and will receive a reward or a punishment in terms of the more or less centered and the more or less broad the new image is. These rewards and punishment will be used for iteratively update which settings must be attached to every class of image. through the continuous interaction with the environment.

Another work will consist on the generation of a more interpretable fuzzy rules set capable of being attached with linguistic labels. The one obtained here has not been shown but we must remark that it has good performance but lacks transparency, what is desirable for a fuzzy rules model. Nevertheless, in the last years several approaches to achieve a trade-off between performance and transparency have arisen [12,4,9] and they will be taken into account in future works.

5 Conclusions and Further Works

This paper proposed a fuzzy rules based model to classify images according to their lighting characteristics. This classifier has been generated using Data Driven Fuzzy Modeling techniques. For this aim 5000 images corresponding to six different positions in a concrete indoor enviroment with controlled ilumination conditions were collected during two weeks and they were manually labeled. Since centered and spread histogram corresponded to correct images, both mean and standard deviation were selected as inputs to the classifier. Afterwards, it has been applied a clustering technique followed by a neuro-fuzzy approach whose adaptive capabilities have been used to tune the parameters of the target fuzzy rules model.

Such classification of images is the first step within a further objetive of generating a fuzzy controller for the automatic calibration of a robot camera. Then, the second step is the application of the Fuzzy Actor-Critic Reinforcement Learning (FACL) [6] to learn which actions (camera settings) must be performed in order to improve the received image.

Another work will consist on the generation of a more interpretable fuzzy rules set capable of being attached with linguistic labels. The one obtained here has not been shown but we must remark that it has good performance but lacks transparency, what is desirable for a fuzzy rules model. Nevertheless, in the last years several approaches to achieve a trade-off between performance and transparency have arisen [12,4,9] and they will be taken into account in future works.

Acknowledgments

This work has been supported by CICYT project DPI2004-07993-C03-02.

References

1. R. Babuška. *Fuzzy Modeling for Control.* Kluwer Academic Publishes, 1998.
2. J. C. Bezdek. *Pattern Recognition with Fuzzy Objective Function Algorithms.* Plenum, New York, USA, 1981.
3. A. F. Gomez-Skarmeta, M. Valdes, F. Jimenez, and J. G. Marn-Blazquez. Approximative fuzzy rules approaches for classification with hybrid-ga techniques. Information Sciences: an International Journal, 136(1-4):193214, 2001. SOURCE Page 9 of 10 4528
4. S. Guillaume. Designing of fuzzy inference systems from data: An interpretability-oriented review. *IEEE Transactions on Fuzzy Sets and Systems*, 9(3):426–443, 2001.
5. J.S. Jang Roger. Anfis: Adaptive-network-based fuzzy inference systems. *IEEE Transactions on Systems, Man, and Cybernetics*, 23(3):665–685, 1993.
6. L. Jouffe. Fuzzy inference systems learning by reinforcement methods: Application to a pig house atmosphere control, 1997.
7. M.W. Powell, S. Sarkar, and D.B. Goldgof. A methodology to extract objetive colors from images. *IEEE Transactions on Systems, Man and Cybernetics-Part B*, 34(5):1964–1978, 2004.
8. Robocup. Robocup. http://www.robocup.org/.
9. M. Setnes and H. Roubos. Ga-fuzzy modeling and classification: Complexity and performance. *IEEE Transactions of Fuzzy Systems*, 8(5):509–522, 2000.
10. M. Sugeno and G. T. Kang. Structure identification of fuzzy model. *Fuzzy Sets and Systems*, 28(1):15–33, 1988.

11. T. Takagi and M. Sugeno. Fuzzy identification of systems and its applications to modeling and control. *IEEE Transactions on Systems, Man and Cybernetic*, 15:116–132, 1985.
12. Y.W Teng and W.J. Wang. Constructing a user friendly ga-based fuzzy system directly from numerical data. *IEEE Transactions on Systems, Man and Cybernetics-Part B*, 34(5):2060–2070, 2004.
13. M. Valdés, A.F. Gómez-Skarmeta, and J.A. Botía. Towards a framework for the specification of hybrid fuzzy modeling. *International Journal of Intelligent Systems*, 20(2):225–252, 2005.

Social Interaction in Robotic Agents Emulating the Mirror Neuron Function

Emilia I. Barakova

Eindhoven University of Technology
P.O.Box 513 5600MB Eindhoven, The Netherlands
e.i.barakova@tue.nl

Abstract. Emergent interactions that are expressed by the movements of two agents are discussed in this paper. The common coding principle is used to show how the mirror neuron system may facilitate interaction behaviour. Synchronization between neuron groups in different structures of the mirror neuron system are in the basis of the interaction behaviour. The robotics experimental setting is used to illustrate the method. The resulting synchronization and turn taking behaviours show the advantages of the mirror neuron paradigm for designing of socially meaningful behaviour.

1 Introduction

Recent neurophysiological, cognitive, and developmental research clearly shows that there are shared representations in the brain between perceived and generated actions, between actions produced by oneself and others (see for instance [20][27][15][16]). These shared representations, conveyed by the mirror neuron system, underlie the process of imitation, social learning, and prediction of the behaviour of conspecifics. Many attempts have been made to model the imitation process, for review see [26] and [17]. However, the imitation that has been modelled so far does not go further than one directional demonstrator-imitator interaction. In this paper we want to make an attempt to show the potential of the mirror neuron paradigm for social interaction, in particular for movement synchronisation, entrainment, and interchangeable turn-taking between two agents.

Entrainment of timing of social interaction has been investigated in multidisciplinary research on conversation. Conversation is an exchange of speech between two or more individuals. Although at first glance it looks like a chaotic process, conversation usually proceeds smoothly, by having the two parties take well timed turns. A number of authors have proposed that the listeners anticipate an upcoming end of a turn by perceiving eye gaze, body movement, or other semantic, syntactic, or prosodic queues from the speaker, for reviews see [8][9]. Conversely, listeners indicate their desire for turn ending. Speech is, in its essence, a motor act and it is likely that the mechanisms of speech and turn taking coevolved, perhaps building on the same preexisting structures and mechanisms for motor expression [29]. In their theoretical study Wilson and Wilson

J. Mira and J.R. Álvarez (Eds.): IWINAC 2007, Part II, LNCS 4528, pp. 389–398, 2007.

[29] argue that turn taking is likely to be successfully modelled by entrainment of endogenous oscillators.

Mutual entrainment of rhythmic activities has been theoretically studied as the basic mechanism of the organization of temporal order by Pavlidis [18]. Endogenous oscillators have been implicated in a range of cognitive processes, including perception, motor control, attention, memory, and consciousness [5].

In a robotic setup turn-taking behaviour is discussed in [6] and [12]. The turn taking behaviour in these studies takes place as a result of interaction of two dynamical recognizers - Elman type of recurrent neural networks that have widely been used to model dynamic systems. The training has been replaced by a genetic algorithm, which aims to produce "genettically different" agents. This will prevent from the low reliability of the interaction process based on neural learning [5].

We base our interaction behaviour on synchronization between neural latices that together simulate the mirror neuron-like functioning. The neuron firing in every lattice of neurons is modelled by an oscillatory model.

This paper is organized as following. In Section 2, we propose the biological background of the mirror neuron model and the common coding paradigm. Section 3 connects the biological modelling to concrete computational framework and shows how it is applied in a robot setting. The experimental setting and results of following and turn taking simulations are shown in Section 4. The discussion (Section 5) summarizes the results and puts this work in perspective.

2 Mirror Neuron System Model for Interagent Interaction

The Common coding paradigm postulates parity between perception and action, i.e. the action and perception arise simultaneously [11]. A core assumption of the Common coding paradigm is that actions are coded in terms of the perceivable effects (i.e. the distal perceptual events) that they should generate. It has the advantages to an information-processing paradigm which is unable to explain perception in many cases related to direct action [3][1]. Common coding paradigm has more solid foundations than Selection paradigm and Gibsons theory of direct perception [7] which fail to explain another group of phenomena like memory and imagination that can certainly originate an action by themselves [3][2].

A growing body of behavioural and neurophysiological studies support the grounding principles of the Common coding paradigm. As first evidence for direct matching between action perception and action execution came the discovery of 'mirror neurons' in the ventral premotor cortex of the macaque monkey [21][22][23]. Mirror neurons fire both when monkey carries out a goal-directed action and when it observes the same action performed by another individual [24], i.e. the perception and the action are likely coded in the same way, by the same structure. More recently, it was found that a subset of these mirror neurons also responds when the final part of a previously seen action is hidden and can only be inferred [28]. Therefore, the observation of an action activates action

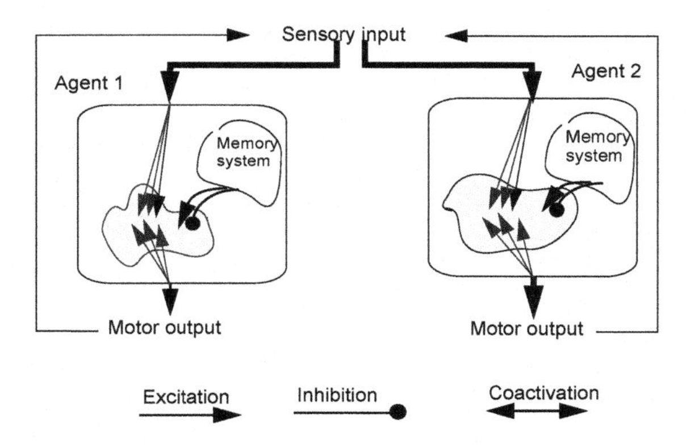

Fig. 1. The common coding theory suggests that sensing and action is related to the activation in the same internal representations. Moreover they can be activated by endogenous factors. Common coding for two agents that share perceptual space as a basis for modelling interaction behaviour.

representations to the degree that the perceived action and the represented action are similar [14]. Specific neurons in this region respond to the representation of an action rather than to the action itself.

It can be inferred that the sensory and the motor activations that represent the same action or intention are related to the activation of the same area in the brain. One such an area is the ventral premotor cortex (PMv). Since the observed, executed, and imagined actions are related to an activation in a common representation, we schematically show this phenomena like the activation from the three events is projected to the common representation (Figure 1, the scheme of the individual agent). Actually, in case of an executed action the activation in the premotor and motor areas occurs in a very short for the behavioural time scale interval, i.e. practically co-occurs. The case of two agents that share perceptual space the common representation for perception and action for each agent will create a basis for an interaction behaviour, as shown in Figure 1.

Actually, there is more than one representational structure that gets active by the same event encountered by the sensory and the motor states. Most of the frontal motor areas receive robust sensory input (visual and somatosensory) from the parietal lobe. This pattern of connectivity supports relatively specialized fronto-parietal area for sensorimotor integration. A posterior area with mirror neuron properties is located in the rostral part of the inferior parietal lobule (IPL). Both areas form the mirror neuron system (MNS) The main visual input to the MNS originates from the posterior sector of the superior temporal sulcus (STS). Together, these three areas form a core circuit for imitation, one of the basic building components of social behaviour. The information flow from the parietal MNS, which is mostly concerned with the motoric description of an action, reaches back to the STS. By macaque, STS and the equivalent of IPL share patchy connections that overlap particularly well with the locations

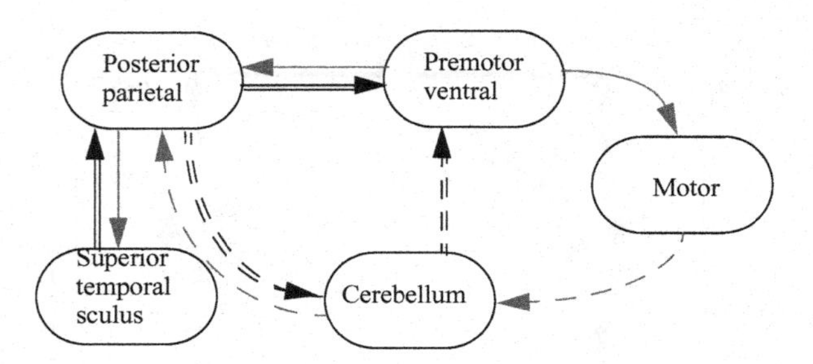

Fig. 2. The information flow by the observation and the imitation of the same action in an individual agent. Double line arrows mark the path of the inverse model that signifies action observation. Single line arrows show the forward path for execution of imitated actions.

in which neurons respond specifically to complex body movements. The STS although considered to be a part of the 'mirror system' [19] do not show any motor activation itself. In spite of lacking mirror properties, STS neurons seem to 'understand' actions quite well, and it is plausible to assume that they send (via IPL) preprocessed signals about actions to the premotor areas that include information about the goal or the meaning of the observed action.

To construct a computational model that can facilitate the imitation and interaction functionality, we have modelled the three interconnected structures with lattices of neurons [4]. The direction of the connectivity between the structures differs while different actions take place. Figure 2 denotes the information flow by the observation and the imitation of the same action. The solid arrows show the part that has been considered in our model for achieving the imitation functionality.

3 Oscillatory Neural Dynamics of the Mirror Neuron System in the Robotics Setting

From the framework proposed in the previous Section becomes apparent that the mirror neuron model that materialises the Common coding paradigm is a useful tool for modelling interactive behaviour. Interagent interaction is initiated from the representation of the movement of each robot within the neural structures of the partner robot. To achieve the imitation functionality and create a model that is suitable for robotics, we have to make some simplification. We base our core scheme for imitation learning on conceptual model of Keisers and Perrett [13], whose experimental work has shown that there are anatomical connections between the macaque analogous of STS, IPL, and PF areas, and therefore a Hebbian learning rule can be applied. The particular network that has been used to simulate the imitation functionality is shown in Figure 3.

In this scheme the role of the STS neurons have the function to transfer the sensory (visual) stimuli and to account for the influence of the inhibitory neurons. For a robotic setting, modelling of the STS area can be reduced to the

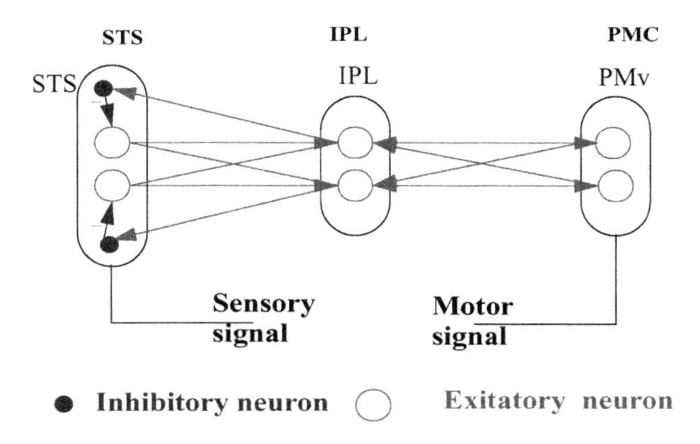

Fig. 3. Core circuit for imitation

influence of the inhibitory neurons. Therefore, the sensory signals project directly to the IPL area which is associated with multisensory integration. The motor information, or the information from the movement of the wheels is co-activated in the simulated PMv area, which has sensorimotor integration functionality. The bidirectional projections between the two areas will insure that both areas represent the sensory and the motor signals.

The embodied implementation of this model is shown in Figure 4. The 8 range sensors of each robot project to the sensory integration area that resembles the functionality of the joined STS-IPL areas. The two wheels project to the sensorimotor integration area, which resembles the PMv, as shown in Figure 4.

Self-organization of rhythmic activity is a fundamental characteristic of biological systems. In addition, rhythmic activities are found in any level of the hierarchical structure, i.e. from the biochemical to the sociobiological level. At neuronal level, single neurons and networks respond with transient oscillations to strong input. The natural frequency, or eigenfrequency of the damped oscillation is a result of two opposing effects, often modelled by the combined effect of executory and inhibitory neurons.

We suggest to use entrainment of endogenous oscillators for modelling turn taking behaviour. The mirror neuron paradigm that allows the behaviour of each robot to be represented in the neuronal structures of its partner makes possible the oscillatory dynamics of the turn taking process to be modelled through the individual agents.

The mutual interaction between two robots has to emerge through self-organizing entrainment of oscillatory neurons. To check this hypothesis, the neurons of each robots mirror system are simulated as oscillators:

$$\theta(t) = \omega t \mod 2\pi \tag{1}$$

The above equation determines the change of rate of the phase with the time. is the the cycle of the limit cycle oscilation. The phase is periodic over the range.

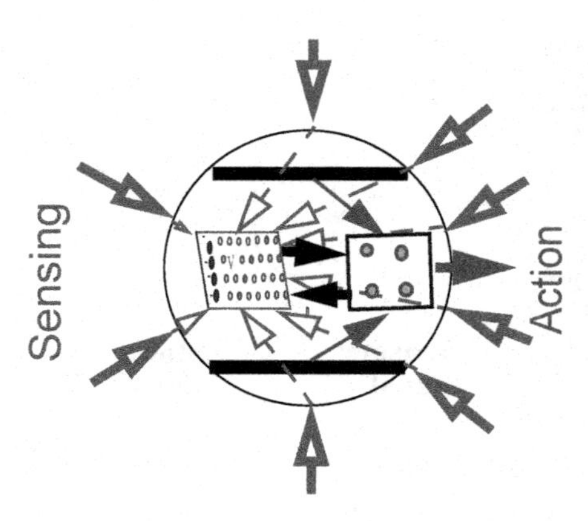

Fig. 4. Robot architecture with mirror circuit

If a synaptic coupling H connects two neurons, their phase equations will be represented in the following way:

$$\frac{d\theta_1}{dt} = \omega_1 + H_1(\theta_2 - \theta_1) \tag{2}$$

$$\frac{d\theta_2}{dt} = \omega_2 + H_2(\theta_1 - \theta_2) \tag{3}$$

where indexes 1 abnd 2 refer to the first and the second neuron respectively. The learning rule for oscillatory networks was used, the original learning rule was shown in [4].

4 Experimental Setting and Results

The experimental setting consists of two parts. In the first part each robot has to build sensory-motor experiences by exploring an environment that consists of a circular arena inhabited by another robot, see Figure 5a. In the second part, interactive turn-taking behaviour emerges, based on the established oscillatory sensorimotor couplings.

Initially we constructed the experimental scenario that performs following behaviour for the training phase, see Figure 5a. Both robots consequently are taking the role of the follower, in order to establish adequate "mirroring" couplings between the lattices that resemble the IPL and PMv areas. Before training, IPL serves as multisensory integration area and PMv is primarily sensorimotor integration area.The hebbian connections between these lattices are modelled in the way that after training both areas will reflect the common sensorimotor representation that is the basis for interaction behaviour:

$$\Delta w_{lk}^{PM-IPL} = \alpha(PM_l - \overline{PM_l})(IPL_k - \overline{IPL_k}) \tag{4}$$

where $\overline{PM_l}$ and $\overline{IPL_k}$ are the average activation values of units l and k over a certain time interval. IPL-PMv synaptic plasticity has the following dynamics:

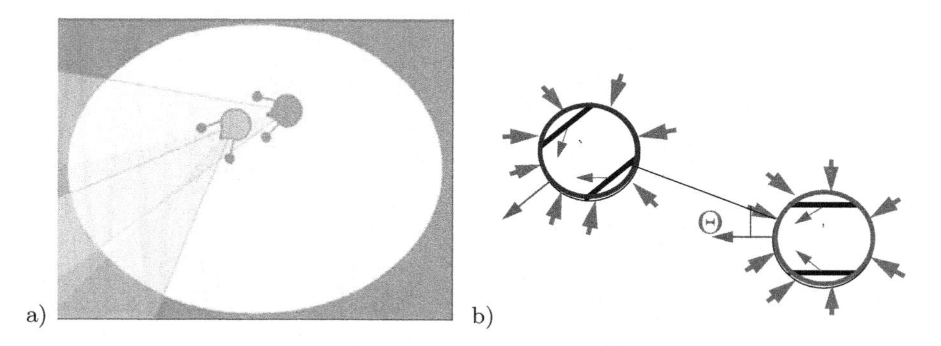

Fig. 5. a) Scenario for following and turn taking behaviours based on the tag game. b) The shortest sensor measurement determines the relative position and direction of the partner robot for the following interaction.

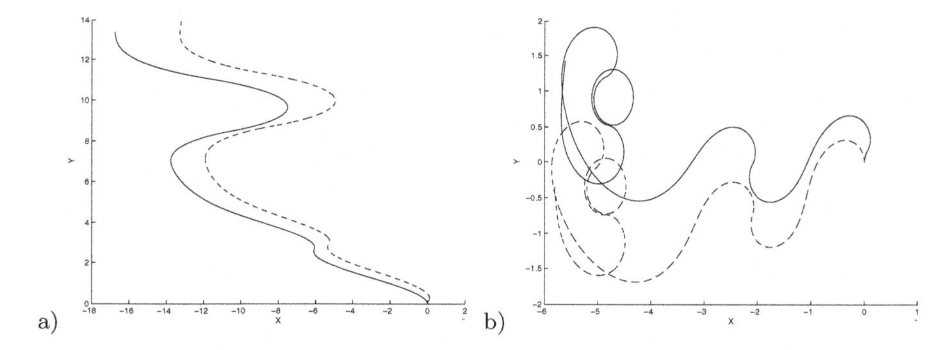

Fig. 6. Movement imitation behaviours

the connection between them is strengthened if both of them are simultaneously active and weakened if the activation of one decreases.

In the initial experiment the robot-follower denotes its shortest distance reading, which signals the presence of the partner robot, as shown in Figure 5b. The placement of the distance sensors defines the relative heading of the partner robot. The robot-follower tends to synchronize its motion direction with the motion direction of the leading robot.

After neurons from the two lattices synchronize, the two simulated robots express a simple form of social behaviour. The leader robot performs movements with different complexity, and the follower (dashed lines) imitates it from its movement perspective, as shown in Figure 6.

At the second part of the experimental scenario, the emergent turn taking is to be shown. The role of the robot, being follower or leader at the present moment depends on which robot is 'within the visual field' of its partner. For the training phase, the tag game is simulated, by which the runner and the tagger functions change between the robots once the tagger reaches the runner.

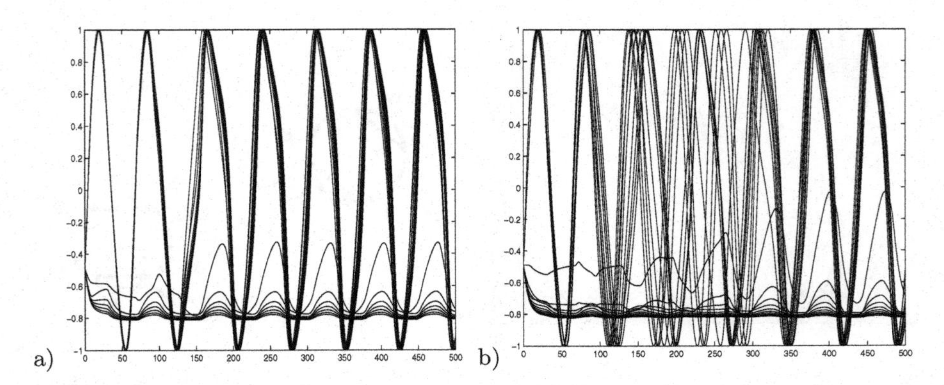

Fig. 7. a) Neural activation during following behaviour. b) A typical desynchronization in the case of turn taking.

After training, the emergent turn taking has to take place which is expressed by symmetry breaking process after a period of synchronization. This way the leading robot can become a follower and later again the lead can be taken over by it. The turn taking, similar as by humans, takes place as a result of some subtle or explicit external stimulation. For the case of the tag game, the external stimulation is usually caused by losing the runner-robot from the perceptual field, caused by reaching the end of the arena or other reason for escape of the runner robot.

Figure 7 shows the neural activation during turn (the left plot) and in the period of turn taking (the plot on the right). The desynchronization of the neurons in central part of the right plot corresponds to the moment of losing the runner robot from the perceptual field. At that period previous follower changes its role to a runner, and vice versa.

5 Discussion

Social interaction has wide spectrum of expressions as synchronous movements, turn taking, gaze sharing, following, imitation and conversation. We have simulated simple interaction behaviours of following and turn taking. In the training phase, the simulated following and the simulated tag game helps to gather examples and establish the sensorimotor couplings between the two robots. In the test runs, there is not an external control that will cause the turn taking behaviour. The turn taking is caused by changing of synchronous firing of the oscillatory neurons. Although external events are initiating turn change, turn taking does not take place only by the same conditions as during the training - turn taking has emergent properties due to nonlinear oscillations and their interaction. This results resemble turn taking in speech: an upcoming end of a turn is anticipated by perceiving eye gaze, body movement, or other queues by the speaker, or indicated in a subtle manner by the listener.

Important questions for designing a movement interactions lies in the respective computational role of each brain area that subserves the internal simulations and shared representations between self and others. We based our model on the simplified mirror neuron network, in which the mirroring functionality is obtained via the selforganization of synchronized neural firing in two robots that share perceptual space. The emergence is an important element, but better understanding of underlying processes and computations will increase the possibilities and reliability of the interaction behaviour. This work is in the process of extensive development.

References

1. Barakova, E.I. and Lourens, T., Event based self-supervised temporal integration for multimodal sensor data, Journal of Integrative Neuroscience 2005 Jun;4(2):265-282.
2. Barakova, E.I. and Lourens, T., Efficient episode encoding for spatial navigation, International Journal of Systems Science, Vol. 36, No. 14, 15 November 2005, 887-895.
3. Barakova, E.I., Social interaction through movement : concepts from perception-action interplay in Feijs, Kyffin, and Young (Eds.), Proceedings of DESFORM conference, 2006, Nov. 25-27.
4. Barakova, E.I. and Yamaguchi Y., Sensorimotor synchronization for action prediction: a mirror neuron model. in the Proc. of 8th BSI Retreat ,Oct. 31- Nov. 1 2005, Tochigi Japan.
5. Barakova, E.I. , Learning Reability: a study on indecisiveness in sample selection , PrintPartners Ipskamp B.V. ISBN: 90-367-0987-3, March 1999.
6. Di Paolo, E.A. , Behavioral coordination, structural congruence and entrainment in a simulation of acoustically coupled agents. Adaptive Behavior 2000, 8:25-46
7. Gibson, J.J., The Senses Considered as Perceptual Systems. Boston: Houghton-Mifflin; 1966. W.J.
8. Ford, C. E., and Thompson, S. A. (1996). Interactional units in conversation. In E. Ochs, E. A. Schegloff, & S. A. Thompson (Eds.),Interaction and grammar (pp. 134-184). NY: Cambridge Univ. Press.
9. Fox Tree, J. E. , Coordinating spontaneous talk. In L. Wheeldon (Ed.), Aspects of language production 2000, (pp. 375-406). Philadelphia: Taylor & Francis.
10. Golubitsky, M. and Stewart I. Bulletin of the american mathematical society, Volume 43, Number 3, July 2006, Pages 305-364.
11. Hommel, B., Musseler J., Aschersleben G, Prinz W: The theory of event coding; a frame-work for perception and action. Behav Brain Sci 2001, 24:849-878.
12. Iizuka, H. and Ikegami, T. (2004). Adaptability and Diversity in Simulated Turn-taking Behaviour, Artificial Life. 10:361-378.
13. Keysers, C., Perrett, D.I., (2004). Demystifying social cognition: a Hebbian perspective. Trends Cogn. Sci. 8, 501-507.
14. Knoblich, G., Flach R: Action identity: evidence from selfrecognition, prediction, and co-ordination. Conscious Cogn 2003, 12:620-632.
15. Meltzoff, A.N., and Moore, M.K. (1983). Newborn infants imitate adult facial gestures. Child Development, 54, 702-709.

16. Meltzoff, A.N., and Moore, M.K. (1989). Imitation in newborn infants: exploring the range of gestures imitated and the underlying mechanisms. Dev. Psychology, 25 (6), 954-962.
17. Oztop, E., Kawato M., and Arbib M., Mirror Neurons and Imitation: A Computationally Guided Review. Neural Networks, 19 (3), 2006, pp 254-271.
18. Pavlidis, T.: Biological oscillators: Their mathematical analysis. New York, London: Academic Press 1973.
19. Rizzolatti, G. and Craighero, L. (2004). The mirror-neuron system. Annual Review of Neuroscience, 27, 169-192.
20. Rizzolatti, G., Fadiga, L., Fogassi, L., and Gallese, V. (2002). From mirror neurons to imitation: Facts and speculations. In A. N. Meltzoff & W.Prinz (Eds.), The imitative mind: Development, evolution and brain bases (pp. 247-266). Cambridge University Press.
21. Rizzolatti, G., Cortical mechanisms subserving object grasping and action recognition: a new view of the cortical motor functions. In The New Cognitive Neuroscience (Gazzaniga, M. ed.), 2000, pp. 539-552, MIT Press.
22. Rizzolatti, G., A unifying view of the basis of social cognition. Trends Cogn. Sci.2004, 8, 396-403.
23. Rizzolatti, G., Localization of grasp representations in humans by PET: 1. Observation versus execution. Exp. Brain Res., 1996, 111, 246-252.
24. Rizzolatti G, Fadiga L, Gallese V, Fogassi L: Premotor cortex and the recognition of motor actions. Brain Res Cogn Brain Res 1996, 3:131-141.
25. Schaal, S., Is imitation learning the route to humanoid robots?, Trends in Cognitive Sciences, 1999, 3, 6, pp.233-242.
26. Schaal, S., Ijspeert, A., and Billard, A.: Computational approaches to motor learningby imitation. Phil. Trans. R. Soc. Lond. B Vol. 358 (2003)537-547.
27. Stoet, G., and Hommel, B., Interaction between feature binding in perception and action. In W. Prinz & B. Hommel (Eds.), Common mechanisms in perception and action: Attention & Performance XIX (pp. 538-552). Oxford: Oxford University Press,2002.
28. Umilta M. A., Kohler E., Gallese V., Fogassi L., Fadiga L., Keysers C., Rizzolatti G.:I know what your are doing: a neurophysiological study. Neuron 2001, 31:155-165.
29. Wilson, M., and Wilson, T.P., Related Articles, Links An oscillator model of the timing of turn-taking.Psychon Bull Rev. 2005 Dec;12(6):957-968.

Integration of Stereoscopic and Perspective Cues for Slant Estimation in Natural and Artificial Systems

Eris Chinellato and Angel P. del Pobil

Robotic Intelligence Lab
Universitat Jaume I, Castellón de la Plana, Spain
eris,pobil@icc.uji.es

Abstract. Within the framework of a model of vision-based robotic grasping inspired on neuroscience data, we deal with the problem of object orientation estimation by analyzing human psychophysical data in order to reproduce them in an artificial setup. A set of ANN is implemented which, on the one hand, allows to replicate some neuroscientific findings and, on the other hand, constitutes a tool for slant estimation that can improve the reliability of artificial vision systems, namely those dedicated to analyze visual data inherent to the interaction robot-environment, such as in grasping actions. The implementation confirms the hypothesis that integration of monocular and binocular data for the extraction of action-related object properties can provide an artificial system with improved pose estimation capabilities.

1 Introduction

The integration between sensory, associative and motor cortex of the human brain in vision-based grasping actions is obtained by coordinating the two visual streams of the human cortex, the action-oriented dorsal stream and the perception-oriented ventral stream. Recent neuroscience findings allowed us to depict the outline of a model of vision-based grasp planning that builds strongly upon primate, and especially human physiology, trying to emulate the dualism and the interaction between the two streams. Our framework, described with more detail in [1], has been conceived to be applied on a robotic setup, and the design of the different brain areas is performed taking into account not only biological plausibility, but also practical issues related to engineering constraints.

In this work, we focus on the section of our framework which models the areas of the brain dedicated to the extraction of basic properties of object features. In particular, we describe issues concerning the use that posterior parietal cortex areas (such as the caudal intraparietal area CIP) make of the visual information coming from retinotopic visual zones as V3. The first considered aspect is the estimation of the distance of the target object, performed by area LIP (lateral intraparietal sulcus) of our cortex. The advantages of expressing and calculating distances in nearness units are discussed. The second aspect is the combination

J. Mira and J.R. Álvarez (Eds.): IWINAC 2007, Part II, LNCS 4528, pp. 399–408, 2007.

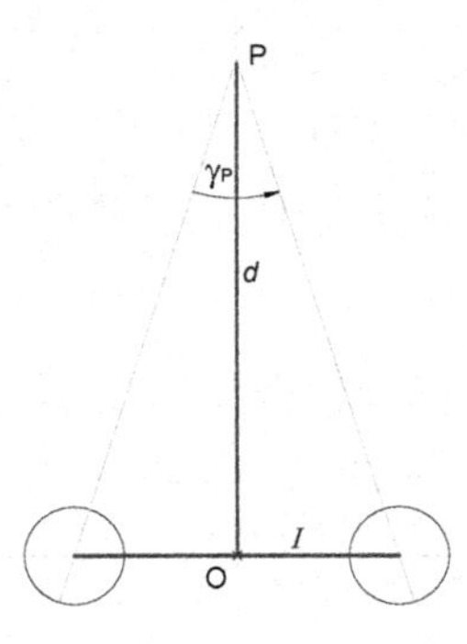

Fig. 1. Vergence angle and distance

of binocular (stereopsis) and monocular (perspective) visual data in order to perform object orientation estimation. A rigorous theoretical analysis for deriving plausible expressions for slant estimation is accompanied by an accurate implementation with a set of artificial neural networks. The behavior of the system in simulated noisy conditions suggests that the model is faithful to the biological reality. Moreover, although we deal with these problems building on real physiological findings, our approach is practical and oriented toward the application to robotic systems. Indeed, as far as we know, no robotic applications have used a combination of the two kinds of visual hints. The neural network simulation shows how and why this solution can be advantageous for robotic applications.

2 Nearness Estimation Through Proprioceptual Data

The distance d of a fixated point P from the viewer can be estimated by either retinal and/or proprioceptive information (accommodation and vergence).

Proprioceptive cues are preferably used when retinal data is not available or considered not reliable, and for short distances. The relation between distance and vergence angle γ_P is simple and depends only on the interocular distance I, which is constant (see Figure 1):

$$I/2 = d \tan(\gamma_P/2) \rightarrow d = \frac{I}{2\tan(\gamma_P/2)} \quad (1)$$

The distance estimator that we designed is based on proprioceptual vergence data. In a robot, an estimation of this proprioceptual data can be obtained in the following way. If a head is available in which the two eyes can rotate independently about a vertical axis, the rotation angle which allows to place a given reference point at the same coordinates in the two images is the required vergence angle. This can be obtained also with only one camera, displacing it sideways by a distance I and then rotating it until the registration of the reference point between the previous and the actual image is obtained. For an efficient calculation, the reference point (and the scene) should be chosen so that the correspondence problem is easy to solve.

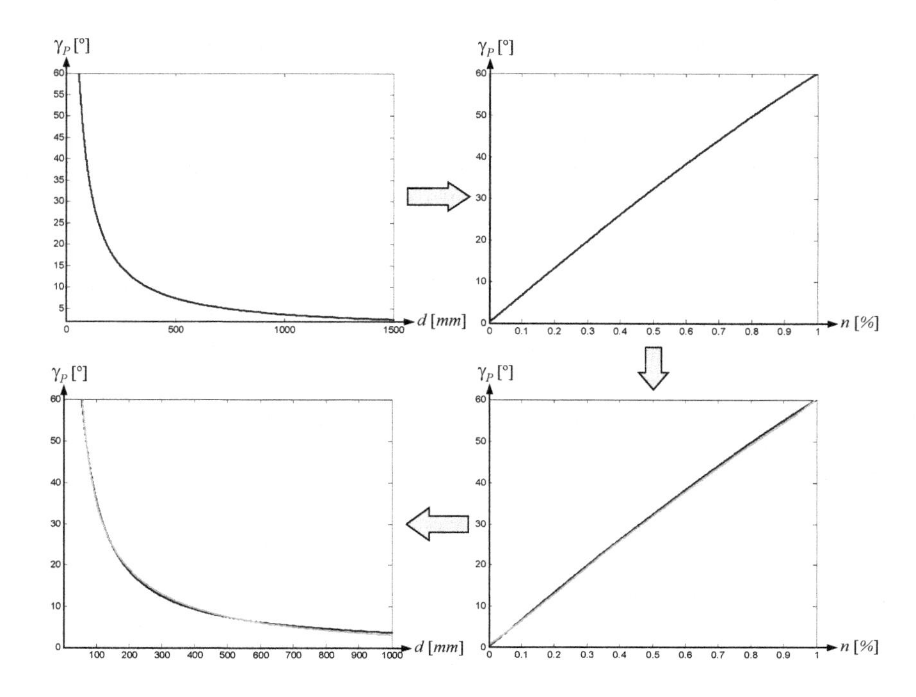

Fig. 2. Nearness to vergence. Dark lines: theoretical equations; light lines: learnt curves.

Neuropsychological experiments [2] suggest that distance estimation is most probably performed in our brain using *nearness* units instead of distance units. Nearness is the reciprocal of distance, so that a point at infinite distance has 0 nearness, and a point at the maximum vergence angle has nearness 1 (or 100%):

$$nearness = 1/d = 2\tan(\gamma_P/2) \tag{2}$$

Following this suggestion, we designed two radial basis function networks for learning the association between vergence and nearness and between vergence and distance. The results can be seen in Figure 2. On the top left, the distance/vergence curve corresponding to Equation 1 is shown. Equation 2 between vergence and nearness is depicted on the top right of the image, and the corresponding learnt curve appears on the bottom right of Figure 2 (lighter curve). The reciprocal of the learnt relation is finally depicted on the bottom left, where it can be compared with the true mathematical relation (they practically coincide). In our simplified example, to obtain similar performances in the estimation of distance, the distance/vergence network requires 11 rbf units, while the nearness/vergence net only requires 4 neurons. This is not surprising due to the nearly linearity of the relation vergence/nearness and considering that, due to evolutionary pressure, our brain often exploits the principle of economy, minimizing the resources required to perform a given task. In our model, and in the related robotic application, object distance is represented in nearness units.

3 From Retinal Angles to Object Orientation

In our approach, we make use of simple visual information for the achievement of a geometric 3D selectivity similar to that observed in neuroscientific and neuropsychological studies. The basic assumption is that different sorts of monocular and binocular data are analyzed and integrated in order to obtain a robust visual representation of the target object [3], composed by its general shape, orientation and position. Concretely, we are now developing a modular computational structure composed of various estimators which make use of proprioceptive and retinal cues in order to obtain the geometrical parameters needed for grasp planning. Our approach differs from related research (e.g. [4]) in that it builds upon retinal data: instead of using pixelated images and projective matrices, our only inputs are retinal angles and proprioceptual eye data. Also, we adopt as center of our coordinate system the cyclopean eye, middle point between the two eyes, as humans do.

In this section we analyze the problem of orientation estimation according to different cues. First, we derive a couple of equations which are neurologically plausible and, at the same time, useful for a practical implementation. Next, we introduce the neural network architecture that we have implemented for the solution of the problem, and finally we describe some results which will allow us to discuss the theoretical and practical implications of the proposed approach.

3.1 Slant Estimation with Stereopsis and Perspective

Object orientation estimation is a very complex problem in robotic vision [5]. Cognitive science experiments showed that this is performed by humans combining the estimators provided by different visual and proprioceptual cues, both binocular (mainly horizontal and gradient disparity cues) and monocular (texture or edge perspective cues) [6]. Our approach for computing the orientation of an object is original in that it pursues estimation reliability through the merging of different estimation methods, as in our brain. For the orientation and basic shape discerning, we rely upon one monocular information source, that is, perspective under the assumption of edge parallelism, and one kind of binocular information: width disparity. In fact, texture perspective and disparity gradient are more complex implementations of the same basic properties. In humans, all these kinds of information are coded by visual areas V3 and V3a [7] and combined in the posterior parietal cortex, most likely by the human homologue of area CIP [8]. Let us first analyze the sort of computation performed by our brain during orientation estimation, in the binocular and in the monocular case.

Stereopsis. In Figure 3(a) a viewing scene is seen from above: object PQ of length l is slanted about a vertical axis with an orientation θ. Its extreme P is the fixation point, placed straight ahead from the cyclopean eye (in this way γ_P corresponds to the vergence angle). All the α angles represent the retinal projections of points P and Q on the left and right eyes, I is the interocular distance, ψ_Q the binocular separation of points P and Q (being $\psi_P = 0$).

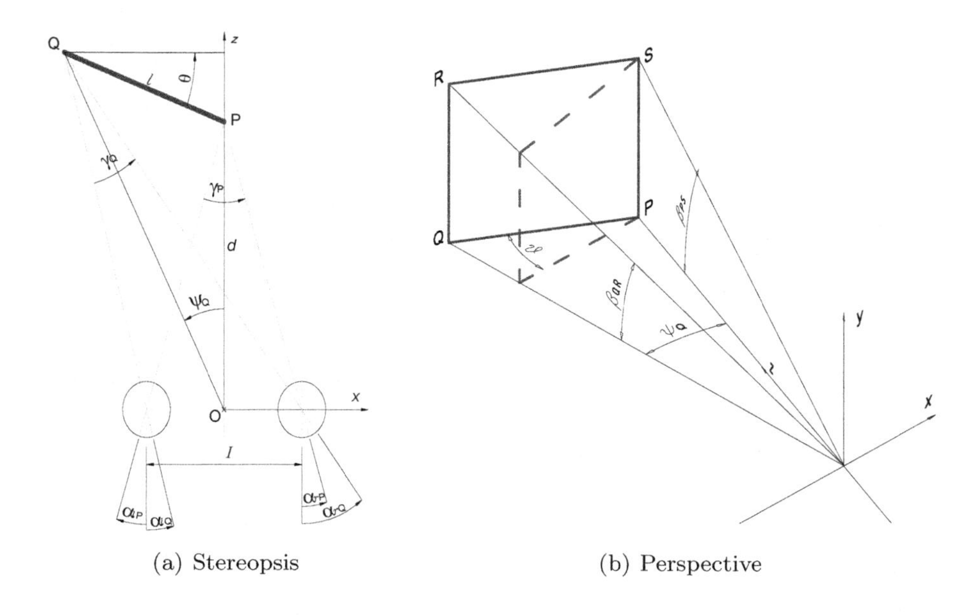

(a) Stereopsis (b) Perspective

Fig. 3. Schemes for deriving slant from stereopsis and perspective

The horizontal slant θ of an object can be computed only from retinal angles using the following expression, which can be derived from the image:

$$\tan\theta = \frac{(\tan\alpha_{rQ} - \tan\alpha_{lQ}) - (\tan\alpha_{rP} - \tan\alpha_{lP})}{\tan\alpha_{lP}\tan\alpha_{rQ} - \tan\alpha_{lQ}\tan\alpha_{rP}} \tag{3}$$

Reminding that P is the fixation point, so that $\alpha_{lP} = -\alpha_{rP} = \gamma_P/2$, we can simplify the equation in this way:

$$\tan\theta = \frac{1}{2\tan(\gamma_P/2)} \cdot \frac{\tan\alpha_{rQ} - \tan\alpha_{lQ} - 2\tan\alpha_{rP}}{(\tan\alpha_{rQ} + \tan\alpha_{lQ})/2} \tag{4}$$

Recalling Equation 2, this relation can be expressed by using only quantities that neuroscientists have observed in our visual brain areas. In this way we obtain a biologically plausible stereoptic estimator for orientation:

$$\hat{\theta}_S = \arctan\frac{relative\ disparity}{nearness \cdot separation} \tag{5}$$

The interpretation of Equation 5 is that the component due to disparity (the fraction relative disparity/separation, which is also called disparity gradient) is modulated by the viewing distance (or nearness), which acts as a gain modulation. Nearness is probably computed from proprioceptual data, as discussed in the previous section. The separation which appears in the formula is binocular, referred to the cyclopean eye. Also, the interocular distance I, which is constant, can be omitted.

Perspective. The slant of an object can be estimated also using only monocular data, as depicted in Figure 3(b), in which the origin of the axes is one of the eyes. To do this, we need to add a dimension to our object, considering it now as a rectangle, and exploit the reasonable assumption of parallelism and equality of opposite edges (PS and RQ in the image). The angles β in the figure represent the vertical retinal angles associated to such edges. The equation which leads from retinal angles to orientation estimation can be derived from the draw, and can be referred entirely to either the left or the right eye:

$$\tan\theta = \frac{\tan\beta_{QR}}{\tan\beta_{PS}\sin\psi_Q} - \frac{1}{\tan\psi_Q} \tag{6}$$

In this case we make use of the monocular separation: $\psi_Q = (\alpha_Q - \alpha_P)/2$.

If we approximate $\sin\psi_Q$ to $\tan\psi_Q$, which is plausible for reasonably small separations, we obtain a formula for a perspective estimation of θ:

$$\hat{\theta}_P = \arctan\frac{vertical\ disparity}{separation} \tag{7}$$

Vertical disparity is the quantity $\frac{\tan\beta_Q}{\tan\beta_P} - 1$ so, again, our estimator depends on a disparity factor and a separation factor.

3.2 Neural Implementation of a Multiple Cue Slant Estimator

We implemented a neural architecture for estimating the orientation θ of a target object. Our framework includes two sets of neural networks, for the stereoscopic estimation and for the monocular estimation based on perspective data.

Neural Network Estimators. The whole framework of our neural network implementation is depicted in Figure 4. Apart for nearness estimation, implemented with the RBF network described in Section 2, all networks are feedforward backpropagation, trained with the Levenberg-Marquardt algorithm. Four neural networks constitute the module for orientation estimation based on stereopsis: they compute nearness, relative disparity and separation (two nets represented as a unique one in the scheme), and the final estimate of θ_S from the outputs of the previous three networks (according to Equation 5). The module for orientation estimation based on perspective makes use of two networks for computing the two components of Equation 7, and a third for the final estimation of θ_P.

Merging the Estimators. Following the human example, our proposal is to compute orientation by combining different estimators. The idea is to use such multiple cue estimator in artificial vision for robotic applications. The merging of the stereoscopic and the perspective estimations has been done at first through a simple average of the two output values:

$$\hat{\theta}_A = \frac{\hat{\theta}_S + \hat{\theta}_P}{2} \tag{8}$$

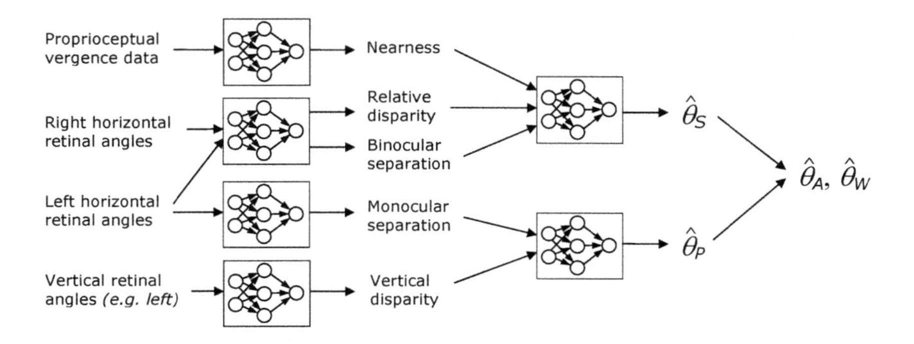

Fig. 4. Scheme of the neural networks architecture

Nevertheless, cognitive science researchers suggest that the different cues are combined in an optimal way, according to a maximum-likelihood estimation process. We thus decided to simulate this sort of combination in an experimental way, making our different estimators "learn" how their reliablity changes in different conditions. According to the literature, the driving factors for the accurateness of orientation estimation are distance and orientation itself (this is not a contradiction: the estimated value can be used as output and, at the same time, as reliability index for the estimation). In fact, although it is known that stereopsis quickly looses its reliability with distance, we are only interested in the near space defined by the arm reaching distance, within which the variation of distance affects the two methods in similar ways. For this reason, we focused our attention more on the effect of orientation and devised a merging method that optimizes the weights given to the two estimators when changing the estimate value of θ.

Cue weighting in humans depends on the reliability of the different estimators, and how our brain can predict this reliability is still a matter of debate. Nevertheless, it has been shown [9] that the different estimates are really weighted through a maximum-likelihood process, so we decided to store in a memory the error pattern obtained with the estimation methods, and generate a joint estimator which is a weighted average of the original ones:

$$\hat{\theta}_W = w_S \hat{\theta}_S + w_P \hat{\theta}_P \tag{9}$$

In the above Equation 9, $\hat{\theta}_S$ and $\hat{\theta}_P$ are the stereoptic and the perspective estimators, w_S and w_P are functions of θ computed in the following way (their behavior is similar to the curves of Figure 5(b):

$$w_S = \frac{SSE_P}{SSE_S + SSE_P}; \qquad w_P = \frac{SSE_S}{SSE_S + SSE_P} \tag{10}$$

where SSE_S and SSE_P are the summed squared errors of stereopsis and perspective respectively.

Table 1. Results with different estimators

Method	**Estimator**	**Error(°)**
Stereopsis	$\hat{\theta}_S$	8.47
Perspective	$\hat{\theta}_P$	5.75
Simple average	$\hat{\theta}_A$	5.24
Weighted average	$\hat{\theta}_W$	4.52

3.3 Results and Discussion

In principle, the neural network implementation allows to achieve any arbitrary precision in the estimation. The study of the reliability of the estimators can thus be done either in the real world or simulating the effect of natural imprecisions introducing stochastic variability in the computation. Before the implementation on a real robotic setup we thus decided to simulate the effect of noise on slant estimation performed with either stereopsis (binocular disparity) and monocular (perspective) methods. With this purpose, we added random noise to all retinal angles which constitute input values for the nets and calculate an error which is the average difference between the estimation and the true value, all over the test space (-80°$< \theta <$80°, 200mm$< d <$700mm).

In table 1 the improvement of joining the two estimators in this way can be observed. For comparison, consider that the nets were trained so that the average error of the two original estimators $\hat{\theta}_S$ and $\hat{\theta}_P$ before the introduction of noise was less than one degree. Although the perspective predictor looks always superior to the stereoscopic one, the contribution of the latter is still very important for improving the final prediction. In fact, the weighted average $\hat{\theta}_W$ improves the $\hat{\theta}_P$ performance of a further 22%, suggesting that the combination of different cues is indeed the best solution for pursuing a reliable estimation. Even the simple average $\hat{\theta}_A$, that can be used a priori without exploiting previous experience, allows to obtain a 10% improvement on $\hat{\theta}_P$.

Considering the effect of distance and orientation on the estimation reliability, experiments with human subjects [9] showed that all estimations become less reliable with increasing distance, though stereoscopic cues are more affected. The effect of orientation is more complex. Excluding very extreme slant values (close to ±90°), perspective methods are more sensitive and precise for pronounced slants, that generate higher differences in vertical disparities. Disparity methods instead favor low slant values, which grant higher binocular disparities. To check if this pattern of behavior could be reproduced in our simulation, we plotted the accurateness of the two estimators and their averages as a function of d and as a function of θ. The graphs we obtained, which can be seen in Figure 5, show that the stereoptic and the perspective estimators behave as predicted by the literature, both worsening with distance (but surely more the binocular method) and showing the different error patterns described above when varying θ. We believe this is a strong argument in support of the appropriateness of our model. Regarding the validity of the cue merging approach, it can be clearly seen how

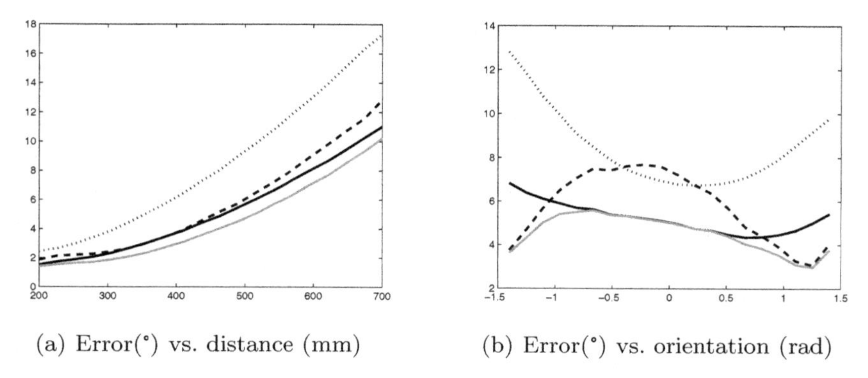

(a) Error(°) vs. distance (mm) (b) Error(°) vs. orientation (rad)

Fig. 5. Estimation errors as a function of distance and orientation. Dotted line: $\hat{\theta}_S$; dashed line: $\hat{\theta}_P$; solid line: $\hat{\theta}_A$; clear line: $\hat{\theta}_W$.

the weighted average estimator $\hat{\theta}_W$ is always superior to both estimators alone, even when their performance is very unequal.

It is probably useful to clarify that we do not claim that our brain uses the formulas explained above to compute orientation from stereopsis and perspective: such formulas were only necessary to train the networks in a simulation. Nevertheless, we believe that the nets themselves are actually modeling the behavior of modules pertaining to our higher visual brain areas. Indeed, our inputs and intermediate results represent variables which are believed to be part of brain processes [7], and this suggests that something very similar to Equations 5 and 7 is probably really computed in our cortex.

Implications for Robotic Vision. Although research on cue integration is not at all new in neuropsychology, we believe that these findings have not been exploited up to their potentialities in computational vision and robotic applications. Orientation and pose estimation are very complex problems in artificial vision, especially when the goal is to develop a reliable robotic application which makes use of visual estimates to interact with the environment, such as in object grasping actions [5].

In the framework of our project, which on the robotic side aims at improving the behavior of a grasping robot within a real environment, we are strongly pursuing methods for increasing the accurateness of estimates, in order to make the robot interact more reliably with its environment. The results obtained so far in this research project encourage to follow the approach presented in this paper, as it looks really promising for help improving the reliability of a real application. The next immediate step of the presented research is the practical experimentation on our robotic platform, which consists of a dexterous three-finger hand mounted on a 7 d.o.g. arm and a stereo vision system.

4 Conclusions

The neural architecture we implemented constitutes an orientation estimator both biologically plausible and practically reliable. We used variables that are available to humans, but also computationally useful for artificial systems (e.g. retinal angles). We could reproduce effects observed in neuropsychological experiments, and are now exploiting them in an artificial set-up.

One of the main limitations of the work at this point is that only horizontal slant is considered. We are now working on a module for estimating binocular orientation disparity, which will allow the extension of the analysis to 3D inclinations. Another limit of our analysis is the reduction to arm-reaching distances. The expected effect at larger distances is that perspective will completely take over stereopsis. This can be understood looking at the distance/vergence graphs of Figure 2. Anyway, for us this aspect has only speculative implications, as our focus is on visual analysis aimed at object reaching and grasping.

Acknowledgments

Support for this research has been provided in part by Fundació Caixa-Castelló (project P1-1B2005-28) and Generalitat Valenciana (project GV05/137). Thanks to professor Oscar Chinellato for his help and advice.

References

1. E. Chinellato, Y. Demiris, and A. P. del Pobil. Studying the human visual cortex for achieving action-perception coordination with robots. In *IASTED Intl. Conf. on Artificial Intelligence and Soft Computing (ASC 2006)*, 2006.
2. J. R. Tresilian and M. Mon-Williams. Getting the measure of vergence weight in nearness perception. *Exp Brain Res*, 132(3):362–368, Jun 2000.
3. K. Tsutsui, M. Jiang, K. Yara, H. Sakata, and M. Taira. Integration of perspective and disparity cues in surface-orientation-selective neurons of area CIP. *J Neurophysiol*, 86(6):2856–2867, Dec 2001.
4. D. G. Jones and J. Malik. Determining three-dimensional shape from orientation and spatial frequency disparities. In *Eur.Conf. on Comp. Vis.*, pages 662–669, 1992.
5. V. Lippiello, B. Siciliano, and L. Villani. 3d pose estimation for robotic applications based on a multi-camera hybrid visual system. In *IEEE Intl. Conf. on Robot. Automat.*, pages 2732–2737, 2006.
6. B. T. Backus, M. S. Banks, R. van Ee, and J. A. Crowell. Horizontal and vertical disparity, eye position, and stereoscopic slant perception. *Vision Res*, 39(6):1143–1170, Mar 1999.
7. A. E. Welchman, A. Deubelius, V. Conrad, H. H. Bülthoff, and Z. Kourtzi. 3D shape perception from combined depth cues in human visual cortex. *Nat Neurosci*, 8(6):820–827, Jun 2005.
8. E. Shikata, F. Hamzei, V. Glauche, R. Knab, C. Dettmers, C. Weiller, and C. Büchel. Surface orientation discrimination activates caudal and anterior intraparietal sulcus in humans: an event-related fMRI study. *J Neurophysiol*, 85(3):1309–1314, 2001.
9. J. M. Hillis, S. J. Watt, M. S. Landy, and M. S. Banks. Slant from texture and disparity cues: optimal cue combination. *J Vis*, 4(12):967–992, Dec 2004.

An Approach to Visual Scenes Matching with Curvilinear Regions

J.M. Pérez-Lorenzo[1], A. Bandera[2], P. Reche-López[1], R. Marfil[2], and R. Vázquez-Martín[2]

[1] Dept. Ing. Telecomunicación, Universidad de Jaén, Linares-23700, Spain
[2] Dept. Tecnología Electrónica, Universidad de Málaga, Málaga-29071, Spain

Abstract. This paper presents a biologically-inspired artificial vision system. The goal of the proposed vision system is to correctly match regions among several images to obtain scenes matching. Based on works that consider that humans perceive visual objects divided in its constituent parts, we assume that a particular type of regions, called curvilinear regions, can be easily detected in digital images. These features are more complex than the basic features that human vision uses in the very first steps in the visual process. We assume that the curvilinear regions can be compared in their complexity to those features analysed by the IT cortex for achieving objects recognition. The approach of our system is similar to other existing methods that also use intermediate complexity features for achieving visual matching. The novelty of our system is the curvilinear features that we use.

1 Introduction

Biological vision and artificial vision are two strongly interconnected fields. Biological vision theories can be helpful for new insights and development of artificial visual systems. At the same time, artificial visual systems allow proving models and theories based on biological experiments. Also, a better theoretical understanding of the computation when processing visual information can supply useful guidelines for empirical studies of the biological processes. Despite of these interrelations, even the best artificial system cannot rival the performance of a few years old child. For example, image matching is still a fundamental aspect of many problems in computer vision, including scene recognition, stereo correspondence and motion tracking. A good starting point to approach this problem can be found in visual theories, which include scientific fields such as biology, psychology and neuroscience.

In artificial vision, image matching is defined as the process of bringing two images into agreement so that corresponding pixels in the two images correspond to the same physical region of the scene. The similarity may be applied to global features derived from the original images. However, this is not the more efficient solution and we think it does not match with the biological principles of human vision. The approach we have used is to analyse first the scenes in order to extract some features and build a scheme to work with these features.

J. Mira and J.R. Álvarez (Eds.): IWINAC 2007, Part II, LNCS 4528, pp. 409–418, 2007.

2 Biological Inspiration

In recent years great advances have been reached in our understanding in the primate visual system. Some experiments show that primate vision works in a hierarchical way. The first stages use simple local features, and the image is subsequently represented in terms of larger and more complex features. In the earliest processing stages, which involve the retina, lateral geniculate nucleus (LGN) and primary visual cortex (V1), the image is represented by simple local features such as center-surround receptive fields and oriented lines and edges. After these steps, moderately complex features are represented in areas V4 and the adjacent region TEO, and finally, partial or complete object views are represented in anterior regions of inferior temporal (IT) cortex. Indeed, there are now great evidences showing that object recognition makes use of neurons in IT that respond to features of intermediate complexity [7][12]. So, a partial answer to objects recognition is to consider that every image can itself be broken down into components and that visual cortical neurons are specialized to detect only a subset of these components. In [12] it is shown by computational analysis and simulations that features of intermediate complexity and partial object views are optimal for visual object classification. Ullman suggests, based on equations of mutual information and entropies, that intermediate complexity features are more informative than very simple ones (detected in V1 receptive fields) or very complex ones. Indeed, it is shown that visual features of intermediate complexity emerge naturally from a coding principle of maximizing the delivered information with respect a class of objects.

In object recognition no much work has solved the question of specificity, that is, which properties of an object are exactly encoded in the neural representation used to recognize that object, and also the problem of feature selection is a fundamental question [12] [4]. Nevertheless, when using intermediate complexity features, these features seem to be very sensitive to particular combinations of local shape, colour, and texture properties [7]. Some researchers have suggested that objects are represented in terms of their constituent parts. Some of the most important ones are Biederman's "Recognition by Components" [2], where objects are formed with an alphabet of simple geometrical shapes called geons, and Marr's "Generalized Cylinders" [9], where these parts are cylinders along the main axes of the object. In both cases, the constituent parts of the object are "intuitive" in the sense that they correspond to the parts in terms of which normal observers commonly understand the objects (e.g. the legs of a table). An alternative approach can be found in [12], who suggested that the intermediate complexity fragments need not correspond to those intuitive parts for the purpose of classifying an object. The work that we present here is more inspired in Marr's theory. In fact, based on the supposition of the generalized cylinders we assume that digital images have got visual regions that can be easily detected thanks to their cylindrical shape, and we call them curvilinear regions. The purpose of our system is not to detect every component described by Marr's theory because it would be a quite complex computational task. However, our experiments have shown that with the detection of some of them, and using them as features for matching and classification stages, the system is able to identify images and scenes with promising results.

Also the above ideas lead to the discussion among "image-based" models, in which objects are represented as collection of viewpoint-specific features and "structural-description" models, in which objects are represented as configurations of 3D volumes. Nowadays, there is still no common agreement among researchers. This issue is extensively discussed in the literature. In [6] a better performance is suggested for 3D shape representations compared with open contour and closed contour models. However, they find also an advantage for using 2D polygons to approximate the object surface shapes, which provides a new framework for the studies of 3D shape representation. In [11] a mixture of both models is suggested to obtain the most appealing aspects of both of them. In our work, although the curvilinear regions are based on the existence of generalized cylinders, we do not build a 3D framework. In fact, our work is more similar to the "image-based" models because we work with the projections of the features, although we transform them in a way that can be interpreted as an extended image-based approach according to [11].

In the computational literature there are other methods that uses intermediate complex features for classifying objects and scenes with good results [7][12][10]. Our work differs mainly from these in the features that we are using, the curvilinear regions.

3 Curvilinear Regions

3.1 Definition

We define our features of intermediate complexity as curvilinear regions represented by a parameter vector $\{a(l), w_l(l), w_r(l)\}_{l=0..L}$, being L the length of the region, $a(l)$ a vector defining the axis between the right and left borders ($b_r(l)$ and $b_l(l)$), and $w_r(l)$ and $w_l(l)$ the widths of the curvilinear region (see Fig. 1). We consider that a region is a curvilinear region if it satisfies the following conditions: i) there must be a geometric similarity around the region axis, ii) ratio between its average width and its total length must be less than a predefined threshold, iii) left and right borders must be locally parallel, iv) the colour along this axis should be homogeneous. The first three properties are geometrical properties. The last one is a restriction colour in order to make the detection easier, as well as it provides the existence of a simple colour descriptor for the region.

3.2 Overview of the Proposed Method

The algorithm for detecting the curvilinear regions can be divided in several steps. Firstly, the original image is segmented into homogeneous colour regions using a pyramid algorithm based on the Bounded Irregular Pyramid (BIP) [8]. The obtained regions comply with the homogeneous colour property, so the geometric properties of every region are checked looking for those regions with the

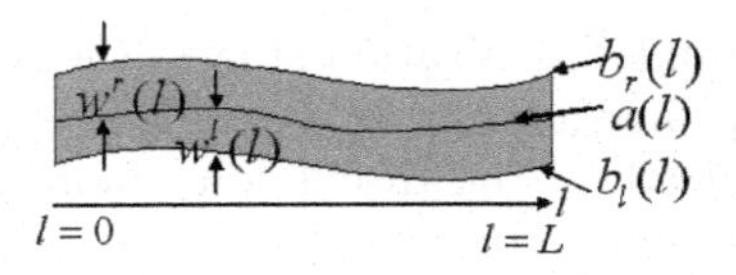

Fig. 1. Definition of curvilinear region. $b_l(l)$,$b_r(l)$: left and right borders, $a(l)$: medial axis, $w_l(l)$,$w_r(l)$: left and right widths.

curvilinear properties (this step requires the extraction of the medial axis for each region). As the result of this analysis step the method obtains the set of the detected curvilinear regions. Then, a normalisation step is applied. If the curvilinear region is enclosed inside an elliptical region whose centre is the centre of mass of the region, this step transforms the elliptical region into a circular reference region of fixed size. These normalised regions are employed as the input of the shape descriptor, which is a contour-based approach to object representation that uses a curvature function for the description of the boundary region. This contour descriptor is invariant to rotation and translation, and partially invariant to noise, scaling and skewing. The recognition stage matches the obtained individual features to a database of features from known scenes using a nearest-neighbour algorithm. This algorithm uses the curvature descriptor, colour descriptor and position of the region in the image. Experimental results show that this approach to scene recognition can match images taken from diffe-rent viewpoints if they present a similar layout, i.e. spatial distribution of curvilinear objects.

3.3 Image Segmentation Based on the Bounded Irregular Pyramid

In the present work we have used a pyramid segmentation algorithm. We have used a framework where the pyramid represents the image at multiple levels of abstraction. In each level the hierarchy presents a set of vertices V_l connected by a set of edges E_l. These edges connect vertices at the same level (intra-level edges) and define also relationships between pyramid levels (inter-level edges), where the vertices at level l connected with one vertex at level $l+1$ are defined as the sons of the vertex at level $l+1$, being this one the parent. The value of each parent is computed from its sons using a reduction function. Using this framework, the local evidence accumulation is achieved by the successive building of level $G_{l+1} = (V_{l+1}, E_{l+1})$ from level $G_l = (V_l, E_l)$. We have used the bounded irregular pyramid (BIP) proposed in [8]. In the irregular pyramids the main problem is that the size is not bounded, but in the BIP this is solved by combining the simplest regular and irregular structures. As we can see in Fig. 2, the segmented images are obtained correctly with this technique.

3.4 Medial Axis Extraction and Skeleton Classification

The skeleton of the region is used for the analysis of the geometric properties that define if a region is curvilinear or not. The skeleton is defined as a subset

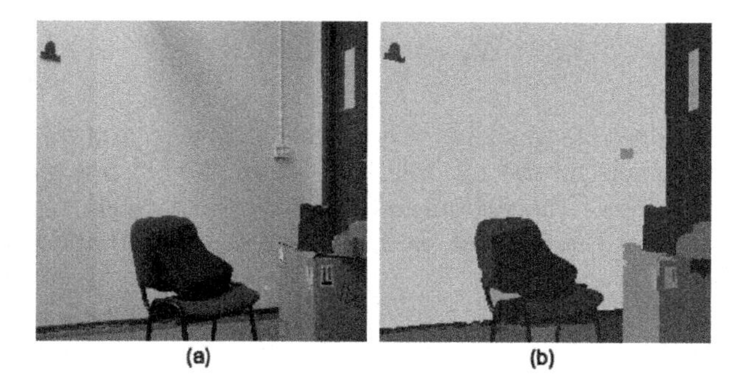

Fig. 2. Segmentation results obtained using the BIP structure: a) original image; b)segmentation results

of pixels that preserve the topological information of the region and it must approximate the medial axis. In this work a distance transformed approach is used for each colour segmented region, where a distance skeleton is a subset of grid points such that every point represents the centre of a maximal disc contained in the given component. For estimating the distance transform of a region we use the algorithm based on the $d8$-distance described in [5] which can approximate the distance transform inside the region in only two steps, so it has got a low computational cost. Those pixels which present a local maximum in the distance transform belong to the distance skeleton, and by choosing them we can obtain a skeleton for each region. These distance skeletons are generally not connected, so we post-process them with morphological operations to obtain connected and smooth skeletons that are used to estimate further geometric properties.

Our method decides which parts of the skeleton belong to a curvilinear region by estimating a set of geometric characteristics: symmetry around the skeleton, ratio between its average width and its total length, and borders parallelism.

Symmetry Around the Skeleton. The method looks for those pixels which comply with the requirement of symmetry around the axis as indicated in 3.1. To describe the algorithm we can define a skeleton as the set of connected pixels $p_s = (i_s, j_s), 0 \leq s \leq N-1$, being N the number of pixels being evaluated of the skeleton. In a first step, the normal vector is calculated for each pixel ps in the skeleton, and the cross-points between the normal and the left and right borders of the region are estimated. If we define p_s^l and p_s^r as these cross-points, then we obtain the triplets $(p_s, p_s^l, p_s^r), 0 \leq s \leq N-1$. The symmetry condition can be defined as:

$$\frac{1}{N}\sum_{s=0}^{N-1}(\Delta w_s - \overline{\Delta w})^2 \leq U_{\Delta w}(1 - e^{-\frac{N^2}{2\sigma^2_{\Delta w}}}) \tag{1}$$

with

$$\Delta w_s = |w_s^l - w_s^r| \tag{2}$$

$$\overline{\Delta w} = \frac{1}{N} \sum_{s=0}^{N-1} \Delta w_s \tag{3}$$

being w_s^l the Euclidean distance between pixels p_s and p_s^l and w_s^r the Euclidean distance between pixels p_s and p_s^r. The left side in Ec. (1) is a term that grows with the asymmetries of the region and the values $U_{\Delta w}$ and $\sigma_{\Delta w}$ in the right side are parameters of the method, we have used $U_{\Delta w} = 10$ and $\sigma_{\Delta w} = \sqrt{50}$ in our experiments.

Ratio. Given a position s in the skeleton, the width w_s of the region is estimated as the Euclidean distance between pixels p_s^l and p_s^r. In order to assure that the ratio between the width and the total length of the region is less than a predefined threshold, the following condition must be satisfied:

$$L_{max} \geq U_w \frac{1}{N} \sum_{s=0}^{N-1} w_s \tag{4}$$

being L_{max} the maximum length that the curvilinear region could have, which is estimated with all the connected pixels of the skeleton. U_w is also a parameter of the method set to 1.5.

Borders Parallelism. To check the borders parallelism requirement we estimate the tangential vectors on the borders at pixels p_s^l and p_s^r. Then, we calculate the angle between those vectors and the normal vector given a position s, obtaining angles α_s^l and α_s^r. We consider that the borders are parallel if the following condition is satisfied:

$$\frac{1}{N} \sum_{s=0}^{N-1} |\alpha_s^l - \alpha_s^r| \leq U_{\Delta\alpha} \tag{5}$$

$U_{\Delta\alpha}$ is a parameter of the method set to 30 degrees.

Classification Algorithm. When the skeleton has been extracted from the distance transform image associated to an object, the algorithm tries to join as many pixels as possible to form a curvilinear skeleton. The algorithm starts in an endpoint of the skeleton and it looks for adding the connected pixels checking if Ec. (1), Ec. (4) and Ec. (5) are true with the new added pixel. If they are true, the pixel is added and the next pixel will be studied. In case that any requirement is not fulfilled, the curvilinear skeleton is finished and a new curvilinear region will begin with the next positive evaluation. When all the pixels have been evaluated inside a region, the curvilinear skeletons with close endpoints are linked. Those parts of the objects with a skeleton evaluated as a curvilinear skeleton are considered curvilinear regions. In our experiments, we demand that these regions must have a minimum length of 10 pixels. Figs. 3.a and 3.b present an experiment and its results.

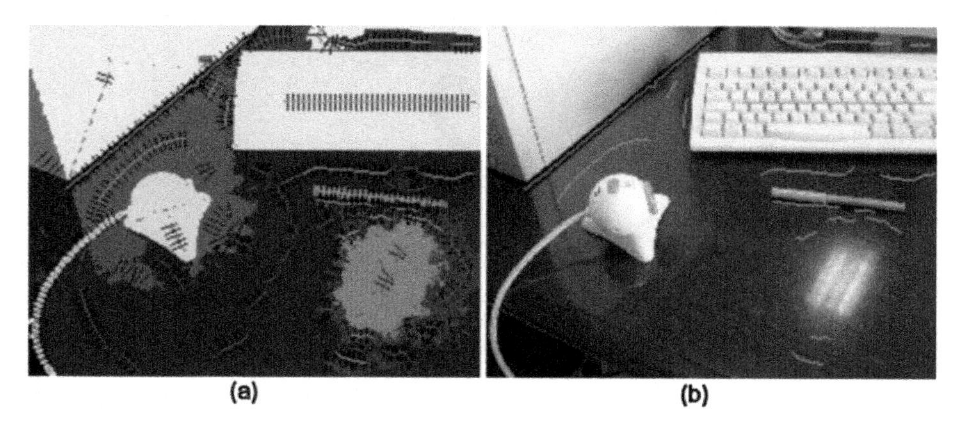

Fig. 3. a) Regions in a segmented image. The extracted skeletons have been drawn (in green colour the skeletons classified as curvilinear and in red colour as not curvilinear). Also some estimated normal vectors (black colour) to the skeletons have been drawn. b) Original image with the detected curvilinear skeletons superimposed. Several interesting objects as the ball pen, keyboard and webcam cable have been detected.

3.5 Normalisation Stage and Shape Description

In the normalisation stage we apply a 2D normalisation before obtaining the descriptor. For this we enclose the region inside an elliptical-shaped region and transform it into a circular region of fixed size (see Fig. 4.a)

The transformed curvilinear regions extracted are characterized using a shape descriptor. We have used the curvature function of the boundary of the region as the shape descriptor, which encodes the shape contour in terms of their local curvature or orientation. We have used an angle-based curvature estimator, where the curve orientation is estimated at each point with respect to a reference direction. In this method, the contour curvature $K(t)$ can be defined as the variation of the curve slope $\psi(t)$ with respect to t, or the inverse of the curvature radius $\rho(t)$:

$$K(t) = \frac{\delta\psi(t)}{\delta t} = \frac{1}{\rho(t)} \tag{6}$$

To extract $K(t)$ we have used an approach based on a k-slope algorithm which estimates the curvature using a k value which is adaptively changed [1]. Fig. 4.c shows an example of a curvature function associated to a shape contour.

4 Experimental Results

The curvilinear regions detected are characterised using a 260-dimensional space whose first two dimensions $(x, y)_i$ are the co-ordinates of the centre of mass of the region, the second two dimensions $(h, s)_i$ are the mean hue and saturation values of the region (HSV colour space), and the other 256 values $fc_{i_{i=1..256}}$

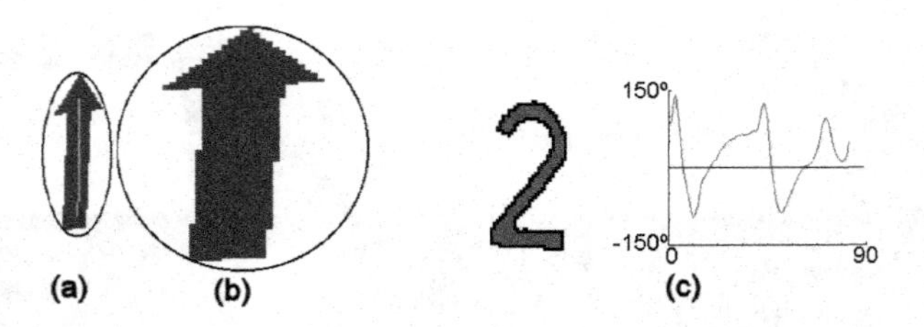

Fig. 4. a),b) Original and normalized curvilinear region; c)Curvilinear region shape and associated curvature function

are the curvature function of the object shape. So, each region is characterised by the shape of the contour, the position in the image and the homogeneous colour inside the region. We decide that two images will be similar if their sets of curvilinear regions are similar. For checking this similarity we define the distance between two curvilinear regions i and j as

$$D(i,j) = \alpha_1\sqrt{(x_i - x_j)^2 + (y_i - y_j)^2} + \alpha_2\sqrt{s_i^2 + s_j^2 - 2s_is_jcos\theta} + \\ +\alpha_3 max\{(fc_i * fc_j)_{k=1..256}\} \quad (7)$$

where θ is equal to $|h_i - h_j|$ if this value is less than π, or equal to $(2\pi - |h_i - h_j|)$ in any other case. Parameters α_i are chosen to define the importance of the position, colour and shape into the distance measure.

Given a query image Q and a dataset of images B_i with the curvilinear regions already characterised, the image matching process firstly extracts the set of N_Q curvilinear regions $\{c_{Q_i}\}_{i=1..N_Q}$ of the query image. The comparison between Q and each image B_i is achieved by comparing each curvilinear region in Q, c_{Q_i}, with all the N_{B_i} curvilinear regions present in B_i, $\{c_{Bi}\}_{i=1...N_{Bi}}$. The most similar region is then selected (if the distance value is less than a threshold) and both curvilinear regions are paired. When all the regions are processed then a similarity value is assigned to the image of the dataset. This similarity value is defined as

$$\lambda = (N_{B_i} - N'_Q + 1)\sum_{i=1}^{N'_Q} D(cQ_i, cBi_j) \quad (8)$$

where N'_Q is the number of paired curvilinear regions. Once this similarity value is computed for each image B_i of the dataset, they are sorted according to it. Those images B_i with a similarity value below a threshold are possible candidates to be matched with the query image Q.

To test this scheme, a database of 40 images obtained in an office-like environment has been created, which are divided into 10 different scenarios. In Fig. 5 we present two example retrievals for this database where the query image is the leftmost image in the rows, and subsequent images are nearest neighbours. In the images, the detected regions for the match among the scenes have been marked. To evaluate the matching performance, a normalised average rank has been used [3]:

$$\overline{R} = \frac{1}{NN_R}(\sum_{i=0}^{N_R-1} R_i - \frac{N_R(N_R-1)}{2}) \tag{9}$$

R_i is the rank at which the ith relevant image is retrieved, N_R is the number of relevant images for a given query, and N is the number of examples in the database. In our experiments the normalised average rank of relevant images present an average value of 0.025 and a standard deviation of 0.001 (if the average value is equal to zero it means a perfect performance)

Fig. 5. Example retrievals for a database of office-like environment images

5 Conclusions and Future Work

We have presented a method which compares scenes based on the matching of detected curvilinear regions. The proposed method is inspired in biological theories which claim that objects can be divided in their constituent parts or regions in order to recognize them.We represent and compare these regions using their colour and position in the scene as well as a set of geometric properties of them. According to the literature, we think that there is an analogy between the curvilinear used features and the intermediate complexity features for the identification of objects in visual processes. These intermediate complexity features seem to be very sensitive to combinations of local shape, colour and texture properties and they are supposed to be used in the process of identifying visual objects. By adding the relative position in the scene to the characterization of the features, we have extended their use not only to object recognition but also to identify the visual scene.

Future work should be focused to further experiments in more scenarios in order to detect more regions in the images, and to study in more depth the possible analogies between the curvilinear regions used in this work and the features used in IT cortex for visual recognition, according to the most recent discoveries in visual neuroscience.

Acknowledgements

This work has been partially granted by FEDER and the Spanish Ministry of Education and Science under Project TEC2006-13883-C04-03 and project TIN2005-01359.

References

1. Bandera, A., Urdiales, C., Arrebola, F., Sandoval, F., "Corner detection by means of adaptively estimated curvature function", *Electronics Letters*, 36(2), 124-126, 2000.
2. Biederman, I., "Recognition-by-components: A theory of human image understanding", *Psychological Review*, 94, 115-147, 1987.
3. Grauman, K., Darrell, T., "Efficient image matching with distributions of local invariant features", *Proc. of IEEE Conf. Computer Vision and Pattern Recognition*, 627-634, 2005.
4. Haushofer, J., Baker, Cl., Kanwisher, N.,"Greater Sensitivity to Convexities than Concavities in Human Lateral Occipital Complex",*Society for Neuroscience Annual Meeting*, 2005.
5. Klette, G., "A comparative discussion of distance transformation and simple deformations in digital image processing", *Machine Graphics & Vision*, 12(2), 235-256, 2003.
6. Leek, E.C., Reppa, I., Arguin, M. "The structure of three-dimensional object representations in human vision: evidence from whole-part matching", *Journal of Experimental Psychology: Human Perception and Performance*, 31(4), 668-684, 2005.
7. Lowe, D., "Towards a computational model for object recognition in IT Cortex". *First IEEE International Workshop on Biologically Motivated Computer Vision*, Seoul, Korea, 20-31, 2000.
8. Marfil, R., Rodriguez, J.A., Bandera, A., Sandoval, F., "Bounded irregular pyramid: a new structure for color image segmentation", *Pattern Recognition*, 37(3), 623-626, 2004.
9. Marr D., Nishihara, H.K., "Representation and recognition of the spatial organization of three-dimensional shapes", *Proceedings of the Royal Society of London B: Biological Sciences*, 200, 269-294, 1978.
10. Matas, J., Chum, O., Urban, M., Pajdla, T.,"Robust wide baseline stereo from maximally stable extremal regions", *Proceedings of the British Machine Vision Conference*, 1, 384-393, 2002.
11. Tarr, M., Bulthof, H., "Image-based object recognition in man, monkey and machine", *Cognition*, 67, 1-20, 1998.
12. Ullman, S.,Vidal-Naquet, M.,Sali E., "Visual features of intermediate complexity and their use in classification", *Nature Neuroscience*, 5, 682-687, 2002.

Supervised dFasArt: A Neuro-fuzzy Dynamic Architecture for Maneuver Detection in Road Vehicle Collision Avoidance Support Systems

Rafael Toledo[1], Miguel Pinzolas[2], and Jose Manuel Cano-Izquierdo[2]

[1] Dept. of Electronics, Computer Technology and Projects
[2] Dept. of Systems Engineering and Automation
Technical University of Cartagena
{Rafael.Toledo,Miguel.Pinzolas,JoseM.Cano}@upct.es

Abstract. A supervised version of dFasArt, a neuronal architecture based method that employs dynamic activation functions determined by fuzzy sets is used for solving support of the problem of inter-vehicles collisions in roads. The dynamic character of dFasArt minimizes problems caused by noise in the sensors and provides stability on the predicted maneuvers. To test the proposed algorithm, several experiments with real data have been carried out, with good results.

Keywords: dFasArt, Collision Avoidance, Maneuver Detection.

1 Introduction

The issue of collision avoidance in roads has been addressed from many different points of view in the current literature. An interesting approach is based on the creation and interpretation of a scene of vehicles in a potentially conflictive situation [1] – [7]. Vehicles exchange pose and geographical information, along with its current maneuver states through ad hoc WLAN networks, in order to determine their roles in the scene [8].

However, the problem of defining appropriate dynamics to represent all possible maneuver states of road vehicles is not simple. To recognize maneuver states, several authors employ multiple model filters [9], [10]. The use of multiple models allows more accurate positioning and noise estimation, and the possibility of recognizing different dynamic states of the vehicle [11]. In order to solve the problem of often unrealistic switches between models, several authors propose the use of interactive multiple model (IMM) filtering, in which maneuver states are formulated as Markovian processes. In works such as [10], the use of an IMM based method combining constant velocity (CV) and constant acceleration (CA) models has been found adequate to this problem. Some other authors like [9] and [12] employ similar assumptions to represent different vehicle dynamics. Nevertheless, despite of the improvements achieved, the tuning of the CA filter noise parameters has been found problematic. For example, in case of highways, typical accelerations do not last long enough to accomplish a transition from CV state to CA, while situations in urban scenarios are completely different [13].

J. Mira and J.R. Álvarez (Eds.): IWINAC 2007, Part II, LNCS 4528, pp. 419–428, 2007.

In our case, market considerations encourage the use of low cost sensors, at the expense of high noise values in the measurements. Therefore, a method capable to interpret very different vehicle dynamics anytime, avoiding noise inconveniences is advisable.

The dFasArt neural architecture can naturally address these issues. Due to its dynamic quality, dFasArt allows taking into account the time span of input data, without the need of keeping buffers of past input or output values. The dynamic characteristic of the activation functions provides a natural way of filtering noise in inputs, while the dynamic evolution of the reset signals allows stability in the predictions.

In this work, a supervised version of dfasArt architecture has been developed and tested for maneuver detection. In the following Section, a brief description of the neural architecture is given. In Section 3, experimental setup is described and obtained results commented. Finally, Section 4 summarizes our conclusions about the work.

2 Supervised dFasArt

FasArt model links the ART architecture with Fuzzy Logic Systems, establishing a relationship between the unit activation function and the membership function of a fuzzy set. On the one hand, this allows interpreting each of the FasArt unit as a fuzzy class defined by the membership-activation function associated to the representing unit. On the other hand, the rules that relate the different classes are determined by the connection weights between the units.

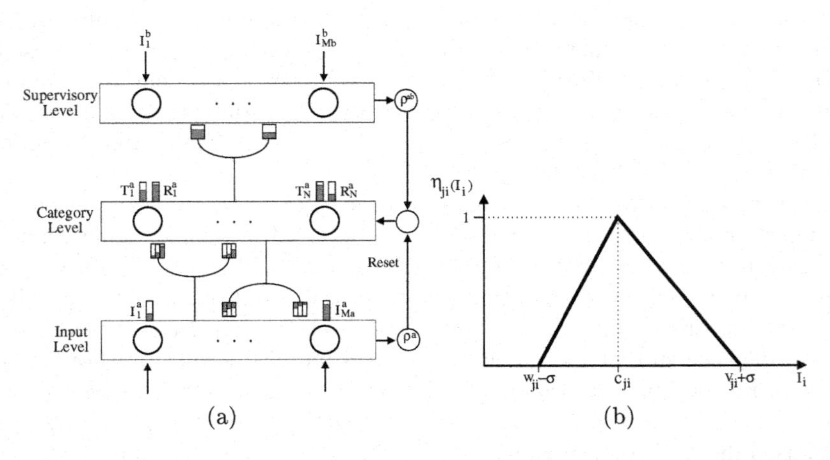

Fig. 1. (a) Supervised dFasArt model arquitecture. (b) Membership-activation function.

Derived from FasArt, dFasArt uses dynamic activation functions, determined by the weights of the unit. These weights can be regarded as the defining parameters of a fuzzy set membership function [14]. In dFasArt, learning is unsupervised and incremental. In this work, a supervised version on the dFasArt algorithm has been developed. This modification follows the ARTMAP philosophy, maintaining the maximum generalization-minimum prediction error principle.

The Supervised dFasArt (SdFasArt) architecture is represented in Fig. 1(a). SdFasArt uses a dynamic activation function determined by the weights of the unit as the membership function of a fuzzy set. The signal activation is calculated as the AND of the activations of each one of the dimensions when a multidimensional signal is considered. This AND is implemented using the product as a T-norm. Hence, the activity T_j of unit j for a M-dimensional input $\boldsymbol{I} = (I_1 \ldots I_M)$ is given by:

$$\frac{dT_j}{dt} = -A_T T_j + B_T \prod_{i=1}^{M} \eta_{ji}(I_i(t))$$

Where η_{ji} is the membership function associated to the *ith-dimension* of unit j, determined by the weights w_{ji}, c_{ji} and v_{ji}, as it is shown in Figure 1(b). The σ parameter determines the fuzziness of the class associated to the unit by:

$$\sigma = \sigma^* |2c_j| + \epsilon$$

The election of the winning unit J is carried out following the winner-takes-all rule:

$$T_J = \max_j \{T_j\}$$

The learning process starts when the winning unit meets a criterion. This criterion is associated to the size of the support of the fuzzy class that would contain the input if this was categorized in the unit. This value is calculated dynamically for each unit according to:

$$\frac{dR_j}{dt} = -A_R R_J + B_R \sum_{i=1}^{M} \left(\frac{\max(v_{Ji}, I_i) - \min(w_{Ji}, I_i)}{|2c_{Ji}| + \epsilon} \right)$$

The R_j value represents a measurement of the change needed on the class associated to the j unit to incorporate the input. To see if the J winning unit can generalize the input, it is compared with the design parameter ρ, so that:

- If:

$$R_J \geq \rho$$

the matching between the input and the weight vector of the unit is good, and the learning task starts.
- If:

$$R_J < \rho$$

there is not enough similarity, so the Reset mechanism is fired. This inhibits the activation of unit J, returning to the election of a new winning unit.

If the Reset mechanism is not fired, then the learning phase is activated and the unit modifies its weights. The *Fast-Commit Slow-Learning* concept is commonly used.

When the winning unit represents a class that had performed some other learning cycle (*committed unit*), the weights are *Slow-Learning* updated according to the equations:

$$\frac{d\boldsymbol{W}}{dt} = -A_W \boldsymbol{W} + B_W \min(\boldsymbol{I}(t), \boldsymbol{W})$$
$$\frac{d\boldsymbol{C}}{dt} = A_C(\boldsymbol{I} - \boldsymbol{C})$$
$$\frac{d\boldsymbol{V}}{dt} = -A_V \boldsymbol{V} + B_V \max(\boldsymbol{I}(t), \boldsymbol{V})$$

For the case of the uncommitted units, the class is initialized with the first categorized value, hence *Fast-Commit*:

$$\boldsymbol{W}_J^{NEW} = \boldsymbol{C}_J^{NEW} = \boldsymbol{V}_J^{NEW} = \boldsymbol{I}$$

Supervision is carried out in the supervisory level, by means of vector $\boldsymbol{I}^b = (I_1^b \dots I_{Mb}^b)$. In this level, for each time instant, $\boldsymbol{I^a}$ actives the corresponding unit. The $\boldsymbol{W_k^{ab}}$ matrix of adaptive weights associates, in a many-to-one mapping, units of the category level to units on the supervisory one. When a unit J is activated for the first time in the category level, weights are adapted by means of a fast-learning process:

$$\boldsymbol{W_J^{ab}} = \boldsymbol{I^b}$$

If unit J is a committed unit, a matching between the membership value to the predicted category and the "crisp" desired value is carried out:

- If:
  $$|\boldsymbol{W_J^{ab}} \wedge \boldsymbol{I^b}| \geq \rho^{ab}|\boldsymbol{I^b}|$$
  So that prediction corroborates supervision.
- If:
  $$|\boldsymbol{W_J^{ab}} \wedge \boldsymbol{I^b}| < \rho^{ab}|\boldsymbol{I^b}|$$
  then matching between prediction and supervision is not strong enough. In this case, the Reset signal is fired, and a new prediction is made.

When no supervision is present, SdFasArt will predict as output the value associated to the weight vector of the winning unit, that is, $\boldsymbol{W_J^{ab}}$.

3 Experimental Tests

3.1 Experimental Setup

For navigation purposes GPS, an odometry captor, INS sensors and GIS maps are installed aboard the vehicle prototype. The use of inertial sensors, accelerometers and gyroscopes, supplies continuous positioning even in cases without GPS coverage, and high frequency measurements [15]. To integrate data coming from different sensors extended Kalman filters (EKF) run a set of different kinematical models as detailed in [15]. Thanks to the MEM (Micro-Electro-Mechanical) technology, low cost inertial sensors can be considered, at the expense of higher measurement noises and low level of performance [16]. For vehicle to vehicle (V2V) communications WLAN ad hoc networks are used [8]. The test vehicle deployment consists of wireless LAN IEEE 802.11b and antenna by Cisco, EGNOS capable GPS and DGPS (differential GPS) sensors by Novatel and Trimble, and MEM based IMUs (Inertial Measurement Unit) by Crossbow and Xsens.

In the measurement collection phase, acceleration and heading turn rate coming from the INS device are gathered (a, ω), while velocity is obtained from the odometry captor installed in the back wheels axis of the vehicle (v). Next, sorting and synchronization processes are performed. In this phase, a spurious detection method based on simplified Nyquist inequation is applied [15]. Finally, these data are used as inputs of the supervised dFasArt algorithm. In this work, although GPS North and East measurements are collected by our system and used for positioning purposes, only odometry and inertial measurements are used by the maneuver detector algorithm. Thus, system performance is not determined by GPS signal outages, very often in built-up environments.

3.2 Vehicle Models

Maneuver states are typically defined by kinematic vehicle models. In order to test the performance of the proposed SdFasArt algorithm, three different vehicle models are used as reference. The outcomes of the algorithm will be then compared with assumed kinematical maneuver truth. The three kinematical models proposed are based on a simplified bicycle model, in which the orientation of velocity and acceleration are assumed to be equal (Fig. 2). Previous results achieved by the authors proved that convenience of this assumption. Three different maneuver states are distinguished: AD (acceleration and deceleration), CR (cruise) and STA (stationary) states.

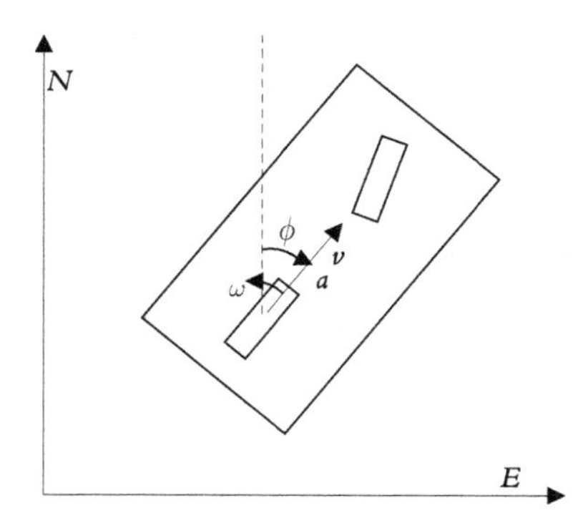

Fig. 2. Simplified bicycle model of the kinematic behavior of the road vehicle. In this model it is assumed that the orientation of velocity and acceleration are defined by heading angle ϕ and rate ω, and referred to East, North (E, N) coordinates.

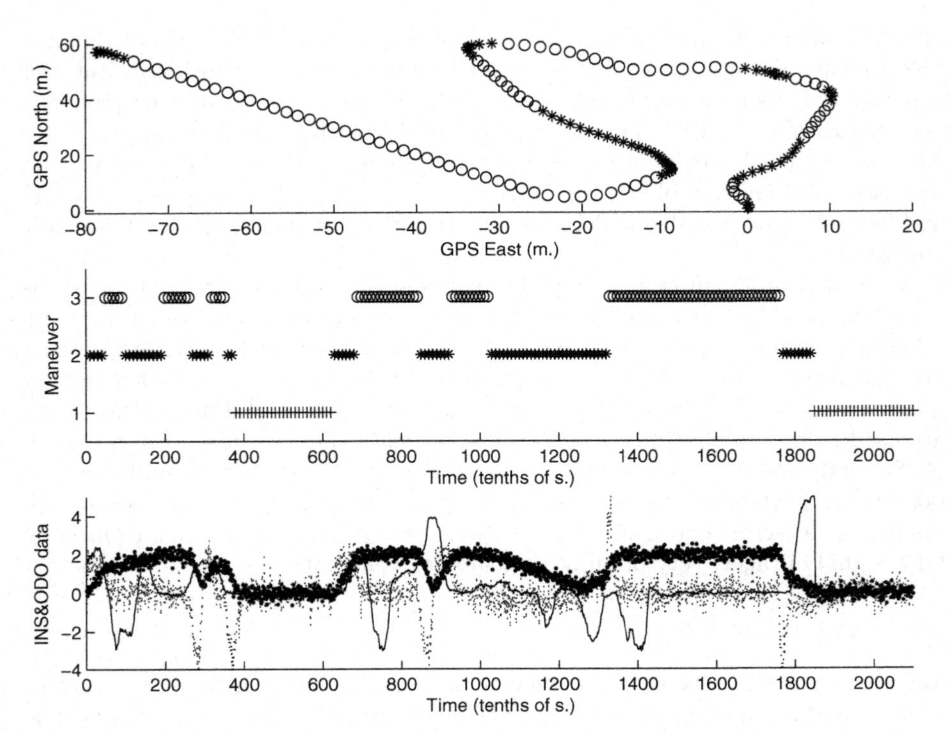

Fig. 3. Circuit used for training. From top to bottom: GPS trajectory; desired maneuver classification: AD (*), CR (o), STA (+); output from the odometry, v (thick line), and INS, a (dotted line), ω (solid line). Real values of a and ω have been scaled to fit in figure limits.

Acceleration/Deceleration (AD). State vector of the acceleration/deceleration model is $\mathbf{x}_{\mathrm{AD}} = (x, y, \phi, v, \omega, a)$, representing east, north, velocity angle, velocity, yaw rate of turn, and the acceleration, in the center of mass of the vehicle. The dynamics of this model are described by

$$\dot{\mathbf{x}}_{\mathrm{AD}} = \begin{bmatrix} (v+at)\cos(\phi) \\ (v+at)\sin(\phi) \\ \omega \\ a \\ 0 \\ 0 \end{bmatrix} + \begin{bmatrix} 0 \\ 0 \\ 0 \\ 0 \\ \eta_{\omega_{\mathrm{AD}}} \\ \eta_{a_{\mathrm{AD}}} \end{bmatrix} \tag{1}$$

where $\eta_{\omega_{\mathrm{AD}}}$ and $\eta_{a_{\mathrm{AD}}}$ are white noise terms representing the errors due to model assumptions of constant acceleration and constant yaw rate.

Cruise (CR). State vector and differential equation of the cruise model are the same as in the AD model (1). As previously commented many authors proposed constant velocity models for cruise maneuver state. However, this has been found

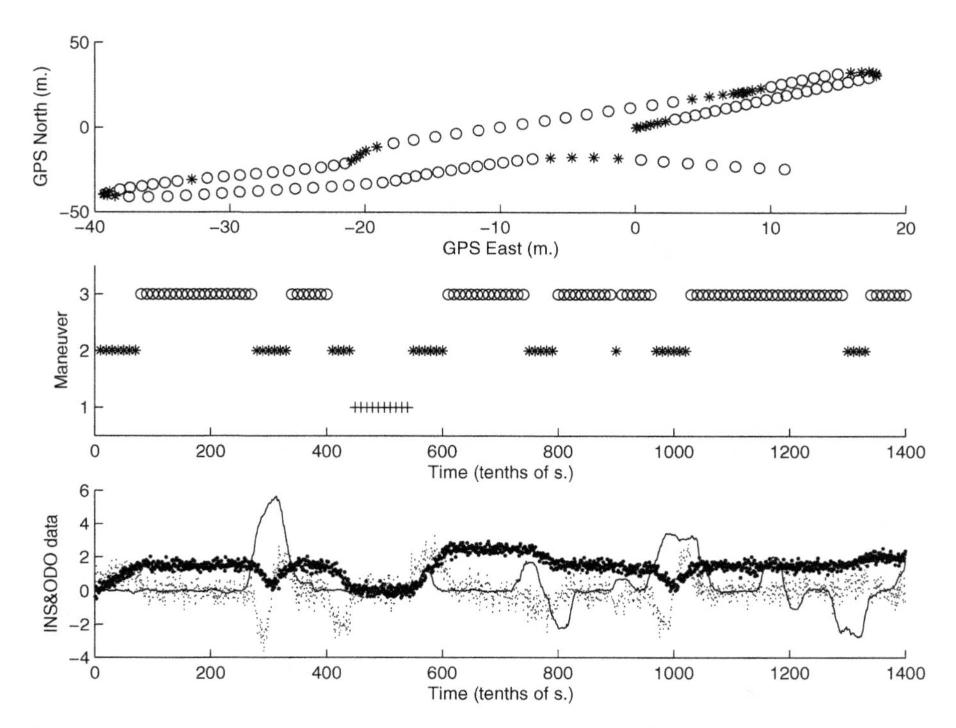

Fig. 4. Circuit used for validation. From top to bottom: GPS trajectory; desired maneuver classification: AD (*), CR (o), STA (+); output from the INS: V (thick line), a (dotted line), ω (solid line). Real values of a and ω have been scaled to fit in the figure limits.

problematic in several cases. In our approach, both AD and CR maneuver states are defined by similar CA models, defining higher noise values for AD state ($\eta_{\omega_{\mathrm{AD}}}, \eta_{a_{\mathrm{AD}}}$) in order to fulfill higher dynamics.

Stationary (S). In this case, state vector is simplified being $v = \omega = a = 0$, and the differential equation

$$\dot{\mathbf{x}}_{\mathrm{S}} = [\ \eta_{x_{\mathrm{S}}}\ \ \eta_{y_{\mathrm{S}}}\ \ \eta_{\phi_{\mathrm{S}}}\ \ 0\ \ 0\ \ 0\ ]',$$

where $\eta_{x_{\mathrm{S}}}$, $\eta_{y_{\mathrm{S}}}$ and $\eta_{\phi_{\mathrm{S}}}$ are white noise terms representing the errors due to the model assumptions. Noise parameters of the models are tuned starting from the sensor specifications.

3.3 Experimental Results

After several experiments to fine-tune the parameters of SdFasArt, values shown in Table 1 have been chosen.

Table 1. Parameters of Supervised dFasArt

Parameter	Value	Description
Aw	0.1	Time constant of weight's dynamic
Av	0.1	Time constant of weight's dynamic
Ac	0.1	Time constant of weight's dynamic
δ	0.2	Minimum fuzziness of the fuzzy categories
α	$1e^{-7}$	Activation value for new classes
RESET	0.2	Reset level
Ar	1.21	Time constant of RESET's dynamic
At	0.98	Time constant of activation's dynamic

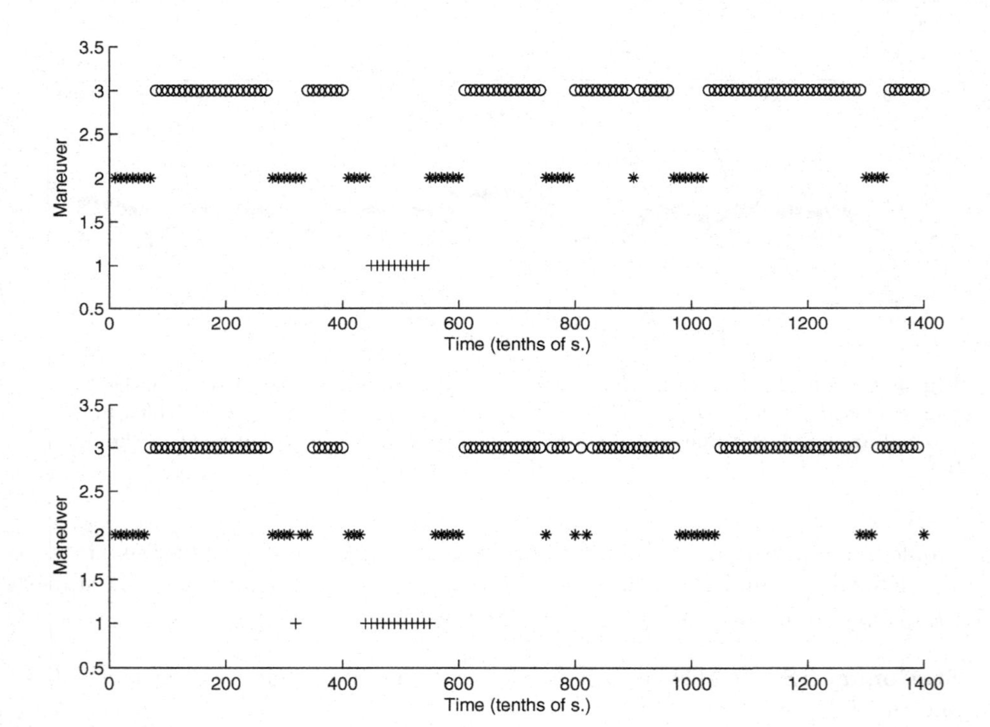

Fig. 5. Results on the validation circuit. Top: desired maneuver classification: AD (*), CR (o), STA (+);Bottom: predicted maneuver classification: AD (*), CR (o), STA (+).

Two different circuits have been employed, one for training (shown in Fig. 3) and the other for validation (see Fig. 4). Results obtained for the validation data are shown in Fig. 5. It can be seen that only a few maneuvers have been misclassified. As referred to the assumed truth, 86.36% of the samples are properly classified. However, manual labeling can only be considered as another proper estimate of the maneuver truth, since there are not unique criteria to classify complex dynamics. Thus, some of the errors can be attributed to errors in the manual labeling for the estimated maneuver truth of the training data. Furthermore, we can appreciate that transitions from STA to CR or from CR to STA maneuver states are never predicted by the algorithm, showing the algorithm prediction consistency.

4 Conclusions

In this work, a Supervised version of dFasArt has been proposed and tested for maneuver detection in road vehicles. The combination of the dynamic character of dFasArt with a supervisory module results in a robust classifier, capable to provide stable outputs in spite of noisy time-varying input data.

The neural architecture has been tested using real data gathered from inertial and odometry sensors mounted on a vehicle, showing good performance and consistent results.

Acknowledgments. We would like to thank the people of the NEUROCOR Group for their support in this work, and the Spanish Ministerio de Fomento and the European Space Agency (ESA) for sponsoring the research activities under the grants FOM/3929/2005 and GIROADS 332599 respectively.

References

1. Kokar M, Matheus C, Letkowski J: Association in Level 2 fusion., SPIE (2004).
2. Ceruti M, Ontology for Level-One Sensor Fusion and Knowledge Discovery. SPIE (2004).
3. Matheus C, Mieczyslaw M. Kokar M, Baclawski K,: A Core Ontology for Situation Awareness. Proceedings of Sixth International Conference on Information Fusion, Cairns, Australia, July (2003), pp. 545–552.
4. Mieczyslaw M. Kokar M , Matheus C, Baclawski K, Letkowski J, Hinman M, Salerno J: Use Cases for Ontologies in Information Fusion. Proceedings of Sixth International Conference on Information Fusion, Cairns, Australia, July (2003), pp. 545–552.
5. Ceruti M, Kaina J: Enhancing Dependability of the Battlefield Single Integrated Picture through Metrics for Modeling and Simulation of Time-Critical Scenarios. Proc. of the IEEE 9th Intl. Workshop on Real-time Dependable Systems, (WORDS 2003F), Oct. (2003).
6. Ceruti M, Kamel M: Preprocessing and Integration of Data from Multiple Sources for Knowledge Discovery. International Journal on Artificial Intelligence Tools, vol. 8, no. 3, June (1999), pp. 159-177.
7. Waltz E, Llinas J: Multisensor Data Fusion. Artech House, Boston (1990).
8. Toledo R, Sotomayor C, Gomez-Skarmeta A: Quadrant: An Architecture Design for Intelligent Vehicle Services in Road Scenarios. Monograph on Advances in Transport Systems Telematics (2006) : 451–460.
9. Huang D, Leung H: EM-IMM based land-vehicle navigation with GPS/INS, Proceedings of the IEE ITSC Conference. Washington DC USA. Oct. (2004). pp. 624–629.

10. Hoffmann C, Dang T: Cheap Joint Probabilistic Data Association Filters in an Interacting Multiple Model Design, Proceedings of the 2006 IEEE-MFI 2006. September 3-6, (2006), Heidelberg, Germany. pp. 197–202.
11. Toledo R, Zamora M, Skarmeta A: A Novel Design of a High Integrity Low Cost Navigation Unit for Road Vehicle Applications, Proceedings of the IEEE-IV 2006, Tokyo, Japan, June (2006) pp.577–582.
12. Barrios C, Himberg H, Motai Y, Sadek A: Multiple Model Framework of Adaptive Extended Kalman Filtering for Predicting Vehicle Location, Proceedings of the IEEE-ITSC 2006, Toronto, Canada, September 17-20, (2006). pp. 1053–1059.
13. Kaempchen N, Weiss K, Shaefer M, Dietmayer K: IMM Object Tracking for High Dynamic Driving Maneuvers. Proceedings of the IEEE-IVS'2004 June (2004) pp. 825 – 830.
14. Cano-Izquierdo J, Almonacid M, Ibarrola J, Pinzolas M: Use of Dynamic Neuro Fuzzy Model dFasArt for Identification of Stationary States in Closed-loop Controlled Systems, Proceedings of EUROFUSE 2007 Jaen, Spain (2007) (in press).
15. Toledo R,: A High Integrity Navigation System for Road Vehicles in Unfriendly Environments Phd. Dissertation. Universidad de Murcia Publishing. (2005).
16. Barshan B, Durrant-Whyte H F,: Inertial Navigation Systems for Mobile Robots. IEEE Internatinal Transactions on Robotics and Automation. June (1995), Vol. II NO. 3: 328–342.

Neuro-fuzzy Based Maneuver Detection for Collision Avoidance in Road Vehicles

M.A. Zamora-Izquierdo[1], R. Toledo-Moreo[2], M. Valdés-Vela[1], and D. Gil-Galván[1]

[1] Univ. Murcia, DIIC, Faculty of Computer Science, 30100 Murcia, Spain
mzamora@um.es, mdvaldes@um.es
http://libra.inf.um.es/ants_vehicles
[2] Technical Univ. of Cartagena, DETCP, Edif. Antigones, 30202 Cartagena, Spain
rafael.toledo@upct.es

Abstract. The issue of collision avoidance in road vehicles has been investigated from many different points of view. An interesting approach for Road Vehicle Collision Assistance Support Systems (RVCASS) is based on the creation of a scene of the vehicles involved in a potentially conflictive traffic situation. This paper proposes a neuro-fuzzy approach for dynamic classification of the vehicles roles in a scene. For that purpose, different maneuver state models for longitudinal movements of road vehicles have been defined, and a prototype has been equipped with INS (Inertial Navigation Systems) and GPS (Global Positioning System) sensors. Trials with real data show the suitability of the proposed neuro-fuzzy approach for solving support to the problem under consideration.

Keywords: Neuro-Fuzzy, FIS, Intelligent Vehicles, Maneuver Detection, Collision Avoidance.

1 Introduction

Intelligent Transportation Systems (ITS) are one of the keys to advance in the development of future societies. ITS-R consists of the applications of telematics to the transportation problem along roads. Navigators, automated vehicles or collision avoidance support systems are a few examples of interesting applications that ITS-R aim [1]. The problem of road vehicles collisions have been intensively reached lately. One interesting approach is based on the situation awareness concept. The issue of situation awareness has been pointed by several authors from the point of view of the artificial intelligence, specially for military purposes [2] – [5]. Most of these authors agree to divide multi-sensor data fusion into four levels of increasing situation complexity. Some other approaches, like the one proposed by University of Melbourne, prefer different architecture schemas not necessarily oriented to military scenes [6]. One of the main aspects of situation awareness is the maneuver recognition of the vehicles in a scene. Very different approaches depending on the sensors used can be found in the current literature. Several authors have been focusing their efforts in the recognition of vehicle

J. Mira and J.R. Álvarez (Eds.): IWINAC 2007, Part II, LNCS 4528, pp. 429–438, 2007.

behaviors by using a set of diverse kinematical models. Each model is developed to represent the vehicle behavior in a particular maneuvering state. In [7] a concrete model is selected according to its dynamic state. In [8], FIR filters are used to detect maneuvers and track targets.

The work presented in this paper is based on a Fuzzy Inference System (FIS) for maneuver classification. The navigation system mounts an onboard equipment (OBE) based on GPS/INS, running several extended Kalman filters (EKF) describing the different maneuver states. Previous papers published by the authors showed the suitability of this proposal for road vehicle navigation, even in unfriendly scenarios, like those with low GPS coverage or changing scenarios [9].

FISs use IF-THEN rules to describe qualitative relationships among variables in the modeled system. Their strength relies on their two fold identity. On the one hand, they look like natural language statements and, therefore, they are suitable for interpretation. On the other hand, they can be regarded as flexible mathematical structures, capable to perform nonlinear mappings between input and output data. Besides, FISs have been widely used to achieve classification tasks [10], given that a fuzzy classifier is only a FIS with crisp and discrete outputs. Rules in a FIS are usually obtained from expert knowledge. However, these kind of models may show poor performance. This can be sorted out by means of methods to build or update fuzzy models from numerical input-output data. This approach is called Data Driven Fuzzy Modeling (DDFM) [11,12]. However, they pay little attention to the transparency of the resulting FISs what is their main drawback. Nevertheless, in the last years several approaches to achieve a trade-off between performance and transparency have arisen [13].

In this work we apply DDFM techniques to build a FIS for classifying road vehicles maneuvers. Concretely, it has been applied a clustering technique followed by a neuro-fuzzy approach in which the adaptive capabilities of neural networks have been used to tune the parameters of the target FIS. As we will see in Section 6, very promising performance results have been obtained. Nevertheless, the current paper must be regarded as a first approximation to the application of fuzzy models to the problem under consideration. Future investigations will be devoted to the generation of more interpretable and accurate FIS for maneuver classification.

2 System Architecture: Quadrant

Quadrant is an architecture for ADAS (Advanced Driver Assistance Systems) applications based on the separation of four layers according the level of abstraction of the fusion performed, paying special attention to its communication framework [14]. First layer (or sensor layer) is in charge of the measurement collection, sorting and synchronization. These measurements are sent to upper layers via the sensor network. Secondly, fusion layer fuses the data coming from the sensors. This layer, oriented to the interpretation of the vehicle behavior to be performed in the following phase, is described by a set of models, representing different dynamic states of the vehicle. The difficulty to find a unique behavior that properly describes all possible maneuvers of a vehicle encourages the definition of multiple models to better describe its movements along roads. Besides, whether the system is able to determine the dynamic behavior of the vehicle, the use of multiple models supplies additional information relative to the maneuver state. Fused data are then supplied to the third layer via ad hoc networks in the vehicle environment. Those networks are supported by WLAN connection availability, thanks to the WIFI PCMCIA card installed in the vehicle computer. Third layer or interpretation layer is in charge of dynamic classification of the vehicle and scene interpretation. The use of the multiple models in the lower layer eases the dynamic classification process, according to the selection of the maneuver states of the vehicle, performed as explained in subsequent Sections. Finally, interpreted data (and any other data coming from previous phases) can be used by the application in order to provide the final service to the user.

A scene will be defined by the information of the vehicle itself, those surrounding, and geographical information of interest in the area (type of road, speed limits, etc.). Each vehicle in the scene is in charge of identifying its maneuver state and reporting it, along with its pose, to the rest of vehicles in the scene (typically a few close enough to be involved and capable to communicate via ad hoc WLAN). On board equipment is described in details in [9].

3 Multiple Model Filtering

The basic idea of using multiple models is based on the fact that a vehicle performs very different maneuvers depending on scenario features. For road vehicles, typical maneuvers in highways differ from those usual in city environments. Thus, a single vehicle model can hardly represent all possible maneuvers, and the use of multiple models, representing different maneuver states and running in parallel is advisable. The output of the multiple model approach is typically the model with highest probability value, or a weighted composite of the individual filters [15]. In our proposal we have distinguished between lateral and longitudinal movements of the vehicle. In this paper we focused on longitudinal movements.

3.1 Vehicle Models

Most common approach found in the current literature for multiple modeling of longitudinal movements, proposes models representing constant acceleration (CA) and constant velocity (CV) maneuver states. The combination of CV and CA models has been analyzed lately in the literature [16]. However, some of these authors found problems with the transition to the CA model, including those who used the IMM method [17]. In this paper, the tuning of the CA noise parameters to avoid often unrealistic switches from one state to the other was found problematic, impoverishing the perception of the situation. In case of highways, typical accelerations and decelerations do not last long enough to accomplish the transition from the CV state, to the CA state. To overcome these difficulties, we propose two different constant acceleration models with different noise parameter adjustments, and a stationary model to consider non-maneuver state. Kinematical model proposed is a simplified bicycle model, in which the orientation of the acceleration and velocity are assumed to be equal. Different tests done proved that this assumption can be done for highway scenarios.

Acceleration/Deceleration (AD). State vector of the acceleration/deceleration model is $\mathbf{x}_{\mathrm{AD}} = (x, y, \phi, v, \omega, a)$, representing east, north, velocity angle, velocity, yaw rate of turn, and the acceleration, in the center of mass of the vehicle. The dynamics of this model are described by

$$\begin{aligned}\dot{\mathbf{x}}_{\mathrm{AD}} = & \left[(v+at)\cos(\phi) \quad (v+at)\sin(\phi) \quad \omega \quad a \quad 0 \quad 0\right]' + \\ & \left[0 \quad 0 \quad 0 \quad 0 \quad \eta_{\omega_{\mathrm{AD}}} \quad \eta_{a_{\mathrm{AD}}}\right]' \end{aligned} \tag{1}$$

where $\eta_{\omega_{\mathrm{AD}}}$ and $\eta_{a_{\mathrm{AD}}}$ are white noise terms representing the errors due to model assumptions of constant acceleration and constant yaw rate.

Cruise (CR). State vector and differential equation of the cruise model are the same as in the AD model (1). As previously commented many authors proposed constant velocity models for cruise maneuver state. However, this has been found problematic in several cases. In our approach, both AD and CR maneuver states are defined by similar CA models, defining higher noise values for AD state ($\eta_{\omega_{\mathrm{AD}}}, \eta_{a_{\mathrm{AD}}}$) in order to fulfill higher dynamics.

Stationary (S). In this case, the vector state is simplified being $v = \omega = a = 0$, and the differential equation

$$\dot{\mathbf{x}}_{\mathrm{S}} = [\, \eta_{x_{\mathrm{S}}} \quad \eta_{y_{\mathrm{S}}} \quad \eta_{\phi_{\mathrm{S}}} \quad 0 \ 0 \ 0\,]',$$

where $\eta_{x_{\mathrm{S}}}$, $\eta_{y_{\mathrm{S}}}$ and $\eta_{\phi_{\mathrm{S}}}$ are white noise terms representing the errors due to the model assumptions. Noise parameters of the models are tuned starting from the sensor specifications. Observations for the AD, CR and S individual filters are GPS north and east values (x_{gps}, y_{gps}), odometry velocity (v_{odo}) and inertial measurements for angular rate (ω_{ins}) and longitudinal acceleration (a_{ins}).

4 TSK Fuzzy Inference System

In this paper we are concerned with FISs of type TSK [18] (see figure 1(a)). In such kind of FISs the rule consequents are usually constant linear functions of the inputs:

IF x_1 is A_{11} AND... AND x_p is A_{1p} THEN $y = a_{10} + a_{11}x_1 + \ldots + a_{1p}x_p$
IF x_1 is A_{21} AND... AND x_p is A_{2p} THEN $y = a_{20} + a_{21}x_1 + \ldots + a_{2p}x_p$
...
IF x_1 is A_{r1} AND... AND x_p is A_{rp} THEN $y = a_{r0} + a_{r1}x_1 + \ldots + a_{rp}x_p$

where r is the number of rules, p is the number of input variables, $x_1, x_2, \ldots, x_p$ are the input variables and y is the output one. $A_{i1}, A_{i2}, \ldots, A_{ip}$ are the antecedent fuzzy sets of i-th rule. They can have an attached linguistic label. The

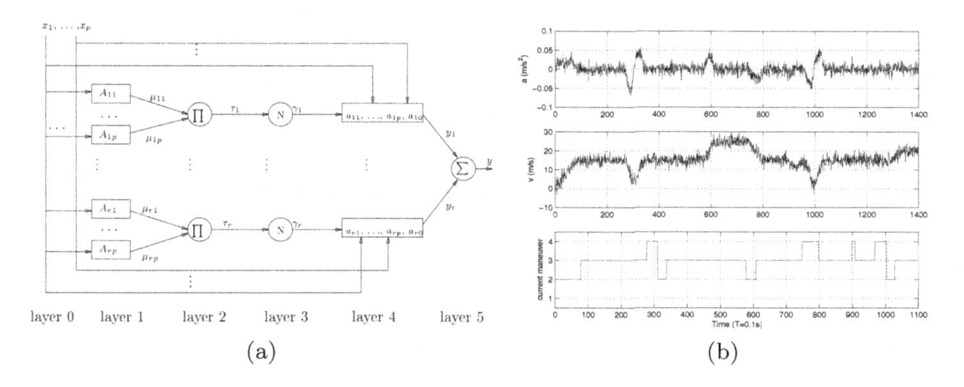

Fig. 1. (a) ANFIS architecture. (b) Training data coming from circuit 4.

fuzzy sets serve as a smooth interface between the numerical nature of input and output variables and linguistic labels.

A type of TSK FIS where every consequent is a crisp value is considered a special case of TSK FIS with zero-order polynomials in the consequents, the so-called Zero-Order TSK FIS. This is more suitable for classification tasks. As we will see in our maneuver classification problem a natural number from 1 to 4 is associated to each maneuver. Nevertheless, first-order TSK FISs (the ones with linear consequents) can also be used.

To obtain the output inferred with a TSK model (of any order), first of all it is necessary to compute the *degree of fulfillment* of the ith rule, denoted as τ_i, with $i = 1, \ldots, r$. This degree of fulfillment is given by the true value of the proposition "x_1 is A_{i1} AND ... AND x_p is A_{ip}", the meaning of which is given through a t-norm as, for instance, the product operator. Therefore, given the input $x_1, \ldots, x_p$, the degree of fulfillment of the i-th rule is computed as:

$$\tau_i = A_{i1}(x_1) \cdot A_{i2}(x_2) \cdot A_{ip}(x_p)$$

Finally, FIS output y is obtained adding partial outputs of each rule, weighted with their respective degrees of fulfillment through the equation:

$$y = \frac{\sum_{i=1}^{r} \tau_i y_i}{\sum_{i=1}^{r} \tau_i}$$

where y_i is the output of the i-th rule:

$$\begin{aligned} y_i &= a_{i0} + a_{i1}x_1 + \ldots + a_{ip}x_p, && \text{in a first-order TSK} \\ y_i &= a_{i0}, && \text{in a zero-order TSK} \end{aligned}$$

Our strategy starts with the selection of the number of rules. Instead of an automatic selection, a suitable number can be chosen according to the distribution of the available input-output data. Therefore, this number is an initial parameter to *a clustering optimization method* resulting in a first approximation to the target FIS. Finally, a *neuro-fuzzy architecture* is fed with the centroids detected. Afterwards, the adaptive capabilities of this network are used in order to adjust FIS parameters.

5 A FIS Maneuver Classification: DDFM Process

In general, three main stages compose a DDFM process: first, the number of rules must be detected (*rules number identification*). After that, a rough approximation to the set of rules is obtained (*rules generation*). Afterward the fuzzy sets involved in the rules are adjusted in the *parameter optimization* stage to better mimic the system behavior.

In this work, instead of the automated selection, a suitable number is chosen manually according to the visual distribution of the available input-output data. Therefore this number is an initial parameter to *a subsequent clustering optimization method* applied to the input-output data. Every cluster can be regarded as a rough fuzzy rule. Hence, their centroids are initial values to the fuzzy sets involved in the FIS which will be tuned using the adaptive capabilities of the *neuro-fuzzy architecture*.

5.1 Rules Generation

Fuzzy C-Means Clustering Method (FCM) [19] is one of the most used methods for optimization of an initial partition of data. Given a set of examples, the algorithm must be provided with a c number a centroids (set beforehand) and an initial fuzzy c-partition of the data. Then, algorithm optimizes the c-partition in such a way that certain *cost function* J is minimized. This function is the weighted within groups sum of squared-errors.

At each step, three operations are successively carried out until the centroids are stable with respect to a given tolerance (ϵ): (1) computation of the centroids (assuming that the membership degrees composing the partition are constants numbers); (2) calculus of the distances from each datum to every centroid (the most commonly used measurement for distance is Euclidean, inducing spherical clusters); (3) update the membership degree of each data point to every cluster (assuming that the centroids remain constant).

The partition of the input-output data obtained by means of this clustering method is considered as first approximation to the target FIS . These clusters centroids are the initial parameters for the fuzzy sets involved in the FIS. Next step is the tuning of those parameters, described in the next Section.

5.2 Parameter Optimization

In order to optimize the parameters, we use a neuro-fuzzy approach. Neuro-fuzzy mechanisms merge the adaptive capabilities of artificial neural networks with the human-readable information representation provided by FISs. Concretely we use the Adaptive Neuro-fuzzy Inference System (ANFIS) [20] that is functionally equivalent to a TSK FIS. It is compound of five layers. Nodes in layer 1 and layer 4 correspond to the antecedent and consequent parameters of the TSK FIS respectively. The ANFIS architecture is shown in the Fig. 1(a).

We have assumed fuzzy sets with gaussian membership functions. Therefore, the mean of the fuzzy sets in the antecedent layers are initialized with the centroids obtained in the previous phase of DDFM. Next, training is carried out during several *epoches* (iterations). Every epoch of the training is compound of a *forward* and *backward pass*. In the forward pass the network is evaluated for every input datum and the rule consequent parameters are identified by means of the *Least-Squares Estimator*. Afterwards, errors for every training datum are calculated and, in the backward pass, error signals are propagated and the rule antecedents modified through backpropagation.

6 Experimental Tests and Results

In order to train and test the proposed system, a prototype vehicle equipped with two IMU (Inertial Measurement Unit) sensors (MT9-B Xsens and Xbow VG-400 IMU units) based on low cost MEM (Micro-Electro-Mechanical) technology has been used. A Trimble DGPS sensor and the odometry of the car were also utilized to obtain the data set. With this information, sensor fusion layer estimates the vehicle pose through different extended Kalman filters, each one for one kinematic model. Real data coming from five different circuits (6908 instances) have been used in the experiments.

Different sets of input data were analyzed, deciding finally to use four inputs to feed the target FIS: two last velocity measurements ($v(k)$ and $v(k-1)$) provided by the odometry of the car, longitudinal acceleration ($a_x(k)$) provided by the IMU unit, and previous output ($o(k-1)$). In these experiments, no angular information is taken into account since we focus exclusively on longitudinal maneuvers. Concretely, four different maneuver states have been considered: stationary (denoted as ST and represented by constant value 1 in a zero-order TSK), acceleration (AC/2), cruise (CR/3) and deceleration (DC/4). Despite the fact that from the point of view of kinematics, acceleration and deceleration can be described by a single model, we make this distinction to check the DDFM process performance for this case.

Input-output data coming from four of the five mentioned circuits were used for training. Fig. 1(b) shows circuit 4 acceleration and velocity measurements, and manual maneuver labeling. Given the data distribution in the input-output space, we decided to set $c = 12$ in the FCM algoritm, that is, 12 clusters and hence 12 rules. The tolerance index ϵ was set to 10^{-4}. Let us call FCS the model of 12 clusters generated by means of the FCM method. Each cluster was projected into each one of the axis obtaining a first approximation of the target FIS (let us call it FCS-p). Afterwards, the ANFIS architecture was fed twice with this rough FIS in order to obtain, respectively, zero and first order TSK FIS. After training twice for 100 epoches, two different classifiers were obtained, TSK-0 and TSK-1.

Final and intermediate FISs were validated using the circuit 5. In Table 1(a) training and validation errors for every FIS are deployed, where RMSE stands for Root Mean Squared Error (RMSE) calculated as

$$RMSE = \sqrt{\frac{\sum_{i=1}^{n} (y_i^t - \hat{y}_i^t)^2}{n}}$$

being y_i^t and $\hat{y}_i^t$ inferred and real outputs for the i-th datum, with n number of data. It must be remarked that training and validation errors were comparable, indicating the absence of overtraining.

It can be observed that TSK-1 presents better performance than TSK-0 due to the more powerful approximation capabilities of the polynomial consequents. On the other side, TSK-0 is more transparent since most of the rule consequents match the maneuver themselves (unfortunately, it cannot be shown due to space limitations). The confusion matrixes of TSK-0 and TSK-1 classifiers are shown in Tables 1(b) and 1(c) respectively, where each column of the matrix represents instances in a inferred maneuver, while each row represents instances in reference maneuver state. Both models classify noticeably well stationary instances while TSK-1 is better than TSK-0 classifying other maneuver states.

Table 1. : (a) RMSE of different FISs. (b) TSK-0 confusion matrix. (c) TSK-1 confusion matrix.

(a)

FIS	RMSE Tr.	RMSE Va.
FCS	0.47	0.83
FCS-p	0.40	0.54
TSK-0	0.24	**0.28**
TSK-1	0.15	**0.24**

(b)

	ST	AC	CR	DC
ST	**499**	28	0	0
AC	0	**231**	3	1
CR	0	50	**857**	6
DC	2	12	0	**409**

(c)

	ST	AC	CR	DC
ST	**499**	1	0	1
AC	1	**306**	6	8
CR	0	3	**854**	7
DC	1	11	0	**400**

In Fig. 2(a) data extracted from the validation circuit are deployed. Fig. 2(b) shows the maneuver inferred by TSK-0 and TSK-1 classifiers versus estimated truth adjusted by manual labeling. It can be observed, for example, that many of errors are concentrated around instant t = 200 tenths of s., where AC points are classified as CR in TSK-1 and DC in TSK-0 due to the large oscillations in $v(t)$.

Once a fuzzy classifier has been obtained, it can be used by the algorithm. In a higher level the system decides which kinematic model will be used, taking into account the output of the fuzzy classifier. With the mentioned four inputs, the system is capable to learn about changes in the velocity fusing information coming from two different sensors. Last output adds information of the current maneuver to the system.

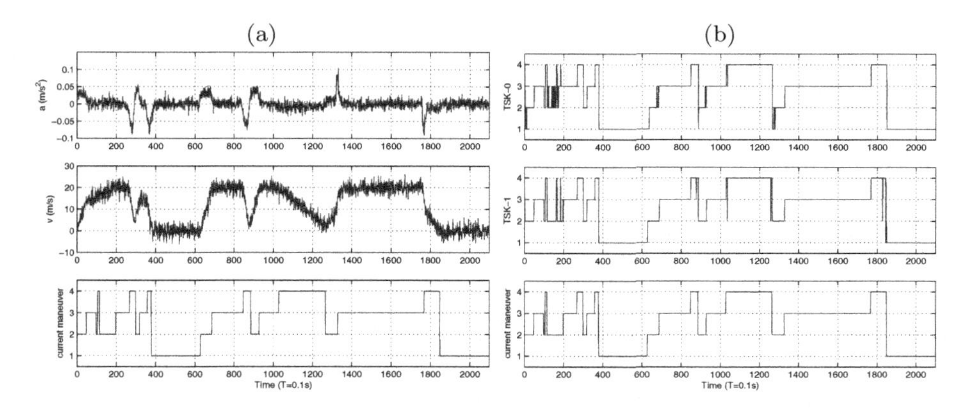

Fig. 2. (a) Circuit 5 used for evaluation. (b) Reference maneuvers and those inferred by TSK-1 and TSK-0 FISs.

7 Conclusions and Future Work

The use of a model-set representing different state maneuvers of a road vehicle provides an appropriate modeling of the vehicle behavior in many different dynamic situations with realistic pose and noise estimates. This paper proposes a fuzzy classifier for state maneuver detection that has been obtained from input-output data by means of a clustering technique and a subsequent neuro-fuzzy approach.

In this first work, promising results have been obtained. Trials performed with real data in a set of road scenarios show the suitability of the proposed system for the problem of detecting different maneuver states for longitudinal movements. Next step is to analyze the system capability for transversal maneuvers, including lane change. Regarding the fuzzy inference system, several improvements can be introduced. In the current experiments the number of fuzzy rules have been fixed a priori. Future works will be focused on the use of automatic methods for the detection of this number directly from data. Further researches will lead us to the generation of more accurate and user-friendly fuzzy rules susceptible of being contrasted with a human expert.

Acknowledgments

The Authors would like to thank the Spanish Ministerio de Fomento and the European Space Agency (ESA) for sponsoring the research activities under the grants FOM/3929/2005 and GIROADS 332599 respectively.

References

1. Skarmeta A., Toledo R., Zamora M.A., Ubeda B., Santa J., Sotomayor C.: Master Plan for ITS in Spain. Ministerio de Fomento Español (2006)

2. Kokar M.M. ,Matheus C.J., Letkowski J.A.: Association in Level 2 fusion, SPIE (2004).
3. Matheus C.J., Kokar M.M., Baclawski K.: A Core Ontology for Situation Awareness. Proceedings of Sixth International Conference on Information Fusion, Cairns, Australia, July (2003), pp. 545–552.
4. Ceruti M.G. and Kamel M.N.: Preprocessing and Integration of Data from Multiple Sources for Knowledge Discovery. International Journal on Artificial Intelligence Tools, vol. 8, no. 3, June (1999), pp. 159-177.
5. Waltz E. and Llinas J.: Multisensor Data Fusion. Artech House, Boston (1990).
6. Dance S., Caelli T., Liu Z.Q.: An architecture for a traffic scene interpretation system. Technical Report 94/12, Dept. of Computer Science. University of Melbourne. (1994).
7. Weiss K., Stueker D., Kirchner A.: Target Modelling and Dynamic Classification for Adaptive Sensor Data Fusion. Proceedings of the IEEE Intelligent Vehicles Symposium 2003 June (2003), pp. 132–137.
8. Hwan Park S., Hyun Kwon W., Oh-Kyu Kwon and P. So Kim: Maneuver Detection and Target Tracking Using State-Space Optimal FIR Filters. Proceedings of the American Control Conference. San Diego, California June (1999). pp. 4253–4257.
9. Toledo R., Zamora M.A. and Gomez-Skarmeta A.F.: A Novel Design of a High Integrity Low Cost Navigation Unit for Road Vehicle Applications, Proceedings of the IEEE-IV 2006, Tokyo, Japan, June (2006) pp.577-582.
10. Gómez-Skarmeta A.F., Valdés M., Jiménez F. and Marín-Blázquez J.G.: Approximative fuzzy rules approaches for classification with hybrid-GA techniques. Information Sciences (2001) vol. 136, no. 1-4, pp.193–214.
11. Babuška R.: Fuzzy Modeling for Control. Kluwer Academic Publishes (1998).
12. Valdés M. Gómez-Skarmeta A.F. and Botía J. A.: Towards a Framework for the Specification of Hybrid Fuzzy Modeling. International Journal of Intelligent Systems (2005), vol 20 no. 2, pp.225–252.
13. Teng Y. W. and Wang W. J., Constructing a User Friendly GA-Based Fuzzy System Directly From Numerical Data. IEEE Transactions on Systems, Man and Cybernetics-Part B (2004), vol.34, no. 5 pp.2060–2070.
14. Toledo R., Sotomayor C. Gomez-Skarmeta A.F., Quadrant: An Architecture Design for Intelligent Vehicle Services in Road Scenarios. Monograph on Advances in Transport Systems Telematics (2006) pp.451-460.
15. Bar-Shalom,Y., and X.R.Li, Multitarget-multisensor tracking: principles and techniques. Storrs, CT. YBS Publishing, (1995).
16. Barrios C., Himberg H., Motai Y., Sadek A.: Multiple Model Framework of Adaptive Extended Kalman Filtering for Predicting Vehicle Location. Proceedings of the IEEE-ITSC 2006, Toronto, Canada, September 17-20, (2006). pp. 1053–1059.
17. Hoffmann C., and Dang T.: Cheap Joint Probabilistic Data Association Filters in an Interacting Multiple Model Design. Proceedings of the 2006 IEEE-MFI 2006. September 3-6, (2006), Heidelberg, Germany. pp. 197–202.
18. Takagi T. and Sugeno M.: Fuzzy Identification of Systems and its Applications to Modeling and Control. IEEE Transactions on Systems, Man and Cybernetic (1985), vol. 15, pp.116–132.
19. Bezdeck J. C.: Pattern Recognition with Fuzzy Objective Function Algorithms. Plenum, New York,(1981).
20. Jang J. S. ANFIS: Adaptive-network-based fuzzy inference systems, IEEE Transactions on Systems, Man, and Cybernetics (1993), vol. 23, no. 3,pps665–685.

The Coevolution of Robot Behavior and Central Action Selection

Fernando Montes-Gonzalez

Departamento de Inteligencia Artificial,
Sebastian Camacho 5, Centro, Xalapa, Ver., Mexico
fmontes@uv.mx
http://www.uv.mx/dia/

Abstract. The evolution of an effective central model of action selection and behavioral modules have already been revised in previous papers. The central model has been set to resolve a foraging task, where specific modules for exploring the environment and for handling the collection and delivery of cylinders have been developed. Evolution has been used to adjust the selection parameters of the model and the neural weights of the exploring behaviors. However, in this paper the focus is on the use of genetic algorithms for coevolving both the selection parameters and the exploring behaviors. The main goal of this study is to reduce the number of decisions made by the human designer.

Keywords: Action Selection, Reactive Robotics, Genetic Algorithms.

1 Introduction

The action selection problem aims to answer the basic question of intelligent systems of what to do next from their sensory perception. However, as Maes has pointed out is also important to consider that the right decision has to be made at the right time [1]. Several researchers in robotics consider important to develop a model able to decide the right time for making the right decision when solving a particular task. Nevertheless, still some researchers are interested in providing a complete solution for the task that is to be solved. Take for instance the work of Nolfi, who proposes the evolution of a can-collection task as a complete solution [2]. Potential problems related to the artificial separation of behavioral modules and to the process fusion of the behaviors support this view; also pointed out by the work of Seth [3]. On the other hand, the works of Kyong-Joong & Sung-Bae [4] and Yamauchi & Randall Beer [5] offer an incremental solution for combining different architectures for solving an entire task. Here, the full integration of evolution with a central action selection model is the main interest.

The model that was employed in the experiments is one of central action selection that uses sensor fusion (CASSF) to build a uniform perception of the world in the form of perceptual variables [6]. The evolution was used in the design of the two exploration behavioral patterns: one for finding walls, and the other for locating cylinders. Then, due to its sequential nature the design of the

J. Mira and J.R. Álvarez (Eds.): IWINAC 2007, Part II, LNCS 4528, pp. 439–448, 2007.

collection and deposit behavioral patterns was carried out as algorithmic routines. However, coevolution was used for both the adjustment of the selection parameters and the design of behavior related to exploring the arena. Therefore, in section 2 a brief background on Genetic Algorithms is provided. Next, section 3 offers an explanation for the use of evolution in the design of the neural behaviors cylinder-seek, and wall-seek. Additionally, an introduction to the handling cylinder behavior, cylinder-pickup and cylinder-deposit, is provided. All these four behavioral patterns will be used in conjunction to solve the foraging task set for the Khepera robot. The selection of a behavior is done by a central selection model namely CASSF that is coevolved with exploration behavior; both the selection model and coevolution are presented in sections 4 and 5. Section 6 offers a description of the results of the use of coevolution in the development of the selection mechanism and neural behavior. Finally, in section 7 a general discussion of the importance of these experiments is presented.

2 Artificial Evolution

Several Evolutionary Algorithms have been considered by the robotics community for modeling survival tasks in semi-controlled environments. Outstanding representatives of the Evolutionary Algorithms are Genetic Algorithms (GAs), Evolution Strategies, Genetic and Evolutionary Programming and Coevolution [7] [8]. However, in this study the use of GAs [9] and Coevolution [10] was preferred. In the existing literature examples of evolutionary algorithms are provided for the modeling of elementary behavior controllers [3] [5] [11]. The use of neural networks with genetic algorithms is a common approach for providing a solution to the modeling of robot behavior [12]. The use of this approach requires the right choice of a topology and the neural weights to control the robot.

The various choices of the weights of the neural controller will span from clumsy to clever individuals. If all possible individuals were to be generated, the plot of their fitness will produce a convoluted space with peaks and valleys. As a result, a gradient ascent search has to be performed in order to find an 'optimal' solution. Hence, is artificial evolution which is guiding this search by evaluating individual fitness and then applying the evolutionary operators. These operators facilitate the selection of more adapted individuals, at every step of the evolution, over those less adapted and ultimately to the rise of the fittest individuals.

The evolutionary operators are applied during the whole evolutionary process. At the beginning of the process an initial population is made of random neural controllers and their fitness evaluated. Then, *Selection* chooses to breed the fittest individuals into the next offspring using crossover and mutation. *Tournament selection* is an instance of selection that chooses to breed new individuals from the winners of local competitions. However, there are some cases where selection occurs by means of agamic reproduction, which inserts intact the fittest individuals into the next offspring (a form of *elitism*). *Crossover* is an operator that takes two individual codifications and swaps their contents around one or several random points along the chromosome. *Mutation* occurs with a probabilistic flip of some of the chromosome bits of the new individuals. Therefore, the genotype encodes information of the neural controller, in fact the weights of the neural net, and the direct encoding is the most employed in evolutionary robotic experiments [12].

The fitness of an individual is evaluated, in the test environment, on the completion of the proposed task. The production of new offspring is halted when a satisfactory solution is found. However, the evaluation of robot fitness is highly time-consuming; as a result, the use of a robot simulator is preferred for the evaluation of candidate solutions. Then, the fittest individuals are transferred to the real robot. The use of an hybrid approach, combining both simulation and real robots [13], minimizes the 'reality gap' between behaviors in simulation and real robots [12].

3 The Design of Behavioral Modules

The development of behavioral modules is achieved by the use of a robot simulator and a physical robot. A commercial robotic simulator offer better support than software offered as a freeware. A table-top robot that is commonly used in evolutionary robotics is the Khepera [14]. A robot like this is equipped with a ring of eight infrared sensors for the detection of close objects. The capabilities of the Khepera can be extended with a gripper for grasping objects. For these experiments the use of a fully functional gripper on both the simulator and the real robot is required. Therefore, for modeling a foraging task the use of Webots [15] is a proper choice.

The definition of this task is as follows, the robot has to take cylinders from the center to the corners of an arena. The task is completed with the use of four different behavioral patterns, two of these for exploring the arena, and two for handling a cylinder. Whereas exploring behavior share the same neural topology, the other two behavioral patterns were programmed as algorithmic routines. A fully connected feedforward multilayer-perceptron with no recurrent connections is employed for neural behavior. The input layer consists of six neurons that are connected to the five neurons of the hidden layer, and then to the two neurons in the output layer. The input neurons are fed with the readings of the six frontal infrared sensors of the Khepera. In order to facilitate the transference of the controller to the robot, sensor inputs are made binary with a collision threshold $th_c = 800$ that can be adjusted to the real sensors. Next, the output of the neural network is scaled to ± 20 values for the DC motors.

The genetic algorithm employs a expanded chromosome that is formed by a first chromosome that directly encodes behavior for seeking a wall, a second chromosome related to the location of a cylinder behavior, and a third chromosome for the selection parameters. The codification of the third chromosome will be further explained in the next section. For the exploring behavior, a chromosome is a vector $\mathbf{ch_x}$ of 40 weights. Random initial values are generated for the vector $\mathbf{ch}_{xi}$, $-K_w < \mathbf{ch}_{xi} < K_w$ with $K_w = 7.5$, and $n = 80$ neural controllers form the initial population G_0. A pair of the fittest individuals of a generation are copied in order to preserve the best solutions so far found (*elitism*). *Tournament selection*, for each of the $(n/2)-1$ local competitions, produces two parents for breeding a new individual using a single random *crossover* point with a probability of 0.5. Next, the entire new population is affected with a *mutation* probability of 0.01. Each of the individuals of the new offspring are evaluated for about 25 seconds in the robot simulator. Although, coevolution is employed in these experiments, individual chromosomes were first evolved as the entire chromosome in order to adjust the collision threshold th_c for the real robot.

Therefore, the first chromosome is defined as behavior for finding a wall (*wall-seek*), which is a form of obstacle avoidance because a wall has to be located without bumping into obstacles. When a wall is found the selection mechanism decides to stop the execution of the behavior. The fitness formula for the obstacle behavior in wall-seek was

$$f_{c1} = \sum_{i=0}^{3000} abs(ls_i)(1-\sqrt{ds_i})(1-max\_ir_i) \qquad (1)$$

where for iteration i: *ls* is the linear speed in both wheels (the absolute value of the sum of the left and right speeds), *ds* is the differential speed on both wheels (a measurement of the angular speed), and *max_ir* is the highest infrared sensor value. The use of a fitness formula like this rewards those fastest individuals who travel on a straight line while avoiding obstacles.

On the other hand, the second chromosome resembles a *cylinder-seek* behavior that is a form of obstacle avoidance using a cylinder-sniffer. For this behavior the robot body has to be positioned right in front of the cylinder if the collection of the can is to occur. The locating cylinder behavior is adjusted through its neural weights, which are directly encoded in the vector $\mathbf{ch}_x$ as described for the previous behavior. The fitness formula for the cylinder-seek behavior was as follows

$$f_{c2} = f_{c1} + K_1 \cdot cnear + K_2 \cdot cfront \qquad (2)$$

A formula such as this select individuals, which avoid obstacles and reach cylinders at different orientations (*cnear*), capable of orienting the robot body in a position where the gripper can be lowered and the cylinder collected (*cfront*). The constants K_1 and K_2, $K_1 < K_2$, are used to reward the right positioning of the robot in front of a cylinder.

The manipulation of the robot gripper requires a fixed number of iterations for clearing the space for lowering the arm, opening the claw, and moving the arm upwards and downwards. Therefore, the behaviors *cylinder-pickup* and *cylinder-deposit* were programmed as algorithmic routines.

4 The Central Action Selection Mechanism

The use of an effective model of central action selection has been presented in previous work [16]. In this model the combination of sensory perception in the robot, and the motor commands of selected behaviors, has been carried out in order to mimic a cylinder-retrieval robot task [6]. The centralized model of action selection with sensor fusion (CASSF) builds a unified perception of the world at every step of the main control loop (Fig. 1). Additionally, the fused perception of the world facilitates the use of the information of multiple non-homogeneous sensors as perceptual variables that can be weighed by the selection mechanism. These perceptual variables are used to calculate the urgency (salience) of a behavioral module to be executed. Furthermore, behavioral modules contribute to the calculation of the salience with a busy-status signal indicating a critical stage where interruption should not occur. Hence, the salience of a behavioral module is calculated by adding relevant sensory information, from the environment, to its own busy status. In turn, the behavior with the highest salience wins the competition and is expressed as motor commands directly sent to the motor wheels and the gripper. In the next paragraph, the computation of the salience is explained.

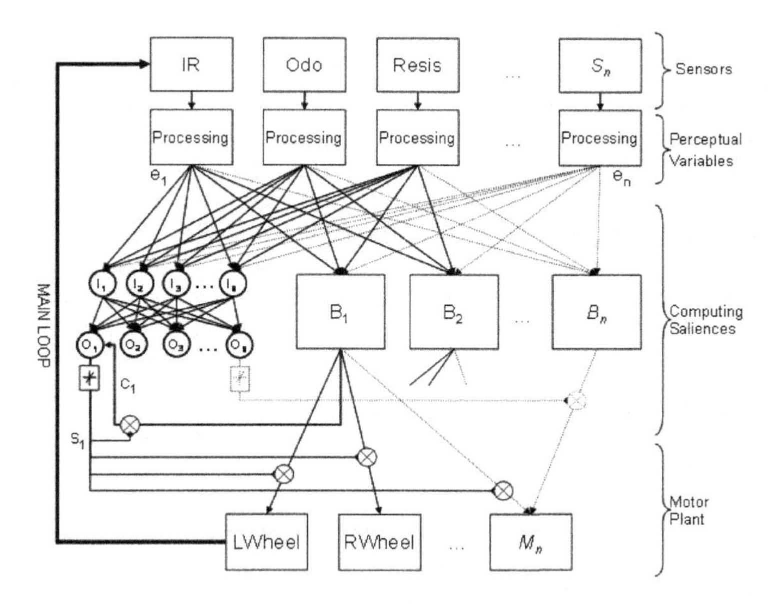

Fig. 1. In the CASFF model, perceptual variables (e_i) form the input to the salience computation of the decision neural network. The output selection of the highest salience (s_i) is gated to the motors of the Khepera. Notice the busy-status signal (c_1) from behavior B_1 to the output neuron.

Firstly, the perceptual variables wall_detector (e_w), gripper_sensor (e_g), cylinder_detector (e_c), and corner_detector (e_r) are coded from the various typical readings of the different sensors. These perceptual variables form the context vector, which is constructed as follows ($\mathbf{e} = [e_w, e_g, e_c, e_r], e_w, e_g, e_c, e_r \in \{1, 0\}$). Secondly, four different behavioral modules return a current busy-status ($\mathbf{c}_i$) indicating that ongoing activities should not be interrupted. Thirdly, the current busy-status vector is formed as next described, $\mathbf{c} = [c_s, c_p, c_w, c_d], c_s, c_p, c_w, c_d \in \{1, 0\}$, for *cylinder-seek*, *cylinder-pickup*, *wall-seek*, and *cylinder-deposit* respectively. Finally, the salience ($\mathbf{s}_i$) or urgency is calculated by adding the weighted busy-status ($c_i \cdot w_b$) to the weighted context vector ($e \cdot [w_j^e]^T$).

It is important to notice that behavioral modules are closely related to specific perception patterns. Most of the time the selection of only one behavioral module occurs. However, overlap may exist and selection has to grant control to the most urgent and relevant action. As a result, a complex pattern emerges and in these experiments the foraging task is solved. The task consists in taking cylinders in the center to the corners of an arena. Then, the centralized model implements a winner-takes-all selection of one of four available behavioral modules. These modules are as follows, *Cylinder-seek* wanders around the arena searching for food while avoiding obstacles, *cylinder-pickup* clears the space for grasping the cylinder, *wall-seek* locates a wall whilst avoiding obstacles, and *cylinder-deposit* lowers and opens an occupied gripper.

Although, the computation of a salience has been described as a vector of the four salience signals from the behavioral modules; this computation can be thought as a decision neural network with an input layer and an output layer both with four neurons (Fig. 1). The output of the neurons make use of the identity function to preserve the original calculation of the salience. The raw sensory information from the Khepera is fed into the neural network in the form of perceptual variables. Next, the input neurons distribute the variables to the output neurons. The behavior that is selected sends a busy signal to the output neurons when its salience is above the salience of the other behaviors. A selected behavior sends a copy of this busy signal to the four output neurons, in turn the four behavioral modules can be selected, thus each of the four modules add four more inputs to the output neurons.

5 Coevolution of CASSF and the Behavioral Modules

In this paper coevolution is employed to subtly adjust the weights of the decision network and exploration behavior. As an initial step, the first two chromosomes of the exploration behaviors were evolved alone in order to adjust the collision threshold for the real robot. It is important to notice that each of the four output neurons of the decision network weighs four perceptual variables plus four different busy signals. Therefore the third chromosome was modeled as a vector $\mathbf{ch_s}$ of 32 weights with initial random values of $\mathbf{ch}_s$, $-K_w < \mathbf{ch}_s < K_w$ with $K_w = 0.75$. As a second step, coevolution was carried out for the three chromosomes, as the entire chromosome, employing the evolutionary operators as described for the development of the individual behavioral patterns. Hence, the fitness formula for the coevolution of the weights of the decision network with cylinder-seek and wall-seek behavior was as follows

$$f_{c3} = (K_1 \cdot fcf) + (K_2 \cdot f_{c2}) + (K_3 \cdot pkfactor) + (K_4 \cdot dpfactor) \quad (3)$$

The coevolution of the weights of the neural network and the exploration behavioral patterns was accomplished using in the fitness formula (f_{c3}) the constants K_1, K_2 , K_3 and K_4 with $K_1 < K_2 < K_3 < K_4$ for the selection of those individuals that find the arena walls and corners (fcf). However, the fitness formula prominently rewards locating cylinders (f_{c2}), their collection inside the arena (*pkfactor*), and their release near the outside walls (*dpfactor*). The average fitness of a population, for over 100 generations, and its maximum individual fitness is shown in Figure 2.

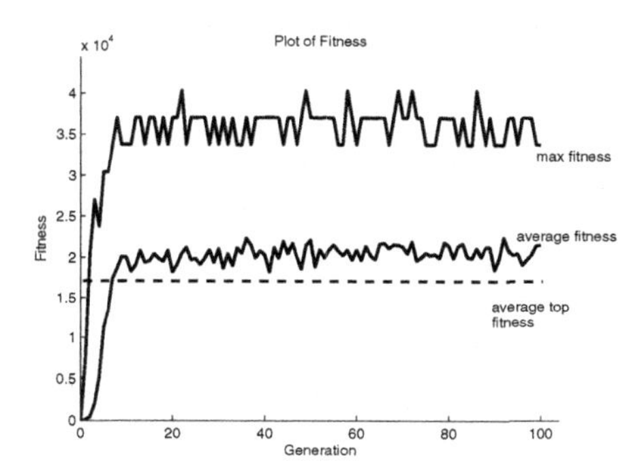

Fig. 2. Fitness is plot across 100 generations. For each generation the highest fitness of one individual was obtained from the averaged fitness of five trials under similar conditions, and the maximum fitness of all the individuals was averaged as a measure of the population fitness. Individuals are more rewarded if they avoid obstacles, collect cylinders, and deposit cylinders close to corners. The evolution is stopped after fitness stabilizes over a value around 3500. For comparison purposes, a horizontal dotted-line indicates the top average fitness of hand-coded selection with evolved behavioral components.

The use of coevolution tends to optimize behavior and their selection in time and in the physical environment. Therefore, it is possible to observe that in attempt to maximize fitness; evolution may disrupt the order of selection as anticipated by the human designer. The latter, has been evidenced in [6]where five behaviors were setup for coevolution. The presence of redundant behavior leads to an evolutionary race, because interference from other behavior may affect the development and selection of a particular behavior. The same phenomenon occur in these experiments.

The cylinder-pickup behavior has been programmed to open the robot claw, even when is already open, to prevent the attempt to fetch a can with a closed gripper. As a result, coevolution overspecialized the selection of this behavior prohibiting the execution of the cylinder-deposit behavioral pattern (as shown in the next section). In order to avoid overspecialization the behavioral modules and the parameters of CASSF were further coevolved at incremental stages using fitness as in 3. As a final step in the coevolution a set of hand-coded parameters for CASSF were chosen to build a seed made of individually evolved behaviors and the selection parameters (Fig. 2). Furthermore, at this step the best of the individuals from the coevolution of behavior and selection parameters was also inserted in the seed. Consequently, a good quality solution was produced as the result of the incremental coevolution of behavior and the selection parameters of CASSF.

6 Experiments and Results

The foraging task was set in an arena with four cylinders as simulated food. In this paper a behavior is considered as the joint product of an agent, environment, and observer. Therefore, a regular grasping-depositing pattern in the foraging task should be the result of the selection of the behavioral types: cylinder-seek, cylinder-pickup, wall-seek, and cylinder-deposit in that order. Collection patterns can be disrupted if for example the cylinder slips from the gripper or a corner is immediately found. Additionally, long search periods may occur if a cylinder is not located. However, there is an additional factor that should be taken into account, which is the fitness of the agent that is solving the foraging task, who finally alters the order in the selection of behavior.

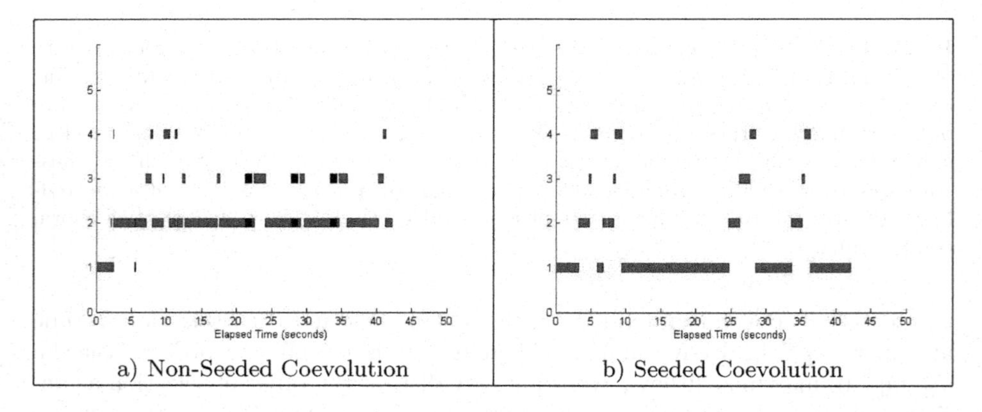

a) Non-Seeded Coevolution b) Seeded Coevolution

Fig. 3. The behaviors are numbered as 1-cylinder-seek, 2-cylinder-pickup, 3-wall-seek, 4-cylinder-deposit and 5-no action selected. A regular grasping-depositing pattern is not easily observed in a). Because of the use of cylinder-pickup to fetch a cylinder already released either in the center of the arena or the inside-walls, the number of deposited cylinders (5 in total) is more than the available cylinders. The latter, explains the reason of the individual presenting a 99 % of the highest fitness. On the other hand in b) four cylinders were properly collected and delivered. However, this individual only scores 83 % of the highest fitness; though, a regular grasping-depositing pattern is shown.

The coevolution of behavior and selection parameters without the use of a seed is presented in Figure 3(a). In contrast to coevolution of the same behavioral patterns and the decision network with the use of a seed (Fig. 3(b)). It is important to notice that the use of a fitness function for coevolution is shaping selection by the optimization of behavior in time and in the physical environment. As seen in Figure 3(a) where a regular pattern occurs during the first ten seconds of the execution of non-seeded coevolution. Then , cylinder-pickup is employed to pick and release cylinders deposited either in the center of the arena or aside the inside-walls. Next, a comparison of a hand-coded network with evolved behavior and a coevolved network with a seed is presented in Figure 4. In this graph is possible to observe an improvement on the fitness of seeded coevolution in comparison to the hand-coded network with evolved behavior.

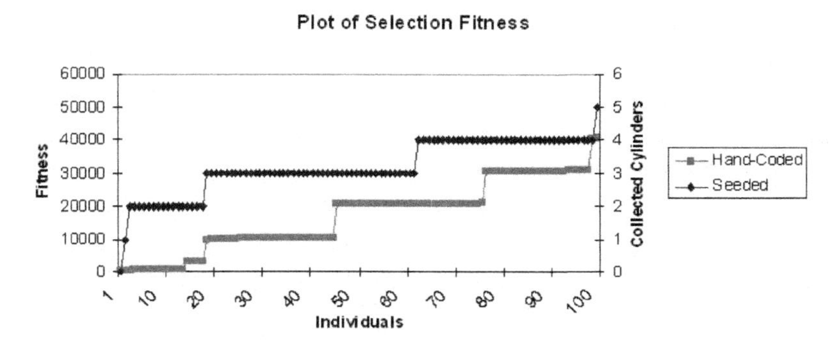

Fig. 4. Plot of the selection fitness of 100 individuals. The secondary Y-axis shows the number of collected cylinders. It can be observed in the graph that seeded coevolution presents the highest fitness. In contrast, the hand-coded network with evolved behavior sometimes fails to collect a cylinder.

7 Conclusions

The coevolution of the central action selection mechanism with neural and programmed behavior was carried out in this study. Later on, both the selection mechanism and neural behavior were further coevolved with a seed made of the initial results of coevolution and a set of hand-coded selection parameters with individually evolved neural behaviors. The experiments presented in this paper provide a insight of the effects of coevolution when evolving behavior that needs to be coupled in a regular pattern. For example, a disruption of the regular selection pattern occurs in an attempt of the selection parameters to increase their fitness as shown in Figure 3(a). However, the use of the seed in the coevolution constrains the proposed solutions to those producing a regular pattern. As a result, maximum fitness of the population fluctuates around a constant level and sometimes even decreases (Fig. 2), yet the individuals complete the foraging task with an optimal fitness (Fig. 4). Finally, the approach here presented aims to reduce the number of decisions made by the human designer when evolving both selection and behavior.

Acknowledgments. This work has been sponsored by CONACyT-MEXICO grant SEP-2004-C01-45726.

References

1. Maes, P.: How to do the right thing. Connection Science Journal Vol. 1, no. 3, 291-323, 1989.
2. Nolfi, S.: Evolving non-trivial behaviors on real robots: A garbage collection robot. Robotics and automation system, vol. 22, 187-98, 1997.
3. Seth, A. K.: Evolving action selection and selective attention without actions, attention, or selection. From animals to animats 5: Proceedings of the Fifth International Conference on the Simulation of Adaptive Behavior, Cambridge, MA. MIT Press, 139-47, 1998.
4. Kyong-Joong, K., Sung-Bae, C.: Robot action selection for higher behaviors with cam-brain modules. Proceedings of the 32nd ISR(International Symposium in Robotics), 2001.
5. Yamauchi, B., Beer, R.: Integrating reactive, sequential, and learning behavior using dynamical neural networks. From Animals to Animats 3, Proceedings of the 3rd International Conference on Simulation of Adaptive Behavior, MIT Press/Bradford Books, 1994.
6. Montes-González, F., Santos Reyes, J., Ríos Figueroa, H.: Integration of Evolution with a Robot Action Selection Model, A. Gelbukh and C. A. Reyes-García (Eds.): MICAI 2006, LNAI 4293, pp. 1160-1170, 2006
7. Bäck, T.: Evolutionary Algorithms in Theory and Practice: Evolution Strategies, Evolutionary Programming, Genetic Algorithms. Oxford University Press, 1996.
8. Santos, J., Duro, R.: Artificial Evolution and Autonomous Robotics (in Spanish). Ra-Ma Editorial, 2005.
9. Holland, J.: Adaptation in Natural and Artificial Systems. University of Michigan Press, Ann Arbor, 1975.
10. Ebner, M.: Coevolution and the red queen effect shape virtual plants. Genetic Programming and Evolvable Machines 7, no. 1, 103-123, 2006.
11. Floreano, D., Mondana F.: Evolution of homing navigation in a real mobile robot. IEEE Transactions on Systems, Man and Cybernetics 26, no. 3, 396-407, 1996.
12. Nolfi, S., Floreano, D.: Evolutionary Robotics. The MIT Press, 2000.
13. Nolfi, S., Floreano D., Miglino, O., Mondada, F.: How to evolve autonomous robots: Different approaches in evolutionary robotics. Proceedings of the International Conference Artificial Life IV, Cambridge MA: MIT Press, 1994.
14. Mondana, F., Franzi, E., Paolo, I.: Mobile robot miniaturisation: A tool for investigating in control algorithms. Presented at the Proceedings of the 3rd International Symposium on Experimental Robotics, Kyoto Japan, 1993.
15. Webots: 'http://www.cyberbotics.com', Commercial Mobile Robot Simulation Software, 2006.
16. Montes González, F., Marín Hernández A., Ríos Figueroa H.: An effective robotic model of action selection. R. Marin et al. (Eds.): CAEPIA 2005, LNAI 4177, 123-32, 2006.

WiSARD and NSP for Robot Global Localization

Paolo Coraggio[1] and Massimo De Gregorio[2]

[1] Dip. Scienze Fisiche, Università degli Studi di Napoli "Federico II", Naples, Italy
pcoraggio@na.infn.it
[2] Istituto di Cibernetica "Eduardo Caianiello", CNR, Pozzuoli (NA), Italy
m.degregorio@cib.na.cnr.it

Abstract. In this paper a hybrid approach for solving a robot global localization problem in an office-like environment is presented. The global localization problem deals with the estimation of the robot position when its initial pose is unknown. The core of this system is formed by a virtual sensor, capable of detecting and classifying the corners in the room in which the robot acts, and an NSP (Neuro Symbolic Processor) control that infers and computes the possible robot locations. In this way, the whole global self localization problem is tackled with a hybrid approach: a classic neurosymbolic hybrid system, composed of a weightless neural network and a BDI agent (it processes the map and build the landmark connections), a neural virtual sensor (for detecting landmarks) and a unified neurosymbolic hybrid system (NSP) devoted to the computation of the robot location on the given map.

1 Introduction

Robot self localization is one of the most important problem to solve in robotic research and it is often addressed as the first property to endow a robotic system to have high level reasoning capacity [1]. The global position problem is the ability, for the robot, to autonomously recognize its pose starting from an unknown position on a given map [2]. It is a very difficult problem needing a lot of computational power, and giving not so much accurate results in terms of robot pose estimation [3]. Several approaches have been successfully proposed during the last years. Markov Localization [4] considers the pose estimation problem as a Markovian process that alternates sensorial readings and motor actions. Monte Carlo Localization [5][6] has the advantage of expressing multimodal probability distributions and of preserving all the plausible hypotheses. Other approaches to the localization problem are based on the detection and recognition of particular (natural or artificial) landmarks of the environment [7][8][9].

In a landmark based approach, the robot position is generally evaluated by means of four different steps: 1) a sensor system for acquiring environmental data; 2) an image processing unit devoted to recognize a priori settled landmarks; 3) a method for establishing the correspondences between the detected

J. Mira and J.R. Álvarez (Eds.): IWINAC 2007, Part II, LNCS 4528, pp. 449–458, 2007.

landmarks and their location on a previously given map; 4) a method for computing the robot position and its error.

In this paper an autonomous mobile robot global localization system is presented. The robot is endowed with a 16 sonar belt array, a camera (the exteroceptive system) and odometric encoders (the proprioceptive system) mounted on the commercial differential drive mobile platform Pioneer 3DX. The robot navigates in an office-like environment so, as distinctive landmarks, the corners between walls represent a good choice. The camera is the sensor devoted to the corner detection (1). The image processing unit is a weightless neural network able to recognize and, above all, to classify the corners in the environment (2). A unified neurosymbolic hybrid system (NSP [10]) maps the detected corners with those on the map (3), and it estimates, during the robot navigation, all the robot plausible locations (4).

In this work, we would like to point out that a hybrid neurosymbolic framework is used to deal with most of the aspects of the global localization problem.

2 System Overview

The robot acts in an office like environment. It can explore rooms with no steps or stairs. A planar 2D map representation of the room completely ensure all the environmental information the robot needs.

The map can be given to the robotic system in two different ways: by means of a bitmap picture (through a file) or by showing the map to the robot (through its camera - see section 4). In the latter case, the map is first processed by an Agent WiSARD like system (a neuro symbolic hybrid system for extracting information from two dimensional geometrical figures) in order to locate and to classify the corners it contains [11] (see section 3).

Once the robotic system receives the map, its navigation starts in a wandering way just looking for a wall or some furniture. When the sonars detect a wall, the robot turns on the right, gets aligned to the wall and keeps on following the wall. In this way, the robots navigates the environment in a clockwise way. During its navigation, if the virtual sensor (WiSARD like system [12]) detects a corner it sends these information to the NSP control module. This module is devoted to determine whether the robot can be localized on the map. In the case that the information (detected corners) are not enough to decide the robot location, the robot keeps going and the virtual sensor try to detect the next corner. Otherwise, the NSP module stops the robot and outputs its location on the map. (The system architecture is sketched in figure 1.)

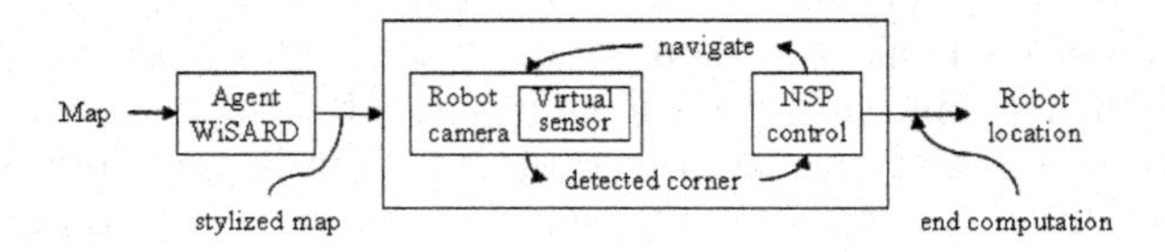

Fig. 1. System architecture

3 Map Input Module

In order to give the system the map in which the robot has to self localize, we decided of taking advantage of a previous system designed for reconstructing and understanding 2D geometrical figures formed by rectangles and triangles [11]. In fact, changing the agent planes devoted to the reconstruction of the mentioned figure, with plans designed to reconstruct and understand a closed polygonal, we got a system capable of understanding a map from an actual picture (fig. 2).

The map has to represent a delimited environment (such as an office) without steps. Furthermore, in the environment are allowed only furniture with well defined shapes (for instance: wardrobe, cabinets, etc.) in order to deal with corner shaped landmarks (such as the cabinet in figure 4).

The neural network (WiSARD like system) has been trained with different instances of corners and adopting the modified training algorithm proposed in [13]. In this way, the system is capable of showing a stylized map formed only by "mental" images of recognized landmarks [14].

Agent WiSARD reproduces the given map labelling all the landmarks (corners) that have been selected and considered belonging to a plausible closed polygonal. In fact, some recognized landmarks are discarded by the agent because their inclusion would not lead to a closed polygonal.

The stylized map with labels, corresponding to the given input, is shown in fig. 2. The same stylized map is used by the system to show the place where the robot localizes itself. The labels have been assigned in clockwise order. Each label is formed by a number and a landmark class: inward or outward corner.

Agent WiSARD creates a list of all the recognized landmarks with their class. This list is passed (with the number of map symmetries) to the control module to generate the devoted neural network (NSP). So doing, the NSP is ready to check, step by step, whether the robot has localized itself on the map (see 5).

The Agent WiSARD calculates the number of symmetries taking into account the clockwise order of the labelled corners. This means that although the map in figure 2 is symmetric from the geometric point of view, it is not symmetric with respect to the paths the robot can follow passing the labelled corners in a clockwise order.

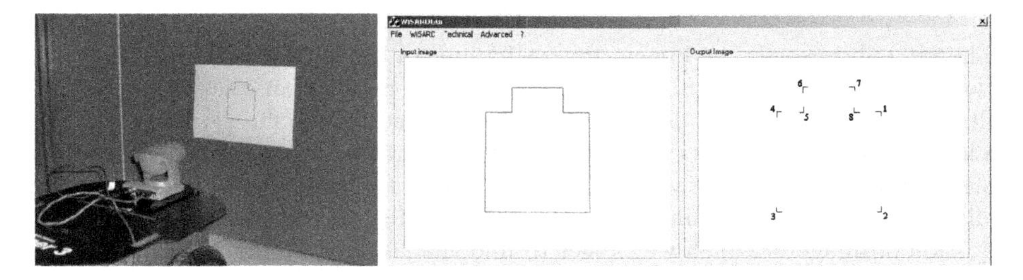

Fig. 2. Map reconstruction. Left: map input. Right: pictorial representation with labels of the reconstructed map.

4 WiSARD as Virtual Sensor

Another WiSARD like systems serves as a virtual sensor. This means that once grabbed a frame from the camera, the virtual sensor processes it to obtain the necessary landmark information. In order to detect and classify the corners, the WiSARD hase been trained on particular images representing the classes of landmarks we expect to detect. The virtual sensor analyzes only the lowest part of the images, by means of a little spot (i.e. a sort of attention window on the image), looking for corners. This choice speeds up the detection process.

4.1 Generating the Training Set

In order to generate the training set, it is necessary to create, for each class of discriminators, different images representing corners seen from different visual angles. This is due to the fact that it is not predictable, neither decidable, the angle from which the robot will see the corner. After a preliminary experimental phase, we found out that the corners were detected at best when the robot stands at about 1.5 meters from the wall. Using a 3D modelling software (in this case Maya), we have generated a set of images representing the same scene but with different angles of view. For each image, examples of corners belonging to the two classes have been extracted and used to create the training sets (fig. 3).

Fig. 3. Part of the scenes used to extract the training set

On this particular scene, a virtual camera, positioned with the same parameters of the real one, analyzed the possible cases. The scene has been rotated step by step (5 degrees step) and a frame has been captured for each new angle. From each frame two types of "relevant" corners have been extracted to form the corresponding training set ("inward" and "outward" corners).

4.2 The Virtual Sensor

The virtual sensor is the image processing unit of the system and it detects and classifies the corners acquired by the robot camera. It is an adapted version of the one used in a previously designed hybrid neurosymbolic system [15]. The adopted WiSARD accepts black and white binary images as input; therefore, a pre elaboration of the camera image is necessary to feed the system. Processing all the image could be computational heavy and, in this case, useless. In fact, the portion of image in which we expect to find the corner (remind that the camera tilt inclination is -15 degrees) is bounded in a particular "strip" of the frame. The image pre processing phase can be summarized in four steps: 1) extracting the strip on the image; 2) converting the strip content from rgb 24 bit format in a grey scale; 3) a trace contour (sobel) filter application; 4) thresholding the strip content to obtain a binary image.

In order to further speed up the detecting process, a little spot (same size of the images used in the training sets) scans on those part of the strip in which there is a certain number of black pixels. So doing, most of the content of the strip is not take into account.

During the spot scanning, the virtual sensors classifies the spot content. Each time the sensor detects a corner, it classifies the corner by putting the corresponding "mental" image on the current position of the spot. In the meanwhile, it communicates to the NSP control module the class of the detected corner.

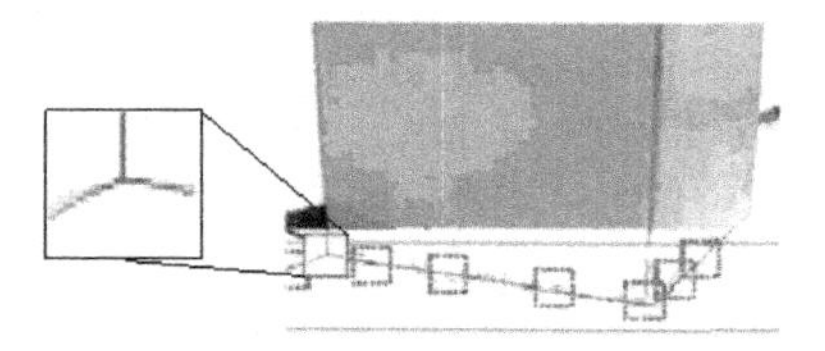

Fig. 4. Spot scanning and "mental" image of "outward" corner class

5 NSP as Control Module

The use of the NSP allow us to have a neural network controlling the robot and, at the same time, the opportunity to implement it on hardware (FPGA) [16]. In this case, we have implemented the self localization strategy using the NSL [17]. Each time the virtual sensor detects a landmark, the NSP runs in order to determine, whether possible, the location of the robot on the map. In such a case, the NSP communicates that the position has been determined. The whole process stops and the system shows the robot position on the virtual map. If more than one position is plausible (this is the case of symmetric maps) the system stops as well and communicates the positions where the robot could be positioned. As reported above, the environment is very simple and just formed by walls, a floor (no steps are allowed) and some furniture. In the case of a symmetric environment and in absence of other landmarks, there is no way for the robot (and even for human beings) to locate itself in a unique manner.

5.1 The Neural Network Architecture

Before to enter the details of the network architecture, we would like to remind the reader some NSL definitions for generating devoted NSPs. As reported in [17], some operators have been defined to automatically generate the logically equivalent NSP. In this paper, we are going to refer to the following operators: $IMPLY$, $ATLEAST$ and $ATMOST$.

The statement "Q is true if P_j is true", where P_j represents the conjunction of all the literals in P_j ($P_j = [P_1, \ldots, P_N]$), will be denoted by the operator $IMPLY(P_j, Q)$, and by the corresponding NSL statement `IMPLY(P[1..N], Q)`. The statement "Q is true if at least h literals belonging to P_j are true", will be represented by means of the operator $ATLEAST(P_j, h, Q)$, and by the corresponding NSL statement `ATLEAST(P[1..N], H, Q)`. Finally, we express that "Q is true if no more than h literals belonging to P_j are true" by means of the operator $ATMOST(P_j, h, Q)$, and by the corresponding NSL statement `ATMOST(P[1..N], H, Q)`.

With these operators and taking advantages of other NSL constructs[1], we can automatically generate the logically equivalent neural networks. Furthermore, we refer to the map of figure 2 in the examples reported from here on.

At the start up, the NSP receives as input from the Agent WiSARD the following information: the list $C = [c_1, \ldots, c_l]$ of corners detected on the 2D map; the list $C^O = [c_1, \ldots, c_m]$ of outward corners; the list $C^I = [c_1, \ldots, c_n]$ of inward corners ($l = m + n$); the number H of possible ambiguities ($H = 1$ for non symmetric maps).

From these information, the corresponding NSP is generated and ready to receive step by step the corners detected by the virtual sensor during the self localization process.

The input layer is formed by $2l$ neurons representing the possible inputs the system can receive during the exploration. In fact, for each label (corner) two neurons are generated: IN^I_j, IN^O_j. IN^I_j represents an inward corner that can be detected at step j, while IN^O_j represents the outward one. Since the system cannot decide in advance what kind of corner is going to detect at step j, both corners (I, O) are represented for each step.

A first layer is generated by the following code and represents the set of possible corners the system could start its exploration (Step 0):

```
FOR J=1 TO M  IMPLY(INO1, CO[J]1)
FOR J=1 TO N  IMPLY(INI1, CI[J]1)
```

Others $l-1$ layers are generated in order to deal with the detected edges after Step 0. The code is the following:

```
FOR K=2 TO L
   FOR J=2 TO M  IMPLY((INOK, PREV(CO[J])K-1), CO[J]K)
   FOR J=2 TO N  IMPLY((INOK, PREV(CI[J])K-1), CI[J]K)
```

The operator $PREV$ is defined as: $PREV(c_i) = c_{i-1}$ for $i = 2 \ldots l$ and $PREV(c_1) = c_l$. This means that the list of corners passed by Agent WiSARD is a circular list representing the corners in clockwise order (see fig. 5).

In the first layer, neurons $c^O_{j,1}$ are actives at Step 0 Time 2 only if IN^O_1 is active at Step 0 Time 1. That is, the system receives as first input an outward corner (IN^O_1). In layer k, neurons $c^O_{j,k}$ are actives if both IN^O_k and $PREV(c^O_j,)_{k-1}$ are actives. This means the neurons threshold in the first layer is $1-\epsilon$ while for the other layers is $2-\epsilon$. Moreover, the input neurons fire on themselves an excitatory impulse. So doing, we trace the overall reasoning carried out by the NSP.

[1] Further NSL constructs are: `FOR`, `IF THEN ELSE`, `WHILE DO` and `REPEAT UNTIL`.

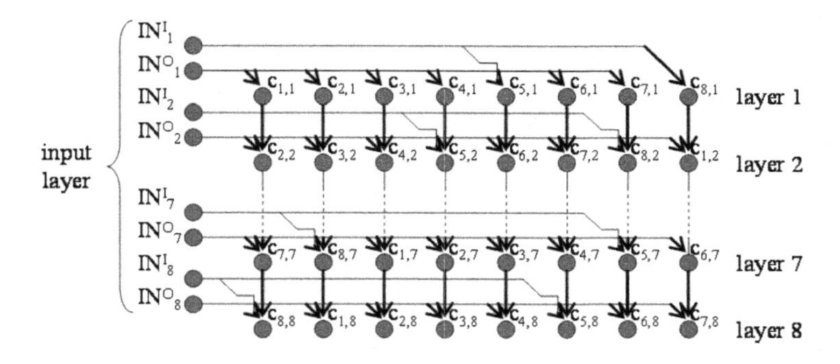

Fig. 5. Input layer and the l network layers

The control subnet, devoted to establish the end of the whole computation, if formed by as many neuron END_j as the number of layers present in the network ($j = 1 \ldots l$) and by a single neuron END that is active when the computation ends. The end computation control is based on the following network property: when the number of active neurons in a layer k is equal to the number of possible ambiguities (H) the system can establish the corner (or the corners, in case of symmetric maps) the robot is looking at with respect to the given map. The following statements are used to generate the control subnet:

```
FOR K=1 to L  ATMOST(C[1..L]K, H, ENDK)
ATMOST(END1..K, 1, END)
IMPLY(END, END_COMP)
```

The neuron END is active when no more than 1 neuron END_k is active, while neuron END_k is active when no more than H neurons of layer k are actives. The last sentence is used to synchronize the outputs. This will be clear in the example of subsection 5.2. Part of the control subnet is sketched in fig. 6.

In order to show the results of the network computation, an output subnet is generated by the following code:

```
FOR K=1 TO L
   FOR J=1 TO L  IMPLY((ENDK,C[J]K), COUT[J]K)
FOR J=1 TO L  ATLEAST(COUT[J]1..K, 1, COUT[J])
```

The neuron $cout_{j,k}$ is active if in the layer k the neurons END_k (that is, the computation is ended in layer k) and $c_{j,k}$ (that is, the neuron corresponding to

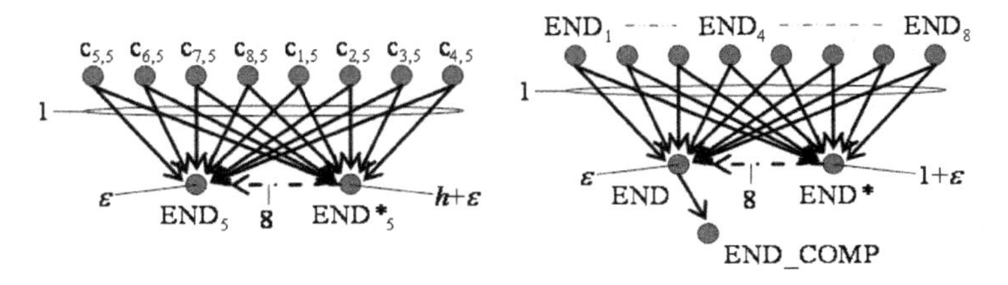

Fig. 6. Left: control on layer 5. Right: end computation control subnet.

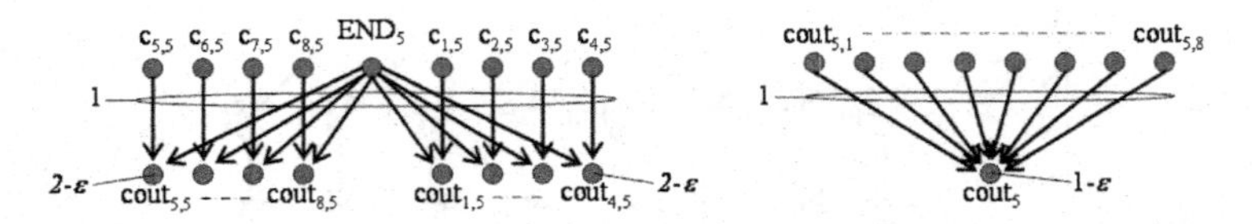

Fig. 7. Left: the generic neuron $cout_{j,k}$ is activated by the corresponding neuron in layer k (that is $c_{j,k}$) and by END_k. Right: subnet for determining whether the robot stopped in corner 5 ($cout_5$ is part of the output layer).

corner label c_j is active in layer k) are both actives. Furthermore, the neuron $cout_j$ is active if at least one of the corresponding $cout_{j,k}$ is active. To sum up, the output layer is formed by the neuron END_COMP and by the set of neurons $cout_{1,...,l}$. Part of the output subnet is reported in fig. 7.

5.2 An Example

Let us consider the map reported in fig. 2, and let suppose that, during the early part of its exploration, the first corner the robot detects is c_2. Since c_2 is an outward corner, the virtual sensor activates neuron IN_1^O as first input for the NSP. From now on, the robot will follow the walls in a clockwise way.

In table 1, the list of active neurons are reported with respect to the detected corner (Step - S) and to the time of activation (Time - T). In the column S, 0 stands for the first detected corner. We remind the reader that the virtual sensor classifies the corner the robot is looking at but cannot establish its position.

In S 1 another outward corner is detected and neuron IN_2^O is activated. One can notice that at T 1 of S 1 the neurons actives at T 3 S 0 are still actives.

At S 3 the robot reaches the corner c_5, and at T 3, for the first time, a neuron END (END_4) is activated instead of an END^* neuron. This means that in layer

Table 1. Sequence of neuron activations during the robot global localization process

S	T	List of active neurons
0	1	IN_1^O
	2	$IN_1^O, c_{1,1}, c_{2,1}, c_{3,1}, c_{4,1}, c_{6,1}$
	3	$IN_1^O, c_{1,1}, c_{2,1}, c_{3,1}, c_{4,1}, c_{6,1}, END_1^*$
1	1	$IN_1^O, c_{1,1}, c_{2,1}, c_{3,1}, c_{4,1}, c_{6,1}, END_1^*$ IN_2^O
	2	$IN_1^O, c_{1,1}, c_{2,1}, c_{3,1}, c_{4,1}, c_{6,1}, END_1^*$ $IN_2^O, c_{2,2}, c_{3,2}, c_{4,2}, c_{7,2}$
	3	$IN_1^O, c_{1,1}, c_{2,1}, c_{3,1}, c_{4,1}, c_{6,1}, END_1^*$ $IN_2^O, c_{2,2}, c_{3,2}, c_{4,2}, c_{7,2}, END_2^*$
2	1	$IN_1^O, c_{1,1}, c_{2,1}, c_{3,1}, c_{4,1}, c_{6,1}, END_1^*$ $IN_2^O, c_{2,2}, c_{3,2}, c_{4,2}, c_{7,2}, END_2^*$ IN_3^O
	2	$IN_1^O, c_{1,1}, c_{2,1}, c_{3,1}, c_{4,1}, c_{6,1}, END_1^*$ $IN_2^O, c_{2,2}, c_{3,2}, c_{4,2}, c_{7,2}, END_2^*$ $IN_3^O, c_{3,3}, c_{4,3}$
	3	$IN_1^O, c_{1,1}, c_{2,1}, c_{3,1}, c_{4,1}, c_{6,1}, END_1^*$ $IN_2^O, c_{2,2}, c_{3,2}, c_{4,2}, c_{7,2}, END_2^*$ $IN_3^O, c_{3,3}, c_{4,3}, END_3^*$
3	1	$IN_1^O, c_{1,1}, c_{2,1}, c_{3,1}, c_{4,1}, c_{6,1}, END_1^*$ $IN_2^O, c_{2,2}, c_{3,2}, c_{4,2}, c_{7,2}, END_2^*$ $IN_3^O, c_{3,3}, c_{4,3}, END_3^*$ IN_4^I
	2	$IN_1^O, c_{1,1}, c_{2,1}, c_{3,1}, c_{4,1}, c_{6,1}, END_1^*$ $IN_2^O, c_{2,2}, c_{3,2}, c_{4,2}, c_{7,2}, END_2^*$ $IN_3^O, c_{3,3}, c_{4,3}, END_3^*$ $IN_4^I, c_{5,4}$
	3	$IN_1^O, c_{1,1}, c_{2,1}, c_{3,1}, c_{4,1}, c_{6,1}, END_1^*$ $IN_2^O, c_{2,2}, c_{3,2}, c_{4,2}, c_{7,2}, END_2^*$ $IN_3^O, c_{3,3}, c_{4,3}, END_3^*$ $IN_4^I, c_{5,4}, END_4$
	4	$IN_1^O, c_{1,1}, c_{2,1}, c_{3,1}, c_{4,1}, c_{6,1}, END_1^*$ $IN_2^O, c_{2,2}, c_{3,2}, c_{4,2}, c_{7,2}, END_2^*$ $IN_3^O, c_{3,3}, c_{4,3}, END_3^*$ $IN_4^I, c_{5,4}, END_4, cout_{5,4}, END$
	5	$IN_1^O, c_{1,1}, c_{2,1}, c_{3,1}, c_{4,1}, c_{6,1}, END_1^*$ $IN_2^O, c_{2,2}, c_{3,2}, c_{4,2}, c_{7,2}, END_2^*$ $IN_3^O, c_{3,3}, c_{4,3}, END_3^*$ $IN_4^I, c_{5,4}, END_4, cout_{5,4}, END, END\_COMP, cout_5$

4, the number of active neurons is equal to H. Since the map is non symmetric, H is equal to 1 and, in fact, the only active neuron in layer 4 is $c_{5,4}$.

The neuron END is activated at T 4 while the output of the network ($cout_5$) will be ready at T 5. In order to have the neurons of the output layer actives at the same time, the neuron END_COMP has been added. In this way, at S 3 and T 5 the network ends its computation and determines where the robot is located ($cout_5$ stands for "the robot is nearby corner c5").

Thanks to this network architecture, we can even trace back all the corners the robot encountered during its navigation. In fact, one can notice that the active neuron $c_{5,4}$ belongs to the "row" of neurons formed by $c_{2,1}$, $c_{3,2}$, $c_{4,3}$. This "row" represents, from $c_{2,1}$ to $c_{5,4}$, in the same order, all the corners encountered by the robot during its navigation.

In order to output all the possible initial corners ($cinit_k$), another output subnet has to be added to the previous one. In particular, the subnet to be added is the one produced by the following NSL code:

```
FOR I=1 to 2*L-1
   FOR J=1 TO I
      IF I < L+1 THEN   IMPLY((C[I]J, ENDJ), CINIT[I-J+1]J)
                 ELSE   IF J>L-1 AND J<L+1 THEN   IMPLY((C[I]J, ENDJ), CINIT[I-J+1]J)
FOR J=1 to L  ATLEAST(CINITJ[1..L], 1, CINITJ)
```

In our example, neuron $cinit_2$ (that stands for "the initial corner was c_2") would have been activated on Time 5 Step 3. We do not need to output the path followed by the robot because having $cinit_2$ and $cout_5$ and being the corners detected in clockwise order, the robot path is unique.

6 Conclusion and Future Works

In this work we presented a hybrid control approach for the global localization problem. The proposed methodology has some interesting peculiarities. Both the map input module and the control module have a common framework. The first one is a classical neurosymbolic hybrid system, the second one is a unified system (NSP). Furthermore, since even the virtual sensor is a WiSARD like system, we can implement most of the system on hardware improving its performances in terms of time response (crucial issue for robot in dynamic environments). Beside this, we have a mechanism that keep track of the robot inference processing. This makes the system capable of "explaining" the given result.

The first experiments we conducted are encouraging and have highlighted the advantages we expected to see. The robot is able to properly localize itself in the sense that indicates the corner in front of which it stands. A further system improvement can be obtained integrating it with a classical probabilistic localization system (e.g. a particle filter). In this way the system (faster and lighter than a particle filter) could process information until robot localize itself on the map, then a particle filter would be applied only to that specific region of the map to compute the precise pose of the robot. In this way, the initial probability density function representing the robot position is no longer a uniform distribution on all the map but is restricted to a small region of the environment.

References

1. Cox, I.J.: Blanche-an experiment in guidance and navigation of an autonomous robot vehicle. Robotics and Automation, IEEE Trans. on **7** (1991) 193–204
2. Borenstein, J., Everett, H.R., Feng, L.: Navigating Mobile Robots: Systems and Techniques. A. K. Peters, Ltd., Natick, MA, USA. (1996)
3. Thrun, S., Burgard, W., Fox, D.: Probabilistic Robotics (Intelligent Robotics and Autonomous Agents). The MIT Press (2005)
4. Simmons, R., Koenig, S.: Probabilistic robot navigation in partially observable environments. In: Proc. of the IJCAI. (1995) 1080–1087
5. Fox, D., Thrun, S., Burgard, W., Dellaert, F.: Particle filters for mobile robot localization (2001)
6. Jensfelt, P., Wijk, O., Austin, D., Andersson, M.: Experiments on augmenting condensation for mobile robot localization. In: IEEE ICRA. (2000) 2518–2524
7. Motomura, A., Matsuoka, T., Hasegawa, T.: Self-localization method using two landmarks and dead reckoning for autonomous mobile soccer robots. In: RoboCup. (2003) 526–533
8. Sim, R., Dudek, G.: Learning and evaluating visual features for pose estimation. In: ICCV (2). (1999) 1217–1222
9. Se, S., Lowe, D., Little, J.: Local and global localization for mobile robots using visual landmarks. In: Proc. of the IEEE/RSJ IROS, Hawaii (2001) 414–420
10. Burattini, E., De Gregorio, M., Ferreira, V.M.G., França, F.M.G.: NSP: A neuro-symbolic processor. In Mira, J., Álvarez, J.R., eds.: IWANN (2). Volume 2687 of Lecture Notes in Computer Science., Springer (2003) 9–16
11. Burattini, E., Coraggio, P., De Gregorio, M.: Agent WiSARD: A hybrid system for reconstructing and understanding two-dimensional geometrical figures. In Abraham, A., Köppen, M., Franke, K., eds.: HIS. Volume 105 of Frontiers in Artificial Intelligence and Applications., IOS Press (2003) 887–896
12. Aleksander, I., Thomas, W., Bowden, P.: WISARD, a radical new step forward in image recognition. Sensor Rev. **4** (1984) 120–124
13. De Gregorio, M.: On the reversibility of multi-discriminator systems. Technical report, Istituto di Cibernetica CNR, Arco Felice (NA) (1997)
14. Burattini, E., De Gregorio, M., Tamburrini, G.: Generation and classification of recall images by neurosymbolic computation. In: Second European Conference on Cognitive Modelling - ECCM98. (1998) 127–134
15. Burattini, E., Coraggio, P., De Gregorio, M., Staffa, M.: Agent WiSARD in a 3D world. In Mira, J., Álvarez, J.R., eds.: IWINAC (2). LNCS 3562, Springer (2005) 272–280
16. Burattini, E., De Gregorio, M., Tamburrini, G.: Neurosymbolic processing: Non-monotonic operators and their FPGA implementation. In: Fourth Brazilian Symposium on Neural Networks, Rio de Janeiro, Brasil. (2000) 93–101
17. Burattini, E., de Francesco, A., De Gregorio, M.: NSL: A neuro-symbolic language for a neuro-symbolic processor (NSP). Int. J. Neural Syst. **13** (2003) 93–101

Design and Implementation of an Adaptive Neuro-controller for Trajectory Tracking of Nonholonomic Wheeled Mobile Robots

Francisco García-Córdova[1], Antonio Guerrero-González[1], and Fulgencio Marín-García[2]

[1] Department of System Engineering and Automation,
[2] Department of Electrical Engineering
Polytechnic University of Cartagena (UPCT),
Campus Muralla del Mar, 30202, Cartagena, Murcia, Spain
{francisco.garcia, antonio.guerrero, pentxo.marin}@upct.es

Abstract. A kinematic adaptive neuro-controller for trajectory tracking of nonholonomic mobile robots is proposed. The kinematic adaptive neuro-controller is a real-time, unsupervised neural network that learns to control a nonholonomic mobile robot in a nonstationary environment, which is termed Self-Organization Direction Mapping Network (SODMN), and combines associative learning and Vector Associative Map (VAM) learning to generate transformations between spatial and velocity coordinates. The transformations are learned in an unsupervised training phase, during which the robot moves as a result of randomly selected wheel velocities. The robot learns the relationship between these velocities and the resulting incremental movements. The neural network requires no knowledge of the geometry of the robot or of the quality, number, or configuration of the robot's sensors. The efficacy of the proposed neural architecture is tested experimentally by a differentially driven mobile robot.

1 Introduction

Several heuristic approaches based on neural networks (NNs) have been proposed for identification and adaptive control of nonlinear dynamic systems. More recently, the efforts have been directed toward the development of control schemes which, besides providing improved performance, can be proved to be stable [1].

In the trajectory-tracking problem, the wheeled mobile robot (WMR) is to follow a prespecified trajectory. Using the kinematic model of WMRs, the trajectory-tracking problem was solved in [2]. Both the local and global tracking problem with exponential convergence have been solved theoretically using time-varying state feedback based on the backstepping technique in [3]. The kinematic model of mobile robots can be described in polar coordinates [4] and stabilization achieved using the backstepping technique. Dynamic feedback linearization has been used for trajectory tracking and posture stabilization of mobile robot systems in chained form [5].

J. Mira and J.R. Álvarez (Eds.): IWINAC 2007, Part II, LNCS 4528, pp. 459–468, 2007.

The approximate linearization of robot equations about the desired reference trajectory has been used for developing trajectory tracking techniques [2], [6] as well as non-linear approaches like the local continuous nonlinear feedback action and the feedback linearization [7], [6]. The backstepping paradigm has been used as well and a number of recursive techniques for trajectory tracking have been proposed [1], [8]. Recent advances tend to focus on the robustness of trajectory tracking techniques with respect to input disturbances and/or parameters variations [9]. In this framework, sliding mode control has been extensively used. Tracking problems for mobile robots not satisfying the nonholonomic constraints have been addressed by a slow manifold method, by using continuous-time variable structure [10] and by a discrete-time sliding mode control technique [11].

Neural networks based controllers have been deeply investigated (see, e.g. [12], [13], [1]), since can be used to learn nonlinear functions, representing direct dynamics, inverse dynamics or any other mapping in the process. In the field of WMRs control, NN are embedded in the closed-loop control system of a WMR and the corresponding NNs-based controllers have shown to effectively deal with unmodelled bounded disturbances and/or unstructured unmodelled dynamics in the mobile robot model [13], [1]. In particular in [1] and [14] a neural network based controller has been developed by combining the feedback velocity control technique and torque controller, using a multilayer feedforward neural network. In this approach the universal approximation property of neural network is used to learn the dynamics of the mobile robot. The resulting control structure and the neural network learning algorithm are very complicated and they require high computational efforts. In the above quoted contributions only numerical simulations studies have been developed. In [13] a simple NN based controller has been proposed by integrating the kinematic control techniques with the universal approximation capabilities of neural networks. In this approach the neural network capabilities to learn unmodelled robot characteristics in a real experimental setup are used for improving the performance of the kinematic control techniques.

The study of autonomous behavior has become an active research area in the field of robotics. Even the simplest organisms are capable of behavioral feats unimaginable for the most sophisticated machines. When an animal has to operate in an unknown environment it must somehow learn to predict the consequences of its own actions. By learning the causality of environmental events, it becomes possible for an animal to predict future and new events [15]. Somehow this learning is possible for organisms in spite of what seem like insurmountable difficulties from a standard engineering viewpoint: noisy sensors, unknown kinematics and dynamics, nonstationary statistics, and so on [16].

In this paper, a real-time, unsupervised neural network controller that can learn to guide a mobile robot towards a target located at an arbitrary location in a 2-D workspace. The robot's movements are controlled by selecting the angular velocity of each of two driving wheels. The neuro-controller we propose is based on existing neural networks of biological sensory-motor control. The neuro-controller requires no information about the robot's structure, and it is resistant to a variety of disturbances. The kinematic adaptive neuro-controller is a Self-Organization Direction Mapping Network (SODMN), uses an associative learning to generate transformations between spatial and velocity coordinates. The efficacy of the proposed neural controller is tested experimentally by a differentially driven mobile robot.

This paper is organized as follows. We first describe the neural control system to mobile robot tracking and approach behaviors in Section II. Experimental results with the proposed scheme for control of a mobile platform are addressed in Section III. Finally, in Section IV, conclusions based on experimental results are given.

2 Architecture of the Neural Control System

Figure 1 illustrates our proposed neural architecture. The trajectory tracking control without obstacles is implemented by the SODMN. The SODMN learns to control the robot through a sequence of spontaneously generated random movements. The random movements enable the neural network to learn the relationship between angular velocities applied at the wheels and the incremental displacement that ensues during a fixed time step. The proposed SODMN combines associative learning and Vector Associative Map (VAM [17]) learning to generate transformations between spatial and velocity coordinates. The nature of the proposed kinematic adaptive neuro-controller is that continuously calculates a vectorial difference between desired and actual velocities, the robot can move to arbitrary distances and angles even though during the initial training phase it has only sampled a small range of displacements. Furthermore, the online error-correcting properties of the proposed architecture endow the controller with many useful properties, such as the ability to reach targets in spite of drastic changes of robot's parameters or other perturbations.

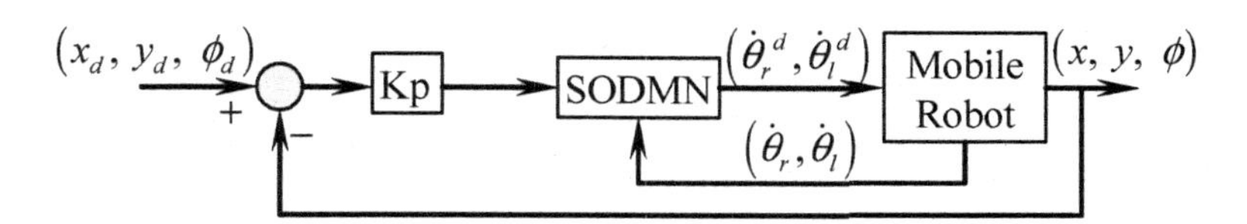

Fig. 1. Neural architecture for reactive and adaptive navigation of a mobile robot

2.1 Self-Organization Direction Mapping Network (SODMN)

At a given set of angular velocities the differential relationship between mobile robot motions in spatial coordinates and angular velocities of wheels is expressed like a linear mapping. This mapping varies with the velocities of wheels. The transformation of spatial directions to wheels' angular velocities is shown in Fig. 2. The spatial error is computed to get a spatial direction vector (DVs). The DVs is transformed by the *direction mapping network* elements V_{ik} to corresponding motor direction vector (DVm). On the other hand, a set of tonically active inhibitory cells which receive broad-based inputs that determine the context of a motor action was implemented as a context field. The context field selects the V_{ik} elements based on the wheels' angular velocities configuration.

A speed-control GO signal acts as a nonspecific multiplicative gate and control the movement's overall speed. The GO signal is a input from a decision center in the brain, and starts at zero before movement and then grows smoothly to a positive value as the movement develops. During the learning, sensed angular velocities of wheels are fed into the DVm and the GO signal is inactive.

Activities of cells of the DVs are represented in the neural network by quantities $(S_1, S_2, ..., S_m)$, while activities of the cells of the motor direction vector (DVm) are represented by quantities $(R_1, R_2, ..., R_n)$. The direction mapping is formed with a field of cells with activities V_{ik}. Each V_{ik} cell receives the complete set of spatial inputs S_j, $j = 1, ..., m$, but connects to only one R_i cell (see Figure 2). The mechanism that is used to ensure weights converge to the correct linear mapping is similar to the VAM learning construction [17]. The direction mapping cells ($\boldsymbol{V} \in \mathbb{R}^{n \times k}$) compute a difference of activity between the spatial and motor direction vectors via feedback from DVm. During learning, this difference drives the adjustment of the weights. During performance, the difference drives DVm activity to the value encoded in the learned mapping.

A context field cell pauses when it recognizes a particular velocity state (i.e., a velocity configuration) on its inputs, and thereby disinhibits its target cells. The target cells (direction mapping cells) are completely shut off when their context cells are active. This is shown in Fig. 2. Each context field cell projects to a set of direction mapping cells, one for each velocity vector component. Each velocity vector component has a set of direction mapping cells associated with it, one for each context. A cell is "off" for a compact region of the velocity space. It is assumed for simplicity that only one context field cell turns "off" at a time. In Figure 2, inactive cells are shown as white disks. The center context field cell is "off" when the angular velocities are in the center region of the velocity space, in this two degree-of-freedom example. The "off" context cell enables a subset of direction mapping cells through the inhibition variable c_k, while "on" context cells disable to the other subsets.

The DVs cell activities, $\boldsymbol{S} \in \mathbb{R}^m$, are driven by the desired spatial direction, $\boldsymbol{xd} \in \mathbb{R}^m$, computed from the difference of the visual target position and the mobile-robot current position

$$\frac{d}{dt}S_j = \delta(xd_j - S_j). \tag{1}$$

where δ is a gain that controls the integration speed rate.

Direction mapping cells with activity V_{ik} compute the difference of the weighted DVs input and the DVm activity. The V_{ik} cell activities are described as

$$\frac{d}{dt}V_{ik} = \alpha(-V_{ik} + c_k(\sum_j z_{jik}S_j - R_i)), \tag{2}$$

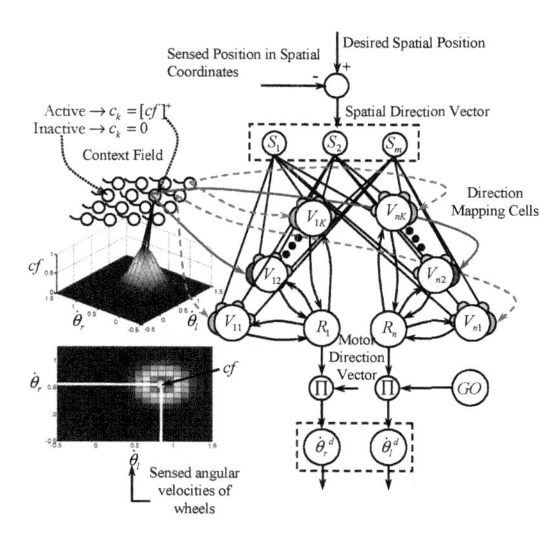

Fig. 2. Self-organization direction mapping network for the trajectory tracking of a mobile robot

where α is a time constant, the coefficients c_k $(k = 1, ..., K)$ represent inhibition from the context field and z_{jik} represents the element of the inverse mapping which multiplies the jth spatial components to contribute to the ith velocity component. Here, the spatial representation contains m-components and n is the number of independent wheels. The kth context field contacts the set $\{V_{ik}, i = 1, ..., n\}$ of direction mapping cells (see Fig. 2). When the context cell is active (modeled as $c_k = 0$), the entire input current to the soma is shunted away such that there remains only activity in the axon hillock, which decays to zero. When the kth context cell shuts off, $c_k = 1$, the V_{ik} receive normal input. The number of cells of the context field is K and it is calculated as

$$K = \left(\prod_{j=1}^{n} D_j \right), \tag{3}$$

where D_j is the number of intervals of the discrete velocity range. Consequently, the total number of the weights, z_T, at the neural network is given by

$$z_T = n.m.K, \tag{4}$$

where m is the number of spatial dimensions.

The motor direction cell activities, $\boldsymbol{R} \in \mathbb{R}^n$, are driven by the V_{ik} during performance and by wheels' rotation velocities $\dot{\theta}_i$ during learning

$$\frac{d}{dt} R_i = \delta \left[(1 - e) \left(\sum_k V_{ik} - R_i \right) + e(\dot{\theta}_i - R_i) \right]. \tag{5}$$

In the learning phase, the endogenous random generator (ERG) circuit is activated, $e = 1$ and the R_i cells are driven to wheels' sensed velocities $\dot{\theta}_i$. During the performance, the ERG circuit is inactive, $e = 0$, and the input is the sum of the V_{ik}, only one of which will be actively processing input.

The Motor Present Direction Vector (PDVm) cell activities ($\dot{\theta}_{id}$) are given by

$$\dot{\theta}_{id} = \alpha(R_i g + e\dot{\theta}_{i_{ERG}}), \tag{6}$$

where g is the sigmoidal GO signal and is described by

$$\begin{aligned} \frac{d}{dt}g^{(1)} &= \varepsilon\left[-g^{(1)} + \left(Cs - g^{(1)}\right)g^{(0)}\right], \\ \frac{d}{dt}g^{(2)} &= \varepsilon\left[-g^{(2)} + \left(Cs - g^{(2)}\right)g^{(1)}\right], \\ g &= g^{(0)}\frac{g^{(2)}}{Cs}. \end{aligned} \tag{7}$$

where $g^{(0)}$ is the step input from a forebrain decision center; ε is a slow integration rate; Cs is the value at which the GO cells are saturated. In the model, wheels' velocities commands are represented by $\dot{\theta}_{id}$, and are given by ERG in the learning phase.

Learning Phase. The learning is obtained by decreasing weights in proportion to the product of the presynaptic and postsynaptic activities. The network can also be redesigned to have only positive activations and weights by using the appropriate push-pull mechanism as in Gaudiano and Grossberg [17]. The training is done by generating random movements, and by using the resulting angular velocities and observed spatial velocities of the mobile robot as training vectors to the direction mapping network. The weights of network are obtained as

$$\frac{d}{dt}z_{jik} = -\eta e V_{ik} S_j. \tag{8}$$

During learning in a particular context, k^{th}, with e=1, we can note that: $S_j \rightarrow xd_j$; $R_i \rightarrow \dot{\theta}_i; \dot{\theta}_i = \dot{\theta}_{id} \rightarrow \dot{\theta}_{i_{ERG}}$; and $V_{ik} \rightarrow \left(\sum_j z_{jik} S_j - R_i\right)$.

Therefore, the learning rule can be obtained by using the gradient-descent algorithm and the equation (8) is modified in discrete form as:

$$z_{jik}(t+1) = z_{jik}(t) + \eta\left(\dot{\theta}_i - \sum_j z_{jik} xd_j\right) xd_j \tag{9}$$

where η is learning rate and is a positive constant gain.

3 Experimental Results

The proposed control algorithm is implemented on a mobile robot from the UPCT named "CHAMAN". The platform has two driving wheels (in the rear)

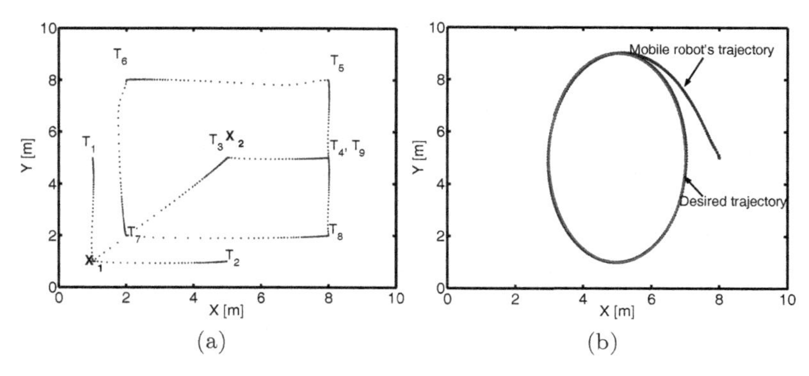

(a) (b)

Fig. 3. Adaptive control by the SODMN. a) Approach behaviors. The symbol X indicates the start of the mobile robot and T_i indicates the desired reach. b) Tracking control of a desired trajectory.

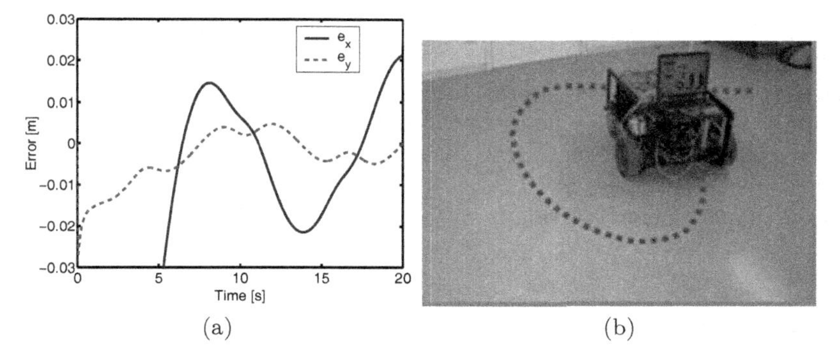

(a) (b)

Fig. 4. Tracking control of the SODMN. a) The trajectory error of the Figure 3b. b) Real-time tracking performance.

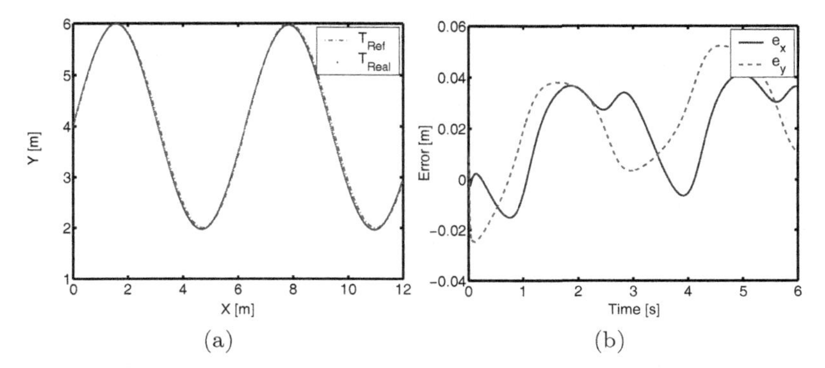

(a) (b)

Fig. 5. Tracking control of a desired trajectory. (a) Tracking trajectory of a sine. (b) Estimated tracking error.

mounted on the same axis and two passive supporting wheels (in front) of free orientation. The two driving wheels are independently driven by two DC-motors to achieve the motion and orientation. The wheels have a radius $r = 18$ cm and are mounted on an axle of length $2R = 22$ cm. The aluminum chassis of the robot measures 102.25×68×44 cm ($L \times W \times h$) and contains, transmission elements, 12-VDC battery, two CCD cameras, and 12 ultrasound sensors. Each motor is equipped with incremental encoder counting 600 pulses/turn and a gear which reduces the speed to 1.25 m/s.

High-level control algorithms (SODMN) are written in VC++ and run with a sampling time of 10 ms on a remote server (a Pentium IV processor). The PC communicates through a serial port with the microcontroller on the robot. Wheel PWM duty cycle commands are sent to the robot and encoder measures and are received for odometric computation. The lower level control layer is in charge of the execution of the high-level velocity commands. It consists of a Texas Instruments TMS320C6701 Digital Signal Processor (DSP). The microcontroller performs three basis tasks: 1) to communicate with the higher-level controller through RS 232; 2) reading encoder counts interrupt driven; and 3) generation of PWM duty cycle.

Figures 3, 4 and 5 show approach behaviors and the tracking of a trajectory by the mobile robot with respect to the reference trajectory. From these figures, it is clear that the tracking of the reference trajectory is very accurate. In the Figure 3, reaches to targets (such as $\times_1$ to T_1, $\times_2$ to T_2 and $\times_1$ to $\times_2$) and a sequence of movement (starting from $\times_2$ to T_4, T_4 to T_5, T_5-T_6, T_6 -T_7, T_7-T_8, T_8-T_4, T_4-$\times_2$) were carried out.

4 Discussion

In present model, appropriate operations are learned in an unsupervised fashion through repeated action-perception cycles by recoding proprioceptive information related to the mobile-robot. The resulting solution has two interesting properties: (a) the required transformation is executed accurately over a large part of the reaching space, although few velocities are actually learned; and (b) properties of single neurons and populations closely resemble those of neurons and populations in parietal and cortical regions [18]. The activity of the population of motor cortical cells which encode movement direction appears to represent the instantaneous velocity of movement [19]. In addition, the preferred directions of individual cells shifts with the movement origin, indicating that the directional coding of motor cortex may be influenced by velocity configuration (in the model is the context field) [20], as is necessary for a Jacobian-based mapping. Correspondence between layers of the network and brain regions can be made tentatively base on anatomical and physiological arguments [18,19]. The representation of DVs could be in posterior parietal cortex (PPC) [21]. Neurons in PPC exhibit activity patterns correlated with the spatial direction of movement [22]. A candidate region for participating in the direction mapping computation is the cerebellum [23]. Also, note that there are certain similarities between the nature of the context field cells in the mobile-robot movement model and the Purkinge cells of the adaptive timing model. Both types of cells are tonically active and allow a response by "pausing" this tonic activity. Thus, the possibility that a context field type of function is performed by cerebellar cortex. The proposed direction mapping model also posits a learning site separate from the context field computation, which might be a cerebellar function. In the model, motor commands were emitted by a layer containing R_i neurons, which contribute to the movement by a displacement along a direction in velocity space. The individual influence of a command neuron is proportional to its discharge level.

4.1 Conclusions

In this paper, a biologically inspired neural network for the spatial reaching tracking has been developed. This neural network is implemented as a kinematic adaptive neuro-controller. The SODMN uses a context field for learning the direction mapping between spatial and angular velocity coordinates. The transformations are learned during an unsupervised training phase, during which the mobile robot moves as result of randomly selected angular velocities of wheels. The performance of this neural network has been successfully demonstrated in experimental results with the trajectory tracking and reaching of a mobile robot. The efficacy of the proposed neural network for reaching and tracking behaviors was tested experimentally by a differentially driven mobile robot.

References

1. Fierro, R., Lewis, F.L.: Control of a nonholonomic mobile robot using neural networks. IEEE Trans. Neural Netw. **9** (1998) 589–600
2. Kanayama, Y., Kimura, Y., Miyazaki, F., Noquchi, T.: A stable tracking control method for an autonomous mobile robot. In: Proc. IEEE Int. Conf. Robotics and Automation. Volume 1., Cincinnati, OH (1990) 384–389
3. Ping, Z., Nijmeijer, H.: Tracking control of mobile robots: A case study in backstepping. Automatica **33** (1997) 1393–1399
4. Oriolo, G., Luca, A.D., Vendittelli, M.: WMR control via dynamic feedback linearization: Design, implementation and experimental validation. IEEE Trans. Control. Syst. Technol. **10** (2002) 835–852
5. Das, T., Kar, I.N.: Design and implementation of an adaptive fuzzy logic-based controller for wheeled mobile robots. IEEE Transactions on Control SystemsTechnology **14** (2006) 501–510
6. De-Luca, A., Oriolo, G., Samson, C.: Feedback control of a nonholonomic car-like robot. In Laumond, J.P., ed.: Robot Motion Planning and Control. Springer, Berlin Heidelberg, New York (1998) 171–253
7. Novel, B.D., Bastin, G., Campion, G.: Control of nonholonomic wheeled mobile robots by state feedback linearization. Int. J. Rob. Res. **14** (1995) 543–559
8. Fukao, T., Nakagawa, H., Adachi, N.: Adaptive tracking control of a nonholonomic mobile robot. IEEE Trans. Robot. Autom. **16** (2000) 609–615
9. Corradini, M., Orlando, G.: Robust tracking control of mobile robots in the presence of uncertainties in the dynamical model. J. Robot. Syst. **18** (2001) 318–323

10. Dixon, W., Dawson, M., Zergeroglu, E.: Tracking and regulation control of a mobile robot system with kinematic disturbance: A variable structure like approach. ASME Trans. J. Dyn. Syst. Meas. Control **122** (2000) 616–623
11. Corradini, M., Leo, T., Orlando, G.: Experimental testing of a discrete-time sliding mode controller for trajectory tracking of a wheeled mobile robot in the presence of skidding effects. J. Robot. Syst. **19** (2002) 177–188
12. Chen, F., Liu, C.: Adaptively controlling nonlinear continuous-time systems using multilayer neural networks. IEEE Trans. Automat. Contr. **39** (1994) 1306–1310
13. Corradini, M., Ippoliti, G., Longhi, S.: Neural networks based control of mobile robots: Development and experimental validation. J. Robot. Syst. **20** (2003) 587–600
14. Lin, S., Goldenberg, A.: Neural-network control of mobile manipulators. IEEE Trans. Neural Netw. **12** (2001) 1121–1133
15. Grossberg, S., Levine, D.: Neural dynamics of attentionally moduled Pavlovian conditioning: Blocking, interstimulus interval, and secondary reinforcement. Applied Optics **26** (1987) 5015–5030
16. Chang, C., Gaudiano, P.: Application of biological learning theories to mobile robot avoidance and approach behaviors. J. Complex Systems **1** (1998) 79–114
17. Gaudiano, P., Grossberg, S.: Vector associative maps: Unsupervised real-time error-based learning and control of movement trajectories. Neural Networks **4** (1991) 147–183
18. Baraduc, P., Guigon, E., Burnod, Y.: Recording arm position to learn visuomotor transformations. Cerebral Cortex **11** (2001) 906–917
19. Georgopoulos, A.P.: Neural coding of the direction of reaching and a comparison with saccadic eye movements. Cold Spring Harbor Symposia in Quantitative Biology **55** (1990) 849–859
20. Caminiti, R., Johnson, P., Urbano, A.: Making arm movements within different parts of space: Dynamic aspects in the primate motor cortex. Journal of Neuroscience **10** (1990) 2039–2058
21. Rondot, P., De-Recondo, J., Dumas, J.: Visuomotor ataxia. Brain **100** (1976) 355–376
22. Lacquaniti, F., Guigon, E., Bianchi, L., Ferraina, S., Caminiti, R.: Representing spatial information for limb movement: Role of area 5 in the monkey. Cerebral Cortex **5** (1995) 391–409
23. Fiala, J.C.: Neural Network Models of Motor Timing and Coordination. PhD thesis, Boston University (1996)

Sharing Gaze Control in a Robotic System*

Daniel Hernandez, Jorge Cabrera, Angel Naranjo, Antonio Dominguez, and Josep Isern

IUSIANI, ULPGC, Spain
dhernandez,jcabrera,adominguez@iusiani.ulpgc.es

Abstract. The use of the vision in humans is a source of inspiration for many research works in robotics. Attention mechanisms has received much of this effort, however, aspects such as gaze control and modular composition of vision capabilities have been much less analyzed. This paper describes the architecture of an active vision system that has been conceived to ease the concurrent utilization of the system by several visual tasks. We describe in detail the functional architecture of the system and provide several solutions to the problem of sharing the visual attention when several visual tasks need to be interleaved. The system's design hides this complexity to client processes that can be designed as if they were exclusive users of the visual system. Besides, software engineering principles for design and integration, often forgotten in this kind of developments, have been considered. Some preliminary results on a real robotic platform are also provided.

1 Introduction

The use of the vision in humans has been a source of inspiration for many research works in robotics. The control of the gaze in an active vision system is usually formulated as a problem of detection of significant points in the image. Under this perspective several aspects such as saliency, bottom-up vs. top-down control, computational modelling, etc, have been analyzed [1][2]. An alternative view considers the shared resource nature of the sensor, transforming the scenario into a management/coordination problem where the control of the gaze must be shared among a set of dynamic concurrent tasks. A good example is the use of the vision in car driving [3], specially when overtaking, changing the lane or braking [4].

In close relation with the aforementioned, but from a more engineering point of view, is the fact that researchers and engineers involved in the development of vision systems have been primarily concerned with the visual capabilities of the system in terms of performance, reliability, knowledge integration, etc. However very little attention has been devoted to the problem of modular composition of vision capabilities in perception-action systems. While new algorithms and techniques add new capabilities and pave the way for new and important challenges,

* This work has been partially supported by Spanish Education Ministry and FEDER (project TIN2004-07087) and Canary Islands Government (project PI2003/160).

J. Mira and J.R. Álvarez (Eds.): IWINAC 2007, Part II, LNCS 4528, pp. 469–478, 2007.

the majority of vision systems are still designed and integrated in a very primitive way according to modern software engineering principles. There are few published results on how to control the visual attention of the system among several tasks that execute concurrently [5].

Posing a simple analogy to clarify these issues, when we execute programs that read/write files that are kept in the hard disk, we aren't normally aware of any contention problem and need not to care if any other process is accessing the same disk at the same time. This is managed by the underlying services, simplifying the writing of programs that can be more easily codified as if they have exclusive access to the device. Putting it on more general terms, we could consider this feature as an "enabling technology" as it eases the development of much more complex programs that build on top of this and other features.

How many of these ideas are currently applied when designing vision systems?. In our opinion, not many. Maybe because designing vision systems has been considered mainly as a research endeavor, much more concerned with other higher level questions, these low level issues tend to be simply ignored. If we translate the former example to the context of vision systems, some important drawbacks can be easily identified.

- Vision systems tend to be monolithic developments. If several visual tasks need to execute concurrently this need to be anticipated since the design stage. If some type of resource arbitration is necessary it is embedded in the code of the related tasks.
- Within this approach it is very difficult to add new visual capabilities that may compete against other tasks for the attention of the system.
- As far as contention situations are dealt with internally and treated by means of ad-hoc solutions, the development of such systems does not produce any reusable technology for coping with these problems.

As a selection of related work, the following three systems can be mentioned. Dickmanns and colleagues [3] have studied the problem of gaze control in the context of their MarVEye Project, where an active multi-camera head is used to drive a car in a highway. In that project, several areas of interest are promoted and ranked by different modules of the control architecture using a measure of information gain. The gaze controller tries to design the gaze trajectory that within 2 seconds can provide the highest gain of information to the system. Several areas of attention can be chained in a gaze trajectory, though in practice this number is never higher than two.

Humanoid robots need to rely in their visual capabilities to perceive the world. Even for simple navigation tasks, a number of tasks (obstacle avoidance, localization, ...) need to operate concurrently to provide the robot with minimal navigation capabilities. Several groups have explored the problem of gaze arbitration in this scenario, both in simulation and with real robots. Seara et al. [6] [7] have experimented with a biped robot that used a combination of two tasks to visually avoid obstacles and localize itself. The decision of where to look next was solved in two stages. Firstly, each task selects its next preferred focus of attention as that providing the largest reduction of incertitude in the robot localization, or in the location of obstacles. In a second stage, a multiagent decision schema, along with a winner-selection society model, was used to finally decide which task was granted the control of gaze. Of the several society models that were evaluated, the best results, as judged by the authors, were obtained by a society that tries to minimize the total "collective unhappiness". Here the concept of unhappiness is derived of loosing the opportunity of reducing incertitude (in self or obstacle localization).

Sprague et al. [8][5] have designed a simulation environment where a biped robot must walk a lane while it picks up litter and avoids obstacles, using vision as the only sensor. These capabilities are implemented as visual behaviors using a reinforcement learning method for discovering the optimal gaze control policy for each task. The lateral position of the robot within the lane and the position of obstacles and litter are modelled by Kalman filters. Every 300 msec, the gaze is given to the task that provides the largest gain in uncertainty reduction.

Similar in spirit to this related research, the motivation of our work is consequently two-fold: contribute to build a vision system more consistent from an engineering point of view, and to take a first step towards systems where the vision becomes integrated in an action context with higher semantic and cognitive level (an "intelligent" way of looking).

In the next sections we will present the proposed architecture, with its design objectives and main components. Some experimental results obtained on a real robot, along with conclusions and intended future development will be described in the last two sections.

2 MTVS

Motivated by the previously stated description of the problem we have designed and implemented MTVS (Multi-Tasking Vision System), a proposal of architecture for active-vision systems in multi-tasking environments. MTVS has been designed to deal with the scheduling of concurrent visual tasks in such a way that resource arbitration is hidden to the user.

2.1 Objectives

More in detail, MTVS pursues the following objectives:

- The assignment of the gaze control to a task is based on a simple scheduler model, so that the behavior can be easily interpreted by an external observer.
- The client tasks are integrated in the system individually with no coordination requirements.
- The set of tasks managed (the activity) can change dynamically.
- The clients should not assume any a priori response time guarantee, though the system can offer high-priority attention modes.
- The system offers services based on a reduced set of visual primitives, in pre-categorical terms.

2.2 Internal Architecture

The figure 1 shows an MTVS application example with two clients. The basic elements making up the vision system architecture are a system server, a task scheduler and the data acquisition subsystem. Basically, the clients connect to the system server to ask for services (visual primitives) with a given configuration. In response, the system server launches both a task-thread, to deal with internal scheduling issues, and a devoted server-thread that will be in charge of the interaction with the external client. The scheduler analyzes the tasks demands under a given scheduling policy and selects one to receive the gaze control (in this example, the client A has been selected). In combination, a second covert scheduler checks for compatibility between tasks to share images among them (FOA's overlapping). The data acquisition subsystem processes the different sensor data streams (images, head pose and robot pose) to generate as accurately as possible time stamps and pose labels for the served images.

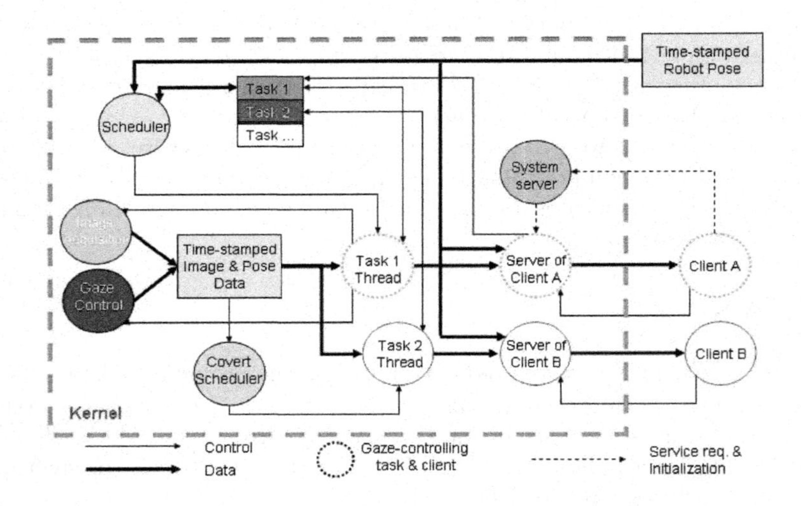

Fig. 1. Control Architecture: example with two clients

2.3 Visual Services

Clients can connect to the vision system and use it through a number of pre-categorical low-level services. The MTVS services are built around basic visual capabilities or primitives that have also been explored by other authors [9]:

WATCH: Capture N images of a 3D point with a given camera configuration.
SCAN: Take N images while the head is moving along a trajectory.
SEARCH: Detect a model pre-categorically in a given image area.
TRACK: Track a model pre-categorically.
NOTIFY: Inform the client about movement, color or other changes.

Except for WATCH, the rest of primitives can be executed discontinuously, allowing for the implementation of interruptible visual tasks.

The clients also regulate their activity in the system by means of the messages they interchange with their devoted server. Currently, the following messages have been defined for a task: creation, suspension, reconfiguration (modify parameters, change priority, commute primitive on success) and annihilation.

2.4 The Scheduler

Several scheduling policies have been implemented and studied inside MTVS. This analysis has considered two main groups of schedulers: time-based and urgency based schedulers.

Time-based Schedulers. Three types of time-based schedulers have been studied: Round-Robin (RR), Earliest Deadline First (EDF) and EDF with priorities (EDFP). The prioritized RR algorithm revealed rapidly as useless in a dynamic and contextual action schema. First, it makes no sense to assign similar time slices to different tasks, and second, the time assigned used for saccadic movements, specially when a slow neck is involved, becomes wasted.

The EDF algorithm yielded a slightly better performance than RR, but was difficult to generalize as visual tasks are not suitable for being modelled as periodic tasks. The best results of this group were obtained by the EDFP algorithm combining critical tasks (strict deadline) with non-critical tasks. Each time a task is considered for execution and not selected its priority is incremented by a certain quantity [10].

Urgency-based Schedulers. The concept of urgency is well correlated with a criteria of loss minimization, as a consequence of the task not receiving the control of the gaze within a time window. This measure can also be put into relation with uncertainty in many visual tasks.

Two schedulers have been studied in this group: lottery [8] and max-urgency. The lottery scheduler is based in a randomized scheme where the probability of a task being selected to obtain the gaze control is directly proportional to its urgency. Every tasks has the possibility of gaining the control of the gaze, but the random unpredictability can sometimes produce undesirable effects.

The max-urgency scheduler substitutes the weighted voting by a direct selection of the task with higher urgency value. This scheme has produced acceptable results provided that the urgency of a task is reduced significantly after gaining the control of the gaze (similar to an inhibition of return mechanism).

3 Implementation

The Vision System can be operated in a number of configurations. In its most simple configuration the system may comprise a simple pan/tilt system and a camera, or a motorized camera, or it can integrate both systems as illustrated in the experiments described later. In this last configuration, a relatively slow pan/tilt system, acting as the neck of the system, carries a motorized camera (SONY EVIG21) equipped with a fast pan/tilt system, that can be considered as the eye of the system. This mechanical system in turn can be used in isolation or it can be installed on a mobile robot.

The vision system runs under the Linux OS as a multithreaded process. Clients run as independent processes. They may share the same host, in which case the communication primitives are implemented using shared memory, or on different hosts connected by a local network. The interface to the image acquisition hardware is based on the Video4Linux2 services so that a large number of cameras and/or digitizers can be supported. For the image processing required by the primitives of the system or at the clients, the OpenCV library is used.

4 Experiments

A set of experiments were carried out to analyze the behavior of MTVS on a real robotic application. We will present in this section some of the results obtained. The basic experimental setup consists of two ActivMedia Pioneer robots, one with the basic configuration and the other mounting an active vision system formed by a Directed Perception PTU (neck) and a motorized Sony EVI-G21 camera (eye).

Two main tasks were combined along the different experiments: target following and obstacle avoidance. The target following task commands the active vision robot (pursuer) to detect and follow a special square target mounted on other robot (leader), trying to keep a predefined constant distance between them. The obstacle avoidance task looks for colored cylinders on the floor, estimating, as exactly as possible their 2D position. Kalman filtering was used to model both target and obstacles positions.

4.1 One-Task-Only Experiments

As a reference for the maximum expected performance some experiments where designed involving only one task.

Experiment 1: Follow Robot Only. In this experiment, the leader robot is commanded to move forward at a constant speed of 200 mm/sec, while the pursuer must try to keep a constant separation of 2 meters. Several tests have been conducted along the main corridor of our lab following a 15 meters straight line path. The pursuer was able to stabilize the robot gap around the desired value with a maximum error of approximately 150 mm.

Experiment 2: Obstacle Avoidance Only. Now the active vision robot is commanded to explore the environment looking for objects (yellow cylinders), trying to reduce their position uncertainty below a predefined threshold. The robot moves straight-line inside a corridor formed by 8 cylinders equally distributed in a zigzag pattern along the path. The figure 2 illustrates the robot path and the different detections for each localized object, including their first (larger) and minimum uncertainty ellipses. The results show how the robot was able to localize all the objects with minimum uncertainty ellipses ranging from 100 to 200 mm in diameter.

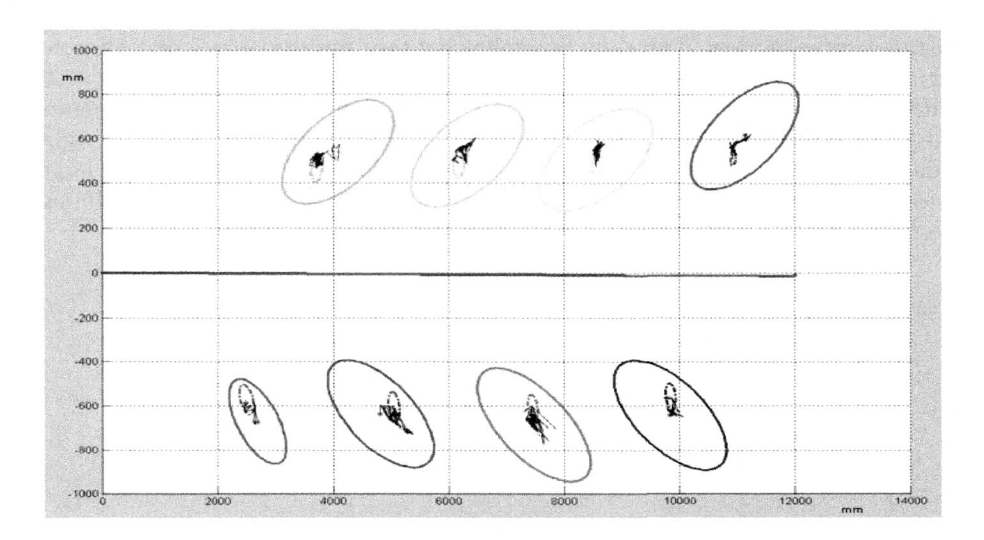

Fig. 2. Obstacle avoidance-Only experiment

4.2 Multiple-Task Experiments

The multiple-task experiments consider an scenario in which each task computes its desired camera configuration and urgency and asks the MTVS scheduler to obtain the gaze control. The scheduler uses this information to select where to look next and how to distribute images. The obstacle avoidance task is extended to classify special configurations of objects as "doors" (two objects aligned perpendicularly to robot initial orientation with a pre-defined separation).

The urgency of the following task is computed as a function of the distance error, the robot velocity and the time. This urgency increases as the distance between the robots separates from the reference, the velocity is high and the elapsed time since the last image was received becomes larger.

The urgency of the obstacle avoidance task is computed separately for three possible focus of attention: front (the urgency increases when the robot moves towards visually unexplored areas), worst estimated object (the urgency increases as the position of a previously detected object is not confirmed with new images), and closest door (the urgency increases with narrow doors).

The first simple multiple-task experiment tries to illustrate the role of the covert scheduler in MTVS. In this case one task has higher priority, so in a context of exclusive camera sensor access the secondary task shows very low performance. Allowing the sharing of images, however, the overall performance can be improved. In experiment 4 a more complex scenario including doors is analyzed.

Experiment 3: Obstacle Avoidance and Robot Following Competing for the Gaze (Following Priority) In this experiment, the control of the gaze is only granted to the avoidance task when both the leader speed and the distance error are low. Typically, the following task performance is not affected significantly, but the avoidance task degrades yielding few objects localization with poor precision. As an example, the upper plot of the figure 3 presents the results of a non sharing run where only half the potential objects (all right sided due to the position of the closest obstacle) have been detected with large uncertainty ellipses. As the lower plot of the figure shows, the covert scheduling permits a much better behavior of the obstacle avoidance task.

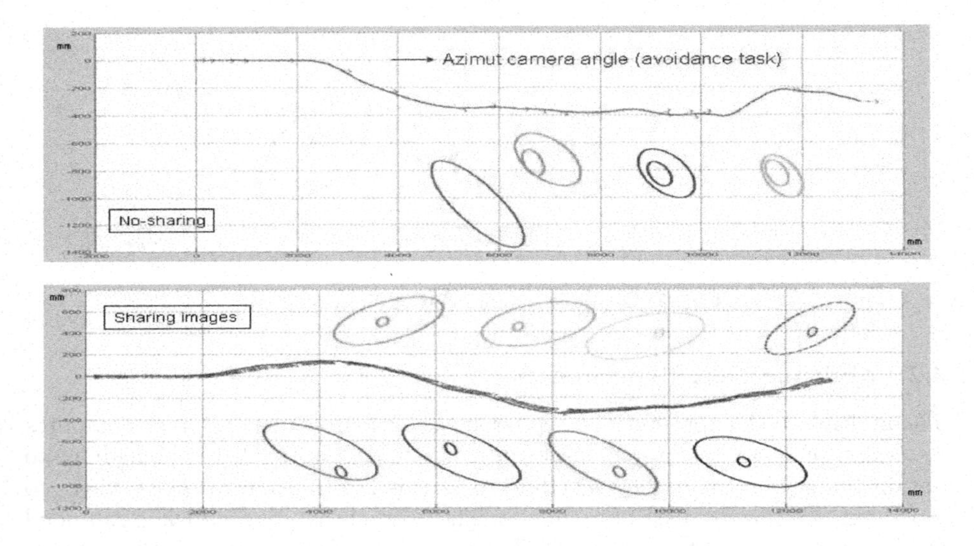

Fig. 3. Follow (priority) and avoidance experiment

Experiment 4: Localize Doors and Robot Following Competing for the Gaze (Narrow and Wide Doors) The configuration of objects used for this experiment consist of a set of four "doors": two narrow type (600 mm width) and two wide type (1500 mm width). All doors are located straight line in front of the robot, the first one (wide) three meters ahead and the rest every 1.5 meters, alternating narrow and wide types. The leader robot is commanded to move at constant speed crossing the doors centered.

The figure 4 illustrates how the camera is pointed to both sides when crossing narrow doors. As a consequence of this behavior, the pursuer robot slows down when approaching a narrow door until the door extremes position have been estimated with the required precision (compare final error ellipses for narrow

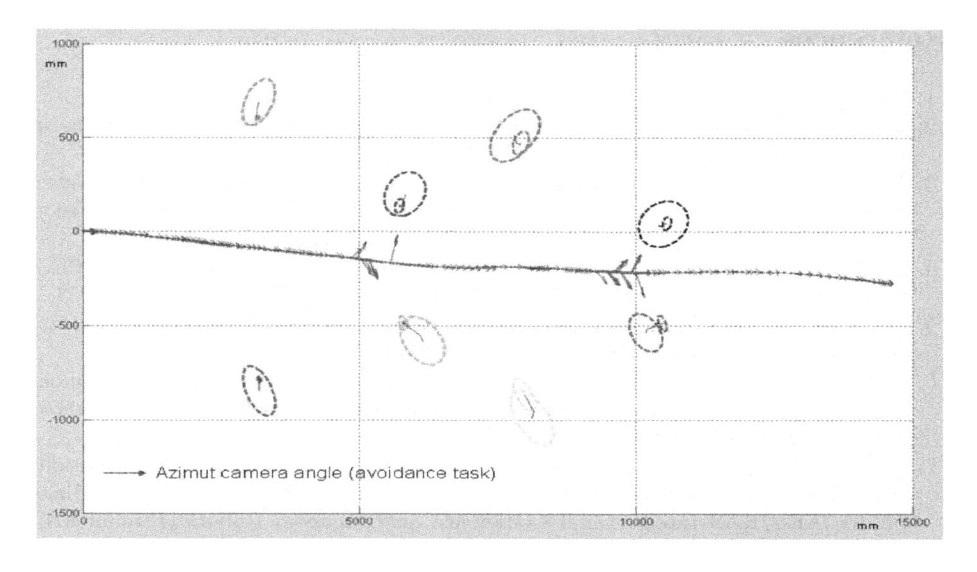

Fig. 4. Narrow and wide doors experiment

and wide doors). After traversing the door, the robot accelerates to recover the desired following distance from the leader.

5 Conclusions and Future Developments

In this paper we propose an open architecture for the integration of concurrent visual tasks. The clients requests are articulated on the basis of a reduced set of services or visual primitives. All the low level control/coordination aspects are hidden to the clients simplifying the programming and allowing for an open and dynamic composition of visual activity from much simpler visual capabilities.

Regarding the gaze control assignation problem, several schedulers have been implemented. The best results are obtained by a contextual scheme governed by urgencies, taking the interaction of the agent with its environment as organization principle instead of temporal frequencies. Usually, a correspondence between urgency and uncertainty about a relevant task element can be established.

The system described in this paper is just a prototype, mainly a proof of concept, and it can be improved in many aspects. The following are just a few of them. We plan to improve the adaptability of the system to different active vision heads (hardware abstraction). A first step is to consider the extension of the system to be applied over a binocular system, where new problems like eye coordination, vergence and accommodation must be tackled. Another issue is the need of an acceptance test for new service requests to avoid overloading the system. Besides, the introduction of homeostasis mechanisms could help to make the system more robust; and the inclusion of bottom-up directed attention (independent movement, e. g.) would allow for new saliency-based primitives.

References

1. Arkin, R., ed.: Behavior-Based Robotics. MIT Press (1998)
2. Itti, L.: Models of bottom-up attention and saliency. In Itti, L., Rees, G., Tsotsos, J.K., eds.: Neurobiology of Attention. Elsevier Academic Press (2005)
3. Pellkoffer, M., Ltzeler, M., Dickmanns, E.: Interaction of perception and gaze control in autonomous vehicles. In: SPIE: Intelligent Robots and Computer Vision XX: Algorithms, Techniques and Active Vision, Newton, USA (2001)
4. Rogers, S.D., Kadar, E.E., Costall, A.: Drivers gaze patterns in braking from three different approaches to a crash barrier. Ecological Psychology, **17** (2005) 39–53
5. Sprague, N., Ballard, D., Robinson, A.: Modeling attention with embodied visual behaviors. ACM Transactions on Applied Perception (2005)
6. F. Seara, J., Lorch, O., Schmidt, G.: Gaze Control for Goal-Oriented Humanoid Walking. In: Proceedings of the IEEE/RAS International Conference on Humanoid Robots (Humanoids), Tokio, Japan (2001) 187–195
7. F. Seara, J., Strobl, K.H., Martin, E., Schmidt, G.: Task-oriented and Situation-dependent Gaze Control for Vision Guided Autonomous Walking. In: Proceedings of the IEEE/RAS International Conference on Humanoid Robots (Humanoids), Munich and Karlsruhe, Germany (2003)
8. Sprague, N., Ballard, D.: Eye movements for reward maximization. In: Advances in Neural Information Processing Systems (Vol. 16). MIT-Press (2003)
9. Christensen, H., Granum, E.: Control of perception. In Crowley, J., Christensen, H., eds.: Vision as Process. Springer-Verlag (1995)
10. Kushleyeva, Y., Salvucci, D.D., Lee, F.J.: Deciding when to switch tasks in time-critical multitasking. Cognitive Systems Research (2005)

An AER-Based Actuator Interface for Controlling an Anthropomorphic Robotic Hand

A. Linares-Barranco[1], A. Jiménez-Fernandez[1], R. Paz-Vicente[1], S. Varona[2], and G. Jiménez[1]

[1] Departamento de Arquitectura y Tecnología de Computadores. Universidad de Sevilla
[2] Lab. of Neurotechnology, Control and Robotics (NEUROCOR). Technical Univ. of Cartagena. Campus Muralla del Mar
alinares@atc.us.es, jl.coronado@upct.es*

Abstract. Bio-Inspired and Neuro-Inspired systems or circuits are a relatively novel approaches to solve real problems by mimicking the biology in its efficient solutions. Robotic also tries to mimic the biology and more particularly the human body structure and efficiency of the muscles, bones, articulations, etc. Address-Event-Representation (AER) is a communication protocol for transferring asynchronous events between VLSI chips, originally developed for neuro-inspired processing systems (for example, image processing). Such systems may consist of a complicated hierarchical structure with many chips that transmit data among them in real time, while performing some processing (for example, convolutions). The information transmitted is a sequence of spikes coded using high speed digital buses. These multi-layer and multi-chip AER systems perform actually not only image processing, but also audio processing, filtering, learning, locomotion, etc. This paper present an AER interface for controlling an anthropomorphic robotic hand with a neuro-inspired system.

1 Introduction

The Address-Event Representation (AER) was proposed by the Mead lab in 1991 [1] for communicating between neuromorphic chips with spikes (Fig. 1). Each time a cell on a sender device generates a spike, it communicates with the array periphery and a digital word representing a code or address for that pixel is placed on the external inter-chip digital bus (the AER bus). Additional handshaking lines (Acknowledge and Request) are used for completing the asynchronous communication. In the receiver chip the spikes are directed to the pixels whose code or address was on the bus. In this way, cells with the same address in the emitter and receiver chips are virtually connected by streams of spikes.

* This email address belongs to J.L. Coronado, the director of the System and Automation Department of the Technical University of Cartagena. His contribution to this paper has been equal to the other authors. He should appear on the author list, but the rules of the congress stipulate a maximum of five recognized authors.

J. Mira and J.R. Álvarez (Eds.): IWINAC 2007, Part II, LNCS 4528, pp. 479–489, 2007.

These spikes can be used to communicate analog information using a rate code, but this is not a requirement. Cells that are more active access the bus more frequently than those less active. Arbitration circuits usually ensure that cells do not simultaneously access the bus. Usually these AER circuits are built using self-timed asynchronous logic by e.g. Boahen [2].

Transmitting the cell addresses allows performing extra operations on the events while they travel from one chip to another. For example the output of a silicon retina can be easily translated, scaled, or rotated by simple mapping operations on the emitted addresses. These mapping can either be lookup-based (using, e.g. an EEPROM) or algorithmic. Furthermore, the events transmitted by one chip can be received by many receiver chips in parallel, by properly handling the asynchronous communication protocol. There is a growing community of AER protocol users for bio-inspired applications in vision, audition systems and robot control, as demonstrated by the success in the last years of the AER group at the Neuromorphic Engineering Workshop series [4]. The goal of this community is to build large multi-chip and multi-layer hierarchically structured systems capable of performing massively-parallel data-driven processing in real time [3].

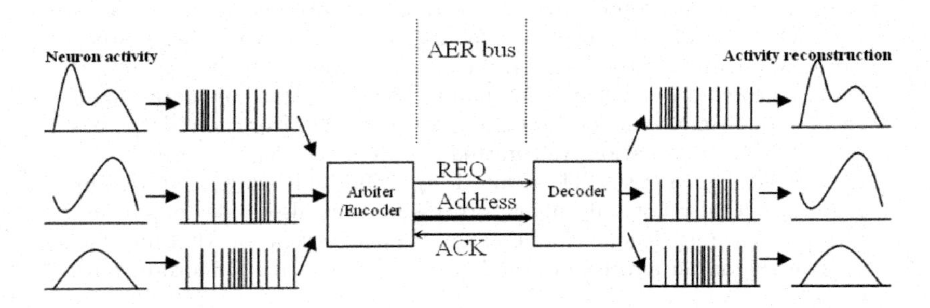

Fig. 1. Rate-coded AER inter-chip communication scheme. REQ is the Request line, ACK is the Acknoledge line a the bus Address hold the neuron code.

The neuromorphic approach of AER can be also applied to actuators, like the muscles in the biology. The success of such systems will strongly depend on the availability of robust and efficient development, debugging and interfacing AER-tools [7][8][9][10][11]. One such tool is a computer interface that allows not only reading a sequence of events with their timestamps, but also reproduces a sequence of events stored in the computer's memory.

From the robotic point of view a very interesting tool will allow the connection between an AER system and a robot. This paper presents, describes and tests the improved AER-Robot [7] interface for this connectivity. This tool is able to interface to actuators and commercial sensors (position, contact, pressure, temperature, ...) to allow movements and feedback allowing a more complex and bio-inspired control of a robot. Depending on the application the interface should be able to: (a) Implement the control algorithm into the interface, using an FPGA or a Microprocessor, or (b) to transmit the sensor information that is receiving to an AER chip or to a computer in AER format, and the opposite, to receive orders format from an AER chip or from a computer, to the actuators in AER, allowing the computer or the AER chip to implement the control algorithm.

Factorization of Length and Tension (FLETE) is a bio-inspired control mechanism for robots that computes not only the position of the robot, but also the rigidity of it. In this case the visual information is insufficient, and another kind of mechanical sensor is needed.

With these interfaces you can control a robotic platform using an AER system to give it orders and obtain other kind of sensory information (pressure, contact, position, etc) into AER format. Furthermore, the AER interface allows debugging the robotic platform if you connect it to the computer using the PCI-AER interface or the mini-USB-AER [10].

In this paper we present an improved AER Interface to connect the AER system to a set of actuators (motors) and sensors, called AER-Robot. The interface has been used to connect a computer to the TUC[1] anthropomorphic robotic hand in order to enable an AER system to control a complex and bio-inspired robot. The hand is driven by an agonist-antagonist opponent system. In order to measure joint position, velocity, and direction of rotation, hall-effect position sensors were integrated at each joint of the fingers. Force sensors are mounted on the curved surface of the fingertips, and on the palm.

In the following sections we describe the anthropomorphic hand, the AER-Robot interface, the PCI-AER interface, the mini-USB-AER interface to the computer for debugging purpose, and the results.

2 Anthropomorphic Robotic Hand

The TUC anthropomorphic robot design hand is based on the biomechanics modelling of the human hand, as well as on designs by manufacturers of robot hands. The hand has three fingers and an opposing thumb, and four degrees of freedom for each finger. The fingers are mounted on a rigid palm. Fig. 2 shows one photography of the TUC Hand, placed over a industrial robot for grasping, reaching and handling tasks.

The design of the multi-jointed finger presents three joints (metacarpophalangeal (MCP), proximal interphalangeal (PIP), and distal interphalangeal (DIP) joints, respectively), where DIP and PIP are coupled. Both the PIP and DIP joints have flexion and extension, and the MCP joint consist of two joints that allow flexion-extension and adduction-abduction motions. In the finger design, the muscle-like actuators are DC motors. Each joint of the finger is actuated through 2 polystyrene tendons, routed through pulleys and driven by DC motors. The joints are moved by an agonist-antagonist opponent system. In order to measure joint position, velocity, and direction of rotation, hall-effect position sensors were integrated at each joint of the fingers. Tactile sensors based

[1] Technical University of Cartagena.

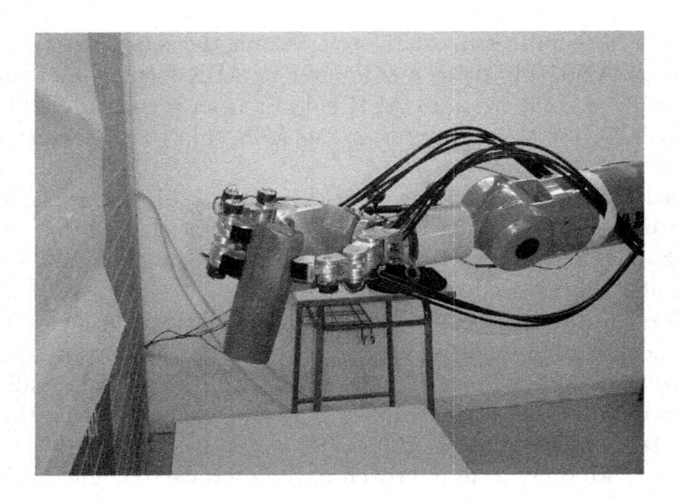

Fig. 2. TUC Robot Hand

on FSR (Force Resistive sensing) technology are mounted on all the joints and on the palm emulating artificial tactile surfaces. The flexibility of these sensors is very suitable for the implementation on the curved surface of the fingertips for precision grasping and manipulations tasks. One two-axis sensor placed on the palm is employed to correct the stability of the gross grasping. It permits the tactile guided for the movement of the wrist of the robot hand-arm [12]. Each one of the fingers that conforms the biomechanical hand is driven by a mechanism constituted by an assembly of pulleys that control the movements of the different phalanges. Each finger is comprised of three articulations with possibility of turn and an additional articulation that permits to reproduce the movement of abduction, besides serving of element of union of the digit with the palm of the hand. The pulleys (on articulations) are driven for a system of cables to way of human tendons. Each articulation possesses two tendons, one flexor and another extensor. The tendon flexor causes the movement of contraction of the articulations while the tendon extensor causes the contrary effect. The mechanical system of actuation arranges of a motor to extend and another to contract the tendon. For control the turn of each articulation, in a synchronized way, the wires remains traction in every instant, and is possible to measure the effort done by the tendons.

3 AER-Robot Interface

This section describes in detail an AER interface to manage actuators and to read analog commercial sensors and convert it to AER format. These actuators are based on DC motors.

The AER-Robot interface can control up to 4 up/down DC motors, each one doted with a two channel encoder. The DC motors are controlled digitally using

Pulse Width Modulation (PWM). AER-Robot can read up to the following sensors: (a) 4 potentiometers for the finger articulations position, (b) 4 contact resistors for the fingertip and the palm object detection, (c) 4 tension sensor for the tendons of the fingers, and (d) 4 current sensor for the power consumption of the motors of the fingers. These sensors information are fundamental for the control algorithms in the hand platform.

The AER-Robot interface has been developed to communicate AER systems with an anthropomorphic robotic hand using two AER buses: one for incoming commands and another for outgoing information of the motors and the sensors. Furthermore, the interface allow the serial connection of several AER-Robot or other AER based components, thanks to other 2 AER busses to (a) pass through the incoming event and (b) to receive events and arbitrate them with the ones produced by itself to send them out. It is based around a Spartan 3 400 FPGA that allows co-processing. This FPGA receives commands through the input AER bus and sends motor and sensor information back. These commands allows to:

- Configure the PWM period that manages all the motors.
- Move a motor attending to PWM intensity and an estimated position through the encoder's information.
- Ask for a motor state.
- Ask for a sensor state.

Fig. 3 shows the block diagram of the circuit of the FPGA, described into VHDL. This circuit is composed by several processes. CMDin receives commands and sends them to the corresponding process. There are 4 independent processes (Motor i) to control de PWM signal to be sent to each motor. There is a process to attend the 16 possible sensors of one finger of the hand (4 potentiometers for articulations positions, 4 contact, 4 hall effect current consumption of the motor and 4 tension tendon sensors). This last process attends to a Cygnal microcontroller that is continuously converting an analog signal to digital from 16 possible analog inputs.

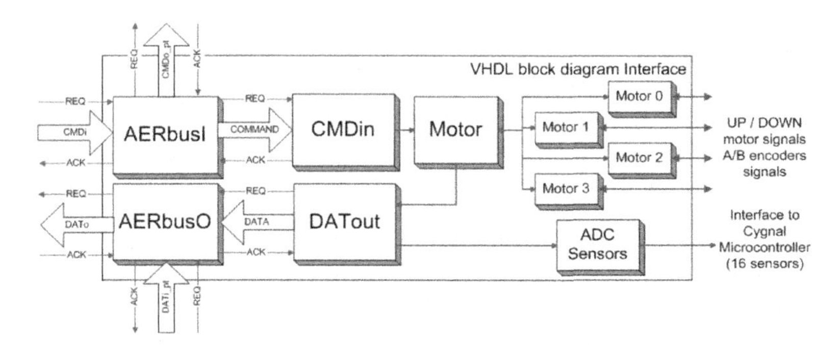

Fig. 3. Circuit block diagram of the FPGA

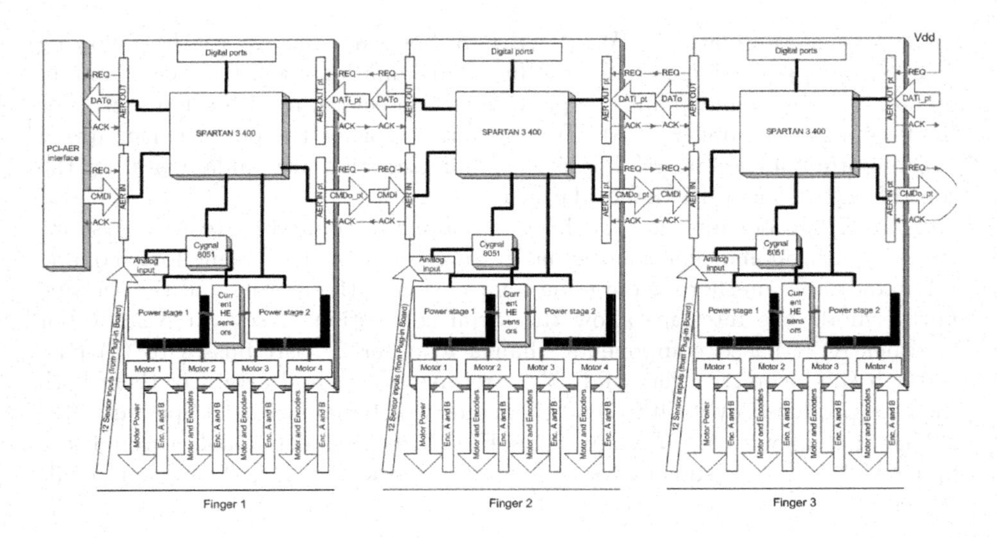

Fig. 4. AER-Robot block diagram interface

These processes work in parallel to allow real-time control of the hand; therefore the interface can receive new commands while it is executing previous ones. There is another process for the AER output bus traffic control (DATout), and another one for the four AER buses control.

For each input command received by the AERin process, the order is sent to the corresponding motor process or sensor process, and then this AERin process is free to attend a new command.

Each motor process is in charge of one motor. If this process receives an order, its motor will go up or down, for a number of encoder pulses and with a programmed intensity.

There is a sensors process that is asked for a value of one the sensors. This process keep updated a 16-address internal RAM memory with the digital value of their sensors, and the digital value is sent to the AERout process when the AERin process received a read-sensor command.

To keep this RAM-table updated, the sensor process communicates with a microcontroller, outside the FPGA, that scan the 16 analog output of the sensors, convert it into 8-bit and send it to the FPGA. The RAM-table is updated every 184μs. Thus, when a command is asking for the value of one sensor, the sensor process doesn't ask it to the microcontroller, but it just has to read it from the internal RAM memory of the FPGA (1 clock cycle).

The microcontroller of the AER-Robot is in charge of 16 sensors of 4 different sets: articulations potentiometers, fingertip and palm contacts, tendon tension and power consumption of the motors.

Fig. 4 shows the block diagram of the interface. The real-time is warranted by the independent process architecture.

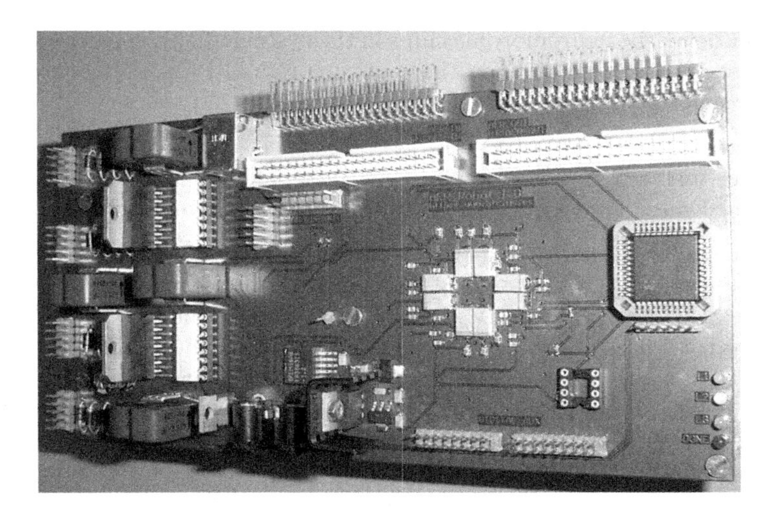

Fig. 5. AER-Robot board photograph

Fig. 5 shows a photograph of the prototype of the AER-Robot Interface PCB. The digital part of the PCB is on the left side. The board has 4 power stages for the 4 motors, 4 Hall Effect sensors for the power consumption measurement of the motors and a Cygnal 80C51F320 microcontroller for the analog to digital conversion (200Ksamples/second and 10-bits) of the sensor measurements, and all the connectors to the Hand and to the AER systems. A small board, called Plug-in board, was connected to the 12 line of the microcontroller to implement the analog adaptation of the output of the different sensor to the voltage needed to convert it into digital by the microcontroller. This plug-in board was composed by 4 x10K amplifiers for the tendon sensors and 8 direct connections to the connectors of the contact sensors and potentiometers sensor installed on the hand,

The AER input bus and AER output bus is connected to a PC using the PCI-AER interface, explained in the next section.

4 PCI-AER Interface

Before the development of our tools the only available PCI-AER interface board was developed by Dante at ISS-Rome (See [4]). This board is very interesting as it embeds all the requirements mentioned above: AER generation, remapping and monitoring. Anyhow its performance is limited to 1Mevent/s approximately. In realistic experiments software overheads reduce this value even further. In many cases these values are acceptable but, currently many address event chips can produce (or accept) much higher spike rates.

As the Computer interfacing elements are mainly a monitoring and testing feature in many address event systems, the instruments used for these proposes should not delay the neuromorphic chips in the system. Thus, speed requirements are at least 10 times higher than those of the original PCI-AER board. Several alternatives are possible to meet these goals:

- Extended PCI buses.
- Bus mastering.
- Hardware based Frame to AER and AER to Frame conversion.

The previously available PCI-AER board uses polled I/O to transfer data to and from the board. This is possibly the main limiting factor on its performance. To increase PCI bus mastering is the only alternative. The hardware and driver architecture of a bus mastering capable board is significantly different, and more complex, than a polling or interrupt based implementation.

A well designed AER system, which produces events only when meaningful information is available, can be very efficient but, an AER monitoring system should be prepared to support the bandwidth levels that can be found in some real systems. These include systems that have not been designed carefully or that are under adjustment. Currently the available spike rates, even in these cases, are far from the value shown above but, some current AER chips may exceed the 40Mevents/s in extreme conditions.

The theoretical maximum PCI32/33 bandwidth is around 133Mbytes/s. This would allow for approximately 44Mevent/s considering 2 bytes per address and two bytes for timing information. Realistic figures in practice are closer to 32Mbyte/s. Thus, in those cases where the required throughput is higher a possible solution is to transmit the received information by hardware based conversion to/from a frame based representation. Although this solution is adequate in many cases, there are circumstances where the developers want to know precisely the timing of each event, thus both alternatives should be preserved.

Implementing AER to Frame conversion is a relatively simple task as it basically requires counting the events over the frame period. Producing AER from a frame representation is not trivial and several conversion methods have been proposed [8][9].

The physical implementation of all the steps is equal. They differ in the VHDL FPGA code and in the operating system dependant driver. This interface is based on SPARTAN-II family. The board is shown in Fig. 6.

Currently a Windows driver that implements bus mastering is available. An API that is compatible, as much as permitted by the different functionality, with that used in the current PCI-AER board has been implemented. MEX files to control the board from MATLAB have also been developed.

The final goal is to transmit an AER sequence to an AER based system (for example a convolution chip) to perform video processing. An adequate sequence of events can be generated by software for testing an AER based system. This sequence of events needs to be sent to the AER based system. For this purpose it is necessary an interface between the computer and the AER bus. This PCI interface has been developed under the European project CAVIAR. The interface, called CAVIAR PIC-AER, has two operation modes that can work in parallel:

Fig. 6. CAVIAR PCI-AER board

A) From PCI to AER
The AER-stream is stored in the computer memory and then it is sent to the AER system through the PCI bus and the OFIFO. This stream is saved in memory using 32 bits for each address event. The sixteen less significant bits represents the address of the pixel that is emitting the event. The sixteen more significant bits represent a time difference from the previous event in clock cycles. The delay between events is specially treated, so a delay in the ACK reception will not cause a distortion in the time distribution of all the events along the time period.

B) From AER to PCI
The AER sequence arrives to the CAVIAR PCI-AER interface through the input AER port. The AER-IN state machine stores the incoming event (16 bits LSbits) into the IFIFO with temporal information. This temporal information (16 bits MSbits) is the number of clock cycles since the last event.

The connection to the PCI bus is done by a VHDL bridge [13] that attends to the Plug & Play protocol of the PCI bus, decodes the access to the base address by the operating system, allows the bus mastering and the interruptions.

5 Results

Both the hand and the AER interfaces have been connected to debug control algorithms algorithm implemented under Matlab.

The Antrophomorphic robotic hand was developed by NEUROCOR group from TUC (Spain). Under the Spanish grant project SAMANTA the hand has been connected to an AER system. One of the objectives of the project was to control the hand using AER vision systems together with other sensor information. Therefore the hand needs to be controlled under the AER protocol. In that project the visual information comes from an AER retina. Thanks to the PCI-AER interface the visual information is sent to a personal computer. A boundary-contour-system feature-control-system (BCS-FCS) algorithm for image processing was implemented under Matlab. With the visual processing results and the hand sensors information, the control algorithm gives orders to the hand, for example to catch an object, by computing not only the visual information, but also the rigidity and fatigue of the muscles.

The results shown in table 1 are the ranges of the parameters that can be programmed in the AER-Robot board. These results have been measured under the following scenario: three AER-Robot interfaces have been connected together using AER busses in a chain configuration. The first AER-Robot interface was connected to a PC through the PCI-AER interface. Therefore, commands where sent from Matlab to the PCI-AER interface into AER format, and then they are sent to the first AER-Robot interface, this one retransmit the command to the second, the second to the third and so on. The AER information that produces a board, this is sent through an AER bus. This one can be connected to the data input AER bus of the previous board to the PCI-AER. Each AER-Robot arbitrates the AER information received from the next board with the information that is produces.

Table 1. Measured ranges for DC motor and Sensors

	Freq min	*Freq max*	*High res.*	
PWM	763 Hz	25 MHz	8-bits	
	Dig. Res.	*Range (volts)*	*Kind*	*Sensor range*
Potentiometers	8-bit	0-3,3v	R	
Contact	8-bit	0-3,3v	R	
Tension	8-bit	2-2,5v	R	0,1mv/V
Hall-Effect current	8-bit	2-3,3v	L	
PCI-AER	*Max Th.*	*Pulse width*		
	6 Mev/s	120 ns		
AER-Robot	3 Mev/s	240 ns		

6 Conclusions and Future Work

An AER to anthropomorphic robotic hand interface that can be connected to an AER system or to a PC through a PCI-AER interface have been presented. The AER-Robot interface is based around a Spartan 3 FPGA that allows it to be configurable and easily modified for other robotic applications based on DC motors, encoders and sensors.

The AER neuro-inspired communication channel is connected with the robot. This implies a neuro-inspired control of the robot. This control is based on visual processing using AER retinas and convolution chips, and neuro-inspired FLETE algorithm in software.

The current interface is able to receive and send 16-bit AER data. Coded under these 16-bit is placed the command or the sensor information. Therefore one event is enough to send a command to a motor or ask for sensor information, then one or two events are sent back with the information required. The future work is focused on the spike based information. In such way, the motor PWM frequency will be sent translating the frequency of one address in the AER bus. Each address corresponds with one motor, and with one sensor in the other way. These VHDL improvements are under development.

Acknowledgements

This work was in part supported by EU project IST-2001-34124 (CAVIAR), Spanish projects TIC-2003-08164-C03-02 (SAMANTA) and SAMANTA II. We also wanted to thank to J.L. Pedreño-Molina, J. Molina-Vilaplana for their contribution in this work with their bio-insipired control algorithms.

References

[1] M. Sivilotti, Wiring Considerations in analog VLSI Systems with Application to Field-Programmable Networks, Ph.D. Thesis, California Institute of Technology, Pasadena CA, 1991.

[2] Kwabena A. Boahen. “Communicating Neuronal Ensembles between Neuromorphic Chips”. Neuromorphic Systems. Kluwer Academic Publishers, Boston 1998.

[3] R. Serrano-Gotarredona et al. AER Building Blocks for Multi-Layer Multi-Chip Neuromorphic Vision Systems. NIPS 2005. Vancouver.

[4] A. Cohen, R. Douglas, C. Koch, T. Sejnowski, S. Shamma, T. Horiuchi, and G. Indiveri, “Report to the National Science Foundation: Workshop on Neuromorphic Engineering”, Telluride, Colorado, USA, June-July 2001. [www.ini.unizh.ch/telluride]

[5] Avis Cohen, et.al., Report on the 2004 Workshop On Neuromorphic Engineering, Telluride, CO. June - July , 2004 [www.ini.unizh.ch/telluride/previous/report04.pdf]

[6] A. Wilen, J.Schade, R. Thornburg. “Introduction to PCI Express”, Intel Press, 2003.

[7] A. Linares-Barranco et al.. “AER Neuro-Inspired interface to Anthropomorphic Robotic Hand”. IEEE World Conference on Computational Intelligence. International Join Conference on Cellular Neural Networks. Vancouver, July-2006.

[8] A. Linares-Barranco, G. Jimenez-Moreno, A. Civit-Ballcels, and B. Linares-Barranco. “On Algorithmic Rate-Coded AER Generation”. *IEEE Transaction on Neural Networks.* May-2006.

[9] A. Linares-Barranco, M. Oster, D. Cascado, G. Jimenez, A. Civit, B. Linares-Barranco. “Inter-Spike-Intervals analysis of AER Poisson like Generator Hardware”. Accepted for publication on Neurocomputing (~December 2007).

[10] R. Paz, F. Gomez-Rodriguez, M. A. Rodriguez, A. Linares-Barranco, G. Jimenez, A. Civit. Test Infrastructure for Address-Event-Representation Communications. IWANN 2005. LNCS 3512. pp 518-526. Springer Verlag.

[11] Address Event Representation tools. [http://www.atc.us.es/AERtools]

[12] J. López-Coronado, J.L. Pedreño-Molina, A. Guerrero-González, P. Gorce. A neural model for visual-tactile-motor integration in robotic reaching and grasping tasks. Robotica, Volume 20, Issue 01, pp. 23-21. Cambridge Press. (January 2002).

[13] R. Paz. “Análisis del bus PCI. Desarrollo de puentes basados en FPGA para placas PCI”. Trabajo de investigación para obtención de suficiencia investigadora. Sevilla, Junio 2003.

Tackling the Error Correcting Code Problem Via the Cooperation of Local-Search-Based Agents

Jhon Edgar Amaya[1], Carlos Cotta[2], and Antonio J. Fernández[2]

[1] Universidad Nacional Experimental del Táchira (UNET)
Laboratorio de Computación de Alto Rendimiento (LCAR), San Cristóbal, Venezuela
jedgar@unet.edu.ve
[2] Dept. Lenguajes y Ciencias de la Computación, ETSI Informática,
University of Málaga, Campus de Teatinos, 29071 - Málaga, Spain
{ccottap,afdez}@lcc.uma.es

Abstract. We consider the problem of designing error correcting codes (ECC), a hard combinatorial optimization problem of relevance in the field of telecommunications. This problem is firstly approached via a battery of local search (LS) methods that are compared and analyzed. Then, we study how to tackle this problem by having a society of interacting autonomous agents where each agent is endowed with a specific (not necessarily unique) strategy for local improvement. Distinct topologies and forms of interaction are analyzed and discussed in the paper. Specifically, it is shown how the election of the LS methods and their combination influences the results. An empirical evaluation shows that agent-based models are promising to solve of this problem.

1 Introduction

Telecommunications undoubtedly constitute one of the most prominent pillars upon which our present society rests. Many of the tasks found in this area can be formulated as combinatorial optimization problems, e.g., assigning frequencies in radio link communications [1], designing telecommunication networks [2], or developing error correcting codes for transmitting messages [3] among others. In this work, we will focus precisely on the latter problem.

Roughly speaking, the development of an error correcting code (ECC) consists of designing a communication scheme for maximizing the reliability of information transmission through a noisy channel. This task admits several formulations. Here, we have considered the case of binary linear-block codes [4]. The design of such codes turns out to be very difficult. There exists no known algorithm for efficiently finding optimal solutions. The utilization of metaheuristic approaches is thus in order.

There have been several attempts to use metaheuristics on the ECC problem (ECCP, see Sect. 2.2). This paper builds on previous approaches, and explores how to solve the ECCP via the cooperation of several algorithmic techniques. To be precise, a set of different LS metaheuristics is used in a first stage to solve the problem, being the results compared and analyzed; then these metaheuristics

J. Mira and J.R. Álvarez (Eds.): IWINAC 2007, Part II, LNCS 4528, pp. 490–500, 2007.

are integrated as autonomous agents in two different agent-based architectures, and deployed on the ECCP. Different versions of these architectures will be used to attack the problem, analyzing the impact of the topology on the results.

2 Background

Before proceeding, let us firstly describe more in depth the ECC problem. Then, we will review previous related work.

2.1 The Error Correcting Code Problem

As discussed in the previous section, an ECC is aimed at maximizing the reliability of message transmissions through a noisy channel. Let us assume messages are expressed in sequences of characters from some alphabet Σ. In the context of the binary linear block codes, we would map each of these characters $ch_i \in \Sigma$ to a sequence of n bits (or code-word) c_i in order to transmit it. Upon reception of an n-bit sequence c, the character encoded could be recovered by looking for the closest –in a Hamming distance sense– valid code-word. Then, if all code-words are separated by at least d bits, any modification of at most $(d-1)/2$ bits in a valid code-word can be easily reverted. Hence large d is sought.

It is possible to increase the value of d by considering larger values of n, but an upper bound of n has to be considered (otherwise messages would be too lengthly, and therefore costly). Thus, we would be interested in maximizing d for a certain alphabet Σ, and a certain value of n. This way, an ECCP instance is fully specified by a pair (M, n), where n is the number of bits in each code-word, and M is the number of code-words. Let $\mathbb{B} = \{0, 1\}$; the solution space for an ECCP instance would comprise all sets $C = \{c_1, \cdots, c_M\}$, $c_i \in \mathbb{B}^n$, i.e., all combinations of M different n-bit sequences. The size of the search space is thus $\binom{2^n}{M}$. Although no known algorithm is available for producing an optimal ECC (i.e., a set of M n-bit code-words with maximal d) in general, the problem has been theoretically studied, and bounds on the attainable values of d for different combinations of n and M have been derived [5].

It is interesting to notice the relation between the ECCP as defined above, and another problem in the realm of physics: finding the lowest energy configuration on M particles in an n-dimensional space. By assimilating particles to code-words, the ECCP can be viewed as distributing M code-words in the corners of a binary n-dimensional space. This connection was used by Dontas and de Jong [6] to define a fitness function (to be maximized) for a genetic algorithm optimizing this problem, i.e.,

$$Fitness(C) = \frac{1}{\sum_{i=1}^{M} \sum_{j=1, i \neq j}^{M} \frac{1}{d_{ij}^2}}, \tag{1}$$

where d_{ij} is the Hamming distance between code-words c_i and c_j. This function is more adequate as a guiding function than a naïve function computing the minimum distance between different code-words in a solution. Although the latter would capture the absolute quality of a solution, it would induce large plateaus in the fitness landscape. This would not be the case for the former function, which is capable of grasping the effects of small changes in a solution.

2.2 Related Work

Many proposals can be found in the scientific literature to deal with the ECCP. For instance, it has been treated with simulated annealing, genetic algorithms, and hybrids thereof, with moderate success (check [7] and the references therein). Recently, in [8], a LS algorithm – called *the Repulsion Algorithm* (RA) – hybridized with a parallel genetic algorithm has been proposed for this problem. The repulsion algorithm was conceived by Birbil and Fang [9], and is based on the repulsion of particles over the surface of a sphere. An experimental study, including comparisons with a pure parallel genetic algorithm, indicated that a significant improvement in efficiency and accuracy can be obtained when the repulsion algorithm is used as LS. The use of scatter search (SS) and memetic algorithms (MAs) to attack the ECCP was also analyzed in [7]. It was shown that SS and MAs are cutting-edge techniques for solving the ECCP, capable of outperforming sophisticated versions of other metaheuristics on this domain, including the parallel genetic algorithm incorporating RA mentioned before. More recently, Blum *et al.* [10] suggested to tackle the ECCP with iterated local search (ILS). An experimental study showed that this proposal is currently a state-of-the-art method for the ECCP.

3 Tackling the ECC Problem Via Local-Search Methods

As explained in Section 2.2, different LS methods have been utilized to solve the ECCP with more or less success. In this section we will describe the deployment of the more successful LS approaches on the ECCP.

3.1 General Issues

There are considerations that must be taken into account regardless of the LS technique used, i.e., the representation, the neighborhood function, etc. Previously to detail the different LS approaches, let us describe these general aspects.

Solution Representation. A candidate solution for an instance (M, n) is a set of M binary code-words of length n represented as a binary matrix C with M rows and n columns. Each row i in C corresponds to a code-word c_i. We also define the *minimum Hamming distance* in C as $d(C) = min\{d_{ij} \mid 1 \leq i, j \leq M\}$.

Neighborhood. The neighborhood relation (unless another one is specified) is based on the 1-Hamming distance neighborhood (i.e., the 1-flip neighborhood) in which a neighbor consists of flipping exactly one position in a solution. This concept of neighborhood is the same as those used in [7,10]. Let C be a candidate solution i.e., an $M \times n$ binary matrix as mentioned above, and let c_{ij} and v_{ij}

(for $1 \leqslant i \leqslant M$ and $1 \leqslant j \leqslant n$) be, respectively, the cell and the value of the cell in row i and column j in matrix C. Then the set of possible moves is $\mathcal{M}(C) = \{(c_{ij}, \overline{v_{ij}}) \mid 1 \leqslant i \leqslant M \wedge 1 \leqslant j \leqslant n \wedge \overline{0} = 1 \wedge \overline{1} = 0\}$.

In the rest of the paper, $C[c_{ij} \leftarrow v]$ denotes the candidate solution C where cell c_{ij} is assigned to v. Then, the *neighborhood for a candidate solution* C is defined as $\mathcal{N}(C) = \{C[c_{ij} \leftarrow v] \mid 1 \leqslant i \leqslant M \wedge 1 \leqslant j \leqslant n \wedge (c_{ij}, v) \in \mathcal{M}(C)\}$.

Fitness Recomputation. The fitness function defined in (1), as pointed out in [7], does not require full re-evaluation, every time a bit is flipped. Assume that we consider a 1-flip that changes the code-word c_i, and let d_i be the minimum distance between code-word c_i and any of the other $M-1$ code-words. We only have to re-compute $\sum_{j=1, i \neq j}^{M} \frac{1}{d_{ij}^2}$ after the 1-flip. Also the Hamming distances do not have to be recomputed from scratch. They can be updated by keeping appropriate data structures. Thus, whenever a configuration is modified, the new fitness can be computed just considering the 1-flip change.

Generation of Initial Solutions. Two methods are considered to generate an initial solution: (1) randomly initializing all the entries in a configuration C, and (2) using an *ad-hoc* constructive heuristic. Regarding this latter possibility, a very interesting proposal is presented in [10]. Basically this intelligent heuristic works as follows: assume that C_i is an optimal code for an instance $(M = 2^{i+1}, n = 2^i)$; then, an optimal code C_{i+1} for the instance $(M = 2^{i+2}, n = 2^{i+1})$ can be obtained by setting

$$C_{i+1} = \begin{pmatrix} C_i & C_i \\ C_i & C_i^f \end{pmatrix}$$

where C_i^f is matrix C_i where each position is flipped. This process starts from the optimal code C_0 for the instance (4,2), and is iterated until both the number of columns of C_k ($k \geqslant 0$) is greater or equal to n, and the number of rows of C_k is greater or equal to M. Doing so, we obtain a solution C_s for the instance $(M' = 2^{k+2}, n' = 2^{k+1})$ with 2^k as minimum Hamming distance (i.e., $d(C_s) = 2^k$). A solution for (M, n) is obtained by removing the last $M' - M$ rows, and the last $n' - n$ columns from C_k.

3.2 Tabu Search (TS)

As it is well-known, tabu-search algorithms are capable of escaping from local optima by allowing down-hill moves. Fig. 1 shows the complete pseudocode of our TS algorithm. The algorithm returns the best solution C_f found. To prevent cycling, a tabu list of movements is kept. The list stores triplets $\langle c_{ij}, v, i \rangle$, where c_{ij} is a cell, $v \in \{0, 1\}$ is a possible value for cell c_{ij}, and i represents the first iteration where cell c_{ij} can be assigned to v again. The tabu tenure, i.e., the number of iterations (c_{ij}, v) stays in the list, is static and equal to 100 (i.e., $\tau = 100$). For a candidate solution C and an iteration k, the set of legal moves is thus defined as

$$\mathcal{M}^+(C, k) = \{(c_{ij}, v) \in \mathcal{M}(C) \mid \neg tabu(c_{ij}, v, k)\}. \tag{2}$$

where $tabu(c_{ij}, v, k)$ holds if the assignment $c_{ij} \leftarrow v$ is tabu at iteration k. The tabu status can be overridden whenever the selected candidate (even if this is tabu) is better. Thus, if C_f is the candidate with the highest $d(C_f)$ found so far, the neighborhood also includes the moves

$$\mathcal{M}^*(C, C_f) = \{(c_{ij}, v) \in \mathcal{M}(C) \mid d(C[c_{ij} \leftarrow v]) > d(C_f)\} \quad (3)$$

```
1.  TS()
2.      C0 ← GenerateInitialSolution(); tabu ← {};
3.      Cf ← C0; C ← C0; k ← 1;
4.      while ¬ timeout do
5.          select randomly (cij, v) ∈ M+(C, k) ∪ M*(C, Cf)
6.          tabu ← tabu ∪ {⟨cij, v, k + τ⟩};
7.          C ← C[cij ← v];
8.          if d(C) > d(Cf) then
9.              Cf ← C;
10.         k ← k + 1;
11.     return Cf;
```

Fig. 1. Pseudocode of the TS algorithm

3.3 Simulated Annealing (SA)

Simulated annealing is another well known method in the field of metaheuristics. It is a computational method that basically mimics the physical process of thermal cooling. In fact it is a variation of the classical Hill climbing method. Fig. 2 shows the complete pseudocode of our SA algorithm.

```
1.  SA()
2.      C0 ← GenerateInitialSolution(); C ← C0; Cf ← C0;
3.      T ← T0; \\select initial temperature
4.      while ¬timeout ∧ ¬(T ⩽ Tmin) do
5.          probabilistically select Ct ∈ N(C);
6.          Δf = d(C) − d(Ct)
7.          τ ← RANDOM([0,1]);
8.          if (Δf < 0) or (τ < e^(−Δf/T)) then
9.              C ← Ct;
10.         if d(Cf) < d(C) then
11.             Cf ← C;
12.         T ← α · T \\ update temperature: geometric decreasing
13.     return Cf
```

Fig. 2. Pseudocode of the SA algorithm

Boltzmann's law is used to determine the probability of acceptation of a candidate solution under a perturbation as result of an energy change ΔE (the fitness increase). The current temperature T modulates this acceptance probability (if T is high, higher energy increases are allowed).

$$P = \begin{cases} 1, & \text{if } \Delta E < 0 \\ e^{-\frac{\Delta E}{k_B T}}, & \text{otherwise} \end{cases}$$

where k_B is Boltzmann's constant (which can be ignored in practice, by considering an appropriate scaling for the temperature). The temperature is decreased from its initial value T_0 to a final value $T_{\min}$ using a geometric cooling schedule. We have considered $T_0 = 100$, $T_{\min} = 0.001$, and $\alpha = 0.998$.

3.4 Iterated Local Search (ILS)

Iterated local search (ILS) is based on the principle of "reconstruction and improvement". This basically means to repeat the following three steps: (1) A perturbation mechanism perturbs at each iteration the current solution. This mechanism basically maximizes the distance between the code-word with minimum average distance to all the other $M - 1$ code-words and another code-word selected at random; (2) a LS method, based on the 1-flip neighborhood, is used to improve this perturbed solution, and (3) a criterion for the acceptance of the improved perturbed solution as new current solution is applied. See the pseudocode shown in Fig. 3, and [10] for more details about the ILS used here.

```
1. ILS ()
2.     C0 ← GenerateInitialSolution(); Cf ← LS(C0);
3.     while ¬ (timeout) do
4.        Cf ← perturbation(Cf);
5.        Cq ← LS(Cf);
6.        C ← AcceptanceCriterion(Cq);
7.     return best C;
```

Fig. 3. Pseudocode of the ILS algorithm

3.5 Hill Climbing (HC) and Variable Neighborhood Search (VNS)

We also consider a standard Hill-Climbing method (HC) with 1-flip neighborhood, and a simple form of VNS algorithm (in which 20 neighbors, in a 1-flip neighborhood, are randomly selected and the best of them is improved locally. This process is repeated until a timeout is reached).

4 Computational Results with LS

We have implemented all the metaheuristics mentioned in previous section in Java and executed them on a PC (Intel Celeron 1.5 GHz 512 MB) under Linux Debian (kernel 2.6.16-2-686). The experiment has been done with the instances chosen in [7,10], i.e., the (M, n) pairs (24,12), (32,16), and (40,20). For each LS $\in$ {HC,ILS,TS,SA,VNS}, we tested two versions of our algorithms: one that uses the heuristic initial solution construction explained in Section 3.1, and another that uses a random initial solution construction instead. We also impose the same time limit as indicated in [10], i.e., 5, 100 and 2000 seconds for the instances $(24, 12)$, $(32, 16)$, and $(40, 20)$ respectively.

Note that our results cannot be directly compared with those presented in [10] because of the following reasons: (1) our computational machine is slower

(1.5 Ghz and 512 MB RAM vs. 3 Ghz and 1 Gb RAM); (2) our algorithms were implemented in Java and not in C (as done in [10]). For normalization purposes we did experiments with our ILS algorithm (note that this is the same as that described in [10]) and our time results are about one order magnitude slower. However, we decide to maintain the same time limits as our purpose is not to obtain new optimums (as these are already known) but to analyze the behavior of two different LS-based agent topologies as shown in next sections.

Fig. 4 shows the results of the experiments (averages of 20 runs). In general, the heuristic initialization-based version of the methods behaves better but, on the smallest problem instance, the random initialization-based versions seems to be slightly better. This means a generalization of one conclusion already pointed out in [10] but only for the ILS method. Also it is not surprising that for the instance (32,16), all the LS methods obtain the optimum with a high success rate (not shown explicitly in the figure) as the constructive heuristic allows to find the optimal solution without a single solution evaluation.

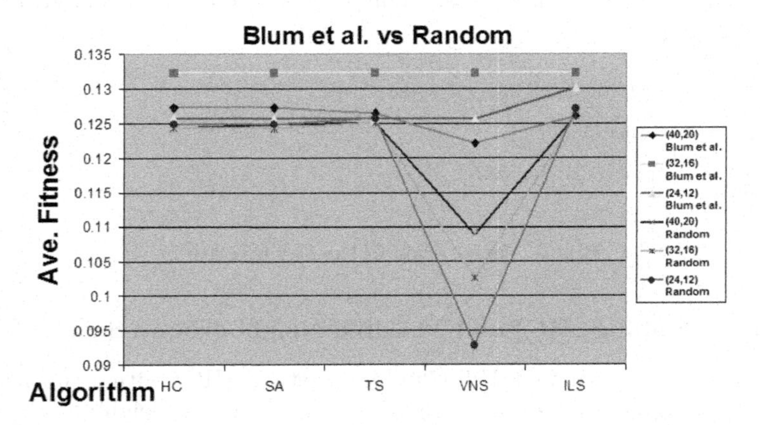

Fig. 4. Average (20 runs) fitness of the different metaheuristics on different ECCP instances. Random = Random generation of initial solutions, Blum *et al.*= generation of initial solutions via the constructive heuristic explained in Section 3.1.

In general, and as expected, ILS performs better than the rest of heuristics, especially in the instance (24,12). However, observe also that all the metaheuristic methods behave very similar in the other instances except VNS (in its two versions) that is significantly worst. We hypothesize the reason in the random selection of the candidate solution because the movement is more dependant on the random factor and not on a heuristic recommendation. To mitigate this effect, we plan to consider for future work neighborhoods relations of variable size, that is to say, a set $\{\mathcal{N}_1(C), \ldots, \mathcal{N}_{k_{max}}(C)\}$ of neighborhood relations ordered according to increasing values of k (i.e., the number of flips that must be simultaneously allowed in a candidate C).

As expected, the heuristic initialization works much better than the random initialization. This is particularly clear in the case of the VNS algorithm.

5 Agent-Based Topologies

So far, we have considered several LS algorithms and have compared their performance as stand-alone techniques. Now, we propose two agent-based architectures in which each agent incorporates a LS mechanism. Let A be an architecture with n agents; each agent a_i $(0 \leqslant i \leqslant n-1)$ in A is driven by one of the metaheuristics described in Sect. 3. In general, we let a_i denote the metaheuristic implemented in the agent i and C_i the current candidate solution handled in agent i.

Ring Topology. The first architecture consists of a ring topology with n agents. Here, the agents are ordered according to some arbitrary criteria and each agent is only connected with its successor in the ordered list. Fig. 5 presents the schematic procedure of the ring-based algorithm for a specific architecture and n agents.

```
1.  RING(A, n) \\
2.     for i ← 0 to n − 1 do
3.        C_i ← GenerateInitialSolution(); \\ As explained in Section 3.1
4.     cyc ← 1; \\ number of cycles
5.     while (cyc ⩽ cyc_max) do
6.        for i ← 0 to n − 1 do
7.           C_i ← a_i(C_i); \\ Agent-specific local improvement
8.        for i ← 0 to n − 1 do \\ Agent cooperation
9.           if d(C_((i+n−1)%n)) > d(C_i) then
10.             C_i ← C_((i%n)−1);
11.       cyc ← cyc + 1;
12.    return C_i (0 ⩽ i ⩽ n − 1) minimizing d(C_i);
```

Fig. 5. Pseudocode of the algorithm of the ring topology

Initially all the agents are "charged" with an initial solution generated as explained in preceding sections (Lines 2-3). Then, the algorithm is executed a specific number of cycles. In each cycle, each agent works independently by improving its internal solution via its corresponding LS method (Lines 6-7). The time dedicated for local improvement in all the agents is the same. Agents coordinate themselves by updating its current best candidate solution with the candidate found by the preceding agent in case of the latter is better.

Broadcast topology. The second architecture consists of a *go with the winners*-like topology, that we call *broadcast topology*, with n agents. As indicated in Fig. 6, the main difference wrt. the algorithm $RING(A, n)$ is that the best global solution at each synchronization point is transmitted to all agents. This means all agents start each cycle (except the first one) in the same local region.

Both $RING(A, n)$ and $BRODCAST(A, n)$ are executed by imposing a time limit of t_{max} (seconds). As a consequence, each cycle cyc takes $t_{cyc} = t_{max}/cyc$ (seconds), and the specific local-improvement method of an agent takes t_{cyc}/n.

```
1.     BROADCAST(A, n) \\
2-7.      same as in algorithm RING(A, n);
8.           for i ← 1 to n − 1 do \\ Agent cooperation
9.              if d(C_i) > d(C_0) then
10.                C_0 ← C_i;
11.          for i ← 1 to n − 1 do
12.                C_i ← C_0;
13-14. same as Lines 11-12 in algorithm RING(A, n);
```

Fig. 6. Pseudocode of the algorithm of the broadcast topology

6 Computational Results

The computational setting for the agent-based algorithms is the same used in Sect. 4. Again, we tested two versions of our algorithms: one using the constructive heuristic of Blum *et al.* as explained in Section 3.1 (these are labelled with the word 'Blum' in the figures shown later), and another using a random initial solution construction instead (labelled with 'Random'). In our experiments $n = 5$, and for both topologies (i.e., ring and broadcast) we consider two different proposals: one in which each agent implements a different metaheuristic in the set {HC,TS,SA,VNS,ILS}, and another one in which all agents implement ILS (i.e., the current state-of-the-art metaheuristic method as mentioned in Sect. 2.2). Again, we impose the same time limits (i.e., t_{max}) as indicated in Section 4, and we consider $cyc = 4$ cycles and $cyc = 8$ cycles.

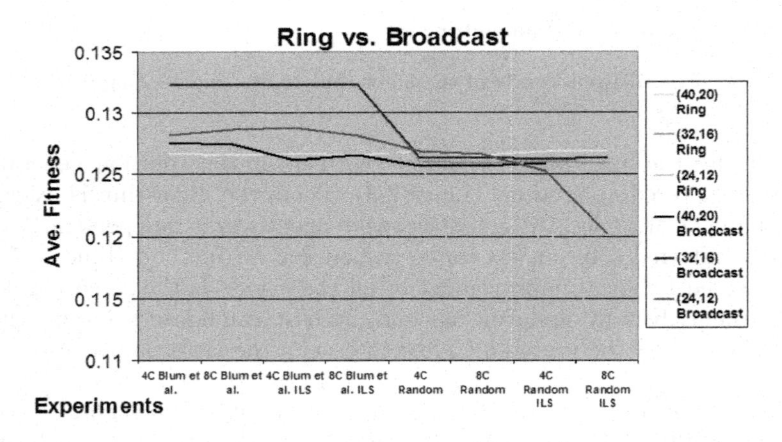

Fig. 7. The graphic displays the average (20 runs) fitness of the different variants of both algorithms $RING(A,n)$ and $BROADCAST(A,n)$ on different ECCP instances. 4C = 4 cycles, 8C = 8 cycles, Random = Random generation of initial solutions, Blum et al.= generation of initial solutions via the constructive heuristic explained in Section 3.1.

Some interesting conclusions can be drawn from results shown in Fig. 7: (1) as expected, algorithms using the constructive heuristic behave better than those based on random initialization; (2) The algorithms seems to be slightly better with a higher number of cycles between synchronization points (although this is not evident in the ILS-only-based model); (3) On the smallest instance, RING outperforms BROADCAST, whereas on larger ones it is the other way around.

7 Conclusions

Agent-based optimization constitutes a very appropriate framework for integrating different search techniques. Each of these techniques has a different view of the search landscape, and by combining the corresponding different exploration patterns, the search can benefit from an increased capability for escaping from local optima. Of course, this capability is more useful whenever the problem tackled poses a challenging optimization task to the individual search algorithms. Otherwise, computational power is diversified in unproductive explorations. In the problem considered in this work, the ECC problem, there exists a killer approach, namely the ILS algorithm of Blum *et al.*. In such a situation, it may be more convenient to combine different instances of the same technique, much like it is done in island-based evolutionary algorithms. Notice however the fact that they are single agents rather than full populations (as in EAs) and this contributes to intensify much more the search. Convergence to near-optimal solutions is thus quicker.

Future work will be directed to extend this agent-based approach to other problems to confirm its usefulness. We plan to include additional techniques in the agent pool, such as population-based algorithms. Finally, we intend to consider smarter mechanisms for exchanging information between agents, not relying exclusively on a fixed interconnection topology.

Acknowledgements. This work is partially supported by Spanish MCyT projects under contracts TIN2004-7943-C04-01 and TIN2005-08818-C04-01

References

1. Dorne, R., Hao, J.: An evolutionary approach for frequency assignment in cellular radio networks. In: 1995 IEEE International Conference on Evolutionary Computation, Perth, Australia, IEEE Press (1995) 539–544
2. Chu, C., Premkumar, G., Chou, H.: Digital data networks design using genetic algorithms. European Journal of Operational Research **127** (2000) 140–158
3. Chen, H., Flann, N., Watson, D.: Parallel genetic simulated annealing: A massively parallel SIMD algorithm. IEEE Transactions on Parallel and Distributed Systems **9** (1998) 126–136
4. Lin, S., Jr., D.C.: Error Control Coding : Fundamentals and Applications. Prentice Hall, Englewood Cliffs, NJ (1983)
5. Agrell, E., Vardy, A., Zeger, K.: A table of upper bounds for binary codes. IEEE Transactions on Information Theory **47** (2001) 3004–3006
6. Dontas, K., Jong, K.D.: Discovery of maximal distance codes using genetic algorithms. In: Proceedings of the Second International IEEE Conference on Tools for Artificial Intelligence, Herndon, VA, IEEE Press (1990) 905–811

7. Cotta, C.: Scatter search and memetic approaches to the error correcting code problem. In Gottlieb, J., Raidl, G.R., eds.: 4th European Conference on Evolutionary Computation in Combinatorial Optimization (EvoCOP 2004). Volume 3004 of Lecture Notes in Computer Science., Coimbra, Portugal, Springer (2004) 51–61
8. Alba, E., Chicano, J.F.: Solving the error correcting code problem with parallel hybrid heuristics. In Haddad, H., Omicini, A., Wainwright, R.L., Liebrock, L.M., eds.: Proceedings of the 2004 ACM Symposium on Applied Computing (SAC), Nicosia, Cyprus, ACM (2004) 985–989
9. Ş. Birbil, Fang, S.C.: An electromagnetism-like mechanism for global optimization. Journal of Global Optimization **25** (2003) 263–282
10. Blum, C., Blesa, M., Roli, A.: Combining ILS with an effective constructive heuristic for the application to error correcting code design. In *et al.*, H.R., ed.: Proceedings of the 6th Metaheuristics International Conference (MIC2005), Universität Wien (2005) 114–119

Strategies for Affect-Controlled Action-Selection in Soar-RL

Eric Hogewoning[1], Joost Broekens[1], Jeroen Eggermont[2], and Ernst G.P. Bovenkamp[2]

[1] Leiden Institute of Advanced Computer Science, Leiden University, P.O. Box 9500, 2300 RA Leiden, The Netherlands
[2] Leiden University Medical Center, Department of Radiology, Division of Image Processing, P.O. Box 9600, 2300 RC Leiden, The Netherlands

Abstract. Reinforcement learning (RL) agents can benefit from adaptive exploration/exploitation behavior, especially in dynamic environments. We focus on regulating this exploration/exploitation behavior by controlling the action-selection mechanism of RL. Inspired by psychological studies which show that affect influences human decision making, we use artificial affect to influence an agent's action-selection. Two existing affective strategies are implemented and, in addition, a new hybrid method that combines both. These strategies are tested on 'maze tasks' in which a RL agent has to find food (rewarded location) in a maze. We use Soar-RL, the new RL-enabled version of Soar, as a model environment. One task tests the ability to quickly adapt to an environmental change, while the other tests the ability to escape a local optimum in order to find the global optimum. We show that artificial affect-controlled action-selection in some cases helps agents to faster adapt to changes in the environment.

1 Introduction

At the core of emotion and mood are states that have certain levels of valence and arousal, i.e, affective states. Valence represents the goodness versus badness of that state, while arousal represents the activity of the organism associated with that state. Affect plays an important role in thinking. Normal affective functioning seems to be necessary for normal cognition [1]. In fact, many cognitive processes (attention, memory) are to some level influenced by affective states [2,3]. Acknowledging the need for affect in human decision making, we investigate how artificial affect can be used to control an artificial agent's equivalent for decision making, i.e., its action-selection mechanism.

Different learning tasks and often even different phases in a single task require different learning-parameters. Tuning these parameters manually is laborious and the algorithms currently available for automatic tuning are often task-specific or need several meta-parameters themselves. Better methods for regulating meta-parameters are needed. In this paper we focus on regulating exploration/exploitation behavior, an important outstanding problem.

J. Mira and J.R. Álvarez (Eds.): IWINAC 2007, Part II, LNCS 4528, pp. 501–510, 2007.

More specifc, we compare diferent methods that use artificial affect to control the greedyness versus randomness of action-selection, thereby influencing exploitation versus exploration respectively. Two existing affect-controlled action-selection strategies are implemented and their performance is compared with that of static action-selection strategies and a new affect-controlled strategy. These strategies are tested on two tasks in which an artificial agent has to find the optimal path to food in a maze. One task tests the agent's capacity to adapt to a change in its environment. The second task tests the agent's ability to escape a local optimum in order to find the global optimum. Affect is operationalized as a measure that keeps track of "how well the agent is doing compared to what it is used to". As such, in this paper we only model the valence part of affect. For a psychological grounding of artifical affect as used in this paper the reader is referred to [4].

In the next sections we first introduce the core components of our approach, after which we discuss our experimental results.

2 Learning, Action-Selection and Soar-RL

The agents in this project are Soar-RL agents. Soar (States, Operators and Results) is a production-rule based architecture [5] that enables the design of cognitive agents, and is used extensively in Cognitive Modelling. Soar agents solve problems based on a cyle of state-perception, proposal of actions, action-selection, and action execution.

Soar has recently been augmented with reinforcement learning. This new version, Soar-RL [6], uses Q-learning [7], which works by estimating the values of state-action pairs. The value $Q(s, a)$ is defined to be the expected discounted sum of future rewards obtained by taking action a from state s and following an optimal policy thereafter. Values learned for actions are called Q-values, and are learned by experience. From the current state s, an action a is selected. This yields an immediate reward r, and arrival at a next state s'. $Q(s, a)$ is updated by a value propagation function [7].

Important to our research is that in each cycle, Soar-RL uses a Boltzmann equation (see Eq. 1) to select an action from a set of possible actions.

$$P(a) = \frac{e^{Q_t(s,a)\cdot\beta}}{\sum_{b=1}^{n} e^{Q_t(s,b)\cdot\beta}} \tag{1}$$

$P(a)$ is the probability of action a being chosen. $Q_t(s, a)$ is the estimated value for performing action a in state s at time t. This equation returns the probability of action a being chosen out of the set of possible actions, based upon an agent's Q-values of those actions and upon the variable β, called *inverse temperature*. The β parameter controls the greediness of the Boltzmann probability distribution. If $\beta \Rightarrow \infty$, the system becomes deterministic and will select the action with the highest estimated value. If $\beta \Rightarrow 0$, each action has the same probability ($\frac{1}{n}$) of being selected. In other words, a low β corresponds to random behavior and a high β to greedy behavior. This β can be adjusted during simulations and is used by our adaptive strategies for regulating the agent's behavior.

Soar-RL creates new actions and updates the estimated values of existing actions as they are explored. Therefore, exploration is needed to construct a good model of the environment. When the agent's model of the environment (its state $\Rightarrow$ action mapping) is accurate enough, the agent should exploit its knowledge by performing greedy behavior in order to maximize the received rewards. Thus, a strategy balancing exploration/exploitation is needed in order to perform tasks efficiently. However, it is hard to decide when an agent should stop exploring and start exploiting, especially in dynamic environments.

3 Affect-Controlled Action-Selection

To address the aforementioned exploration/exploitation tradeoff problem, we investigate whether artificial affect can be used to control the β parameter. For more detail on the relation between artificial affect and natural affect see [4][8].

3.1 Schweighofer and Doya's Method

Schweighofer and Doya [9] proposed that emotion should not just be considered 'emergency behavioral routines', but a highly important component of learning: emotion can be considered a metalearning system. Doya argued that a mapping exists between RL parameters and neuromodulators [10]. The β parameter is regulated by a search strategy. A random amount of noise is added to the β and the newly obtained β is tested. If the resulting behavior proves to perform better, then the β is adjusted in the direction of the noise. If, for example, positive noise yields higher rewards, then the β is increased.

In Schweighofer and Doya's model (referred to as SD), β is governed by the following set of equations:

$$\beta(t) = e^{\kappa} + \sigma(t) \tag{2}$$

β is used in the Boltzmann distribution to determine the amount of exploration, σ is a Guassian noise source term with mean 0 and variance ν. A new noise value is drawn every N steps, with N $\gg$ 1.

$$\Delta\kappa = \mu \cdot (\bar{r}(t) - \bar{\bar{r}}(t)) \cdot \sigma(t-1) \tag{3}$$

$\bar{r}(t)$ is the short-term average reward and $\bar{\bar{r}}(t)$ is the long-term average reward. μ is a learning rate.

$$\Delta\bar{r}(t) = \frac{1}{\tau_1} \cdot (r(t) - \bar{r}(t-1)) \tag{4}$$

$$\Delta\bar{\bar{r}}(t) = \frac{1}{\tau_2} \cdot (\bar{r}(t) - \bar{\bar{r}}(t-1)) \tag{5}$$

τ_1 and τ_2 are time constants for respectively the short-term and the long-term. The term $(\bar{r}(t) - \bar{\bar{r}}(t))$ in equation 4 represents valence: positive when doing better than expected, negative when doing worse than expected.

3.2 Broekens and Verbeek's Method

Where Schweighofer and Doya use a search-based method, Broekens and Verbeek [8] use a direct relation between affect and exploration. High valence results in exploitation, while low valence leads to exploration. As a measure for valence, the difference between short and long term average rewards is used. This method (referred to as BV) is not a search algorithm and does not attempt to find an optimal value for β, but tries to balance exploration/exploitation by responding to changes in the environment.

The following equations are used to govern the agent's exploration behavior:

$$e_p = (\bar{r}(t) - (\bar{\bar{r}}\ (t) - f \cdot \sigma_{ltar}))/2 \cdot f \cdot \sigma_{ltar} \tag{6}$$

e_p is a measure for valence and σ_{ltar} is the long-term variance of $\bar{r}(t)$. $\bar{r}(t)$ and $\bar{\bar{r}}\ (t)$ are again short-term and long-term running averages and are computed in the same way as in SD.

$$\beta = e_p \cdot (\beta_{max} - \beta_{min}) + \beta_{min} \tag{7}$$

Thus, the agent's valence (on a scale from 0 to 1) directly translates to the amount of exploration/exploitation (on a scale from β_{min} to β_{max}).

3.3 Hybrid-χ^2 Method

Both methods described above have their own strengths and drawbacks. BV has the ability to respond quickly to sudden changes in the environment. However, it is bound to a fixed range of values, and β will always converge to the center of that range when the environment stabilizes. SD, on the other hand, is able to cope with a broader value range, and β converges to the optimal[1] value in this range. The downside of this method is that it has trouble responding to sudden changes (see results section). To overcome these drawbacks, we propose a new method, The Hybrid-χ^2 method. This method combines both methods and uses the environmental stability to balance the contribution of SD and BV to the actual β used. The more substantial the changes in the environment, the more influence BV has. In stable environments, SD gets most influence. A heuristic for detecting environmental change is the reward distribution. If we assume two equally long, consecutive reward histories, the difference in reward distribution between these histories is a measure for the stability of the environment. A large difference indicates substantial changes[2], and vice versa.

This difference is computed with the aid of the statistical χ^2 test. It measures whether two sets of numbers are significantly different, and returns a value between 0 (different) and 1 (equal). In our case, we compare the long term rewards

[1] Possibly a local optimum.

[2] Or a situation in which the agent is heavily exploring, in which case BV should also have the most influence.

with the short term rewards and compute the significance of the differences. Using χ^2, β is computed as follows:

$$\beta(t) = \chi^2 \cdot \beta_{SD} + (1 - \chi^2) \cdot \beta_{BV} \quad (8)$$

The value for β is restricted to the interval $[\beta_{min} \ldots \beta_{max}]$ because BV directly couples valence to the amount of exploration. Some flexibility was added to this interval by using the, β found by SD to influence β_{max}. If SD's proposed β is different from the current value, the maximum value of the range will be adjusted according to:

$$\beta_{max} = \beta_{max} + \chi^2 \cdot (\beta_{SD} - \beta(t-1)) \quad (9)$$

4 Experiments

We tested the methods described in the previous section on two different maze tasks. We focused on the resulting exploration/exploitation behavior of these methods, not exclusively on their learning performance. In addition to the affective methods, we tested Soar-RL's own performance for several fixed β's without the addition of affect-controlled action-selection. We refer to 'unaffected' Soar-RL as the *static method.*

In both tasks a Soar-RL agent needs to find food in a maze. The vision of the agent is limited to one tile in all directions. For both mazes, this vision is large enough to satisfy the Markov property[3]. The agent has to learn this behavior purely from the rewards it receives.

4.1 Cue-Inversion and Candy Task

In the maze of Figure 1a, the agent has to learn to go to A if the light is *on* and to B if the the light is *off*, using the shortest path. A reward of +1 is given if the agent reaches the correct goal state. -1 is the reward for going to the wrong goal state or for walking into a wall. The agent is set back to the starting position when an end-state has been reached.

After 3000 steps, after the agent has learned this maze quite well, we switch the rewards of the two goal states; so now the agent has to go to A when the light is *off* to get the +1 reward and to B when the light is *on*. The agent has to unlearn its previous behavior and learn a new behavior. This task is constructed to be similar to psychological cue inversion tasks (e.g., [11]) and tests the agent's capacity to adapt to a sudden change in the environment. One run equals 8000 steps in this maze.

Figure 1b shows the maze used for the Candy task (see also [4]). Close to the starting position of the agent is some candy (end state B). The agent receives a reward of 0.5 for grabbing this candy. A few steps further is some real food (A), rewarded with +5. The reward per step is higher for going after this food

[3] No knowledge of the history of states is required to determine the future behavior of the environment.

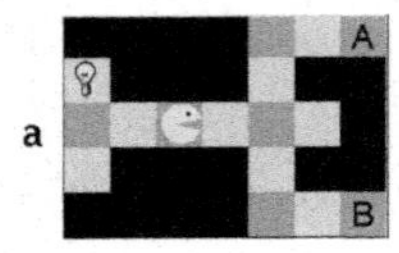

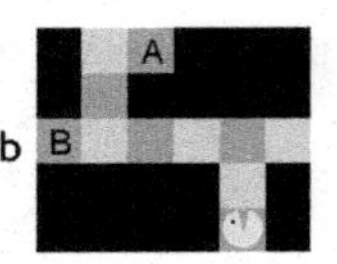

Fig. 1. a: Maze used in Cue-inversion task. b: Maze used in Candy task.

and it is the agent's goal to find this real food, instead of the candy. Whenever the agent finds the candy or the real food, it is set back to the starting position. This task tests the agent's ability to escape a local optimum. Contrary to the other task, this task does not have any switches, but it does however require flexible behavior: the agent needs to switch from initial exploration to exploiting the local optimum (the candy) and then revert to exploration to find the global optimum (the real food) and exploit that afterwards. In this task a run is aborted at t=25000. At this point it was clear how different methods perform.

4.2 Simulation Settings

Simulations are performed using a short term moving average of 50 rewards and a long term moving average of 400 short term averages. In initial experiments, these values proved to be useful for determining whether 'we are doing better than we used to do'. The two other parameters used by Q-learning (the learning rate (α) and a discount factor (γ)), are set at the default values of Soar-RL: $\alpha = 0.4$ and $\gamma = 0.9$. Results are aggregated over 50 runs. This proved to be enough to reproduce the same results when repeating the experiments. Every 30 steps, the average values of β and the received rewards are computed and printed.

5 Results

BV performs best on the cue inversion task (Figure 2). It is faster than the three other methods both for initial learning and for relearning after the cue inversion. The impact of the cue-inversion on β is visible in Figure 2B. The agent appears to benefit from the exploration phase after the inversion. SD performs poorly, even worse than the control method. SD does not seem to respond to the switch, but slowly increases β over time instead. This is an average β and the actual situation is somewhat more complex. We measured a high diversity in results: the standard deviation near the switch is 8.1, but even at step=6000 it is still 5.6. Data of individual runs seem to suggest that there are runs that increase rapidly to the maximum value of β=30 and there are some that cannot escape low β values. These high β runs do not compensate for the lower rewards obtained by low β runs.

The characteristics of the β curves of the Hybrid-χ^2 method are quite similar to those of BV: β quickly decreases after the inversion and it slowly converges to a center value. SD's influence in determining the β_{max} is not noticeable. All

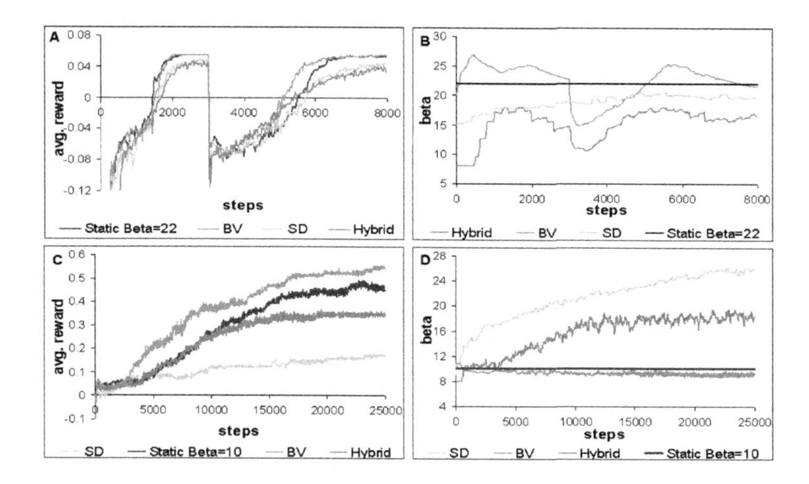

Fig. 2. Graphs A and C display the average reward per time step on the Cue-inversion task (A) and the Candy task (C). Graphs B and D show the corresponding value for β. The inversion occurs at t=3000 in the cue-inversion task.

runs seem to stay well within their initial ranges and no trend can be detected besides the one that pulls the β to the center value.

It is clear from Figure 2C and 2D that SD performs poorly on the Candy task. The β increases when the agent encounters the candy and, as a result, the agent exploits the candy afterwards.

BV again clearly shows that it cannot search for an optimal β, it merely responds to the rewards received: after an initial exploitation burst, β converges to the center of the β range. Although the optimal center value for BV is close to the optimal value (β=10) for the static method, BV's 'noisy' β value performs slightly better than the static β value.

Again, BV's component in the hybrid method has the upper hand. The β's are drawn to the center value (β=15). It is not fair to compare the rewards of the Hybrid-χ^2 method directly with those of BV, as the center values for the hybrid runs were set higher than those of BV's simulations. As mentioned above, SD cannot compensate by decreasing β, as SD mainly attempts to exploit the candy by increasing the β even further.

6 Discussion and Future Work

For an elaborate discussion of related work (e.g., the work on Soar and emotion by Marinier [12], and the work on affect and problem solving by Belavkin [13]) we would like to refer the reader to [14]. In this paper, we concentrate on comparing the three methods discussed above. For more detail on BV, we refer to [4,8].

SD performed poorly on the given tasks. In the Candy task this is probably due to the agent's quick exploitation. To exploit the local optimum, a very high β is beneficial. The random noise used as seed for the search process is not enough to trigger exploration. A similar problem can occur with a very low β: an agent can get trapped in explorative behavior when changes to an already low β do not change the agent's behavior enough to lead to significantly different rewards. Further increasing the noise, to counter these problems, would change SD into a random search method, which is not the intention of SD.

On the switch task, SD performed worse than expected, given that successful use was reported on another task [10]. In that task, an agent repeatedly reacts to a stimulus by pressing one of two buttons, one leading to a small or large positive reward, the other to a small or large loss. The problem is such that the optimal behavior is to take a number of small losses but receive a large reward later. In contrast taking a number of small rewards in the beginning results in a large loss in the end. After a number of trials, the rewards are changed and the agent has to switch from a 2-step planning situation to one where it must loose 7 times to get the large reward at step 8. On this task, they found a strong response to the switch and noticed the agent relearning the new scenario quickly, whereas our results do not show a clear response to a change in the enviroment. A possible explanation for this is that in Schweighofer and Doya's scenario, there are no walls or other none-goal states that give big negative rewards. The behavior the agent showed before the switch is the worst possible thing to do after the switch and thus exploration will be beneficial to the agent in terms of reward, i.e., exploration is encouraged. In contrast, in our task higher exploration also results in more bumping into a wall and can therefore be discouraged by SD.

SD seems to be more suitable for fine-tuning parameters in a stable environment, but its use as an adaptive control method is limited.

In the switch task, the β guided by BV improves learning performance, and reacts to the switch. It qualifies for a useful affect-based method for steering the β parameter. The opposite is true for the β in the Candy task. The method guides β only when the agent is learning to find the candy. After that, the β gets very noisy. The β fluctuates around its center value, only limited by the value range imposed on β. Granted, learning performance is better than the best static method, but not due to meaningfully balancing exploration and exploitation, as is the method's goal. As such, it does not qualify for a good adaptive method in Candy-like tasks. The main cause for the instability is the standard deviation of the long term average reward (σ_{ltar}), which is meant to normalize valence, but when the environment stabilizes, this term approaches 0. As a result, all minor differences between $\bar{\bar{r}}$ and $\bar{r}$ are magnified to extreme values. A possible solution to this problem is the introduction of some noise in the reward history, such that σ_{ltar} never approaches 0.

It is clear that the hybrid method does not provide the desired behavior. The influence of BV on SD's measurements is probably one of the main reasons for this. SD measures the influence of the added noise over a long period and then checks whether the added noise was beneficial to the agent's performance. With BV influencing the β and thus the performance within this period, SD's method cannot accurately measure the influence of the noise. The result is that the two methods are in each others' way instead of cooperating. Updating BV's β only when SD's β is updated (i.e., *not* influence β during SD's evaluation of its effect on reward) is not an option; it dramatically reduces BV's capacity to quickly react to changes in the environment.

Another important reason for the failing of the hybrid method is the incorrect assumption that BV stabilizes in a static environment. Because of large fluctuations of BV, the two reward histories keep having different reward distributions, and thus SD will never have much influence. It would be interesting to test the performance of the hybrid method with the addition of noise as describe above.

As Soar-RL agents attempt to maximize reward over time, our performance measure is the average reward the agent received per step: the agent's *Quality of Life (QoL)*. Nonetheless, it would be interesting to also measure the 'time to goal' to obtain more information on the agent's actual behavior.

We only focussed on regulating exploration/exploitation through the β parameter. The methods in this paper might also be used to control other RL parameters ([10] used their method to control α and γ as well). The χ^2 test in the hybrid method could be used to control, e.g., α.

Better hybrid methods might be constructed by merging a search strategy and a directional approach by another balancing algorithm (instead of χ^2). An alternative would be to let one strategy search the window in which parameters can vary, while the second determines the exact position within this window.

7 Conclusions

We conclude that Soar-RL agents can use artificial affect to directly control the amount of exploration [4] by coupling valence to the β in the Boltzmann distribution used in action-selection. Experiments show that such agents adapt better to a sudden change in the environment, as compared to agents that use an amount of exploration that is either static or determined by an affect-based search strategy (e.g., as used in [10]). These results are compatible with those presented in [4], indicating that this specific benefit of affective control of exploration is not tied to a particular learning architecture.

On the other hand, affect-based search enables convergence of this β to a (local) optimal value, while the directed affect-based control method converges to a "middle" β value in a range of values. This requires careful setting of the value range. The downside of the affect-based search method is that it cannot guarantee to escape a local optimum (as in the Candy task), and that it cannot react to sudden changes in the environment.

We conclude that the affect-based search method examined in this paper is suitable for fine-tuning parameters in static environments, but that its use as an adaptive control method is limited. We further conclude that directed exploration-control is suitable for fast reaction to changes in the environment, but has little use in static environments. Currently, our hybrid method that attempts to merge positive elements of both does not behave as intended.

Artificial affect can make a useful contribution to controlling an agent's exploration/exploitation, but there is much work to be done and a better understanding of the interplay between human learning and affect is needed.

Acknowledgments

We wish to thank S. Nason and J. Laird for providing us with the experimental Soar-RL software. This research is supported by the Netherlands Organization for Scientific Research (NWO, grant 612.066.408).

References

1. Damasio, A.R.: Descartes' error: Emotion, reason, and the human brain. Gosset/Putnam Press, New York (1994)
2. Craig, S.D., Graesser, A.C., Sullins, J., Gholson, B.: Affect and learning: An exploratory look into the role of affect in learning with autotutor. Journal of Educational Media **29** (2004) 241–250
3. Ashby, F.G., Isen, A.M., Turken, U.: A neuropsychological theory of positive affect and its influence on cognition. Psychological Review **106** (1999) 529–550
4. Broekens, J., Kosters, W.A., Verbeek, F.J.: On affect and self-adaptation: Potential benefits of valence-controlled action-selection, submitted to IWINAC'2007. (2007)
5. Newell, A.: Unified Theories of Cognition. Harvard University Press, Cambridge, Massachusetts (1990)
6. Nason, S., Laird, J.: Soar-RL, integrating reinforcement learning with Soar. Cognitive Systems Research **6** (2005) 51–59
7. Sutton, R., Barto, A.: Reinforcement learning, an introduction. MIT Press, Cambridge, Massachusetts (1998)
8. Broekens, J., Verbeek, F.J.: Simulation, emotion and information processing: Computational investigations of regulative role of pleasure in adaptive behaviour. In: Proceedings of the Workshop on Modeling Natural Action Selection, Edinburgh, UK (2005) 166–173
9. Doya, K.: Metalearning, neuromodulation and emotion. Affective Minds **1** (2000) 101–104
10. Schweighofer, N., Doya, K.: Meta-learning in reinforcement learning. Neural Networks **16** (2003) 5–9
11. Dreisbach, G., Goschke, T.: How positive affect modulates cognitive control: Reduced perseveration at the cost of increased distractibility. Journal of Experimental Psychology **30** (2004) 343–353
12. Marinier, R., Laird, J.: Towards a comprehensive computational model of emotion and feelings. In: Proceedings of the Sixth International Conference on Cognitive Modeling. (2004) 172–177
13. Belavkin, R.V.: On relation between emotion and entropy. In: Proceedings of the AISB'04 Symposium on Emotion, Cognition and Affective Computing, Leeds, UK (2004) 1–8
14. Hogewoning, E.: Strategies for affect-controlled action-selection in soar-rl. Technical Report 07-01, Leiden Institute of Advanced Computer Science, Leiden University (2007)

An Agent-Based Decision Support System for Ecological-Medical Situation Analysis

Marina V. Sokolova[1,2] and Antonio Fernández-Caballero[1]

[1] Universidad de Castilla-La Mancha, Escuela Politécnica Superior de Albacete & Instituto de Investigación en Informática de Albacete, 02071-Albacete, Spain
{marina,caballer}@dsi.uclm.es
[2] Kursk State Technical University, Kursk, ul.50 Let Oktyabrya, 305040, Russia

Abstract. This paper presents an architecture of an agent-based decision support system (ADSS) for ecological-medical situation assessment. The system receives statistical information in form of direct and indirect pollution indicator values. The ultimate goal of the modeled multi-agent system (MAS) is to evaluate the impact of the exposure to pollutants in population health. The proposed ADSS interacts with humans in real-time "what-if" scenarios, providing the user with evidence for optimal decision making. A detailed description of all the agents and their BDI (beliefs, desires, intentions) cards is presented.

1 Introduction

During our lives we are open to environmental air, water, food and soil contaminants and pay progress with our health. Our research serves to the purpose of the evaluation of population exposures caused by indoor and outdoor pollution. The population exposures permit assessing the impact of environmental pollution into population morbidity in general and the contribution of each contaminant in particular by means of an agent-based decision support system (ADSS). Different research institutes and organizations confirm the ecological contribution to illness and death in adults and children [1]. The World Health Organization (WHO) declared the necessity of broader analysis of environmental risk factors over public health, with a strong emphasis on mortality [2].

To date there are some known applications of multi-agent systems (MAS) for situation assessment, including environment and public health. For example, in a recent work [3] the agent-based model for estimating residential water demand was described; the agent-based approach was used to create a system as support in clinical management [4]; data mining techniques for knowledge discovery within MAS was used for early diagnostics to early intervention to developmentally-delayed children [5]. In [6] the MAS paradigm was introduced with the objective of revealing correlations between human health and environmental stress factors (traffic activity, meteorological data, and noise monitoring information) using a wide range of data mining (DM) methods, including regression analysis, neural networks, ANOVA, and others.

J. Mira and J.R. Álvarez (Eds.): IWINAC 2007, Part II, LNCS 4528, pp. 511–520, 2007.

2 The ADSS Functionality and Design Characteristics

We suggest implementing an agent-oriented software system, dedicated to ecological / medical situation estimation. The system receives retrospective statistical information in form of direct indicator values - water pollution, solar radiation - and in form of indirect indicator values - types and number of vehicles used, energy used annually and energy conserved, types and quantity of used fuel, etc. [7]. The indirect indicators are utilized in accordance with ISO 14031 "Environmental Performance Evaluation" standard in order to estimate air and soil pollution [8]. The population exposure is registered as number of morbidity cases, with respect to International Statistical Classification of Diseases and Related Health Problems, 10th review (ICD-10) [9].

The system is aimed to fulfill the enumeration of general goals, subdivided into plans and actions, in MAS terminology, as shown in Fig. 1. This enumeration includes a set of scenarios starting from meta-ontology creation and finishing

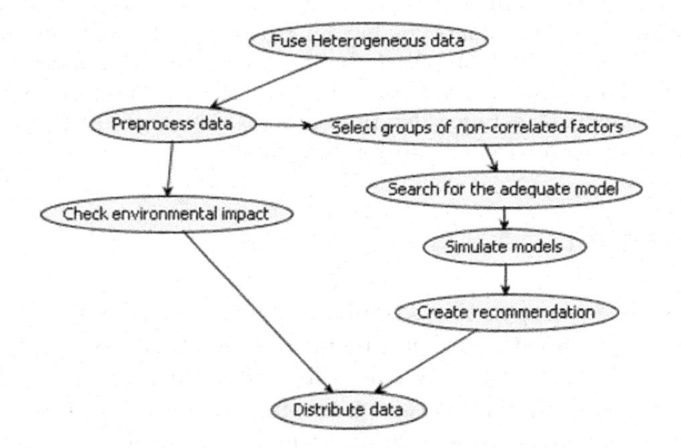

Fig. 1. The ADSS Goals Overview diagram

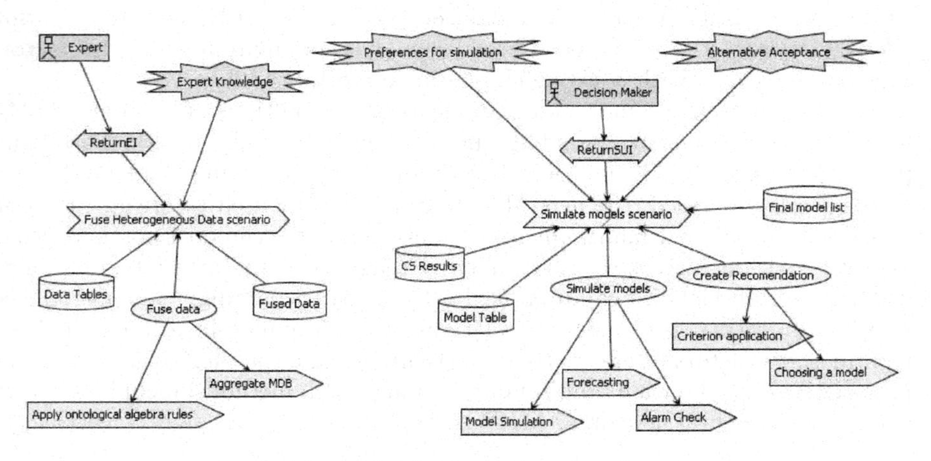

Fig. 2. The ADSS Analysis Overview diagram

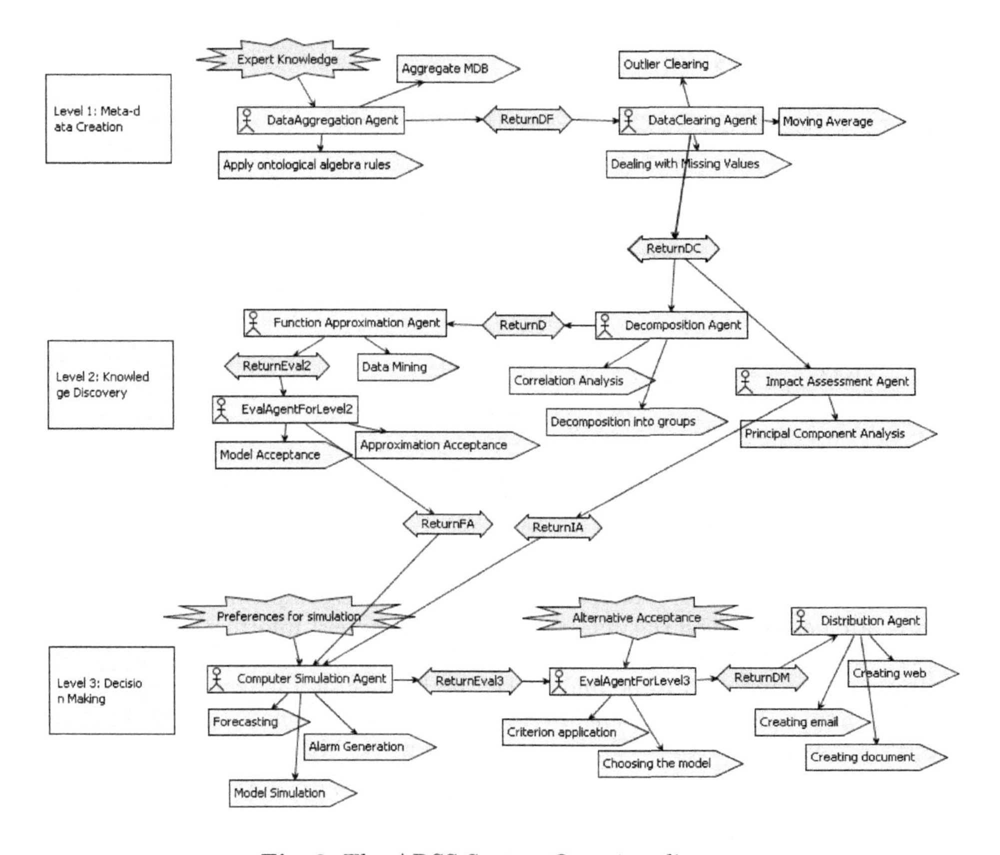

Fig. 3. The ADSS System Overview diagram

with decision making and data distribution. The ADSS design has entirely been modeled by means of *Prometheus* methodology [10] through *Prometheus Design Tool* [11]. The tool provides the possibility of checking the consistency of the created system and of generating a skeleton code for *JACK Intelligent Agents* [12] development tool, as well as design reports in HTML.

The process of creatiing a multi-system architecture, in accordance with the *Prometheus* methodology, consists of three phases, which are:

1. *System specification*, aimed to the identification of multi-agent system entities, such as actors, system goals, scenarios, actions, percepts and roles.
2. *High-level (architectural) design*, which is centered in the description of agent-role coupling, general system structure and interaction protocols.
3. *Detailed design*, in which each agent is described in detail in terms of capabilities, events, plans and data.

The goals drawn in Fig. 1 repeat the main points of a traditional decision-making process, which includes the following steps: (1) problem definition, (2)

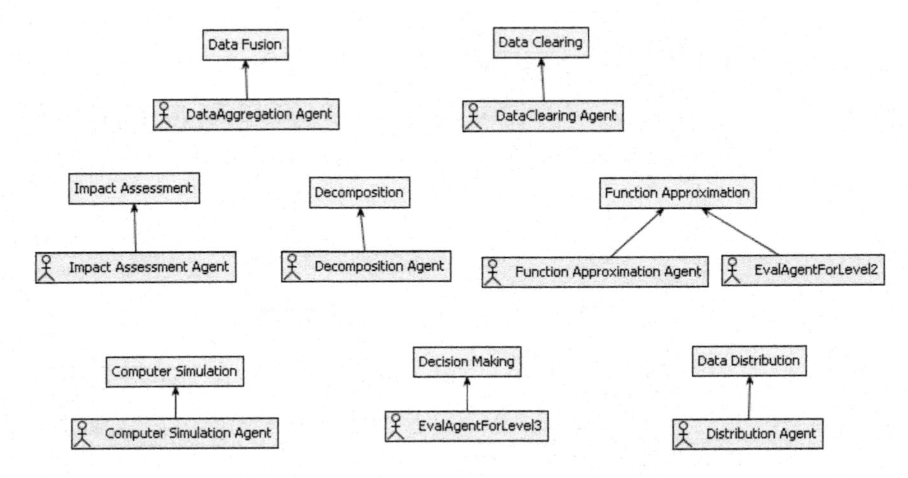

Fig. 4. The ADSS Agent-Roles Coupling diagram

information gathering, (3) alternative actions identification, (4) alternatives evaluation, (5) best alternative selection, and, (6) alternative implementation. Phases 1 and 2 are performed on the initial step, when the expert information and initial retrospective data is gathered. Phases 3 to 5 are solved by means of the ADSS, and phase 6 is supposed to be undertaken by the decision maker.

Being implemented by means of the *Prometheus Design Tool*, the Analysis Overview Diagram of the ADSS enables seeing the high-level view composed of external actors, key scenarios and actions (see Fig. 2). The proposed ADSS presupposes communication with two actors. One actor is named as "Expert" and it embodies the external entity which possesses the information about the problem area - in more detail, it includes the knowledge of the domain of interest represented as an ontology - and delivers it through protocol *ReturnEI* to the ADSS. The second actor, named "Decision Maker", is involved in an interactive process of decision making and choosing the optimal alternative. This actor communicates with agents by message passing through protocol *ReturnSUI*, stating the model, simulation values, prediction periods, levels of variable change, etc. It accepts the best alternative in accordance with its believes and the ADSS recommendation model.

The initial analysis of the system results in obtaining and describing the system roles and protocols. The architecture of the ADSS is offered in Fig. 3. There, the proposed system is logically and functionally divided into three layers; the first is dedicated to meta-data creation (information fusion), the second is aimed to knowledge discovery (data mining), and the third layer provides real-time generation of alternative scenarios for decision making. Fig. 3 also provides a look on connections between agents with correspondent interactions and undertaken sets of actions. This view gives a sufficient understanding of the entire system design.

The system resembles a typical organizational structure, as the agents are strictly dedicated to work with the stated sets of data sources, solve the particular tasks and are triggered when all the necessary conditions are fulfilled and there are positive messages from previously executed agents [13]. The system includes a set of roles, correlated with the main system functions and a set of agents related to each role (see Fig. 4). Actually, mostly every agent is associated to one role; only in case of "Function Approximation" role, there are two agents, one for data mining, and the other one for validation.

3 Description of the ADSS Agents

Each type of agent has its own sphere of awareness (competence), belonging to some agent type. Agents possess different characteristics and use various techniques, as summarized in detail in Table 1, where agent types are given in columns, and the tasks they solve are labeled by a "+" sign. We have accepted the following abbreviations: DBA - *DB Handling agents*, AA - *Analysis agents*, EA - *Evaluation agents*, SA - *Simulation agents*, DA - *Distribution agents*. The types of agents are described in detail in the following section of the article. We also represent for every agent its BDI (beliefs, desires, intentions) card, which was adapted from [14].

Table 1. Task Delegation in the ADSS

Level	Role	Task	DBA	AA	EA	SA	DA
Level 1	Data Fusion	DB Transformation	+	-	-	-	-
		Meta-data Fusion	+	-	-	-	-
	Data Clearing	Anomalies Elimination	+	-	-	-	-
		Missing Values Treatment	+	-	-	-	-
		Smoothing	+	-	-	-	-
Level 2	Decomposition	Correlation Analysis	-	+	-	-	-
		Subsets Selection	-	+	-	-	-
	Function Approximation	Data Mining	-	+	-	-	-
		Model Evaluation	-	-	+	-	-
		Approximation Acceptance	-	-	+	-	-
	Impact Assessment	Principal Component Analysis	-	+	-	-	-
Level 3	Computer Simulation	Simulation	-	-	-	+	-
		Forecasting	-	-	-	+	-
		Alarm Check	-	-	-	+	-
	Decision Making	Criterion Calculation	-	-	+	-	-
		Model Selection	-	-	+	-	-
	Distribution	Creation of Documents	-	-	-	-	+
		Creation of Web Source	-	-	-	-	+
		Creation of E-mails	-	-	-	-	+

3.1 DB Handling Agent Type

The *DB Handling agent* type contains two agents, the *Data Aggregation agent* and the *Data Clearing agent*. The difference between their competence areas has been introduced in their names; the first one is oriented to data fusion, and the second one is focused on data preprocessing.

Table 2. BDI card for the *Data Aggregation agent*

Desires	Create
Pre-condition	Expert information is received
Beliefs	Problem ontology
Post-condition	Termination message
Collaborator	-
Intentions	Search data in data sources according to ontology
	Fuse data into meta-schema
	Send a termination message to the *Data Clearing agent*
Type	*DB Handling agent*

Table 3. BDI card for the *Data Clearing agent*

Desires	Replace data
Pre-condition	Message from *Data Aggregation agent*
Beliefs	Meta-DB
Post-condition	Termination message
Collaborator	-
Intentions	Eliminate outliers and anomalies
	Treat missing values
	Smooth time series
	Send termination message
Type	*DB Handling agent*

The Data Aggregation agent. The *Data Aggregation agent* (DAA) is responsible for initial information reception and meta-data creation. The agent plays an essential role as it has to realize the plan of important tasks, including: (1) reception of expert information and initial data, (2) scanning of data sources for selecting the values related to identical hierarchical and semantical levels of the meta-ontology, (3) pooling homogeneous data into a meta-data structure. The DAA communicates with the *Data Clearing agent* through communication protocol *ReturnDF*, sending a message that confirms meta-data creation and triggers data preprocessing.

The Data Clearing agent. The *Data Clearing agent* (DCA) performs all data preprocessing procedures, including outliers and anomalies detection, dealing with missing values, smoothing, normalization, etc. The final meta-data consists of sequences of ordered indicator values measured at equal time intervals (time-series).

3.2 Analysis Agent Type

The Impact Assessment agent. The single agent that belongs to this type, *Impact Assessment agent* (IAA), solves the principal component analysis (PCA) procedure for every type of disease and the totality of pollutant indicators. The outcomes are interpreted as the determination of the most influencing subsets of pollutants for every nosology. It is message-triggered from the DCA and after being executed it creates a file with the results of its work. The IAA delivers messages to *Simulation agent* through protocol *ReturnIA*.

The Decomposition agent. The main task of the *Decomposition agent* (DA) is to separate the totality of factors (pollution indicators) and output variables

Table 4. BDI card for the *Impact Assessment agent*

Desires	Execute
Pre-condition	Message from *Data Clearing agent*
Beliefs	Meta-DB
Post-condition	Termination message
Collaborator	-
Intentions	Execute PCA procedure
Type	*Analysis agent*

Table 5. BDI card for the *Decomposition agent*

Desires	Execute
Pre-condition	Message from *Data Clearing agent*
Beliefs	Meta-DB
Post-condition	Termination message
Collaborator	-
Intentions	Executes procedures of correlation analysis
	Select subsets for every output variable
	Send termination message
Type	*Analysis agent*

(morbidity) into subsets (commonly intersected) with respect to every output variable (nosology). Decomposition is aimed to select for every output variable those factors that (1) do not correlate significantly with the output variable - in order to avoid inter-correlation, or, (2) do not correlate significantly between themselves - in order to avoid multi-colinearity. The DA creates a correlation matrix and analysis it. The *Decomposition agent* creates an output source used then by *Simulation agent* and *Distribution agent.*

The Function Approximation agent. The *Function Approximation agent* (FAA) carries out a core ability: it explores unknown relationships, trends and connections in data, received from the DA agent. The FAA is competent in applying data mining (DM) techniques, including theory-driven methods (linear, non linear regression, ANOVA, etc.) as well as data-driven methods (neural networks, decision trees, rules, etc.). Taking into account the nature of metadata base, we state the following requirements to DM algorithms: robustness, ability to work with missing data values, and ability to deal with short data sets (as data are usually being registered once a year; so, in better case we can operate with data sets including up to 50 values). As DM techniques satisfy these requirements, they will be used.

3.3 Evaluation Agent Type

As shown before in Table 1, the *Evaluation agent* type is involved in two roles: "Function Approximation" and "Decision Making", which also belong to different levels. In role "Function Approximation" the *Evaluation agent* is thought to calculate statistical performance characteristics for the model, created by the FAA, and the model's approximation abilities. To do it, we approximate initial data by the models, compare the real and the approximated data sets and evaluate the

Table 6. BDI card for the *Function Approximation agent*

Desires	Execute
Pre-condition	Message from *Decomposition agent*
Beliefs	Meta-DB, Grouping Table
Post-condition	Termination message
Collaborator	-
Intentions	Apply Data Mining technique
	Send a message to *Evaluation agent*
Type	*Analysis agent*

Table 7. BDI card for the *Evaluation agent for level 2*

Desires	Execute
Pre-condition	Message from the *Function Approximation agent*
Beliefs	Meta-DB, Models table
Post-condition	Termination message
Collaborator	-
Intentions	Model evaluation
	Model approximation acceptance
	Send a termination message
Type	*Evaluation agent*

Table 8. BDI card for the *Evaluation agent for level 3*

Desires	Execute
Pre-condition	Message from the *Function Approximation agent*
Beliefs	CS results table, Final model list
Post-condition	Termination message
Collaborator	*Computer Simulation agent*
Intentions	Game theory criteria application
	Best model selection
	Send a termination message
Type	*Evaluation agent*

performance statistics. In role "Decision Making" the *Evaluation agent* applies decision making theory criteria (Bayes, minimax, Hurwicz, and so on).

3.4 Simulation Agent Type

The Computer Simulation agent. Actually, there is one agent that belongs to this type, namely the *Computer Simulation agent* (CSA), which works with models created by the FAA and accepted by an *Evaluation agent.* The simulation process involves the interaction with the user. During this process the CSA learns the user´s preferences: the model to be simulated, the factors the user is interested in analyzing, the forecast period, etc. The CSA simulates a model within capability "Simulation", compares the outputs with the permitted and hazardous levels within capability "Alarm Check" and shows the results to the user in order to be revised and accepted, or refined and repeated. This creates a decision support system, which helps the user in decision making by providing several alternatives.

Table 9. BDI card for the *Computer Simulation agent*

Desires	Execute
Pre-condition	Message from the *Evaluation agent for level 2* and from *Impact Assessment agent*
Beliefs	Models table, Alarm levels
Post-condition	Termination message
Collaborator	-
Intentions	Simulation
	Checking the permitted levels
	Send a termination message
Type	*Simulation agent*

3.5 Distribution Agent Type

The *Data Distribution agent* (DDA) emphasizes in data visualization and delivering to the user. The revealed information is transformed by the DDA into multiple understandable forms. These tasks are performed by the *Data Distribution agent*, which operates by combining textual and graphical descriptions of recommendations. This information jointly with computer simulation results for chosen models (including forecast and alarm check) are delivered. This includes impact assessment outcomes (which discover the exposure of urban pollution on the morbidity of people), correlation results (which reveal dependencies between pollutants and different types of diseases), and, optionally, those models recommended by the system to be accepted as optimal for regional development and situation correction.

Table 10. BDI card for the *Data Distribution agent*

Desires	Create
Pre-condition	Message from *Evaluation agent for level 3*
Beliefs	Final model list, Correlation table, PCA results
Post-condition	Termination message
Collaborator	-
Intentions	Create textual document Create web-source Create and send e-mail
Type	*Distribution agent*

4 Conclusions

The agent-based decision support system (ADSS) introduced in this paper provides all the necessary steps for standard managerial decision procedure by utilizing intelligent agents. The levels of the system architecture, logically and functionally connected, have been presented. Real-time interaction with the user provides a range of possibilities in choosing one course of action from among several alternatives, which are generated by a system through directed data mining and computer simulation. The system is aimed to regular usage for adequate and effective management by responsible municipal and state government authorities.

The proposed ADSS architecture may be generalized and applied in closely related areas, such as economics, sociology, public health, etc. Some specific DM techniques, for example, econometric system dynamic models, can be used in these kinds of applications.

Acknowledgements

Marina V. Sokolova is the recipient of a Postdoctoral Scholarship (Becas MAE) awarded by the Agencia Española de Cooperación Internacional of the Spanish Ministerio de Asuntos Exteriores y de Cooperación.

References

1. Kaiser R., Romieu I., Medina S., Schwartz J., Krzyzanowski M., Künzli N. (2004) Air pollution attributable postneonatal infant mortality in U.S. metropolitan areas: A risk assessment study. Environmental Health: A Global Access Science Source, vol.3, no. 1, p. 4.
2. World Health Organization. http://www.who.int/en/
3. Athanasiadis, I.N., Mentes, A.K., Mitkas, P.A., Mylopoulos, Y.A. (2005). A hybrid agent-based model for estimating residential water demand. Simulation, vol. 81, no. 3, pp. 175-187.
4. Foster D., McGregor C., Samir E.S. (2006) A survey of agent-based intelligent decision support systems to support clinical management and research. http://www.diee.unica.it/biomed05/pdf/W22-102.pdf.
5. Chang, C.L. (2006). A study of applying data mining to early intervention for developmentally-delayed children. Expert Systems with Applications. In press.
6. Chen, H., Bell, M. (2002). Instrumented city database analysts using multi-agents. Transportation Research, Part C, vol. 10, pp. 419-432.
7. Sokolova, M.V., Fernández-Caballero, A. (2007). A multi-agent architecture for environmental impact assessment: Information fusion, data mining and decision making. 9th International Conference on Enterprise Information Systems, ICEIS 2007.
8. ISO 14031:1999. Environmental management - Environmental performance - Guidelines. http://www.iso.org/
9. International Classification of Diseases (ICD). http://www.who.int/classifications/icd/en/
10. Padgham, L., Winikoff, M. (2004). Developing Intelligent Agent Systems: A Practical Guide. John Wiley and Sons.
11. Prometheus Design Tool (PDT). http://www.cs.rmit.edu.au/agents/pdt/
12. Jack Intelligent Agents. http://www.agent-software.com/shared/home/
13. Weiss, G. (2000). Multiagent Systems: A Modern Approach to Distributed Artificial Intelligence. The MIT Press.
14. Jo, C.H., Einhorn, J.M. (2005) A BDI agent-based software process. Journal of Object Technology, vol. 4, no. 9, pp. 101-121.

A Meta-ontological Framework for Multi-agent Systems Design

Marina V. Sokolova[1,2] and Antonio Fernández-Caballero[1]

[1] University of Castilla-La Mancha, Polytechnical Superior School of Albacete, Campus Universitario s/n, 02071-Albacete, Spain
{marina,caballer}@dsi.uclm.es
[2] Kursk State Technical University, Kursk, ul.50 Let Oktyabrya, 305040, Russia

Abstract. The paper introduces an approach to using a meta-ontology framework for complex multi-agent systems design, and illustrates it in an application related to ecological-medical issues. The described shared ontology is pooled from private sub-ontologies, which represent a problem area ontology, an agent ontology, a task ontology, an ontology of interactions, and the multi-agent system architecture ontology.

1 Introduction

The aim of this work is to apply an ontological analysis for complex situation descriptions which should be modelled as multi-agent systems (MAS) in a natural manner. The example used is that of a complex problem that includes a set of factors, which are grouped into indicators of environmental issues in a city caused by traffic, industrial activity and man-made sphere pollution, and indicators of public health represented through morbidity. In spite of numerous and interesting works in this sphere, this problem continues to be actual, as new techniques and tools that can be utilized for the examination appear, and new facts and knowledge about the problem are discovered.

Therefore, the main practical objective of the paper is the creation of an agent-based system for state situation assessment, monitoring the environment pollution and following the corresponding changes in public health, and generating a set of alternatives for successful and sustainable situation management. The MAS paradigm helps reducing the complexity of such a system and pooling the optimal solutions produced by autonomous and semiautonomous entities (agents) into a general strategy or plan [1]-[2].

When analyzing the problem of environmental impact upon population health [3], we come to the conclusion that this task has to be seen in relation to the studied region as a root stable entity. The region, in turn, is characterized by some environmental situation (resulted from industrial development, transport activity, water, air and soil pollution, etc.) and some health indicators, which are represented by morbidity and the number of diseased people grouped into classes endogeneous and exogeneous diseases.

The main step in organizing the terminological and informational foundation for further analysis and usage (MAS creation, simulation, alarm awareness, etc.)

J. Mira and J.R. Álvarez (Eds.): IWINAC 2007, Part II, LNCS 4528, pp. 521–530, 2007.

supposes describing a distributed meta-ontology framework. This step is a basic one and states the initial quality of further study processes and correct treatment of the concepts.

According to Guarino et al. [4], an ontology can be understood as an intentional semantic structure which encodes the implicit rules constraining the structure of a piece of reality. There are a number of approaches to ontology creation, mostly induced by the specificity of the domain of interest and the nature of the tasks to solve (e.g. [5]), from which we can induce and convert to our aims an algorithm of distributed ontology creation:

1. Situation description in natural language.
2. Vocabulary creation (extraction of concepts describing the situation).
3. Taxonomy creation.
4. Distributed meta-ontology structure creation.
5. Domain of interest ontology statement.
6. Description of tasks to solve and creation of the respective private ontology.
7. Description of MAS roles, agents and creation of the system architecture ontology.
8. Description of agent ontology.
9. Agent environment ontology statement by specifying interaction and communication protocols.
10. Ontologies mapping.
11. Data Bases filling for a MAS.
12. Data Sources delivering to agents.

To look briefly the steps of a given algorithms, it is worth noting that step 1 - problem description - serves for a better understanding the aims of the research and structure of the functionality of the situation. This initial analysis helps defining concretely the problem at hand and recovering the concepts, their characteristics and relations to examine. On this stage expert information, which is supplemented by statistical data and multimedia references related to the problem, is used. The consequentially following task (2) is the creation of a vocabulary, which includes the necessary and sufficient information about the concepts. The further step 3 consists in adding a set of relations (including hierarchical ones) between the concepts to a vocabulary, which results into a taxonomy. As in our work we use the inductive method of ontology creation, then, on step 4 we determine the general structure of the meta-ontology and extract the main functionally and semantically separated components. On steps 5 to 8 we create private ontologies for the extracted components of the meta-ontology, namely domain of interest, MAS architecture, tasks, agents and interactions. At steps 9 and 10 the private ontologies are mapped together. Finally, we fill data bases for a MAS (11) and deliver the real data to agents (12). In the following part of the article the distributed meta-ontology and the private ontologies, as well as the mapping procedure, are described in detail.

2 Description of the Ontological Basis of the Multi-agent Architecture

It is well-known that ontology creation is based on expert knowledge about the problem area and the developer's experience and understanding. Though, the more generalized approach implies extracting the main group of concepts, semantically and functionally connected. As it is shown in [6], a typical ontology for a MAS includes the following models: domain of interest, aims and tasks, agents, interaction, and environment. Having accepted this model of distributed meta-ontology creation, the structure shown in Fig. 1 is proposed as a framework for meta-ontological MAS design:

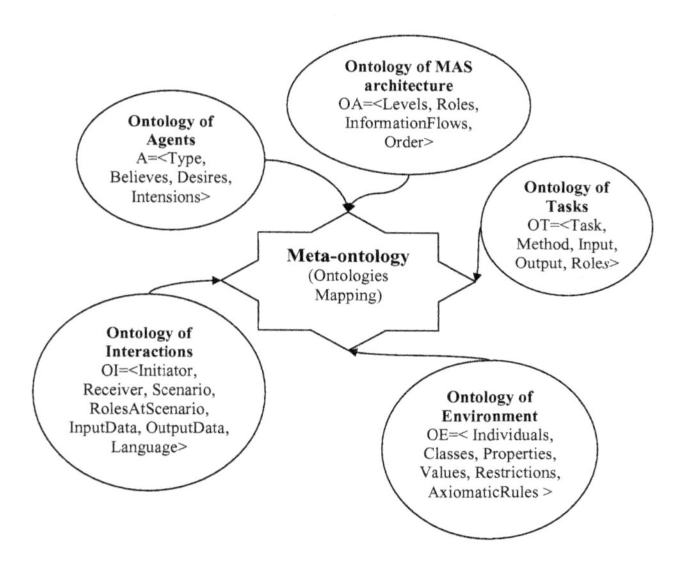

Fig. 1. The components of the distributed meta-ontology

This meta-ontology model specifies the private ontologies and gives opportunities to generalize knowledge about the MAS and the problem area. In the following subsections the focus is set on the components of the proposed meta-ontology.

2.1 The Domain of Interest Ontology

If defining the ontology O in terms of algebraic system, we have the following three attributes:

$$O = (C, R, \Omega) \tag{1}$$

where C is a set of concepts, R a set of relations between the concepts, and Ω a set of rules. Formula (1) proposes that the ontology for the domain of interest (or the problem ontology) may be describe by offering proper meanings to C, R and Ω.

As we have used the ontology editing software Protégé [7], after widening the formula for our system, we get the following specialization for equation (1) :

$$OE =< Individuals, Classes, Properties, Values, Restrictions, AxiomaticRules > \quad (2)$$

Individuals are entities representing regions under study. These have certain mortality and pollution levels; concretely, in our current research, the possibilities for region are Spain, Castilla-La Mancha, Albacete, etc.

Classes are interpreted as "sets containing individuals", and are organized in a taxonomy in accordance with the hierarchical superclass-subclass relations.

Properties are binary relations on individuals, which enable asserting facts about classes and individuals and can be functional, inverse functional, symmetric, or transitive. The properties are used in restrictions and in axioms. *Values* contains the values that can be assigned to individuals. *Restrictions* state the permitted and extreme ranges. Generally speaking, *Restrictions* impose constraints on the properties of the classes. *AxiomaticRules* use restrictions, boolean algebra and some other concepts such as general classes to create properties and class axioms.

Let us take a look at the *Classes* with respect to our domain of interest. The general illustration to our understanding of the domain of interest includes regions, which are characterized with some environmental pollution and human health level. The accent is made on regions, which are represented by instances as Toledo, Albacete, etc. The ontology for the dimension "Regions, pollution, health population indicators", as created in Protégé 3.2, is represented in Fig. 1 and described as:

$$Morbidity = M^t = < m_j, m_{j,k}^{t,g,ag} >$$

where m is the set of nosologies, $m \in M$, t represents the time of registration (year), $j = 1 \dots |M|$, k stands for a general class of disease (endogeneous or exogeneous) $k = 1, 2$, g is the gender, and ag stands for the age.

The superclass-subclass relations are stated through the indexes, as it is shown for theclass *Morbidity* in the Fig. 2.

$$Pollution = P_t = < p_i, pp_{i,j}, pp_{i,j}^t >,$$

where p represents the set of main pollutants, $p \in P$, pp is a sub-pollutant from class p, $pp \in PP$, $PP \subseteq P$, t is again the time of registration (year), $i = 1 \dots |P|$, $j = 1 \dots |PP|$. The *Environment* class is represented in similar manner as the class *Morbidity*, as it shown in Fig.3.

As stated previously, the *Morbidity* class includes two subclasses of diseases: endogeneous and exogeneous, which are detailed into nosologies in accordance with the International Classification of Diseases [9]. The *Environment* class includes the following performance indicators: water pollution, dangerous wastes, transport activity, and industrial activity parameters revealing dangerous emissions during energy life-cycles (use of energy, and so on).

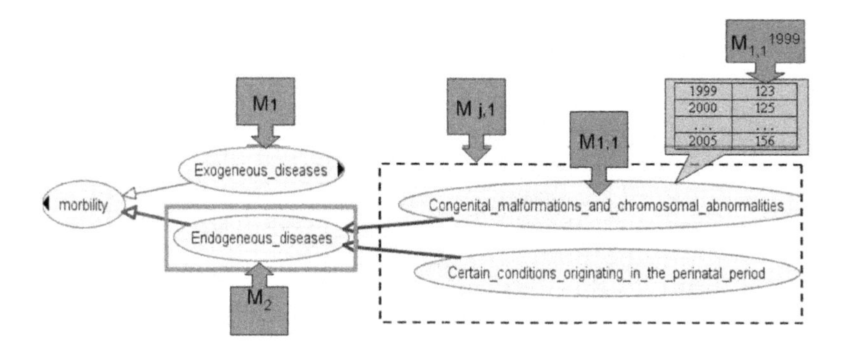

Fig. 2. Illustration of *Morbidity* class representation

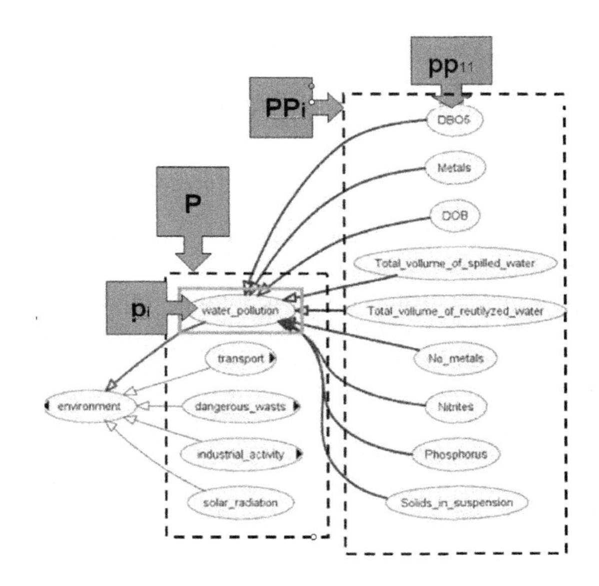

Fig. 3. Illustration of *Environment* class representation

2.2 The MAS Architecture Ontology

The initial analysis of the system was carried out with the Gaia methodology [10] and resulted in revealing and describing the system roles and protocols. The ontology for MAS architecture is stated as:

$$OA =< Levels, Roles, InformationFlows, Orders > \tag{3}$$

where *Levels* correspond to logical levels of the MAS (see Fig. 1), *Roles* is a set of determined roles, *InformationFlows* is a set of the corresponding input and

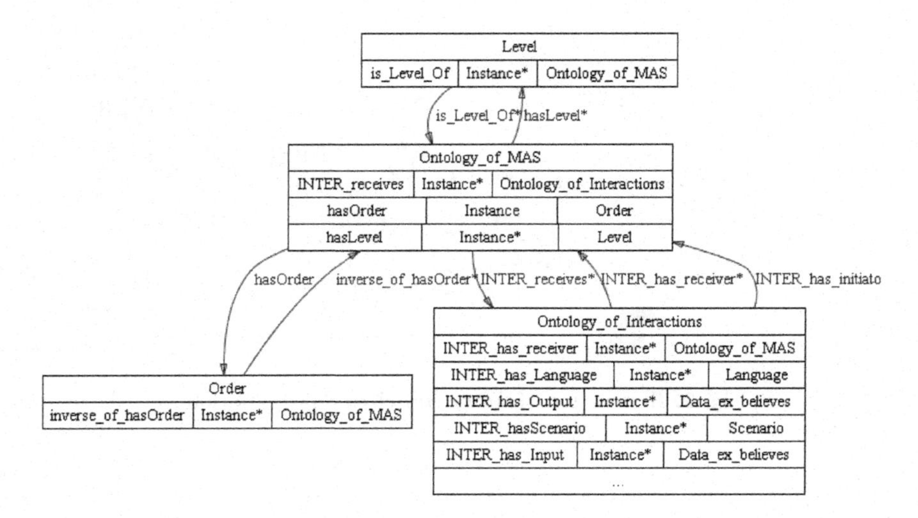

Fig. 4. The MAS Ontology

output information, represented by protocols. Lastly, the set *Order* determines the sequence of execution for every role.

The system consists of three levels. The first level is aimed for meta-data creation, the second one is responsible for hidden knowledge discovering, and the third level provides real-time decision support making, data distribution and visualization. This architecture satisfies all the required criteria to decision support systems as it includes the necessary procedures and functions.

In Fig. 4 there is an ontology created in accordance with the given formal description (3), which presents also its connection with the Interaction ontology (see 2.5), which determines protocols.

The first level is named "Information fusion" and it acquires data from diverse sources and preprocess the initial information to be ready for further analysis. The second layer is named "Data Mining" and there are three roles at this level, dedicated to knowledge recover through modelling, and calculation impact of various pollutants upon human health. The third level, "Decision Making", carries out a set of procedures including model evaluation, computer simulation, decision making and forecasting, based on the models created on the previous level. The main function of this level is to provide a user - actually, a person who makes decision - with the possibility to run online real-time "what - if" scenarios.

The end-user, that is to say the person making decisions, interacts with the MAS through a System-User Interaction protocol, which is responsible for human-computer interaction. The user chooses the indicator he wants to examine and initiates a computer simulation.

2.3 The Tasks Ontology

In order to fulfil the assigned aims, the MAS have to realize the set of tasks and subtasks. The task ontology is represented by the following components:

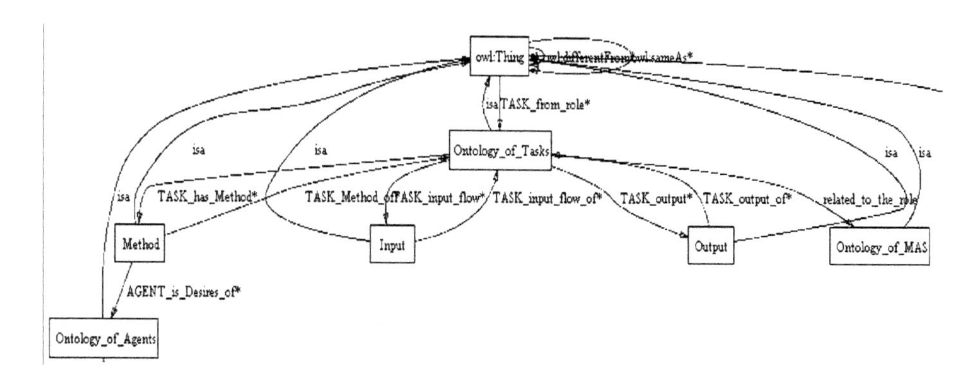

Fig. 5. The Task Ontology

$$OT =< Task, Method, Input, Output, Role > \tag{4}$$

where *Task* is a set of tasks to be solved in the MAS, and *Method* is a set of activities related to the concrete task, *Input* and *Output* are input and output data flows, *Role* is a set of roles, which utilize task.

In Fig. 5 there is a private ontology created in Protégé, which demonstrates the formal model of the task ontology. The component "Role" is inherited from the Ontology of MAS. The tasks are shared and can be accomplished independently, in accordance with an order, which is inherited from the MAS architecture ontology through the *Role* component. The tasks delegation is being delivered for every type of agent. Actually, all the types of agents solve particular tasks and have determined responsibilities. This fact let us relate our system to organizational MAS [12], which is strictly organized and does not require any kind of control agents.

2.4 The Agent Ontology

In our approach we model by BDI agents; their architecture consists of three data structures: *Beliefs*, *Desires* and *Intentions* (which include a plan library). The *Beliefs* are usually represented as facts or in form of information files, data bases, and correspond to the information the agent has about its environment. *Desires* are actions or goals that the agent wants to achieve, and *Intentions* are the desires that the agent chooses under the given circumstances. Intentions are realized in form of actions, which are formed in a plan library, which consists of sequences of steps the agent can execute to achieve its goals. Actually, *Intentions* is a subset of *Desires*. Hence, we describe every agent as a composition of the following components:

$$Agent =< Beliefs, Desires, Intentions, Type >, \tag{5}$$

Every agent has a detailed description in accordance with the given ontology, which is offered in a form of BDI cards, in which the pre-conditions and

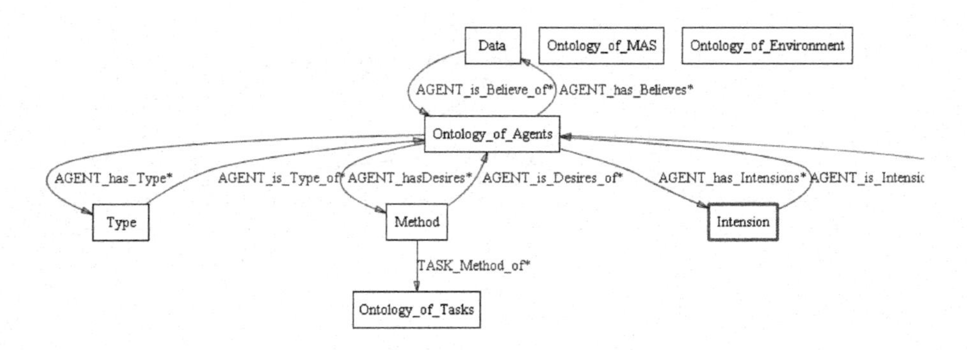

Fig. 6. The Agent Ontology

post-conditions of agent execution, explaining necessary conditions and resources for the agent successful execution, are stated. There is also a collaborator, in case if there are two or more agents needed to solve the task.

The Agent Ontology created is represented in Fig. 6 with general details that include its components. The Beliefs are referenced to Data ontology class, which determine information data resources, necessary for every agent. Desired include methods, stored in the Task Ontology, Intensions call Desires, necessary for every activity or task specification.

The component *Type* determines the common function of an agent and indicates its relation to a certain level of a MAS, and has a value in the range *DB Handling agents, Analysing agents, Evaluation agents, Simulation agents, Distribution agents. DBHandling agents* deal with raw initial data sources, which may be distributed and heterogeneous. Agents of this type fuse the information, clear fused meta-data from outlets, noise, deal with missing values, etc. *Analyzing agents* realize statistical analysis of meta-data and organize and execute data mining procedures (function approximation, classification, etc.). *Evaluation agents* provide model evaluation; they are called for model acceptance and decision making. *Simulation agents* organize and execute model simulation, alarm check and forecast. *Distribution agents* deliver actual information for end-users in the form of electronic documents, files, e-mails.

2.5 The Interactions Ontology

The interactions between the agents include an initiator and a receiver, a scenario and the roles taken by the interacting agents, the input and output information and a common communication language.

The private ontology is setup as:

$$OI = < Initiator, Receiver, Scenario, RolesAtScenario, InputData, OutputData, Language > \quad (6)$$

Actually, as *Initiator* and *Receiver* we use roles, which are delegated to split the information and deliver to proper agents. A *Scenarios* corresponds to a protocol. *RolesAtScenario* is a set of roles the agents play during the interaction,

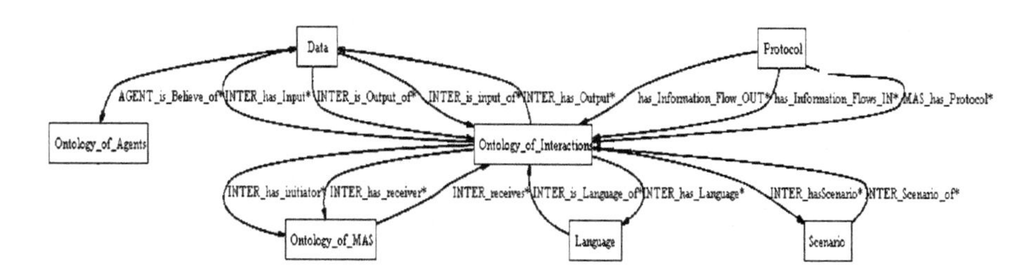

Fig. 7. The Interactions ontology

InputData and *OutputData* are represented by informational resources, read and created, respectively. *Language* determines the communication language.

The graphical representation of the protocol (depicted in Fig. 7) shows the main components, their properties and connections with other sub-ontologies.

2.6 The Distributed Meta-ontology

The Distributed Meta-Ontology is obtained as a result of private ontologies mapping, and is pooled by their common use and execution. This is achieved at step 10 of the algorithm proposed in the Introduction section. The shared ontological dimension, filled with the data, provides agents with correct addressing to proper concepts and synchronizes the MAS functionality.

The problems that appear at this stage are mostly concerned with data heterogeneity. Indeed, data might be stored in different sources, represented by various identifiers and be measured unequally. These procedures can be solved by different methods and are marked as a future step of our work.

3 Conclusions

Ontology creation may be viewed as a crucial step in MAS design as it determines the system knowledge area and potential capabilities. In this article a model of distributed meta-ontology has been proposed that serves as a framework for MAS design. Its components - private ontologies - have been described in extensive with respect to application area and in terms of used semantics. Our future work will be dedicated to the problem of sharing data in the MAS through information fusion.

Acknowledgements

Marina V. Sokolova is the recipient of a Postdoctoral Scholarship (Becas MAE) awarded by the Agencia Española de Cooperación Internacional of the Spanish Ministerio de Asuntos Exteriores y de Cooperación.

References

1. Bradshaw, J.M. (1997). Software Agents. The MIT Press.
2. López-Jaquero, V., Montero, F., González, P., Fernández-Caballero, A. (2005). A multi-agent system architecture for the adaptation of user interfaces. International Central and Eastern European Conference on Multi-Agent Systems, CEEMAS 2005. Lecture Notes in Artificial Intelligence, 3690, pp. 583-586.
3. Sokolova, M.V., Fernández-Caballero, A. (2007). A multi-agent architecture for environmental impact assessment: Information fusion, data mining and decision making. 9th International Conference on Enterprise Information Systems, ICEIS 2007.
4. Guarino, N., Giaretta, P. (1995), Ontologies and knowledge bases: Towards a terminological clarification. In: Mars, N.J.I. (ed.), Towards Very Large Knowledge Bases, IOS Press 1995, pp. 25-32.
5. Samoilov, V., Gorodetsky, V. (2005). Ontology issue in multi-agent distributed learning. Autonomous Intelligent SAystems: Agents and Data Mining, AIS-ADM 2005. Lecture Notes in Computer Science, 3505, pp. 215-230.
6. Gómez-Sanz, J., Pavon, J. (2003). Agent oriented software engineering with INGENIAS. International Central and Eastern European Conference on Multi-Agent Systems, CEEMAS 2003. Lecture Notes in Computer Science, 2691, pp. 394-403.
7. Protégé. http://protege.stanford.edu/
8. OWL Web Ontology Language. (2004). http://www.w3.org/TR/owl-features/
9. International Classification of Diseases (ICD). http://www.who.int/classifications/icd/en/
10. Wooldridge, M., Jennings, N.R., Kinny, D. (2000). The Gaia Methodology for Agent-Oriented Analysis and Design. Journal of Autonomous Agents and Multi-Agent Systems, 3, 285-312.
11. ISO 14031:1999. Environmental management - Environmental performance evaluation - Guidelines
12. Weiss, G. (2000). Multi-agent Systems: A Modern Approach to Distributed Artificial Intelligence. The MIT Press.
13. Georgeff, M., Pell, B., Pollack, M., Tambe, M., Wooldridge, M. (1998). The Belief-Desire-Intention model of agency. Intelligent Agents V: Agent Theories, Architectures, and Languages, ATAL'98. Lecture Notes in Computer Science, 1555, pp. 1-10.

Design of an Agent-Based System for Passenger Transportation Using PASSI

Claudio Cubillos, Sandra Gaete, and Broderick Crawford

Pontificia Universidad Católica de Valparaíso, Escuela de Ingeniería Informática,
Av. Brasil 2241, Valparaíso, Chile
claudio.cubillos@ucv.cl, sandra.gaete@gmail.com,
broderick.crawford@ucv.cl

Abstract. This work presents the experience on designing a multiagent system devoted to the transportation of passengers using the PASSI methodology. The agent system is in charge of the planning and scheduling of passenger trips using the contract-net protocol as base coordination mechanism. It also allows the events' processing caused by changes to the original plan due to vehicle failures or delays, detours and traffic jams, cancellations or passenger no-show. The system has been modeled with the PASSI Toolkit (PTK) and implemented over Jade.

1 Introduction

Multiagent systems can be seen as a natural evolution from Distributed Artificial Intelligence (DAI) and Distributed Computing (DC), the first one providing a more anthropomorphic vision of agents and its societies. In fact agents are defined as autonomous entities capable of flexible behavior denoted by reactiveness, pro-activeness and social ability [9]. In a multiagent system (MAS) diverse agents communicate and coordinate generating synergy to pursue a common goal. Hence as modeling artifact, agent-based systems borrow their key characteristics from us,humans, and our societies.

This higher level of abstraction has allowed agents to tackle the increasing complexity of nowadays open software systems where integration, transparency and interoperation among heterogeneous components are a must. Under such an scenario, the agent paradigm has leveraged as an important modeling abstraction, in areas such as web and grid service, peer to peer and ambient intelligence architectures just to mention some cases.

For this technology to get more mature and widespread, the use of agent-oriented software engineering (AOSE) methodologies and tools are a key factor of success. Hence, the present work describes the design of a multiagent system using a particular AOSE methodology called PASSI [2]. The chosen domain for the system corresponds to the passenger transportation under a flexible scenario in which transport requests coming from clients must be processed on-demand and where changes can occur due to changes in the environment(e.g. traffic jams, detours), problems with the service (e.g. vehicle delay or break down) or client eventualities (e.g. delay or no-show at pickup, trip cancellation).

J. Mira and J.R. Álvarez (Eds.): IWINAC 2007, Part II, LNCS 4528, pp. 531–540, 2007.

2 Related Work

In literature, the passenger transportation problem can be found under different names such as dial-a-ride problem (DARP), on-demand, online or flexible passenger transportation and Demand-Responsive transportation Service (DRTS). Most research focuses in the inherent planning-scheduling algorithm rather than in the software architecture either under the object-oriented or agent-oriented paradigm.

Research tackling this problem tends to use a distributed market-based philosophy based in the Contract-Net Protocol (CNP) (see [3][5]). The MARS System [5] and the TeleTruck [3] approach use the Extended Contract-Net Protocol (ECNP) with Simulated Trading improvement for dealing with dynamics and uncertainty in a transportation scheduling problem. Agent-based systems are presented in [10] and [14]. All of them make use of the CNP for the assignment of client's rides. In addition, [10] uses a stochastic post-optimization phase to improve the result initially obtained. In [14] is presented the Provisional Agreement Protocol (PAP), based on the ECNP and de-commitment.

Although the above works mention the agent approach for designing the overall system, no formal models or AOSE methodology is used mainly due to a lack of maturity that AOSE methodologies presented at that time. At present, there is an increasing need in using the existing methodologies in order to get experience and refine them. Therefore, the aim of this work is twofold; on one side, to provide an experience in the design of multiagent system using the PASSI methodology and on the other, to present the design of an agent-based passenger transportation system, all of the above to show the richness of the agent technology to model systems under complex scenarios.

3 The Passenger Transportation Service

The problem we are treating consists of transport requests coming from a set of clients which should be satisfied by a heterogeneous fleet. The system should process the trips as they come and provide an answer back (bus number, pickup and delivery times) in a timely way. Service requests have to be assigned to vehicles and scheduled according to time restrictions. A restriction exists about the maximum number of passengers to carry (capacity). Transport requests commonly specify a pick-up and delivery place. They also indicate time windows, that is, time intervals within which the client has to be picked-up at the origin node and delivered at the destination node.

In practice, the passenger transportation system we are tackling considers the transport requests coming from different types of clients and that should be satisfied by a heterogeneous fleet of vehicles, composed by buses, minivans, vehicles for disabled people and shuttles, among others. These vehicles are characterized by different properties, but in general they have in common: a limited passenger's capacity, availability time-periods along the day and an area of geographic coverage. As well, there are additional characteristics that vehicles have but like the types of seats, WC, air conditioning or complementary services like Bar and bicycle rack among others. All of these characteristics, that are specific of this transportation scenario, are tackled by the domain ontology which is presented detailed in Section 5.4.

4 PASSI Methodology

PASSI is a step-by-step methodology for designing and developing multi-agent societies. Its name stands for a Process for Agent Societies Specification and Implementation. PASSI integrates design models and concepts from both OO software engineering and artificial intelligence approaches using the UML notation. The design process with PASSI is supported by PTK (PASSI ToolKit [7]) to be used as an add-in for Rational Rose. Figure 1 shows the PASSI methodology, which is made up of five models plus twelve steps in the process of building multi-agent. These are briefly described in the following. Please refer to [2] for a more detailed description.

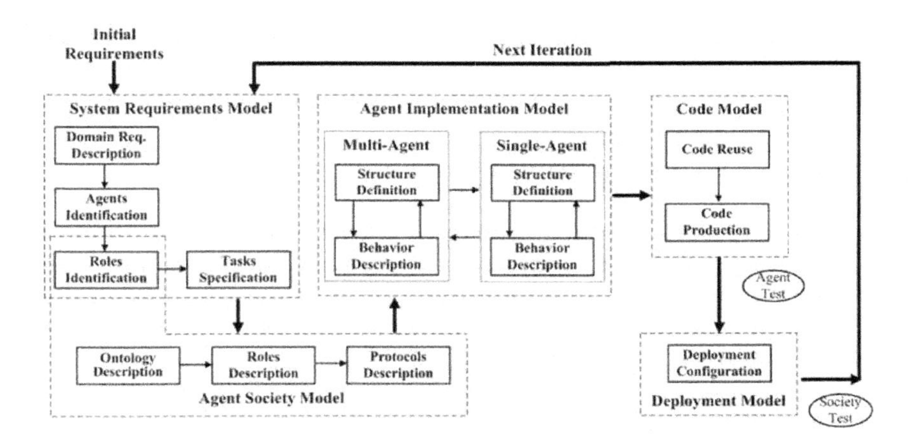

Fig. 1. The PASSI methodology

- System Requirements Model. Corresponds to an anthropomorphic model of the system requirements in terms of agency and purpose. It involves 4 steps: 1) a Domain Description (D.D.), which provides a functional description of the system using conventional use-case diagrams, 2) an Agent Identification (A.Id.), leveraging the separation of responsibility concerns into agents, represented as UML packages, 3) a Role Identification (R.Id.), consisting in use of sequence diagrams to explore each agent's responsibilities through role-specific scenarios and 4) a Task Specification (T.Sp.), detailing through activity diagrams the capabilities of each agent.
- Agent Society Model. Considers the social interactions and dependencies among the agents involved. It considers 3 additional steps: 1) an Ontology Description (O.D.), using class diagrams and OCL constraints to describe the

knowledge ascribed to individual agents and the pragmatics of their interactions, 2) a Role Description (R.D.), using class diagrams to show distinct roles played by agents, the tasks involved by their roles, communication capabilities and inter-agent dependencies and 3) a Protocol Description (P.D.), making use of sequence diagrams to specify the grammar of each pragmatic communication protocol in terms of speech-act performatives.
- Agent Implementation Model. Provides the solution architecture in terms of classes and methods and considers: 1) an Agent Structure Definition (A.S.D.), which uses conventional class diagrams to describe the structure of solution agent classes and 2) an Agent Behavior Description (A.B.D.), considering activity diagrams or statecharts to describe the behavior of individual agents.
- Code Model. Considers a model of the solution at code level, requiring: 1) a Code Reuse Library (C.R.), that is, a library of class and activity diagrams with associated reusable code and 2) a Code Completion Baseline (C.C.) consisting in the Source code of the target system.
- Deployment Model. Describes a model of the distribution of the parts of the system across hardware processing units, and their migration between processing units. It involves the Deployment Configuration (D.C.), which makes use of deployment diagrams to describe the allocation of agents to the available processing units and any constraints on migration and mobility.

5 The Agent Architecture

The multiagent transportation system stands over the Jade Agent Platform[2], which provides a full environment for agents to work. In the following subsections, the agent architecture is described in terms of the PASSI methodology phases. Due to space restrictions, the description will focus on the System Requirements and Agent Society models.

5.1 Agent Identification (A.Id.)

In this step the use cases capturing the system requirements are grouped together to conform an agent. The diagram in Fig. 2 shows the identified use cases for the transport systems and the leveraged agents.

Client agents play an interface role, providing the means for the end user (requester) to generate a trip request. They are also in charge of capturing the description of the desired transportation service through a *Trip Request Profile*. It also provides a channel for client-to-system events' communication (e.g. a delay or change on the agreed service) and for system-to-client events' communication (e.g. a traffic jam, or vehicle break down).

Trip-Request agent's main role is to represent the client and his decisions concerning the transportation request, residing on a device with more processing power. This agent is responsible of having the client's request fulfilled and of communicating him about the result. It is also responsible for events pre-processing and managing the negotiation with the Planner.

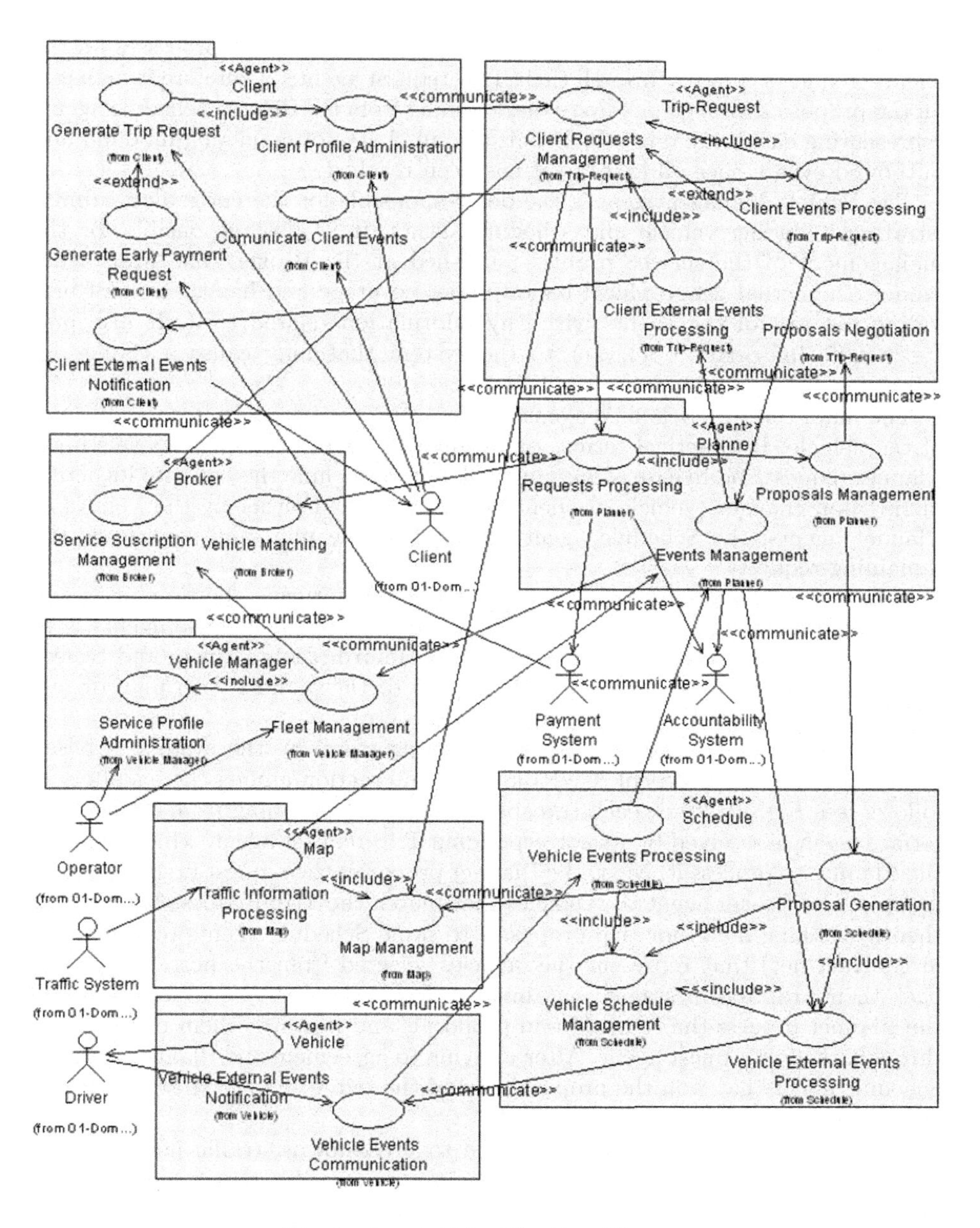

Fig. 2. Agents' Identification Diagram

The Broker's main role is to know which transportation services are available and their characteristics. In addition is able to analyze those service characteristics upon planner request. It provides a publish/subscribe infrastructure that allows vehicles to enter or leave the system freely and allows clients to query the system for available transport services.

The Planner is a key agent in the system's architecture. It processes all the clients' requests coming through their Trip-request agents. Therefore it manages all the proposals for a given trip request coming from the diverse Schedule agents representing each available vehicle.It is also in charge of managing inbound and outbound events once an agreement has been reached.

The Vehicle Manager agent is the one responsible for the entire fleet administration including vehicle and schedule agents. It is also responsible for the management of the service profiles published at the Broker. The Map agent models the actual geographical region under coverage and has the role of providing the rest of the agents with any information related to it. It also process traffic information relevant to the system that may cause a change on schedules.

The main role of a Schedule agent is to manage the trip plan (work-schedule) of the vehicle. In practical terms, the agent will have to make proposals upon Planner request and in case of winning will have to include the trip into its actual plan. Upon changes (vehicle or client events) informed either by the Vehicle or Planner agents, the Schedule agent will update the plan and reschedule the remaining requests.

Vehicle agents are interface agents in charge of capturing the properties and status of the real vehicle. The most important role of the Vehicle agents is to monitor the real vehicle while in service and inform their Drivers and Schedule agents about any differences with respect to the planned trip plan or any eventualities (e.g client no show, delay, detour, etc).

The routing and scheduling functionality provided by the system is based on the contract-net protocol (CNP) [8]. The interaction among the agents is as follows (see Fig. 2): First, each transportation request coming from a Client interface agent is received by its corresponding Trip-request agent, which requests the Planner to process it. Next, the Planner processes the request first by obtaining from the Broker agent the vehicles that match the Trip Request Profile, and then by making a call for trip-proposals to some Schedule agents (call for bids in contract-net) that represent the vehicles selected from the fleet. They send back their proposal messages containing a *TripProposalProfile* description and the Planner process the received trip proposals and forwards them to the client through its Trip-request agent. After arriving to agreement the Planner tells the Schedule agent that won the proposal to add the trip to its actual schedule and tells the others their proposal rejection.

Upon differences in the planning (due to breakdowns, traffic jam, etc) the Schedule agent re-plans. In the case of having an infeasible trip request (mainly due to the time-window restrictions), it informs the Planner agent about the situation. The Planner makes a call for trip-proposals to try reallocating the request in other available vehicle. In any case, the result is informed to the corresponding Trip-request agent, which will inform the client about the change. This change may imply a different vehicle processing the trip only or also a delay or an anticipation of the pickup and delivery times defined previously.

5.2 Roles Identification (R.Id.)

Roles Identification consists in exploring all the possible paths of the preceding Agents' Identification Diagram in Fig. 2. In fact, each "communicate" relationship among two agents can be in one or more scenarios showing interacting agents working together to achieve a certain desired system behavior.

As example, the following Fig. 3 shows the scenario in which the Driver actor communicates the system that the passenger did not show at the pickup point. Each object in the diagram is described following the $\langle role \rangle$:$\langle agent \rangle$ convention. Therefore, this scenario involves the Driver and Accountability System actors plus the vehicle, Schedule and Planner agents.

The scenario starts with the Driver communicating the absence of the passenger at the pickup point through the Vehicle (agent) interface. This last one requests its Schedule agent to process the absence, providing the trip details. This request is processed by the Events' Processing role of the Schedule. This role performs two tasks: the first is the corresponding trip deletion from the actual schedule while the second is the communication of such event to the Planner. The first part is carried out by the schedule but through another role, the Schedule Management. Finally the Planner informs the rest of the system (the Accountability System in this case) about the event(*Passenger Absence Confirmation* as defined in the ontology) in order to update its corresponding trip-service registry.

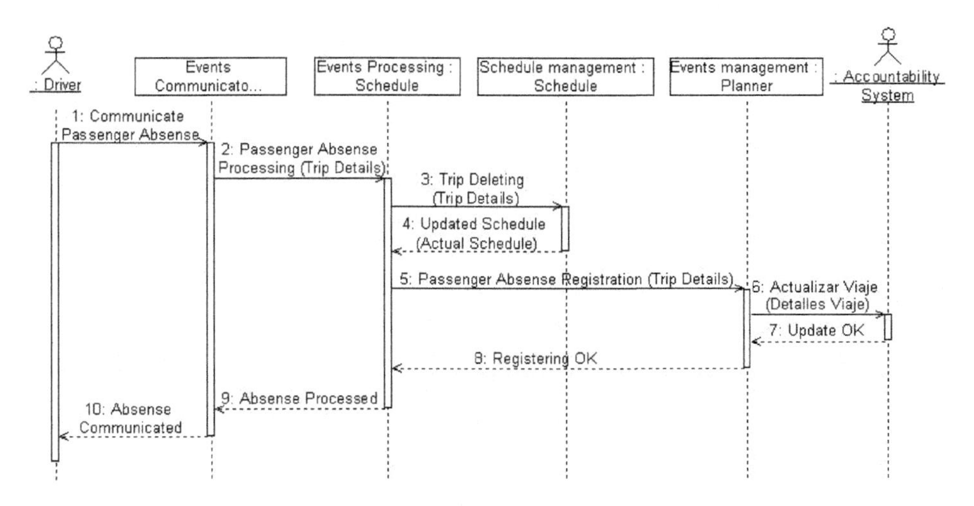

Fig. 3. Roles Identification for the "Driver Communicates Passenger Absence at Pickup Point" scenario

5.3 Task Identification (T.Sp.)

In this phase the scope is to focus on each agent's behavior, decomposing it into tasks, which usually capture some functionality that forms a logical unit of work and generating cohesion. Therefore for each agent an activity diagram is developed, containing what that agent is capable of along the diverse roles it performs. In general, an agent will be requiring one task for handling each incoming and outgoing message.

As example, tasks of the Vehicle agent are depicted in Fig. 4. The diagram shows five tasks that constitute the Vehicle agent capabilities. The *ObtainEventData* task handles Driver messages which notify events. This one calls the *SendEventNotification* task which performs the event forwarding. The message is processed by the *ReceiveVehicleEventNotification* task of the Schedule agent. The next couple of tasks are related to the passenger confirm scenario (as opposed to a client no-show). When confirming a passenger at the pickup point, the Driver will need to have displayed on its interface (vehicle agent) the scheduled trip-passenger list (performed by the *ShowPassengersSchedule* task) and will confirm the picking of the corresponding passenger (processed by *SendPassengerConfirmation*). Finally, the last task *ShowUpdatedScheduleDetails* acknowledges the Driver about the last changes in the vehicle schedule.

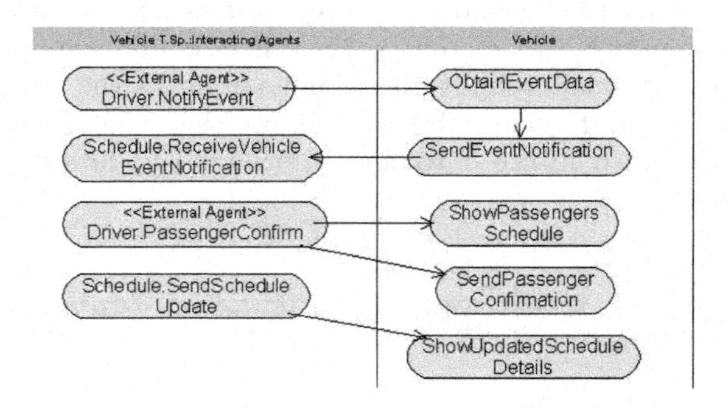

Fig. 4. Tasks of the Vehicle agent

5.4 Ontology Description (O.D.)

In this step the agent society is described from an ontological perspective, providing them with a common knowledge of the domain, and thus, possibilitating communication. The following Fig. 5 shows a portion of the Domain Ontology Description regarding the concepts, while leaving away the actions and predicates.

In the upper-left corner we can identify the *Event* concept hierarchy, containing all the symbols used by agents to understand, without ambiguity, the eventualities that can happen after a trip has been contracted and scheduled: traffic, client or vehicle delay or cancellation, detour, vehicle break down and the presence or absence confirmation of the passenger (actually used in the scenario of Fig. 3 above). On the right is possible to see the service *Requester* and *Provider* concepts, corresponding to the Client and Vehicle respectively. Both have an *Utility Function* consisting on a list of *Utility Properties.* On its turn an utility property is the specification of a *Property* plus its weight in the utility function (e.g. ClientWaitingTime:1, ClientExcessTravelTime:0.5 for a given requester or client).

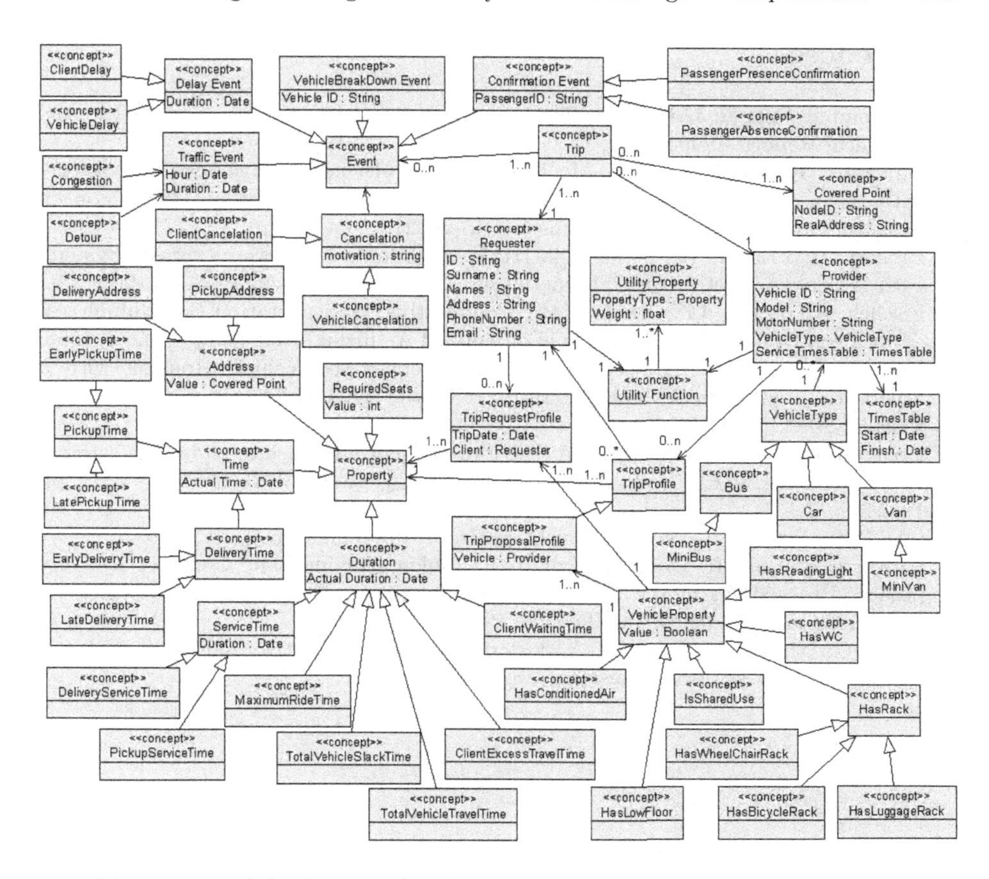

Fig. 5. Part of the Domain Ontology Description considering concepts only

In lower-left area, the *Property* hierarchy is depicted, describing all the characteristics needed to specify a trip service, either when requesting it (*TripRequestProfile*), when making proposals to clients (through *TripProposalProfile*) or once the service has been contracted and scheduled (using a *TripProfile* to describe it). Finally in the lower-right part, the *VehicleType* and *VehicleProperty* hierarchies are shown. These are used to further specify each vehicle's characteristics. Such information is published in the Broker and used by the Planner in order to perform the matchmaking between the required transport service and the available vehicles.

6 Conclusions

The design of an agent-based software architecture devoted to passenger transportation was described. The PASSI methodology use allowed an appropriate level of specification along its diverse phases. In addition, the specification of a domain ontology allows a clear understanding of the knowledge model agents must understand in order to interact within the society, thus enabling the system openness.

References

1. Bellifemine, F. et al: JADE - A FIPA Compliant Agent Framework. C SELT Internal Technical Report, 1999.
2. Burrafato, P., and Cossentino, M.: Designing a multiagent solution for a bookstore with the passi methodology. In Fourth International Bi-Conference Workshop on AgentOriented Information Systems (AOIS-2002).
3. Brckert, H; Fischer, K.; et al.: TeleTruck: A Holonic Fleet Management System. 14th European Meeting on Cybernetics and Systems Research, 1998, pp. 695–700.
4. Fischer, K.; Mller, J.P.; Pischel, M.: Cooperative Transportation Scheduling: An application Domain for DAI. Journal of Applied Artificial Intelligence, Vol. 10, 1996.
5. Kohout, R; Erol, K. Robert C. In-Time Agent-Based Vehicle Routing with a Stochastic Improvement Heuristic. In Proc. Of the AAAI/IAAI Int. Conf. Orlando, Florida, 1999, pp. 864–869.
6. Perugini, D.; Lambert, D.; et al.: A distributed agent approach to global transportation scheduling. IEEE/ WIC Int. Conf. on Intelligent Agent Technology, 2003, pp 18–24.
7. PASSI Toolkit (PTK) Available at http://sourceforge.net/projects/ptk
8. Smith, R. G. and R. Davis: Distributed Problem Solving: The Contract Net Approach. Proceedings of the 2nd National Conference of the Canadian Society for Computational Studies of Intelligence. 1978.
9. Weiss, G.: Multiagent Systems: A Modern Approach to Distributed Artificial Intelligence. MIT Press, Massachusetts, USA. 1999.

The INGENIAS Methodology for Advanced Surveillance Systems Modelling

José M. Gascueña[1] and Antonio Fernández-Caballero[1,2]

[1] Instituto de Investigación en Informática de Albacete (I3A)
[2] Escuela Politécnica Superior de Albacete
Universidad de Castilla-La Mancha, 02071-Albacete, Spain
{jmanuel,caballer}@dsi.uclm.es

Abstract. The use of surveillance systems has grown exponentially during the last decade. Moreover, the agency paradigm has shown to be suitable for the design and development of complex systems such as surveillance systems. They provide autonomy, reactivity, social ability and pro-activeness to carry out surveillance tasks in a semi-automatic way, collaborating with users in a more effective manner. Agents provide coordination mechanisms, solve conflicts, and determine through negotiation processes the more appropriate distribution of the surveillance tasks. Existent agent-based surveillance systems do not really use agent-based methodologies to develop them. In this paper, our experience for modelling advanced surveillance systems using the INGENIAS methodology is described.

1 Introduction

The use of surveillance systems has grown exponentially during the last decade, and has been applied in many different environments [7]. These systems are focused on (1) monitoring the environment, (2) informing of detected pre-alarm situations, and (3) performing the most suitable actions, in collaboration with the users, when a real alert situation is confirmed. Nowadays, there is also an increasing interest in researchers in integrating some capabilities to interpret scenes, that is to say, to understand dynamic situations and events that appear in the global scene. The purpose is to inform of the presence of certain previously defined situations, which go one step further than simply detecting and tracking the objects. This is precisely called advanced surveillance.

On the other hand, one important aim of multi-agent systems (MAS) is to provide a high-level abstraction software engineering paradigm for designing and developing complex software systems [14]. Advanced surveillance systems are precisely a good example of it, as they include elements that coordinate the execution of the activities for solving sufficiently complex tasks and achieving the outlined goals. There are proposals for several methodologies that allow facing the development of complex systems. Remember, for example, MAS-CommonKADS, ZEUS, Gaia, MaSE, or Prometheus, among others. In this article we describe our experience for modelling an advanced surveillance system using the INGENIAS methodology [9].

J. Mira and J.R. Álvarez (Eds.): IWINAC 2007, Part II, LNCS 4528, pp. 541–550, 2007.

2 Multi-agents in Surveillance Systems

Franklin and Graesser [4] state: *"an autonomous agent is a system situated within and a part of an environment that senses that environment and acts on it, over time, in pursuit of its own agenda and so as to effect what it senses in the future"*. According to this definition, any agent should satisfy four properties: autonomy, social ability, reactivity and pro-activeness. Usually one single agent does not solve problems by itself. Generally the cooperation between various agents is essential to tackle a complex problem more easily. In this kind of process probably the agents ask and provide data, request to carry out actions, enter into negotiations, etc.

A MAS is formed by a set of agents possessing the previously cited properties. Thus, in a visual surveillance system, the software that manages the operation of a camera group might easily be modelled as a multi-agent system because it satisfies the properties. For instance, each camera could be associated to an agent that controls it, that is to say, it is monitoring what is happening in its field of view (autonomy), it reacts to changes that take place in the scene (reactivity), it may communicate with agents associated to others cameras to obtain additional information about a common target (social ability), and it may rotate the camera with the aim of looking for new objects (pro-activeness). In this example it is necessary to coordinate a set of cameras in order to solve conflicts that may appear, and to determine through negotiation processes the more appropriate distribution of tasks aimed to locate and track at any moment the targets. These problems have widely been studied in the area of agents; this is another powerful reason that justifies their use in surveillance systems. This example can be extended to even more complex multi-sensorial surveillance systems [8], which also include non-visual sensors (e.g., temperature, infrared or biometric sensors) and cameras mounted on moving robots. Consider also that in order to control a surveillance system a decentralized architecture is preferred to a centralized one, as this allows to obtain a more robust, scalable, flexible and fault tolerant system. Multi-agent systems are a good solution to face more easily the complex problem of the distribution because they provide a natural way of solving problems that are inherently distributed [14].

Due to all these reasons, in the last years there have been several efforts to create agent-based visual surveillance systems. There is the tracking system proposed in [12], where each moving object equipped with a Global Positioning System (GPS) is tracked by an agent. Also in [13] a group of active vision agents, whose fields of view may be overlapped, have the capability to track in real time multiple simultaneously moving objects, whereas the traffic surveillance system Monitorix uses non overlapped cameras [1]. A framework to understand the dynamics of a scene have also been introduced so far. There is the proposal by Remagino and colleagues [10] based on the cooperation between agents, where persons and vehicles appear entering or leaving an area; and the architecture shown in [2] uses wireless visual sensors, region agents searching for objects of interest (persons, vehicles and animals) and object agents tracking them to determine if the scene contains a threat pattern, and performing a appropriate action. On the other hand, robots and other sensors can also be used for monitoring and security tasks. For example, robot and sensors controlled by agents have been used for fire detection [11], [6] or in rescue operations [5]. Our critique to the approaches cited previously is that there is a lack of methodological development of the multi-agent systems. Our concern is that using a methodology is absolutely useful as it offers solutions on techniques, guidelines, and tools necessary during the life cycle stages of any system under development.

3 MAS Description Using the INGENIAS Methodology

Our proposal is based on INGENIAS methodology for several reasons: (a) it is based in a previously tested MESSAGE methodology [3], (b) it is largely documented [9], (c) useful tools that support the analysis, design and code generation of based-agent software are included in the INGENIAS Development Kit, (d) it uses a model-driven approach that facilitates the independence from the implementation platform, (e) it is not oriented towards a particular agent platform, and (f) it has been validated in real applications (e.g. information access system, tourist guide services using mobile devices, scheduling of a ceramic tile factory production process, or knowledge management). The models created using INGENIAS are based on five meta-models which define different views and concepts to describe the agency: organization, agent, goals and tasks, interaction, and environment. Some elements of the agency may appear in different meta-models. This repetition of entities across different views induces dependencies among meta-models.

3.1 System Requirements

In our approach the system does not replace the human personnel, but it has to serve as his complement to jointly perform surveillance and security tasks. The user of a surveillance system should be informed of relevant events; this way, a semi-automatic system helps in the arduous task of permanently looking at the screens. Figure 1 shows the functionalities offered to the security guards: watching the observed environment and informing about the interactions produced

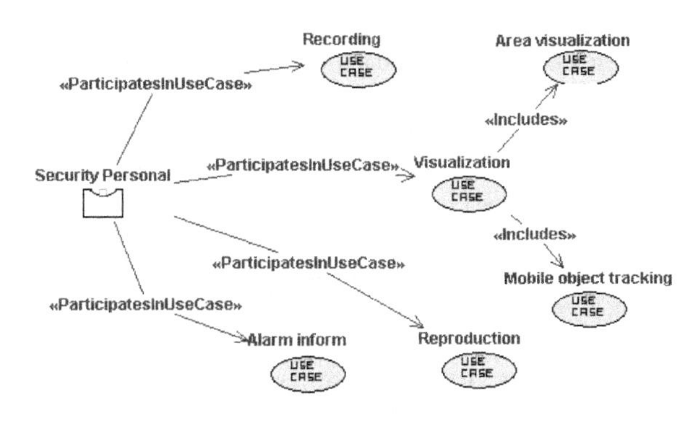

Fig. 1. Functional requirements for the security guard

between monitored targets (*area visualization* use case), tracking the selected targets (*mobile object tracking* use case), recording video sequences (*recording* use case) to reproduce later on (*reproduction* use case), as well informing about detected alarm situations (*alarm inform* use case). After identifying the system functional requirements we have performed the decomposition into agents in order to carry out the surveillance system functions.

3.2 Organization Model

In a multi-agent system modelled in INGENIAS the organization model is the equivalent to the system architecture. It defines how the agents are grouped, the system functionality and which restrictions are imposed to the agents' behaviour. From a structural point of view, the organization is a set of entities with relationship of aggregation and inheritance. It defines a schema where agents, resources, tasks and goals may exist. In this model, groups may be used to decompose the organization, plans, and workflows to establish the way the resources are assigned, which tasks are necessary to achieve a goal, and who has the responsibility of carrying them out.

Thus, the scheme of the organization shown in figure 2 describes the global system architecture. Notice that agents belonging to groups that gather information of the monitored place (*CameraGroup*, *RobotGroup*, *OtherSensorGroup* groups) are distinguished. These agents send the information to agents included in other groups that have the functionality of showing information to the user about what is happening in the monitored environment (*DeviceMobileGroup*, *BuildingCentral*, *PrincipalCentral* groups). The surveillance system, represented as organization *Surveillance System*, has the goals to monitor the environment, to inform about suspicious events, to interpret what is happening in a scene and to handle alarms that are fired (all represented in form of circles).

The Camera group incorporates a series of different agents. An Image agent (*AImage*) captures the image of the environment, does a pre-processing and sends the results to the Blob agent (*ABlob*). This one is in charge of identifying image regions that correspond to objects of interest (vehicles, people, animals, and so on). An Object agent (*AObject*) is responsible for organizing the information associated to the objects of interest and to classify them. An Activity agent (*AActivity*) describes the activities of the objects identified in the scene, whereas a Social agent (*ASocial*) describes the interaction between objects. A Camera Manager agent (*AManager*) communicates with the previous agents, and may establish a communication with a mobile device, a Building Alarm System or Central Alarm System agents (*APDA*, *AMPhone*, *ABCentral*, *APCentral*). Moreover, in the Robot group there is an agent that controls the robot (*ARobot*) whose goals are to obtain a vision of the objects of interest from a different perspective, and to accede to zones that are inaccessible to fixed cameras or that are dangerous for the humans. Finally, the administration of the information provided by non visual sensors is carried out by the corresponding associated agents (*temperature*, *volumetric* agents, among others).

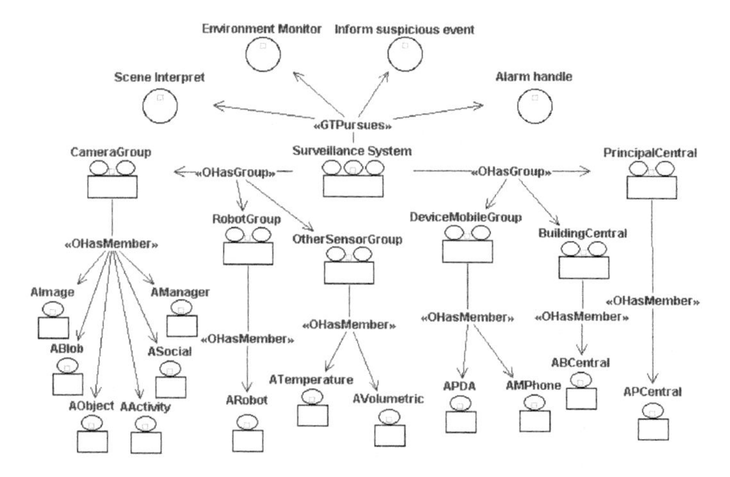

Fig. 2. Representation of the surveillance system organization

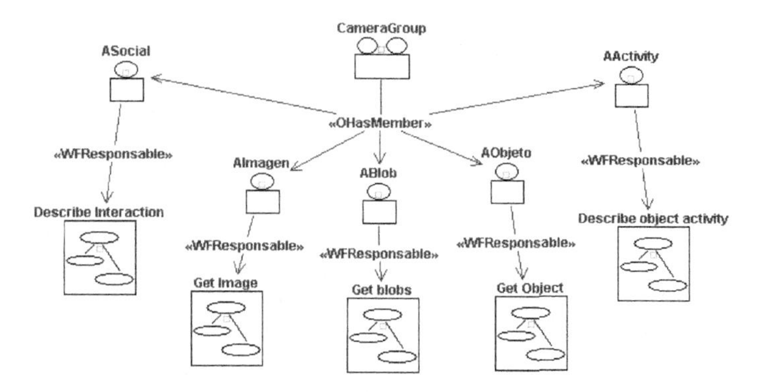

Fig. 3. Plans of the agents belonging to Camera group

The security guard may be equipped with a mobile device (PDA, mobile phone) wirelessly connected to the surveillance environment. This way he can be informed while he freely walks around the monitored indoor or outdoor zone. Therefore, we consider an agent that is working in a mobile device receiving information from the manager, the robot or other sensor agents. Moreover, this agent interacts with the security guard to alert him of detected suspicious events, to send images or any other required information.

In the INGENIAS-based organization model, plans realizing tasks can also be included. As an example, figure 3 shows some plans for the agents belonging to the Camera group. Plans define associations among tasks and general information about their execution. Thus, for instance, consider figure 4 where the tasks execution sequence associated with plan *"Get blobs"* is shown. Its goal is to segment the input image, represented by a circle, in order to obtain the blobs (moving patches) that appear in an image. It is decomposed into two

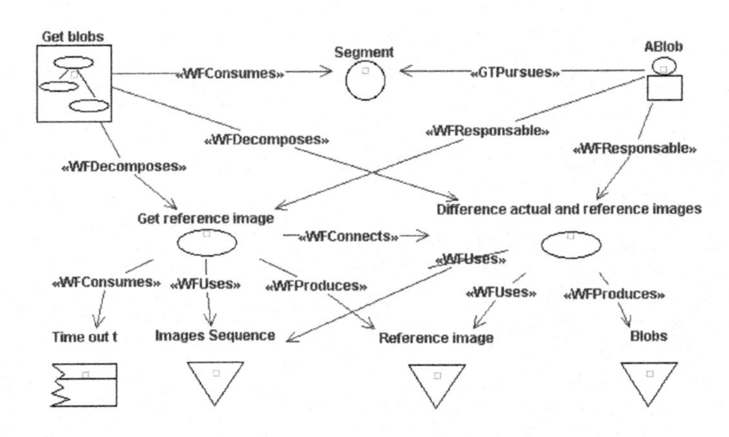

Fig. 4. Representation of plan *Get blobs*

tasks, and a *Blob* agent is responsible for its execution. Task *"Get reference image"* requires an image sequence (resource represented as a triangle) to produce a reference image that is consumed by task *"Difference among current and reference image"* to produce blobs (resource *blobs*) extracted from the current image. The reference image is gotten each certain time; this is indicated by event *Time out t*.

3.3 Agent Model

In INGENIAS, particular agents are defined in the agent model, excluding the interactions with other agents. This model is centred in defining the tasks to execute and the goals to pursue, that is to say, the agent functionality (associations of plans, tasks, goals and roles with agents), and it defines which mechanisms ensure the tasks execution within the decided parameters, that is to say, its control design (type of control, mental state specification and its evolution).

For each agent modelled previously in the organization model, an agent model describing it is created. Figure 5 shows a graphical representation of the Image agent model. It is responsible for acquiring images and pre-processing them (*get image* and *image pre-processing* goals). It possesses the capability to carry out tasks in order to achieve these goals. Moreover, it has a mental state manager (*MSManagerAImage*) with the aim of developing the mental state evolution through operations "create, destroy, modify and monitor agent knowledge"; and a mental state processor (*MSProcessorAImage*) to make decisions of which task has to be executed (agent control). It contains an initial mental state representing the camera initial position (fact *Camera initial position*). Finally, through internal application *APICamera* it can turn left, right, up, or down its associated physical camera, and apply zoom in order to modify its field of view.

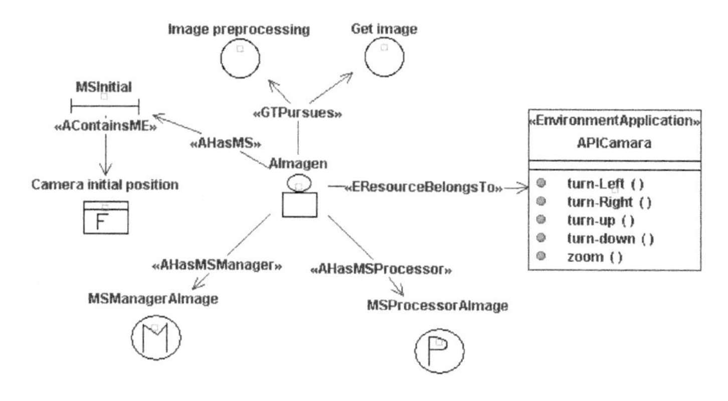

Fig. 5. AImage agent model

3.4 Goals and Tasks Model

The goals and tasks model has the purpose of gathering the motivations of the multi-agent system, to define actions identified in the organization, interactions or agents models, and how these actions affect his responsible. It also expresses which consequences have to execute tasks and why they should be executed. Diagrams from this view can be used to explain how the mental state processor and manager work.

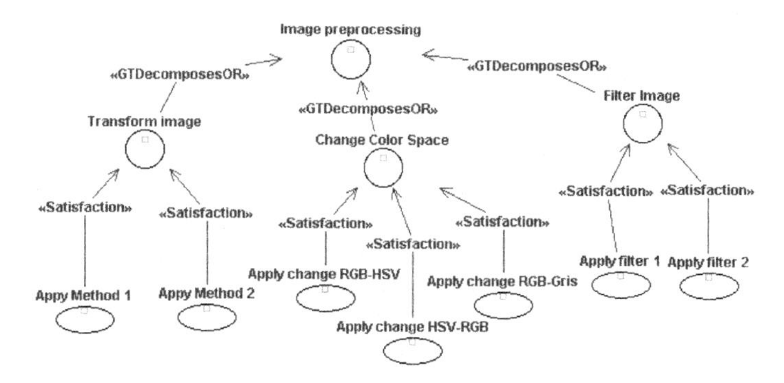

Fig. 6. Goals and tasks model

For example, goal *"Image pre-processing"* may be decomposed into three goals as shown in figure 6: (1) Transform image, (2) Filter image, and (3) Change colour space. It disposes of several tasks to achieve each sub-goal, but not all of them have always to be executed; this depends on the scenario where the camera is placed and the proper camera characteristics, among other features.

3.5 Interaction Model

In the interaction model there are elements such as agents, roles, goals, interactions and interaction units. It offers details about the way the agents communicate

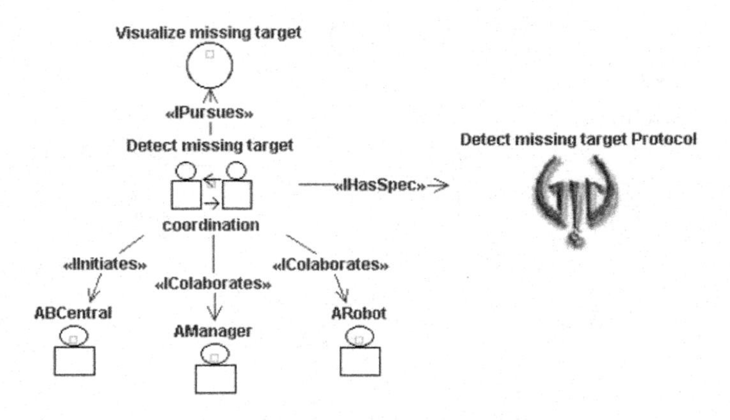

Fig. 7. Interaction *"Detect target missing"*

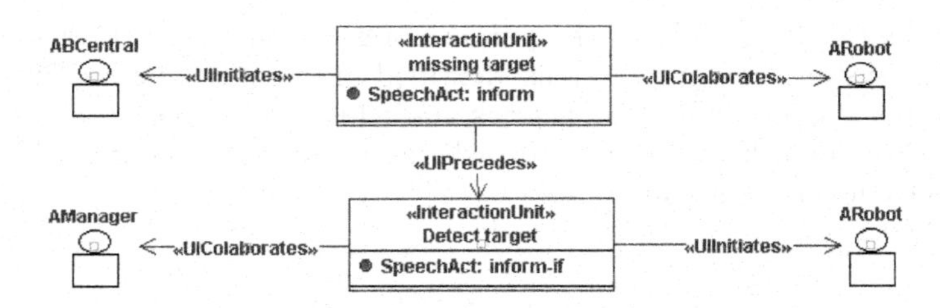

Fig. 8. Protocol *"Detect missing target"*

and coordinate. The interaction model addresses the exchange of information or requests between agents, or between agents and users. Indeed, some tasks require the collaboration of other agents to carry them out. Interaction units are performed in the interactions where there is one initiator and several collaborators. Moreover, the participation of the actors in the interaction and the existence of the interaction itself are justified through goals.

For example, figure 7 shows the interaction *"Detect missing target"*. In a certain moment a possible target can be outside the field of view of the installed cameras; so this fact is communicated to a Robot agent. The Robot agent cooperates with the Camera Manager agent responsible of controlling the camera installed on the robot in order to detect and visualize the target again. In first place, figure 8 describes protocol *"Detect missing target"*, where a Central Building agent (*ABCentral*) requests the Robot agent to start navigating in order to detect the missing target (*missing target* unit interaction). Next, the Robot agent asks the Camera Manager agent if it has to visualize the target looked for (*detected target* unit interaction). Through UIPrecedes relation we indicate that the *missing target* unit interaction precedes the *detected target* unit interaction.

3.6 Environment Model

The environment model addresses what surrounds the system modelled in INGENIAS. The environment only includes agents, resources (e.g., required number of processes or threads, number of necessary connections with data bases) and applications whose main utility is to express perceptions and actions of the agents. Agents act on the environment by invoking the methods or procedures defined in the applications, and they define their perceptions indicating which events are produced by the applications they listen to.

A part of the environment model of a surveillance system is shown in figure 9. The Robot agent is associated an external application named *RobotApplication*. This application controls the robot by providing operations drive straight, left, right, specifying to advance forward, back, at a given speed, etc. In the same way, sensors are associated some Application Programming Interfaces (API) that can be used by agents to carry out their tasks.

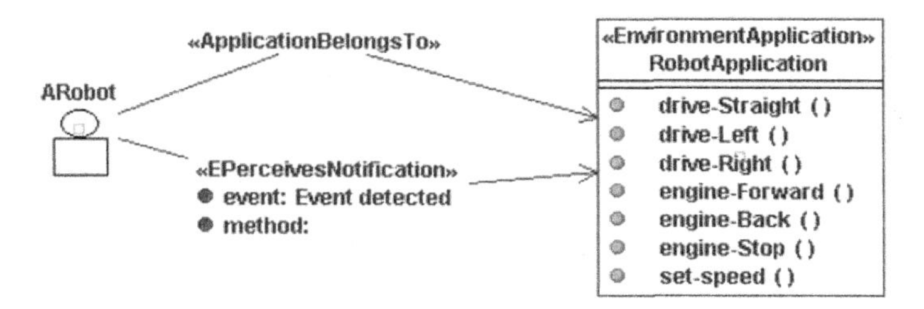

Fig. 9. Environment model

4 Conclusions

Agents offer characteristics such as autonomy to reach a given objective, communication between agents, and capabilities to react before changes in the environment, as well as to take the initiative to perform certain tasks. Therefore, agent-based surveillance systems are essential to assist users in a semi-automatic way in target detection and monitoring tasks. Multi-agent systems provide the necessary foundations to reduce the complexity, to increase the flexibility, scalability and the necessary tolerance to failures in a surveillance system, and they are a natural way of solving distributed problems.

To date, the modelling of agent-based surveillance systems is usually carried out empirically, without using a methodology for the development of the agents. We have studied and proven the adequacy of using the INGENIAS methodology for the analysis and design of such a complex application as an advanced surveillance system.

Acknowledgments

This work is supported in part by the Junta de Comunidades de Castilla-La Mancha PBI06-0099 grant and the Spanish TIN2004-07661-C02-02 grant. The authors are thankful to Dr. Juan Pavón and Dr. Jorge Gómez-Sanz of the Universidad Complutense de Madrid, Spain, for answering to all their queries concerning the INGENIAS methodology.

References

1. Abreu, B., Botelho, L., Cavallaro, A., Douxchamps, D., Ebrahimi, T., Figueiredo, P., Macq, B., Mory, B., Nunes, L., Orri, J., Trigueiros, M.J., Violante A.: Video-based multi-agent traffic surveillance system. Proceedings of the IIEE Intelligent Vehicles Symposium, IV2000 (2000) 457-462.
2. Aguilar-Ponce, R., Kumar, A, Tecpanectl-Xihuitl, J.L., and Magdy Bayoumi.: A network of sensors based framework for automated visual surveillance. Journal of Networks and Computer Applications (2007) to appear.
3. Caire, G., Leal, F., Chainho, P., Evans, R., Garijo, F., Gómez-Sanz, J.J., Pavón, J., Kearny, P., Stark, J., Massonet, P.: Agent-oriented analysis using MESSAGE/UML. Lecture Notes in Computer Science 2222 (2001) 119-135.
4. Franklin, S., Graesser, A.: Is it an agent, or just a program?: A taxonomy for autonomous agents. Lecture Notes in Computer Science 1193 (1996) 21-35.
5. Farinelli, A., Grisetti, G., Iocchi, L., Lo Cascio, S., Nardi, D.: Design and evaluation of multi agent systems for rescue operations. Proceedings of the International Conference on Intelligent Robots and Systems, IROS 2003 (2003) 3138-3143.
6. Fok, Ch.L., Roman, G.C., Lu, Ch.: Agilla: A Mobile Agent Middleware for Sensor Networks. Technical Report, Washington University in St. Louis, WUCSE-2006-16 (2006).
7. Patricio, M.A., Carbó, J., Pérez, O., García, J., Molina, J.M.: Multi-agent framework in visual sensor networks. EURASIP Journal on Advances in Signal Processing (2007) Article ID 98639.
8. Pavón J., Gómez-Sanz J., Fernández-Caballero A. & Valencia-Jiménez J.J. Development of intelligent multi-sensor surveillance systems with agents. Submitted, Robotics and Autonomous Systems (2007).
9. Pavón, J., Gómez-Sanz, J.J., Fuentes, R. The INGENIAS methodology and tools. In: Agent-Oriented Methodologies. Idea Group Publishing (2005).
10. Remagnino, P., Shihab, A.I., Jones, G.A.: Distributed intelligence for multi-camera visual surveillance. Pattern Recognition 37:4 (2004) 675-689.
11. Roman-Ballesteros, I., Pfeiffer, C.F.: A framework for cooperative multi-robot surveillance tasks. Proceedings of the Electronics, Robotics and Automotive Mechanics Conference, CERMA'06 (2006) 163-170.
12. Shakshuki E., Wang, Y.: Using agent-based approach to tracking moving objects. Proceedings of the 17th International Conference on Advanced Information Networking and Applications, ANIA'03 (2003) 578-481.
13. Ukita, N., Matsuyama, T.: Real-time cooperative multi-target tracking by communicating active vision agents. Computer Vision and Image Understanding 97:2 (2005) 137-179.
14. Weiss, G.: Multiagent Systems: A Modern Approach to Distribuited Artificial Intelligence. The MIT Press (2000).

Propos: A Dynamic Web Tool for Managing Possibilistic and Probabilistic Temporal Constraint Networks

Francisco Guil[1], Ivan Gomez[1], Jose M. Juarez[2], and Roque Marin[2]

[1] Departamento de Lenguajes y Computacion
Universidad de Almeria
04120 Almeria
francisco.guil@ual.es, igf286@alboran.ual.es
[2] Dept. Ingenieria de la Informacion y las Comunicaciones
Universidad de Murcia
30071 Espinardo (Murcia)
{jmjuarez, roque}@dif.um.es

Abstract. In this paper we present *Propos*, a novel dynamic Web tool for managing probabilistic and possibilistic temporal constraint networks. These networks are a special sort of temporal constraint satisfaction problems, useful for representing and reasoning with uncertain temporal relations between temporal points. They can be applied as an effective formalism in a very different kind of domains. *Propos* has been developed using dynamic Web technology for mainly, make easier the complex process of interpreting, evaluating, and finally, extracting useful knowledge from the network. It can be viewed as a novel knowledge acquisition tool and, for practical purposes, it will be used for work with temporal patterns extracted from temporal data mining processes.

1 Introduction

Representing and reasoning about time is an important area of Artificial Intelligent (AI). Temporal formalism are applied in a lot of different tasks, like temporal planning, diagnosis or prognosis, realized in different dynamic domains like Web, agriculture or medicine. Temporal points and intervals are the main ontological primitives used by temporal formalism. In this paper, we are centered in models dealing with temporal points-based events.

Many temporal representation and reasoning approaches assume that precise and certain temporal information is available, and very few works give support for situations in which we are dealing with imperfect temporal information. Uncertainty is one kind of imperfect information, which arises from the lack of information about the state of the world. As we can see in [2], the topic of handling uncertainty in temporal knowledge was considered as one of the major, newly, and growing area of temporal representation and reasoning. Dealing with this topic, we want to highlight two different works which proposed models for representing and reasoning with uncertain temporal relations. The first

J. Mira and J.R. Álvarez (Eds.): IWINAC 2007, Part II, LNCS 4528, pp. 551–560, 2007.

one, proposed by V. Ryabov and S. Puuronen [13], defined a model where the probabilistic approach is used to deal with uncertain temporal relations between temporal points. Based in this work, in [9], A. HadjAli, D. Dubois, and H. Prade presented the same model but with the proposal of using the possibility theory as a mathematical tool for dealing with uncertain temporal relations. In the works, the authors specified the way in which temporal information can be represented and the way in which we can reason with them. Both, the probabilistic and the possibilistic model belong to a special class of model called binary temporal constraint networks [3].

Temporal constraint satisfaction is an information technology useful for representing and answering queries about the times of events and the temporal relations about them. In our case, information is represented as temporal constraint network, and constraints represent the uncertain temporal relations between temporal points. Usually, the expert build the networks practically starting from scratch. They combine binary pieces of information obtaining a global model which represents the knowledge about the domain. In our case, we proposed a method for building this kind of model using temporal data mining techniques [8]. In particular, the main goal of the proposed method is to obtain an enumeration of temporal constraint networks that better summarizes the temporal information existing in the dataset.

In this work we present a dynamic Web-based tool for representing and managing the mined temporal constraint networks (probabilistic and possibilistic). This tool will assist the expert in the knowledge discovery process, in particular, in the visualization, interpretation, and evaluation of the patterns extracted from temporal data mining processes, tasks necessaries for complete the main goal, discover useful knowledge.

The remainder of the paper is organized as follows. Section 2 describes briefly our proposal for temporal data mining and its place in the global Knowledge Discovery in Databases Process. The Section 3 describes the Probabilistic and Possibilistic Temporal Constraint Networks Model. Section 4 presents the *Propos* tool and shows the basic functionality via a practical example. Conclusions and future work are finally drawn in Section 5.

2 Temporal Data Mining

Data mining is an essential step in the process of knowledge discovery in databases that consists of applying data analysis and discovery algorithms that produce a particular enumeration of structures over the data [5]. There are two types of structures: models and patterns. So, we can talk about local and global methods in data mining [11].

Since the problem of mining association rules was introduced by *Agrawal* in [1], a large amount of work has been done in several directions, including improvement of the *Apriori* algorithm, mining generalized, multi-level, or quantitative association rules, mining weighted association rules, fuzzy association rules mining, constraint-based rule mining, efficient long patterns mining, maintenance of the discovered association rules, etc. Temporal data mining can be viewed as an extension of this work.

Temporal data mining is an important extension of data mining techniques. Temporal data mining can be defined as the activity of looking for interesting correlations (or patterns) in large sets of temporal data accumulated for other purposes. It has the capability of mining activity, inferring associations of contextual and temporal proximity, that could also indicate a cause-effect association. This important kind of knowledge can be overlooked when the temporal component is ignored or treated as a simple numeric attribute [12]. In non-temporal data mining techniques, there are usually two different tasks: the description of the characteristics of the database (or analysis of the data), and the prediction of the evolution of the population. However, in temporal data mining this distinction is less appropriate, because the evolution of the population is already incorporated in the temporal properties of the analyzed data.

We can found in the literature a large quantity of temporal data mining techniques. We want to highlight some of the most representative ones. So, we can talk about sequential pattern mining, episodes in event sequences, temporal association rules mining, discovering calendar-based temporal association rules, patterns with multiple granularities mining, and cyclic association rules mining. However, there is an important form of temporal associations which are useful but could not be discovered with this techniques. These are the inter-transaction associations presented in [10].

In [6] we presented an algorithm, named $TSET$, based on the inter-transactional framework for mining frequent sequences from several kind of datasets, mainly transactional and relational datasets. The improvement of the proposed solution was the use of a unique structure to store all frequent sequences. The data structure used is the well-known set-enumeration tree, widely used in the data mining area, in which the temporal semantic is incorporated. The result is a set of frequent sequences describing partially the dataset. This set forms a potential base of temporal information that, after the experts analysis, can be very useful to obtain valuable knowledge. However, the overwhelming number of discovered frequent sequences may make such task absolutely impossible in practice. This problem can be viewed as a second-order data mining problem, which consists in the necessity of obtaining a more understandable and useful sort of knowledge from a huge volume of temporal associations resulting after the data mining process.

In [8], we proposed a method for building a special model of temporal network formed by a set of uncertain relations amongst temporal points. The temporal model, proposed by HadjAli, Dubois and Prade in [9], is based on the Possibility Theory as expressive tool for the representation and management of uncertainty in point-based temporal relations. The uncertainty is represented by a vector describing three possibility values, expressing the relative plausibility of the three basic relations between two temporal points, that is, "before", "at the same time" and "after". Thus, the authors define the basic operations (inversion, composition, combination and negation) that allow to infer new temporal information and to propagate uncertainty in a possibilistic way.

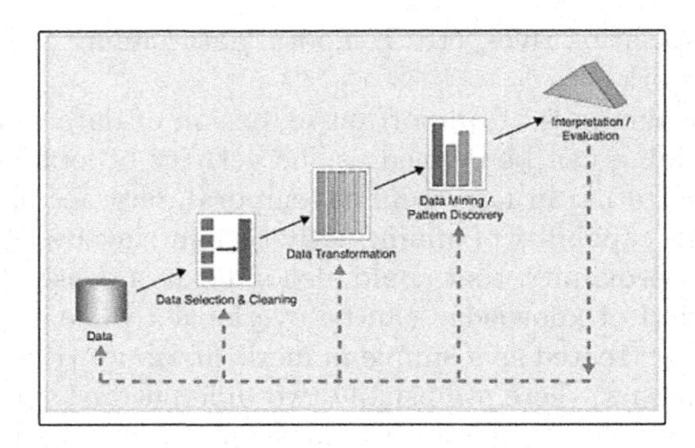

Fig. 1. The Knowledge Discovery Process

Another related work is presented in [13], where the authors proposed a model for representing and reasoning with uncertain temporal relations using the probabilistic approach to deal with uncertainty. The building of this kind of networks from databases is currently in the experimentation step.

As we can see in Figure 1, the next step in the global Knowledge Discovery in Databases is the interpretation/visualization and evaluation of the patterns extracted in the temporal data mining process. In this case, the pattern is a temporal constraint network, and it can be viewed as a graph abstract data type, a very difficult structure to deal with. Moreover, the operations for reasoning involves very complex calculus. A simple addition implies a lot of operations for reconstructing the network. Another problem we find in this step is the understanding about how the temporal model works in order to infer new knowledge. The necessity of a tool to deal with this type of patterns is clear.

In the next section we describe deeper the sort of temporal pattern we mined using temporal data mining techniques.

3 The Temporal Patterns Model

Both models, the Possibilistic Temporal Constraint Networks [9] and the Probabilistic Temporal Constraint Networks [13], belong to a special sort of binary temporal constraint networks used to represent and to infer (to reason) with uncertain temporal relations between point-based events. The main difference between them is the approach to deal with temporal uncertainty. In the former, the authors proposed the Possibility Theory [4], whereas in the later, the authors proposed the well-know probabilistic approach. The possibilistic model appeared in the literature as an improvement of the probabilistic one in the sense that classical probability theory can not model ignorance (even partial ignorance) in a natural way. In [4], the authors show how the possibility theory copes with the situation of complete ignorance in a non-biased way, that is, without making any prior assumption. However, in this paper we do not want to argue what of the two approaches is better to deal with temporal uncertainty. For us, both models are designed with different applicability, depending of the problem we want to deal with. It will be the responsibility of the user to choose the appropriate theory for a particular problem.

Independently of the chosen approach, both models (in the sequel we call them Temporal Patterns Model) belong to the general temporal constraint networks model, and therefore, can be modeled as a tuple $< X, D, C >$, where X is a set of variables, D the set of variable domains and C the set of constraints restricting the values that the variables can simultaneously take. In particular, the variables represent point-based events, and the constraints represent uncertain relations between temporal points. The constraint is represented as a vector with three possibility/probability values denoting the plausibility/probability of the three basic relations: "$<$" (before), "$=$" (at the same time), and "$>$" (after). Given this three basic relations, an uncertain relation between two temporal points is expressed as any possible disjunction of these basic relations:

$$\begin{array}{l} \leq \Longleftrightarrow < or = \\ \geq \Longleftrightarrow > or = \\ \neq \Longleftrightarrow < or > \\ ? \Longleftrightarrow <, =, or > \end{array}$$

The last case represents *total ignorance*, that is, any of the three basic relations is possible. In the probabilistic model, this case is impossible to represent. Let a and b two temporal points, and r_{ab} the uncertain relation between them. In the probabilistic model,

$$r_{ab} = (P^{<}_{ab}, P^{=}_{ab}, P^{>}_{ab})$$

with $P^{<}_{ab} + P^{=}_{ab} + P^{>}_{ab} = 1.0$, where $P^{<}_{ab}$ (respectively $P^{=}_{ab}, P^{>}_{ab}$) is the probability of $a < b$ (respectively $a = b, a > b$). In the possibilistic model,

$$r_{ab} = (\Pi^{<}_{ab}, \Pi^{=}_{ab}, \Pi^{>}_{ab})$$

with $\Pi^{<}_{ab} + \Pi^{=}_{ab} + \Pi^{>}_{ab} = 1.0$, where $\Pi^{<}_{ab}$ (respectively $\Pi^{=}_{ab}, \Pi^{>}_{ab}$) is the possibility of $a < b$ (respectively $a = b, a > b$). Here, and using the duality between possibility and necessity, namely $N(A) = 1 - \Pi(A^c)$, where A^c is the complement of A, we can derive the possibility and necessity degrees of each basic relations and their disjunctions.

Both models complete the proposal with the reasoning mechanism that, as we can see in the Figure 2, includes inversion, composition, addition, and negations operations. This operations define the rules that enable us to infer new temporal information and to propagate uncertainty in a probabilistic/possibilistic way. These rules complete the definition of a model for representing and reasoning with temporal uncertainty, handled in the settings of point algebra.

$$\begin{array}{ll} inversion & \Longleftrightarrow \widetilde{r}_{ab} = r_{ba} \\ composition & \Longleftrightarrow r_{ac} = r_{ab} \otimes r_{bc} \\ combination & \Longleftrightarrow r_{ab} = r1_{ab} \oplus r2_{ab} \\ negation & \Longleftrightarrow \neg \end{array}$$

Fig. 2. Operations for reasoning with uncertain temporal relations

4 The *Propos* Tool

In this section, we present *Propos*, the Web dynamic tool designed for managing temporal patterns. In order to show both the interface and the functionality, we will follow an example obtained after a temporal data mining process over Diagnostic Evolution Database. The example is a very representative temporal possibilistic pattern extracted in practical experience in an Intensive Care Unit (ICU) [7].

The pattern describes patients to whom physicians diagnosed at income day: d_6 (Acute Miocardial Infarction of the Pared Inferior), d_7 (haemorrage complications), and d_{169} (Acute Miocardial Infarction). Also, d'_{169} is diagnosed again the third day of stay in the ICU. The uncertain relation that the temporal mining process extract are:

$$\Pi_{d_6,d_7} = (0.44, 1, 0.44), \quad \Pi_{d_6,d_{169}} = (0.7, 1, 0.7)$$

$$\Pi_{d_7,d_{169}} = (0.7, 1, 0.7), \quad \Pi_{d_{169},d'_{169}} = (1, 0.87, 0.87)$$

In ICU domains, as well as the final diagnosis (like other hospital services), there are *evolutive diagnoses* that state the diagnostic hypotheses. These hypotheses are daily made by physicians during patient's stay at the ICU service. Furthermore, they can be considered high-level medical information since it is obtained from physician's knowledge and medical observations (like EKGs, tests, or nursing care data).

Despite the importance of other clinical information within the health record, such as treatments or demographic data, we consider in our experiment that the evolution of these diagnosis are a good representation of patient problems and the discovery of temporal pattern diagnosis could be useful in many AI systems for temporal diagnosis or prognosis.

In our experiment, each patient was represented in the database by a temporal sequence of diagnoses (temporal points) and the data mining process results were frequent temporal patterns (or frequent sequences) of diagnosis evolution. In the analysis of this data, different parameters had been empirically stated ($maxspan$ = 24 , and *support* value = 3, 5, 9) depending of the dataset of 144 patients.

Let us see how *Propos* works. Once *Propos* is started, and after the authentication process, we can create a new possibilistic project using the dymanic forms that we can see in Figure 3. With the project loaded, we can start to work within the network. First we have to accomplish is to create all the temporal points (nodes) of the network. By clicking "Add Node" in the control panel we will get a new temporal point in the desktop. The node properties can be modified opening the "Node Properties Inspector" clicking twice on the node. Once we got the nodes created, it's time to connect them creating relations and bringing life to the network. For creating a new relation, first we have to select, by clicking once on a node, two nodes and then click the "add link" button to establish the relation. After create the relation, we have to modify the possibility vector values, and it is done with the "Link Properties" inspector. In Figure 4 we can see the global network formed by binary pieces of temporal information.

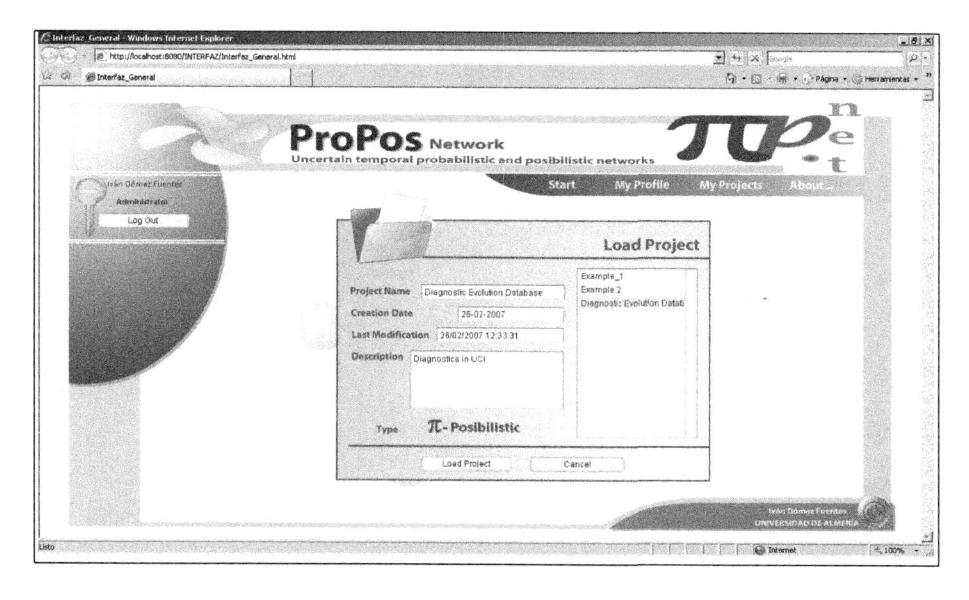

Fig. 3. Loading a project

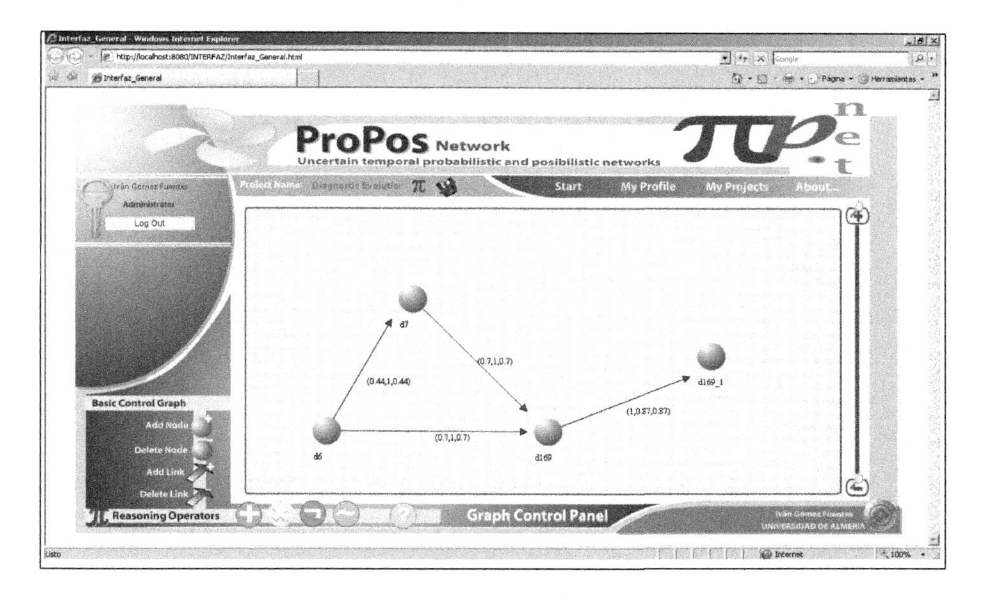

Fig. 4. The complete network

After the network has been created, we can use the tool as a Knowledge Acquisition tool. We can interact with it (combining two or more relations, adding new uncertain relations to the network, and so on) and we can obtain

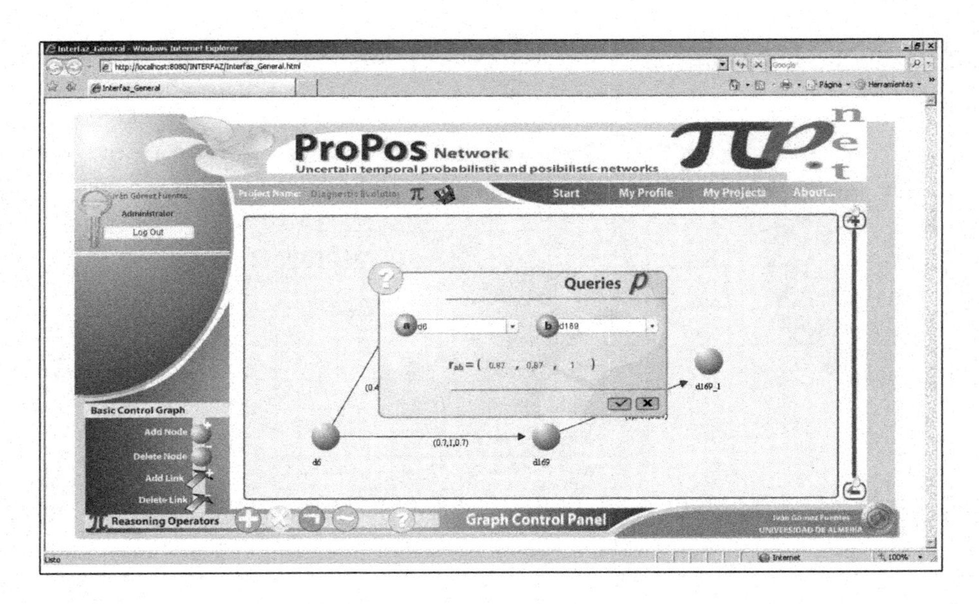

Fig. 5. Queries form

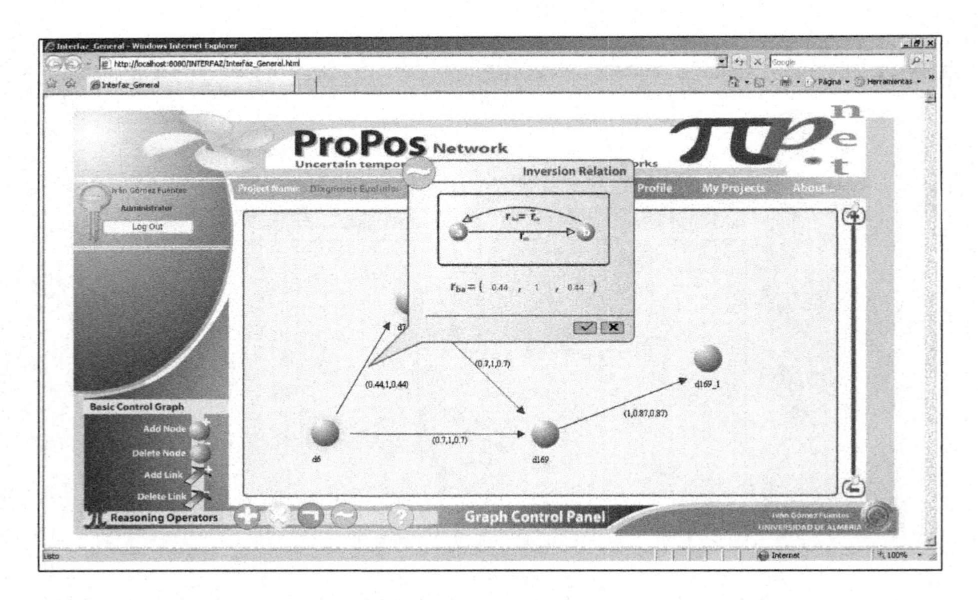

Fig. 6. Inversion operation form

new knowledge from it by opening the "Queries Form", located in the "Graph Control Panel" and represented by the symbol "?". When we open this form, the network is recombined in order to obtain its minimal representation, where every node is connected with the others. We can ask for two nodes and obtain the possibility values associated. We can see in the Figure 5 the "Queries Form" asking for the possibility degrees between the nodes d_6 and d_{169}. In this case, $r_{d_6,d_{169}} = (0.87, 0.87, 1)$.

When editing networks we can use the "Reasoning Operators" to avoid calculate possibility vectors when we get new information. In Figure 6 we can see "Inversion Operator Form":

5 Conclusions and Future Work

In this paper, we have presented a novel tool for representing and managing temporal constraint networks formed by uncertain temporal relations between point-based events. The tool allows to represent the uncertainty from two different point of view, the possibilistic and the probabilistic one. We propose the use od dynamic Web technology in order to obtain better mechanisms for representing the networks and understanding how they work they infer new knowledge. The Web-based technology will also allow us to use the tool from very different areas of the world.

As future work, we propose first to integrate the tool with an existing information system designed for clinical guidelines managing. The basic idea is to use *Propos* as an assistant for representing and reasoning with mined temporal patterns in a visual way. Secondly, we propose to extend the tool for representing uncertain temporal relations between temporal intervals.

Acknowledgments

This work is supported in part by MEC grant TIN2006-15460-C04-01 and MEC grant TIN2004-05694.

References

1. R. Agrawal, T. Imielinski, and A. N. Swami. Mining association rules between sets of items in large databases. In P. Buneman and S. Jajodia, editors, *Proc. of the ACM SIGMOD Int. Conf. on Management of Data, Washington, D.C., May 26-28, 1993*, pages 207–216. ACM Press, 1993.
2. L. Chittaro and A. Montanari. Temporal representation and reasoning in artificial intelligence: Issues and approaches. In *Annals of Mathematics and Artificial Intelligence*, number 28, pages 47–106. IEEE Computer Society, 2000.
3. R. Dechter, I. Meiri, and J. Pearl. Temporal constraint networks. *Artificial Intelligence*, 1-3(49):61–95, 1991.
4. D. Dubois and H. Prade. *Possibility Theory*. Plenum Press, 1988.
5. U. Fayyad, G. Piatetky-Shapiro, and P. Smyth. From data mining to knowledge discovery in databases. *AIMagazine*, 17(3):37–54, 1996.
6. F. Guil, A. Bosch, and R. Marín. TSET: An algorithm for mining frequent temporal patterns. In *Proc. of the First Int. Workshop on Knowledge Discovery in Data Streams, in conjunction with ECML/PKDD 2004*, pages 65–74, 2004.

7. F. Guil, J. M. Juárez, and R. Marín. Mining possibilistic temporal constraint networks: A case study in diagnostic evolution at intensive care units. In *Intelligen Data Anlisis in Biomedicine and Pharmacology (IDAMAP 2006)*, 2006.
8. F. Guil and R. Marín. Extracting uncertain temporal relations from mined frequent sequences. In *Proc. of the 13th Int. Symposium on Temporal Representation and Reasoning (TIME 2006)*, pages 152–159, 2006.
9. A. HadjAli, D. Dubois, and H. Prade. A possibility theory-based approach for handling of uncertain relations between temporal points. In *11th International Symposium on Temporal Representation and Reasoning (TIME 2004)*, pages 36–43. IEEE Computer Society, 2004.
10. H. Lu, L. Feng, and J. Han. Beyond intra-transaction association analysis: Mining multi-dimensional inter-transaction association rules. *ACM Transactions on Information Systems (TOIS)*, 18(4):423–454, 2000.
11. H. Mannila. Local and global methods in data mining: Basic techniques and open problems. In P. Widmayer, F. Triguero, R. Morales, M. Hennessey, S. Eidenbenz, and R. Conejo, editors, *In Proc. of the 29th Int. Colloquium on Automata, Languages and Programming (ICALP 2002), Malaga, Spain, July 8-13, 2002*, volume 2380 of *Lecture Notes in Computer Science*, pages 57–68. Springer, 2002.
12. J. F. Roddick and M. Spiliopoulou. A survey of temporal knowledge discovery paradigms and methods. *IEEE Transactions on Knowledge and Data Engineering*, 14(4):750–767, 2002.
13. V. Ryabov and S. Puuronen. Probabilistic reasoning about uncertain relations between temporal points. In *8th International Symposium on Temporal Representation and Reasoning (TIME 2001)*, pages 1530–1511. IEEE Computer Society, 2001.

BIRD: Biomedical Information Integration and Discovery with Semantic Web Services*

Juan Miguel Gomez[1], Mariano Rico[2], Francisco García-Sánchez[3], Ying Liu[4], and Marília Terra de Mello[5]

[1] Universidad Carlos III de Madrid
juanmiguel.gomez@uc3m.es
[2] Universidad Autónoma de Madrid
Mariano.Rico@uam.es
[3] Universidad de Murcia
frgarcia@um.es
[4] University of Texas at Dallas (UTD)
Ying.liu@utdallas.edu
[5] Universidade Federal do RioGrande do Sul
mtmello@inf.ufrgs.br

Abstract. Biomedical research is now information intensive; the volume and diversity of new data sources challenges current database technologies. The development and tuning of database technologies for biology and medicine will maintain and accelerate the current pace for innovation and discovery. New promising application fields such as the Semantic Web and Semantic Web Services can leverage the potential of biomedical information integration and discovery, facing the problem of semantic heterogeneity of biomedical information sources in a variety of storage and data formats widely distributed both across the Internet and within individual organizations. In this paper, we present BIRD, a fully-fledged biomedical information integration solution that combines natural language analysis and semantically-empowered techniques to ascertain how the user needs can be best fit. Our approach is backed with a proof-of-concept implementation where the breakthrough and efficiency of integrating the biomedical publications database PubMed, the Database of Interacting Proteins (DIP) and the Munich Information Center for Protein Sequences (MIPS) has been tested.

1 Introduction

Integration and exchange of data within and among organizations is a universally recognized need in bioinformatics and genomics research. By far the most obvious

* This work is founded by the Ministry of Science and Technology of Spain under the project DAWIS (TIC2002-04050-C02-01) and Arcadia (TIC2002-1948). We also thank the Spanish Ministry for Science and Education through the projects CIT-380000-2005-1 and TSI2004-06475-C02. The third author is supported by the Seneca Foundation through the FPI Program.

J. Mira and J.R. Álvarez (Eds.): IWINAC 2007, Part II, LNCS 4528, pp. 561–570, 2007.

frustration of a life scientist today is the extreme difficulty in putting together information available from multiple distinct sources. A commonly noted obstacle for integration efforts in bioinformatics is that relevant information is widely distributed, both across the Internet and within individual organizations, and is found in a variety of storage formats, both traditional relational databases and non-traditional sources (e.g. text data sources in semi-structured text files or XML, and the result of analytic applications such as gene-finding application or homology searches).

Arguably the most critical need in biomedical data integration is to overcome semantic heterogeneity i.e. to identify objects in different databases that represent the same or related biological objects (genes, proteins, etc) and to resolve the differences in database structures or schemas, among the related objects. Such data integration is technically difficult for several reasons. First, the technologies on which different databases are based may differ and do not interoperate smoothly. Standards for cross-database communication allow the databases (and their users) to exchange information. Secondly, the precise naming conventions for many scientific concepts (such as individual genes, proteins or drugs) in fast developing fields are often inconsistent, and so mappings are required between different vocabularies. Third, the precise underlying biological model for the data may be different (scientists view things differently) and so to integrate these data requires a common model of the concepts that are relevant and their allowable relations. This reason is particularly crucial because unstated assumptions may lead to improper use of information, on the surface, appears to be valid. Fourth, as our understanding of a particular domain improves, not only will data change, but even database structures will evolve. Any users of the data source, including in particular any data integrators must be able to manage such data source evolution.

Since the current Web is an environment primarily developed for human users, the need of adding semantics to the Web becomes more critical as organizations rely on the service-oriented architecture paradigms to expose functionality and data sources by means of Web Services. The Semantic Web is about adding machine-understandable and machine-processable metadata to Web resources through its key-enabling technology: ontologies [1]. Ontologies are a formal, explicit and shared specification of a conceptualization [2]. The breakthrough of adding semantics to Web Services leads to the Semantic Web Services (SWS) paradigm [3], which offers the possibility of ascertain which services could best fit the wishes and fulfill the goals of the user. SWS can be discovered, located and accessed since they provide formal means of leveraging different vocabularies and terminologies and foster mediation. However, the problem of bridging the gap between the current Web, primarily designed for human users whose intentions are expressed in natural language, and the formalization of those wishes remains. Potential users might deter from using SWS, since its underlying formalization and unease of use hampers its use from rich user-interaction perspective.

Hence, we present in this paper our work on the Biomedical Information and Integration Discovery with Semantic Web Services (BIRD) platform, which fosters the intelligent interaction between natural language user intentions and the existing Semantic Web Services available. Our contribution is an overall solution, based on a fully-fledged architecture and proof-of-concept implementation that transforms the user intentions into semantically-empowered goals. These goals are then used to find the most appropriate services by accessing their semantic description. Next, the services are invoked in the correct order and, eventually, the results returned to the user.

The remainder of this paper is organized as follows. Section 2 describes the Biomedical Information and Integration Discovery with Semantic Web Services (BIRD) platform for Semantic Web services. Section 3 presents the proof-of-concept implementation based in a real world scenario in which the integration of the biomedical publications database PubMed, the Database of Interacting Proteins (DIP) and the Munich Information Center for Protein Sequences (MIPS) has been tested. Finally, conclusions and related work are discussed in Section 4.

2 BIRD: Biomedical Information Integration Discovery

BIRD is a two-faced software agent designed to interact with and human beings as a gateway or a man-in-the-middle towards Semantic Web Services. The main goal of the system is to help users express their needs in terms of information retrieval and achieve both biological data and analytical functionality integration by means of SWS. BIRD allows users to either state their needs via natural language or go through a list of the most important terms, extracted from the Gene Ontology[1] (GO). For this, BIRD makes use of ontology-driven data mining. This implies that it firstly captures and gathers which are the terms the user would like to search (e.g. Gene A, Protein Y) by using as a reference the aforementioned terms of the GO. Secondly, it builds up a "lightweight ontology" i.e. a very simple graph made of the relationships of those terms. The lightweight ontology is then managed as the formal "goal" the system has to achieve on behalf of its user. A "goal" in SWS technology refers to the aim a user expects to fulfill by the use of the service. When BIRD has inferred the goal derived from the users' wishes, it starts looking for the necessary services to achieve the objective. With that purpose, BIRD queries the repository of services semantic descriptions. Once the appropriate services have been located, BIRD orchestrates them and starts their execution. Finally, BIRD retrieves the outcome resulting of the integration of the applications being accessed (e.g. all the biomedical, biological publications and medical databases) and presents it back to the user.

Fig. 1 depicts the main components of BIRD. The core component is the Manager, which supervises all the process and act as intermediary among the other components. The GUI is placed between the user and the Manager. The users have two possibilities: they can either introduce a text in natural language or use the ontology-guided tool (but this option would require further work which is not envisaged in this section), which assists them in expressing their

[1] See `http://www.geneontology.org/`

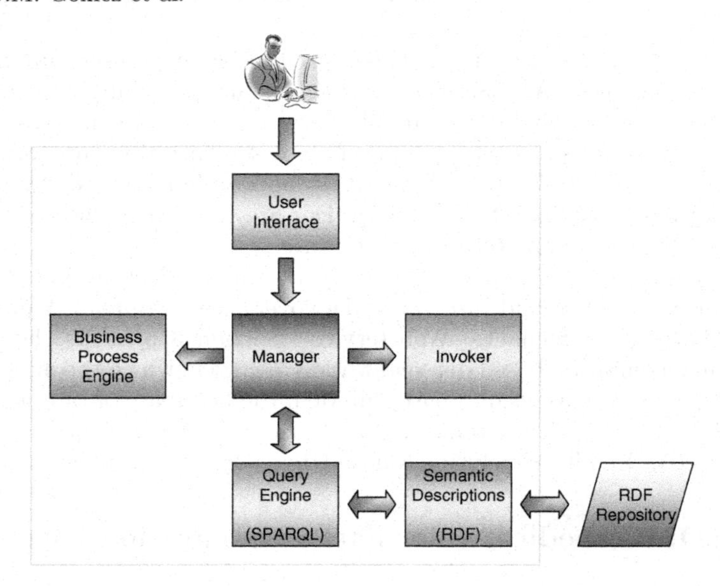

Fig. 1. The BIRD Architecture

goals. On the other hand, both the Query Engine and the Semantic Descriptions are responsible for finding and selecting the most suitable services. Then, the Business Process Engine determines how the orchestration should take place and the Invoker invokes the services one after the other. In the following subsections, a concise description of each of these components will be presented.

2.1 Manager

As the main component of our architecture, the Manager coordinates all processes among the various components of the architecture. It follows the execution semantics explained in the previous paragraph and ends the BIRD execution once the invoker has returned its results and the corresponding information is forwarded to the end-user.

2.2 Semantic Descriptions and RDF Repository

Semantic descriptions of the information resources connected to BIRD are currently paving the way to provide the automation and dynamic discovery of information resources. In our particular case, these semantic descriptions are SWS descriptions.

In our architecture, the user can use a two-fold approach to access these services: they can be described and annotated either with a semantic web language, such as OWL, or these descriptions can be found in a SWS goal repository.

Particularly, RDF use is highly advisable given the good results of the SPARQL language in its querying.

2.3 Query Engine

The Query Engine component searches and locates the most accurate biomedical resource to be accessed. A pivotal concept that enables the location of these ERP systems is usually known as discovery, what implies the fact of locating a semantic description of a resource that may have been previously unknown and that meets certain functional criteria.

In BIRD, discovery takes place querying the RDF descriptions of the multiple biomedical information systems interconnected. This query is performed using the SPARQL[2] language. The results of the query are a given number of SWS. These SWS will be invoked by the Invoker.

2.4 Business Process Engine

The Business Process Engine specifies the order or choreography of the invocations of BIRD, regarding how the different sources to be accessed or queried, should be addressed. This implies an order for temporal constraints and also a particular "workflow" or execution of such queries. The Business Process Engine also raises a challenge when a particular choreography provides more added-value than a given one and optimization techniques must be applied.

2.5 Invoker

This component simply uses the grounding of the Web Services technology to invoke and interact with the particular Semantic Web Service selected by the Query Engine.

2.6 User Interface

The users have two possibilities: they can either introduce a text in natural language or use the ontology-guided tool (but this option would require further work which is not envisaged in this section), which assists them in expressing their goals.

As it is shown in Fig. 2, for example, a biomedical researcher could write in plain natural language: "Give me information about the gene ASB1 and its effects on the anthranilate protein". This would revert into the triggering of Natural Language Processing (NLP) techniques to express the most important concepts of the query to configure a "lightweight ontology".

A more in-deep rundown on the case study is attained in the next section where the different components of the BIRD architecture fit in the picture of a value-added scientific case study.

---

[2] See `http://www.w3.org/TR/rdf-sparql-query/`

Fig. 2. BIRD Frontpage

3 Using BIRD for Biomedical Information Integration

In this section, we subsequently present a real-world based use case scenario in order to validate the appropriateness of the architecture presented above and show the advantages provided by BIRD from the user perspective. First, the three data sources that are being integrated are put forward. Then, some insights on the ontology utilized for integrating the disparate data sources are offered. Thirdly, Web Services developed that are capable of showing off some parts of the functionality available in the biomedical Web portals along with their semantic annotation are depicted. Finally, a typical example of how a common user might use the application is shown.

3.1 Data and Functionality Sources

Firstly, the PubMed[3] database, a free search engine offered by the United States National Library of Medicine (as part of the Entrez information retrieval system). The inclusion of an article in PubMed does not endorse that article contents and the service allows searching the MEDLINE database. MEDLINE covers over 4,800 journals published and also offers access to citations to articles that are out-of-scope (e.g., covering plate tectonics or astrophysics) from certain MEDLINE journals, primarily general science and general chemistry journals, for which the life sciences articles are indexed for MEDLINE, in-process citations which provide a record for an article before it is indexed and added to MEDLINE

[3] See `http://pubmed.gov/`

or converted to out-of-scope status and citations that precede the date that a journal was selected for MEDLINE indexing (when supplied electronically by the publisher).

Secondly, the Database of Interacting Proteins (DIP)[4] database catalogs experimentally determined interactions between proteins. It combines information from a variety of sources to create a single, consistent set of protein-protein interactions. The data stored within the DIP database were curated, both, manually by expert curators and also automatically using computational approaches that utilize the knowledge about the protein-protein interaction networks extracted from the most reliable, core subset of the DIP data.

Finally, the Munich Information Center for Protein Sequences (MIPS)[5] database was selected, through its endpoint MPact. MPact[6] provides a common access point to interaction resources at MIPS. It is designed to support for both downloading and uploading data related to protein interaction. It provides the user with intuitive query forms to quickly retrieve the interactions of interest. Graphical representations allow an easy navigation through the protein interaction networks.

3.2 The Integrating Ontology

An ontology of biological terminology provides a model of biological concepts that can be used to form a semantic framework for data storage, retrieval and analysis tasks. As it was aforementioned, the Gene Ontology (GO) [4] is the common reference model employed in BIRD.

GO is an ontology that describes attributes of gene products. GO offers a way to resolve the semantic heterogeneity of the annotations of genetic products in diverse databases: annotations of different databases are linked to the same GO term. The main GO components are the terms and their relations. Each term has a sole identifier apart from the name of the term. GO is divided into three independent ontologies: molecular functions, biological processes, and cellular components. Molecular functions describe the basic molecular role of gene products. Each biological process is comprised of different molecular functions and describes its role at conceptual level. The cellular component ontology represents the structure of the eukaryotic cell.

3.3 Semantic Web Services for Integration

In order to test BIRD, we have developed several Web Services that expose some of the functions provided by the three biomedical sources accessed. Each of these services has been subsequently annotated with semantic content using the OWL-S approach. For this purpose the OWL-S Editor [5], a plugin for Protégé that allows users to semantically annotate services, has been applied. So, for example, the service that provides global access to the MIPS database consist

[4] See `http://dip.doe-mbi.ucla.edu/`

[5] See `http://mips.gsf.de/services/ppi`

[6] `http://mips.gsf.de/genre/proj/mpact/`

```
<wsdl:definitions targetNamespace="http://klt.inf.um.es/">
  <wsdl:documentation> MPact Service </wsdl:documentation>
+ <wsdl:types></wsdl:types>
+ <wsdl:message name="getDescriptionMessage"></wsdl:message>
+ <wsdl:message name="getDescriptionResponseMessage"></wsdl:message>
+ <wsdl:message name="getInteractionMessage"></wsdl:message>
+ <wsdl:message name="getInteractionResponseMessage"></wsdl:message>
+ <wsdl:portType name="MPact_WebServicePortType"></wsdl:portType>
- <wsdl:binding name="MPact_WebServiceSOAP11Binding" type="axis2:MPact_WebServicePortType">
    <soap:binding transport="http://schemas.xmlsoap.org/soap/http" style="document"/>
  + <wsdl:operation name="getDescription"></wsdl:operation>
  + <wsdl:operation name="getInteraction"></wsdl:operation>
  </wsdl:binding>
+ <wsdl:binding name="MPact_WebServiceSOAP12Binding" type="axis2:MPact_WebServicePortType"></wsdl:binding>
+ <wsdl:binding name="MPact_WebServiceHttpBinding" type="axis2:MPact_WebServicePortType"></wsdl:binding>
+ <wsdl:service name="MPact_WebService"></wsdl:service>
</wsdl:definitions>
```

Fig. 3. MPact service WSDL file

```
<j.1:Service rdf:ID="getDescriptionService">
  <j.1:describedBy rdf:resource="#getDescriptionProcess"/>
- <j.1:presents>
  - <j.0:Profile rdf:ID="getDescriptionProfile">
      <j.1:presentedBy rdf:resource="#getDescriptionService"/>
    - <j.0:hasInput>
      + <j.2:Input rdf:ID="term"></j.2:Input>
      </j.0:hasInput>
      <j.0:hasOutput rdf:resource="#description"/>
    + <j.0:textDescription rdf:datatype="http://www.w3.org/2001/XMLSchema#string"></j.0:textDescription>
      <j.0:serviceName rdf:datatype="http://www.w3.org/2001/XMLSchema#string">getDescription</j.0:serviceName>
    </j.0:Profile>
  </j.1:presents>
+ <j.1:supports></j.1:supports>
</j.1:Service>
```

Fig. 4. A portion of the OWL-S file for the MPact service

of two methods, namely, getDescription and getInteraction, which return the description and the interactions of a protein respectively given its name. An extract of the WSDL file that describes the capability of the service is shown in Fig. 3. Finally, a piece of the OWL-S semantic description of the service is presented in Fig. 4.

3.4 Use Case Scenario

In this section we present a use case scenario (see Fig. 5) in which a user, hereafter Alice, aims to get some information about a particular gene. The first thing Alice must do is to type the application URL on her standard Web browser. Then, she expresses her query through a sentence in natural language as shown in Fig. 2. Once she had clearly stated her goal, she presses the "Send" button and BIRD starts to process the query. Firstly, it translates the natural language sentence into a lightweight ontology by means of a NLP tool. In the second place, BIRD searches for those services that can achieve the goal. This is done by matching the goal lightweight ontology with the semantic description of the services' capabilities. Next, the services are ordered by the Business Process Engine and invoked. Finally, the results are integrated and returned to the user.

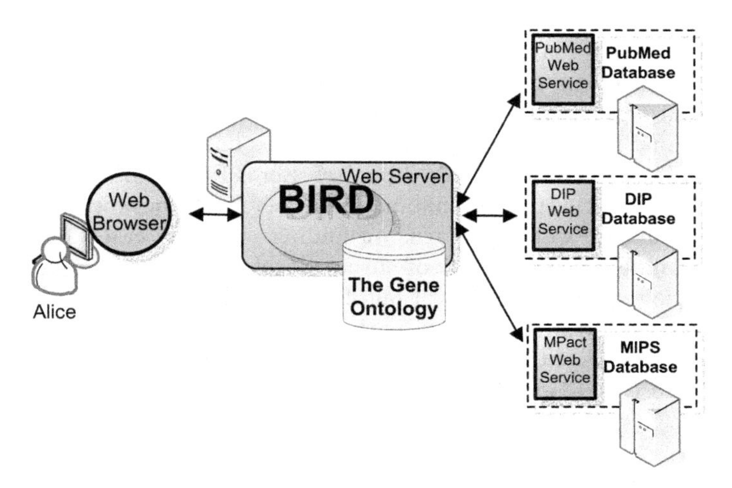

Fig. 5. Use case scenario

Our experience shows that users prefer guided typing when they begin using BIRD. This is because free typing becomes frustrating for novel users when BIRD is not able to match user' goals with goals in the repository. According to our experience, when users acquire enough experience they move to free typing. However, guided typing is not so easy when the number of options increase. Critical mass in goals repositories and more available services hamper the efficiency of the selection criteria. A potential solution is dividing the list in a hierarchical manner in order to avoid hoarding of an excessive number of elements to the user.

4 Conclusions and Related Work

The increasing volume and diversity of information in biomedical research is demanding new approaches for data integration in this domain. Semantic Web Services along with Semantic Web technologies can leverage the potential of biomedical information integration and discovery, facing the problem of semantic heterogeneity of biomedical information sources. In this paper we have proposed a solution and a proof-of-concept implementation to integrate both biological data and analytical functionality from disparate sources. Semantic Web Services provide access to these sources and BIRD executes them according to the user wishes. The forthcomings of our approach are mainly two, namely: striving for a rich and ease-of-use interaction to understand the user goals and the ability to interact with the existing Semantic Web Services.

Several projects are running at the moment aiming at developing a reference architecture for data and functionality integration in this application field. This is the purpose of the BioMOBY project[7] . A system for providing interoperability

[7] See `http://biomoby.org`

between biological data hosts and analytical services is being built within the scope of this project. This system defines an ontology-based messaging standard through which a client might be able to automatically discover and interact with task-appropriate biological data and analytical service providers, without requiring manual manipulation of data formats as data flows from one provider to the next. The Semantic MOBY [8] project stems from BioMOBY. It intends to provide an infrastructure for enabling interoperability between Web Services.

Finally, our future work will focus on finding more use cases and real-world scenarios to validate the efficiency of our approach and determine the feasibility of the semantic match of lightweight ontologies extracted from natural language text and ontologies defining goals in particular domains. This work is related to existing efforts about ontology merging and alignment. Future version of BIRD will be oriented towards that direction.

References

1. Berners-Lee, T., Hendler, J., Lassila, O.: The semantic web. Scientific American **284**(5) (2001) 28–37
2. Borst, W.: Construction of Engineering Ontologies for Knowledge Sharing and Reuse. Centre for Telematics and Information Technology (1997)
3. Fensel, D., Bussler, C.: The Web Service Modeling Framework WSMF. Electronic Commerce Research and Applications **1**(2) (2002) 113–137
4. Ashburner, M., Ball, C., Blake, J., Botstein, D., Butler, H., Cherry, J., Davis, A., Dolinski, K., Dwight, S., Eppig, J., et al.: Gene ontology: tool for the unification of biology. The Gene Ontology Consortium. Nat Genet **25**(1) (2000) 25–9
5. Elenius, D., Denker, G., Martin, D., Gilham, F., Khouri, J., Sadaati, S., Senanayake, R.: The OWL-S editor-a development tool for semantic web services. Proc. 2nd European Semantic Web Conf (ESWC2005) **3532** (2005) 78–92

[8] See `http://semanticmoby.org`

Neural Networks to Predict Schooling Failure/Success

María Angélica Pinninghoff Junemann, Pedro Antonio Salcedo Lagos, and Ricardo Contreras Arriagada

Informatics Engineering and Computer Science Department
Research and Educational Informatics Department
Universidad de Concepción, Chile
{mpinning,psalcedo,rcontrer}@udec.cl

Abstract. This paper depicts an already developed experience in search for a predictable mechanism with respect to the future performance of a student considering the numerous factors that influence in its failure/success. The use of different neural networks configurations in conjunction with a large data volume on top of detailed attributes consideration for each student makes for an adequate base for the results obtained to be analyzed. The idea behind this paper is to arrange a mechanism that allows us to estimate before hand taking into consideration data from the student in reference to family, social and wealth surroundings for the student future performance identifying those factors that favors the tendency to failure or success.

Keywords: Prediction, Schooling Performance, Neural Networks.

1 Introduction

The development of learning processes involves packages of complexes activities where as the success is not guaranteed. Such is the case that great efforts have been internationally devoted to make this teaching process be more effective identifying failures at early stages and weaknesses that can be remedied. An important conclusion is that the learning and performance of the studies is influenced by the interaction of different factors that are coupled to individual features, family and of the teaching institutions involved.

It seems clear that the socioeconomic and cultural features of the family affect the students' performance since they appear as determinant in the development of the student well before his/her entrance to the educational system and during the whole academic experience. In another stage, schooling performance is also affected by what happens within the classroom apart from features of the classmates and teachers.

Judging from this, being able to identify related factors associated to the impact about the educational process to an individual student could become determinant to the actual decision taking of corrective actions. Besides the above this paper depicts a comparable analysis about different neural networks architectures, which gravitate around the same database that considers a group of

J. Mira and J.R. Álvarez (Eds.): IWINAC 2007, Part II, LNCS 4528, pp. 571–579, 2007.

a bit more than 20000 students derived from PISA (Program for International Student Assessment) [10]. 46 variables were considered including among others the family wealth condition, school discipline environment, the number of permanent teachers at the specific school, location, size and type of school.

This article is arranged as follows: First section is made up of the present introduction; second section discuss some related work. On third section, factors related to the learning process are analyzed. Section four describes some general aspects related to neural networks, while section five points out in a summarized manner different architectures considered in this experience and section six shows the final results attained. Finally, section seven is the one showing the conclusions for the work.

2 Related Work

All data utilized in this paper comes from the PISA Program, mentioned in the first section. This program evaluates the abilities and knowledge in Reading, Math and Sciences of 15 years old students besides of gathering data from the educational and family environment. This database considers information from 42 different countries [11,12]. PISA strives to be a follow up mechanism of the students' performance in reading, math and sciences, for which every three years, ability and knowledge evaluations in these three areas are carried out.

Neural networks have been used in a variety of different areas that encompass medical diagnosis to the behavior prediction of specific human groups. There is a very interesting book [1], that presents some developments from leading experts and scientists working in health, biomedicine, biomedical engineering and computing areas. Applications include cardiology, electromyography, electroencephalography, therapeutic drug monitoring, among others. Areas include earthquake prediction, as shown in the work of Liu [8].

A work that calls for attention is the one developed by Karamouzis [5], which calls for a form of behavior prediction in teenagers' involved in serious crime in the USA; this work tries to identify risk factors using neural networks, perceptron type in particular, to influence over these factors in order to diminish recurrence risks. Although the student population considered in this work is extremely low, 166 cases in total of which 120 are used for training and do not allow to gather general conclusions, this work shows the interest in determining critical variables as a first step to modify some undesirable behavior.

The array of predictive applications that uses neural networks can be also depicted on the work of Chakarida [3], designed to provide information to assist in the reduction of fatalities, injuries and economic losses caused by car crashes.

Something closer to the work shown in this article is the one presented by Linstrom [7]. This is a proposal that applies artificial neural network to predict successful or unsuccessful completion of special education programming for students diagnosed with serious emotional disturbance. In line with this is also found the work of Wu [14], which uses neural networks to identify and diagnose learning disability problems even though this¡work still requires a greater degree of development to make it more precise.

An interesting work is the one where Cripps [4] presents the use of neural networks in order to predict academic performance over a large students population covering data collected over a number of years, even though it is referred to a specific group of data from a single institution; this makes a difference with our presentation. Another interesting work is the proposal of Nebot [9], that presents fuzzy models applied to an e-learning environment.

Wangs' work [13] presents a feedforward, backpropagation neural network as a first step towards an intelligent problem selection agent for the SQL-Tutor Intelligent Tutoring System. It firstly predicts the number of errors the student will make on a set of problems, and then in the second stage decides on a suitable problem for the student. However this is a very specific application.

Lastly it is convenient to cite Bekele's work [2], a research aimed at investigating the use of Bayesian networks for predicting performance of a student based on values of some identified attributes, although this is a particular domain and considers only 8 attributes for a student.

3 Relevant Factors in the Learning Process

A student academic performance related to the PISA study, considers a significant amount of variables. It can determine the most important factors assess its influence on a theme that must be taken into account for minimize the risk to failure during the learning process. From this study it can be concluded that representative variables from the socioeconomic and cultural aspects have a great impact in the performance of the students.

Students originating from wealthier families attain better performance than those originating in families with lesser incomes. The average socioeconomic incoming of the school attendance is also a factor on the students' performance. This effect gets greater since students' from a lower socioeconomic level usually attend schools where the average student is of the same characteristics. PISA results show that students' from a wealthier background obtain better marks on the tests, at the same time these students' are incline to attend schools with better facilities, more resources and well motivated teachers.

Nonetheless, there is not a single factor alone to explain why some schools or countries get better results, even though there are more relevant elements than others. The most important is the socioeconomic level of the students', parents' education, degree of relationship between teachers' and students', availability of learning tools and cultural assets at home, school environment, expectations and teachers' opinion of each student, motivation and degree of compromise of the teachers', students, identity with their schools, school's autonomy and the parents' participation in the learning process.

4 General Aspects

The idea behind this work is to operate with neural networks over a significant amount of data focused on the student achievement separating these students' in two groups: the successful group, those who obtain the approval marks on the subjects considered by the PISA study and the failure group, those who obtain marks lower than the minimum for approval. This way exiting from the implemented neural networks will indicate *Success* or *Failure*.

A large amount of prototypes was developed for this work, setting as common element the learning mechanism, and based on previous experiences it was decided to work with supervised learning networks, since they provided better results in the general performance. In the supervised learning the network can learn from the input and the error (the difference between the output gotten from the network and the expected output).

The necessary elements are the input, the expected output, the error definition and a learning general rule. The network good performance must be indicated assigning a small value to the error which is typically defined as a cost function.

The general rule of learning defines a systematic way to modify the values assigned to weights in order to reduce costs to a minimum. The most known of these rules is backpropagation. At the same time this mechanism is the principal error propagation method that provides the commercial application employed (*Neurosolutions*).

The error propagation is based on the approximation technique of descent gradients employed by the LMS algorithm (Least Mean Squared Error), also called Delta Rule [6], one of the most used training mechanisms for neural networks.

Neurosolutions considers two families of components: *Backprop*, which is responsible to propagate the error backwards in the neural network, besides to calculate gradients of weights; the second component, *Gradientsearch*, uses these gradients to update weights in the network aiming to find the global minimums (optimum values).

The weight updating rule considers different parameters and is given by the following expression (a detailed explanation can be found in documentation of the software used):

$$\Delta\overline{w} = f(\overline{x}, \overline{y}, \overline{w}, \nabla\overline{w})$$

where, $\overline{x}$: is an input vector; ;
$\overline{y}$: is an output vector;
$\overline{w}$: is a current weights vector;
$\nabla\overline{w}$: gradients for current weights.

From the available data different blocks are defined which will be used for training and validation of the network. These blocks are as follows:

Training data. Training is the process by which the network free parameters such as weights are adjusted to optimum values for which this block must have an adequate number of cases and sufficient in order to discover patterns that allows making a classification in the problem domain.

Data for cross validation. Periodically, while the network is being trained, performance of it is tested with a particular data set, specifically selected. During this test, weights are not modified, and its objective is to determine the moment at which the networks starts to degrade due to overtraining.

Data for testing. Is the data set selected for testing performance of the network; once the training process is finished, weights are frozen, the the test set is handed over to the network and the output is compared to the expected output in order to get an evaluation of the performance quality.

5 The Neural Networks Architectures Considered

From the neural network theory and from some prototypes developed, it does not seem to be any reason to utilize architectures that consider more than two hidden layers; in fact experimenting with more than two hidden layers only increases response times (about 200%) with no better output results. Different networks are tested with a different amount of neurons per layer. Activation functions are modified on different networks on the hidden layers as well as on the output.

The step size affects the speed of the algorithm convergence. This speed is controlled with the learning rate α. The idea is to use an adequate value in order to avoid oscillation in the network, if α is too large and in order to prevent a great number of iterative steps which would imply a lesser value.

The number of training cycles (epochs) was set at 1000; for cases that the learning curve shows an unwanted behavior this training can be stopped before reaching this value. Percentages utilized in all cases for the data sets are shown in the following table:

Table 1. Data percentages

Data Sets	Percentages	Number of Cases
Training	55%	11690
Cross Validation	30%	6375
Testing	15%	3177

Architectures considered grouped into families are as follows:

- Network type: Perceptron
  Hidden layers: 1 or 2
  Activation function: Hyperbolic tangent, Sigmoid, Linear
  Gradient type: Momentum (from 0.7 to 0.8)
  Step size hidden layer: 0.1 to 1.0
  Numbers of neurons on hidden layer: from 8 to 30
  Step size on output layer: from 0.01 to 0.1
- Network type: Global feed forward
  Hidden layers: 1 or 2

Activation function: Hyperbolic tangent
Gradient type: Momentum (0.8)
Step size on hidden layer: 1
Number of neurons on hidden layer: from 8 to 16
Step size on output layer: 0.1

- Network type: Recurrent (partial)
  Hidden layers: 1 or 2
  Activation function: Hyperbolic tangent
  Gradient type: Momentum (0.8)
  Step size on hidden layer: 0.1
  Number of neurons on hidden layer: from 16 to 24
  Step size on output layer: 0.1
- Network type: Modular
  Hidden layers: 1 (in two levels)
  Activation function: Hyperbolic tangent
  Gradient type: Momentum (0.8)
  Step size on hidden layer: 0.1
  Number of neurons on hidden layer: 8 per level
  Step size on output layer: 0.1

6 Results

In the perceptron hidden layer case, training reaches its peak before completion of the 1000 cycles of training and percentage of error is slightly over 22%. Adding to the number of neurons from 8 to 12 this error diminishes hence continuing adding more neurons to the hidden layer until this number reaches 30; although

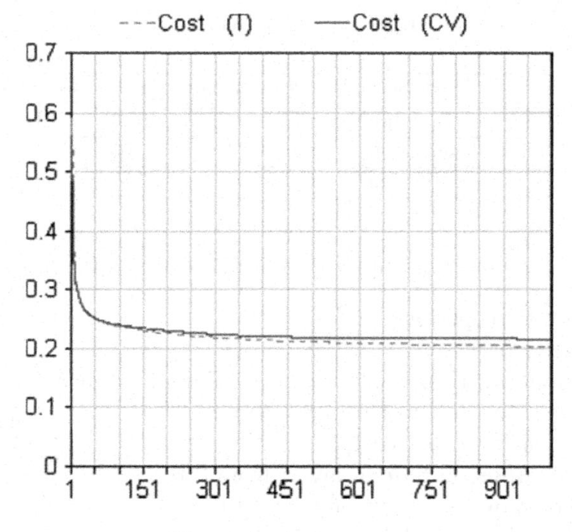

Fig. 1. Learning curves

Training

	Success	Failure
Success	82.290413	17.709583
Failure	17.688236	82.311760

Testing

	Success	Failure
Success	82.453911	17.546091
Failure	19.950434	80.049568

Fig. 2. Classification matrixes

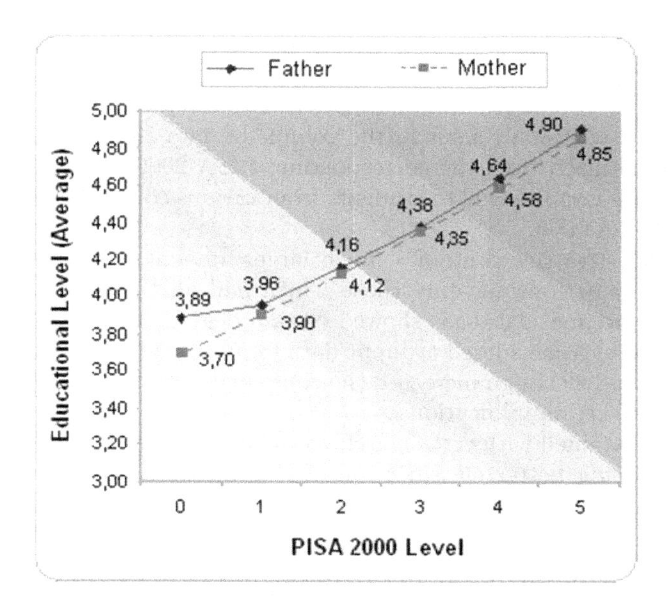

Fig. 3. Parents' influence

in this situation the network shows an erratic behavior, thus indicating that increasing this parameter must be done carefully. The best behavior for this architecture is attained with 16 neurons on the hidden layer.

In the feedforward network case, the results do not vary significantly with respect to the ones attained with the perceptron network, although the computer time increases. Using a recurrent network, results showed an error increase of almost 50% with respect to perceptron network even though the model holds a good generalization, detected by a minimum difference between the training data and the cross validation. In the case of modular net the error is also larger than the perceptron case in around 2%.

The best behavior is presented by the perceptron with a hidden layer containing 16 neurons. The learning curve is the one shown in figure 1, while percentage classification matrixes are shown in figure 2. In both matrixes the explanation is the same, on the first row successful cases are shown correctly as successful cases (first column) then the successful cases that the network shows as failures (second column). The second row interpretation is similar.

Even though the analysis proves a well defined difference between father and mother, results are coherent with international studies. Thus, the mothers' educational level has a greater influence in the academic performance than the fathers' educational level. Tendency of the sample considered establishes the students' greater level of performance corresponds to a greater average level of education of the parents. This is represented in figure 3.

7 Conclusions

An important conclusion relates to the volume of data considered in this work. The database utilized from the corresponding PISA 2000 evaluation contains a great amount of data referred to students from various countries which conforms a high variable domain.

The original database contain a much larger amount of records, but it was filtered in order to consider only those which had all the information (lots of records in the original database showed certain degree of incompleteness). The amount of variables considered as input data for each student is also large, using 46 attributes; and although increased the complexity of the computer processing it avoided arbitrary simplifications.

The results attained with certain architectures are better if the neural network is capable of doing better classification of the input set based on the already defined academic performance. In this experience, the best result were obtained from architectures perceptron, that are used as a base for the analysis trying to identify more influential variables in the output of the neural net; in other words those who determine which factors of the input set have a direct relationship with the students' performance.

The analysis allowed identifying a specific influence in each input set which served to categorize in line of importance those variables to focus into detailed factors which proved to have a larger sensibility with respect to the output. And also the country variable is determinant in the results, which were expected, it is interesting to illustrate the results attained when considering parents' education.

In this particular case we must recognize the great complexity and almost null possibility to revert the effects from these indexes on the students since almost nothing can be done to improve the educational level of the parents. That's the reason to consider a large amount of variables and select those who are relevant and at the same time allow certain degree of intervention in order to positively modify the students' performance.

Acknowledgements. This work is supported by Project DIUC 205.164.001-1.0, University of Concepción.

References

1. Begg, Rezaul; Kamruzzaman, Joarder; Sarkar, Ruhul (Eds.). *Neural Network in Healthcare: Potential and Challenges.* Idea Group Publishing, 2006.
2. Bekele, R.; Manzel, W.*A Bayesian Approach to Predict Performance of a Student (BAPPS): A Case with Ethiopian Students.* The IASTED Conference Artificial Intelligence and Applications, AIA 2005, Innsbruck, Austria, February 2005.
3. Chakarida, Nukoolkit; Hui-Chuan Chen; David Brown. *A Data Transformation Technique for Car Injury Prediction.* 39th ACM-SE, Athens, Georgia, USA, 2001.
4. Cripps, A. *Using Artificial Neural Nets to Predict Academic Performance.* Proceedings of the ACM Symposium on Applied Computing. Philadelphia, PA, USA. February 1996.
5. Karamouzis, S.T.; Katsiyannis, A.; Archwamety, T. *An Application of Neural Networks for Predicting Juvenile Recidivism.* The IASTED Conference on Artificial Intelligence and Applications. Innsbruck, Austria, February 2006.
6. Kroese, B.; van der Smagt, P. *An Introduction to Neural Networks.* University of Amsterdam, 1996.
7. Linstrom, Kristopher; Boye, John. *A Neural Network Prediction Model for a Psychiatric Application.* Proceedings of the Sixth International Conference on Computational Intelligence and Multimedia Application (ICCIMA'05). IEEE Computer Society, Washington, DC, USA, 2005.
8. Tian-Yu Liu, Guo-Zheng Li, Yue Liu, Gengfeng Wu, Wei Wang. *Estimation of the Future Earthquake Situation by Using Neural Networks Ensemble.* Advances in Neural Networks. Lecture Notes in Computer Science, Vol. 3973. Springer Verlag, 2006.
9. Nebot, A.; Castro, F.; Mugica, F.; Vellido, A. *Identification of Fuzzy Models to Predict Students Performance in a E-Learning Environment.* The IASTED Conference on Web-based Education. Puerto Vallarta, Mexico, January 2006.
10. OECD. *Knowledge and Skills for Life. First Results from PISA 2000.* Unesco 2001.
11. OECD. *PISA 2000 Technical Report.* Unesco 2001.
12. OECD. *Education at a Glance, OECD Indicators.* Unesco 2002.
13. Wang, T.; Mitrovic, A. *Using Neural Networks to Predict Student's Performance.* International Conference on Computers in Education, ICCE'02. Auckland, New Zealand, December 2002.
14. Wu, Tung-Kuang; Meng, Ying-Ru, Huang, Shian-Chang. *Application of Artificial Neural Network to the identification of Students with Learning Disabilities.* Proceedings of the 2006 International Conference on Artificial Intelligence, ICAI 2006, Las Vegas, Nevada, USA, June 2006.

Application of Genetic Algorithms for Microwave Oven Design: Power Efficiency Optimization

Juan Monzó-Cabrera[1], Alejandro Díaz-Morcillo[1], Elsa Domínguez-Tortajada[2], and Antonio Lozano-Guerrero[1]

[1] Depto. Tecnologías de la Información y las Comunicaciones, Universidad Politécnica de Cartagena, Cuartel de Antiguones, 30202 Cartagena, Spain
juan.monzo@upct.es
[2] Colegio Oficial de Ingenieros de Telecomunicación, Comunidad Valenciana, Avda. Jacinto Benavente, 12, 1B, 46005 Valencia, Spain

Abstract. In this work we present power efficiency optimization for microwave ovens by means of genetic algorithms (GA). Two kind of microwave applicators are analyzed in this case: cylindrical and rectangular ones. In the first case, optimization of the oven uses cavity dimensions, waveguide feeding location and polarization as design parameters. In the second case, waveguide slots are used to feed the rectangular multimode cavity. The slots' dimension, position and angle in the waveguide are optimized by the genetic algorithm in order to achieve the best power efficiency. All simulations are carried out by the CST Microwave Studio^{TM}, a Finite Integration Technique (FIT) commercial software capable of solving electromagnetic (EM) structures in three dimensions whereas GA are implemented in Matlab^{TM}. The obtained results show that combination of both the EM software and GA provides a very powerful tool for microwave oven design and optimization.

1 Introduction

Industrial microwave heating devices need to be very efficient in order to conform to ever-demanding economic requirements. This feature is particularly important since traditional industrial heating technologies employ energy sources at much lower cost than that of electricity. Unfortunately for the designer, multimode cavities show a much more complex impedance behavior than single-mode applicators. Due to this, there are very few works that deal with the efficiency optimization for this kind of devices.

Multimode microwave-heating cavities that contain high permittivity dielectrics may be adapted by several means. Conventional ones require the use of irises and tuning screws acting as triple stubs [1], [2]. These elements, however, need to be finely tuned and tweaked to provide an adequate match to load and frequency conditions. Thus, several authors have tried to avoid their use by means of other approaches such as the use of dielectric moulds [3]. This alternative leads to several practical problems such as heat migration, accurate

J. Mira and J.R. Álvarez (Eds.): IWINAC 2007, Part II, LNCS 4528, pp. 580–588, 2007.

mechanical specifications or the need to manufacture materials with very precise dielectric constants. Therefore, other authors have developed other technique that consists of placing the sample at an optimal location within the cavity [4].

In this work two different design alternatives are studied in order to obtain good load matching levels for two different kinds of microwave applicators. In the first case, a comprehensive design is carried out over a multimode cylindrical cavity partially loaded with a rectangular dielectric sample so that the reflections are minimized. In the second case, a rectangular multimode applicator is fed by means of a slotted waveguide. The slots placed at the lower wall of the feeding waveguide are allowed to change their dimensions, position and angle in order to obtain the best adaptation.

2 Theoretical Background: Efficiency and Genetic Algorithms

2.1 Power Efficiency

In this work we will use two different multimode cavities, both of them being fed through a single waveguide port. Reflections at this feeding port are usually characterized and measured by the so-called reflection coefficient ρ or the S_{11} scattering parameter

$$\rho = S_{11} = b/a. \tag{1}$$

In this case b represents the normalized reflected wave at the feeding port and a the incident one. From this definition, the power efficiency η can be written as

$$\eta = 1 - |S_{11}|^2 . \tag{2}$$

Therefore, the power efficiency can be related to the S_{11} scattering parameter which is readily provided by most commercial EM software. Usually S_{11} strongly varies as a function of frequency and consequently efficiency design must take into account this fact. Because of this, efficiency has been optimized for the whole 2.4-2.5 GHz Industrial Scientific and Medical (ISM) band.

2.2 Genetic Algorithms

In order to design and optimize power efficiency in the proposed multimode microwave ovens, genetic algorithms (GA) have been chosen as the optimization strategy in this work for several reasons: they provide global optimization and, therefore, can avoid local minima; they can handle multiple optimization parameters and non-continuous functions and have been used successfully in previous works [5], [6].

The implementation of the GA configuration can be found in previous works [5], [7] and is not reproduced here for the sake of conciseness. Basically, GA are used to obtain the optimum design in terms of power efficiency, which implies obtaining the dimensions and feeding characteristics in the case of the cylindrical applicator and the slot structure for the waveguide feeding in the case of the rectangular oven. A combination of these design parameters constitutes an individual that is assessed by using the evaluation function (f):

$$f = 2 - \left(\frac{\sum_{k=1}^{M} |S_{11}(k)|}{M} \right) = 2 - \left| \overline{S}_{11} \right| \ . \tag{3}$$

In (3), M represents the number of frequency points used in the S_{11} simulation, 1001 in this case. In this way, f evaluates the frequency mean value of the S_{11} magnitude. As it can be observed f is positively defined since $\left|\overline{S}_{11}\right|$ can never exceed 1.

GA have been implemented in Matlab^{TM} as a Toolbox whereas CST Microwave Studio^{TM} has been used to evaluate the fitness of each individual since it provides S_{11} as a function of frequency. Communication between both commercial applications was accomplished by means of text files. The used GA configuration always tries to maximize the evaluation function, which in this case is accomplished when S_{11} is minimized.

For the cylindrical cavity, the GA were configured in the following way: 5 individuals per generation, 10 generations, 5 arithmetic crossovers and 2 non-uniform mutations per generation. For the cubic applicator, however, crossovers per generation were reduced to 2 and mutations to 1.

3 Simulation Scenarios

3.1 Cylindrical Cavity: Scenario A

Fig. 1 shows the considered scenario for the cylindrical multimode cavity (Scenario A). In Fig. 1a the sample location and dimensions are depicted. As it can be perceived, sample is centered within the cavity and placed on a cylindrical PTFE tray that completely fills the section of the cavity. PTFE tray thickness is 0.5 cm and is located 6 cm upon the lowest wall of the cavity whereas sample dimensions are 5.58x0.92x4.35cm^3 along the x, y and z axis, respectively.

Fig. 1b shows the design parameters of the cylindrical cavity, which correspond to the genes of each individual: R is the cavity radius, H is the cavity height, h is the distance of the waveguide-feeding center to the bottom of the cavity, θ represents angle of the longest waveguide dimension with respect to the bottom of the cavity and φ is the angle of the center of the waveguide with respect to x axis.

Due to limitations of CST Microwave Studio^{TM} ,θ can only be chosen from two values: 0 and 90 whereas φ can only be assigned the angle: 0, 90, 180 or 270 . Additionally, the variation ranges for the rest of design parameters are: $H \in$[20,40] (cm), $R \in$[10,20] (cm) and $h \in$[11,35.5] (cm). The sample used in simulations was an epoxy resin with relative electric permittivity equal to $\epsilon_r^* = 3.78 - j0.92$ and the assumed PTFE relative permittivity value was $\epsilon_r^* = 2.1 - j0.0003$.

3.2 Rectangular Cavity Fed by a Slotted Waveguide: Scenario B

Scenario B consists of a 24x24x24cm^3 cubic applicator such as the one depicted in Fig. 2 and is fed by means of a slotted waveguide that is excited with its

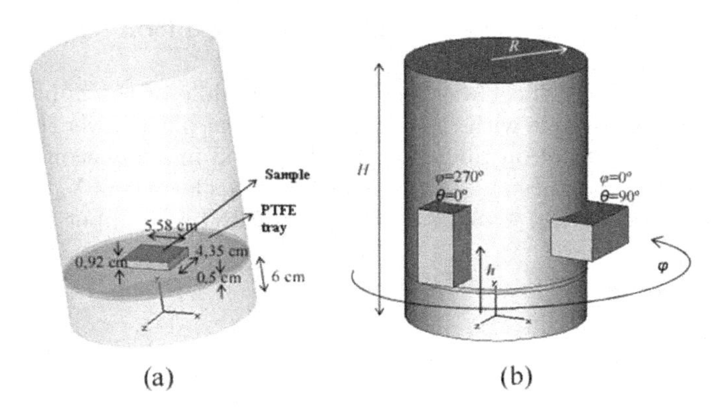

Fig. 1. (a) Sample, tray dimensions and location (b) Design parameters for efficiency optimization

fundamental mode TE_{10}. In Fig. 2, the depicted slot configuration is equally spaced and is not optimized. It will be used only for comparison purposes.

As it can be perceived from Fig. 2, the waveguide is placed on the upper side of the microwave oven. We considered four slots in the waveguide. The parameters to be optimized are in this case: dimensions, position, angle of the slots in the $x - z$ plane (ψ) and the thickness of the upper wall. Slot superposition was

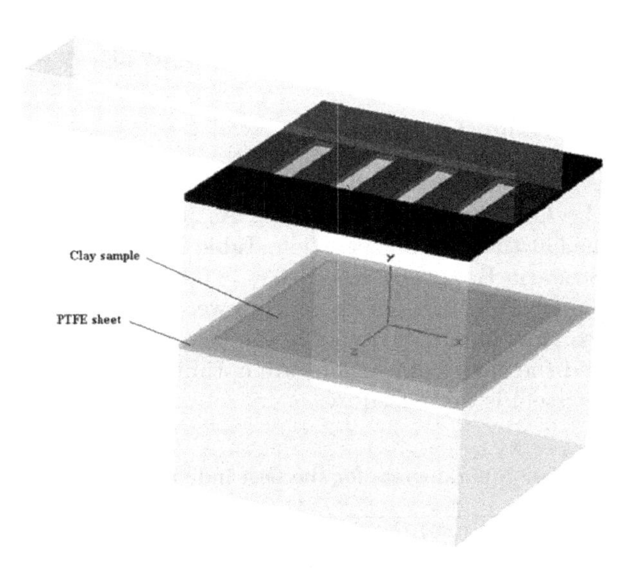

Fig. 2. Rectangular cavity with slots equally spaced at the upper wall. Non-optimized design.

allowed and consequently the slots could intersect and form bigger ones during the optimization process.

A 1 cm thick PTFE sheet was placed at the middle of the applicator and a 20x0.3x20cm^3 clay sample with $\epsilon_r^* = 26.64 - j5.45$ was placed over the sheet. The allowed parameter variation range for the slots used in simulations was: length (l_s) $\in$[0.3,4] (cm), width (w_s) $\in$[0.3,11.5] (cm), x-axis location (X_s) $\in$[-11.5,11.5] (cm), z-axis location (Z_s) $\in$[-3.9,3.9] (cm), $\psi \in$[0,180] (°), upper wall thickness (t_w) $\in$[0.1,0.9] (cm).

4 Results

The GA algorithms and the CST Microwave Studio^{TM} have been applied to scenarios A and B in order to obtain good optimum matching and efficiency values across the 2.4-2.5 GHz ISM band.

4.1 Optimized Design for Scenario A

Fig. 3 depicts the optimization evolution for scenario A versus the generation number. Instead of showing evaluation function (f) we show the average absolute value for S_{11}, which must be minimized. Table 1 shows the optimized parameters for Scenario A and its efficiency. From obtained data one can appreciate that the best solution is found at the last generation achieving a very low average magnitude value for S_{11} (0.067) and an average efficiency in the ISM band of 99.55%. Additionally, Fig. 4 represents S_{11} magnitude in the 2.445-2.455 GHz frequency range showing a minimum value around 2.4775 GHz and very low S_{11} values even for magnetron deviations of ±5 MHz from their nominal frequency (2.45 GHz).

4.2 Optimized Design for Scenario B

Fig. 5 represents the evolution of all individuals' S_{11} magnitude in each generation during the GA optimization process. As it can be observed the best solution is obtained at the fourth generation. Table 2 writes down the optimized parameters for Scenario B. In this case, $|\overline{S}_{11}| = 0.382$ and the efficiency value ($\eta = 85.4\%$) although is a high one, is much lower than that obtained in Scenario A. The reason for this can be found in the fact that in Scenario A both the feeding system and the cavity dimensions are optimized whereas for Scenario B only the feeding system is optimized.

Table 1. Design parameters for the best individual in Scenario A

h(cm)	H(cm)	R(cm)	φ(°)	θ(°)	$\|S_{11}(k)\|$	η
16.25	35.25	10.15	180	0	0.067	99.95

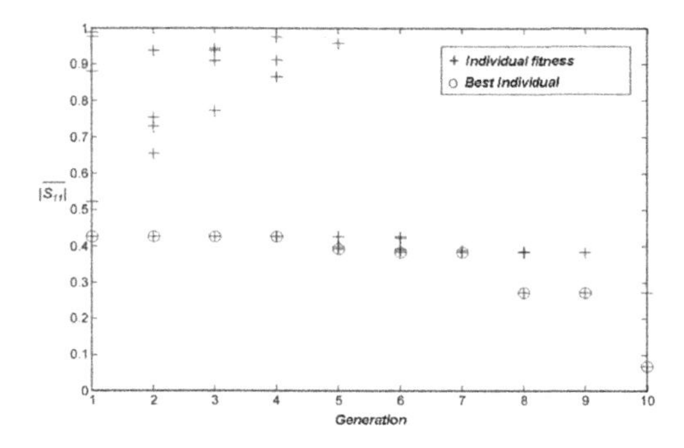

Fig. 3. Evolution for Scenario A optimization

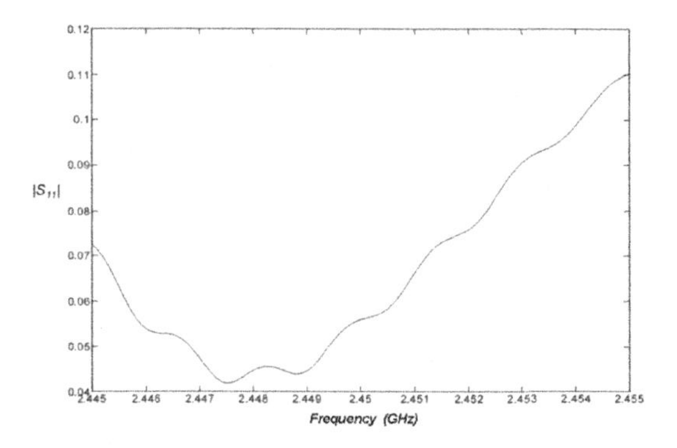

Fig. 4. Frequency response for the optimum cylindrical cavity design

Table 2. Design parameters for the best individual in Scenario B

Slot number	l_s(cm)	w_s(cm)	X_s(cm)	Z_s(cm)	ψ(rad)	t_w(cm)
1	3.016	10.26	2.245	3.748	1.003	
2	2.087	6.052	-2.195	-1.474	0.6451	
3	2.672	6.512	-11.59	2.969	0.9681	0.4611
4	2.542	1.758	1.466	-0.1826	1.197	

In Fig. 6 the frequency behavior for both the optimized and non optimized (Fig. 2) designs of Scenario B can be found. It is obvious from this data that optimization provides much less reflections (lower S_{11} magnitude) and therefore better power efficiency. Additionally, the design is especially interesting for 2.45-2.47 GHz band since it provides the best efficiency at this frequency range.

The designs obtained by means of GA are depicted in Fig. 7 for both Scenario A and B. The geometry of both designs differs greatly. In the case of the cylindrical microwave oven (Fig. 7a) the design and mechanization is simple whereas for the rectangular applicator (Fig. 7b) the waveguide slot is quite irregular and has floating pieces of metal which could not be physically implemented.

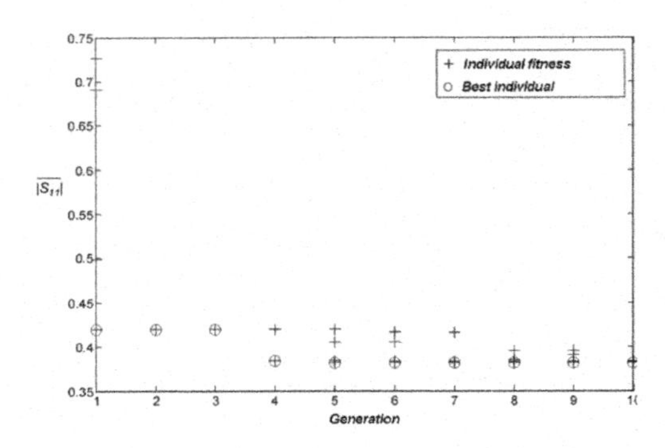

Fig. 5. Evolution for Scenario B optimization

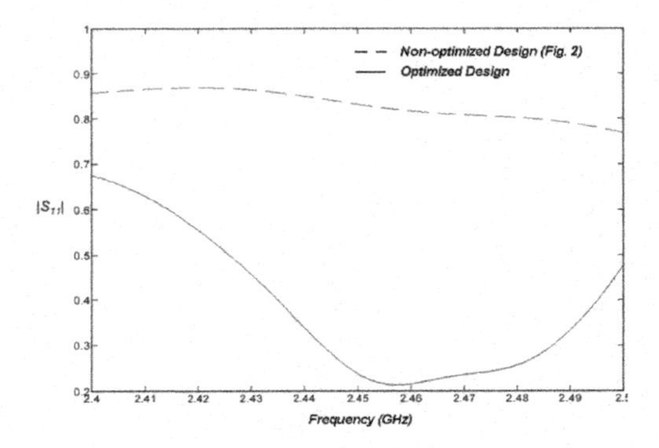

Fig. 6. Rectangular cavity frequency response for the optimum slotted waveguide design

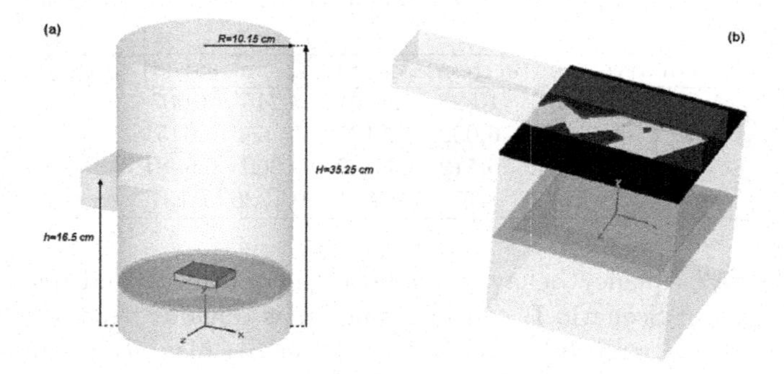

Fig. 7. Schemes for the optimized cylindrical (a) and rectangular (b) microwave oven designs

Finally, the electric field distribution at the sample, which is an indicator of the microwave oven heating quality [7], is shown for both applicators in Fig. 8. In Fig. 8a it can be appreciated that the electric field pattern across the epoxy

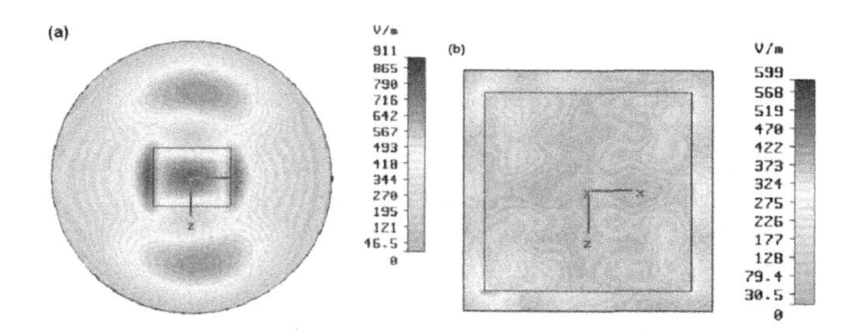

Fig. 8. Electric field distributions at the sample for (a) Scenario A and (b) Scenario B

sample in the cylindrical cavity is quite uniform although the sample borders show less field intensity. In Fig. 8b the electric field at the clay sample within the rectangular oven is depicted. In this case, the electric field is very uniform along the sample.

Therefore, although it was not the main objective of this work, the designed applicators provide uniform heating which is also a very important feature of this kind of devices [7].

5 Conclusions

A new design strategy based on the use of GA and the CST Microwave Studio™ commercial EM simulator has been presented and applied to obtain high efficiency microwave ovens. Two different methods have been applied. In the first case, the GA have varied all the permitted parameters that constituted the cylindrical cavity. In the second one, only the waveguide slots that fed the rectangular applicator were allowed to vary. Both approaches have obtained very good results. In the case of the cylindrical applicator an average efficiency of 99.55% has been obtained across the ISM band. The optimization of waveguide slots position and orientation has achieved an average efficiency value of 85.4% in the ISM frequency range, which is lower than the efficiency obtained in the cylindrical case.

From obtained results it seems logical to conclude that the more freedom parameters are given to the GA, the better solution they are capable to provide since GA have shown that are able to manage complex optimization processes.

Finally, although no experimental validation has been provided in this work, it must be remarked that this procedure and the CST Microwave Studio have been validated successfully in previous works with experimental tests [6], [7].

Acknowledgments. This work was supported in part by Fundación Séneca under project reference 00700/PPC/04.

References

1. Metaxas, A.C., Meredith, R.J.: Industrial Microwave Heating. Ed. Peter Peregrinus London. (1981)
2. Meredith, R.J.: Engineer's Handbook of Industrial Microwave Heating (Power & Energy Series). Institution of Electrical Engineers (1998)
3. Domínguez-Tortajada, E., Monzó-Cabrera, J.,Díaz Morcillo, A.: Load Matching in microwave-heating applicators by means of genetic algorithms optimization of dielectric multilayer estructures. Microwave and Optical Technologie Letters **47** (2005) 426–460
4. Requena-Pérez, M.E.,Monzó-Cabrera, J.,Pedreo-Molina, J.L., Díaz-Morcillo, A.: Multimode Cavity Efficiency Optimization by Optimum Load Location - Experimental Approach. Microwave Theory and Techniques (2005) 2114-2120
5. Monzó-Cabrera, J., Escalante, J., Díaz-Morcillo, A., Martínez-González, A., Sánchez-Hernández, D.: Load Matching in Multimode Microwave-heating Applicators based on the use of Dielectric-layer Moulding with Commercial Materials. Microwave and Optical Technologie Letters **41** (2004) 414–417
6. Requena-Pérez, M.E., Albero-Ortiz, A., Monzó-Cabrera, J., Díaz Morcillo, A.: Genetic Algorithms and Gradient Descent Optimization Methods for Accurate Inverse Permittivity Measurement. IEEE Transactions on Microwave Theory and Techniques **54** (2006) 615-624
7. Domínguez-Tortajada, E., Monzó-Cabrera, J., Díaz-Morcillo, A.: Uniform Electric Field Distribution in Microwave Heating Applicators by means of Genetic Algorithms Optimization of Dielectric Multilayer Structures. IEEE Transactions on Microwave Theory and Techniques **55** (2007) 85–91

Application of Neural Networks to Atmospheric Pollutants Remote Sensing

Esteban García-Cuesta[1], Susana Briz[2], Isabel Fernández-Gómez[1], and Antonio J. de Castro[1]

[1] LIR - Laboratorio de Sensores y Teledetección IR, Departamento de Física, Universidad Carlos III de Madrid
[2] Departamento de Ciencias, Universidad Europea de Madrid
egc@fis.uc3m.es, susana.briz@fis.cie.uem.es, ifgomez@fis.uc3m.es, decastro@fis.uc3m.es

Abstract. Infrared remote sensing is an extended technique to measure "in situ" atmospheric pollutant gas concentration. However, retrieval of concentrations from the absorbance spectra provided by technique is not a straightforward problem. In this work the use of artificial neural networks to analyze infrared absorbance spectra is proposed. A summary of classical retrieval codes is presented, highlighting advantages and important drawbacks that arise when these methods are applied to spectral analysis. As an alternative, a neural network retrieval approach is suggested, based on a multi layer perceptron. This approach has been focused to the retrieval of carbon monoxide concentration, because of the great environmental importance of this gas. Absorption overlapping of atmospheric gases such as carbon dioxide, nitrous oxide or water vapour is one the most important problem in the retrieval process. The training dataset has been generated with special care to overcome this aspect and guarantee a successful training phase. Results obtained from the ANN method are very promising. However, high retrieval errors have been found when ANN method is applied to experimental spectra. This fact reveals the need of a deep study of the instrumental parameters to be included in the model.

1 Introduction

One of the most important problems of the current society is atmospheric pollution. The importance of this issue has led the European Union to design European environmental policies addressed to control and minimize pollution levels. The Air Quality and Atmospheric Emissions Programme is one of the main activity lines of the Environmental European Agency, created in 1990 in order to coordinate, analyze and use strategically the environmental information, and provided with a strong subvention (26,9 million of € for 2005 financial year)[1].

There are many methods to analyze gaseous atmospheric pollutants. Traditionally, extractive or 'in situ' methods have been used for control and measurement of pollutants. These measurements methods are selective, sensible, accredited and accepted. However, they present some issues that may be avoided by using remote sensing techniques [2,3]. For example: a) the need to take a gas sample involves interfering and

J. Mira and J.R. Álvarez (Eds.): IWINAC 2007, Part II, LNCS 4528, pp. 589–598, 2007.

modifying the gas conditions and b) the measurements, taken in a very specific point, are not always representative of the area under study.

One of the main advantages of remote sensing is that it is a non intrusive method that does not require the collection of samples, avoiding any alteration in the analyzed gas. In addition, it allows the measurement through an integrated path of the chemical composition of the atmosphere in an area. More precisely, the proposed remote sensing technique, Open-Path FTIR Spectroscopy, allows the simultaneous measurement with the same instrument of all the atmospheric gases that present absorption in the infrared (IR) region of the electromagnetic spectrum. It is also applicable to a great variety of gas detection problems [4]. For all these advantages, this is a very promising technique that can contribute significantly to the environmental monitoring.

However, even though Open-Path FTIR is a well-established technique (EPA and VDI[1] have published guidelines for its use), some discrepancies with extractive methods have been noticed sometimes. In previous works [5], it has been demonstrate that these occasional problems can be due to the analysis programs and not to the Open-Path technique itself. Most of these programs are based on Classical-Least-Squared or Partial-Least-Squared procedures. Alternately, methods based on non linear least-squares spectral fitting programs are also used. The objective of this work is to design an analysis method based on artificial neural networks that overcomes the disadvantages of the analysis methods used previously.

The Open-Path instrument consist of a source of infrared energy and a spectroradiometer that analyzes the infrared energy coming to the instrument The distribution of the energy as a function of the wavelength (or more commonly as a function of the wavenumber[2]) is called the spectrum of energy. Fig.1 illustrates a diagram of a typical experimental set-up for atmospheric gas measurements.

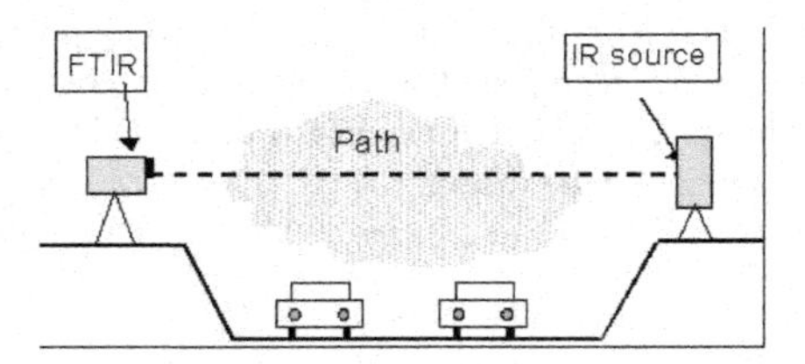

Fig. 1. Typical experimental setup for FTIR open-path atmospheric gas measurement

Pollutants along the optical path absorb infrared energy, but only in particular wavenumbers. This characteristic allows detecting if a gas is in the atmosphere just looking at the spectrum: if the energy corresponding with the characteristic wavenumber of a gas has been absorbed it can be inferred that this gas is in the path between the infrared source and the FTIR. In Fig.2 a transmittance spectrum with the absorption of typical gases is shown.

[1] U.S. Environmental Protection Agency and Verein Deutscher Ingenieure.

[2] Relationship between wavenumbers (ν) expressed in cm^{-1} and Wavelength (λ) expressed in (μm): ν=$10^{-4}/\lambda$.

The spectral resolution of this spectrum is high enough to resolve fine spectral lines of gases such as carbon monoxide (CO) and water vapour (H_2O), (see Fig.2). In order to retrieve the concentration from spectra, analysis programmes are needed. In that way, the Open-Path FTIR system is able to detect and quantify gases in the path.

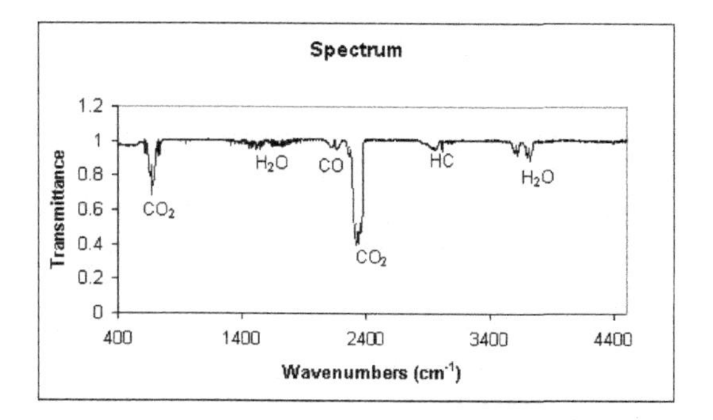

Fig. 2. Example of a transmittance spectrum showing absorption bands of atmospheric gases

To obtain the value of concentration, the analysis codes mentioned before use the transmittance or the absorbance spectra. However, these spectra are not measured directly by the Open-Path system. The transmittance spectrum is obtained by dividing a spectrum measured with pollutants in the path (measurement spectrum) by other one which is intended to be measured without these gases (background spectrum) but in the same experimental conditions. In Fig.3 a measurement and background spectra are shown.

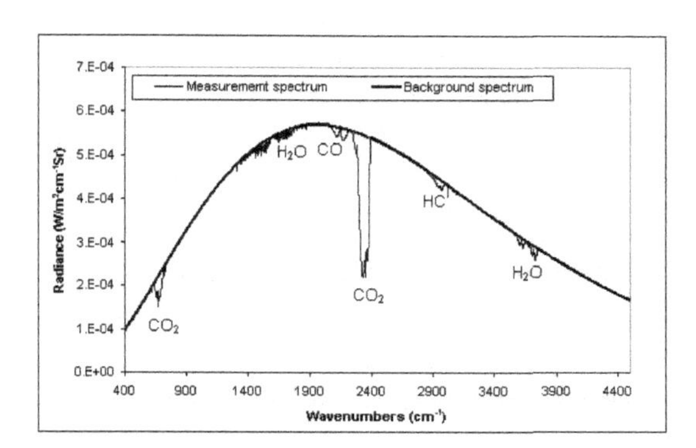

Fig. 3. Example of background an measurement spectra used to obtain the transmittance spectrum

An important problem that analysis programs have to work out is the interference between lines of different gases. Although each gas has characteristic lines, sometimes their lines can overlap. Then, analysis programs have to resolve which is the contribution to the absorption lines of each gas in order to retrieve their concentration, and this is not a straightforward task.

Nevertheless for many applications the analysis of transmittance spectra gives accurate values for gases concentrations in real time. However, in some occasions is impossible to measure an experimental pure background spectrum, because there are always gases in the atmosphere such as carbon dioxide (CO_2), water vapour (H_2O), carbon monoxide (CO), etc. In this case, the experimental background spectrum can entail occasional failures in the retrieved concentration. In fact, this is one of the main problems of this technique. Sometimes, this problem is overcomed by generating a 'synthetic' background spectrum that is, fitting the baseline of the experimental spectrum to a curve such as the background spectrum shown in Fig.3. However, this process can be a time-consuming task because commercial analysis programs do not usually offer this possibility and the advantage of obtaining values at real time would be lost.

On the other hand, although all these programs run automatically, it is necessary a previous work to define appropriate analysis methods that can work out the overlapping lines problem. This procedure requires the supervision of a skilled specialist, and this is another disadvantage of these methods.

Neural networks techniques can be used as an approach to solve problems of experimental data fitting. In the context of approximation of nonlinear maps, the architecture most widely used is the multi perceptron layer (MPL). In previous works, neural networks also have been used successfully in the development of computationally efficient inversion methods for satellite data and for geophysical applications [6,7] and with promising results to retrieve physical parameters using infrared spectra for ground based remote sensing [8].

The main purpose of using neural network in this study is to obtain a mapping between the transmittance spectra and gas concentrations. The MLP will provide a unique and global method to retrieve the four molecules concentration which is being studied in this work from transmittance spectra. One of the main advantages over other methods is its speed and that it is unnecessary to build a new model for new observations once the MLP has been trained. In next sections classical analysis methods will be explained briefly. Then they will be applied to a test data set and results will be discussed. Afterwards, the neural network-based method will be applied to the same data set and the results will be compared to the classical ones.

2 Classical Methods to Obtain CO Concentration from Open-Path FTIR Spectra

Most of the commercial software use classical-least-squared algorithms to retrieve the unknown concentrations from absorbance spectra. These codes work in real time and give suitable results. However, sometimes this procedure can generate offset errors in the retrieved concentrations, mainly for low values. These errors can be attributed to the appearance of absorption lines of investigated gases in the background spectrum. In addition, the method has to be defined and validated. This is a delicate and time consuming phase that has to be developed under an expert supervision.

Line-by-line approach is a non commercial software used as an alternative to CLS method, and it gives satisfactory results because they solve the problem associated to the absorption lines in the background spectrum by using a synthetic background spectrum.

The main problem is that these algorithms do not work in real time, and need a skilled operator.

Experimental open-path FTIR spectra will be analyzed with both methods to compare the results whit the proposed artificial neural networks process.

2.1 CLS Method

Classical or partial least-squared methods are the most standard multicomponent quantitative methods to analyze infrared spectra. They fit the experimental absorbance spectrum to a set of calibration spectra, that is, a set of spectra of known concentration stored in a spectral library.

Commercial software based on these algorithms needs the definition by the user of the so called analytical method. For this, it is necessary to provide: which gases present active infrared absorption band in the spectral region of interest, a set of calibrated spectra, for each component to be analyzed, a definition of the spectral window to be used in the analysis, trying to avoid the overlapping between bands, and finally it is essential a validation procedure to verify the quality of the retrievals, because not all the methods retrieve the same concentration.In this work a specific method has been designed to optimize the retrieval of carbon monoxide concentrations. The study conditions are represented in Table 1.

Table 1. Description of the method designed for the retrieval of carbon monoxide

CO spectral region absorption band	Spectral region of calibrated spectra	IR active gases in this region	Dataset conditions	Spectral resolution
$(2050,2250)\text{cm}^{-1}$	$(2000,2300)\text{cm}^{-1}$	CO_2, N_2O, H_2O	300K, 100m path	$0.5\ \text{cm}^{-1}$

In this case, the calibrated spectra have been generated by using the well known HITRAN[9] spectroscopic database. The most important interference event to take into account in the analysis is the water vapour lines, because they are all over the spectral region, as can be seen in Fig.4. Therefore small spectral regions for the carbon monoxide band have been chosen avoiding the CO_2, N_2O and H_2O overlapping.

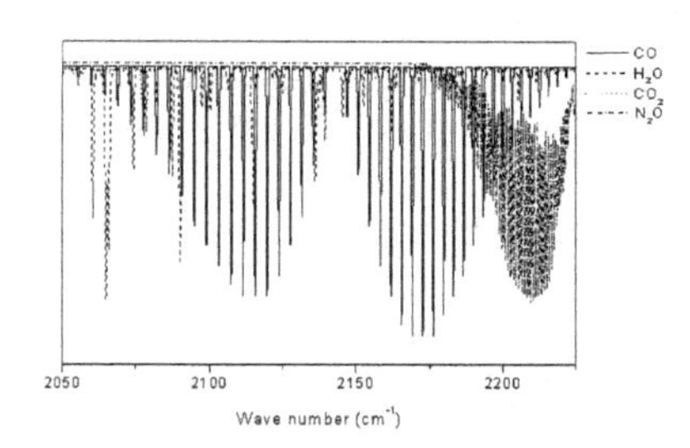

Fig. 4. Overlapping in simulated transmittance spectra of different gases in the CO absorption band

In order to validate the method, a set of spectra has been generated in the same conditions of the dataset, considering a broad range of CO concentrations, from background conditions to polluted scenarios. For the other gases, concentrations have been selected from the atmospheric model 1976 US Standard Atmosphere provided by HITRAN, covering wet/dry conditions with high/low values of the considered gases. Results are shown in Fig.5.

From the results of the validation it is deduced that the proposed analytical method is very suitable to retrieve carbon monoxide concentration from FTIR spectrum in a very wide range of atmospheric conditions. However, it is observed that there is a strong discrepancy between nominal and retrieved values of CO concentration when these values are very low (non-polluted scenarios). This discrepancy is associated to the fact that the background reference spectrum presents CO absorption that is eliminated when division between spectra is performed to obtain the absorbance spectra. This effect decreases when CO concentrations in measurement spectra are higher. However, the error will be higher if the background reference spectrum is acquired in a more polluted atmosphere.

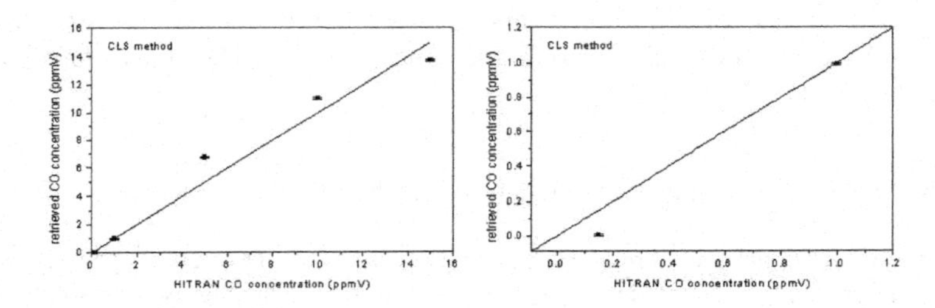

Fig. 5. (Left)Comparison of retrieved and nominal CO concentrations with CLS method. (Right) Detail of the graphic for the lower concentration scale.

2.2 Line-by-Line Methods

SFIT code is a line-by-line method based in a non-linear least-square fitting program. It was originally used to analyze solar absorption spectra, and occasionally it has been used to analyze open-path spectra. In order to know the concentrations of an experimental spectrum, this program generates itself a set of synthetic spectra from an initial set of gas concentrations, using spectral information from HITRAN96 database. Then the program modifies the initial gas concentrations so that the synthetic spectrum matches the experimental spectrum. The best fitting gives the concentration of the experimental spectrum.

The problem associated with the background spectra in the CLS methods, does not exists in line-by-line methods because it is not necessary to provide a background reference spectrum to the fitting procedure. The 100% transmission level is obtained by fitting the baseline slope of the single beam spectrum.

The most important parameter in this method is the spectral fitting range, which is decisive in the results. From the CO validation process, it was deduced that (2080, 2142)cm^{-1} is the spectral fitting range that gives the best results. This spectral region prevents the CO_2 and N_2O overlapping and allows considering the unavoidable water vapour interference Fig.4.

The same data set analyzed by the CLS method, has been analyzed by the line-by-line program, with the same conditions but in the new spectral range, and the results are represented are equivalent to those shown in Fig.5, with several minor improvements in the retrieval of low CO concentrations.

Conclusions are that line-by-line method works out the overlapping with other gases for all considered concentrations, because his standard deviations are very small, similar to the CLS ones. For low concentrations, the retrieval error is smaller than in the CLS case, but it increases lineally for high values.

In both cases, the retrieval errors are not homogeneous. Another disadvantage is the need of a specialist to define analysis methods during the validation phase.

3 Neural Network Based Analysis Method: A Proposal

The neural network method is base in a learning phase in which the global model is generated. Thus, the generation of a good set of samples to train the neural network is one of the most important steps. These samples should cover as much as it can all the possibilities of the problem to be solved.

For the purposes of this study a data set of 10000 different samples has been generated. This samples are associated to synthetic transmittance spectra generated with a specific computer code developed at University Carlos III (CASIMIR)[10] based on HITRAN[9]. All this spectra have been generated at high spectral resolution (0.05 cm^{-1}) and smoothed to the desired resolution of 0.482 cm^{-1}. This resolution has been selected for maintaining the relationship between the discrete wavelengths obtained on experimental measurements using a spectroradiometer and the ones which have been obtained with computer code CASIMIR.

The data set cover a broad concentration range for each molecule at a constant temperature of 300 K and a path length of 100 m. The concentration ranges covered by each molecule and the incremental step used to generate the data set are shown in Table.2.

Table 2. Data set generation parameters

Gas	MSE Concentration range (ppmV)	Step (ppmV)
CO_2	330 – 500	17
CO	0.15 – 15	1.45
N_2O	0.30 – 0.35	0.005
H_2O	4316 – 25600	2158

A total of 1246 variables at a resolution of 0.482 cm^{-1} have been used as inputs. These variables are associated to different wave numbers which - as has been explained in previous methodologies - are related to different physical properties of each molecule. A previous selection of variables could be done trying to exploit this fact.

But since the MLP performance can ignore those irrelevant inputs, a previous analysis of the data has not been done in this work.

3.1 Neural Network Architecture

Let y be the physical forward model output (the radiative transfer equation RTE in this case) then $Y^n = y(x)$, where n are the different observations. Note that y and x are vectors which represents multiple observations and parameters data. At this point the goal is to obtain the x physical variables once an observation Y^n is given. The input layer receives all the observation variables corresponding to the physical properties x. Thus the estimated value of x will be $\hat{x} = mw(y^n)$ where mw is the neural network m with w parameters learnt during the training process. The goal is to minimize the euclidean distance $D_e(y(\hat{x}), y^n)$, and this is achieved during the learning process adjusting the weights of the net with backpropagation algorithm.

The architecture selected has been an MLP with one hidden layer. The number of neurons in this hidden layer has been changed in order to find the architecture which best fits the expected output. The results for these proofs are shown in Table3.

Table 3. Concentration error of CO molecule

Hidden neurons	MSE Train	MSE Test	Mean error CO (ppm)	Standard Deviation
10	0.004557	0.004607	0.673598	0.466015
20	0.000235	0.000325	0.053307	0.050109
30	0.000215	0.000237	0.012437	0.010191
40	0.000165	0.000174	0.023160	0.018051

This process of try and error, is necessary since they are not a unique way to know what is the best number of hidden neurons, and a validation method is needed to find the best configuration. Furthermore, although as more neurons are included a better fit to the learning dataset is obtained, this could not be a good criterion to select the MLP configurations because a problem of overfitting can arise. When this happens the generalization is lost in benefit of the specification. Thus, the criterion to select the number of hidden neurons has been the smallest generalization error.

The number of outputs is given by the number of molecules to recover, in this study are four molecules: CO_2, CO, N_2O and H_2O. For clarification, because the main point in this study is to recover the CO concentration the other molecules retrievals has been omitted in the tables. To validate the MLP performance synthetic and experimental tests have been done. The results obtained by the MLP using synthetic data fit almost perfectly the real ones. In the Fig.6 is showed how the values obtained by MLP (marks over the line) fit the ones expected (solid line), and its low standard deviation (width of the marks). The validation has been done also with experimental data but the results have not been as good, having errors up to 50%.

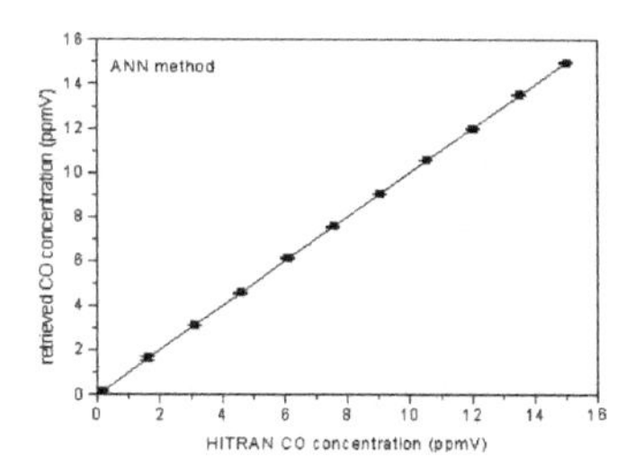

Fig. 6. Results given by the MLP using for validation synthetic data

4 Conclusions

Open-Path FTIR spectroscopy is a remote sensing technique widely used to measure atmospheric gas concentrations. This experimental technique, as well as the classical methods used to retrieve concentration values has been described in the first sections of the paper, highlighting advantages and drawbacks of these classical methods.

An analysis method based on artificial neural network architecture has been proposed and designed to keep the advantages of classical methods and overcoming the main drawbacks. Special care has been taken to generate the dataset. In this design, a wide range of of low and high carbon monoxide concentrations has been used to cover different polluted scenarios. Overlapping of CO absorption band with bands due to other gases has been identified as an important problem and consequently different concentration (and realistic) ranges of water vapor, nitrous oxide and carbon dioxide has been taken into account to in the data set generation.

The results obtained in the framework of this work are very promising. Carbon monoxide concentrations have been retrieved with accuracy in a wide range of values, and for very different concentrations for the interference gases. This result can be achieved taking into account that a supervised phase is only needed during the training of the network. After this training process, the method is able to provide results automatically in real time. Other important advantage over classical methods concerns the retrieval errors, that keep a high homogeneity over all the concentration ranges under study.

Experimental measurements of carbon monoxide concentrations have been performed by using a commercial FTIR spectroradiometer working in Open-Path configuration. Experimental spectra have been selected to be analyzed by both CLS and ANN methods. However, retrieval errors up to 50% have been observed when ANN method is applied to these experimental spectra. We think these errors are due mainly to the instrumental response of the spectroradiometer that slighly modifies the spectral shape, spectral location and height of the CO absorption peaks. Many experimental factors (apodization, resolution, truncation, experimental noise) have influence in this spectral response.

Summarizing the analysis method based in ANN architecture presents clear advantages over classical methods in order to obtain in real time accurate values of pollutant gases, avoiding the need of expertise supervision (except in the initial training phase), but work has been done to include instrumental effects in the model.

Acknowledgments

The authors wish to thank Samuel Rodríguez for his support developing computer codes used in this work. They also wish to acknowledge Spanish Ministry of Education and Science for financial support under projects TRA2005-08892-C02-01 and TRA2005-08892-C02-02.

References

1. Informe sobre las cuentas anuales de la agencia europea de medio ambiente. Environmental European Agency: Diario oficial de la Unión Europea (2005)
2. Sigrist, M.W.: Air monitoring by spectroscopy techniques. John Wiley and Sons, New York (1994)
3. Melendez, J.: Medida de gases contaminantes en la atmósfera. Universidad Carlos III de Madrid, Leganes (Spain) (1994)
4. Schafer, K. Heland, J., Haak, A.: Measurements of atmospheric trace gases by emission and absortion spectroscopy with ftir. Berichte der Bunsengesellschaft fr Physikalische Chemie 99 **3** (1995) 405–411
5. Briz, S., de Castro, A.J., Diez, S., Lopez, F., Schafer, K.: Remote sensing by open-path ftir spectroscopy. comparison of different analysis techniques applied to ozone and carbon monoxide detection. Journal of quantitative spectroscopy and radiative transfer. **103** (2007) 314–330
6. Chevalier, F., J.-J, M., Chéruy, F., Scott, N. Q. J. R. Meteorol. Soc. **126** (2000) 761–776
7. Aires, F.: Surface and atmospheric temperature retrieval with the high resolution interferometer iasi. Proc. Am. Meteorol. Soc. 98 (1998) 182–186
8. García-Cuesta, E., de Castro, A., Galván, I.M.: Spectral high resolution feature selection for retrieval of combustion temperature profiles. Lect. Notes Comput. SC **4224** (2006) 754–762
9. Rothman, L.S.: The hitran molecular spectroscopic database: edition of 2000 including updates through 2001. J. Quant. Spectrosc. Radiat. Transfer (2003)
10. García-Cuesta, E.: CASIMIR: Cálculos Atmosféricos y Simulación de la Transmitancia en el Infrarrojo. University Carlos III L/PFC 01781, Madrid (in Spanish) (2003)

Air Pollutant Level Estimation Applying a Self-organizing Neural Network

J.M Barron-Adame[1], J.A. Herrera Delgado[1], M.G. Cortina-Januchs[1], D. Andina[2], and A. Vega-Corona[1]

[1] Universidad de Guanajuato
Facultad de Ingeniería Mecánica, Eléctrica y Electrónica.
Salamanca, Guanajuato. México
badamem,herreraj,januchs,tono@salamanca.ugto.mx
[2] Universidad Politécnica de Madrid
Departamento de Señales, Sistemas y Radiocomunicaciones, E.T.S.I.
Telecomunicación, Madrid, Spain
d.andina@upm.es

Abstract. This paper presents a novel Neural Network application in order to estimate Air Pollutant Levels. The application considers both Pollutant concentrations and Meteorological variables. In order to compute the Air Pollutant Level the method considers three important stages. In first stage, A process to validate data information and built a threedimensional Information Feature Vector with Pollutant concentrations and both wind speed and wind direction meteorological variables is developed. The information Feature Vector is orderly like a time series to estimate the Air Pollutant Level. In second stage, considering the behavior space knowledge a priori about pollutant and meteorological variables distribution a threedimensional Representative Vector is built in order to reduces the computational cost in Neural Network training process. In last stage, a Neural Network is designed and trained with the Threedimensional Representative Vector, then using the Threedimensional Information Feature Vector the Air Pollutant Level is estimated. This paper considers a real time series from an Automatic Environmental Monitoring Network from Salamanca, Guanajuato, Mexico, and therefore in this proposal a real Air Pollutant Level is also estimated.

Keywords: SOM Neural Networks, Air Pollutant Level, Pattern Recognition.

1 Introduction

Air pollution is one of the most important environmental problems. Pollution is caused by both natural and man-made sources. Major man-made sources of ambient air pollution include industries, automobiles, domestic activities and power generation [1]. Air pollution has both acute and chronic effects on human health. Health effects range anywhere from minor eyes irritations and the upper

J. Mira and J.R. Álvarez (Eds.): IWINAC 2007, Part II, LNCS 4528, pp. 599–607, 2007.

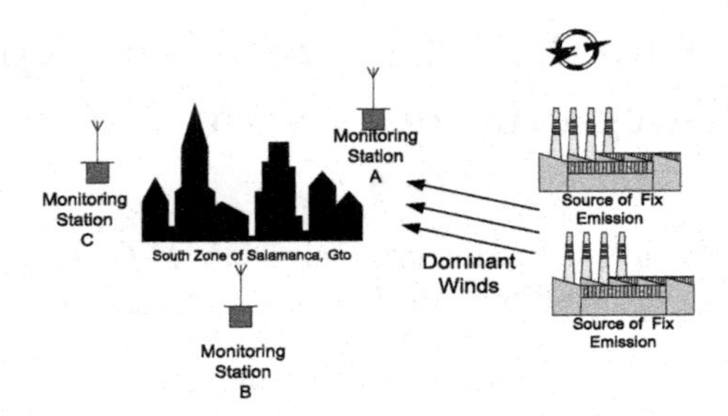

Fig. 1. Environmental Monitoring Network distribution

respiratory system to chronic respiratory disease, heart disease, lung cancer, and death.

Nowadays, many countries make big efforts to minimize air pollution [2,3,4]. In polluted countries like Mexico a continuous monitoring of Air Quality to measure pollutant concentrations to reduce possible negative effects in population health is necessary. A special case with great pollution is Salamanca, Guanajuato in Mexico. Salamanca city is catalogued as one of the most polluted cities in Mexico [5]. The main causes of pollution in Salamanca are due to fixed emission sources such as Chemical Industry and Electricity Generation, being Sulphur Dioxide (SO_2) (measured in Parts Per Billion, (PPB)), and Particulate Matter less than 10 micrometers in diameter PM_{10} (measured in micrometers, (μm)) the most important Air pollutants. This article focuses the analysis on PM_{10} concentration.

In 1999, an Automatic Environmental Monitoring Network (AEMN) was established in Salamanca, this AEMN provided time series about criteria pollutant [6] among other meteorological variables. Figure 1 shows the AEMN distribution. In an effort to fight pollution on the region, in July 2005, the Environmental Contingency Program was launched [7], with the purpose to protect population health, especially of the vulnerable groups. This program contemplates the urgent and immediate reduction of SO_2 and PM_{10} emissions when measurements of these pollutants register levels above those established by Health Authorities. To accomplish it, 3 phases were established: *Pre-contingency, Phase I Contingency* and *Phase II Contingency* for SO_2, PM_{10} particles and for a combination of both [5,8].

1.1 Particulate Matter

Particulate Matter (PM) consists of solid or liquid aerosol particles suspended in the air and has a diverse chemical composition related to its sources. Under normal ambient conditions of sampling and analysis, particulate matter exists almost exclusively in solid phase but can include liquid aerosols such as the heavier components of diesel combustion products and nitric acid. Some particles are emitted directly into the air from a variety of sources that are either natural or related to human activity. Natural sources include bushfires, dust storms, pollens and sea spray. Those related to human activity include motor vehicle emissions, industrial processes, unpaved roads and wood heaters. PM is commonly designated as $PM_{2.5}$ or PM_{10}, refereing to particles with aerodynamic diameters less than 2.5 μm and 10 μm, respectively.

The statistical correlation between high levels of inhalable particulate matter and increased mortality has been widely reported. The particles in the $PM_{2.5}$ and PM_{10} fractions can be inhaled into the lungs, causing damage to the alveolar tissues and inducing various health problems [9,10]. The adverse effects may range from the irritation of the lung tissues resulting in coughing to severe respiratory problems for individuals with asthma or heart disease. The mechanic details of how the constituents of the PM induce adverse health effects are currently areas of intense scientific research. The polycyclic aromatic hydrocarbons and heavy metals present in the $PM_{2.5}$ and PM_{10} samples have been studied extensively with regard to their roles in inducing toxicity effects [11].

1.2 Artificial Neural Network

Artificial Neural Networks (ANN) are biologically inspired networks based on the neuron organization and decision making process in the human brain [12]. In other words, it is the mathematical model of the human nervous system.

SOM Neural Network. *The Self-organizing Map (SOM9)* [13] basically provides a form of cluster analysis by producing a mapping of high-dimensional input data $\mathbf{x}$, $\mathbf{x} \in \Re^n$ onto a usually bidimensional output space while preserving the topological relationships between the input data items as faithfully as possible. It consists of a set of units, which are arranged in some topology where the most common choice is a two dimensional grid [14,15].

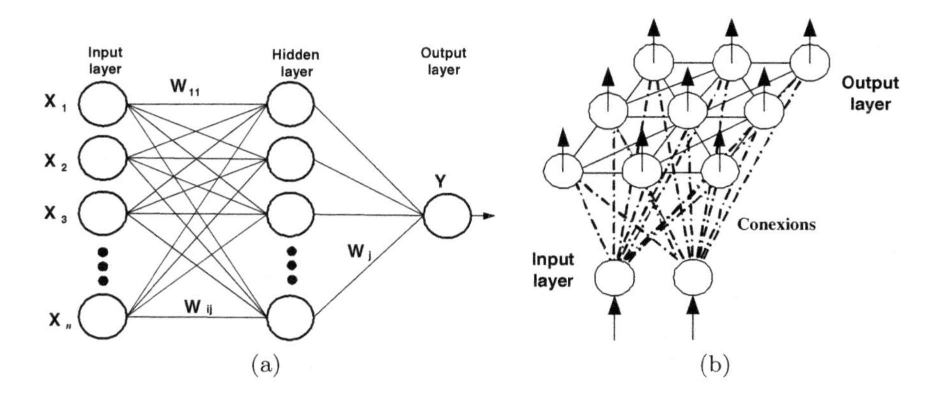

Fig. 2. Basic Representation for (a) Neural Network, (b) Self-organazin Map Neural Network

Each of the units i is assigned a weight vector m_i of the same dimension as the input data, $m_i \in \Re^n$. In the initial setup of the model prior to training, the weight vectors are filled with random values.

During each learning step, the unit c with the highest activity level, the winner c with respect to a randomly selected input pattern x, is adapted in a way that it will exhibit an even higher activity level at future presentations of that specific input pattern. Commonly, the activity level of a unit is based on the Euclidean distance between the input pattern and that unit's weight vector. The unit showing the lowest Euclidean distance between its weight vector and the presented input vector is selected as winner. Hence, the selection or the winner c may be written as

$$c : \|x - m_c\| = min_i \|x - m_i\| \tag{1}$$

Adaptation takes place at each learning iteration and is performed as a gradual reduction of the difference between the respective components of the input vector and the weight vector. The amount of adaptation is guided by a learning-rate α that is gradually decreasing in the course of time. As an extension to standard competitive learning, units in a time-varying and gradually decreasing neighborhood around the winner are adapted, too. This strategy enables the formation of large clusters in the beginning and fine-grained input discrimination towards the end of the learning process. In combining these principles of self-organizing map training, we may write the learning rule as given in expression 2. We make use of a discrete time notation with t denoting the current learning iteration. The other variables of this expression are α representing the time-varying learning-rate, h_{ci} representing the time-varying neighborhood-kernel, x representing the currently presented input pattern, and m_i denoting the weight vector assigned to unit i.

$$m_i(t+1) = m_i(t) + \alpha(t) h_{ci}[x(t) - m_i(t)] \tag{2}$$

2 Database and Variables Definition

2.1 Database

In this research, a real and historical time series database from the AEMN has been used. Data series of the months from January to June in the year 2006 have been analyzed. These time series consist of a total of 260,000 threedimensional patterns of pollutants and meteorological variables.

3 Variables Definition

Variables definition consists of normalized pollutants concentration values PM_{10} and normalized meteorological values (Wind Speed and Wind Direction). In Table 1, variables are defined in order to build a Threedimensional Representative Feature Vector $\mathbf{x}_j$ and to define a pattern set $\mathbf{X}_* = \{\mathbf{x_1}, \mathbf{x_2}, .., \mathbf{x_j}, .., \mathbf{x_n}\}$.

Table 1. Variables Definition: $\mathbf{PM}_{10}$ Concentration, **WS**; Wind Speed, **WD**; Wind Direction

	$\mathbf{X}_{\mathbf{PM_{10}}}$
Variables x_i	$\mathbf{x}_j$
x_1	PM_{10}
x_2	WS
x_3	WD

Let $\mathbf{X}_{\mathbf{PM_{10}}}$ be a particles concentration set, thus their corresponding pattern is defined as $\mathbf{x_j} = \{x_1, x_2, x_3\}$. Figure 3 shows Thredimensional Representative Feature Vector (TRFV) representation.

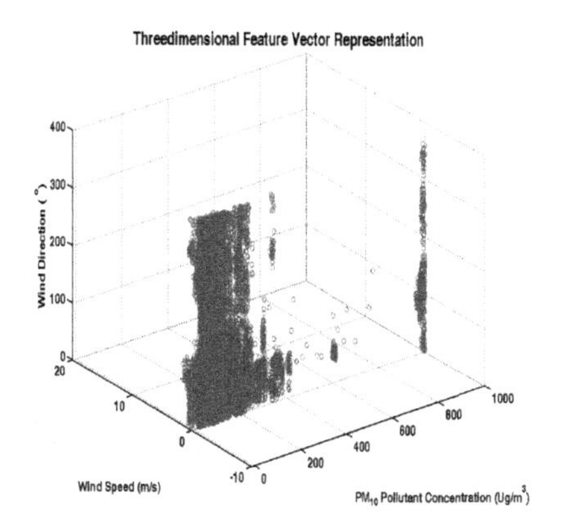

Fig. 3. Pollutant concentration and booth Wind Speed and Wind Direction environmental variables

4 Proposed Methodology

Figure 4 shows the flow diagram of the methodology that was followed for the classification process. This process has three steps which will be explained in details:

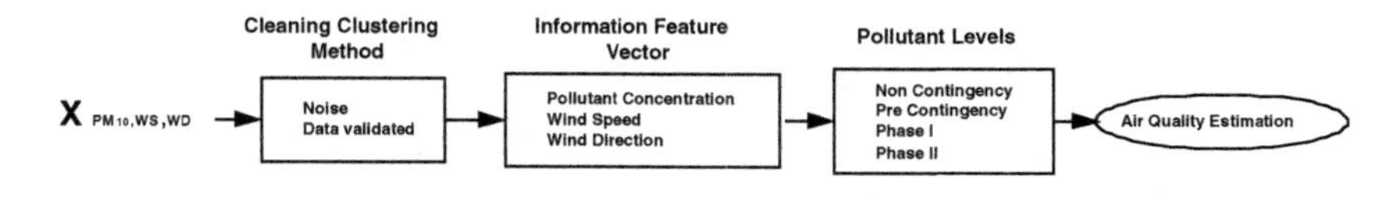

Fig. 4. Proposed Methodology

- The Cleaning Clustering Method
- The Threedimensional Representative Feature Vector Method
- The Clustering Classification Method

Cleaning Clustering Method. This method uses a SOM Neural Network to classify the patterns to be analyzed. Because time series have noise, we trained the SOM Neural Network using time series in two classes, the first class is noisy data and the second class is validated data. Validated data will be used in the Clustering Classification Method, meanwhile noisy data will be deleted. Figure 5 shows a bidimensional noise representation.

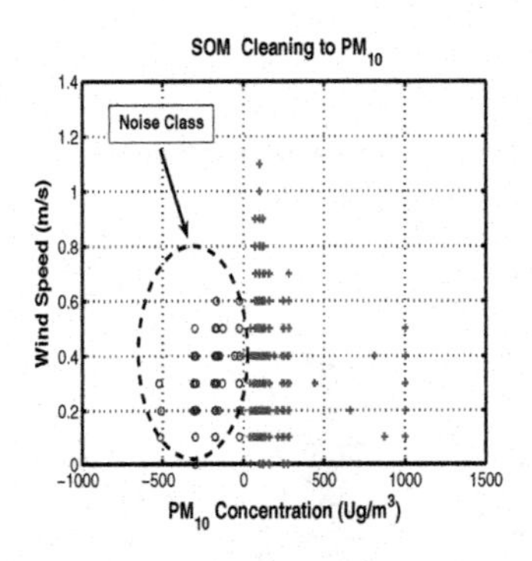

Fig. 5. Bidimensional noise representation

Threedimensional Representative Feature Vector Method. Because of data nature of the variables, it is possible to know the variables limits. In this section a Threedimensional Representative Feature Vector (TRFV) is built [16], with the centers of the involved variables, these centers are for *Non Contingency, Pre-Contingency, Phase I Contingency* and *Phase II Contingency.* This TRFV was used to train a new SOM Neural Network with a [4 1 1] line topology, and try to locate the SOM Neural Neural node in the appropriate contingency level (see Table 2).

Table 2. Contingency Concern Levels and Pattern centers vector

Contingency Level Center	
Contingency Levels	Pattern Center PM_{10}-WS-WD
Non Contingency	75, 10, 180
Pre-Contingency	200.5, 10, 180
Phase I Contingency	304.5, 10, 180
Phase II Contingency	450, 10, 180

Clustering Classification Method. after the SOM Neural Network has been trained with a [4 1 1] line topology [16, 17] in order to have four clusters and therefore four prototypes according to contingency levels, we proceed to group the new patterns set in four classes according to the contingency levels.

5 Experimental Results

Figure 6 shows the SOM Neural Network trained with the TRFV shown in Table 7. Figure 8 shows a real time series for PM_{10} concentrations analyzed every minute. They are classified according to the process followed by our methodology. The meteorological variables used were Wind Speed and Wind Direction, creating 129,600 three dimensional pattern vectors $\mathbf{x}_i$, for pollutant XPM_{10}, as it is shown in Table 1. Pollutant concentrations and meteorological variables are provided by the AEMN from Salamanca. In the clustering method, four clusters have been performed from a new feature vector.

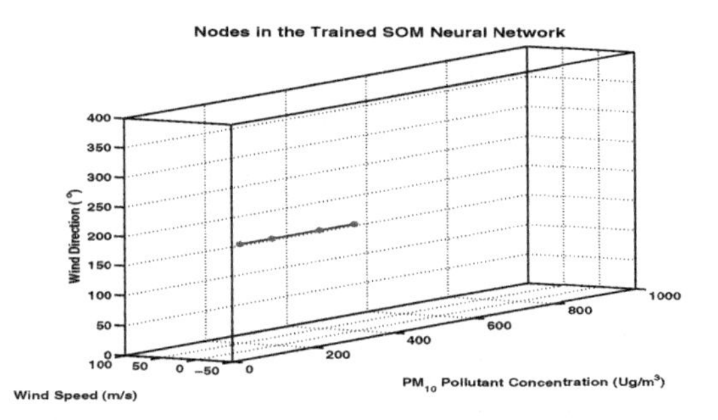

Fig. 6. Nodes in line topology for a trained SOM Neural Network

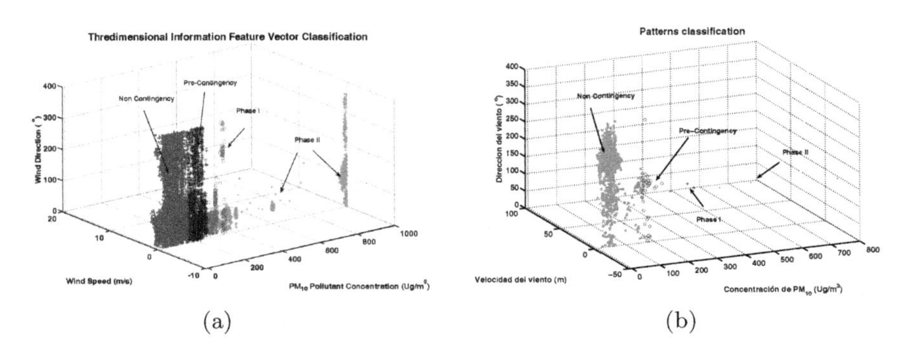

(a) (b)

Fig. 7. Patterns classification with a line topology SOM NN, for (a) one month, (b) one week

In Table 3 the error percent obtained with the proposed methodology is shown for both January 2006 and January 1-7 in 2006. We can observe that the error depends on the number of data. Noisy pattern is an inconsistent element in the time series and it is caused by wind blasts. Noisy elements can cause bad estimation, so with this method a better estimate is obtained.

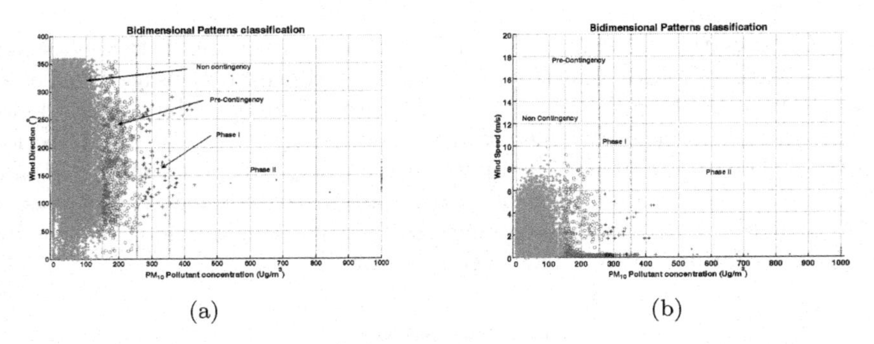

(a) (b)

Fig. 8. Bidimensional representation for Patterns classification, for (a) PM - WS, (b) PM - WS

Table 3. Error percent using a SOM Neural Network trained with the TRFV

% Error		
	Data analyzed	% Error $PM_{10} - WS - WD$
January, 2006	43200	0.45
January 1-7, 2006	10080	0.32

6 Conclusions

In this work, a time series was obtained from the Automatic Environmental Monitoring Network (AEMN) in Salamanca, Mexico. This time series contain noisy data and valid data with a SOM Neural Network this time series was clustered in two classes, the class with noisy data was deleted and the valid data was using to train a new SOM Neural Network. This SOM classify the valid data according to the Air Contingency Levels. This methodology presented good results, because the Contingency Levels are known, allowing to create a Representative Feature Vector for each level. Thus, less patterns are required to train a SOM Neural Network. The classification error depends only on the number of data. Our method produces errors that are less than 1.

References

1. Environmed Research Inc. Alpha Nutrition. Problem Air Pollution. http://www.nutramed.com/environment/particles.htm. 2004.
2. (In Spanish): Sistema de Información Medioambiental. La contaminación atmosférica. *Madrid, España*, Febrero 2007.
3. Chinese Ambassy. China so2 emissions reductions policies. February 2007.
4. Department for Environment Food, Rural Affairs, and the Devolved Administrations. *UK Air Quality Archive*, February 2007.
5. (In Spanish): Sistema de información ambiental, SIMA. Red automática de monitoreo atmosférico, http://www.sima.com.mx/. 2004.
6. United States Environmental Protection Agency. Risk assessment for toxic air pollutants: A citizens guide-epa 450/3-90-024. air risk information support center (md-13). March 1991.
7. (In Spanish): Gobierno del Estado de Guanajuato. Programa de contingencias ambientales atmosféricas del municipio de salamanca, http://www.guanajuato.gob.mx/iee/contingencias.pdf. Abril 2004.
8. (In Spanish): Instituto de Ecologia del Estado de Guanajuato. Programa de contingencias ambientales atmosféricas. 2005.
9. Wordl Health Organization. Health aspect of air pollution with particulate matter, ozone and nitrogen dioxide. January 2003.
10. Secretariat of Commission for Environmental Cooperation. Continental pollutant pathway. communication and public outreach deppartment of cec secretariat. 1997.
11. Ngee-Sing Chong; Kavitha Sivaramakrishnan; Marion Well and Kathy Jones. Characrerization of inhalable particulate matter in ambient air by scanning electron macroscopy and energy-dispersive x-rau analysis. 3:145 164, 2002.
12. R. Setiono and Huan Liu. Neural-network feature selector. *IEEE Transactions on Neural Networks*, 8(3):654 – 662, May 1997.
13. T. Kohonen. The self organization map. *Proceedings. IEEE*, 78(9):1464 – 1480, September 1990.
14. James A. Freman and David M. Skacura. Neural networks algorithms, application and programming techniques. 1991.
15. S.Haykin. *Neural Networks.* 2nd edition edition, 1999.
16. Jain A.K. John Wiley and Sons Inc. Pattern recognition. pages 1052 – 1063, 1988.
17. MatLab. Math Works. Neural networks toolbox. page September, 2005.

Application of Genetic Algorithms in the Determination of Dielectric Properties of Materials at Microwave Frequencies

Alejandro Díaz-Morcillo[1], Juan Monzó-Cabrera[1], María E. Requena-Pérez[2], and Antonio Lozano-Guerrero[1]

[1] Universidad Politécnica de Cartagena, Plaza del Hospital, 1. 30202 Cartagena, Spain
alejandro.diaz@upct.es
[2] Consejería de Educación y Cultura de la Región de Murcia C/ Villaleal, 2. Edificio Centro. 30001 Murcia, Spain

Abstract. In this paper the application of an evolutionary procedure based on genetic algorithms for obtaining the dielectric properties of arbitrary shaped, homogeneous or inhomogeneous materials is presented. The optimization procedure matches the measured and simulated scattering parameters of a waveguide setup that contains the sample under study. Depending on the geometry of the sample, analytic or numerical (2D or 3D) electromagnetic simulations must be carried out in order to obtain the simulated scattering parameters for a set of electric permittivities. Results for different polymeric and biological materials are presented with similar uncertainties than conventional direct methods, with the advantage that this new technique can deal with non-canonical and heterogeneous samples.

1 Introduction

An accurate and reliable determination of the electric permittivity of a dielectric material is of key importance in order to know its behavior under the influence of electromagnetic fields. The dielectric characterization of materials at microwave frequencies (300MHz - 300GHz) has undergone an important advance in the last fifty years, with the development of new measurement techniques based, mainly, on transmission lines, as reflection or transmission-reflection methods, and resonators [1]. These direct methods are quick, but they are only valid for macroscopically homogeneous materials which, moreover, can be shaped in canonical geometries such as rectangular prisms or circular cylinders.

When the material under study has an irregular geometry and does not admit mechanical treatment (for instance, in a non-destructive quality control) these conventional methods fail in obtaining its electric permittivity. Besides, if the material is non-homogeneous they only can obtain an effective permittivity as a whole. In these cases it is necessary to develop alternative techniques that can deal with arbitrary shape or non-homogeneous samples. This paper introduces

J. Mira and J.R. Álvarez (Eds.): IWINAC 2007, Part II, LNCS 4528, pp. 608–616, 2007.

a procedure of reconstruction of the permittivity of materials in an arbitrary sample from the scattering parameters obtained in a waveguide setup which includes the sample under study. In this inverse measurement an excitation is applied to the waveguide by means of a vector network analyzer and the behavior of the material or materials under study is obtained by means of the scattering parameters of the two-port network. The key in this procedure is the knowledge of the relationship between the measured parameters (scattering parameters) and the desired physical parameter, i.e., the electric permittivity.

Unfortunately, in this case that relationship is not straightforward and, therefore, it is necessary to introduce a new element in the measurement process: the optimization. This optimization can be realized with a search algorithm, in this case a genetic algorithm, which minimizes an error function related with the relationship between scattering parameters and permittivity. In our problem, the computation of the error function requires a computer simulation of the waveguide setup and a comparison of the simulated scattering parameters with the measured ones. Fig. 1 shows this philosophy.

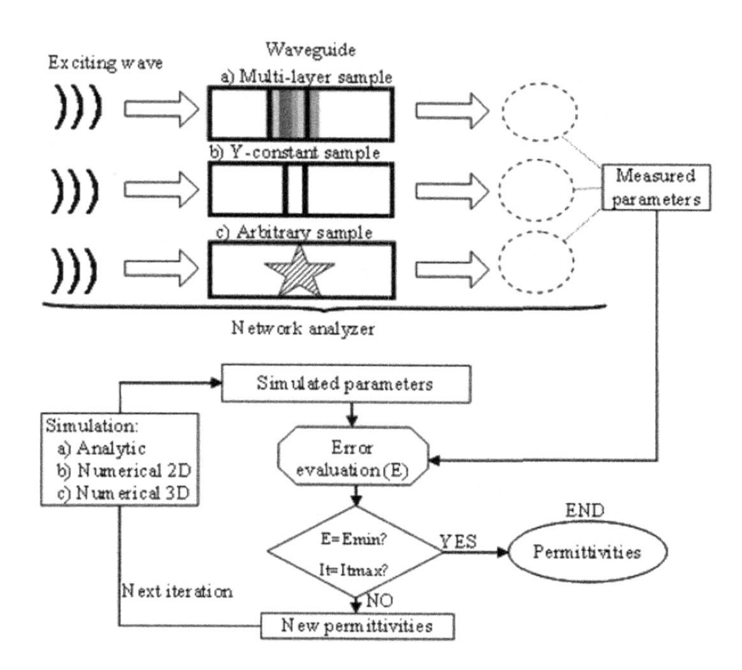

Fig. 1. Flow chart for inverse measurement of permittivities in a waveguide

2 Genetic Algorithm

Genetic algorithms are robust search procedures based on the principles of natural selection and evolution. They are particular effective in finding a global minimum and in last years they have been successfully used in the electromagnetic community [2]. In this work, the objective is obtaining the complex permittivity which minimizes the difference the measured and the simulated scattering parameters of the waveguide setup which contains the sample.

The genetic algorithm used here is based on a Matlab^{TM} implementation [3] and shown in Fig. 2. The individuals are the candidates for the solution of optimization problem. In this case, their genes are the complex permittivities of the different materials contained in the sample. The evaluation of these individuals is realized with the fitness function

$$f = p\sum_{i=1}^{2}\sum_{j=1}^{2}\sqrt{\left(\sum_{k=1}^{N}\left(\left|S_{ij}^{m}(k)\right| - \left|S_{ij}^{s}(k)\right|\right)^2\right)} +$$

$$(1-p)\sum_{i=1}^{2}\sum_{j=1}^{2}\sqrt{\left(\sum_{k=1}^{N}\left(\theta\left(S_{ij}^{m}(k)\right) - \theta\left(S_{ij}^{s}(k)\right)\right)^2\right)}\ , \qquad (1)$$

where $S_{ij}^{m}(k)$ are the scattering parameter measured with the network analyzer, $S_{ij}^{s}(k)$ are those obtained with electromagnetic simulation, and N is the number of frequencies where this parameters are evaluated. The weighting factor $p \in [0, 1]$ allows to give more importance to modulus or phase in the fitness function.

In the first step, the parameters of the genetic algorithm (number of individuals per generation, number of generations, crossover and mutation probabilities) and the range of variation of the solution, that is, the range of variation for the dielectric constant $\epsilon^{'}$ and the loss factor $\epsilon^{''}$, are assigned. After the initial generation is randomly generated, its individuals are evaluated, that is, the electromagnetic problem is solved in the range of frequencies under study for each individual and its fitness value is obtained by (1). Depending on the results of the evaluation different parents are selected for crossover, the children are evaluated and substitutes the parents. Next, depending on the mutation probability different individuals are mutated. Finally, as this implementation is elitist, the best individual from the parent generation is added to the new one. This procedure is repeated until a number of generations is reached.

The main characteristics of the genetic algorithm are summarized in Table 1.

Table 1. Characteristics of the genetic algorithm

Characteristic	Type
Selection	Geometric normalized, Roulette, Tournament
Crossover	Arithmetic, Simple
Mutation	Uniform, Non-uniform
Fitness function	Simple (one objective)
Improvements	Elitist strategy, Weighting fitness
End condition	Number of generations

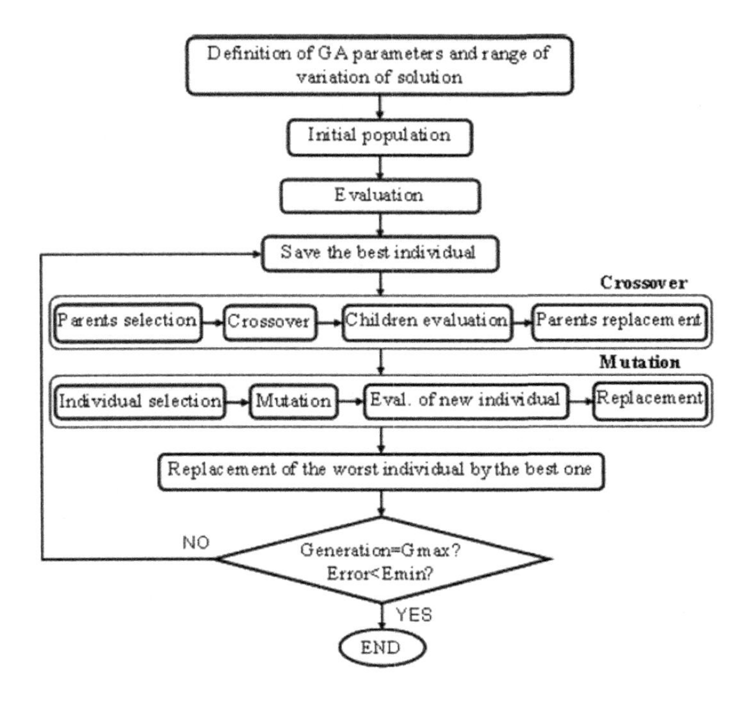

Fig. 2. Genetic algorithm

3 Different Simulation Strategies

As shown in Fig. 1, depending on the characteristics of the sample (geometry, composition) different simulation techniques can be used. The optimization procedure governed by the genetic algorithm needs to carry out a deep search in the space of solution, which leads to a great number of generations and individuals per generation. Each individual will require an electromagnetic simulation. The computational cost of the overall optimization procedure is determined mainly by the simulation technique employed in the evaluation of individuals. If t_s is the time for an electromagnetic simulation, m the number of individuals per generation, g the number of generations, n_c the number of crossovers per generation and $n_m$the number of mutations per generation, the overall optimization time is

$$t = m \cdot t_s \cdot [1 + (2n_c + n_m) \cdot (g - 1)] \ . \quad (2)$$

Therefore it is of key importance to employ a simulation method with a minimal computational cost.

3.1 Transmission Line Theory

Prism samples filling completely the section of the waveguide permits the use of the purely analytical transmission line theory, where the waveguide setup is analyzed as a cascade of transmission lines with different characteristic impedance, as Fig. 3 depicts. In this figure Z^{v}_{TE10} and $\gamma^{v}_{\mathrm{TE10}}$ are the characteristic impedance (TE_{10} impedance) and the propagation constant, respectively, and Z^{Mi}_{TE10}, $\gamma^{Mi}_{\mathrm{TE10}}$ refer to material i of the multilayer sample.

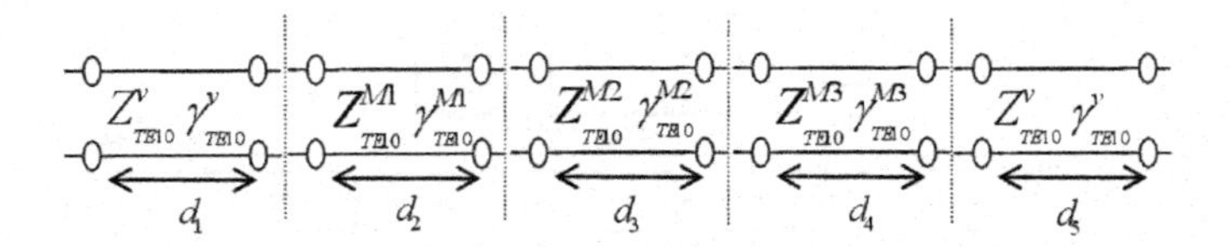

Fig. 3. Transmission line cascade for a three-layer sample

From the scattering matrix of the different sections of transmission line and the scattering matrix of the discontinuities of characteristic impedance it is straightforward [4] and extremely quick to obtain the overall scattering parameters. This approach provides the fastest way for evaluating individuals, but it is constrained to single or multi-layer samples which fill completely the waveguide section.

3.2 Two-Dimensional Numerical Method

For samples with a more complex geometry it is necessary to employ numerical methods for solving the waveguide problem. This electromagnetic problem is defined by the vector wave equation

$$\nabla \times \mu^{-1} \nabla \times \boldsymbol{E} - \omega^2 \epsilon \boldsymbol{E} = 0 \ , \tag{3}$$

where $\boldsymbol{E}$ is the electric field, ω is the angular frequency, μ is the magnetic permeability (a tensor in general) and ϵ is the electric permittivity.

The boundary conditions are perfect electric conductor at the waveguide walls and excitation and matching conditions at ports of the waveguide. Once the electric field is obtained in the domain of the problem, the computation of the scattering parameters is straightforward [5].

When the sample under study have a constant shape over the y axis of the waveguide, the variation of the electric field is known (constant for TE_{10} mode) and it is possible to solve the waveguide problem with a reduction to a 2D domain (x and z axis). In this work the 2D Finite Element Method implemented in the Matlab^{TM} Pdetool [6] was employed. The reduction of unknowns from 3D to 2D cut down on computational cost.

3.3 Three-Dimensional Numerical Method

In the more general case, that is, completely arbitrary geometry , it is necessary to use a 3D full-wave electromagnetic simulator. There are a great number of commercial tools, based on different numerical method as Finite Difference or Finite Elements, that can be used for this purpose. In this work the CST Microwave Studio™ [7], based on the Finite Integration technique, is employed. For both 2D and 3D electromagnetic simulators an accurate modeling of the sample is of key importance for obtaining accurate results in the measured permittivities.

4 Results

For the sake of conciseness only results for analytic and 3D-numerical techniques are shown in Fig. 4, 5, 7 and 8. A three-layer sample (10mm of Polyester, 5.25mm of Styrofoam and 11mm of PTFE) was measured in the range 2 - 3 GHz with the analytic technique. Table 2 shows the final permittivities obtained and the error as regards to the values in literature for these materials. The error for Polyester and Styrofoam was in the range of direct methods (<5%). Nevertheless the result for PTFE was slightly worse. The evolution of the fitness function throughout the procedure in Fig. 4 shows a quick convergence which finds an stable result in the 33rd generation, which is reached in 79 seconds. An additional measurement of the quality of the final results is the comparison between the simulated and the measured scattering parameters. A good matching can be observed in Fig. 5.

Table 2. Permittivities for 3-layer sample

Material	Polyester	Styrofoam	PTFE
ϵ'	3.648	1.59	1.94
ϵ''	0.0676	0.0182	1.7e-5
Error (%)	1.3	2.9	7.6

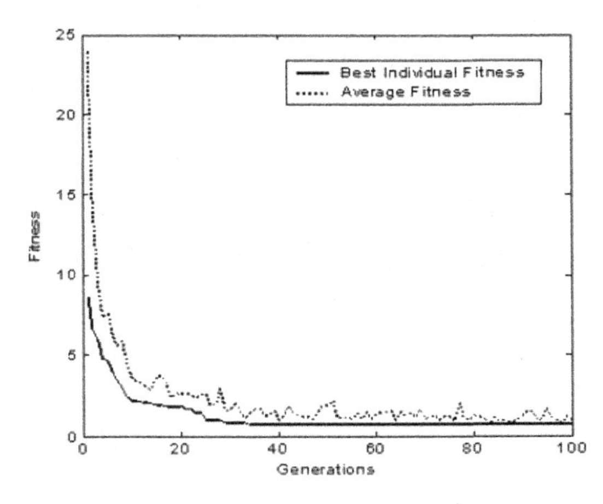

Fig. 4. Evolution of fitness function in the GA for the 3-layer sample

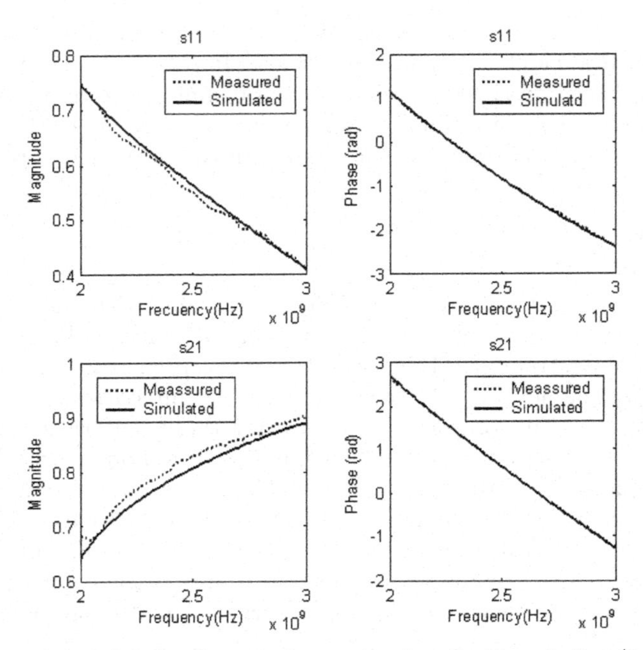

Fig. 5. Measured and simulated scattering parameters for the solution (3-layer sample)

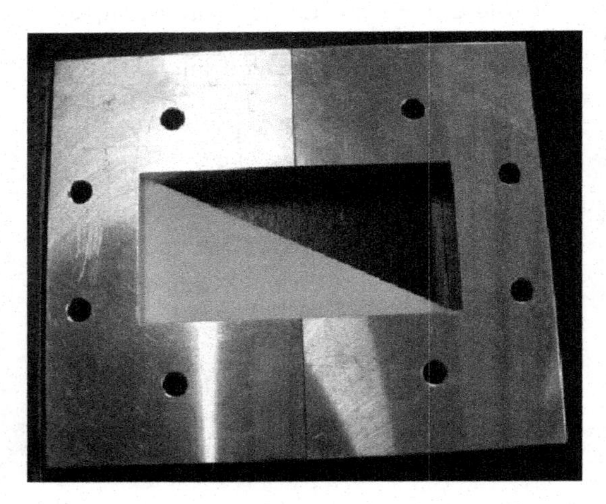

Fig. 6. Triangular sample of PTFE

A triangular sample of PTFE (Fig. 6) was used for testing the behavior of the 3D-numeric technique in the range 2.2 - 3.3 GHz. The convergence in this case is slower and Fig. 7 indicates that the fitness can even improve with a bigger number of generations. Anyway, the comparison in magnitude and phase of

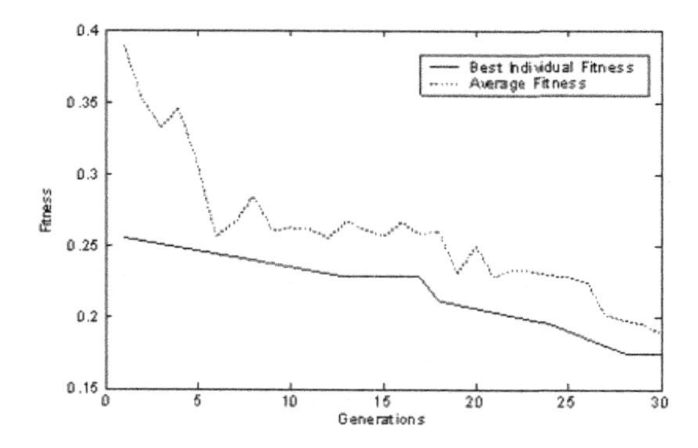

Fig. 7. Evolution of fitness function in the GA for the triangular PTFE sample

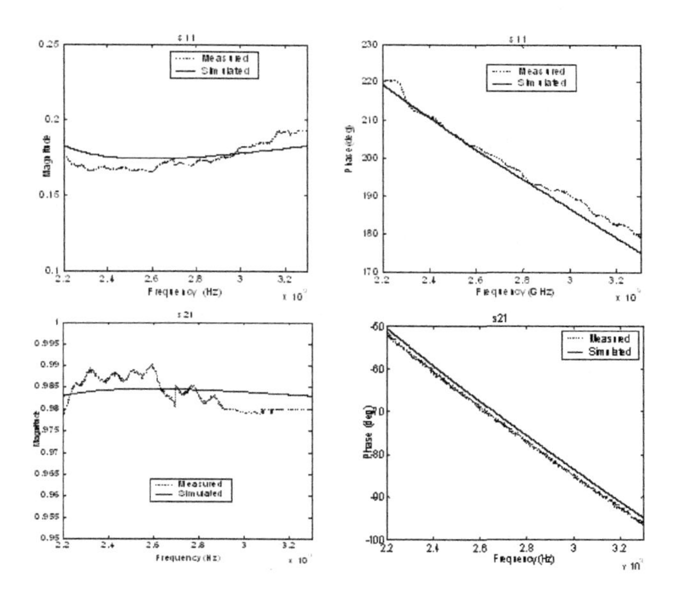

Fig. 8. Measured and simulated scattering parameters for the solution (triangular PTFE)

measured scattering parameters and simulated ones for the final solution (Fig. 8) shows again a good matching. In this case both the PTFE and the air permittivity were assumed unknown. The final solution was $\epsilon^{'} = 2.1036$, $\epsilon^{''} = 7e-4$ for PTFE and $\epsilon^{'} = 1.0024$, $\epsilon^{''} = 1.9e-4$, which yields to a relative error of 0.17% and 0.24%, respectively. The time needed for this result was 9835 seconds, which points out the high computational cost of the 3D-numerical technique.

5 Conclusions

The results presented in this paper show that it is possible to measure the permittivity of non-homogeneous materials or samples with arbitrary geometry by means of a genetic algorithm with an error. Depending on the characteristics of the sample, different simulation techniques can be employed, taking into account that more general techniques (3D-numerical) will require much more time for obtaining the solution. Normally the permittivity of the material is unknown and it is not possible to obtain an error in the solution obtained. In this case, the convergence of the genetic algorithm as well as the comparison between the measured and the simulated scattering parameters can give information about the quality of the final solution.

Acknowledgments. This work was supported in part by Fundación Séneca under project reference 00700/PPC/04.

References

1. Chen, L.F., Ong, C.K., Neo, C.P., Varadan, V.V., Varadan, V.K.: Microwave Electronics. Measurement and Material Characterization. John Wiley & Sons, Ltd., England (2004)
2. Ramat-Samii, Y., Michielssen, E.: Electromagnetic Optimization by Genetic Algorithms. John Wiley & Sons, Inc. Canada (1999)
3. Houck, C.R., Joines, J.A., Kay, M.G.: A Genetic Algorithm for Function Optimization: a MATLAB Implementation. The Mathworks, Natick, MA. NCSU-IE TR (1995) 95–09
4. Requena-Pérez, M.E.: Desarrollo de un Método Inverso para la Determinacin de las Propiedades Dieléctricas de los Materiales. Dissertation (in Spanish). Universidad Politécnica de Cartagena, Spain (2006)
5. Requena-Pérez, M.E., Albero-Ortiz, A., Monzó-Cabrera, J., Díaz Morcillo, A.: Genetic Algorithms and Gradient Descent Optimization Methods for Accurate Inverse Permittivity Measurement. IEEE Transactions on Microwave Theory and Techniques **54** (2006) 615–624
6. Partial Differential Equation Toolbox, user's guide. www.mathworks.com
7. HF Design and Analysis Manual, Version 5, CST Microwave Studio, Darmstadt, Germany (2004)

Putting Artificial Intelligence Techniques into a Concept Map to Build Educational Tools

Denysde Medina, Natalia Martínez, Zoila Zenaida García,
María del Carmen Chávez, and María Matilde García Lorenzo

Department of Computer Science, Central University of Las Villas, Cuba
{denysde, natalia, zgarcia, mchavez, mmgarcia}@uclv.edu.cu

Abstract. When a tutoring aims to guide students in the teaching/ learning process, it needs to know what knowledge the student has and what goals the student is currently trying to achieve. The Bayesian framework offers a number of techniques for inferring individual's knowledge state from evidence of mastery of concepts or skills. Using Bayesian networks, we have devised the probabilistic student models for MacBay, a tutoring system that is an authoring tool. MacBay's models provide prediction of student's action during teaching/learning process. We combined the Concept Maps and the Bayesian networks in order to obtain a Concept Map with intelligent behavior, where "intelligence" is considered as the capacity to adapt the interaction to its user's specific needs. In this paper we describe the way in which we do this combination and inference process.

1 Introduction

One of the key elements that distinguishes Intelligent Tutoring Systems (ITS) from more traditional educational systems is their ability to interpret student actions to maintain a model of student reasoning and learning- the student model [1]. Like user models for non-ITS software, the student model allows an ITS to adapt the interaction to its user's specific needs. However, unlike non-ITS software, the ultimate goal of an ITS's interventions is to help its users learn a target instructional domain.

In the whole world one way to represent organized knowledge in Educational area are the Concept Maps (Cmaps). They are graphical representations of knowledge that are comprised of concepts and the relationships between them. We define a concept as a perceived regularity in events or objects, or a record of events or objects, designated by a label. Concepts are usually enclosed in circles or boxes, and relationships between concepts are indicated by connecting lines that link them together. Words on the linking line specify the relationship between the concepts. Another characteristic of Cmaps is that the concepts are represented in a hierarchical fashion with the most inclusive, most general concepts at the top of the map and the more specific, less general concepts arranged below. The hierarchical structure for a particular domain of knowledge also depends on the context in which that knowledge is being applied or considered [2].

J. Mira and J.R. Álvarez (Eds.): IWINAC 2007, Part II, LNCS 4528, pp. 617–627, 2007.

With the advent of small, powerful computers and GUI interfaces, modeling tools based on Bayesian networks (BNs) are seeing frequent use in real-world applications, including diagnosis, forecasting, automated vision, sensor fusion and manufacturing control. In our case we are going to apply them in education [3].

The structures of a Cmap are similar to a BN. Hence, this paper attempts to develop a Bayesian model that will be applied to the Student Model's in order to guide the student in the teaching/learning process and carry out the process of diagnosis. The appearance of this Bayesian network is like a Cmap with intelligent behavior, where "the intelligence" is considered as the capacity to adapt the interaction to its user's specific needs.

2 Bayesian Networks

BNs are powerful tools both for graphically representing the relationships among a set of variables and for dealing with uncertainties in expert systems see [4] and [5] for an introduction to this field. A key problem in BNs is evidence propagation that is, obtaining the posterior distributions of variables when some evidence is observed. Several efficient methods for propagation of evidence in BNs have been proposed in recent years. Exact methods exploit the independence structure contained in the network to efficiently propagate uncertainty see [6], [7] and [8]. Stochastic simulation constitutes an interesting alternative in highly connected networks, where exact algorithms may become inefficient [9]. Recently, search-based approximation algorithms, which search for high probability configurations through a space of possible values, have emerged as an alternative to the above methods in special cases as, for example, in Bayesian networks with extreme probabilities.

In real learning problems, however, we are typically interested in looking for relationships among a large number of variables. The BN is a representation suited to this task. It is a graphical model that efficiently encodes the joint probability distribution (physical or Bayesian) for a large set of variables [10].

A BN for a set of variables X = {X1, , Xn} consists of (1) a network structure S that encodes a set of conditional independence assertions about variables in X, and (2) a set P of local probability distributions associated with each variable. Together, these components define the joint probability distribution for X. The network structure S is a directed acyclic graph. The nodes in S are in one-to-one correspondence with the variables X. We use Xi to denote both the variable and its corresponding node, and Pai to denote the parents of node Xi in S as well as the variables corresponding to those parents. The lack of possible arcs in S encodes conditional independencies. In particular, given structure S, the joint probability distribution for X is given by

$$p(x) = \Pi_{i=1}^{n}(px_i|pa_i) \tag{1}$$

The local probability distributions P are the distributions corresponding to the terms in the product of 1. Consequently, the pair (S; P) encodes the joint distribution p(x).

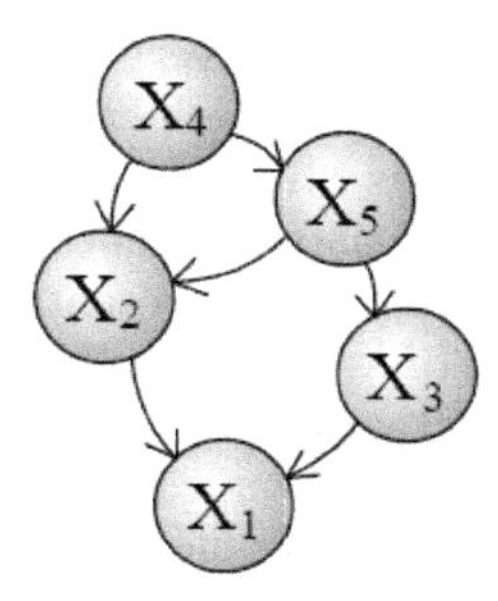

Fig. 1. Shows an example BN structure (without depicting the probability distributions)

The power of the BNs is that once specified the structure of the network and the information available are specified, it is possible to make any type of inference. It is possible to make predictive inferences (if the student knows the concept C, which is the probability that he responds to the question P1 correctly?) or adductive (if the student has answered incorrectly the question P1, which is the probability that he knows the concept C). In this form, a same node can be information source or prediction object. These inferences are made applying algorithms of probabilities propagation [4] that have been developed specifically to reach this goal.

For a general introduction to bayesian network theory see [10].

3 Concepts Maps

Cmaps were developed in the 1970's by Joe Novak (1984) and his research team at Cornell University as a means to help determine how students advanced in their understanding of Science.

The structure of a Cmap is dependent on its context. Consequently, maps having similar concepts can vary from one context to another and are highly idiosyncratic. The strength of Cmaps lies in their ability to measure a particular person's knowledge about a given topic in a specific context. Thereby, Cmaps constructed by different persons on the same topic are necessarily different, as each represents its creator's personal knowledge. Similarly, we cannot refer to the correct Cmap about topic, as there can be many different representations of the topic that are correct [11].

Cmaps have particular characteristics that make them amenable to smart tools. These include:

1. Cmaps have structure: By definition, more general concepts are presented at the top with more specific concepts at the bottom. Other structural information, e.g. the number of ingoing and outgoing links of a concept, may provide additional information regarding a concept's role in the map.

2. Cmaps are based on propositions: every two concepts with their linking phrase forms a "unit of meaning". This propositional structure distinguishes Cmaps from other tools such as Mind Mapping and The Brain, and provides semantics to the relationships between concepts.
3. Cmaps have a context: A Cmap is a representation of a person's understanding of a particular domain of knowledge. As such, all concepts and linking phrases are to be interpreted within that context.
   In addition, in well constructed Cmaps:
4. Concepts and linking phrases are as short as possible, possibly single words.
5. Every two concepts joined by a linking phrase form a standalone proposition. That is, the proposition can be read independently of the map and can still "make sense".
6. The structure is hierarchical and the root node of the map is a good representative of the topic of the map

4 Bayesian Networks and Concept Maps in MacBay

Although the BNs are powerful in the diagnosis problems, their application in the student model is not very frequent in comparison with the great number of developed systems. We consider that the cause is that the application of the BNs demands much more effort than the application of other models of approximate reasoning (certainty factors, fuzzy logic) and the development of heuristic for define and update the student model [12]. This additional effort is mainly caused by two factors:

1. Specification of the network (structure and parameters). Specification of a BN demands a careful study of the variables that take part in the system and the relations of causal influence among them. In addition, once the network has been defined we have to estimate the conditional probabilities that are normally a quite great number of parameters difficult to consider.
2. Difficulty to implement the algorithms of propagation of probabilities, which besides being more or less complex are very expensive computationally.

4.1 Model Specifications and Motivation

It is possible to appreciate that the Cmaps and the BN can be represented like an acyclic graph. This structural similarity and the own characteristics of the Cmap to facilitate the teaching/learning process were the cause that motivated the authoresses of this paper to design this model that combines the BN and the Cmap in order to obtain a tutoring system to improve the teaching/learning process.

To obtain a BN from data requires a learning process that is divided in two stages: the structural learning and the parametric learning [5]. The first stage consists in obtaining the structure of the BN, the relations of dependency and independence between the involved variables. The second stage has the purpose of obtaining the required prior and conditional probabilities from a given structure. The structure of the BN is equivalent to the Cmap, because only those concepts (nodes of the Cmap) that influence in the elaboration of the student model can be considered variables (nodes of the BN). See in the Figure 1 an example of a Cmap and in Figure 2 the structure of the corresponding BN. The second stage will be obtained through a questionnaire that allows catching the cognitive and affective state of the student. We use this information to calculate the prior and conditional probabilities.

With this model we facilitate the specification of the BN. First we define a Cmap (its structural part) and later with the experience of the professor we obtain the tables of conditional probabilities(its parametric part), diminishing the additional effort to implement a BN, since this is one of the factors that causes its little use to modeled this kind of problem.

4.2 Calculation of Prior and Conditional Probabilities from the Concept Map

The professor who develops the Cmap, must formulate an initial questionnaire, that has to be able to catch the student's cognitive state, turning it in a customized Cmap, where the student "navigate" in an oriented way according to his knowledge and not in a free way like in the traditional Cmap. And precisely this characteristic allows the Cmap to adapt the interaction to its user's specific needs.

The variables (features) that take part in the calculation of the prior and conditional probabilities are:

1. The evaluation of the questions that conform the questionnaire
2. The results of the test that the pedagogical assistant makes (see Topic 4.3).

The number of variables (features) conforms a table of 2n combinations, where n is a natural number that can be considerably great; which constitutes a difficulty whose solution could reduce the space of initial representation, so that if there are superfluous variables, it is analyzed if they stay or not, according to its importance from the methodological point of view.

An alternative solution to the selection of problem variables is the use of the set of typical testors [13]. The typical testors are the number of variables (attributes or features) that describe the objects (questionnaires) that affect a problem in a significant way. In the proposed model the LEX [14] algorithm is applied to obtain the typical testors.

4.3 Pedagogical Assistant

The affective characteristics are obtained through a pedagogical assistant in order to obtain a bridge of connection between affectivity and cognition. Traditionally there has been an almost absolute separation between the cognitive and affective-motivational aspects when studying their influence in the process of learning. Some authors have centered their studies in the cognitive aspects and have forgotten the others, or in the contrary form. At the present time, there is an increasing interest in studying both types of components in an integrated form. "It is possible to affirm that learning is characterized simultaneously as a cognitive and motivational process". Consequently, in the improvement of the academic yield it is appropriate to consider both the cognitive and motivational aspects. In the process of learning it is essential "to be able" to make it, which makes reference to the capacities, the knowledge, the strategies, and the necessary skills (cognitive component); but in addition, it is necessary "to want" to do it, to have the sufficient disposition, intention and motivation (motivational component).

For this component a system was developed based on rules that collaborate with the learning. It allows to know the mood of the student, if he is motivated or not and if he understood all that he has been taught. Thus the tutoring system will be able to change the strategy that is being used to learn or motivate and reconstruct the way that the student studies.

4.4 Example of Application of the Proposed Model

The subject of Binary Tree (BT) in the course of Data Structures in the studies to obtain a professional title in Computer Science includes Binary Heaps (BH), concept that has been selected to show in a simple way the use of the proposed model.

Figure 2 shows a Cmap of BH. As it can be observed a BH is defined as a Balanced Binary Tree (BBT), which is as well a BT that can be perfect or not. In addition, the BH has a specific structure and can be implemented with arrays or connected nodes list.

Figure 3 shows the structure of the BN associated to the Cmap of Figure 2. Figure 4 shows the parametric part equivalent to the probability distribution associated to each concept present in the network. The initial tables associated to each node of the BN, are constructed with the professors' experiences on the educational behavior of the students in the subject that is being analyzed.

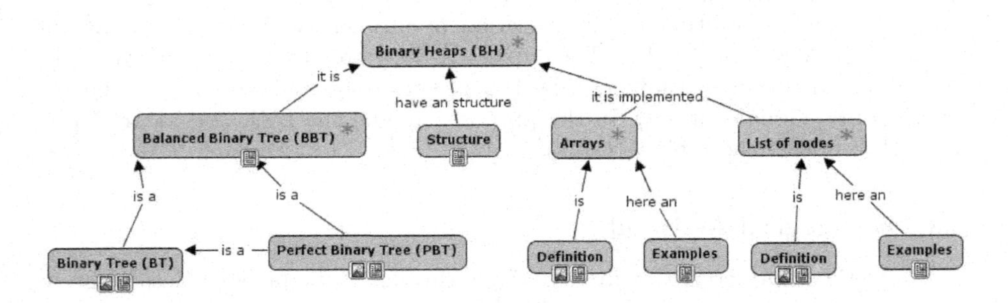

Fig. 2. Example of a Cmap to learn BH

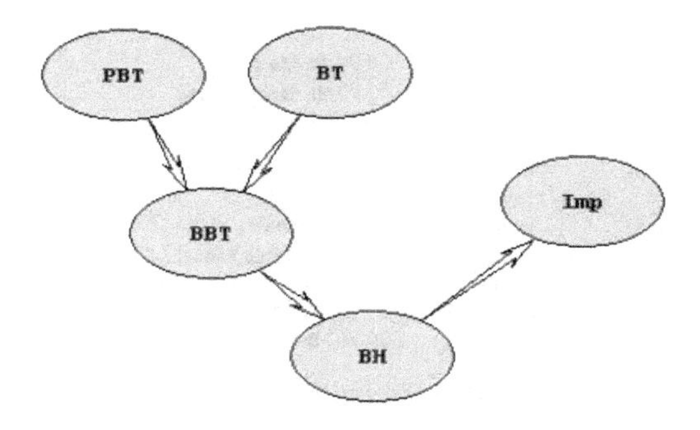

Fig. 3. BN associated to the Cmap of Figure 2

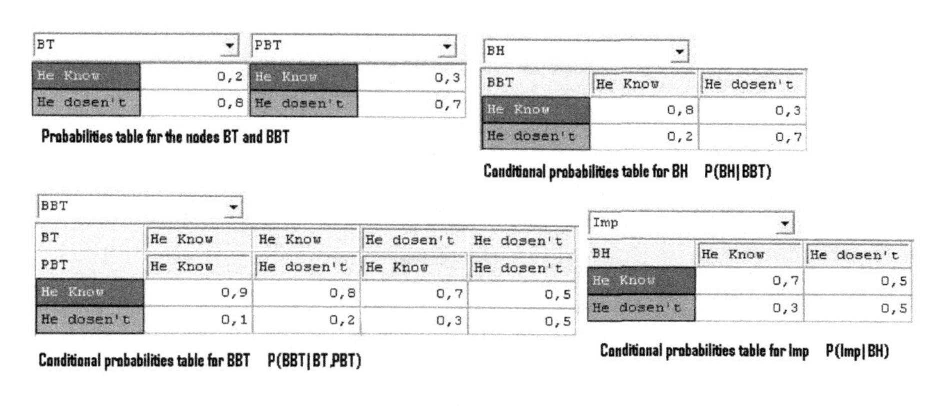

BT	
He Know	0,2
He dosen't	0,8

PBT	
He Know	0,3
He dosen't	0,7

Probabilities table for the nodes BT and BBT

BH		
BBT	He Know	He dosen't
He Know	0,8	0,3
He dosen't	0,2	0,7

Conditional probabilities table for BH P(BH|BBT)

BBT				
BT	He Know	He Know	He dosen't	He dosen't
PBT	He Know	He dosen't	He Know	He dosen't
He Know	0,9	0,8	0,7	0,5
He dosen't	0,1	0,2	0,3	0,5

Conditional probabilities table for BBT P(BBT|BT,PBT)

Imp		
BH	He Know	He dosen't
He Know	0,7	0,5
He dosen't	0,3	0,5

Conditional probabilities table for Imp P(Imp|BH)

Fig. 4. Tables of probabilities associated to the BN of Figure 3

In the initial model of these tables, that is a simple example, we only consider the probability that the student dominates (knows) the subject or not. Finally Figure 4 shows the tables after an inference is made, when the student responds a questionnaire elaborated by the professor.

In Figure 2, the nodes of the Cmap that have associate questionnaire have been marked with an asterisk (*), where the professor considers opportune to catch the cognitive state of the student to guide his navigation by the Cmap. According to the answers to the questionnaire, the BN variables can be valorized. This is equivalent to making an inference, which allows us to determine the probability that the student know certain contents. And based on this the navigation in the Cmap is qualified.

Figure 5 and Figure 6 show the form in which the inference process is made. In Figure 5 the initial probabilities are shown according to the algorithm of propagation of knowledge. In Figure 5 we propagate the case of a student that according to the answer to the made questionnaire has the evidence that he knows BT and PBT subjects. We can conclude that the student can learn balanced binary trees with greater probability (0.9).

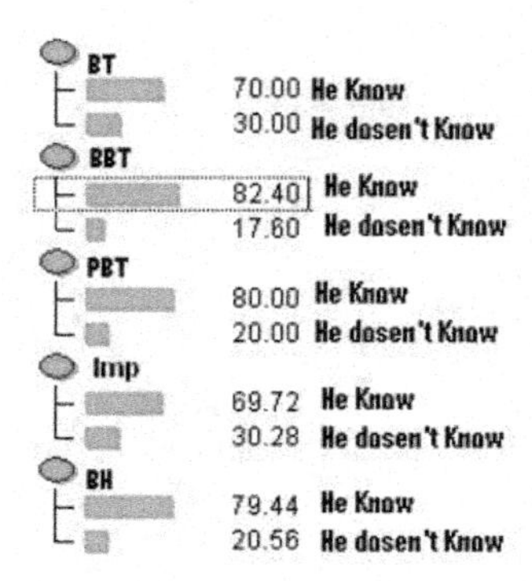

Fig. 5. Initial model from the probabilistic point of view

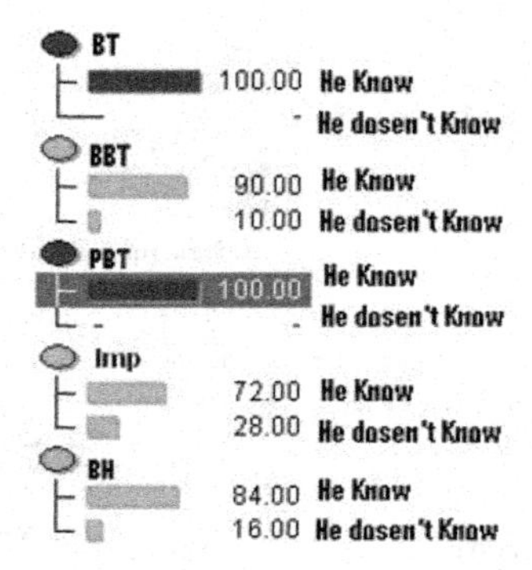

Fig. 6. Model after an inference is made

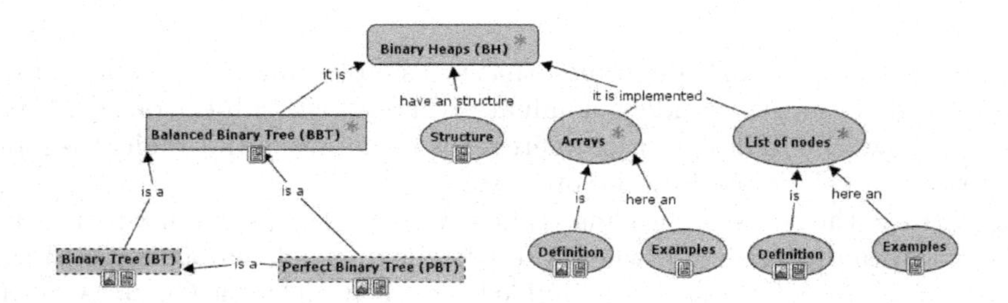

Fig. 7. Example of a Cmap that results after a questionnaire is made

The structure of BN allows the professor to know that if the student doesn't know a thematic when he makes a questionnaire it is evidence will be that he doesn't know the topic. In this case the evidence is "he doesn't know", that is equivalent to say that the probability of the evidence is equal to 0. In consequence, this branch in a Cmap is available.

The Figure 6 shows an example of the Cmap that corresponds with the model of one student that answers a questionnaire, where the nodes with square form mean that he must study this contents and the nodes with elliptical form mean that he doesn't know these topics. The nodes with square form, including the nodes with square form with broken lines in its border, are the only nodes that the student can access. The student could study the contents that the nodes with elliptical form contain if he knows all the contents that the nodes with square form contain. The system knows that across a questionnaire.

5 Results

In order to evaluate the proposed model, we have used simulated students. The use of simulated students has the following advantages: (a) the total control of the conditions of the evaluation questionnaires is possible; (b) it allows to compare the obtained results with the real results, since it is impossible to have the true cognitive state of a real student and (c) it allows to evaluate the techniques before being used with people.

The operation of a simulated student is the following: be C1,, Cn the concepts of the Cmap associated to one subject that is intended to be evaluated. We defined a simulated student that knows the 100% of the concepts C1, , Cn, where the set of well-known concepts is generated in random way. In this way we got simulated students of the same level but whose group of well-known concepts is different. Once the simulated student has been generated we use the net to calculate the probabilities of responding correctly to each one of the questions in the questionnaires. This probability will be used in order to simulate the behavior of the student in the following way: suppose that the probability of responding correctly the question P is p. If the questionnaire has the question P we use the probability p to determine the membership degree of this question to one linguistic label of the fuzzy sets of qualifications that was defined by an expert (in this case a professor) (see [15]). For example, the linguistic label of the fuzzy sets of qualifications could be: "The question was answered correctly" and "The question was answer incorrectly".

For the validation of the model 40 students of six different types were generated (taken in consideration the group of well-known concepts that they have). This made 240 students. Each simulated student answered 4 questionnaires with 10 questions, this made 40 questions.

When we applied our model we obtained the results that we shown in Figure 8.

We could qualify the outputs gotten as very good. Without a doubt it is due to the theoretical consistency of the utilized model.

The efficacy of the evaluator process depend, to a extent on the efficacy of the diagnosis process. The measurement of the errors for topic varies between 0.0264 and 0.0530, with very small typical deviations. In a decimal scale, the error would vary between two and five tenth, which seems acceptable given that the model admits that students without knowledge give the correct answer and students with all the necessary knowledge may fail in answering a question.

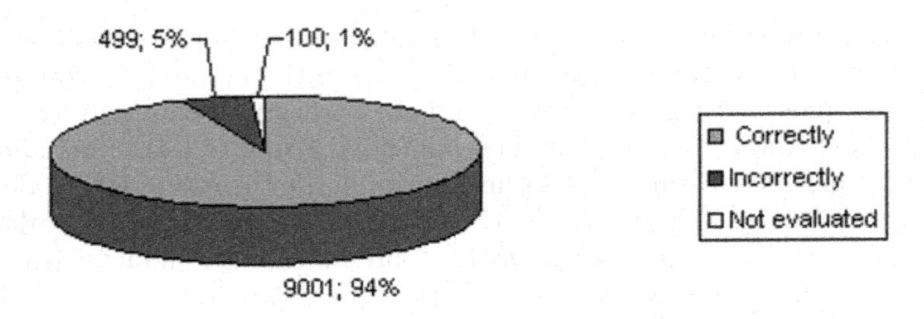

Fig. 8. The results that were obtained when the MacBay model was applied

6 Conclusions

In this paper we propose a new form of implementing the student model, where we combine the facilities of the Cmaps and the BNs capacities. As a result of this combination we obtain a nice way of representing knowledge and a powerful inference tool that allows our system to have the capacity to adapt the interaction to its user's specific needs.

These ideas have been implemented in MacBay. This ITS has been applied in the elaboration of many teaching/learning systems and we have obtained good results.

Acknowledgments

The authors would like to thanks VLIR (Vlaamse InterUniversitaire Raad, Flemish Interuniversity Council, Belgium) for supporting this work under the IUC Program VLIR-UCLV.

References

1. Shute, V. J. and Psotka, J. (1996). Intelligent Tutoring Systems: Past, Present and Future. Handbook of Research on Educational Communications and Technology. Jonassen, D., (Ed.), Scholastic Publications.
2. Alberto J Cañas: A Summary of Literature Pertaining to the Use of Concept Mapping Techniques and Technologies for Education and Performance Support. The Institute for Human and Machine Cognition. July 2003
3. Eugene Charniak. Bayesian Networks without Tears. AI Magazine http://www.aaai.org. 1991.
4. Castillo, E., Gutiérrez, J. M., and Hadi, A. S. (1997), Expert Systems and Probabilistic Network Models, Springer-Verlag, New York.
5. Pearl, J. (1988), Probabilistic Reasoning in Intelligent Systems: Networks of Plausible Inference, Morgan Kaufmann, San Mateo, CA.
6. Kim, J. H. and Pearl, J. (1983), "A Computation Model for Causal and Diagnostic Reasoning in Inference Systems," in Proceedings of the 8th International Joint Conference on AI, Los Angeles, 190-193.

7. Lauritzen, S. L. and Spiegelhalter, D. J. (1988), "Local Computations with Probabilities on Graphical Structures and Their Application to Expert Systems," Journal of the Royal Statistical Society (B), 50, 157-224.
8. Jensen, F. V., Olesen, K. G., and Andersen, S. K. (1990), "An Algebra of Bayesian Belief Universes for Knowledge-Based Systems," Networks, 20, 637-659
9. Shachter, R. D. and Peot, M. A. (1990a), "Simulation Approaches to General Probabilistic Inference on Belief Networks," in Uncertainty in Artificial Intelligence 5, Machine Intelligence and Pattern Recognition Series, 10 (Henrion et al. Eds.), North Holland, Amsterdam, 221-231.
10. David Heckerman. A Tutorial on Learning with Bayesian Networks. Technical Report MSR-TR-95-06, Microsoft Research, March, 1995.
11. Alberto J Cañas and Marco Carvalho. Concept Maps and AI: an Unlikely Marriage? The Institute for Human and Machine Cognition. July 2003.
12. Millán Eva y José Luis Pérez-de-la-Cruz. Un algoritmo de diagnóstico para modelado del alumno basado en test adaptativos y redes bayesianas. http://www.lcc.uma.es/~eva/doc/materiales/millane.pdf
13. Ruiz J."Modelos matemáticos para el reconocimiento de patrones". Edit UCLV 1993.
14. Santiesteban Alganza Yovanis y Aurora Pons Porrata. REVISTA CIENCIAS MATEMATICAS Vol. 21, No. 1, 2003. LEX: UN NUEVO ALGORITMO PARA EL CALCULO. DE LOS TESTORES TIPICOS.
15. Medina Sotolongo, D; Martinez Sánchez, N; León Espinosa, M; Garcia Valdivia, Z. A fuzzy Bayesian model for intelligent teaching learning system. In Proceedings of the International Symposium on Fuzzy and Rough Sets, Cuba. December 2006. ISBN 959-250-308-7

A Preliminary Neural Model for Movement Direction Recognition Based on Biologically Plausible Plasticity Rules

Eduardo A. Kinto[1], Emílio Del Moral Hernandez[1], Alexis Marcano[2], and Javier Ropero Peláez[3]

[1] Laboratório de Sistemas Integráveis. São Paulo University
eakinto@lsi.usp.br, emilio_del_moral@ieee.org
[2] Group for Automation and Soft Computing, Technical Univ. of Madrid
a.marcano@gc.ssr.upm.es
[3] Laboratório de Microeletrônica. São Paulo University
fjavier@usp.br

Abstract. In this work we implement a neural architecture for recognizing the direction of movement using neural properties that are consistent with biological findings like intrinsic plasticity and synaptic metaplasticity. The network architecture has two memory layers and two competitive layers. This un-supervised neural network is able to identify the direction of movement of an object, being a promising network for object tracking, hand-written and speech recognition.

Keywords: intrinsic plasticity, synaptic metaplasticity, un-supervised neural network, movement direction identification.

1 Introduction

Cinematographic perception is a concept that has been recently coined by several brain researchers [10,11]. According to them, the brain is able to capture reality as it is done in cinematography, frame by frame, one frame each 100 msec. This point of view agrees with our previous works [3,7], in which we propose that the thalamus captures one frame of reality each 100 msec. At the end of the 100msec. interval, each frame is sent to the cortex [3] where it is linked to other frames producing a sequence. Shorter sequences are linked in lower cortical layers. These sequences are further linked into longer sequences in higher cortical layers. We propose that the linkage of frames in each layer of the cortex is compatible with competitive processes taking place in each layer. To test this possibility we created an architecture composed of several layers, each one of them consisting in a competitive Self Organizing Feature Map (SOFM) neural network that was applied to recognition of trajectories of persons inside a building [4] and to predictive maintenance [7]. Apparently this architecture was able to combine sequential and competitive learning for identifying sequences (trajectories).

J. Mira and J.R. Álvarez (Eds.): IWINAC 2007, Part II, LNCS 4528, pp. 628–636, 2007.

One important characteristic of the architecture is that the neurons' output in each SOFM was maintained during a while, and slowly vanishes. This characteristic is also present in pyramidal neurons of the cerebral cortex [9]. Pyramidal neurons trigger a burst of action potentials, in which the first action potentials are close together whereas the last action are sparse, thereby creating a "comet tail" effect.

Our intention was to gradually substitute artificial elements in that neural architecture by more biological elements. In this paper we substitute the SOFM network that was used in the previous work, by a competitive network with biologically realistic neurons. In this biologically realistic neural network, the winning "responsibility" in the competitive layer is distributed (by means of a biological property called intrinsic plasticity), so that each neuron in the competitive layer has its own chance to win, avoiding that a single neuron wins the whole time. Intrinsic plasticity [1], [2] consists in the shifting of the activation function leftwards or rightwards in the cases of a lowly or highly activated neuron respectively. For a lowly activated neuron the leftward gradual shifting of the activation function makes this neuron to increment its probability of firing in the future. In the case of a highly activated neuron, intrinsic plasticity leads the neuron to reduce its firing probability in the future by shifting its activation function rightwards.

We also use neurons in which synapses are not only reinforced or depressed in the same condition biological neurons, but also have the property of metaplasticity [5].

With all these biological elements we have created a multilayered network and tested preliminarily its capacity of discriminate, without supervision, whether a stimulus moves rightwards or leftwards in the visual field. The results of the network are discussed, encouraging us to apply this architecture to more demanding tasks like handwriting and speech recognition.

This paper is organized as follows. In section two we present some biological findings regarding synaptic plasticity, metaplasticity and intrinsic plasticity. In section three the network architecture that uses the homeostatic property is presented and commented. In section four, the results of the simulation are shown and, finally, in section five results are discussed.

2 Synaptic Plasticity

Plasticity is related to the capacity of changing. This term is applied either to weight changes at synapses or to changes in the basal firing frequency of neurons. In the following section we will introduce these two types of plasticity, synaptic plasticity and intrinsic plasticity.

2.1 Synaptic Plasticity

Synaptic plasticity is a crucial concept to understand how information is stored in complex neural systems. Understanding synaptic plasticity it is possible to visualize how learning is shaped by experience. Donald Hebb proposes in 1949 that: When neuron A excites neuron B in a short time interval, some metabolic changes take place in synapses so that cell A contribution to the cell B postsynaptic potential becomes more significant. Hebbian plasticity is therefore referred to any kind of synaptic plasticity that is consistent with Donald Hebb statement.

We will use the following probabilistic Hebbian rule, usually called pre-synaptic rule, for modeling synaptic plasticity:

$$w_{AB} = P(B/A) = \frac{n(A \cap B)}{n(A)}. \quad (1)$$

Where:

- A and B represents pre-synaptic and postsynaptic action potentials respectively.

- W_{AB} represents the synapse weight between neurons A and B.

- P(B/A) is a conditional probability between a presynaptic neuron A and a postsynaptic neuron B. This probabilistic rule evaluates how many times presynaptic and postsynaptic action potentials coincide given that a pre-synaptic action potential has taken place.

The reason of using the pre-synaptic learning rule is that it captures important properties of real synapses like synaptic directionality, potentiation and depression intervals and synaptic metaplasticity.

2.2 Intrinsic Plasticity

Synaptic plasticity may not handle some situations of "synapse saturation" due to the correlation-based rules. A "pitfall" of Hebbian-like synaptic learning arises when the average input activity is too low (never fires) or too high (easily saturated). Homeostasis is one of accepted explanations to deal with this situation. Homeostasis is the mechanism through which living organisms maintain its internal equilibrium beyond tolerable limits for proper functioning. Our body fluids that keep us alive must be in equilibrium, that is, we have a built-in physiological mechanism that regulates our internal state due to alteration in temperature, molecules and ion concentration. The concept of homeostasis was introduced by the French Claude Bernard in 1865 (in his *Introduction to Experimental Medicine*). Claude Bernard is considered the father of modern physiology. Another great physiologist in the field of homeostasis is the North American Walter Bradford Cannon (1932), who coined the word *homeostasis*. Cannon formalized the mechanism of equilibrium.

Intrinsic plasticity (IP) is related to the intrinsic electrical properties of individual neurons. For the case of homeostatic regulation, IP describes the cellular mechanism for handling variations on neuron input. If the input is too low, we will hardly see a postsynaptic action potential and if the input is too high, the postsynaptic action potential will be saturated. So, shifting the action potential curve (see figure 1, input-output curves) to the input average region will enable better postsynaptic response.

Hebbian plasticity does not account for input fluctuations. During our life, the average value of input activity to a cell may change, and also the average input activity (sensory information) may change even in a simple situation like eyes opened and eyes closed.

Desai et al. [1,2] showed that neurons that had been prevented from spiking for 48 hours increased their response to current injection. They become easier to fire.

In this work, the intrinsic plasticity is being represented (according to [8]) by the sigmoid displacement (shifting) in the activation function (see figure 1). This displacement of the plasticity curve depends on the initial weight value.

The activation function can be expressed by:

$$y = \frac{1}{1 + e^{-A(x-l)}} \tag{2}$$

Where A represents one constant of the sigmoid to be adjusted and l é the displacement constant given by:

$$l_t = \frac{\xi \cdot O_{t-1} + l_{t-1}}{\xi + 1} \tag{3}$$

ξ is the constant that controls the sigmoid displacement.
O is the ANN output in the instant t-1

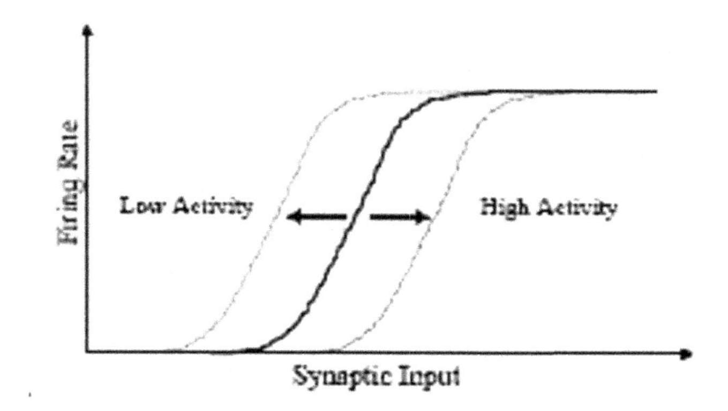

Fig. 1. Intrinsic plasticity. The shifting or displacement of the activation function is rightwards or leftwards depending on the synaptic input of the neuron. If the input is high most of the time, the activation function shifts to right, and if the input is usually low the activation function shifts to left.

3 Computational Simulations

In this section we propose a neural network architecture for allowing movement direction recognition. Matlab (a commercial software for simulation purposes) was used for the development of aforementioned architecture and the whole simulation was done on a simple Pentium 4 home desktop computer.

3.1 Network Topology

Figure 2 represents the architecture of the neural network developed in this paper. It is composed by one input layer with 4 neurons, which maps input data into a memory layer. The memory layer main function is to store past states by a fading mechanism similar to that of pyramidal neurons in the cerebral cortex. As was explained at the introduction, pyramidal neurons fire in a "comet tail" manner: when triggering a burst of action potentials, the initial action potentials are close-together but, at the end of the burst, the action potentials are wide apart. We simulate this behavior by connecting each neurons of the memory layer to itself with a fixed weight value (being the weight value less than one). To the right of the memory layer, there is third layer with 3 neurons. This layer performs dimensionality reduction through a competitive process. Therefore, we have two memory layers and two competitive layers, the latter with only 2 neurons. With this architecture we expect that when the movement of input patterns is in one direction, neuron 15 will fire, whereas when the movement is in the opposite direction, neuron 16 will fire.

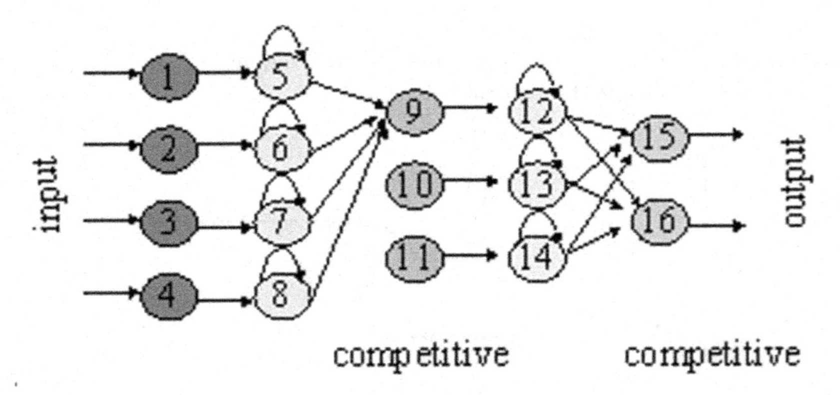

Fig. 2. Movement direction identification – Architecture of artificial neural network with Hebbian pre-synaptic learning

The architecture of the ANN presented on the figure 2 above can be represented in matrix notation as shown in table 1. The matrix M is a topological representation where each line corresponds to a neuron (so we have 16 lines) and each column represents whether the neuron A (defined by the line number) receives input from other neuron B. The diagonal line of this square Matrix corresponds to self-input of each neuron.

Let us focus on the fifth line, **1 0 0 0 1 0 0 0 0 0 0 0 0 0 0 0**. for explaining how the fifth neuron is connected.. The ones in the first and fifth position mean that neuron 5 receives its inputs from neuron 1 and from itself.

Only weights arriving to competitive layers undergo synaptic plasticity. Or, in other words, only the synaptic weights of neurons 9, 10, 11, 15 and 16 are modified. The lack of synaptic weights in some neurons of the model is not

Table 1. Matrix representation of the proposed architecture

```
M = [0 0 0 0 0 0 0 0 0 0 0 0 0 0 0 0;
0 0 0 0 0 0 0 0 0 0 0 0 0 0 0 0;
0 0 0 0 0 0 0 0 0 0 0 0 0 0 0 0;
0 0 0 0 0 0 0 0 0 0 0 0 0 0 0 0;
1 0 0 0 1 0 0 0 0 0 0 0 0 0 0 0;
0 1 0 0 0 1 0 0 0 0 0 0 0 0 0 0;
0 0 1 0 0 0 1 0 0 0 0 0 0 0 0 0;
0 0 0 1 0 0 0 1 0 0 0 0 0 0 0 0;
0 0 0 0 1 1 1 1 0 0 0 0 0 0 0 0;
0 0 0 0 1 1 1 1 0 0 0 0 0 0 0 0;
0 0 0 0 1 1 1 1 0 0 0 0 0 0 0 0;
0 0 0 0 0 0 0 0 1 0 0 1 0 0 0 0;
0 0 0 0 0 0 0 0 0 1 0 0 1 0 0 0;
0 0 0 0 0 0 0 0 0 0 1 0 0 1 0 0;
0 0 0 0 0 0 0 0 0 0 0 1 1 1 0 0;
0 0 0 0 0 0 0 0 0 0 0 1 1 1 0 0];
```

incompatible with biology: in the nervous system there are also neurons that lack modifiable synaptic weights.

For example, the network input layer, do not have modifiable weights, their weights are always set to one. Another type of connections lacking modifiable weights is the feedback connection of memory layers. Arbitrarily we have set the weights of these connections to 0.4

For dealing with modifiable and not modifiable weights we split the connectivity matrix M into three connectivity matrices MA, MP and MF .

Therefore, the weight matrix can be expressed by:

$$w = \left(\frac{num}{den}\right) * MA + MP + 0.4 * MF \tag{4}$$

Where **num** is a numerator matrix randomly initialized at the beginning with dimension equals to M and **den** is a denominator matrix of dimension M with all values equal to one.

MA is a matrix of zeros with dimension M. Only points (line and column) representing neurons 9, 10, 11, 15 and 16 have a unit value. It represents a connectivity matrix in which only neurons with modifiable weights participates.

MP is a matrix of zeros with dimension M. Only points representing neurons 5, 6, 7, 8, 12, 13 and 14 have a unit value. It represents the one to one connections from memory to competitive layers.

MF is a matrix of zeros with dimension M. Only points representing neurons 5, 6, 7, 8, 12, 13 and 14 and representing the "self-feedback" have a unit value. It represents the memory layer retaining changes for some number of iterations. The constant of smoothness is 0.4 (empirically found).

Equation 5 shows how the activation function shifting rate used on the activation function (eq. 3) is decreased over iterations. This decrease is dependent on the number of iterations set in the beginning, the greater is iteration the lower is shifting speed.

$$\xi = \frac{\xi_c}{It} \tag{5}$$

Where:
ξ is a variable that controls the sigmoid displacement (used on equation 3)
ξ_c is some initial constant value
It is the current iteration
This equation represents an annealing process for stabilizing the weights.

3.2 Network Simulation

The sequence of input patterns represents the direction of movement.

Movement from left to right:	Movement from right to left:
1 0 0 0	0 0 0 **1**
0 **1** 0 0	0 0 **1** 0
0 0 **1** 0	0 **1** 0 0
0 0 0 **1**	**1** 0 0 0

On every iteration we can compute the matrix weight using the equation 4. The activation of each neuron is computed as follow.

$$a = w * o. \tag{6}$$

Where w is a matrix and o is a vector composed by the outputs of all neurons in the previous iteration.

New output value is obtained using equation 2. Equation 3 is used to adjust the shifting rate.

The stopping criterion is based on number of iterations. See pseudo-code.

Initialize O_T, with zeros
While condition is true
Adjust O_T
  update epsilon, according to equation 5
  initialize W, according to equation 4
  calculate the net input, according to equation 6
  update shifting, according to equation 3
  calculate and adjust O_{T+1}, according to equation 2
End while

4 Results

The results of the simulation confirmed the expected behavior. We can see in figures 3 and 4 the one at the input layer going in one direction or another;

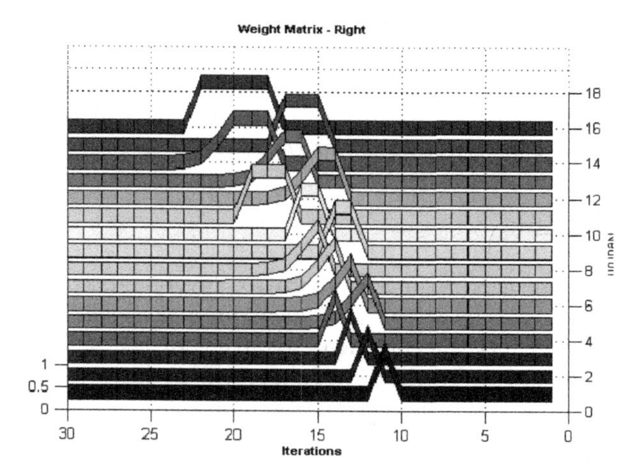

Fig. 3. Movement identification to right. The first four lines (corresponding to neurons 1, 2, 3 and 4) receive inputs (value 1) following a temporal sequence showing a movement to the right direction (see figure 2). Note that after neuron 1 has received a unit value as input, in the following time step the memory layer retains values from the previous time. Also note that neurons smooth their value toward zero as expected. On the competitive layer, only one neuron can be active at each time.

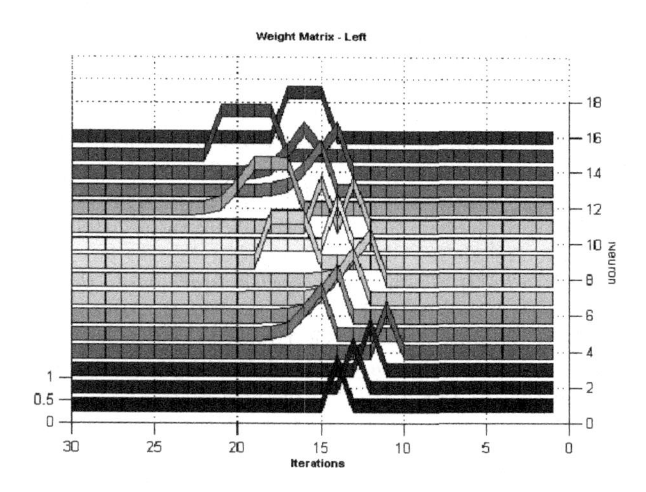

Fig. 4. Identification of the movement to the left. We have similar conditions as in the experiments in the figure 3, but changing movement direction.

the memory layer delaying a response for about 3 unit time, and the competitive layer performing the dimensionality reduction, allowing only one "winner" neuron for each time step. At the end of the training, the network "learned" to identify movement direction. However, since the rightward displacement of input patterns was always followed by the opposite movement, the network has also learned to predict the direction of following patterns. In this sense the winning neuron of the last layer indicating direction is always followed by another winning neuron that predicts a reversal direction in future iterations.

It takes about 10.000 iterations for training the network.

5 Conclusions

The neural architecture with 2 competitive layer and 2 memory layer implemented in this work, based on biological findings such as plasticity (eq. 1), metaplasticity and intrinsic plasticity (shifting of the activation function, see eq. 2 and 3), showed to be effective for movement direction identification. Displacing a unit value over the input layer represents movement. After training, this self-organizing network "learned" the direction of input moving patterns: when the direction is rightwards one neuron in the output layer fires, and when the direction is leftwards, the another neuron fires.

References

1. Desai, N. S., Rutherford, L. C. and Turrigiano, G., (1999) *Plasticity in the intrinsic excitability of cortical pyramidal neurons*, Nature Neuroscience, 2(6):515-520.
2. Desai, N.,(2003) *Homeostatic plasticity in the CNS: synaptic and intrinsic forms*, Journal of Physiology. 97:391-402.
3. Peláez, F. J. R., (1997) *Plato's theory of ideas revisited*, (Special Issue) Neural Networks. 7(10): 1269-1288.
4. Peláez, F. J. R., Simões, M. G., (1999a), *A neural network based intruder detection system*, In: Anais do 4o Simpósio Brasileiro de Automação Inteligente. Sao Paulo. pages 377-381
5. Peláez, F. J. R., Simões, M. G., (1999b), *A computational model of synaptic metaplasticity*,In: Proceedings of the IJCNN 99 International Joint Conference of Artificial Neural Networks. Washington D.C.
6. Peláez, F. J. R., (2000), *Towards a neural network based therapy for hallucinatory disorders*, (Special Issue) Neural Networks. 2000(13): 1047-1061.
7. Peláez, F. J. R., Aguiar, M. A., Destro, R. C., Kovács, Z. L., Simões, M. G. (2001), *Predictive Maintenance Oriented Neural Network System(PREMON)*, In: Proceedings of the IECON 2001 Denver (Colorado)
8. Pelaéz, J.R, & Piqueira, J. R. C., (2006) *Biological clues for up-to-date artificial neurons"*, In: Diego Andina; Duc Truong Phan. (Org.). Computational Inteligence: for Engineering and Manufacturing. 1 ed. Berlin: Springer-Verlag, 2006, v. 1, p. 1-19.
9. Shepherd, G.M.,(1998) *The synaptic organization of the brain*,Oxford University Press. New York. Pag.437.
10. VanReullen, R. & Koch, C.,(2003) *Is perception discrete or continuous?*,Trends in Cognitive Sciences. 7(5):207-213.
11. VanReullen, R., Reddy, L. & Koch, C., (2005) *Attention-driven discrete sampling of motor perception*, Proceedings of the National Academy of Sciences. 102(14):5292-5296

Classification of Biomedical Signals Using a Haar 4 Wavelet Transform and a Hamming Neural Network

Orlando José Arévalo Acosta and Matilde Santos Peñas

University Complutense of Madrid, Faculty of Computer Science,
Madrid, 28040, Spain
ojareval@fis.ucm.es

Abstract. This contribution consists on the application of a hybrid technique of signals digital processing and artificial intelligence, to classify two kinds of biomedical spectra, normal brain and meningioma tumor. Each signal is processed to extract the relevant information within the range of interest. Then, a Haar 4 wavelet transform is applied to reduce the size of the spectrum without loosing its main features. This signal approximation is coded in a binary set which keeps the frequencies that could have representative amplitude peaks of each signal. The coding is input in a recursive Hamming neural network previously trained, which is able to classify it by comparing it with patterns. The results of the classification are shown for a group of signals that corresponds to human brain tissue. The advantages and disadvantages of the implemented method are discussed.

Keywords: Neural networks, wavelets, classification, biomedical signals.

1 Introduction

As it has been proved in many applications, Artificial Intelligence techniques developed so far have found in Medicine an interesting implementation field, specifically for automation of medical diagnosis. This is the aim of the following contribution, where previous works have already explored the possibility of using heuristic algorithms and computational resources in the characterization of this kind of biomedical signals [2]. In this paper, the identification of brain tumors of the human tissue, obtained by cranial surgery from patients with two different medical conditions (normal brain and meningioma brain tumor) is under study.

Different strategies have been applied [5,6,10,12,18] to deal with this problem of pattern recognition (linear discriminant analysis, statistical correlation, other kinds of neural networks, cluster analysis, support vector machines, etc.). The final classification is strongly influenced by the measurement conditions [17] and by the parameters of the classifier. In many cases they are based in specialized medical knowledge [12,13,16,19], but in some of other cases a feed forward multi-layer neural network was implemented and non medical knowledge was used to classify any sample of the database [2]. Nevertheless, the use

J. Mira and J.R. Álvarez (Eds.): IWINAC 2007, Part II, LNCS 4528, pp. 637–646, 2007.

of a different network structure could improve the expectations concerning the performance obtained by the other approaches. The medical diagnosis is faced using a recursive bi-layer neural network, with four nodes, and under supervised learning. This structure has been selected due to the fact it provides a better adaptation to the task [7].

The signals under analysis are biomedical spectra which were obtained in vitro, using the well known magnetic resonance spectroscopy method ^{1}H [13]. These spectra have a resolution of 16384 samples for the real component, being uniformly distributed in a determined range. Previously to the application of the neural network, another computational tool is applied to extract the relevant information. Because the data are influenced by the conditions in which the samples were taken, the signals need to be pre-processed. Moreover, the spectra are large and complex. Therefore the application of some compression technique is required in order to reduce the size of the spectra and to obtain the main features while filtering the noise. Wavelet Transforms (WT) are then used as they differ form the traditional Fourier techniques by the way in which they represent the information in the time-frequency plane. In particular, they are suitable for the analysis of non-stationary signals [9].

2 First Signal Processing

Nuclear Magnetic Resonance (NMR) has provided a great help in the knowledge of the different pathologies and also to establish the relations between the different lesions. This particular aspect has a great importance in tumor pathology [4,13,16]. It is not an invasive method and it is able to determine, qualitatively and quantitatively, a great variety of metabolites in each tissue, giving significant information. Nevertheless, its diagnostic application has been limited due to the fact that the in vivo ^{1}H NMR spectra only offers small accuracy, and because of the difficulties founded in the quantification of the metabolites.

Most of the limitations could be overcome if extractions of tumor biopsies are available, by applying the in vitro technique [15]. This method has been used to obtain the spectra that will be analyzed in this paper. Samples from normal brain and different tumor tissues were obtained after craniotomy. The preparation and characterization of biopsies are described in the literature[16]. The spectra obtained by this way have thousand of data. In fact, each spectrum has 16384 samples taking into account only the real part. The spectrum of each tumor represents the intensity, which is proportional to the concentration of protons in the tissue (y axis) vs. the distance in parts per million (ppm) (x axis), i.e. at a particular resonance frequency. This frequency depends on the magnetic field, it is not an absolute value but a ratio, and then the chemical displacement δ is defined as in equation 1,

$$\delta = \left(\frac{\Delta\nu}{\nu_{\text{obs}}} \right) \times 10^{6} \, [\text{ppm}] \quad , \tag{1}$$

where ν_{obs} is the frequency of resonance in the magnetic field (8 Tesla) and $\Delta\nu$ is the difference between:

– the point we are on and the reference point (0 ppm) in Hz, and
– the frequency of observation.

Before the signal processing, some of the spectra presented different resolution (8kb, 16kb, and 32kb). Therefore, the first step was to adjust all the signals to the same size (16384 kb), applying zero padding when the number of samples needed to be enlarged. Due to this, the interval between points is different for each signal, in spite of the fact that the interval is constant for every particular signal. The size of the mentioned interval ΔP (in ppm) was computed by the absolute value of the difference between the final P_f and the initial P_i points of the spectrum, dividing it by the total number of points minus one, as it is shown in equation 2. Most of the spectra presented values of P_i and P_f between approximately 10ppm and 1ppm respectively.

$$\delta P = \frac{|P_f - P_i|}{(16384 - 1)} \ . \tag{2}$$

Once the size of the interval between samples for each signal has been obtained, we proceeded to extract the amplitudes in the region of interest, which covers the range from 0.8 ppm up to 4.2 ppm, according to the information provided by the medical experts. In that interval, the characteristic peaks of the two kinds of spectra, normal brain and meningioma tumor, will be analyzed.

The calibration of the intensity for the signals was also different depending on how they were obtained. For that reason, a normalization procedure of $I(\delta)$ was applied, as expressed in equation 3 [20].

$$I_{\text{norm}}(\delta) = \frac{I(\delta)}{I_{\max}} \ . \tag{3}$$

Then, the signals have been normalized in both, the resonance intensity and the number of samples of the spectra. The data are thus prepared to be classified. In figure 1 it is shown a spectrum after this processing, where it is possible to observe that the number of points of each spectrum is now 4655.

3 Coding the Signals

3.1 Reducing the Signal's Size by Using Wavelets

Since a large input vector requires a more complex network, with many elements that must be configured and adjusted, the wavelet approximation is convenient and even necessary as it provides a most suitable and reduced representation of the signals. The use of the Discrete Wavelets Transform (DWT) makes possible to reach a desired decomposition level preserving the signal information [1]. The redundant information is minimized and so the computational load is substantially cut down [8].

In this application, after trying different wavelet transforms at different levels, the Haar 4 transform was selected [11]. It reduces the signals to 291 points. The

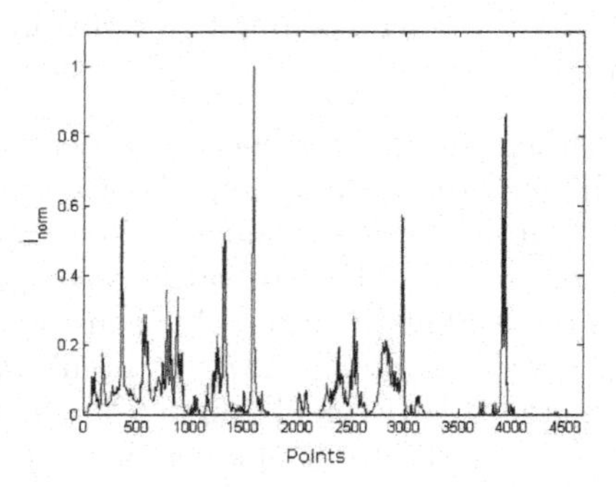

Fig. 1. Normalized normal brain spectrum

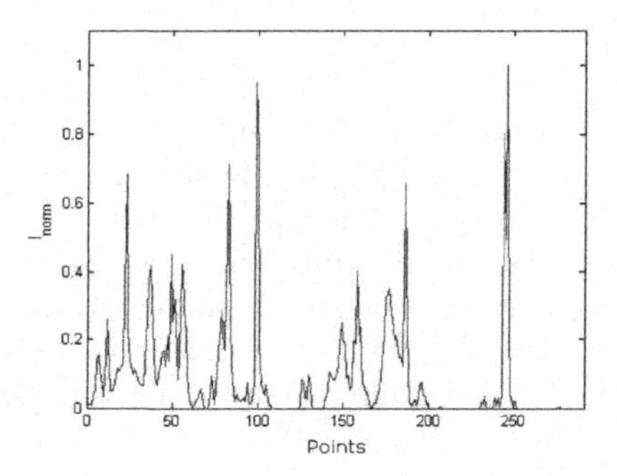

Fig. 2. Normal brain spectrum processed by a Haar 4 wavelet transform

results of Section 5 prove that this is the best configuration for these signals, as other decomposition levels were also tried [2]. Figure 2 shows the approximation of the signal of figure 1 after applying wavelet Haar 4.

It is possible to see that the representative peaks of the signal are kept almost without any loss. This is due to the main property of the Haar transform, which preserves the features of abrupt changes in signals [8,21]. As in the original biomedical spectra, the peaks are well represented, with a relative small standard deviation regarding the intensities. The coding we are going to apply to the spectrum is based on the profile of the signals, specifically on the peaks they present. This will help to convert the signals into a binary format in order to feed the sensorial layers of the Hamming neural network.

3.2 Binary Coding of the Approximated Signals

Before proceeding to classify the signals, the last step is to find out the best representation for the inputs of the classification system. As the important aspect of the biomedical signals is the profile, specifically the peaks, a discretizing method is applied to the approximations of the signals. First, a threshold is fixed in order to discriminate the existence of a representative peak. This threshold is defined according to an estimation of the peaks shape observed in the spectra, with an exponential function given by $\exp(-(x - x_o)^2/\sigma^2)$ [18]. In most of the peaks, the σ^2 value is relatively small regarding the intensity, because the range of the chemical displacement is large, as it runs from 0,8 ppm up to 4,2 ppm. It is then expected that the coding assigns ones to the points of the spectrum with amplitude over the 35% of the signal's highest peak, and a zeroes to the points which have intensity below this threshold. The problem of the shifted peaks, in the different samples of the same type of spectrum, is mitigated when implementing this methodology.

Then, we proceed to code the patterns that will be learned by the neural network. The net will compare unknown signals to the patterns and will identify them. Figure 3 shows the coding of two different signals, both of them randomly chosen between all the samples of their corresponding groups, normal brain or meningioma tumor. These data can be considered as vectors with 291 binary components, which will be input to the net's sensorial layer.

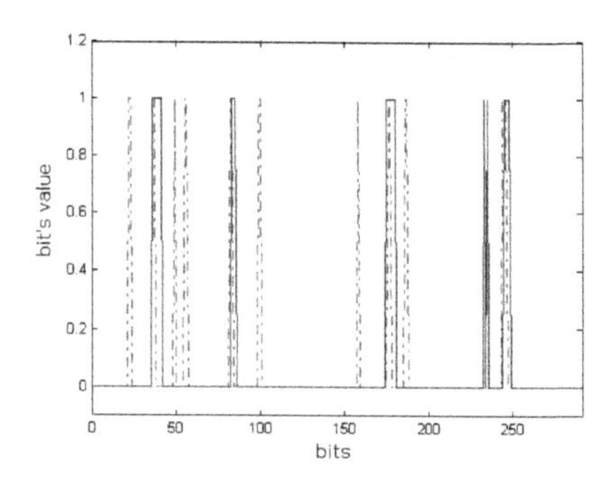

Fig. 3. Binary coding for a normal brain (dashed line) and a meningioma (solid line)

4 Signals Classification

Once the size of the spectrum has been reduced, preserving its relevant information in the range of interest, and has also been conveniently coded, this knowledge representation is delivered to the sensorial layer of a Hamming neural network in order to classify the signals into one of the sets under consideration. The design of the networks consists of two layers: a feed-forward lower layer and a recursive upper layer. This has been proved as one of the most adequate structures to solve classification problems with minimum error [7]. For training, coded signals of 291 binary components (0,1) are presented to the net. It generates the representative pattern of each cluster. When a coded unknown signal is shown to the net it computes the Hamming distance between that signal and each one of the already memorized samples. The Hamming distance is defined by the number of bits of the input signal that do not match the respective ones of each sample

[3]. Therefore, the classification is done by selecting the class which gives the minimum value of the distance.

4.1 Layout of the Neural Network

The structure of this supervised learning network consists of two layers, as it is shown in figure 4. The lower one is in charge of the computation of the concordance punctuation, and the upper one has the task of selecting the right answer, using the inhibition effect imposed by the output nodes. Both layers have two neurons as this number is directly related to the total number of classes under consideration.

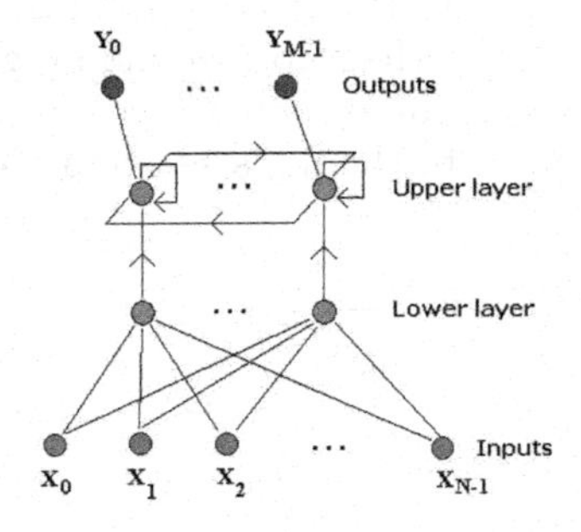

Fig. 4. Hamming net. N=291 inputs (0,1); M=2 classes (Normal Brain, Meningioma).

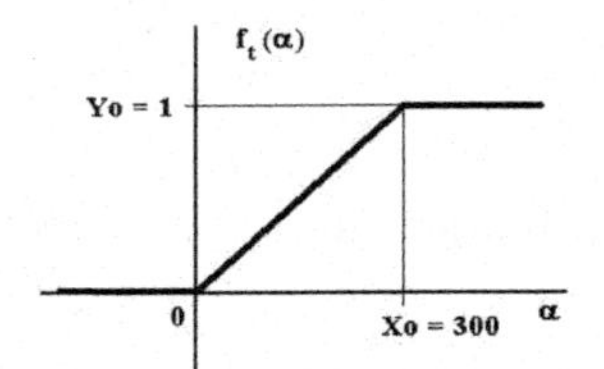

Fig. 5. Activation function of the neural network

The neural network algorithm gives hints of the recommended values for the configuration parameters in order to guarantee the convergence [14]. The implemented activation function is a saturated ramp, which reaches a value of 1, as it is shown in figure 5.

4.2 The Hamming Neural Network Algorithm

The first step when configuring the network is to initialize the weights of the connections and the offsets of each node, taking into account that these values will remain constant along the iteration time t. It is needed to remark that it is a dynamic network. For the lower layer the initialization is given by equation 4.

$$w_{ij} = \frac{x_i^j}{2}, \ \theta_j = \frac{N}{2} \ / \ 0 \leq i \leq (N-1) \ \wedge \ 0 \leq j \leq (M-1) \ . \tag{4}$$

For the upper layer the initialization is done as in equation 5.

$$t_{kl} = \{+1 \ \forall \ k = l; \ -\epsilon \ \forall \ k \neq l\} \ / \ 0 \leq k, l \leq (M-1) \ \wedge \ \epsilon < \frac{1}{M} \ . \tag{5}$$

In equations 4 and 5, w_{ij} means the weight of the connection between the ith input and the jth node of the lower layer; θ_j is the corresponding node's activation threshold; N is the number of binary inputs, and M the number of outputs or classes to be identified. The weighting of the connection between node k and node l in the upper layer is given by t_{kl}, and all the thresholds in this output layers are initially set to zero. The notation of x_i^j represents the ith element of the jth sample that is supplied to train the network.

The second step is to present an unknown signal to the neural network and to compute the initial value of the iteration in t. In equation 6, $\mu_j(0)$ is the output of the jth node in the upper layer, for the initial time $t = 0$; x_i is the ith element of the unknown signal, and f_t is the activation function already shown in figure 5.

$$\mu_j(0) = f_t \left(\sum_{i=0}^{N-1} w_{ij} x_i + \theta_j \right) \ / \ 0 \leq j \leq (M-1) \ . \tag{6}$$

From now on it is assumed that the maximum input for this function never causes saturation on its output. The last step is the iteration in t at the upper layer of the network that it is given by equation 7. The process continues until the convergence is obtained when only the output of one node remains positive.

$$\mu_j(t+1) = f_t \left(\mu_j(t) - \epsilon \sum_{k \neq j} \mu_k(t) \right) \ / \ 0 \leq j, k \leq (M-1) \ . \tag{7}$$

The convergence and the identification of the right node are guaranteed by values of $\epsilon < (1/M)$ [7]. The number t of iterations in this case is around 10.

5 Implementations and Results

The available signals represent a very reduced set of samples to make neither a final assessment nor a detailed statistical study of the developed method in the medical diagnosis field, but it helps to show how this computational tool can be applied to this field, and how could it be improved in order to get significant results if more data were available. The set of normal brain signals consists of 15 spectra, and 29 for the meningioma tumors class. The selection of training

Table 1. Results of the classification with the Hamming network

Signal Class	Normal Brain	Meningioma
Correctly classified	13 (92.8%)	26 (92.8%)
Wrongly classified	1 (7.2%)	2 (7.2%)
Total classified	14 (100.0%)	28 (100.0%)

samples for both classes was random, extracting only one pattern signal from each class. The groups of signals that have been considered for the classification applying a Haar 4 WT with their corresponding results are shown in table 1.

It is needed to remark that, due to the supervised learning method, the adjustments of the network parameters such as the activation function, the maximum iteration time, the inhibition factor, and even the threshold for the coding of the peaks were easily set. It also has to be mentioned that the Haar transform decomposition level was modified to observe the influence on the identification results. For instance, table 2 shows the results when applying the Haar 5 transform to the same group of spectra. It is possible to observe that there is a higher percentage of errors in the classification of the meningioma tumors. In this way, it has been also discovered that when using higher decomposition levels of the Haar transform the success percentage in the classification of the meningioma tumor signals decreases even more. This fact suggests that more reduction implies the loss of essential information in the tumor signal.

Table 2. Classification with the Hamming network and Haar 5 wavelet transforms

Signal Class	Normal Brain	Meningioma
Correctly classified	13 (92.8%)	23 (82.1%)
Wrongly classified	1 (7.2%)	5 (17.9%)
Total classified	14 (100.0%)	28 (100.0%)

6 Conclusions and Future Work

The process of signal knowledge extraction, normalization, approximation by a Haar 4 wavelet transform, coding, and classification with a Hamming neural network has been successfully automated. The original data are obtained from a text file that contains the magnetic resonance spectroscopy ^{1}H of brain human tissue. The code was developed in the MATLAB®language, version 6.5, Release 13, running in a PC with an Intel Pentium III processor at 800 MHz and 256 MB RAM, under Windows XP Professional. The computing time required for this task was about 7 seconds for every new signal.

The percentage of success in the classification was the 92.8% for both classes (normal brain and meningioma tumor). That means the combination of Haar

4 wavelet transform as a preprocessing tool and a recursive Hamming neural network for classification can help for medical diagnosis. This percentage of success is slightly better than the one obtained by a feed forward neural network implemented in a prior work [2], where a 91.0% of correct classification between normal brain and meningioma tumors was achieved in spite of a more complex and slower nodal structure.

Nevertheless, a multi-class recognition system should be considered to make the classification tool more general. In order to approach this multi-class identification, different strategies could be taken into account. For example, if the priority is to detect anomalies in the behavior of the signals, and there are not interests in a specific identification, this tool could still be useful.

A cascade classification scheme or a dual classification tree between normal brain spectra and all the different kinds of brain tumor could be proposed to face this problem, giving more importance to the identification of any pathology than to the type of the pathology itself.

Acknowledgments

We appreciate the collaboration of the members of the Neurosurgery Service at University Hospital of La Paz (Community of Madrid), and thanks to the Biomedical Research Institute (CSIC), who have provided us with the signals and data used in this work.

References

1. DeVore R., and Lucier B., (1992) “Wavelets”, Acta Numerica, Iserles, Ed., Cambridge University Press, v. 1, 1-56.
2. Farías G., and Santos M., (2006) “A computational fusion of wavelets and neural networks in a classifier for biomedical applications”, IeCCS e-Conference.
3. Gallaguer R., (1968) “Information theory and reliable communication”, John Wiley & Sons, U.S.A.
4. García-Martín M., Hérigault G., Rémy C., Farion R., Ballesteros P., Coles J., and Cerdán S., (2001) “Mapping Extracellular pH in rat Brain Gliomas in vivo by H Magnetic Resonance Spectroscopic Imaging: Comparison with maps of metabolites”. Cancer Research 61: 6524-6531.
5. Hagberg G., (1998) “From magnetic resonance spectroscopy to classification tumors”. A review of pattern recognition methods. NMR Biomed. 11: 148-156.
6. Howells S., Maxwell R., and Griffiths J., (1992) “An investigation of tumor ^{1}H NMR spectra by pattern recognition”. NMR Biomed. 5: 59-64.
7. Lippmann R., (1987) “An introduction to computing with neural networks”, IEEE ASSP Magazine, 4-22.
8. Mallat S., (1989) “A theory for Multi-resolution Signal Decomposition: The Wavelet representation”. IEEE trans. on Pattern Analysis and Machine Intelligence. 11(7): 674-693.
9. Mallat S., (2001) “A Wavelet Tour of signal Processing”, 2nd ed., Academic Press, San Diego.

10. Martínez-Pérez I., Maxwell R., Howells S., Van Der Bogaart A., Mazucco R., Griffiths J., and Arús C., (1995) "Pattern recognition analysis of ^{1}H NMR spectra from human brain tumours biopsies". Proc. Soc. Magn. Reson., 3rd Annual Meeting, Abstract P1709.
11. MATLAB®, (1989) "Wavelet Toolbox Users Guide", The Math Works, Inc., U.S.A.
12. Maxwell R., Martínez-Pérez I., Cerdán S., Cabañas M, Arús C., Moreno A., Capdevila A., Ferrer E., Bartomeus F., Aparicio A., Conesa G., Roda J., Carceller F., Pascual J., Howells S., Mazzuco R., and Griffiths J., (1998) "Pattern Recognition Analysis oh ^{1}H NMR Spectra from Perchloric Acid Extracts of Human Brain Tumor Biopsies". Magn. Reson. Med. 39: 869-877.
13. Pascual J., Carceller F., Cerdán S., and Roda J., (1998) "Diagnóstico diferencial de tumores cerebrales *in vitro* por espectroscopía de resonancia magnética de protón. Método de los cocientes espectrales". Neurocirugía (Santiago) 9: 4-10.
14. Passino K., (2001) "Intelligent control: An overview of techniques", Chapter in T. Samad, Ed., Perspectives in control: New concepts and applications, IEEE Press, U.S.A.
15. Peeling J., and Sutherland G., (1992) "High-resolution ^{1}H NMR spectroscopy studies of extracts of human cerebral neoplasm". Magn. Reson. Med. 24: 123-136.
16. Roda J., Pascual J., Carceller F., and González-Llanos F., (2000) "Nonhistological Diagnosis of Human Cerebral Tumors by ^{1}H Magnetic Resonance Spectroscopy and Amino Acid Analysis". Clinical Cancer Research 6: 3983-3993.
17. Somorjai R., Dolenko B., Nikulin A., Pizzi N., Scarth G., Zhilkin P., Halliday W., Fewer D., Hill N., Ross I., West M., Smith I., Donnelly S., Kuesel A., and Bière K., (1996) "Classification of ^{1}H MR spectra of human brain neoplasms: the influence of preprocessing and computerized consensus diagnosis on classification accuracy". J. Magn. Reson. Imaging 6: 437-444.
18. Tate A., (1997) "Statistical pattern recognition for the analysis of biomedical magnetic resonance spectra". J. Magn. Resonance Anal. 3: 63-78.
19. Tate A., Griffiths J., Martínez-Pérez I., Moreno A., Barba I., Cabañas M., Watson D., Alonso J., Bartumeus F., Isamat F., Ferrer I., Vila F., Ferrer E., Capdevilla A., and Arús C., (1998) "Towards a method for automated classification of ^{1}H MRS spectra from brain tumours". NMR Biomed. 11: 177-191.
20. (1985) "The fundamentals of signal analysis", Application Note 243, Hewlett Packard Co., U.S.A.
21. http://www.wavelet.org/wavelet/index.html

Author Index

Lecture Notes in Computer Science

For information about Vols. 1–4445

please contact your bookseller or Springer

Vol. 4543: A.K. Bandara, M. Burgess (Eds.), Inter-Domain Management. XII, 237 pages. 2007.

Vol. 4542: P. Sawyer, B. Paech, P. Heymans (Eds.), Requirements Engineering: Foundation for Software Quality. IX, 384 pages. 2007.

Vol. 4541: T. Okadome, T. Yamazaki, M. Makhtari (Eds.), Pervasive Computing for Quality of Life Enhancemanet. IX, 248 pages. 2007.

Vol. 4539: N.H. Bshouty, C. Gentile (Eds.), Learning Theory. XII, 634 pages. 2007. (Sublibrary LNAI).

Vol. 4538: F. Escolano, M. Vento (Eds.), Graph-Based Representations in Pattern Recognition. XII, 416 pages. 2007.

Vol. 4537: K.C.-C. Chang, W. Wang, L. Chen, C.A. Ellis, C.-H. Hsu, A.C. Tsoi, H. Wang (Eds.), Advances in Web and Network Technologies, and Information Management. XXIII, 707 pages. 2007.

Vol. 4534: I. Tomkos, F. Neri, J. Solé Pareta, X. Masip Bruin, S. Sánchez Lopez (Eds.), Optical Network Design and Modeling. XI, 460 pages. 2007.

Vol. 4531: J. Indulska, K. Raymond (Eds.), Distributed Applications and Interoperable Systems. XI, 337 pages. 2007.

Vol. 4530: D.H. Akehurst, R. Vogel, R.F. Paige (Eds.), Model Driven Architecture- Foundations and Applications. X, 219 pages. 2007.

Vol. 4529: P. Melin, O. Castillo, L.T. Aguilar, J. Kacprzyk, W. Pedrycz (Eds.), Foundations of Fuzzy Logic and Soft Computing. XIX, 830 pages. 2007. (Sublibrary LNAI).

Vol. 4528: J. Mira, J.R. Álvarez (Eds.), Nature Inspired Problem-Solving Methods in Knowledge Engineering, Part II. XXII, 650 pages. 2007.

Vol. 4527: J. Mira, J.R. Álvarez (Eds.), Bio-inspired Modeling of Cognitive Tasks, Part I. XXII, 630 pages. 2007.

Vol. 4526: M. Malek, M. Reitenspieß, A. van Moorsel (Eds.), Service Availability. X, 155 pages. 2007.

Vol. 4525: C. Demetrescu (Ed.), Experimental Algorithms. XIII, 448 pages. 2007.

Vol. 4524: M. Marchiori, J.Z. Pan, C.d.S. Marie (Eds.), Web Reasoning and Rule Systems. XI, 382 pages. 2007.

Vol. 4523: Y.-H. Lee, H.-N. Kim, J. Kim, Y. Park, L.T. Yang, S.W. Kim (Eds.), Embedded Software and Systems. XIX, 829 pages. 2007.

Vol. 4522: B.K. Ersbøll, K.S. Pedersen (Eds.), Image Analysis. XVIII, 989 pages. 2007.

Vol. 4521: J. Katz, M. Yung (Eds.), Applied Cryptography and Network Security. XIII, 498 pages. 2007.

Vol. 4519: E. Franconi, M. Kifer, W. May (Eds.), The Semantic Web: Research and Applications. XVIII, 830 pages. 2007.

Vol. 4517: F. Boavida, E. Monteiro, S. Mascolo, Y. Koucheryavy (Eds.), Wired/Wireless Internet Communications. XIV, 382 pages. 2007.

Vol. 4516: L. Mason, T. Drwiega, J. Yan (Eds.), Managing Traffic Performance in Converged Networks. XXIII, 1191 pages. 2007.

Vol. 4515: M. Naor (Ed.), Advances in Cryptology - EUROCRYPT 2007. XIII, 591 pages. 2007.

Vol. 4514: S.N. Artemov, A. Nerode (Eds.), Logical Foundations of Computer Science. XI, 513 pages. 2007.

Vol. 4513: M. Fischetti, D.P. Williamson (Eds.), Integer Programming and Combinatorial Optimization. IX, 500 pages. 2007.

Vol. 4510: P. Van Hentenryck, L. Wolsey (Eds.), Integration of AI and OR Techniques in Constraint Programming for Combinatorial Optimization Problems. X, 391 pages. 2007.

Vol. 4509: Z. Kobti, D. Wu (Eds.), Advances in Artificial Intelligence. XII, 552 pages. 2007. (Sublibrary LNAI).

Vol. 4508: M.-Y. Kao, X.-Y. Li (Eds.), Algorithmic Aspects in Information and Management. VIII, 428 pages. 2007.

Vol. 4507: F. Sandoval, A. Prieto, J. Cabestany, M. Graña (Eds.), Computational and Ambient Intelligence. XXVI, 1167 pages. 2007.

Vol. 4506: D. Zeng, I. Gotham, K. Komatsu, C. Lynch, M. Thurmond, D. Madigan, B. Lober, J. Kvach, H. Chen (Eds.), Intelligence and Security Informatics: Biosurveillance. XI, 234 pages. 2007.

Vol. 4505: G. Dong, X. Lin, W. Wang, Y. Yang, J.X. Yu (Eds.), Advances in Data and Web Management. XXII, 896 pages. 2007.

Vol. 4504: J. Huang, R. Kowalczyk, Z. Maamar, D. Martin, I. Müller, S. Stoutenburg, K.P. Sycara (Eds.), Service-Oriented Computing: Agents, Semantics, and Engineering. X, 175 pages. 2007.

Vol. 4501: J. Marques-Silva, K.A. Sakallah (Eds.), Theory and Applications of Satisfiability Testing – SAT 2007. XI, 384 pages. 2007.

Vol. 4500: N. Streitz, A. Kameas, I. Mavrommati (Eds.), The Disappearing Computer. XVIII, 304 pages. 2007.

Vol. 4497: S.B. Cooper, B. Löwe, A. Sorbi (Eds.), Computation and Logic in the Real World. XVIII, 826 pages. 2007.

Vol. 4496: N.T. Nguyen, A. Grzech, R.J. Howlett, L.C. Jain (Eds.), Agent and Multi-Agent Systems: Technologies and Applications. XXI, 1046 pages. 2007. (Sublibrary LNAI).

Vol. 4495: J. Krogstie, A. Opdahl, G. Sindre (Eds.), Advanced Information Systems Engineering. XVI, 606 pages. 2007.

Vol. 4494: H. Jin, O.F. Rana, Y. Pan, V.K. Prasanna (Eds.), Algorithms and Architectures for Parallel Processing. XIV, 508 pages. 2007.

Vol. 4493: D. Liu, S. Fei, Z. Hou, H. Zhang, C. Sun (Eds.), Advances in Neural Networks – ISNN 2007, Part III. XXVI, 1215 pages. 2007.

Vol. 4492: D. Liu, S. Fei, Z. Hou, H. Zhang, C. Sun (Eds.), Advances in Neural Networks – ISNN 2007, Part II. XXVII, 1321 pages. 2007.

Vol. 4491: D. Liu, S. Fei, Z.-G. Hou, H. Zhang, C. Sun (Eds.), Advances in Neural Networks – ISNN 2007, Part I. LIV, 1365 pages. 2007.

Vol. 4490: Y. Shi, G.D. van Albada, J. Dongarra, P.M.A. Sloot (Eds.), Computational Science – ICCS 2007, Part IV. XXXVII, 1211 pages. 2007.

Vol. 4489: Y. Shi, G.D. van Albada, J. Dongarra, P.M.A. Sloot (Eds.), Computational Science – ICCS 2007, Part III. XXXVII, 1257 pages. 2007.

Vol. 4488: Y. Shi, G.D. van Albada, J. Dongarra, P.M.A. Sloot (Eds.), Computational Science – ICCS 2007, Part II. XXXV, 1251 pages. 2007.

Vol. 4487: Y. Shi, G.D. van Albada, J. Dongarra, P.M.A. Sloot (Eds.), Computational Science – ICCS 2007, Part I. LXXXI, 1275 pages. 2007.

Vol. 4486: M. Bernardo, J. Hillston (Eds.), Formal Methods for Performance Evaluation. VII, 469 pages. 2007.

Vol. 4485: F. Sgallari, A. Murli, N. Paragios (Eds.), Scale Space and Variational Methods in Computer Vision. XV, 931 pages. 2007.

Vol. 4484: J.-Y. Cai, S.B. Cooper, H. Zhu (Eds.), Theory and Applications of Models of Computation. XIII, 772 pages. 2007.

Vol. 4483: C. Baral, G. Brewka, J. Schlipf (Eds.), Logic Programming and Nonmonotonic Reasoning. IX, 327 pages. 2007. (Sublibrary LNAI).

Vol. 4482: A. An, J. Stefanowski, S. Ramanna, C.J. Butz, W. Pedrycz, G. Wang (Eds.), Rough Sets, Fuzzy Sets, Data Mining and Granular Computing. XIV, 585 pages. 2007. (Sublibrary LNAI).

Vol. 4481: J. Yao, P. Lingras, W.-Z. Wu, M. Szczuka, N.J. Cercone, D. Ślęzak (Eds.), Rough Sets and Knowledge Technology. XIV, 576 pages. 2007. (Sublibrary LNAI).

Vol. 4480: A. LaMarca, M. Langheinrich, K.N. Truong (Eds.), Pervasive Computing. XIII, 369 pages. 2007.

Vol. 4479: I.F. Akyildiz, R. Sivakumar, E. Ekici, J.C.d. Oliveira, J. McNair (Eds.), NETWORKING 2007. Ad Hoc and Sensor Networks, Wireless Networks, Next Generation Internet. XXVII, 1252 pages. 2007.

Vol. 4478: J. Martí, J.M. Benedí, A.M. Mendonça, J. Serrat (Eds.), Pattern Recognition and Image Analysis, Part II. XXVII, 657 pages. 2007.

Vol. 4477: J. Martí, J.M. Benedí, A.M. Mendonça, J. Serrat (Eds.), Pattern Recognition and Image Analysis, Part I. XXVII, 625 pages. 2007.

Vol. 4476: V. Gorodetsky, C. Zhang, V.A. Skormin, L. Cao (Eds.), Autonomous Intelligent Systems: Multi-Agents and Data Mining. XIII, 323 pages. 2007. (Sublibrary LNAI).

Vol. 4475: P. Crescenzi, G. Prencipe, G. Pucci (Eds.), Fun with Algorithms. X, 273 pages. 2007.

Vol. 4474: G. Prencipe, S. Zaks (Eds.), Structural Information and Communication Complexity. XI, 342 pages. 2007.

Vol. 4472: M. Haindl, J. Kittler, F. Roli (Eds.), Multiple Classifier Systems. XI, 524 pages. 2007.

Vol. 4471: P. Cesar, K. Chorianopoulos, J.F. Jensen (Eds.), Interactive TV: a Shared Experience. XIII, 236 pages. 2007.

Vol. 4470: Q. Wang, D. Pfahl, D.M. Raffo (Eds.), Software Process Dynamics and Agility. XI, 346 pages. 2007.

Vol. 4468: M.M. Bonsangue, E.B. Johnsen (Eds.), Formal Methods for Open Object-Based Distributed Systems. X, 317 pages. 2007.

Vol. 4467: A.L. Murphy, J. Vitek (Eds.), Coordination Models and Languages. X, 325 pages. 2007.

Vol. 4466: F.B. Sachse, G. Seemann (Eds.), Functional Imaging and Modeling of the Heart. XV, 486 pages. 2007.

Vol. 4465: T. Chahed, B. Tuffin (Eds.), Network Control and Optimization. XIII, 305 pages. 2007.

Vol. 4464: E. Dawson, D.S. Wong (Eds.), Information Security Practice and Experience. XIII, 361 pages. 2007.

Vol. 4463: I. Măndoiu, A. Zelikovsky (Eds.), Bioinformatics Research and Applications. XV, 653 pages. 2007. (Sublibrary LNBI).

Vol. 4462: D. Sauveron, K. Markantonakis, A. Bilas, J.-J. Quisquater (Eds.), Information Security Theory and Practices. XII, 255 pages. 2007.

Vol. 4459: C. Cérin, K.-C. Li (Eds.), Advances in Grid and Pervasive Computing. XVI, 759 pages. 2007.

Vol. 4453: T. Speed, H. Huang (Eds.), Research in Computational Molecular Biology. XVI, 550 pages. 2007. (Sublibrary LNBI).

Vol. 4452: M. Fasli, O. Shehory (Eds.), Agent-Mediated Electronic Commerce. VIII, 249 pages. 2007. (Sublibrary LNAI).

Vol. 4451: T.S. Huang, A. Nijholt, M. Pantic, A. Pentland (Eds.), Artifical Intelligence for Human Computing. XVI, 359 pages. 2007. (Sublibrary LNAI).

Vol. 4450: T. Okamoto, X. Wang (Eds.), Public Key Cryptography – PKC 2007. XIII, 491 pages. 2007.

Vol. 4448: M. Giacobini et al. (Ed.), Applications of Evolutionary Computing. XXIII, 755 pages. 2007.

Vol. 4447: E. Marchiori, J.H. Moore, J.C. Rajapakse (Eds.), Evolutionary Computation,Machine Learning and Data Mining in Bioinformatics. XI, 302 pages. 2007.

Vol. 4446: C. Cotta, J. van Hemert (Eds.), Evolutionary Computation in Combinatorial Optimization. XII, 241 pages. 2007.